SOLUTIONS MANUAL

Joseph Topich
Virginia Commonwealth University

CHEMISTRY

John McMurry
Cornell University

Robert C. Fay
Cornell University

Prentice Hall, Upper Saddle River, New Jersey 07458

Production Editor: Tina M. Trautz
Special Projects Manager: Barbara A. Murray
Editor-in-Chief: Paul Corey
Associate Editor: Mary P. Hornby
Production Coordinator: Ben Smith

Simon & Schuster/A Viacom Company
Upper Saddle River, New Jersey 07458

Printed in the United States of America

10 9 8 7 6 5 4 3 2 1

ISBN: 0-13-350497-2

PRENTICE-HALL INTERNATIONAL (UK) LIMITED, LONDON
PRENTICE-HALL OF AUSTRALIA PTY. LIMITED, SYDNEY
PRENTICE-HALL CANADA INC. TORONTO
PRENTICE-HALL HISPANOAMERICANA, S.A., MEXICO
PRENTICE-HALL OF INDIA PRIVATE LIMITED, NEW DELHI
PRENTICE-HALL OF JAPAN, INC., TOKYO
SIMON & SCHUSTER ASIA PTE. LTD., SINGAPORE
EDITORA PRENTICE-HALL DO BRASIL, LTDA., RIO DE JANEIRO

Contents

Preface

The principles of chemistry are learned through a combination of different experiences. These include reading about them from your textbook, hearing about them from your general chemistry instructor, seeing them in a laboratory experiment, and most importantly, understanding them by using chemistry principles to solve chemistry problems.

The Complete Solutions Manual to accompany CHEMISTRY by McMurry and Fay contains the solutions to all in-chapter, end-of-chapter, and understanding key concept problems. To be successful in chemistry you must be proficient in setting up and solving chemistry problems. To develop your chemistry problem solving skills, you should always attempt to set up and solve a problem before consulting this solutions manual. Read each problem carefully to see what it is asking and to see what information it contains. Then call upon your chemistry background to attempt a solution for the problem. Look at your answer, and units where applicable, to see if they appear reasonable. Only after this is done should you consult the solutions manual. If you fall into the habit of looking at the solutions manual before you set up a problem, you will not develop your own problem solving skills.

During the many years that I have taught general chemistry, students have voiced to me on many occasions their frustrations in setting up and solving problems and then consulting a solutions manual only to find something completely different because of an error in the manual. I have worked to ensure that the solutions in this manual are as error free as possible. Solutions have been double-checked and in many cases triple-checked. Students must realize that small differences in numerical answers between their results and those in the Complete Solutions Manual can result because of rounding and significant figure differences. Students must also be aware of the fact that there may be more than one legitimate set up for a problem in some cases. Learn to use the Complete Solutions Manual to your advantage, not to your disadvantage.

I would like to thank John McMurry and Robert Fay for the opportunity to contribute to their CHEMISTRY package. I also want to thank them for their comments and input as I worked on this solutions manual. I also want to acknowledge accuracy checkers, Anina Carter, John DeKorte, Ray Fort, Art Meyer, and Barbara Mowery. Finally, I want to thank in a special way my wife, Ruth, and our children, Joey and Judy, for their constant support and encouragement as I worked on this project.

Joseph Topich
Virginia Commonwealth University

Chapter 1 – Chemistry: Matter and Measurement

1.1 (a) Cu (b) Pt (c) Pu

1.2 (a) silver (b) rhodium (c) rhenium

(d) cesium (e) argon (f) arsenic

1.3 (a) Ti, metal (b) Te, semimetal (c) Se, nonmetal

(d) Sc, metal (e) At, semimetal (f) Ar, nonmetal

1.4 (a) The decimal point must be shifted ten places to the right so the exponent is –10. The result is 3.72×10^{-10} m.

(b) The decimal point must be shifted eleven places to the left so the exponent is 11. The result is 1.5×10^{11} m.

1.5 (a) microgram (b) decimeter (c) picosecond

(d) kiloampere (e) millimole

1.6 $$°C = \frac{5}{9} \times (°F - 32) = \frac{5}{9} \times (1474 - 32) = 801.1°C$$

$$K = °C + 273.15 = 801.1 + 273.15 = 1074.2\ K$$

1.7 (a) $K = °C + 273.15 = -78 + 273.15 = 195.15\ K = 195\ K$

(b) $°F = (\frac{9}{5} \times °C) + 32 = (\frac{9}{5} \times 158) + 32 = 316.4°F = 316°F$

(c) $°C = K - 273.15 = 375 - 273.15 = 101.85°C = 102°C$

$$°F = (\frac{9}{5} \times °C) + 32 = (\frac{9}{5} \times 101.85) + 32 = 215.33°F = 215°F$$

1.8 $°C = \frac{5}{9} \times (°F - 32)$

Set °C = °F: $°C = \frac{5}{9} \times (°C - 32)$

Solve for °C: $°C \times \frac{9}{5} = °C - 32$

$(°C \times \frac{9}{5}) - °C = -32$

$°C \times \frac{4}{5} = -32$

$°C = \frac{5}{4}(-32) = -40°C$

The Celsius and Farenheit scales "cross" at –40°C.

1.9 The actual mass of the bottle and the acetone = 38.0015 g + 0.7791 g = 38.7806 g. The measured values are 38.7798 g, 38.7795 g, and 38.7801 g. These values are both close to each other and close to the actual mass. Therefore the results are both precise and accurate.

1.10 (a) 76.600 kg has 5 significant figures because zeros at the end of a number and after the decimal point are always significant.

(b) 4.502 00 x 10^3 g has 6 significant figures because zeros in the middle of a number are significant and zeros at the the end of a number and after the decimal point are always significant.

(c) 3000 nm has 1, 2, 3, or 4 significant figures because zeros at the end of a number and before the decimal point may or may not be significant.

(d) 0.003 00 mL has 3 significant figure because zeros at the beginning of a number are not significant and zeros at the end of a number and after the decimal point are always significant.

(e) 18 students has an infinite number of significant figures since this is an exact number.

1.11 (a) Since the digit to be dropped (the second 4) is less than 5, round down. The result is 3.774 L.

(b) Since the digit to be dropped (0) is less than 5, round down. The result is 255 K.

(c) Since the digit to be dropped is equal to 5 with nothing following and the preceding digit is even, round down. The result is 55.26 kg.

1.12 (a)

$$\begin{array}{r} 24.567\ \ \ \ \ \text{g} \\ +\ \ 0.044\ 78\ \text{g} \\ \hline 24.611\ 78\ \text{g} \end{array}$$

This result should be expressed with 3 decimal places. Since the digit to be dropped (7) is greater than 5, round up. The result is 24.612 g (5 significant figures).

(b) 4.6742 g / 0.003 71 L = 1259.89 g/L

0.003 71 has only 3 significant figures so the result of the division should have only 3 significant figures. Since the digit to be dropped (first 9) is greater than 5, round up. The result is 1260 g/L (3 significant figures).

(c)

$$\begin{array}{lr} & 0.378\ \ \text{mL} \\ + & 42.3\ \ \ \ \ \ \text{mL} \\ - & 1.5833\ \text{mL} \\ \hline & 41.0947\ \text{mL} \end{array}$$

This result should be expressed with 1 decimal place. Since the digit to be dropped (9) is greater than 5, round up. The result is 41.1 mL (3 significant figures).

1.13 (a) °F ≈ 2 x °C if °C is large. The melting point of gold ≈ 2000°F.

(b) $r = d/2 = 3 \times 10^{-6}$ m $= 3 \times 10^{-4}$ cm; $h = 2 \times 10^{-6}$ m $= 2 \times 10^{-4}$ cm

$$\text{volume} = \pi r^2 h \approx 3r^2h \approx 3(3 \times 10^{-4}\ \text{cm})^2(2 \times 10^{-4}\ \text{cm}) \approx 5 \times 10^{-11}\ \text{cm}^3$$

1.14 1 mi = 1760 yd and 1m = 1.0936 yd

$$26\ \text{mi} \times \frac{1760\ \text{yd}}{1\ \text{mi}} = 45{,}760\ \text{yd}$$

marathon in yards = 45,760 yd + 385 yd = 46,145 yd

marathon in meters = $46{,}145 \text{ yd} \times \dfrac{1 \text{ m}}{1.0936 \text{ yd}} = 42{,}196 \text{ m}$

1.15 1 carat = 200 mg = 200×10^{-3} g = 0.200 g

Mass of Hope Diamond in grams = $44.4 \text{ carats} \times \dfrac{0.200 \text{ g}}{1 \text{ carat}} = 8.88 \text{ g}$

1 ounce = 28.35 g

Mass of Hope Diamond in ounces = $8.88 \text{ g} \times \dfrac{1 \text{ ounce}}{28.35 \text{ g}} = 0.313 \text{ ounces}$

1.16 1100 μg = 1100×10^{-6} g

$$\frac{1100 \times 10^{-6} \text{ g}}{2300 \text{ people}} = 4.8 \times 10^{-7} \text{ g/person}$$

1.17 $d = \dfrac{m}{V} = \dfrac{27.43 \text{ g}}{12.40 \text{ cm}^3} = 2.212 \text{ g/cm}^3$

1.18 volume = $9.37 \text{ g} \times \dfrac{1 \text{ mL}}{1.483 \text{ g}} = 6.32 \text{ mL}$

Understanding Key Concepts

1.

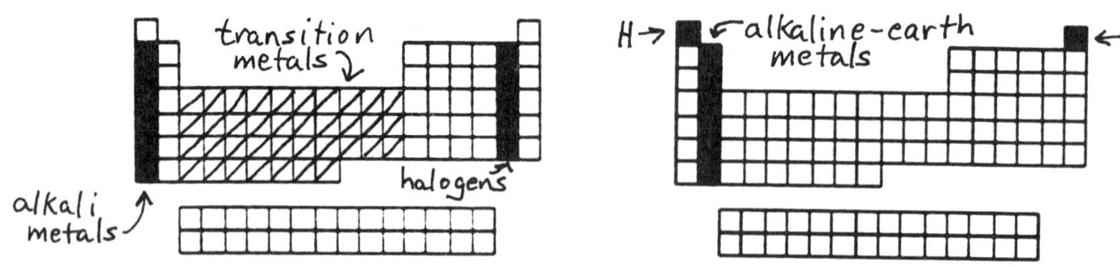

2.

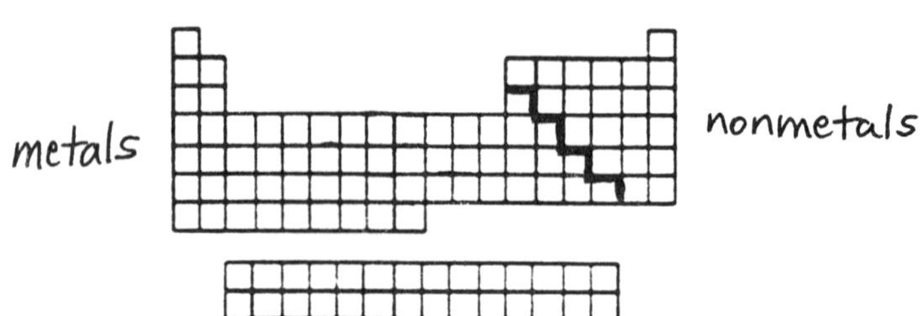

3. red – gas; blue – 42; green – sodium

4. (a) darts are clustered together (good precision) but are away from the bullseye (poor accuracy).

   (b) darts are clustered together (good precision) and hit the bullseye (good accuracy).

   (c) darts are scattered (poor precision) and are away from the bullseye (poor accuracy).

5. Weight = 135 lb – 85 lb = 50 lb

   Mass does not depend on gravity. Mass = 135 lb = 61.2 kg

   If the antigravity machine was set for 165 lb of antigravity force, the person would fly into space.

Additional Problems

Elements and the Periodic Table

1.19 109 elements are presently known. About 90 elements occur naturally.

1.20 The rows are called periods, and the columns are called groups.

1.21 There are 18 groups in the periodic table. They are labeled as follows:

1A, 2A, 3B, 4B, 5B, 6B, 7B, 8B (3 groups), 1B, 2B, 3A, 4A, 5A, 6A, 7A, 8A

1.22 Elements within a group have similar chemical properties.

1.23

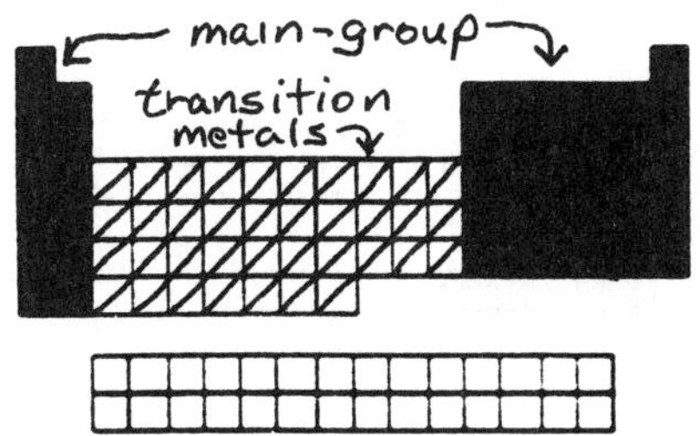

1.24

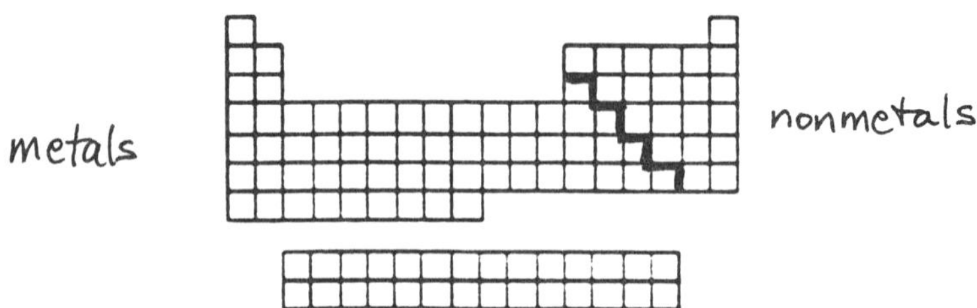

1.25 A semimetal is an element with properties that fall between those of metals and nonmetals.

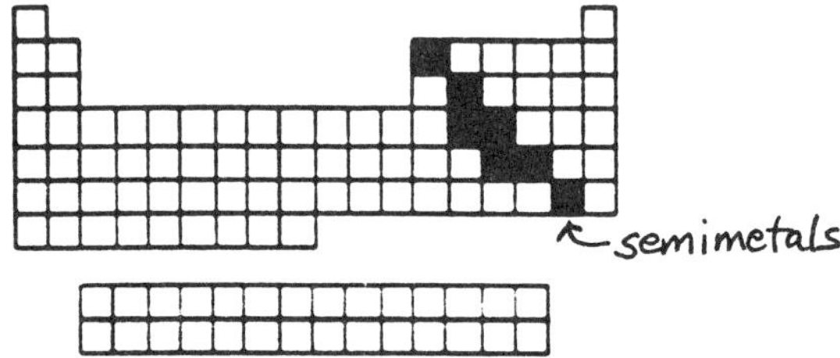

1.26 (a) The alkali metals are shiny, soft, low-melting metals that react rapidly with water to form products that are alkaline.

(b) The noble gases are gases of very low reactivity.

(c) The halogens are nonmetallic and corrosive. They are found in nature only in combination with other elements.

1.27 Li, Na, K, Rb, and Cs

1.28 F, Cl, Br, and I

1.29 He, Ne, Ar, Kr, Xe, and Rn

1.30 (a) americium, Am (b) germanium, Ge (c) technetium, Tc

(d) arsenic, As (e) cadmium, Cd (f) iridium, Ir

1.31 (a) Te, tellurium (b) Re, rhenium (c) Be, beryllium

(d) Ar, argon (e) Pu, plutonium

1.32 (a) Tin is Sn: Ti is titanium.

(b) Manganese is Mn: Mg is magnesium.

(c) Potassium is K: Po is polonium.

(d) The symbol for helium is He.

1.33 (a) The symbol for carbon is C. (b) The symbol for sodium is Na.

(c) The symbol for nitrogen is N. (d) The symbol for chlorine is Cl.

1.34 Accurate measurement is crucial in science because, if our experiments are to be reproducible, we must be able to describe fully the substances we are working with — their amounts, sizes, temperatures, etc.

1.35 A physical quantity contains both a number and a unit label. The physical quantity length would contain a number and a unit label such as meter.

1.36 (a) kg (b) m (c) K (d) m^3

1.37 There are only seven fundamental (base) SI units for scientific measurement. A derived SI unit is some combination of two or more base SI units.

	<u>example</u>
base SI unit	mass, kg
derived SI unit	density, kg/m^3

1.38 (a) kilo, k (b) micro, μ (c) giga, G

(d) pico, p (e) centi, c

1.39 Mass measures the amount of matter in an object, whereas weight measures the pull of gravity on an object by the earth or other celestial body.

1.40 A Celsius degree is larger than a Farenheit degree by a factor of $\frac{9}{5}$.

1.41 A kelvin and Celsius degree are the same.

1.42 The volume of a cubic decimeter (dm^3) and a liter (L) are the same.

1.43 It is necessary to round off the results of numerical calculations so that the results are reported with the correct number of significant figures.

1.44 (a) and (b) are exact because they are obtained by counting. (c) is not exact because it results from a measurement.

1.45

$$\begin{array}{r} 4.8673\ \text{g} \\ -\ 4.8\ \text{g} \\ \hline 0.0673\ \text{g} \end{array}$$

The result should contain only 1 decimal place. Since the digit to be dropped (6) is greater than 5, round up. The result is 0.1 g.

1.46 cL is centiliter (10^{-2} L)

Units and Significant Figures

1.47 (a) deciliter (10^{-1} L)   (b) decimeter (10^{-1} m)

(c) micrometer (10^{-6} m)   (d) nanoliter (10^{-9} L)

1.48 1 mg = 1×10^{-3} g and 1 pg = 1×10^{-12} g

$$\frac{1 \times 10^{-3}\ \text{g}}{1\ \text{mg}} \times \frac{1\ \text{pg}}{1 \times 10^{-12}\ \text{g}} = 1 \times 10^{9}\ \text{pg/mg}$$

35 ng = 35×10^{-9} g

$$\frac{35 \times 10^{-9}\ \text{g}}{35\ \text{ng}} \times \frac{1\ \text{pg}}{1 \times 10^{-12}\ \text{g}} = 3.5 \times 10^{4}\ \text{pg/35 ng}$$

1.49 $1\ \mu L = 10^{-6}\ L$ number of µL per L = $1\ L \times \dfrac{1\ \mu L}{10^{-6}\ L} = 10^{6}\ \mu L$

$20\ mL = 20 \times 10^{-3}\ L = 2.0 \times 10^{-2}\ L$

number of µL per 20 mL = $2.0 \times 10^{-2}\ L \times \dfrac{1\ \mu L}{10^{-6}\ L} = 2 \times 10^{4}\ \mu L$

1.50 (a) $5\ pm = 5 \times 10^{-12}\ m$

$$5 \times 10^{-12}\ m \times \frac{100\ cm}{1\ m} = 5 \times 10^{-10}\ cm$$

$$5 \times 10^{-12}\ m \times \frac{1\ nm}{1 \times 10^{-9}\ m} = 5 \times 10^{-3}\ nm$$

(b) $$8.5\ cm^3 \times \left(\frac{1\ m}{100\ cm}\right)^3 = 8.5 \times 10^{-6}\ m^3$$

$$8.5\ cm^3 \times \left(\frac{10\ mm}{1\ cm}\right)^3 = 8.5 \times 10^{3}\ mm^3$$

(c) $$65.2\ mg \times \frac{1 \times 10^{-3}\ g}{1\ mg} = 0.0652\ g$$

$$65.2\ mg \times \frac{1 \times 10^{-3}\ g}{1\ mg} \times \frac{1\ pg}{1 \times 10^{-12}\ g} = 6.52 \times 10^{10}\ pg$$

1.51 (a) A liter is just slightly larger than a quart.

(b) A mile is about twice as long as a kilometer.

(c) An ounce is about 30 times larger than a gram.

(d) An inch is about 2.5 times larger than a centimeter.

1.52 (a) 35.0445 g has 6 significant figures because zeros in the middle of a number are significant.

(b) 59.0001 cm has 6 significant figures because zeros in the middle of a number are significant.

(c) 0.030 03 kg has 4 significant figures because zeros at the beginning of a number are not significant and zeros in the middle of a number are significant.

(d) 0.004 50 m has 3 significant figures because zeros at the beginning of a number are not significant and zeros at the end of a number and after the decimal point are always significant.

(e) 67,000 m^2 has 2, 3, 4, or 5 significant figures because zeros at the end of a number and before the decimal point may or may not be significant.

(f) 3.8200 x 10^3 L has 5 significant figures because zeros at the end of a number and after the decimal point are always significant.

1.53 To convert 3,666,500 m^3 to scientific notation, move the decimal point 6 places to the left and include an exponent of 10^6. The result is 3.6665 x 10^6.

1.54 (a) To convert 453.32 mg to scientific notation, move the decimal point 2 places to the left and include an exponent of 10^2. The result is 4.5332 x 10^2 mg.

(b) To convert 0.000 042 1 mL to scientific notation, move the decimal point 5 places to the right and include an exponent of 10^{-5}. The result is 4.21 x 10^{-5} mL.

(c) To convert 667,000 g to scientific notation, move the decimal point 5 places to the left and include an exponent of 10^5. The result is 6.67 x 10^5 g.

1.55 (a) Since the exponent is a negative 3, move the decimal point 3 places to the left to get 0.003 221 mm.

(b) Since the exponent is a positive 5, move the decimal point 5 places to the right to get 894,000 m.

(c) Since the exponent is a negative 12, move the decimal point 12 places to the left to get 0.000 000 000 001 350 82 m^3.

1.56 Since the digit to be dropped (3) is less than 5, round down. The result to 4 significant figures is 7926 mi or 7.926 x 10^3 mi.

Since the digit to be dropped (2) is less than 5, round down. The result to 2 significant figures is 7900 mi or 7.9×10^3 mi.

1.57 (a) Since the digit to be dropped (0) is less than 5, round down. The result is 3.567×10^4 or 35,670 m (4 significant figures).

Since the digit to be dropped (the second 6) is greater than 5, round up. The result is 35,670.1 m (6 significant figures).

(b) Since the digit to be dropped is 5 with nonzero digits following, round up. The result is 69 g (2 significant figures).

Since the digit to be dropped (0) is less than 5, round down. The result is 68.5 g (3 significant figures).

(c) Since the digit to be dropped is 5 with nothing following and the preceding digit is odd, round up. The result is 5.00×10^3 cm (3 significant figures).

(d) Since the digit to be dropped is 5 with nothing following and the preceding digit is even, round down. The result is 2.3098×10^{-4} kg (5 significant figures).

1.58 (a) $4.884 \times 2.05 = 10.012$

The result should contain only 3 significant figures because 2.05 contains 3 significant figures (the smaller number of significant figures of the two). Since the digit to be dropped (1) is less than 5, round down. The result is 10.0.

(b) $94.61 / 3.7 = 25.57$

The result should contain only 2 significant figures because 3.7 contains 2 significant figures (the smaller number of significant figures of the two). Since the digit to be dropped (second 5) is 5 with nonzero digits following, round up. The result is 26.

(c) $3.7 / 94.61 = 0.0391$

The result should contain only 2 significant figures because 3.7 contains 2 significant figures (the smaller number of significant figures of the two). Since the digit to be dropped (1) is less than 5, round down. The result is 0.039.

(d)

$$\begin{array}{r} 5502.3 \\ 24 \\ +\quad 0.01 \\ \hline 5526.31 \end{array}$$

This result should be expressed with no decimal places. Since the digit to be dropped (3) is less than 5, round down. The result is 5526.

(e)

$$\begin{array}{r} 86.3 \\ +\ 1.42 \\ -\ 0.09 \\ \hline 87.63 \end{array}$$

This result should be expressed with only 1 decimal place. Since the digit to be dropped (3) is less than 5, round down. The result is 87.6.

(f) 5.7 x 2.31 = 13.167

The result should contain only 2 significant figures because 5.7 contains 2 significant figures (the smaller number of significant figures of the two). Since the digit to be dropped (second 1) is less than 5, round down. The result is 13.

1.59 (a) $\frac{3.41 - 0.23}{5.2333} \times 0.205 = \frac{3.18}{5.2333} \times 0.205 = 0.12457 = 0.125.$

Complete the subtraction first. The result has 2 decimal places and 3 significant figures. The result of the multiplication and division must have 3 significant figures. Since the digit to be dropped is 5 with nonzero digits following, round up.

(b) $\frac{5.556 \times 2.3}{4.223 - 0.08} = \frac{5.556 \times 2.3}{4.14} = 3.09 = 3.1.$

Complete the subtraction first. The result has 2 decimal places and 3 significant figures. The result of the multiplication and division must have 2 significant figures. Since the digit to be dropped (9) is greater than 5, round up.

Unit Conversions

1.60 (a) $0.25 \text{ lb} \times \frac{453.59 \text{ g}}{1 \text{ lb}} = 113.4 \text{ g} = 110 \text{ g}$

(b) $1454 \text{ ft} \times \frac{12 \text{ in}}{1 \text{ ft}} \times \frac{2.54 \text{ cm}}{1 \text{ in}} \times \frac{1 \text{ m}}{100 \text{ cm}} = 443.2 \text{ m}$

(c) $2{,}941{,}526 \text{ mi}^2 \times \left(\frac{1.6093 \text{ km}}{1 \text{ mi}}\right)^2 \times \left(\frac{1000 \text{ m}}{1 \text{ km}}\right)^2 = 7.6181 \times 10^{12} \text{ m}^2$

1.61 (a) $5.4 \text{ in} \times \frac{2.54 \text{ cm}}{1 \text{ in}} \times \frac{1 \text{ m}}{100 \text{ cm}} = 0.14 \text{ m}$

(b) $66.31 \text{ lb} \times \frac{1 \text{ kg}}{2.2046 \text{ lb}} = 30.08 \text{ kg}$

(c) $0.5521 \text{ gal} \times \frac{3.7854 \text{ L}}{1 \text{ gal}} \times \frac{1 \times 10^{-3} \text{ m}^3}{1 \text{ L}} = 2.090 \times 10^{-3} \text{ m}^3$

(d) $65 \frac{\text{mi}}{\text{h}} \times \frac{1.6093 \text{ km}}{1 \text{ mi}} \times \frac{1000 \text{ m}}{1 \text{ km}} \times \frac{1 \text{ h}}{60 \text{ min}} \times \frac{1 \text{ min}}{60 \text{ s}} = 29 \frac{\text{m}}{\text{s}}$

(e) $978.3 \text{ yd}^3 \times \frac{1 \text{ m}}{1.0936 \text{ yd}} \times \frac{1 \text{ m}}{1.0936 \text{ yd}} \times \frac{1 \text{ m}}{1.0936 \text{ yd}} = 748.0 \text{ m}^3$

1.62 (a) $1 \text{ acre–ft} \times \frac{1 \text{ mi}^2}{640 \text{ acres}} \times \left(\frac{5280 \text{ ft}}{1 \text{ mi}}\right)^2 = 43{,}560 \text{ ft}^3$

(b) $116 \text{ mi}^3 \times \left(\frac{5280 \text{ ft}}{1 \text{ mi}}\right)^3 \times \frac{1 \text{ acre–ft}}{43{,}560 \text{ ft}^3} = 3.92 \times 10^8 \text{ acre–ft}$

1.63 (a) $18.6 \text{ hands} \times \frac{1/3 \text{ ft}}{1 \text{ hand}} \times \frac{12 \text{ in}}{1 \text{ ft}} \times \frac{2.54 \text{ cm}}{1 \text{ in}} = 189 \text{ cm}$

(b) $(6 \times 2.5 \times 15) \text{ hands}^3 \times \left(\frac{1/3 \text{ ft}}{1 \text{ hand}}\right)^3 \times \left(\frac{12 \text{ in}}{1 \text{ ft}}\right)^3 \times \left(\frac{2.54 \text{ cm}}{1 \text{ in}}\right)^3 \times \left(\frac{1 \text{ m}}{100 \text{ cm}}\right)^3 = 0.2 \text{ m}^3$

1.64 (a) $\frac{200 \text{ mg}}{100 \text{ mL}} \times \frac{1000 \text{ mL}}{1 \text{ L}} = 2000 \text{ mg/L}$

(b) $\frac{200 \text{ mg}}{100 \text{ mL}} \times \frac{1 \times 10^{-3} \text{ g}}{1 \text{ mg}} \times \frac{1 \ \mu\text{g}}{1 \times 10^{-6} \text{ g}} = 2000 \ \mu\text{g/mL}$

(c) $\frac{200\text{ mg}}{100\text{ mL}} \times \frac{1 \times 10^{-3}\text{ g}}{1\text{ mg}} \times \frac{1000\text{ mL}}{1\text{ L}} = 2\text{ g/L}$

(d) $\frac{200\text{ mg}}{100\text{ mL}} \times \frac{1 \times 10^{-3}\text{ g}}{1\text{ mg}} \times \frac{1000\text{ mL}}{1\text{ L}} \times \frac{1\text{ ng}}{1 \times 10^{-9}\text{ g}} \times \frac{1 \times 10^{-6}\text{ L}}{1\ \mu\text{L}} = 2000\text{ ng}/\mu\text{L}$

1.65 2 g/L x 5 L = 10 g

1.66 $8.65\text{ stones} \times \frac{14\text{ lb}}{1\text{ stone}} = 121\text{ lb}$

1.67 $55\ \frac{\text{mi}}{\text{h}} \times \frac{5280\text{ ft}}{1\text{ mi}} \times \frac{12\text{ in}}{1\text{ ft}} \times \frac{2.54\text{ cm}}{1\text{ in}} \times \frac{1\text{ h}}{3600\text{ s}} \times \frac{2.5 \times 10^{-4}\text{ s}}{1\text{ shake}} = 0.61\ \frac{\text{cm}}{\text{shake}}$

1.68 $160\text{ lb} \times \frac{1\text{ kg}}{2.2046\text{ lb}} = 72.6\text{ kg}$

$72.6\text{ kg} \times \frac{20\text{ mg}}{1\text{ kg}} = 1452\text{ mg} = 1500\text{ mg}$

Temperature

1.69 $^\circ\text{F} = (\frac{9}{5} \times {}^\circ\text{C}) + 32$

$^\circ\text{F} = (\frac{9}{5} \times 39.9\ {}^\circ\text{C}) + 32 = 103.8^\circ\text{F}$

$^\circ\text{F} = (\frac{9}{5} \times 22.2^\circ\text{C}) + 32 = 72.0^\circ\text{F}$

1.70 For Hg: mp is $\left[\frac{9}{5} \times (-38.87)\right] + 32 = -37.97^\circ\text{F}$

For Ga: mp is $\left[\frac{9}{5} \times (29.78)\right] + 32 = 85.60^\circ\text{F}$

For Br_2: mp is $\left[\frac{9}{5} \times (-7.2)\right] + 32 = 19.0°F$

1.71 $°C = \frac{5}{9} \times (°F - 32) = \frac{5}{9} \times (6170 - 32) = 3410°C$

$K = °C + 273.15 = 3410 + 273.15 = 3683.15\ K = 3683\ K$

1.72			
Ethanol boiling point	78.5°C	173.3°F	200°E
Ethanol melting point	–117.3°C	–179.1°F	0°E

(a) $\frac{200°E}{[78.5°C - (-117.3°C)]} = \frac{200°E}{195.8°C} = 1.021\ °E/°C$

(b) $\frac{200°E}{[173.3°F - (-179.1°F)]} = \frac{200°E}{352.4°F} = 0.5675\ °E/°F$

(c) $°E = \frac{200}{195.8} \times (°C + 117.3)$

H_2O melting point = 0°C; $°E = \frac{200}{195.8} \times (0 + 117.3) = 119.8°E$

H_2O boiling point = 100°C; $°E = \frac{200}{195.8} \times (100 + 117.3) = 220.0°E$

(d) $°E = \frac{200}{352.4} \times (°F + 179.1) = \frac{200}{352.4} \times (98.6 + 179.1) = 157.6°E$

(e) $°F = \left(°E \times \frac{352.4}{200}\right) - 179.1 = \left(130 \times \frac{352.4}{200}\right) - 179.1 = 50.0°F$

Since the outside temperature is 50.0°F, I would wear a sweater or light jacket.

1.73			
NH_3 boiling point	–33.4°C	–28.1°F	100°A
NH_3 melting point	–77.7°C	–107.9°F	0°A

(a) $\dfrac{100°A}{[-33.4 - (-77.7°C)]} = \dfrac{100°A}{44.3°C} = 2.26\ °A/°C$

(b) $\dfrac{100°A}{[-28.1 - (-107.9°F)]} = \dfrac{100°A}{79.8°F} = 1.25\ °A/°F$

(c) $°A = \dfrac{100}{44.3} \times (°C + 77.7)$

H_2O melting point = 0°C; $°A = \dfrac{100}{44.3} \times (0 + 77.7) = 175°A$

H_2O boiling point = 100°C; $°A = \dfrac{100}{44.3} \times (100 + 77.7) = 401°A$

(d) $°A = \dfrac{100}{79.8} \times (°F + 107.9) = \dfrac{100}{79.8} \times (98.6 + 107.9) = 259°A$

Density

1.74 The intensive physical property that relates the mass of an object to its volume is called density. Density equals mass divided by volume. A more dense substance appears heavier than a less dense substance because the more dense substance contains more mass per unit volume.

1.75 $250\ mg \times \dfrac{1 \times 10^{-3}\ g}{1\ mg} = 0.25\ g$; $V = 0.25\ g \times \dfrac{1\ cm^3}{1.40\ g} = 0.18\ cm^3$

$500\ lb \times \dfrac{453.59\ g}{1\ lb} = 226{,}795\ g$; $V = 226{,}795\ g \times \dfrac{1\ cm^3}{1.40\ g} = 161{,}996\ cm^3 = 162{,}000\ cm^3$

1.76 For H_2: $V = 1.0078\ g \times \dfrac{1\ L}{0.0899\ g} = 11.2\ L$

For Cl_2: $V = 35.45\ g \times \dfrac{1\ L}{3.214\ g} = 11.03\ L$

1.77 $d = \frac{m}{V} = \frac{220.9\text{ g}}{(0.50 \times 1.55 \times 25.00)\text{ cm}^3} = 11.4\frac{\text{g}}{\text{cm}^3} = 11\frac{\text{g}}{\text{cm}^3}$

1.78 d = 2.40 mm = 0.240 cm

r = d/2 = 0.120 cm and $V = \pi r^2 h$

$$d = \frac{m}{V} = \frac{0.3624\text{ g}}{(3.1416)(0.120\text{ cm})^2(15.0\text{ cm})} = 0.534\text{ g/cm}^3$$

1.79 $d = \frac{m}{V} = \frac{8.763\text{ g}}{(28.76 - 25.00)\text{ mL}} = \frac{8.763\text{ g}}{3.76\text{ mL}} = 2.331\frac{\text{g}}{\text{cm}^3} = 2.33\frac{\text{g}}{\text{cm}^3}$

1.80 The explosion was caused by a chemical property. Na reacts violently with H_2O.

General Problems

1.81 An interpretation of experimental data that explains the result is called an hypothesis. The hypothesis can be used to make more predictions and to suggest more experiments until a consistent explanation, or theory, of known observations is arrived at.

1.82 Physical properties are those characteristics like color, mass, melting point, and temperature that can be determined without changing the chemical makeup of the sample. Chemical properties are those that do change the chemical makeup of the sample.

1.83 Color, melting point, mass, and solubility (see Table 1.5).

1.84 (a) selenium, Se (b) rhenium, Re (c) cobalt, Co (d) rhodium, Rh

1.85 (a) Se, selenium (b) Sb, antimony (c) Ba, barium (d) Cs, cesium

(e) Np, neptunium

1.86 (a) M (b) n (c) da (d) h

1.87 (a) cL, centiliter (b) m^2, square meter (c) K, kelvin

(d) nm, nanometer

1.88 (a) 105.331 mg has 6 significant figures because zeros in the middle of a number are significant.

(b) 4050 km has 3 or 4 significant figures because zeros in the middle of a number are significant and zeros at the end of a number and before the decimal point may or may not be significant.

(c) 405.0 km has 4 significant figures because zeros in the middle of a number are significant and zeros at the end of a number and after the decimal point are always significant.

(d) 0.022 107 kg has 5 significant figures because zeros at the beginning of a number are not significant and zeros in the middle of a number are significant.

1.89 NaCl melting point = 1074 K

$°C = K - 273.15 = 1074 - 273.15 = 800.85°C = 801°C$

$$°F = (\frac{9}{5} \times °C) + 32 = (\frac{9}{5} \times 800.85) + 32 = 1473.53°F = 1474°F$$

NaCl boiling point = 1686 K

$°C = K - 273.15 = 1686 - 273.15 = 1412.85°C = 1413°C$

$$°F = (\frac{9}{5} \times °C) + 32 = (\frac{9}{5} \times 1412.85) + 32 = 2575.13°F = 2575°F$$

1.90 $$°F = \left(\frac{9}{5} \times °C\right) + 32 = \left[\frac{9}{5} \times (-38.9)\right] + 32 = -38.0°F$$

1.91 $$V = 112.5 \text{ g} \times \frac{1 \text{ mL}}{1.4832 \text{ g}} = 75.85 \text{ mL}$$

1.92 $15.28\ \frac{\text{lb}}{\text{gal}} \times \frac{453.59\ \text{g}}{1\ \text{lb}} \times \frac{1\ \text{gal}}{3.7854\ \text{L}} \times \frac{1\ \text{L}}{1000\ \text{mL}} = 1.831\ \text{g/mL}$

1.93 $80.31 \times 10^{9}\ \text{lb} \times \frac{453.59\ \text{g}}{1\ \text{lb}} = 3.643 \times 10^{13}\ \text{g}$

$V = 3.643 \times 10^{13}\ \text{g} \times \frac{1\ \text{mL}}{1.831\ \text{g}} \times \frac{1\ \text{L}}{1000\ \text{mL}} = 1.990 \times 10^{10}\ \text{L}$

1.94 $0.22\ \text{in} \times \frac{2.54\ \text{cm}}{1\ \text{in}} \times \frac{10\ \text{mm}}{1\ \text{cm}} = 5.6\ \text{mm}$

2.1 First, find the S:O ratio in each compound.

Substance 1: S:O mass ratio = (6.00 g S) / (5.99 g O) = 1.00

Substance 2: S:O mass ratio = (8.60 g S) / (12.88 g O) = 0.668

$$\frac{\text{S:O mass ratio in substance 1}}{\text{S:O mass ratio in substance 2}} = \frac{1.00}{0.668} = 1.50 = \frac{3}{2}$$

2.2 $0.0002 \text{ in} \times \frac{2.54 \text{ cm}}{1 \text{ in}} \times \frac{1 \text{ Au atom}}{2.9 \times 10^{-8} \text{ cm}} = 2 \times 10^4 \text{ Au atoms}$

2.3

$$1 \times 10^{19} \text{ C atoms} \times \frac{1.5 \times 10^{-10} \text{ m}}{\text{C atom}} \times \frac{1 \text{ km}}{1000 \text{ m}} \times \frac{1 \text{ time}}{40{,}075 \text{ km}} = 37.4 \text{ times} \approx 40 \text{ times}$$

2.4 $^{75}_{34}Se$ has 34 protons, 34 electrons, and (75 – 34) = 41 neutrons.

2.5 $^{35}_{17}Cl$ has (35 – 17) = 18 neutrons. $^{37}_{17}Cl$ has (37 – 17) = 20 neutrons.

2.6 (a) $^{15}_{7}N$ has 7 protons, 7 electrons, and (15 – 7) = 8 neutrons.

(b) $^{131}_{53}I$ has 53 protons, 53 electrons, and (131 – 53) = 78 neutrons.

(c) $^{242}_{94}Pu$ has 94 protons, 94 electrons, and (242 – 94) = 148 neutrons.

2.7 The element with 47 protons is Ag. The mass number is the sum of the protons and the neutrons, 47 + 62 = 109. The isotope symbol is $^{109}_{47}Ag$.

2.8 atomic weight = [0.6917 x 62.94 amu] + [0.3083 x 64.93 amu] = 63.55 amu

2.9 $2.15 \text{ g} \times \frac{1 \text{ amu}}{1.6605 \times 10^{-24} \text{ g}} \times \frac{1 \text{ Cu}}{63.55 \text{ amu}} = 2.04 \times 10^{22} \text{ Cu atoms}$

2.10

```
      H   H
      |   |
  H — C — N — H
      |
      H
```

2.11 (a) LiBr is composed of a metal (Li) and nonmetal (Br) and is ionic.

(b) $SiCl_4$ is composed of only nonmetals and is molecular.

(c) BF_3 is composed of only nonmetals and is molecular.

(d) CaO is composed of a metal (Ca) and nonmetal (O) and is ionic.

2.12 (a) HF is an acid. In water, HF dissaciates to produce $H^+(aq)$.

(b) $Ca(OH)_2$ is a base. In water, $Ca(OH)_2$ dissociates to produce $OH^-(aq)$.

(c) LiOH is a base. In water, LiOH dissociates to produce $OH^-(aq)$.

(d) HCN is an acid. In water, HCN dissociates to produce $H^+(aq)$.

2.13 (a) CsF, cesium fluoride (b) K_2O, potassium oxide

(c) CuO, copper(II) oxide (d) BaS, barium sulfide

(e) $BeBr_2$, beryllium bromide

2.14 (a) vanadium(III) chloride, VCl_3 (b) manganese(IV) oxide, MnO_2

(c) copper(II) sulfide, CuS (d) aluminum oxide, Al_2O_3

2.15 (a) NCl_3, nitrogen trichloride (b) P_4O_6, tetraphosphorus hexoxide

(c) S_2F_2, disulfur difluoride

2.16 (a) disulfur dichloride, S_2Cl_2 (b) iodine monochloride, ICl

(c) nitrogen triiodide, NI_3

2.17 (a) $Ca(ClO)_2$, calcium hypochlorite

(b) $Ag_2S_2O_3$, silver(I) thiosulfate or silver thiosulfate

(c) NaH_2PO_4, sodium dihydrogen phosphate

(d) $Sn(NO_3)_2$, tin(II) nitrate

(e) $Pb(C_2H_3O_2)_4$, lead(IV) acetate

2.18 (a) lithium phosphate, Li_3PO_4

(b) magnesium hydrogen sulfate, $Mg(HSO_4)_2$

(c) manganese(II) nitrate, $Mn(NO_3)_2$

(d) chromium(III) sulfate, $Cr_2(SO_4)_3$

2.19 (a) HIO_4, periodic acid (b) $HBrO_2$, bromous acid (c) H_2CrO_4, chromic acid

Understanding Key Concepts

1. Drawing (a) depicts a collection of SO_2 molecules.

2. H_2O

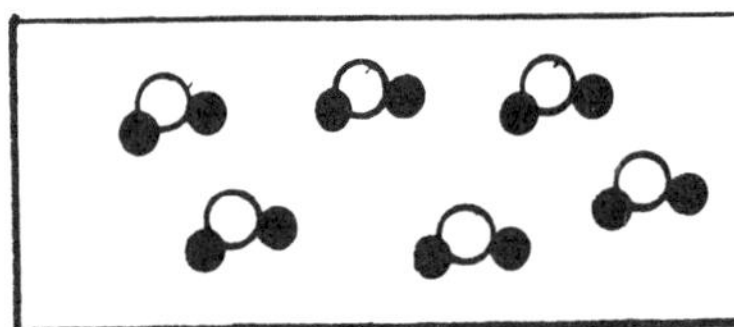

H_2O_2

3. A Na atom has 11 protons and 11 electrons [drawing (b)].

A Ca^{2+} ion has 20 protons and 18 electrons [drawing (c)].

A F^- ion has 9 protons and 10 electrons [drawing (a)].

4.

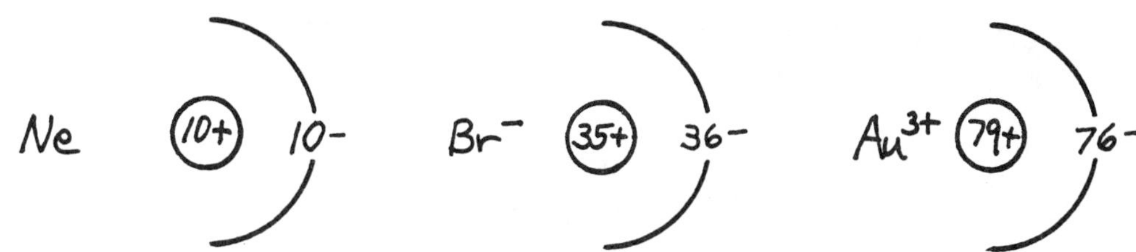

5. In order to obey the law of mass conservation the correct drawing must have the same number of red and yellow spheres as in drawing (a). The correct drawing is (d).

6. (a) gallium, Ga (b) chromium, Cr (c) aluminum, Al

7. The subscript giving the atomic number of an atom is often left off of an isotope symbol because one can readily look up the atomic number from the periodic table knowing the element symbol.

8. Te has isotopes with more neutrons than the isotopes of I.

9.

```
      H     H
      |     |
  H — C — O — C — H
      |     |
      H     H
```

10. Alanine, $C_3H_7NO_2$; Ethylene glycol, $C_2H_6O_2$; Acetic acid, $C_2H_4O_2$

Additional Problems

Atomic Theory

2.20 Different samples of the same compound always contain the same mass ratio of elements.

2.21 In a chemical reaction, no mass is lost or gained.

2.22 The law of mass conservation in terms of Dalton's atomic theory is that chemical reactions only rearrange the way that atoms are combined; the atoms themselves are not changed.

The law of definite proportions in terms of Dalton's atomic theory is that the chemical combination of elements to make different substances occurs when atoms join together in small, whole-number ratios.

2.23 The law of multiple proportions states that if two elements combine in different ways to form different substances, the mass ratios are small, whole-number multiples of each other. This is very similar to Dalton's statement that the chemical combination of elements to make different substances occurs when atoms join together in small, whole-number ratios.

2.24 First, find the C:H ratio in each compound.

Benzene: C:H mass ratio = (4.61 g C) / (0.39 g H) = 12

Ethane: C:H mass ratio (4.00 g C) / (1.00 g H) = 4.0

Ethylene: C:H mass ratio = (4.29 g C) / (0.71 g H) = 6.0

$$\frac{\text{C:H mass ratio in benzene}}{\text{C:H mass ratio in ethane}} = \frac{12}{4.0} = \frac{3}{1}$$

$$\frac{\text{C:H mass ratio in benzene}}{\text{C:H mass ratio in ethylene}} = \frac{12}{6.0} = \frac{2}{1}$$

$$\frac{\text{C:H mass ratio in ethylene}}{\text{C:H mass ratio in ethane}} = \frac{6.0}{4.0} = \frac{3}{2}$$

2.25 First, find the C:O ratio in each compound.

Carbon suboxide: C:O mass ratio = (1.32 g C) / (1.18 g O) = 1.12

Carbon dioxide: C:O mass ratio = (12.00 g C) / (32.00 g O) = 0.375

$$\frac{\text{C:O mass ratio in carbon suboxide}}{\text{C:O mass ratio in carbon dioxide}} = \frac{1.12}{0.375} = \frac{3}{1}$$

2.26 (a) For benzene:

$$4.61 \text{ g x } \frac{1 \text{ amu}}{1.6605 \times 10^{-24} \text{ g}} \text{ x } \frac{1 \text{ C atom}}{12.011 \text{ amu}} = 2.31 \times 10^{23} \text{ C atoms}$$

$$0.39 \text{ g x } \frac{1 \text{ amu}}{1.6605 \times 10^{-24} \text{ g}} \text{ x } \frac{1 \text{ H atom}}{1.008 \text{ amu}} = 2.3 \times 10^{23} \text{ H atoms}$$

$$\frac{C}{H} = \frac{2.31 \times 10^{23} \text{ C atoms}}{2.3 \times 10^{23} \text{ H atoms}} = \frac{1 \text{ C}}{1 \text{ H}}$$

A possible formula for benzene is CH.

For ethane:

$$4.00 \text{ g x } \frac{1 \text{ amu}}{1.6605 \times 10^{-24} \text{ g}} \text{ x } \frac{1 \text{ C atom}}{12.011 \text{ amu}} = 2.01 \times 10^{23} \text{ C atoms}$$

$$1.00 \text{ g x } \frac{1 \text{ amu}}{1.6605 \times 10^{-24} \text{ g}} \text{ x } \frac{1 \text{ H atom}}{1.008 \text{ amu}} = 5.97 \times 10^{23} \text{ H atoms}$$

$$\frac{C}{H} = \frac{2.01 \times 10^{23} \text{ C atoms}}{5.97 \times 10^{23} \text{ H atoms}} = \frac{1 \text{ C}}{3 \text{ H}}$$

A possible formula for ethane is CH_3.

For ethylene:

$$4.29 \text{ g x } \frac{1 \text{ amu}}{1.6605 \times 10^{-24} \text{ g}} \text{ x } \frac{1 \text{ C atom}}{12.011 \text{ amu}} = 2.15 \times 10^{23} \text{ C atoms}$$

$$0.71 \text{ g x } \frac{1 \text{ amu}}{1.6605 \times 10^{-24} \text{ g}} \text{ x } \frac{1 \text{ H atom}}{1.008 \text{ amu}} = 4.2 \times 10^{23} \text{ H atoms}$$

$$\frac{C}{H} = \frac{2.15 \times 10^{23} \text{ C atoms}}{4.2 \times 10^{23} \text{ H atoms}} = \frac{1 \text{ C}}{2 \text{ H}}$$

A possible formula for ethylene is CH_2.

(b) The results in part (a) give the smallest whole-number ratio of C to H for benzene, ethane, and ethylene, and these ratios are consistent with their modern formulas.

2.27 $1.32\text{ g} \times \dfrac{1\text{ amu}}{1.6605 \times 10^{-24}\text{ g}} \times \dfrac{1\text{ C atom}}{12.011\text{ amu}} = 6.62 \times 10^{22}\text{ C atoms}$

$1.18\text{ g} \times \dfrac{1\text{ amu}}{1.6605 \times 10^{-24}\text{ g}} \times \dfrac{1\text{ O atom}}{15.9994\text{ amu}} = 4.44 \times 10^{22}\text{ O atoms}$

$\dfrac{C}{O} = \dfrac{6.62 \times 10^{22}\text{ C atoms}}{4.44 \times 10^{22}\text{ O atoms}} = \dfrac{1.5\text{ C}}{1\text{ O}}$;

therefore the formula for carbon suboxide is $C_{1.5}O$ or C_3O_2.

2.28 (a) $(1.67 \times 10^{-24}\ \dfrac{\text{g}}{\text{H atom}})(6.02 \times 10^{23}\text{ H atoms}) = 1.01\text{ g}$

This result is numerically equal to the atomic weight of H in grams.

(b) $(26.558 \times 10^{-24}\ \dfrac{\text{g}}{\text{O atom}})(6.02 \times 10^{23}\text{ O atoms}) = 16.0\text{ g}$

This result is numerically equal to the atomic weight of O in grams.

(c) The mass of 6.02×10^{23} atoms is its atomic weight expressed in grams. If the atomic weight of an element is Z, then 6.02×10^{23} atoms of this element weighs Z grams.

2.29 Assume a 1.00 g sample of the binary compound of zinc and sulfur.

$0.671 \times 1.00\text{ g} = 0.671\text{ g Zn}$; $0.329 \times 1.00\text{ g} = 0.329\text{ g S}$

$0.671\text{ g} \times \dfrac{1\text{ amu}}{1.6605 \times 10^{-24}\text{ g}} \times \dfrac{1\text{ Zn atom}}{65.39\text{ amu}} = 6.18 \times 10^{21}\text{ Zn atoms}$

$$0.329 \text{ g} \times \frac{1 \text{ amu}}{1.6605 \times 10^{-24} \text{ g}} \times \frac{1 \text{ S atom}}{32.066 \text{ amu}} = 6.18 \times 10^{21} \text{ S atoms}$$

$$\frac{\text{Zn}}{\text{S}} = \frac{6.18 \times 10^{21} \text{ Zn atoms}}{6.18 \times 10^{21} \text{ S atoms}} = \frac{1 \text{ Zn}}{1 \text{ S}}; \text{ therefore the formula is ZnS.}$$

2.30 Assume a 1.000 g sample of one of the binary compounds.

0.3104 x 1.000 g = 0.3104 g Ti; 0.6896 x 1.000 g = 0.6896 g Cl

$$0.3104 \text{ g} \times \frac{1 \text{ amu}}{1.6605 \times 10^{-24} \text{ g}} \times \frac{1 \text{ Ti atom}}{47.88 \text{ amu}} = 3.90 \times 10^{21} \text{ Ti atoms}$$

$$0.6896 \text{ g} \times \frac{1 \text{ amu}}{1.6605 \times 10^{-24} \text{ g}} \times \frac{1 \text{ Cl atom}}{35.453 \text{ amu}} = 1.17 \times 10^{22} \text{ Cl atoms}$$

$$\frac{\text{Cl}}{\text{Ti}} = \frac{1.17 \times 10^{22}}{3.90 \times 10^{21}} = \frac{3}{1}$$

Assume a 1.000 g sample of the other binary compound.

0.2524 x 1.000 g = 0.2524 g Ti; 0.7476 x 1.000 g = 0.7476 g Cl

$$0.2524 \text{ g} \times \frac{1 \text{ amu}}{1.6605 \times 10^{-24} \text{ g}} \times \frac{1 \text{ Ti atom}}{47.88 \text{ amu}} = 3.17 \times 10^{21} \text{ Ti atoms}$$

$$0.7476 \text{ g} \times \frac{1 \text{ amu}}{1.6605 \times 10^{-24} \text{ g}} \times \frac{1 \text{ Cl atom}}{35.453 \text{ amu}} = 1.27 \times 10^{22} \text{ Cl atoms}$$

$$\frac{\text{Cl}}{\text{Ti}} = \frac{1.27 \times 10^{22}}{3.17 \times 10^{21}} = \frac{4}{1}$$

2.31 A cathode-ray tube is a glass tube from which the air has been removed and in which two thin pieces of metal called electrodes have been sealed. When a sufficiently high voltage is applied across the electrodes, an electric current flows through the tube from the negatively charged electrode (the cathode) to the positively charged electrode (the anode). If the tube is not fully evacuated but still contains a small amount of air or other gas, the flowing current is visible as a glow called a cathode ray.

Elements and Atoms

2.32 proton, 1.007 276 amu, +1

neutron, 1.008 664 amu, 0

electron, $5.485\ 799 \times 10^{-4}$ amu, −1

2.33 The atomic number is equal to the number of protons.

The mass number is equal to the sum of the number of protons and the number of neutrons.

2.34 The atomic number is equal to the number of protons.

The atomic weight is the weighted average mass (in amu) of the various isotopes for a particular element.

2.35 Atoms of the same element that have different numbers of neutrons are called isotopes.

2.36 (a) carbon, C (b) argon, Ar (c) vanadium, V

2.37 The mass number is equal to the sum of the number of protons and the number of neutrons for a particular isotope.

For ${}^{14}_{6}C$, mass number = 6 protons + 8 neutrons = 14.

For ${}^{14}_{7}N$, mass number = 7 protons + 7 neutrons = 14.

2.38 ${}^{137}_{55}Cs$

2.39 (a) ${}^{220}_{86}Rn$ (b) ${}^{210}_{84}Po$ (c) ${}^{197}_{79}Au$

2.40 (a) ${}^{140}_{58}Ce$ (b) ${}^{60}_{27}Co$

2.41 (a) $^{15}_{7}N$, 7 protons, 7 electrons, (15 – 7) = 8 neutrons

(b) $^{60}_{27}Co$, 27 protons, 27 electrons, (60 – 27) = 33 neutrons

(c) $^{131}_{53}I$, 53 protons, 53 electrons, (131 – 53) = 78 neutrons

2.42 (a) ^{27}Al, 13 protons and (27 – 13) = 14 neutrons

(b) ^{32}S, 16 protons and (32 – 16) = 16 neutrons

(c) ^{64}Zn, 30 protons and (64 – 30) = 34 neutrons

(d) ^{207}Pb, 82 protons and (207 – 82) = 125 neutrons

2.43 (a) $^{24}_{12}Mg$, magnesium (b) $^{58}_{28}Ni$, nickel

2.44 (a) $^{202}_{80}Hg$, mercury (b) $^{195}_{78}Pt$, platinum

2.45 [0.199 x 10.0129 amu] + [0.801 x 11.009 31 amu] = 10.8 amu for B

2.46 [0.5184 x 106.9051 amu] + [0.4816 x 108.9048 amu] = 107.9 amu for Ag

2.47 24.305 amu = [0.7899 x 23.985 amu] + [0.1000 x 24.986 amu] + [0.1101 x Z]

Solve for Z. Z = 25.982 (^{26}Mg)

2.48 $$\text{mass of 1 C atom} = \frac{12.011 \text{ amu}}{1 \text{ C atoms}} \times \frac{1.6605 \times 10^{-24} \text{ g}}{1 \text{ amu}} = 2.00 \times 10^{-23} \text{ g/C atom}$$

$$\text{number of C atoms} = \frac{1 \times 10^{-5} \text{ g}}{2.00 \times 10^{-23} \text{ g/C atom}} = 5 \times 10^{17} \text{ C atoms}$$

$$\text{time} = \frac{5 \times 10^{17} \text{ C atoms}}{2 \text{ C atoms/s}} = 2.5 \times 10^{17} \text{ s}$$

Compounds and Mixtures, Molecules and Ions

2.49 A compound is a pure substance with constant composition. A mixture is formed when two or more substances are mixed together in some random proportion without chemically changing the individual substances in the mixture. Sucrose ($C_{12}H_{22}O_{11}$) is a compound. Sucrose dissolved in water is a mixture.

2.50 Homogeneous mixtures, called solutions, have a constant composition throughout. Heterogeneous mixtures have regions with differing compositions. Sucrose dissolved in water is a homogeneous mixture. Water and oil is a heterogeneous mixture.

2.51 (a) muddy water, heterogeneous mixture

(b) concrete, heterogeneous mixture

(c) house paint, homogeneous mixture

(d) a soft drink, homogeneous mixture
(heterogeneous mixture if it contains CO_2 bubbles)

2.52 An atom is the smallest particle that retains the chemical properties of an element. A molecule is matter that results when two or more atoms are joined by covalent bonds. H and O are atoms, H_2O is a water molecule.

2.53 A molecule is the unit of matter that results when two or more atoms are joined by covalent bonds. An ion results when an atom gains or loses electrons. CH_4 is a methane molecule. Na^+ is the sodium cation.

2.54 A covalent bond results when two atoms share several (usually two) of their electrons. An ionic bond results from a complete transfer of one or more electrons from one atom to another. The C—H bonds in methane (CH_4) are covalent bonds. The bond in NaCl (Na^+Cl^-) is an ionic bond.

2.55 Element symbols are composed of one or two letters. If the element symbol is two letters, the first letter is uppercase and the second is lowercase. CO stands for carbon and oxygen in carbon monoxide.

2.56 (a) The formula of ammonia is NH_3.

(b) The ionic solid of potassium chloride has the formula KCl.

(c) Cl^- is an anion.

(d) CH_4 is a neutral molecule.

2.57 (a) Be^{2+}, 4 protons and 2 electrons (b) Rb^+, 37 protons and 36 electrons

(c) Se^{2-}, 34 protons and 36 electrons (d) Au^{3+}, 79 protons and 76 electrons

2.58 (a) A 2+ cation that has 36 electrons must have 38 protons. X = Sr.

(b) A 1– anion that has 36 electrons must have 35 protons. X = Br.

2.59 C_3H_8O

2.60

```
    H   H   H   H
    |   |   |   |
H — C — C — C — C — H
    |   |   |   |
    H   H   H   H
```

2.61

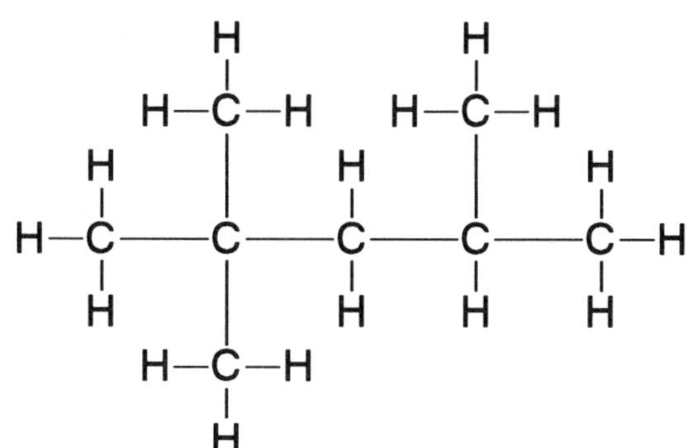

Acids and Bases

2.62 (a) HI, acid (b) CsOH, base (c) H_3PO_4, acid

(d) $Ba(OH)_2$, base (e) H_2CO_3, acid

2.63 $HI(aq) \rightarrow H^+(aq) + I^-(aq)$; the anion is I^-

$H_3PO_4(aq) \rightarrow H^+(aq) + H_2PO_4^-(aq)$; the anion is $H_2PO_4^-$

$H_2CO_3(aq) \rightarrow H^+(aq) + HCO_3^-(aq)$; the anion is HCO_3^-

2.64 (b) $CsOH(aq) \rightarrow Cs^+(aq) + OH^-(aq)$; the cation is Cs^+

(d) $Ba(OH)_2(aq) \rightarrow Ba^{2+}(aq) + 2\ OH^-(aq)$; the cation is Ba^{2+}

Naming Compounds

2.65 (a) KCl (b) $SnBr_2$ (c) CaO (d) $BaCl_2$ (e) AlH_3

2.66 (a) $Ca(C_2H_3O_2)_2$ (b) $Fe(CN)_2$ (c) $Na_2Cr_2O_7$

(d) $Cr_2(SO_4)_3$ (e) $Hg(ClO_4)_2$

2.67 (a) barium ion (b) cesium ion (c) vanadium(III) ion

(d) hydrogen carbonate ion (e) ammonium ion (f) nickel(II) ion

(g) nitrite ion (h) chlorite ion (i) manganese(II) ion

(j) perchlorate ion

2.68 (a) carbon tetrachloride (b) silicon dioxide

(c) dinitrogen monoxide (d) dinitrogen trioxide

2.70 (a) Zn^{2+} (b) Fe^{3+} (c) Ti^{4+} (d) Sn^{2+}

(e) Hg_2^{2+} (f) Mn^{4+} (g) K^+ (h) Cu^{2+}

2.71 (a) zinc(II) cyanide (b) iron(III) nitrite (c) titanium(IV) sulfate

(d) tin(II) phosphate (e) mercury(I) sulfide (f) manganese(IV) oxide

(g) potassium periodate (h) copper(II) acetate

2.72 (a) magnesium sulfite (b) cobalt(II) nitrate

(c) manganese(II) hydrogen carbonate (d) zinc(II) chromate

(e) barium sulfate (f) potassium permanganate

(g) aluminum sulfate (h) lithium chlorate

2.73 (a) Na^+ and SO_4^{2-}; therefore the formula is Na_2SO_4

(b) Ba^{2+} and PO_4^{3-}; therefore the formula is $Ba_3(PO_4)_2$

(c) Ga^{3+} and SO_4^{2-}; therefore the formula is $Ga_2(SO_4)_3$

2.74 (a) Na_2O_2 (b) $AlBr_3$ (c) $Cr_2(SO_4)_3$

2.75 (a) sodium sulfate (b) barium phosphate (c) gallium sulfate

General Problems

2.76 atomic weight = [0.205 x 69.924 amu] + [0.274 x 71.922 amu]
+ [0.078 x 72.923 amu] + [0.365 x 73.921 amu]
+ [0.078 x 75.921 amu] = 72.6 amu

2.77 (a) sodium bromate (b) phosphoric acid

(c) phosphorous acid (d) vanadium(V) oxide

2.78 (a) $Ca(HSO_4)_2$ (b) SnO (c) $Ru(NO_3)_3$

(d) $(NH_4)_2CO_3$ (e) HI (f) $Be_3(PO_4)_2$

2.79 For NH_3, $(2.34\text{ g N})\left(\frac{3 \times 1.0079\text{ amu H}}{14.0067\text{ amu N}}\right) = 0.505\text{ g H}.$

For N_2H_4, $(2.34\text{ g N})\left(\frac{4 \times 1.0079\text{ amu H}}{2 \times 14.0067\text{ amu N}}\right) = 0.337\text{ g H}.$

2.80 $\text{g N} = (1.575\text{ g H})\left(\frac{3.670\text{ g N}}{0.5275\text{ g H}}\right) = 10.96\text{ g N}$

From Problem 2.79:

for NH_3, $\frac{\text{g N}}{\text{g H}} = \frac{2.34\text{ g N}}{0.505\text{ g H}} = 4.63$

for N_2H_4, $\frac{\text{g N}}{\text{g H}} = \frac{2.34\text{ g N}}{0.337\text{ g H}} = 6.94$

for compound X, $\frac{\text{g N}}{\text{g H}} = \frac{10.96\text{ g N}}{1.575\text{ g H}} = 6.96$; X is N_2H_4,

2.81 TeO_4^{2-}, tellurate; TeO_3^{2-}, tellurite.

TeO_4^{2-} and TeO_3^{2-} are analogous to SO_4^{2-} and SO_3^{2-}.

2.82 H_2TeO_4, telluric acid; H_2TeO_3, tellurous acid

2.83 (a) I^- (b) Au^{3+} (c) Kr

2.84 $\frac{12.0000\text{ amu}}{15.9994\text{ amu}} = \frac{X}{16.0000\text{ amu}}$; X = 12.0005 amu for ^{12}C prior to 1961.

2.85 $\frac{39.9626\text{ amu}}{15.9994\text{ amu}} = \frac{X}{16.0000\text{ amu}}$; X = 39.9641 amu for ^{40}Ca prior to 1961.

Chapter 3 – Formulas, Equations, and Moles

3.1 $2\ KClO_3 \rightarrow 2\ KCl + 3\ O_2$

3.2 (a) $C_6H_{12}O_6 \rightarrow 2\ C_2H_6O + 2\ CO_2$

(b) $4\ Fe + 3\ O_2 \rightarrow 2\ Fe_2O_3$

(c) $4\ NH_3 + Cl_2 \rightarrow N_2H_4 + 2\ NH_4Cl$

3.3 (a) Fe_2O_3: 2(55.85) + 3(16.00) = 159.7 amu

(b) H_2SO_4: 2(1.01) + 1(32.07) + 4(16.00) = 98.1 amu

(c) $C_6H_8O_7$: 6(12.01) + 8(1.01) + 7(16.00) = 192.1 amu

(d) $C_{16}H_{18}N_2O_4S$:

16(12.01) + 18(1.01) + 2(14.01) + 4(16.00) + 1(32.07) = 334.4 amu

3.4 $Fe_2O_3(s) + 3\ CO(g) \rightarrow 2\ Fe(s) + 3\ CO_2(g)$

$$0.5\text{ mol } Fe_2O_3 \times \frac{3\text{ mol CO}}{1\text{ mol } Fe_2O_3} = 1.5\text{ mol CO}$$

3.5 Mass of 1 HCl molecule

$$= \left(36.5\ \frac{\text{amu}}{\text{molecule}}\right)\left(1.6605 \times 10^{-24}\ \frac{\text{g}}{\text{amu}}\right) = 6.06 \times 10^{-23}\text{ g/molecule}$$

$$\text{Avogadro's Number} = \left(\frac{36.5\text{ g/mol}}{6.06 \times 10^{-23}\text{ g/molecule}}\right) = 6.02 \times 10^{23}\text{ molecules/mole}$$

3.6 $C_9H_8O_4$, 180.2 amu; 500 mg = 500×10^{-3} g = 0.500 g

$$0.500\text{ g} \times \frac{1\text{ mol}}{180.2\text{ g}} = 2.77 \times 10^{-3}\text{ mol aspirin}$$

$$2.77 \times 10^{-3}\text{ mol} \times \frac{6.02 \times 10^{23}\text{ molecules}}{1\text{ mol}} = 1.67 \times 10^{21}\text{ aspirin molecules}$$

3.7 salicylic acid, $C_7H_6O_3$, 138.1 amu; acetic anhydride, $C_4H_6O_3$, 102.1 amu

aspirin, $C_9H_8O_4$, 180.2 amu; acetic acid, $C_2H_4O_2$, 60.1 amu

$$4.5\text{ g } C_7H_6O_3 \times \frac{1\text{ mol } C_7H_6O_3}{138.1\text{ g } C_7H_6O_3} \times \frac{1\text{ mol } C_4H_6O_3}{1\text{ mol } C_7H_6O_3} \times \frac{102.1\text{ g } C_4H_6O_3}{1\text{ mol } C_4H_6O_3} = 3.3\text{ g } C_4H_6O_3$$

$$4.5\text{ g } C_7H_6O_3 \times \frac{1\text{ mol } C_7H_6O_3}{138.1\text{ g } C_7H_6O_3} \times \frac{1\text{ mol } C_9H_8O_4}{1\text{ mol } C_7H_6O_3} \times \frac{180.2\text{ g } C_9H_8O_4}{1\text{ mol } C_9H_8O_4} = 5.9\text{ g } C_9H_8O_4$$

$$4.5\text{ g } C_7H_6O_3 \times \frac{1\text{ mol } C_7H_6O_3}{138.1\text{ g } C_7H_6O_3} \times \frac{1\text{ mol } C_2H_4O_2}{1\text{ mol } C_7H_6O_3} \times \frac{60.1\text{ g } C_2H_4O_2}{1\text{ mol } C_2H_4O_2} = 2.0\text{ g } C_2H_4O_2$$

3.8 C_2H_4, 28.1 amu; C_2H_6O, 46.1 amu

$$4.6\text{ g } C_2H_4 \times \frac{1\text{ mol } C_2H_4}{28.1\text{ g } C_2H_4} \times \frac{1\text{ mol } C_2H_6O}{1\text{ mol } C_2H_4} \times \frac{46.1\text{ g } C_2H_6O}{1\text{ mol } C_2H_6O} = 7.5\text{ g } C_2H_6O$$

(theoretical yield)

$$\text{Percent yield} = \frac{\text{Actual yield}}{\text{Theoretical yield}} \times 100\% = \frac{4.7\text{ g}}{7.5\text{ g}} \times 100\% = 63\%$$

3.9 CH_4, 16.04 amu; CH_2Cl_2, 84.93 amu; 1.85 kg = 1850 g

$$1850\text{ g } CH_4 \times \frac{1\text{ mol } CH_4}{16.04\text{ g } CH_4} \times \frac{1\text{ mol } CH_2Cl_2}{1\text{ mol } CH_4} \times \frac{84.93\text{ g } CH_2Cl_2}{1\text{ mol } CH_2Cl_2} = 9800\text{ g } CH_2Cl_2$$

(theoretical yield)

Actual yield = (9800 g)(0.431) = 4220 g CH_2Cl_2

3.10 $C_3H_8 + 5\ O_2 \rightarrow 3\ CO_2 + 4\ H_2O$

C_3H_8, 44.1 amu; O_2, 32.0 amu

$$65\text{ g } O_2 \times \frac{1\text{ mol } O_2}{32.0\text{ g } O_2} \times \frac{1\text{ mol } C_3H_8}{5\text{ mol } O_2} \times \frac{44.1\text{ g } C_3H_8}{1\text{ mol } C_3H_8} = 18\text{ g } C_3H_8 \text{ reacted}$$

mass C_3H_8 remaining = 700 g – 18 g = 682 g C_3H_8 remaining

3.11 Li_2O, 29.9 amu: 65 kg = 65,000 g; H_2O, 18.0 amu: 80 kg = 80,000 g

$$65{,}000 \text{ g } Li_2O \times \frac{1 \text{ mol } Li_2O}{29.9 \text{ g } Li_2O} = 2.2 \times 10^3 \text{ mol } Li_2O$$

$$80{,}000 \text{ g } H_2O \times \frac{1 \text{ mol } H_2O}{18.0 \text{ g } H_2O} = 4.4 \times 10^3 \text{ mol } H_2O$$

The reaction stoichiometry between Li_2O and H_2O is one to one. There are twice as many moles of H_2O as there are moles of Li_2O. Therefore, Li_2O is the limiting reactant.

$(4.4 \times 10^3 \text{ mol} - 2.2 \times 10^3 \text{ mol}) = 2.2 \times 10^3 \text{ mol } H_2O$ remaining

$$2.2 \times 10^3 \text{ mol } H_2O \times \frac{18.0 \text{ g } H_2O}{1 \text{ mol } H_2O} = 40{,}000 \text{ g } H_2O$$

40,000 g H_2O = 40 kg H_2O

3.12 LiOH, 23.9 amu; CO_2, 44.0 amu

$$500 \text{ g LiOH} \times \frac{1 \text{ mol LiOH}}{23.9 \text{ g LiOH}} \times \frac{1 \text{ mol } CO_2}{1 \text{ mol LiOH}} \times \frac{44.0 \text{ g } CO_2}{1 \text{ mol } CO_2} = 921 \text{ g } CO_2$$

3.13 (a) 125 mL = 0.125 L; (0.20 mol/L)(0.125 L) = 0.025 mol $NaHCO_3$

(b) 650 mL = 0.650 L; (2.50 mol/L)(0.650 L) = 1.62 mol H_2SO_4

3.14 (a) NaOH, 40.0 amu; 500 mL = 0.500 L

$$1.25 \frac{\text{mol NaOH}}{\text{L}} \times 0.500 \text{ L} \times \frac{40.0 \text{ g NaOH}}{1 \text{ mol NaOH}} = 25.0 \text{ g NaOH}$$

(b) $C_6H_{12}O_6$, 180.2 amu

$$0.250 \frac{\text{mol } C_6H_{12}O_6}{\text{L}} \times 1.50 \text{ L} \times \frac{180.2 \text{ g } C_6H_{12}O_6}{1 \text{ mol } C_6H_{12}O_6} = 67.6 \text{ g } C_6H_{12}O_6$$

3.15 $C_6H_{12}O_6$, 180.2 amu;

$$25.0 \text{ g } C_6H_{12}O_6 \text{ x } \frac{1 \text{ mol } C_6H_{12}O_6}{180.2 \text{ g } C_6H_{12}O_6} = 0.1387 \text{ mol } C_6H_{12}O_6$$

$$0.1387 \text{ mol x } \frac{1 \text{ L}}{0.20 \text{ mol}} = 0.69 \text{ L}; \;\; 0.69 \text{ L} = 690 \text{ mL}$$

3.16 $C_{27}H_{46}O$, 386.7 amu; 750 mL = 0.750 L

$$0.005 \frac{\text{mol } C_{27}H_{46}O}{\text{L}} \text{ x } 0.750 \text{ L x } \frac{386.7 \text{ g } C_{27}H_{46}O}{1 \text{ mol } C_{27}H_{46}O} = 1 \text{ g } C_{27}H_{46}O$$

3.17 $M_i \text{ x } V_i = M_f \text{ x } V_f$; $\quad M_f = \frac{M_i \text{ x } V_i}{V_f} = \frac{3.50 \text{ M x } 75.0 \text{ mL}}{400.0 \text{ mL}} = 0.656 \text{ M}$

3.18 $M_i \text{ x } V_i = M_f \text{ x } V_f$; $\quad V_i = \frac{M_f \text{ x } V_f}{M_i} = \frac{0.500 \text{ M x } 250 \text{ mL}}{18.0 \text{ M}} = 6.94 \text{ mL}$

Dilute 6.94 mL of 18.0 M H_2SO_4 with enough water to make 250 mL of solution. The resulting solution will be 0.500 M H_2SO_4.

3.19 50.0 mL = 0.0500 L; (0.100 mol/L)(0.0500 L) = 5.00×10^{-3} mol NaOH

$$5.00 \text{ x } 10^{-3} \text{ mol NaOH x } \frac{1 \text{ mol } H_2SO_4}{2 \text{ mol NaOH}} = 2.50 \text{ x } 10^{-3} \text{ mol } H_2SO_4$$

$$\text{volume} = 2.50 \text{ x } 10^{-3} \text{ mol x } \frac{1 \text{ L}}{0.250 \text{ mol}} = 0.0100 \text{ L}; \;\; 0.0100 \text{ L} = 10.0 \text{ mL } H_2SO_4$$

3.20 $HNO_3(aq) + KOH(aq) \rightarrow KNO_3(aq) + H_2O(l)$

25.0 mL = 0.0250 L and 68.5 mL = 0.0685 L

$$0.150 \frac{\text{mol KOH}}{\text{L}} \text{ x } 0.0250 \text{ L x } \frac{1 \text{ mol } HNO_3}{1 \text{ mol KOH}} = 3.75 \text{ x } 10^{-3} \text{ mol } HNO_3$$

$$HNO_3 \text{ molarity} = \frac{3.75 \times 10^{-3}\text{ mol}}{0.0685\text{ L}} = 5.47 \times 10^{-2}\text{ M}$$

3.21 From the reaction stoichiometry, moles NaOH = moles $HC_2H_3O_2$

(0.200 mol/L)(0.0947 L) = 0.018 94 mol NaOH = 0.018 94 mol $HC_2H_3O_2$

$$\text{molarity} = \frac{0.018\ 94\text{ mol}}{0.0250\text{ L}} = 0.758\text{ M}$$

3.22 For dimethylhydrazine, $C_2H_8N_2$, divide each subscript by 2 to obtain the empirical formula. The empirical formula is CH_4N.

$C_2H_8N_2$, 60.1 amu or 60.1 g/mol

$$\%C = \frac{2 \times 12.0\text{ g}}{60.1\text{ g}} \times 100\% = 39.9\%$$

$$\%H = \frac{8 \times 1.01\text{ g}}{60.1\text{ g}} \times 100\% = 13.4\%$$

$$\%N = \frac{2 \times 14.0\text{ g}}{60.1\text{ g}} \times 100\% = 46.6\%$$

3.23 Assume a 100.0 g sample. From the percent composition data, a 100.0 g sample contains 14.25 g C, 56.93 g O, and 28.83 g Mg.

$$14.25\text{ g C} \times \frac{1\text{ mol C}}{12.0\text{ g C}} = 1.19\text{ mol C}$$

$$56.93\text{ g O} \times \frac{1\text{ mol O}}{16.0\text{ g O}} = 3.56\text{ mol O}$$

$$28.83\text{ g Mg} \times \frac{1\text{ mol Mg}}{24.3\text{ g Mg}} = 1.19\text{ mol Mg}$$

$Mg_{1.19}C_{1.19}O_{3.56}$; divide each subscript by the smallest, 1.19.

$Mg_{1.19\,/\,1.19}C_{1.19\,/\,1.19}O_{3.56\,/\,1.19}$

The empirical formula is $MgCO_3$.

3.24 $1.174 \text{ g } H_2O \times \frac{1 \text{ mol } H_2O}{18.0 \text{ g } H_2O} \times \frac{2 \text{ mol H}}{1 \text{ mol } H_2O} = 0.130 \text{ mol H}$

$1.910 \text{ g } CO_2 \times \frac{1 \text{ mol } CO_2}{44.0 \text{ g } CO_2} \times \frac{1 \text{ mol C}}{1 \text{ mol } CO_2} = 0.0434 \text{ mol C}$

$0.130 \text{ mol H} \times \frac{1.01 \text{ g H}}{1 \text{ mol H}} = 0.131 \text{ g H}$

$0.0434 \text{ mol C} \times \frac{12.0 \text{ g C}}{1 \text{ mol C}} = 0.521 \text{ g C}$

1.00 g – (0.131 g + 0.521 g) = 0.35 g O

moles O = $0.35 \text{ g O} \times \frac{1 \text{ mol O}}{16.0 \text{ g O}} = 0.022 \text{ mol O}$

$C_{0.0434}H_{0.130}O_{0.022}$; divide each subscript by the smallest, 0.022.

$C_{0.0434 / 0.022}H_{0.130 / 0.022}O_{0.022 / 0.022}$

$C_{1.97}H_{5.91}O$

The empirical formula is C_2H_6O.

3.25 (a) empirical formula is CH, 13 amu; molecular weight = 78 amu

78 amu/13 amu = 6; molecular formula = $C_{(6 \times 1)}H_{(6 \times 1)} = C_6H_6$

(b) empirical formula is CH, 13 amu; molecular weight = 26 amu

26 amu/13 amu = 2; molecular formula = $C_{(2 \times 1)}H_{(2 \times 1)} = C_2H_2$

(c) empirical formula is CH_2O, 30 amu; molecular weight = 60 amu

60 amu/30 amu = 2; molecular formula = $C_{(2 \times 1)}H_{(2 \times 2)}O_{(2 \times 1)} = C_2H_4O_2$

3.26 (a) Assume a 100.0 g sample. From the percent composition data, a 100.0 g sample contains 21.86 g H and 78.14 g B.

$$21.86 \text{ g H} \times \frac{1 \text{ mol H}}{1.01 \text{ g H}} = 21.6 \text{ mol H}$$

$$78.14 \text{ g B} \times \frac{1 \text{ mol B}}{10.8 \text{ g B}} = 7.24 \text{ mol B}$$

$B_{7.24}H_{21.6}$; divide each subscript by the smaller, 7.24.

$B_{7.24 / 7.24}H_{21.6 / 7.24}$

The empirical formula is BH_3, 13.8 amu.

27.7 amu / 13.8 amu = 2; molecular formula = $B_{(2 \times 1)}H_{(2 \times 3)} = B_2H_6$.

(b) Assume a 100.0 g sample. From the percent composition data, a 100.0 g sample contains 6.71 g H, 40.00 g C, and 53.28 g O.

$$6.71 \text{ g H} \times \frac{1 \text{ mol H}}{1.01 \text{ g H}} = 6.64 \text{ mol H}$$

$$40.00 \text{ g C} \times \frac{1 \text{ mol C}}{12.0 \text{ g C}} = 3.33 \text{ mol C}$$

$$53.28 \text{ g O} \times \frac{1 \text{ mol O}}{16.0 \text{ g O}} = 3.33 \text{ mol O}$$

$C_{3.33}H_{6.64}O_{3.33}$; divide each subscript by the smallest, 3.33.

$C_{3.33 / 3.33}H_{6.64 / 3.33}O_{3.33 / 3.33}$

The empirical formula is CH_2O, 30.0 amu

90.08 amu / 30.0 amu = 3; molecular formula = $C_{(3 \times 1)}H_{(3 \times 2)}O_{(3 \times 1)} = C_3H_6O_3$

Understanding Key Concepts

1. The concentration of a solution is cut in half when the volume is doubled. This is best represented by box (b).

2. (c) $2\text{ A} + B_2 \rightarrow A_2B_2$

3. $C_2H_4 + 3\ O_2 \rightarrow 2\ CO_2 + 2\ H_2O$

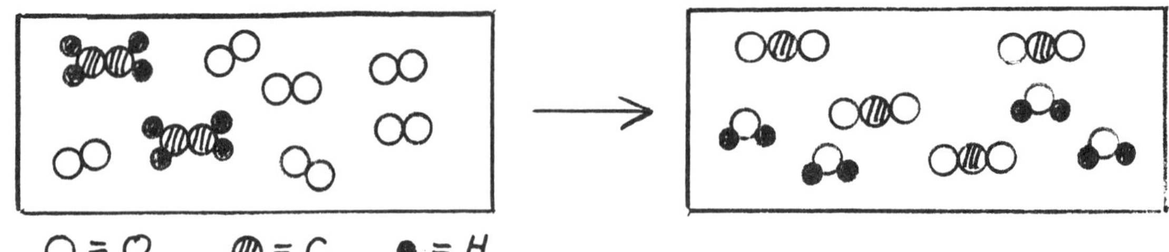

4. Reactants, box (d) and products, box (c).

5. $C_5H_{11}NO_2S$

   5(12.01) + 11(1.01) + 1(14.01) + 2(16.00) + 1(32.07) = 149.2 amu

ADDITIONAL PROBLEMS — Balancing Equations

3.27 (a) not balanced (b) balanced (c) not balanced

3.28 (a) $2\ Al + Fe_2O_3 \rightarrow Al_2O_3 + 2\ Fe$

(c) $4\ Au + 8\ NaCN + O_2 + 2\ H_2O \rightarrow 4\ NaAu(CN)_2 + 4\ NaOH$

3.29 Adding a subscript of "2" to the oxygen in H_2O would change water to a different compound, hydrogen peroxide. This is not allowed.

3.30 (a) $Mg + 2\ HNO_3 \rightarrow H_2 + Mg(NO_3)_2$

(b) $CaC_2 + 2\ H_2O \rightarrow Ca(OH)_2 + C_2H_2$

(c) $3\ O_2 + 2\ S \rightarrow 2\ SO_3$

(d) $UO_2 + 4\ HF \rightarrow UF_4 + 2\ H_2O$

3.31 (a) $2\ NH_4NO_3 \rightarrow 2\ N_2 + O_2 + 4\ H_2O$

(b) $C_2H_6O + O_2 \rightarrow C_2H_4O_2 + H_2O$

(c) $C_2H_8N_2 + 2\ N_2O_4 \rightarrow 3\ N_2 + 2\ CO_2 + 4\ H_2C$

Molecular Weights and Moles

3.32 Hg_2Cl_2: 2(200.59) + 2(35.45) = 472.1 amu

$C_4H_8O_2$: 4(12.01) + 8(1.01) + 2(16.00) = 88.1 amu

CF_2Cl_2: 1(12.01) + 2(19.00) + 2(35.45) = 120.9 amu

3.33 One mole equals the atomic weight or molecular weight in grams.

(a) Ti, 47.88 g (b) Br_2, 159.81 g (c) Hg, 200.59 g (d) H_2O, 18.02 g

3.34 (a) $1.00 \text{ g Cr} \times \dfrac{1 \text{ mol Cr}}{52.0 \text{ g Cr}} = 0.0192 \text{ mol Cr}$

(b) $1.00 \text{ g } Cl_2 \times \dfrac{1 \text{ mol } Cl_2}{70.9 \text{ g } Cl_2} = 0.0141 \text{ mol } Cl_2$

(c) $1.00 \text{ g Au} \times \dfrac{1 \text{ mol Au}}{197.0 \text{ g Au}} = 0.005\ 08 \text{ mol Au}$

(d) $1.00 \text{ g } NH_3 \times \dfrac{1 \text{ mol } NH_3}{17.0 \text{ g } NH_3} = 0.0588 \text{ mol } NH_3$

3.35 There are 2 ions per each formula unit of NaCl.

(2.5 mol)(2 mol ions/mol) = 5.0 mol ions

3.36 There are 2 K^+ ions per each formula unit of K_2SO_4.

$$1.45 \text{ mol } K_2SO_4 \times \frac{2 \text{ mol } K^+}{1 \text{ mol } K_2SO_4} = 2.90 \text{ mol } K^+$$

3.37 There are 3 F^- anions per each formula unit of AlF_3.

$$35.6 \text{ g } AlF_3 \times \frac{1 \text{ mol } AlF_3}{84.0 \text{ g } AlF_3} \times \frac{3 \text{ mol anions}}{1 \text{ mol } AlF_3} = 1.27 \text{ mol } F^-$$

3.38 Molar mass = $\frac{3.28\text{ g}}{0.0275\text{ mol}}$ = 119 g/mol

Molecular weight = 119 amu

3.39 Molar mass = $\frac{221.6\text{ g}}{0.5731\text{ mol}}$ = 386.7 g/mol; The molecular weight is 386.7 amu.

3.40 $FeSO_4$, 151.9 amu; 300 mg = 0.300 g

$$0.300\text{ g }FeSO_4 \times \frac{1\text{ mol }FeSO_4}{151.9\text{ g }FeSO_4} = 1.97 \times 10^{-3}\text{ mol }FeSO_4$$

$$1.97 \times 10^{-3}\text{ mol }FeSO_4 \times \frac{6.02 \times 10^{23}\text{ Fe atoms}}{1\text{ mol }FeSO_4} = 1.19 \times 10^{21}\text{ Fe atoms}$$

3.41 $0.0001\text{ g C} \times \frac{1\text{ mol C}}{12.0\text{ g C}} \times \frac{6.02 \times 10^{23}\text{ C atoms}}{1\text{ mol C}} = 5 \times 10^{18}\text{ C atoms}$

3.42 (a) $Be_3Al_2Si_6O_{18}$

3(9.01) + 2(26.98) + 6(28.09) + 18(16.00) = 537.5 amu

(b) $C_{17}H_{19}NO_3$

17(12.01) + 19(1.01) + 1(14.01) + 3(16.00) = 285.4 amu

(c) $Na_2Cr_2O_7$

2(22.99) + 2(52.00) + 7(16.00) = 262.0 amu

(d) $NaC_5H_8NO_4$

1(22.99) + 5(12.01) + 8(1.01) + 1(14.01) + 4(16.00) = 169.1 amu

3.43 $C_8H_{10}N_4O_2$, 194.2 amu; 125 mg = 0.125 g

$$0.125\text{ g caffeine} \times \frac{1\text{ mol caffeine}}{194.2\text{ g caffeine}} = 6.44 \times 10^{-4}\text{ mol caffeine}$$

$$0.125 \text{ g caffeine} \times \frac{1 \text{ mol caffeine}}{194.2 \text{ g caffeine}} \times \frac{6.022 \times 10^{23} \text{ molecules}}{1 \text{ mol}} =$$

3.88×10^{20} caffeine molecules

3.44 $$45 \frac{\text{g}}{\text{egg}} \times \frac{6.02 \times 10^{23} \text{ eggs}}{1 \text{ mol eggs}} = 2.7 \times 10^{25} \text{ g/mol of eggs}$$

3.45 (a) $$1.0 \text{ g Li} \times \frac{1 \text{ mol Li}}{6.94 \text{ g Li}} = 0.14 \text{ mol Li}$$

(b) $$1.0 \text{ g Au} \times \frac{1 \text{ mol Au}}{197.0 \text{ g Au}} = 0.0051 \text{ mol Au}$$

(c) penicillin G: $C_{16}H_{17}N_2O_4SK$, 372.5 amu

$$1.0 \text{ g} \times \frac{1 \text{ mol penicillin G}}{372.5 \text{ g penicillin G}} = 2.7 \times 10^{-3} \text{ mol penicillin G}$$

3.46 (a) $$0.0015 \text{ mol Na} \times \frac{23.0 \text{ g Na}}{1 \text{ mol Na}} = 0.034 \text{ g Na}$$

(b) $$0.0015 \text{ mol Pb} \times \frac{207.2 \text{ g Pb}}{1 \text{ mol Pb}} = 0.31 \text{ g Pb}$$

(c) $C_{16}H_{13}ClN_2O$, 284.7 amu

$$0.0015 \text{ mol diazepam} \times \frac{284.7 \text{ g diazepam}}{1 \text{ mol diazepam}} = 0.43 \text{ g diazepam}$$

Stoichiometry Calculations

3.47 TiO_2, 79.88 amu

$$100 \text{ kg Ti} \times \frac{79.88 \text{ kg } TiO_2}{47.88 \text{ kg Ti}} = 167 \text{ kg } TiO_2$$

3.48 Fe_2O_3, 159.7 amu

$$\% \text{ Fe} = \frac{2(55.85 \text{ g}) \text{ Fe}}{159.7 \text{ g } Fe_2O_3} \times 100\% = 69.94\%$$

mass Fe = (0.6994)(105 kg) = 73.4 kg

3.49 (a) $2\ Fe_2O_3 + 3\ C \rightarrow 4\ Fe + 3\ CO_2$

(b) Fe_2O_3, 159.7 amu

$$525 \text{ g } Fe_2O_3 \times \frac{1 \text{ mol } Fe_2O_3}{159.7 \text{ g } Fe_2O_3} \times \frac{3 \text{ mol C}}{2 \text{ mol } Fe_2O_3} = 4.93 \text{ mol C}$$

(c) $4.93 \text{ mol C} \times \frac{12.01 \text{ g C}}{1 \text{ mol C}} = 59.2 \text{ g C}$

3.50 (a) $Fe_2O_3 + 3\ CO \rightarrow 2\ Fe + 3\ CO_2$

(b) Fe_2O_3, 159.7 amu; CO, 28.01 amu

$$3.02 \text{ g } Fe_2O_3 \times \frac{1 \text{ mol } Fe_2O_3}{159.7 \text{ g } Fe_2O_3} \times \frac{3 \text{ mol CO}}{1 \text{ mol } Fe_2O_3} \times \frac{28.01 \text{ g CO}}{1 \text{ mol CO}} = 1.59 \text{ g CO}$$

(c) $1.68 \text{ mol } Fe_2O_3 \times \frac{3 \text{ mol CO}}{1 \text{ mol } Fe_2O_3} \times \frac{28.01 \text{ g CO}}{1 \text{ mol CO}} = 141 \text{ g CO}$

3.51 (a) $2\ Mg + O_2 \rightarrow 2\ MgO$

(b) Mg, 24.30 amu; O_2, 32.00 amu; MgO, 40.30 amu

$$25.0 \text{ g Mg} \times \frac{1 \text{ mol Mg}}{24.30 \text{ g Mg}} \times \frac{1 \text{ mol } O_2}{2 \text{ mol Mg}} \times \frac{32.00 \text{ g } O_2}{1 \text{ mol } O_2} = 16.5 \text{ g } O_2$$

$$25.0 \text{ g Mg} \times \frac{1 \text{ mol Mg}}{24.30 \text{ g Mg}} \times \frac{2 \text{ mol MgO}}{2 \text{ mol Mg}} \times \frac{40.30 \text{ g MgO}}{1 \text{ mol MgO}} = 41.5 \text{ g MgO}$$

(c) $25.0\ g\ O_2 \times \frac{1\ mol\ O_2}{32.00\ g\ O_2} \times \frac{2\ mol\ Mg}{1\ mol\ O_2} \times \frac{24.30\ g\ Mg}{1\ mol\ Mg} = 38.0\ g\ Mg$

$25.0\ g\ O_2 \times \frac{1\ mol\ O_2}{32.00\ g\ O_2} \times \frac{2\ mol\ MgO}{1\ mol\ O_2} \times \frac{40.30\ g\ MgO}{1\ mol\ MgO} = 63.0\ g\ MgO$

3.52 $C_2H_4 + H_2O \rightarrow C_2H_6O$

C_2H_4, 28.05 amu; H_2O, 18.02 amu; C_2H_6O, 46.07 amu

(a) $0.133\ mol\ H_2O \times \frac{1\ mol\ C_2H_4}{1\ mol\ H_2O} \times \frac{28.05\ g\ C_2H_4}{1\ mol\ C_2H_4} = 3.73\ g\ C_2H_4$

$0.133\ mol\ H_2O \times \frac{1\ mol\ C_2H_6O}{1\ mol\ H_2O} \times \frac{46.07\ g\ C_2H_6O}{1\ mol\ C_2H_6O} = 6.13\ g\ C_2H_6O$

(b) $0.371\ mol\ C_2H_4 \times \frac{1\ mol\ H_2O}{1\ mol\ C_2H_4} \times \frac{18.02\ g\ H_2O}{1\ mol\ H_2O} = 6.69\ g\ H_2O$

$0.371\ mol\ C_2H_4 \times \frac{1\ mol\ C_2H_6O}{1\ mol\ C_2H_4} \times \frac{46.07\ g\ C_2H_6O}{1\ mol\ C_2H_6O} = 17.1\ g\ C_2H_6O$

3.53 (a) $2\ HgO \rightarrow 2\ Hg + O_2$

(b) HgO, 216.6 amu; Hg, 200.6 amu; O_2, 32.0 amu

$45.5\ g\ HgO \times \frac{1\ mol\ HgO}{216.6\ g\ HgO} \times \frac{2\ mol\ Hg}{2\ mol\ HgO} \times \frac{200.6\ g\ Hg}{1\ mol\ Hg} = 42.1\ g\ Hg$

$45.5\ g\ HgO \times \frac{1\ mol\ HgO}{216.6\ g\ HgO} \times \frac{1\ mol\ O_2}{2\ mol\ HgO} \times \frac{32.00\ g\ O_2}{1\ mol\ O_2} = 3.36\ g\ O_2$

(c) $33.3\ g\ O_2 \times \frac{1\ mol\ O_2}{32.00\ g\ O_2} \times \frac{2\ mol\ HgO}{1\ mol\ O_2} \times \frac{216.6\ g\ HgO}{1\ mol\ HgO} = 451\ g\ HgO$

3.54 $2.00 \text{ g Ag} \times \dfrac{1 \text{ mol Ag}}{107.9 \text{ g Ag}} = 0.0185 \text{ mol Ag}$

$0.657 \text{ g Cl} \times \dfrac{1 \text{ mol Cl}}{35.45 \text{ g Cl}} = 0.0185 \text{ mol Cl}$

$Ag_{0.0185}Cl_{0.0185}$

Both subscripts are the same. The empirical formula is AgCl.

3.55 $5.0 \text{ g Al} \times \dfrac{1 \text{ mol Al}}{27.0 \text{ g Al}} = 0.19 \text{ mol Al};$ $4.45 \text{ g O} \times \dfrac{1 \text{ mol O}}{16.0 \text{ g O}} = 0.28 \text{ mol O}$

$Al_{0.19}O_{0.28}$; divide both subscripts by the smaller, 0.19.

$Al_{0.19\,/\,0.19}O_{0.28\,/\,0.19}$

$Al_1O_{1.5}$; multiply both subscripts by 2 to obtain integers.

The empirical formula is Al_2O_3.

3.56 5.60 kg = 5600 g; $TiCl_4$, 189.7 amu; TiO_2, 79.88 amu

$5600 \text{ g } TiCl_4 \times \dfrac{1 \text{ mol } TiCl_4}{189.7 \text{ g } TiCl_4} \times \dfrac{1 \text{ mol } TiO_2}{1 \text{ mol } TiCl_4} \times \dfrac{79.88 \text{ g } TiO_2}{1 \text{ mol } TiO_2}$

$= 2358 \text{ g } TiO_2 = 2.36 \text{ kg } TiO_2$

Limiting Reactants and Reaction Yield

3.57 $3.44 \text{ mol } N_2 \times \dfrac{3 \text{ mol } H_2}{1 \text{ mol } N_2} = 10.3 \text{ mol } H_2$ required.

Since there is only 1.39 mol H_2, H_2 is the limiting reactant.

$1.39 \text{ mol } H_2 \times \dfrac{2 \text{ mol } NH_3}{3 \text{ mol } H_2} \times \dfrac{17.03 \text{ g } NH_3}{1 \text{ mol } NH_3} = 15.8 \text{ g } NH_3$

$$1.39 \text{ mol } H_2 \times \frac{1 \text{ mol } N_2}{3 \text{ mol } H_2} \times \frac{28.01 \text{ g } N_2}{1 \text{ mol } N_2} = 13.0 \text{ g } N_2 \text{ reacted}$$

$$3.44 \text{ mol } N_2 \times \frac{28.01 \text{ g } N_2}{1 \text{ mol } N_2} = 96.3 \text{ g } N_2 \text{ initially}$$

(96.3 g – 13.0 g) = 83.3 g N_2 left over

3.58 H_2, 2.016 amu; Cl_2, 70.91 amu; HCl 36.46 amu

$$3.56 \text{ g } H_2 \times \frac{1 \text{ mol } H_2}{2.016 \text{ g } H_2} = 1.77 \text{ mol } H_2$$

$$8.94 \text{ g } Cl_2 \times \frac{1 \text{ mol } Cl_2}{70.91 \text{ g } Cl_2} = 0.126 \text{ mol } Cl_2$$

Since the reaction stoichiometry between H_2 and Cl_2 is one to one, Cl_2 is the limiting reactant.

$$0.126 \text{ mol } Cl_2 \times \frac{2 \text{ mol HCl}}{1 \text{ mol } Cl_2} \times \frac{36.46 \text{ g HCl}}{1 \text{ mol HCl}} = 9.19 \text{ g HCl}$$

3.59 C_2H_4, 28.05 amu; Cl_2, 70.91 amu; $C_2H_4Cl_2$, 98.96 amu

$$15.4 \text{ g } C_2H_4 \times \frac{1 \text{ mol } C_2H_4}{28.05 \text{ g } C_2H_4} = 0.549 \text{ mol } C_2H_4$$

$$3.74 \text{ g } Cl_2 \times \frac{1 \text{ mol } Cl_2}{70.91 \text{ g } Cl_2} = 0.0527 \text{ mol } Cl_2$$

Since the reaction stoichiometry between C_2H_4 and Cl_2 is one to one, Cl_2 is the limiting reactant.

$$0.0527 \text{ mol } Cl_2 \times \frac{1 \text{ mol } C_2H_4Cl_2}{1 \text{ mol } Cl_2} \times \frac{98.96 \text{ g } C_2H_4Cl_2}{1 \text{ mol } C_2H_4Cl_2} = 5.22 \text{ g } C_2H_4Cl_2$$

3.60 (a) NaCl, 58.44 amu; $AgNO_3$, 169.9 amu; AgCl, 143.3 amu; $NaNO_3$, 85.00 amu

$NaCl + AgNO_3 \rightarrow AgCl + NaNO_3$

$$1.3 \text{ g NaCl} \times \frac{1 \text{ mol NaCl}}{58.44 \text{ g NaCl}} = 0.0222 \text{ mol NaCl}$$

$$3.5 \text{ g AgNO}_3 \times \frac{1 \text{ mol AgNO}_3}{169.9 \text{ g AgNO}_3} = 0.0206 \text{ mol AgNO}_3$$

Since the reaction stoichiometry between NaCl and $AgNO_3$ is one to one, $AgNO_3$ is the limiting reactant.

$$0.0206 \text{ mol AgNO}_3 \times \frac{1 \text{ mol AgCl}}{1 \text{ mol AgNO}_3} \times \frac{143.3 \text{ g AgCl}}{1 \text{ mol AgCl}} = 3.0 \text{ g AgCl}$$

$$0.0206 \text{ mol AgNO}_3 \times \frac{1 \text{ mol NaNO}_3}{1 \text{ mol AgNO}_3} \times \frac{85.00 \text{ g NaNO}_3}{1 \text{ mol NaNO}_3} = 1.8 \text{ g NaNO}_3$$

(b) $BaCl_2$, 208.2 amu; H_2SO_4, 98.08 amu; $BaSO_4$, 233.4 amu; HCl, 36.46 amu

$BaCl_2 + H_2SO_4 \rightarrow BaSO_4 + 2\ HCl$

$$2.65 \text{ g BaCl}_2 \times \frac{1 \text{ mol BaCl}_2}{208.2 \text{ g BaCl}_2} = 0.0127 \text{ mol BaCl}_2$$

$$6.78 \text{ g H}_2\text{SO}_4 \times \frac{1 \text{ mol H}_2\text{SO}_4}{98.08 \text{ g H}_2\text{SO}_4} = 0.0691 \text{ mol H}_2\text{SO}_4$$

Since the reaction stoichiometry between $BaCl_2$ and H_2SO_4 is one to one, $BaCl_2$ is the limiting reactant.

$$0.0127 \text{ mol BaCl}_2 \times \frac{1 \text{ mol BaSO}_4}{1 \text{ mol BaCl}_2} \times \frac{233.4 \text{ g BaSO}_4}{1 \text{ mol BaSO}_4} = 2.96 \text{ g BaSO}_4$$

$$0.0127 \text{ mol BaCl}_2 \times \frac{2 \text{ mol HCl}}{1 \text{ mol BaCl}_2} \times \frac{36.46 \text{ g HCl}}{1 \text{ mol HCl}} = 0.926 \text{ g HCl}$$

3.61 Results from Problem 3.60 are used here.

(a) $0.0206 \text{ mol } AgNO_3 \times \frac{1 \text{ mol NaCl}}{1 \text{ mol } AgNO_3} \times \frac{58.44 \text{ g NaCl}}{1 \text{ mol NaCl}} = 1.2 \text{ g NaCl reacted}$

(1.3 g – 1.2 g) = 0.1 g NaCl left over

(b) $0.0127 \text{ mol } BaCl_2 \times \frac{1 \text{ mol } H_2SO_4}{1 \text{ mol } BaCl_2} \times \frac{98.1 \text{ g } H_2SO_4}{1 \text{ mol } H_2SO_4} = 1.25 \text{ g } H_2SO_4 \text{ reacted}$

(6.78 g – 1.25 g) = 5.53 g H_2SO_4 left over

3.62 $CaCO_3$, 100.1 amu; HCl, 36.46 amu

$$CaCO_3 + 2\,HCl \rightarrow CaCl_2 + H_2O + CO_2$$

$$2.35 \text{ g } CaCO_3 \times \frac{1 \text{ mol } CaCO_3}{100.1 \text{ g } CaCO_3} = 0.0235 \text{ mol } CaCO_3$$

$$2.35 \text{ g HCl} \times \frac{1 \text{ mol HCl}}{36.46 \text{ g HCl}} = 0.0645 \text{ mol HCl}$$

The reaction stoichiometry is 1 mole of $CaCO_3$ for every 2 moles of HCl. For 0.0235 mol $CaCO_3$, we only need 2(0.0235 mol) = 0.0470 mol HCl. We have 0.0645 mol HCl, therefore $CaCO_3$ is the limiting reactant.

$$0.0235 \text{ mol } CaCO_3 \times \frac{1 \text{ mol } CO_2}{1 \text{ mol } CaCO_3} \times \frac{22.4 \text{ L}}{1 \text{ mol } CO_2} = 0.526 \text{ L } CO_2$$

3.63 $C_2H_4O_2 + C_5H_{12}O \rightarrow C_7H_{14}O_2 + H_2O$

$C_2H_4O_2$, 60.05 amu; $C_5H_{12}O$, 88.15 amu; $C_7H_{14}O_2$, 130.19 amu

$$1.87 \text{ g } C_2H_4O_2 \times \frac{1 \text{ mol } C_2H_4O_2}{60.05 \text{ g } C_2H_4O_2} = 0.0311 \text{ mol } C_2H_4O_2$$

$$2.31 \text{ g } C_5H_{12}O \times \frac{1 \text{ mol } C_5H_{12}O}{88.15 \text{ g } C_5H_{12}O} = 0.0262 \text{ mol } C_5H_{12}O$$

Since the reaction stoichiometry between $C_2H_4O_2$ and $C_5H_{12}O$ is one to one, isopentyl alcohol ($C_5H_{12}O$) is the limiting reactant.

$$0.0262 \text{ mol } C_5H_{12}O \times \frac{1 \text{ mol } C_7H_{14}O_2}{1 \text{ mol } C_5H_{12}O} \times \frac{130.19 \text{ g } C_7H_{14}O_2}{1 \text{ mol } C_7H_{14}O_2} = 3.41 \text{ g } C_7H_{14}O_2$$

3.41 g $C_7H_{14}O_2$ is the theoretical yield.

$$\% \text{ Yield} = \frac{\text{Actual yield}}{\text{Theoretical yield}} \times 100\% = \frac{2.96 \text{ g}}{3.41 \text{ g}} \times 100\% = 86.8\%$$

3.64 $$3.58 \text{ g } C_2H_4O_2 \times \frac{1 \text{ mol } C_2H_4O_2}{60.05 \text{ g } C_2H_4O_2} = 0.0596 \text{ mol } C_2H_4O_2$$

$$4.75 \text{ g } C_5H_{12}O \times \frac{1 \text{ mol } C_5H_{12}O}{88.15 \text{ g } C_5H_{12}O} = 0.0539 \text{ mol } C_5H_{12}O$$

Since the reaction stoichiometry between $C_2H_4O_2$ and $C_5H_{12}O$ is one to one, isopentyl alcohol ($C_5H_{12}O$) is the limiting reactant.

$$0.0539 \text{ mol } C_5H_{12}O \times \frac{1 \text{ mol } C_7H_{14}O_2}{1 \text{ mol } C_5H_{12}O} \times \frac{130.19 \text{ g } C_7H_{14}O_2}{1 \text{ mol } C_7H_{14}O_2} = 7.02 \text{ g } C_7H_{14}O_2$$

7.02 g $C_7H_{14}O_2$ is the theoretical yield.

Actual yield = (7.02 g)(0.45) = 3.2 g

3.65 $K_2PtCl_4 + 2\ NH_3 \rightarrow 2\ KCl + Pt(NH_3)_2Cl_2$

K_2PtCl_4, 415.1 amu; NH_3, 17.03 amu; $Pt(NH_3)_2Cl_2$, 300.0 amu

$$55.8 \text{ g } K_2PtCl_4 \times \frac{1 \text{ mol } K_2PtCl_4}{415.1 \text{ g } K_2PtCl_4} = 0.134 \text{ mol } K_2PtCl_4$$

$$35.6 \text{ g } NH_3 \times \frac{1 \text{ mol } NH_3}{17.03 \text{ g } NH_3} = 2.09 \text{ mol } NH_3$$

Only 2(0.134) = 0.268 mol NH_3 are needed to react with 0.134 mol K_2PtCl_4. Therefore, the NH_3 is in excess and K_2PtCl_4 is the limiting reactant.

$$0.134 \text{ mol } K_2PtCl_4 \times \frac{1 \text{ mol } Pt(NH_3)_2Cl_2}{1 \text{ mol } K_2PtCl_4} \times \frac{300.0 \text{ g } Pt(NH_3)_2Cl_2}{1 \text{ mol } Pt(NH_3)_2Cl_2} = 40.2 \text{ g } Pt(NH_3)_2Cl_2$$

40.2 g $Pt(NH_3)_2Cl_2$ is the theoretical yield.

Actual yield = (40.2 g)(0.95) = 38 g $Pt(NH_3)_2Cl_2$

Molarity, Solution Stoichiometry, Dilution, and Titration

3.66 (a) 35.0 mL = 0.0350 L

$$1.200 \frac{\text{mol } HNO_3}{L} \times 0.0350 \text{ L} = 0.0420 \text{ mol } HNO_3$$

(b) 175 mL = 0.175 L

$$0.67 \frac{\text{mol } C_6H_{12}O_6}{L} \times 0.175 \text{ L} = 0.12 \text{ mol } C_6H_{12}O_6$$

3.67 (a) C_2H_6O, 46.07 amu; 250 mL = 0.250 L

$$0.600 \frac{\text{mol } C_2H_6O}{L} \times 0.250 \text{ L} = 0.150 \text{ mol } C_2H_6O$$

(0.150 mol)(46.07 g/mol) = 6.91 g C_2H_6O

(b) $B(OH)_3$, 61.83 amu; 167 mL = 0.167 L

$$0.200 \frac{\text{mol } B(OH)_3}{L} \times 0.167 \text{ L} = 0.0334 \text{ mol } B(OH)_3$$

(0.0334 mol)(61.83 g/mol) = 2.07 g $B(OH)_3$

3.68 $BaCl_2$, 208.2 amu

$$15.0 \text{ g } BaCl_2 \times \frac{1 \text{ mol } BaCl_2}{208.2 \text{ g } BaCl_2} = 0.0720 \text{ mol } BaCl_2$$

$$0.0720\ \text{mol} \times \frac{1\ \text{L}}{0.45\ \text{mol}} = 0.16\ \text{L}; \quad 0.16\ \text{L} = 160\ \text{mL}$$

3.69 NaCl, 58.4 amu; 400 mg = 0.400 g; 100 mL = 0.100 L

$$0.400\ \text{g NaCl} \times \frac{1\ \text{mol NaCl}}{58.4\ \text{g NaCl}} = 0.006\ 85\ \text{mol NaCl}$$

$$\text{Molarity} = \frac{0.006\ 85\ \text{mol}}{0.100\ \text{L}} = 0.0685\ \text{M}$$

3.70 $C_6H_{12}O_6$, 180.2 amu; 90 mg = 0.090 g; 100 mL = 0.100 L

$$0.090\ \text{g}\ C_6H_{12}O_6 \times \frac{1\ \text{mol}\ C_6H_{12}O_6}{180.2\ \text{g}\ C_6H_{12}O_6} = 0.000\ 50\ \text{mol}\ C_6H_{12}O_6$$

$$\text{Molarity} = \frac{0.000\ 50\ \text{mol}}{0.100\ \text{L}} = 0.0050\ \text{M} = 5.0 \times 10^{-3}\ \text{M}$$

3.71 NaCl, 58.4 amu; KCl, 74.6 amu; $CaCl_2$, 111.0 amu; 500 mL = 0.500 L

$$4.30\ \text{g NaCl} \times \frac{1\ \text{mol NaCl}}{58.4\ \text{g NaCl}} = 0.0736\ \text{mol NaCl}$$

$$0.150\ \text{g KCl} \times \frac{1\ \text{mol KCl}}{74.6\ \text{g KCl}} = 0.002\ 01\ \text{mol KCl}$$

$$0.165\ \text{g}\ CaCl_2 \times \frac{1\ \text{mol}\ CaCl_2}{111.0\ \text{g}\ CaCl_2} = 0.001\ 49\ \text{mol}\ CaCl_2$$

$$0.0736\ \text{mol} + 0.002\ 01\ \text{mol} + 2(0.001\ 49\ \text{mol}) = 0.0786\ \text{mol}\ Cl^-$$

$$Na^+\ \text{molarity} = \frac{0.0736\ \text{mol}}{0.500\ \text{L}} = 0.147\ \text{M}$$

$$Ca^{2+}\ \text{molarity} = \frac{0.001\ 49\ \text{mol}}{0.500\ \text{L}} = 0.002\ 98\ \text{M}$$

$$K^+ \text{ molarity} = \frac{0.002\ 01 \text{ mol}}{0.500 \text{ L}} = 0.004\ 02 \text{ M}$$

$$Cl^- \text{ molarity} = \frac{0.0786 \text{ mol}}{0.500 \text{ L}} = 0.157 \text{ M}$$

3.72 $M_f \times V_f = M_i \times V_i$; $\quad M_f = \dfrac{M_i \times V_i}{V_f} = \dfrac{12.0 \text{ M} \times 35.7 \text{ mL}}{250 \text{ mL}} = 1.71 \text{ M HCl}$

3.73 $2\ HBr(aq) + K_2CO_3(aq) \rightarrow 2\ KBr(aq) + CO_2(g) + H_2O(l)$

K_2CO_3, 138.2 amu; 450 mL = 0.450 L

$$0.500 \frac{\text{mol HBr}}{\text{L}} \times 0.450 \text{ L} = 0.225 \text{ mol HBr}$$

$$0.225 \text{ mol HBr} \times \frac{1 \text{ mol } K_2CO_3}{2 \text{ mol HBr}} \times \frac{138.2 \text{ g } K_2CO_3}{1 \text{ mol } K_2CO_3} = 15.5 \text{ g } K_2CO_3$$

3.74 $C_2H_2O_4$, 90.04 amu; 400 mL = 0.400 L

$$12.0 \text{ g } C_2H_2O_4 \times \frac{1 \text{ mol } C_2H_2O_4}{90.04 \text{ g } C_2H_2O_4} = 0.133 \text{ mol } C_2H_2O_4$$

$$\text{Molarity} = \frac{0.133 \text{ mol}}{0.400 \text{ L}} = 0.333 \text{ M } C_2H_2O_4$$

3.75 $C_2H_2O_4(aq) + 2\ KOH(aq) \rightarrow K_2C_2O_4(aq) + 2\ H_2O(l)$

From Problem 3.74, $C_2H_2O_4$ molarity = 0.333 M

$$0.333 \frac{\text{mol } C_2H_2O_4}{\text{L}} \times 0.0250 \text{ L} = 0.008\ 32 \text{ mol } C_2H_2O_4$$

$$0.008\ 32 \text{ mol } C_2H_2O_4 \times \frac{2 \text{ mol KOH}}{1 \text{ mol } C_2H_2O_4} = 0.0166 \text{ mol } C_2H_2O_4$$

$$0.0166 \text{ mol} \times \frac{1 \text{ L}}{0.100 \text{ mol}} = 0.166 \text{ L};\ 0.166 \text{ L} = 166 \text{ mL}$$

3.76 $C_2H_2O_4$, 90.04 amu

$$3.225\ g\ C_2H_2O_4 \times \frac{1\ mol\ C_2H_2O_4}{90.04\ g\ C_2H_2O_4} \times \frac{2\ mol\ KMnO_4}{5\ mol\ C_2H_2O_4} = 0.0143\ mol\ KMnO_4$$

$$0.0143\ mol \times \frac{1\ L}{0.250\ mol} = 0.0572\ L$$

3.77 $3.045\ g\ Cu \times \frac{1\ mol\ Cu}{63.546\ g\ Cu} = 0.047\ 92\ mol\ Cu$; 50.0 mL = 0.0500 L

$$Cu(NO_3)_2\ molarity = \frac{0.047\ 92\ mol}{0.0500\ L} = 0.958\ M$$

Formulas and Elemental Analysis

3.78 CH_4N_2O, 60.1 amu

$$\%\,C = \frac{12.0\ g\ C}{60.1\ g} \times 100\% = 20.0\%$$

$$\%\,H = \frac{4 \times 1.01\ g\ H}{60.1\ g} \times 100\% = 6.72\%$$

$$\%\,N = \frac{2 \times 14.0\ g\ N}{60.1\ g} \times 100\% = 46.6\%$$

$$\%\,O = \frac{16.0\ g\ O}{60.1\ g} \times 100\% = 26.6\%$$

3.79 (a) $Cu_2(OH)_2CO_3$, 221.1 amu

$$\%\ Cu = \frac{2 \times 63.5\ g\ Cu}{221.1\ g} \times 100\% = 57.4\%$$

$$\%\ O = \frac{5 \times 16.0\ g\ O}{221.1\ g} \times 100\% = 36.2\%$$

$$\% \text{ C} = \frac{12.0 \text{ g C}}{221.1 \text{ g}} \times 100\% = 5.43\%$$

$$\% \text{ H} = \frac{2 \times 1.01 \text{ g H}}{221.1 \text{ g}} \times 100\% = 0.91\%$$

(b) $C_8H_9NO_2$, 151.2 amu

$$\% \text{ C} = \frac{8 \times 12.0 \text{ g C}}{151.2 \text{ g}} \times 100\% = 63.5\%$$

$$\% \text{ H} = \frac{9 \times 1.01 \text{ g H}}{151.2 \text{ g}} \times 100\% = 6.01\%$$

$$\% \text{ N} = \frac{14.0 \text{ g N}}{151.2 \text{ g}} \times 100\% = 9.26\%$$

$$\% \text{ O} = \frac{2 \times 16.0 \text{ g O}}{151.2 \text{ g}} \times 100\% = 21.2\%$$

(c) $Fe_4[Fe(CN)_6]_3$, 859.2 amu

$$\% \text{ Fe} = \frac{7 \times 55.85 \text{ g Fe}}{859.2 \text{ g}} \times 100\% = 45.5\%$$

$$\% \text{ C} = \frac{18 \times 12.01 \text{ g C}}{859.2 \text{ g}} \times 100\% = 25.2\%$$

$$\% \text{ N} = \frac{18 \times 14.01 \text{ g N}}{859.2 \text{ g}} \times 100\% = 29.3\%$$

3.80 Assume a 100.0 g sample. From the percent composition data, a 100.0 g sample contains 24.25 g F and 75.75 g Sn.

$$24.25 \text{ g F} \times \frac{1 \text{ mol F}}{19.00 \text{ g F}} = 1.276 \text{ mol F}$$

$$75.75 \text{ g Sn} \times \frac{1 \text{ mol Sn}}{118.7 \text{ g Sn}} = 0.6382 \text{ mol Sn}$$

$Sn_{0.6382}F_{1.276}$; divide each subscript by the smaller, 0.6382.

$Sn_{0.6382\,/\,0.6382}F_{1.276\,/\,0.6382}$

The empirical formula is SnF_2.

3.81 (a) Assume a 100.0 g sample of ibuprofen. From the percent composition data, a 100.0 g sample contains 75.69 g C, 15.51 g O, and 8.80 g H.

$$75.69 \text{ g C} \times \frac{1 \text{ mol C}}{12.01 \text{ g C}} = 6.302 \text{ mol C}$$

$$15.51 \text{ g O} \times \frac{1 \text{ mol O}}{16.00 \text{ g O}} = 0.9694 \text{ mol O}$$

$$8.80 \text{ g H} \times \frac{1 \text{ mol H}}{1.01 \text{ g H}} = 8.71 \text{ mol H}$$

$C_{6.302}H_{8.71}O_{0.9694}$, divide each subscript by the smallest, 0.9694.

$C_{6.302\,/\,0.9694}H_{8.71\,/\,0.9694}O_{0.9694\,/\,0.9694}$

$C_{6.5}H_9O$; multiply each subscript by 2 to obtain integers.

The empirical formula is $C_{13}H_{18}O_2$.

(b) Assume a 100.0 g sample of tetraethyllead. From the percent composition data, a 100.0 g sample contains 29.71 g C, 6.23 g H, and 64.06 g Pb.

$$29.71 \text{ g C} \times \frac{1 \text{ mol C}}{12.01 \text{ g C}} = 2.474 \text{ mol C}$$

$$6.23 \text{ g H} \times \frac{1 \text{ mol H}}{1.01 \text{ g H}} = 6.17 \text{ mol H}$$

$$64.06 \text{ g Pb} \times \frac{1 \text{ mol Pb}}{207.2 \text{ g Pb}} = 0.3092 \text{ mol Pb}$$

$Pb_{0.3092}C_{2.474}H_{6.17}$; divide each subscript by the smallest, 0.3092.

$Pb_{0.3092\,/\,0.3092}C_{2.474\,/\,0.3092}H_{6.17\,/\,0.3092}$

The empirical formula is PbC_8H_{20}.

(c) Assume a 100.0 g sample of zircon. From the percent composition data, a 100.0 g sample contains 34.91 g O, 15.32 g Si, and 49.76 g Zr.

$$34.91 \text{ g O} \times \frac{1 \text{ mol O}}{16.00 \text{ g O}} = 2.182 \text{ mol O}$$

$$15.32 \text{ g Si} \times \frac{1 \text{ mol Si}}{28.09 \text{ g Si}} = 0.5454 \text{ mol Si}$$

$$49.76 \text{ g Zr} \times \frac{1 \text{ mol Zr}}{91.22 \text{ g Zr}} = 0.5455 \text{ mol Zr}$$

$Zr_{0.5455}Si_{0.5454}O_{2.182}$; divide each subscript by the smallest, 0.5454.

$Zr_{0.5455\,/\,0.5454}Si_{0.5454\,/\,0.5454}O_{2.182\,/\,0.5454}$

The empirical formula is $ZrSiO_4$.

3.82 Toluene contains only C and H.

152.5 mg = 0.1525 g and 35.67 mg = 0.035 67 g

$$0.1525 \text{ g } CO_2 \times \frac{1 \text{ mol } CO_2}{44.01 \text{ g } CO_2} \times \frac{1 \text{ mol C}}{1 \text{ mol } CO_2} = 0.003\ 465 \text{ mol C}$$

$$0.035\ 67 \text{ g } H_2O \times \frac{1 \text{ mol } H_2O}{18.02 \text{ g } H_2O} \times \frac{2 \text{ mol H}}{1 \text{ mol } H_2O} = 0.003\ 959 \text{ mol H}$$

$C_{0.003465}H_{0.003959}$; divide each subscript by the smaller, 0.003 465.

$C_{0.003465\,/\,0.003465}H_{0.003959\,/\,0.003465}$

$CH_{1.14}$; multiply each subscript by 7 to obtain integers.

The empirical formula is C_7H_8.

3.83 5.024 mg = 0.005 024 g; 13.90 mg = 0.013 90 g; 6.048 mg = 0.006 048 g

$$0.013\ 90 \text{ g } CO_2 \times \frac{1 \text{ mol } CO_2}{44.01 \text{ g } CO_2} \times \frac{1 \text{ mol C}}{1 \text{ mol } CO_2} = 3.158 \times 10^{-4} \text{ mol C}$$

$$0.006\ 048\ \text{g}\ H_2O \times \frac{1\ \text{mol}\ H_2O}{18.02\ \text{g}\ H_2O} \times \frac{2\ \text{mol H}}{1\ \text{mol}\ H_2O} = 6.713 \times 10^{-4}\ \text{mol H}$$

$$3.158 \times 10^{-4}\ \text{mol C} \times \frac{12.01\ \text{g C}}{1\ \text{mol C}} = 0.003\ 793\ \text{g C}$$

$$6.713 \times 10^{-4}\ \text{mol H} \times \frac{1.008\ \text{g H}}{1\ \text{mol H}} = 0.000\ 676\ 7\ \text{g H}$$

mass N = 0.005 024 g – (0.003 793 g + 0.000 676 7 g) = 0.000 554 g N

$$0.000\ 554\ \text{g N} \times \frac{1\ \text{mol N}}{14.01\ \text{g N}} = 3.95 \times 10^{-5}\ \text{mol N}$$

Scale each mol quantity to eliminate exponents.

$C_{3.158}H_{6.713}N_{0.395}$; divide each subscript by the smallest, 0.395.

$C_{3.158\,/\,0.395}H_{6.713\,/\,0.395}N_{0.395\,/\,0.395}$

The empirical formula is $C_8H_{17}N$.

3.84 23.46 mg = 0.023 46 g; 20.42 mg = 0.02042 g; 33.27 mg = 0.033 27 g

$$0.033\ 27\ \text{g}\ CO_2 \times \frac{1\ \text{mol}\ CO_2}{44.01\ \text{g}\ CO_2} \times \frac{1\ \text{mol C}}{1\ \text{mol}\ CO_2} = 7.560 \times 10^{-4}\ \text{mol C}$$

$$0.020\ 42\ \text{g}\ H_2O \times \frac{1\ \text{mol}\ H_2O}{18.02\ \text{g}\ H_2O} \times \frac{2\ \text{mol H}}{1\ \text{mol}\ H_2O} = 2.266 \times 10^{-3}\ \text{mol H}$$

$$7.560 \times 10^{-4}\ \text{mol C} \times \frac{12.01\ \text{g C}}{1\ \text{mol C}} = 0.009\ 080\ \text{g C}$$

$$2.266 \times 10^{-3}\ \text{mol H} \times \frac{1.008\ \text{g H}}{1\ \text{mol H}} = 0.002\ 284\ \text{g H}$$

mass O = 0.023 46 g – (0.009 080 g + 0.002 284 g) = 0.012 10 g O

$$0.012\ 10\ \text{g O} \times \frac{1\ \text{mol O}}{16.00\ \text{g O}} = 7.563 \times 10^{-4}\ \text{mol O}$$

Scale each mol quantity to eliminate exponents.

$C_{0.7560}H_{2.266}O_{0.7563}$; divide each subscript by the smallest, 0.7560.

$C_{0.7560\,/\,0.7560}H_{2.266\,/\,0.7560}O_{0.7563\,/\,0.7560}$

The empirical formula is CH_3O, 31.0 amu.

62.0 amu / 31.0 amu = 2; molecular formula = $C_{(2 \times 1)}H_{(2 \times 3)}O_{(2 \times 1)} = C_2H_6O_2$

3.85 High resolution mass spectrometry is capable of measuring the mass of a particular isotopic composition of the molecule.

3.86 Let X equal the molecular weight of cytochrome c.

$$0.0043 = \frac{55.847 \text{ amu}}{X}; \qquad X = \frac{55.847 \text{ amu}}{0.0043} = 13{,}000 \text{ amu}$$

3.87 Let X equal the molecular weight of nitrogenase.

$$0.000\ 872 = \frac{2 \times 95.94 \text{ amu}}{X}; \qquad X = \frac{2 \times 95.94 \text{ amu}}{0.000\ 872} = 220{,}000 \text{ amu}$$

3.88 Let X equal the molecular weight of disilane.

$$0.9028 = \frac{2 \times 28.09 \text{ amu}}{X}$$

$$X = \frac{2 \times 28.09 \text{ amu}}{0.9028} = 62.23 \text{ amu}$$

62.23 amu – 2(Si atomic wt) = 62.23 amu – 2(28.09 amu) = 6.05 amu

6.05 amu is the total mass of H atoms.

$$6.05 \text{ amu} \times \frac{1 \text{ H atom}}{1.01 \text{ amu}} = 6 \text{ H atoms}$$

Disilane is Si_2H_6.

3.89 Let X equal the molecular weight of MS_2.

$$0.4006 = \frac{2 \text{ x } 32.07 \text{ amu}}{X}$$

$$X = \frac{2 \text{ x } 32.07 \text{ amu}}{0.4006} = 160.1 \text{ amu}$$

Atomic weight of M = 160.1 amu – 2(S atomic wt)

= 160.1 amu – 2(32.07 amu) = 95.96 amu

M is Mo.

General Problems

3.90 (a) $C_6H_{12}O_6$, 180.2 amu

$$\%C = \frac{6 \text{ x } 12.01 \text{ g C}}{180.2 \text{ g}} \text{ x } 100\% = 39.99\%$$

$$\%H = \frac{12 \text{ x } 1.008 \text{ g H}}{180.2 \text{ g}} \text{ x } 100\% = 6.71\%$$

$$\%O = \frac{6 \text{ x } 16.00 \text{ g O}}{180.2 \text{ g}} \text{ x } 100\% = 53.27\%$$

(b) H_2SO_4, 98.08 amu

$$\%H = \frac{2 \text{ x } 1.008 \text{ g H}}{98.08 \text{ g}} \text{ x } 100\% = 2.06\%$$

$$\%S = \frac{32.07 \text{ g S}}{98.08 \text{ g}} \text{ x } 100\% = 32.70\%$$

$$\%O = \frac{4 \text{ x } 16.00 \text{ g O}}{98.08 \text{ g}} \text{ x } 100\% = 65.25\%$$

(c) $KMnO_4$, 158.0 amu

$$\%\,K = \frac{39.10\ g\ K}{158.0\ g} \times 100\% = 24.75\%$$

$$\%\,Mn = \frac{54.94\ g\ Mn}{158.0\ g} \times 100\% = 34.77\%$$

$$\%\,O = \frac{4 \times 16.00\ g\ O}{158.0\ g} \times 100\% = 40.51\%$$

(d) $C_7H_5NO_3S$, 183.2 amu

$$\%\,C = \frac{7 \times 12.01\ g\ C}{183.2\ g} \times 100\% = 45.89\%$$

$$\%\,H = \frac{5 \times 1.008\ g\ H}{183.2\ g} \times 100\% = 2.75\%$$

$$\%\,N = \frac{14.01\ g\ N}{183.2\ g} \times 100\% = 7.65\%$$

$$\%\,O = \frac{3 \times 16.00\ g\ O}{183.2\ g} \times 100\% = 26.20\%$$

$$\%\,S = \frac{32.07\ g\ S}{183.2\ g} \times 100\% = 17.51\%$$

3.91 (a) Assume a 100.0 g sample of aspirin. From the percent composition data, a 100.0 g sample contains 60.00 g C, 35.52 g O, and 4.48 g H.

$$60.00\ g\ C \times \frac{1\ mol\ C}{12.01\ g\ C} = 5.00\ mol\ C$$

$$35.52\ g\ O \times \frac{1\ mol\ O}{16.00\ g\ O} = 2.22\ mol\ O$$

$$4.48\ g\ H \times \frac{1\ mol\ H}{1.01\ g\ H} = 4.44\ mol\ H$$

$C_{5.00}H_{4.44}O_{2.22}$; divide each subscript by the smallest, 2.22.

$C_{5.00 / 2.22}H_{4.44 / 2.22}O_{2.22 / 2.22}$

$C_{2.25}H_2O_1$; multiply each subscript by 4 to obtain integers.

The empirical formula is $C_9H_8O_4$.

(b) Assume a 100.0 g sample of ilmenite. From the percent composition data, a 100.0 g sample contains 31.63 g O, 31.56 g Ti, and 36.81 g Fe.

$$31.63\text{ g O} \times \frac{1\text{ mol O}}{16.00\text{ g O}} = 1.98\text{ mol O}$$

$$31.56\text{ g Ti} \times \frac{1\text{ mol Ti}}{47.88\text{ g Ti}} = 0.659\text{ mol Ti}$$

$$36.81\text{ g Fe} \times \frac{1\text{ mol Fe}}{55.85\text{ g Fe}} = 0.659\text{ mol Fe}$$

$Fe_{0.659}Ti_{0.659}O_{1.98}$; divide each subscript by the smallest, 0.659.

$Fe_{0.659 / 0.659}Ti_{0.659 / 0.659}O_{1.98 / 0.659}$

The empirical formula is $FeTiO_3$.

(c) Assume a 100.0 g sample of sodium thiosulfate. From the percent composition data, a 100.0 g sample contains 30.36 g O, 29.08 g Na, and 40.56 g S.

$$30.36\text{ g O} \times \frac{1\text{ mol O}}{16.00\text{ g O}} = 1.90\text{ mol O}$$

$$29.08\text{ g Na} \times \frac{1\text{ mol Na}}{22.99\text{ g Na}} = 1.26\text{ mol Na}$$

$$40.56\text{ g S} \times \frac{1\text{ mol S}}{32.07\text{ g S}} = 1.26\text{ mol S}$$

$Na_{1.26}S_{1.26}O_{1.90}$; divide each subscript by the smallest, 1.26.

$Na_{1.26 / 1.26}S_{1.26 / 1.26}O_{1.90 / 1.26}$

$NaSO_{1.5}$; multiply each subscript by 2 to obtain integers.

The empirical formula is $Na_2S_2O_3$.

3.92 (a) $SiCl_4 + 2\ H_2O \rightarrow SiO_2 + 4\ HCl$

(b) $P_2O_5 + 3\ H_2O \rightarrow 2\ H_3PO_4$

(c) $CaCN_2 + 3\ H_2O \rightarrow CaCO_3 + 2\ NH_3$

(d) $3\ NO_2 + H_2O \rightarrow 2\ HNO_3 + NO$

3.93 Molar mass = $\frac{221.6\ g}{0.5731\ mol}$ = 386.7 g/mol; molecular weight = 386.7 amu

3.94 (a) $C_{10}H_{18}O$, 10(12.01) + 18(1.01) + 1 (16.00) = 154.3 amu

(b) $Na_2B_4O_7$, 2(22.99) + 4(10.81) + 7(16.00) = 201.2 amu

4.1 (a) precipitation (b) redox (c) acid-base neutralization

4.2 $FeBr_3$ contains 3 Br^- ions.

The molar concentration of Br^- ions = 3 x 0.255 M = 0.765 M

4.3 (a) Ionic equation:

$2\ Ag^+(aq) + 2\ NO_3^-(aq) + 2\ Na^+(aq) + CrO_4^{2-}(aq) \rightarrow$
$Ag_2CrO_4(s) + 2\ Na^+(aq) + 2\ NO_3^-(aq)$

Delete spectator ions from the ionic equation to get the net ionic equation.

Net ionic equation: $2\ Ag^+(aq) + CrO_4^{2-}(aq) \rightarrow Ag_2CrO_4(s)$

(b) Ionic equation:

$2\ H^+(aq) + SO_4^{2-}(aq) + MgCO_3(s) \rightarrow H_2O(l) + CO_2(g) + Mg^{2+}(aq) + SO_4^{2-}(aq)$

Delete spectator ions from the ionic equation to get the net ionic equation.

Net ionic equation: $2\ H^+(aq) + MgCO_3(s) \rightarrow H_2O(l) + CO_2(g) + Mg^{2+}(aq)$

4.4 (a) $CdCO_3$, insoluble (b) MgO, insoluble (c) Na_2S, soluble

(d) $PbSO_4$, insoluble (e) $(NH_4)_3PO_4$, soluble (f) $HgCl_2$, soluble

4.5 (a) Ionic equation:

$Ni^{2+}(aq) + 2\ Cl^-(aq) + 2\ NH_4^+(aq) + S^{2-}(aq) \rightarrow NiS(s) + 2\ NH_4^+(aq) + 2\ Cl^-(aq)$

Delete spectator ions from the ionic equation to get the net ionic equation.

Net ionic equation: $Ni^{2+}(aq) + S^{2-}(aq) \rightarrow NiS(s)$

(b) Ionic equation:

$2\ Na^+(aq) + CrO_4^{2-}(aq) + Pb^{2+}(aq) + 2\ NO_3^-(aq) \rightarrow$
$PbCrO_4(s) + 2\ Na^+(aq) + 2\ NO_3^-(aq)$

Delete spectator ions from the ionic equation to get the net ionic equation.

Net ionic equation: $Pb^{2+}(aq) + CrO_4^{2-}(aq) \rightarrow PbCrO_4(s)$

(c) Ionic equation:

$$2\ Ag^+(aq) + 2\ ClO_4^-(aq) + Ca^{2+}(aq) + 2\ Br^-(aq) \rightarrow 2\ AgBr(s) + Ca^{2+}(aq) + 2\ ClO_4^-(aq)$$

Delete spectator ions from the ionic equation, and reduce coefficients to get the net ionic equation.

Net ionic equation: $Ag^+(aq) + Br^-(aq) \rightarrow AgBr(s)$

4.6 $3\ CaCl_2(aq) + 2\ Na_3PO_4(aq) \rightarrow Ca_3(PO_4)_2(s) + 6\ NaCl(aq)$

Ionic equation:

$$3\ Ca^{2+}(aq) + 6\ Cl^-(aq) + 6\ Na^+(aq) + 2\ PO_4^{3-}(aq) \rightarrow Ca_3(PO_4)_2(s) + 6\ Na^+(aq) + 6\ Cl^-(aq)$$

Delete spectator ions from the ionic equation to get the net ionic equation.

Net ionic equation: $3\ Ca^{2+}(aq) + 2\ PO_4^{3-}(aq) \rightarrow Ca_3(PO_4)_2(s)$

4.7 (a) Ionic equation:

$$2\ Cs^+(aq) + 2\ OH^-(aq) + 2\ H^+(aq) + SO_4^{2-}(aq) \rightarrow 2\ Cs^+(aq) + SO_4^{2-}(aq) + 2\ H_2O(l)$$

Delete spectator ions from the ionic equation, and reduce coefficients to get the net ionic equation.

Net ionic equation: $H^+(aq) + OH^-(aq) \rightarrow H_2O(l)$

(b) Ionic equation:

$$Ca^{2+}(aq) + 2\ OH^-(aq) + 2\ CH_3COOH(aq) \rightarrow Ca^{2+}(aq) + 2\ CH_3COO^-(aq) + 2\ H_2O(l)$$

Delete spectator ions from the ionic equation, and reduce coefficients to get the net ionic equation.

Net ionic equation: $CH_3COOH(aq) + OH^-(aq) \rightarrow CH_3COO^-(aq) + H_2O(l)$

4.8 (a) $SnCl_4$: Cl −1, Sn +4

(b) CrO_3: O −2, Cr +6

(c) $VOCl_3$: O −2, Cl −1, V +5

(d) V_2O_3: O −2, V +3

(e) HNO_3: O −2, H +1, N +5

(f) $FeSO_4$: O −2, S +6, Fe +2

4.9 $2\ Cu^{2+}(aq) + 4\ I^-(aq) \rightarrow 2\ CuI(s) + I_2(aq)$

Oxidation numbers:

Cu^{2+} +2; I^- −1; CuI: Cu +1, I −1; I_2: 0

Oxidizing agent (oxidation number decreases), Cu^{2+}

Reducing agent (oxidation number decreases) , I^-

4.10 (a) $SnO_2(s) + 2\ C(s) \rightarrow Sn(s) + 2\ CO(g)$

C is oxidized (its oxidation number increases from 0 to +2). C is the reducing agent.

The Sn in SnO_2 is reduced (its oxidation number decreases from +4 to 0). SnO_2 is the oxidizing agent.

(b) $Sn^{2+}(aq) + 2\ Fe^{3+}(aq) \rightarrow Sn^{4+}(aq) + 2\ Fe^{2+}(aq)$

Sn^{2+} is oxidized (its oxidation number increases from +2 to +4). Sn^{2+} is the reducing agent.

Fe^{3+} is reduced (its oxidation number decreases from +3 to +2). Fe^{3+} is the oxidizing agent.

4.11 (a) Pt is below H in the activity series, therefore NO REACTION.

(b) Mg is below Ca in the activity series, therefore NO REACTION.

4.12 $Cr_2O_7^{2-}(aq) + I^-(aq) \rightarrow Cr^{3+}(aq) + IO_3^-(aq)$

$Cr_2O_7^{2-}(aq) + I^-(aq) \rightarrow 2\ Cr^{3+}(aq) + IO_3^-(aq)$

+6 (Cr), −1 (I), +3 (Cr), +5 (I)

2(+6)= +12 ; 2(+3)= +6

lose $6e^-$ (I: −1 → +5)

gain $6e^-$ (Cr: +12 → +6)

$8\ H^+(aq) + Cr_2O_7^{2-}(aq) + I^-(aq) \rightarrow 2\ Cr^{3+}(aq) + IO_3^-(aq) + 4\ H_2O(l)$

4.13 $MnO_4^-(aq) + Br^-(aq) \rightarrow MnO_2(s) + BrO_3^-(aq)$

+7 (Mn), −1 (Br), +4 (Mn), +5 (Br)

gain 3 e^- (Mn: +7 → +4)

lose 6 e^- (Br: −1 → +5)

$2\ MnO_4^-(aq) + Br^-(aq) \rightarrow 2\ MnO_2(s) + BrO_3^-(aq)$

$2\ H^+(aq) + 2\ MnO_4^-(aq) + Br^-(aq) \rightarrow 2\ MnO_2(s) + BrO_3^-(aq) + H_2O(l)$

$2\ H^+(aq) + 2\ OH^-(aq) + 2\ MnO_4^-(aq) + Br^-(aq) \rightarrow$
$2\ MnO_2(s) + BrO_3^-(aq) + H_2O(l) + 2\ OH^-(aq)$

$H_2O(l) + 2\ MnO_4^-(aq) + Br^-(aq) \rightarrow 2\ MnO_2(s) + BrO_3^-(aq) + 2\ OH^-(aq)$

4.14 (a) $MnO_4^-(aq) \rightarrow MnO_2(s)$ (reduction)

$IO_3^-(aq) \rightarrow IO_4^-(aq)$ (oxidation)

(b) $NO_3^-(aq) \rightarrow NO_2(g)$ (reduction)

$SO_2(aq) \rightarrow SO_4^{2-}(aq)$ (oxidation)

4.15 $NO_3^-(aq) + Cu(s) \rightarrow NO(g) + Cu^{2+}(aq)$

$[Cu(s) \rightarrow Cu^{2+}(aq) + 2\ e^-] \times 3$ (oxidation half reaction)

$NO_3^-(aq) \rightarrow NO(g)$

$NO_3^-(aq) \rightarrow NO(g) + 2\ H_2O(l)$

$4\ H^+(aq) + NO_3^-(aq) \rightarrow NO(g) + 2\ H_2O(l)$

$[3\ e^- + 4\ H^+(aq) + NO_3^-(aq) \rightarrow NO(g) + 2\ H_2O(l)] \times 2$
(reduction half reaction)

Combine the two half reactions.

$2\ NO_3^-(aq) + 8\ H^+(aq) + 3\ Cu(s) \rightarrow 3\ Cu^{2+}(aq) + 2\ NO(g) + 4\ H_2O(l)$

4.16 $Fe(OH)_2(s) + O_2(g) \rightarrow Fe(OH)_3(s)$

$[Fe(OH)_2(s) + OH^-(aq) \rightarrow Fe(OH)_3(s) + e^-] \times 4$ (oxidation half reaction)

$O_2(g) \rightarrow 2\ H_2O(l)$

$4\ H^+(aq) + O_2(g) \rightarrow 2\ H_2O(l)$

$4\ e^- + 4\ H^+(aq) + O_2(g) \rightarrow 2\ H_2O(l)$

$4\ e^- + 4\ H^+(aq) + 4\ OH^-(aq) + O_2(g) \rightarrow 2\ H_2O(l) + 4\ OH^-(aq)$

$4\ e^- + 4\ H_2O(l) + O_2(g) \rightarrow 2\ H_2O(l) + 4\ OH^-(aq)$

$4\ e^- + 2\ H_2O(l) + O_2(g) \rightarrow 4\ OH^-(aq)$ (reduction half reaction)

Combine the two half reactions.

$4\ Fe(OH)_2(s) + 4\ OH^-(aq) + 2\ H_2O(l) + O_2(g) \rightarrow$
$4\ Fe(OH)_3(s) + 4\ OH^-(aq)$

$4\ Fe(OH)_2(s) + 2\ H_2O(l) + O_2(g) \rightarrow 4\ Fe(OH)_3(s)$

4.17 31.50 mL = 0.031 50 L; 10.00 mL = 0.010 00 L

$$0.031\ 50\ \text{L} \times \frac{0.105\ \text{mol}\ BrO_3^-}{1\ \text{L}} \times \frac{6\ \text{mol}\ Fe^{2+}}{1\ \text{mol}\ BrO_3^-} = 1.98 \times 10^{-2}\ \text{mol}\ Fe^{2+}$$

$$\text{molarity} = \frac{1.98 \times 10^{-2}\ \text{mol}\ Fe^{2+}}{0.010\ 00\ \text{L}} = 1.98\ \text{M}\ Fe^{2+}\ \text{solution}$$

Understanding Key Concepts

1. (a) $Ba^{2+}(aq) + SO_4^{2-}(aq) \rightarrow BaSO_4(s)$

   (b) $2\ H^+(aq) + CO_3^{2-}(aq) \rightarrow CO_2(g) + H_2O(l)$

2. Dissolve the solid in water. Set up a conductivity apparatus similar to that shown in Figure 4.1. If the bulb glows brightly, the substance is a strong electrolyte. If the bulb glows dimly, the substance is a weak electrolyte. If the bulb does not glow at all, the substance is a nonelectrolyte.

3. "Any element higher in the activity series will react with the ion of any element lower in the activity series."

   $C + B^+ \rightarrow C^+ + B$; therefore C is higher than B.

   $A^+ + D \rightarrow$ no reaction; therefore A is higher than D.

   $C^+ + A \rightarrow$ no reaction; therefore C is higher than A.

   $D + B^+ \rightarrow D^+ + B$; therefore D is higher than B.

   The net result is C > A > D > B

4. (a) The reaction, $A^+ + C \rightarrow A + C^+$, will occur since C is above A in the activity series.

   (b) The reaction, $A^+ + B \rightarrow A + B^+$, will not occur since B is below A in the activity series.

5. (a) 2 $Na^+(aq)$ + $CO_3^{2-}(aq)$ does not form a precipitate. This is represented by box (1).

(b) $Ba^{2+}(aq) + CrO_4^{2-}(aq) \rightarrow BaCrO_4(s)$. This is represented by box (2).

(c) $2\ Ag^+(aq) + SO_4^{2-}(aq) \rightarrow Ag_2SO_4(s)$. This is represented by box (3).

Additional Problems

Aqueous Reactions and Net Ionic Equations

4.18 (a) precipitation (b) redox (c) acid-base neutralization

4.19 (a) redox (b) precipitation (c) acid-base neutralization

4.20 (a) Ionic equation:

$$Hg^{2+}(aq) + 2\ NO_3^-(aq) + 2\ Na^+(aq) + 2\ I^-(aq) \rightarrow 2\ Na^+(aq) + 2\ NO_3^-(aq) + HgI_2(s)$$

Delete spectator ions from the ionic equation to get the net ionic equation.

Net ionic equation: $Hg^{2+}(aq) + 2\ I^-(aq) \rightarrow HgI_2(s)$

(b) $2\ HgO(s) \xrightarrow{\text{heat}} 2\ Hg(l) + O_2(g)$

(c) Ionic equation:

$$H_3PO_4(aq) + 3\ K^+(aq) + 3\ OH^-(aq) \rightarrow 3\ K^+(aq) + PO_4^{3-}(aq) + 3\ H_2O(l)$$

Delete spectator ions from the ionic equation to get the net ionic equation.

Net ionic equation: $H_3PO_4(aq) + 3\ OH^-(aq) \rightarrow PO_4^{3-}(aq) + 3\ H_2O(l)$

4.21 (a) $S_8(s) + 8\ O_2(g) \rightarrow 8\ SO_2(g)$

(b) $Ni^{2+}(aq) + 2\ Cl^-(aq) + 2\ Na^+(aq) + S^{2-}(aq) \rightarrow NiS(s) + 2\ Na^+(aq) + 2\ Cl^-(aq)$

Net ionic equation: $Ni^{2+}(aq) + S^{2-}(aq) \rightarrow NiS(s)$

(b) $2\ CH_3COOH(aq) + Ba^{2+}(aq) + 2\ OH^-(aq) \rightarrow$
$2\ CH_3COO^-(aq) + Ba^{2+}(aq) + 2\ H_2O(l)$

Net ionic equation: $CH_3COOH(aq) + OH^-(aq) \rightarrow CH_3COO^-(aq) + H_2O(l)$

4.22 (a) $2\ Na(s) + 2\ H_2O(l) \rightarrow 2\ Na^+(aq) + 2\ OH^-(aq) + H_2(g)$

(b) $AgNO_3(aq) + HCl(aq) \rightarrow AgCl(s) + HNO_3(aq)$

Ionic equation:

$Ag^+(aq) + NO_3^-(aq) + H^+(aq) + Cl^-(aq) \rightarrow AgCl(s) + H^+(aq) + NO_3^-(aq)$

Delete spectator ions from the ionic equation to get the net ionic equation.

Net ionic equation: $Ag^+(aq) + Cl^-(aq) \rightarrow AgCl(s)$

4.23 A strong electrolyte is a compound that completely dissociates into its ions when placed in water. When a weak electrolyte is placed in water it dissociates only slightly into its ions.

4.24 (a) HBr, strong electrolyte

(b) HF, weak electrolyte

(c) $NaClO_4$, strong electrolyte

(d) $(NH_4)_2CO_3$, strong electrolyte

(e) NH_3, weak electrolyte

4.25 (a) K_2CO_3 contains 3 ions (2 K^+ and 1 CO_3^{2-}).

The molar concentration of ions = 3 x 0.750 M = 2.25 M

(b) $AlCl_3$ contains 4 ions (1 Al^{3+} and 3 Cl^-).

The molar concentration of ions = 4 x 0.355 M = 1.42 M

4.26 (a) CH_3OH is a nonelectrolyte. The ion concentration from CH_3OH is zero.

(b) $HClO_4$ is a strong acid.

$HClO_4(aq) \rightarrow H^+(aq) + ClO_4^-(aq)$

In solution, there are 2 moles of ions per mole of $HClO_4$.

The molar concentration of ions = 2 × 0.225 M = 0.450 M

4.27 H_2O is polar and a good H^+ acceptor. It allows the polar HCl to dissociate into ions in aqueous solution: $HCl + H_2O \rightarrow H_3O^+ + Cl^-$

$CHCl_3$ is not very polar and not a H^+ acceptor and does not allow the polar HCl to dissociate into ions.

Precipitation Reactions and Solubility Rules

4.28 (a) Ag_2O, insoluble (b) $Ba(NO_3)_2$, soluble

(c) $SnCO_3$, insoluble (d) Fe_2O_3, insoluble

4.29 (a) ZnS, insoluble (b) $Au_2(CO_3)_3$, insoluble

(c) $PbCl_2$, insoluble (soluble in hot water) (d) MnO_2, insoluble

4.30 (a) No precipitate will form.

(b) $FeCl_2(aq) + 2\ KOH(aq) \rightarrow Fe(OH)_2(s) + 2\ KCl(aq)$

(c) No precipitate will form.

4.31 (a) $MnCl_2(aq) + Na_2S(aq) \rightarrow MnS(s) + 2\ NaCl(aq)$

(b) No precipitate will form.

(d) $3\ Hg(NO_3)_2(aq) + 2\ Na_3PO_4(aq) \rightarrow Hg_3(PO_4)_2(s) + 6\ NaNO_3(aq)$

4.32 (a) $Pb(NO_3)_2(aq) + Na_2SO_4(aq) \rightarrow PbSO_4(s) + 2\ NaNO_3(aq)$

(b) $3\ MgCl_2(aq) + 2\ K_3PO_4(aq) \rightarrow Mg_3(PO_4)_2(s) + 6\ KCl(aq)$

(c) $ZnSO_4(aq) + Na_2CrO_4(aq) \rightarrow ZnCrO_4(s) + Na_2SO_4(aq)$

4.33 (a) $AlCl_3(aq) + 3\ NaOH(aq) \rightarrow Al(OH)_3(s) + 3\ NaCl(aq)$

(b) $Fe(NO_3)_2(aq) + Na_2S(aq) \rightarrow FeS(s) + 2\ NaNO_3(aq)$

(c) $CoSO_4(aq) + K_2CO_3(aq) \rightarrow CoCO_3(s) + K_2SO_4(aq)$

4.34 Add HCl(aq); it will selectively precipitate AgCl(s)

4.35 Add $Na_2SO_4(aq)$; it will selectively precipitate $BaSO_4(s)$.

4.36 Ag^+ is eliminated because it would have precipitated as AgCl(s), Ba^{2+} is eliminated because it would have precipitated as $BaSO_4(s)$. The solution might contain Cs^+ and/or NH_4^+. Neither of these will precipitate with OH^-, SO_4^{2-}, or Cl^-.

4.37 Cl^- is eliminated because it would have precipitated as AgCl(s). OH^- is eliminated because it would have precipitated as either AgOH(s) or $Cu(OH)_2(s)$. SO_4^{2-} is eliminated because it would have precipitated as $BaSO_4(s)$. The solution might contain NO_3^- because all nitrates are soluble.

Acids, Bases, and Neutralization Reactions

4.38 In water a strong acid completely dissociates into $H^+(aq)$ and the aqueous anion of the acid. A weak acid dissociates only a few percent into $H^+(aq)$ and the aqueous anion of the acid.

In water a strong base completely dissociates into $OH^-(aq)$ and the aqueous cation of the base. A weak base reacts only slightly with water to produce a small amount of $OH^-(aq)$.

4.39 Add the solution to an active metal, such as magnesium. Bubbles of H_2 gas indicate the presence of an acid.

4.40 We use a forward-and-backward arrow to show the dissociation of a weak acid or weak base in aqueous solution to indicate the equilibrium between reactants and products.

4.41 (a) $2\ H^+(aq) + 2\ ClO_4^-(aq) + Ca^{2+}(aq) + 2\ OH^-(aq) \rightarrow$
$Ca^{2+}(aq) + 2\ ClO_4^-(aq) + 2\ H_2O(l)$

(b) $CH_3COOH(aq) + Na^+(aq) + OH^-(aq) \rightarrow CH_3COO^-(aq) + Na^+(aq) + H_2O(l)$

4.42 (a) $2\ HF(aq) + Ca^{2+}(aq) + 2\ OH^-(aq) \rightarrow Ca^{2+}(aq) + 2\ F^-(aq) + 2\ H_2O(l)$

(b) $Mg(OH)_2(s) + 2\ H^+(aq) + 2\ NO_3^-(aq) \rightarrow$
$Mg^{2+}(aq) + 2\ NO_3^-(aq) + 2\ H_2O(l)$

4.43 (a) $LiOH(aq) + HI(aq) \rightarrow LiI(aq) + H_2O(l)$

Ionic equation:

$Li^+(aq) + OH^-(aq) + H^+(aq) + I^-(aq) \rightarrow Li^+(aq) + I^-(aq) + H_2O(l)$

Delete spectator ions from the ionic equation to get the net ionic equation.

Net ionic equation: $H^+(aq) + OH^-(aq) \rightarrow H_2O(l)$

(b) $2\ HBr(aq) + Ca(OH)_2(aq) \rightarrow CaBr_2(aq) + 2\ H_2O(l)$

Ionic equation:

$2\ H^+(aq) + 2\ Br^-(aq) + Ca^{2+}(aq) + 2\ OH^-(aq) \rightarrow$
$Ca^{2+}(aq) + 2\ Br^-(aq) + 2\ H_2O(l)$

Delete spectator ions from the ionic equation to get the net ionic equation.

Net ionic equation: $H^+(aq) + OH^-(aq) \rightarrow H_2O(l)$

4.44 (a) $2\ Fe(OH)_3(s) + 3\ H_2SO_4(aq) \rightarrow Fe_2(SO_4)_3(aq) + 6\ H_2O(l)$

Ionic equation and net ionic equation are the same.

$2\ Fe(OH)_3(s) + 3\ H^+(aq) + 3\ HSO_4^-(aq) \rightarrow 2\ Fe^{3+}(aq) + 3\ SO_4^{2-}(aq) + 6\ H_2O(l)$

(b) $HClO_3(aq) + NaOH(aq) \rightarrow NaClO_3(aq) + H_2O(l)$

Ionic equation:

$HClO_3(aq) + Na^+(aq) + OH^-(aq) \rightarrow Na^+(aq) + ClO_3^-(aq) + H_2O(l)$

Delete spectator ions from the ionic equation to get the net ionic equation.

Net Ionic equation: $HClO_3(aq) + OH^-(aq) \rightarrow ClO_3^-(aq) + H_2O(l)$

Redox Reactions and Oxidation Numbers

4.45 The best reducing agents are at the bottom left of the periodic table. The best oxidizing agents are at the top right of the periodic table (excluding inert gases).

4.46 The most easily reduced elements in the periodic table are in the top-right corner, excluding group 8A.

The most easily oxidized elements in the periodic table are in the bottom-left corner.

4.47 (a) An oxidizing agent gains electrons.

(b) A reducing agent loses electrons.

(c) A substance undergoing oxidation loses electrons.

(d) A substance undergoing reduction gains electrons.

4.48 (a) In a redox reaction, the oxidation number decreases for an oxidizing agent.

(b) In a redox reaction, the oxidation number increases for a reducing agent.

(c) In a redox reaction, the oxidation number increases for a substance undergoing oxidation.

(d) In a redox reaction, the oxidation number decreases for a substance undergoing reduction.

4.49 (a) NO_2 O −2, N +4

(b) SO_3 O −2, S +6

(c) $COCl_2$ O −2, Cl −1, C +4

(d) CH_2Cl_2 Cl −1, H +1, C 0

(e) $KClO_3$ O −2, K +1, Cl +5

(f) HNO_3 O −2, H +1, N +5

4.50 (a) $VOCl_3$ O −2, Cl −1, V +5

(b) $CuSO_4$ O −2, S +6, Cu +2

(c) CH_2O O −2, H +1, C 0

(d) Mn_2O_7 O −2, Mn +7

(e) OsO_4 O −2, Os +8

(f) H_2PtCl_6 Cl −1, H +1, Pt +4

4.51 (a) ClO_3^- O −2, Cl +5 (b) SO_3^{2-} O −2, S +4

(c) $C_2O_4^{2-}$ O −2, C +3 (d) NO_2^- O −2, N +3

(e) BrO^- O −2, Br +1

4.52 (a) $Cr(OH)_4^-$ O −2, H +1, Cr +3

(b) $S_2O_3^{2-}$ O −2, S +2

(c) NO_3^- O −2, N +5

(d) MnO_4^{2-} O −2, Mn +6

(e) HPO_4^{2-} O −2, H +1, P +5

4.53 (a) $Ca(s) + Sn^{2+}(aq) \rightarrow Ca^{2+}(aq) + Sn(s)$

Ca(s) is oxidized (oxidation number increases from 0 to +2).

Sn^{2+}(aq) is reduced (oxidation number decreases from +2 to 0).

(b) $ICl(s) + H_2O(l) \rightarrow HCl(aq) + HOI(aq)$

No oxidation numbers change. The reaction is not a redox reaction.

4.54 (a) $Si(s) + 2\ Cl_2(g) \rightarrow SiCl_4(l)$

Si(s) is oxidized (oxidation number increases from 0 to +4).

$Cl_2(g)$ is reduced (oxidation number decreases from 0 to –1).

(b) $Cl_2(g) + 2\ NaBr(aq) \rightarrow Br_2(aq) + 2\ NaCl(aq)$

$Br^-(aq)$ is oxidized (oxidation number increases from –1 to 0).

$Cl_2(g)$ is reduced (oxidation number decreases from 0 to –1).

4.55 (a) Zn is below Na^+; therefore no reaction.

(b) Pt is below H^+; therefore no reaction.

(c) Au is below Ag^+; therefore no reaction.

(d) Ag is above Au^{3+}; the reaction is $Au^{3+}(aq) + 3\ Ag(s) \rightarrow 3\ Ag^+(aq) + Au(s)$

4.56 "Any element higher in the activity series will react with the ion of any element lower in the activity series."

$A + B^+ \rightarrow A^+ + B$; therefore A is higher than B.

$C^+ + D \rightarrow$ no reaction; therefore C is higher than D.

$B + D^+ \rightarrow B^+ + D$; therefore B is higher than D.

$B + C^+ \rightarrow B^+ + C$; therefore B is higher than C.

The net result is A > B > C > D.

4.57 (a) C is below A^+; therefore no reaction.

(b) D is below A^+; therefore no reaction.

Balancing Redox Reactions

4.58 (a) N oxidation number decreases from +5 to +2; reduction.

(b) Zn oxidation number increases from 0 to +2; oxidation.

(c) Ti oxidation number increases from +3 to +4; oxidation.

(d) Sn oxidation number decreases from +4 to +2; reduction.

4.59 (a) O oxidation number decreases from 0 to –2; reduction.

(b) O oxidation number increases from –1 to 0; oxidation.

(c) Mn oxidation number decreases from +7 to +6; reduction.

(d) C oxidation number increases from –2 to 0; oxidation.

4.60 (a) $NO_3^-(aq) \rightarrow NO(g)$

$NO_3^-(aq) \rightarrow NO(g) + 2\ H_2O(l)$

$4\ H^+(aq) + NO_3^-(aq) \rightarrow NO(g) + 2\ H_2O(l)$

$3\ e^- + 4\ H^+(aq) + NO_3^-(aq) \rightarrow NO(g) + 2\ H_2O(l)$

(b) $Zn(s) \rightarrow Zn^{2+}(aq) + 2\ e^-$

(c) $Ti^{3+}(aq) \rightarrow TiO_2(s)$

$Ti^{3+}(aq) + 2\ H_2O(l) \rightarrow TiO_2(s)$

$Ti^{3+}(aq) + 2\ H_2O(l) \rightarrow TiO_2(s) + 4\ H^+(aq)$

$Ti^{3+}(aq) + 2\ H_2O(l) \rightarrow TiO_2(s) + 4\ H^+(aq) + e^-$

(d) $Sn^{4+}(aq) + 2\ e^- \rightarrow Sn^{2+}(aq)$

4.61 (a) $O_2(g) \rightarrow OH^-(aq)$

$O_2(g) \rightarrow OH^-(aq) + H_2O(l)$

$3\ H^+(aq) + O_2(g) \rightarrow OH^-(aq) + H_2O(l)$

$$3\ H^+(aq) + 3\ OH^-(aq) + O_2(g) \rightarrow 4\ OH^-(aq) + H_2O(l)$$

$$3\ H_2O(l) + O_2(g) \rightarrow 4\ OH^-(aq) + H_2O(l)$$

$$4\ e^- + 2\ H_2O(l) + O_2(g) \rightarrow 4\ OH^-(aq)$$

(b) $H_2O_2(aq) \rightarrow O_2(g)$

$$H_2O_2(aq) \rightarrow O_2(g) + 2\ H^+(aq)$$

$$2\ OH^-(aq) + H_2O_2(aq) \rightarrow O_2(g) + 2\ H^+(aq) + 2\ OH^-(aq)$$

$$2\ OH^-(aq) + H_2O_2(aq) \rightarrow O_2(g) + 2\ H_2O(l) + 2\ e^-$$

(c) $MnO_4^-(aq) \rightarrow MnO_4^{2-}(aq)$

$$MnO_4^-(aq) + e^- \rightarrow MnO_4^{2-}(aq)$$

(d) $CH_3OH(aq) \rightarrow CH_2O(aq)$

$$CH_3OH(aq) \rightarrow CH_2O(aq) + 2\ H^+(aq)$$

$$CH_3OH(aq)\ 2\ OH^-(aq) \rightarrow CH_2O(aq) + 2\ H^+(aq) + 2\ OH^-(aq)$$

$$CH_3OH(aq)\ 2\ OH^-(aq) \rightarrow CH_2O(aq) + 2\ H_2O(l)$$

$$CH_3OH(aq)\ 2\ OH^-(aq) \rightarrow CH_2O(aq) + 2\ H_2O(l) + 2\ e^-$$

4.62 (a) $Te(s) + NO_3^-(aq) \rightarrow TeO_2(s) + NO(g)$

oxidation: $Te(s) \rightarrow TeO_2(s)$

reduction: $NO_3^-(aq) \rightarrow NO(g)$

(b) $H_2O_2(aq) + Fe^{2+}(aq) \rightarrow Fe^{3+}(aq) + H_2O(l)$

oxidation: $Fe^{2+}(aq) \rightarrow Fe^{3+}(aq)$

reduction: $H_2O_2(aq) \rightarrow H_2O(l)$

4.63 (a) $Mn(s) + NO_3^-(aq) \rightarrow Mn^{2+}(aq) + NO_2(g)$

oxidation: $Mn(s) \rightarrow Mn^{2+}(aq)$

reduction: $NO_3^-(aq) \rightarrow NO_2(g)$

(b) $Mn^{3+}(aq) \rightarrow MnO_2(s) + Mn^{2+}(aq)$

oxidation: $Mn^{3+}(aq) \rightarrow MnO_2(s)$

reduction: $Mn^{3+}(aq) \rightarrow Mn^{2+}(aq)$

4.64 (a) $Cr_2O_7^{2-}(aq) \rightarrow Cr^{3+}(aq)$

$Cr_2O_7^{2-}(aq) \rightarrow 2\ Cr^{3+}(aq)$

$Cr_2O_7^{2-}(aq) \rightarrow 2\ Cr^{3+}(aq) + 7\ H_2O(l)$

$14\ H^+(aq) + Cr_2O_7^{2-}(aq) \rightarrow 2\ Cr^{3+}(aq) + 7\ H_2O(l)$

$14\ H^+(aq) + Cr_2O_7^{2-}(aq) + 6\ e^- \rightarrow 2\ Cr^{3+}(aq) + 7\ H_2O(l)$

(b) $CrO_4^{2-}(aq) \rightarrow Cr(OH)_4^-(aq)$

$4\ H^+(aq) + CrO_4^{2-}(aq) \rightarrow Cr(OH)_4^-(aq)$

$4\ H^+(aq) + 4\ OH^-(aq) + CrO_4^{2-}(aq) \rightarrow Cr(OH)_4^-(aq) + 4\ OH^-(aq)$

$4\ H_2O(l) + CrO_4^{2-}(aq) \rightarrow Cr(OH)_4^-(aq) + 4\ OH^-(aq)$

$4\ H_2O(l) + CrO_4^{2-}(aq) + 3\ e^- \rightarrow Cr(OH)_4^-(aq) + 4\ OH^-(aq)$

(c) $Bi^{3+}(aq) \rightarrow BiO_3^-(aq)$

$Bi^{3+}(aq) + 3\ H_2O(l) \rightarrow BiO_3^-(aq)$

$Bi^{3+}(aq) + 3\ H_2O(l) \rightarrow BiO_3^-(aq) + 6\ H^+(aq)$

$Bi^{3+}(aq) + 3\ H_2O(l) + 6\ OH^-(aq) \rightarrow BiO_3^-(aq) + 6\ H^+(aq) + 6\ OH^-(aq)$

$Bi^{3+}(aq) + 3\ H_2O(l) + 6\ OH^-(aq) \rightarrow BiO_3^-(aq) + 6\ H_2O(l)$

$Bi^{3+}(aq) + 6\ OH^-(aq) \rightarrow BiO_3^-(aq) + 3\ H_2O(l)$

$Bi^{3+}(aq) + 6\ OH^-(aq) \rightarrow BiO_3^-(aq) + 3\ H_2O(l) + 2\ e^-$

(d) $ClO^-(aq) \rightarrow Cl^-(aq)$

$ClO^-(aq) \rightarrow Cl^-(aq) + H_2O(l)$

$2\ H^+(aq) + ClO^-(aq) \rightarrow Cl^-(aq) + H_2O(l)$

$2\ H^+(aq) + 2\ OH^-(aq) + ClO^-(aq) \rightarrow Cl^-(aq) + H_2O(l) + 2\ OH^-(aq)$

$2\ H_2O(aq) + ClO^-(aq) \rightarrow Cl^-(aq) + H_2O(l) + 2\ OH^-(aq)$

$H_2O(aq) + ClO^-(aq) \rightarrow Cl^-(aq) + 2\ OH^-(aq)$

$H_2O(aq) + ClO^-(aq) + 2\ e^- \rightarrow Cl^-(aq) + 2\ OH^-(aq)$

4.65 (a) $VO^{2+}(aq) \rightarrow V^{3+}(aq)$

$VO^{2+}(aq) \rightarrow V^{3+}(aq) + H_2O(l)$

$2\ H^+(aq) + VO^{2+}(aq) \rightarrow V^{3+}(aq) + H_2O(l)$

$2\ H^+(aq) + VO^{2+}(aq) + e^- \rightarrow V^{3+}(aq) + H_2O(l)$

(b) $Ni(OH)_2(s) \rightarrow Ni_2O_3(s)$

$2\ Ni(OH)_2(s) \rightarrow Ni_2O_3(s) + H_2O(l)$

$2\ Ni(OH)_2(s) \rightarrow Ni_2O_3(s) + H_2O(l) + 2\ H^+(aq)$

$2\ Ni(OH)_2(s) + 2\ OH^-(aq) \rightarrow Ni_2O_3(s) + H_2O(l) + 2\ H^+(aq) + 2\ OH^-(aq)$

$2\ Ni(OH)_2(s) + 2\ OH^-(aq) \rightarrow Ni_2O_3(s) + 3\ H_2O(l) + 2\ e^-$

(c) $NO_3^-(aq) \rightarrow NO_2(g)$

$NO_3^-(aq) \rightarrow NO_2(g) + H_2O(l)$

$2\ H^+(aq) + NO_3^-(aq) \rightarrow NO_2(g) + H_2O(l)$

$2\ H^+(aq)\ +\ NO_3^-(aq)\ +\ e^-\ \rightarrow\ NO_2(g)\ +\ H_2O(l)$

(d) $Br_2(aq) \rightarrow BrO_3^-(aq)$

$Br_2(aq) \rightarrow 2\ BrO_3^-(aq)$

$Br_2(aq) + 6\ H_2O(l) \rightarrow 2\ BrO_3^-(aq)$

$Br_2(aq) + 6\ H_2O(l) \rightarrow 2\ BrO_3^-(aq) + 12\ H^+(aq)$

$Br_2(aq) + 6\ H_2O(l) + 12\ OH^-(aq) \rightarrow 2\ BrO_3^-(aq) + 12\ H^+(aq) + 12\ OH^-(aq)$

$Br_2(aq) + 6\ H_2O(l) + 12\ OH^-(aq) \rightarrow 2\ BrO_3^-(aq) + 12\ H_2O(l)$

$Br_2(aq) + 12\ OH^-(aq) \rightarrow 2\ BrO_3^-(aq) + 6\ H_2O(l) + 10\ e^-$

4.66 (a) $MnO_4^-(aq) \rightarrow MnO_2(s)$

$MnO_4^-(aq) \rightarrow MnO_2(s) + 2\ H_2O(l)$

$4\ H^+(aq) + MnO_4^-(aq) \rightarrow MnO_2(s) + 2\ H_2O(l)$

$[4\ H^+(aq) + MnO_4^-(aq) + 3\ e^- \rightarrow MnO_2(s) + 2\ H_2O(l)\ ] \times 2$
(reduction half reaction)

$IO_3^-(aq) \rightarrow IO_4^-(aq)$

$H_2O(l) + IO_3^-(aq) \rightarrow IO_4^-(aq)$

$H_2O(l) + IO_3^-(aq) \rightarrow IO_4^-(aq) + 2\ H^+(aq)$

$[H_2O(l) + IO_3^-(aq) \rightarrow IO_4^-(aq) + 2\ H^+(aq) + 2\ e^-] \times 3$
(oxidation half reaction)

Combine the two half reactions.

$8\ H^+(aq) + 3\ H_2O(l) + 2\ MnO_4^-(aq) + 3\ IO_3^-(aq) \rightarrow$
$6\ H^+(aq) + 4\ H_2O(l) + 2\ MnO_2(s) + 3\ IO_4^-(aq)$

$2\ H^+(aq) + 2\ MnO_4^-(aq) + 3\ IO_3^-(aq) \rightarrow 2\ MnO_2(s) + 3\ IO_4^-(aq) + H_2O(l)$

$2\ H^+(aq) + 2\ OH^-(aq) + 2\ MnO_4^-(aq) + 3\ IO_3^-(aq) \rightarrow$
$2\ MnO_2(s) + 3\ IO_4^-(aq) + H_2O(l) + 2\ OH^-(aq)$

$2\ H_2O(l) + 2\ MnO_4^-(aq) + 3\ IO_3^-(aq) \rightarrow$
$2\ MnO_2(s) + 3\ IO_4^-(aq) + H_2O(l) + 2\ OH^-(aq)$

$H_2O(l) + 2\ MnO_4^-(aq) + 3\ IO_3^-(aq) \rightarrow$
$2\ MnO_2(s) + 3\ IO_4^-(aq) + 2\ OH^-(aq)$

(b) $Cu(OH)_2(s) \rightarrow Cu(s)$

$Cu(OH)_2(s) \rightarrow Cu(s) + 2\ H_2O(l)$

$2\ H^+(aq) + Cu(OH)_2(s) \rightarrow Cu(s) + 2\ H_2O(l)$

$[2\ H^+(aq) + Cu(OH)_2(s) + 2\ e^- \rightarrow Cu(s) + 2\ H_2O(l)] \times 2$
(reduction half reaction)

$N_2H_4(aq) \rightarrow N_2(g)$

$N_2H_4(aq) \rightarrow N_2(g) + 4\ H^+(aq)$

$N_2H_4(aq) \rightarrow N_2(g) + 4\ H^+(aq) + 4\ e^-$ (oxidation half reaction)

Combine the two half reactions.

$4\ H^+(aq) + 2\ Cu(OH)_2(s) + N_2H_4(aq) \rightarrow$
$2\ Cu(s) + 4\ H_2O(l) + N_2(g) + 4\ H^+(aq)$

$2\ Cu(OH)_2(s) + N_2H_4(aq) \rightarrow 2\ Cu(s) + 4\ H_2O(l) + N_2(g)$

(c) $Fe(OH)_2(s) \rightarrow Fe(OH)_3(s)$

$Fe(OH)_2(s) + H_2O(l) \rightarrow Fe(OH)_3(s)$

$Fe(OH)_2(s) + H_2O(l) \rightarrow Fe(OH)_3(s) + H^+(aq)$

$[Fe(OH)_2(s) + H_2O(l) \rightarrow Fe(OH)_3(s) + H^+(aq) + e^-] \times 3$
(oxidation half reaction)

$CrO_4^{2-}(aq) \rightarrow Cr(OH)_4^-(aq)$

$4\ H^+(aq) + CrO_4^{2-}(aq) \rightarrow Cr(OH)_4^-(aq)$

$4\ H^+(aq) + CrO_4^{2-}(aq) + 3\ e^- \rightarrow Cr(OH)_4^-(aq)$
(reduction half reaction)

Combine the two half reactions.

$3\ Fe(OH)_2(s) + 3\ H_2O(l) + 4\ H^+(aq) + CrO_4^{2-}(aq) \rightarrow$
$3\ Fe(OH)_3(s) + 3\ H^+(aq) + Cr(OH)_4^-(aq)$

$3\ Fe(OH)_2(s) + 3\ H_2O(l) + H^+(aq) + CrO_4^{2-}(aq) \rightarrow$
$3\ Fe(OH)_3(s) + Cr(OH)_4^-(aq)$

$3\ Fe(OH)_2(s) + 3\ H_2O(l) + H^+(aq) + OH^-(aq) + CrO_4^{2-}(aq) \rightarrow$
$3\ Fe(OH)_3(s) + Cr(OH)_4^-(aq) + OH^-(aq)$

$3\ Fe(OH)_2(s) + 4\ H_2O(l) + CrO_4^{2-}(aq) \rightarrow$
$3\ Fe(OH)_3(s) + Cr(OH)_4^-(aq) + OH^-(aq)$

(d) $ClO_4^-(aq) \rightarrow ClO_2^-(aq)$

$ClO_4^-(aq) \rightarrow ClO_2^-(aq) + 2\ H_2O(l)$

$4\ H^+(aq) + ClO_4^-(aq) \rightarrow ClO_2^-(aq) + 2\ H_2O(l)$

$4\ H^+(aq) + ClO_4^-(aq) + 4\ e^- \rightarrow ClO_2^-(aq) + 2\ H_2O(l)$
(reduction half reaction)

$H_2O_2(aq) \rightarrow O_2(g)$

$H_2O_2(aq) \rightarrow O_2(g) + 2\ H^+(aq)$

$[H_2O_2(aq) \rightarrow O_2(g) + 2\ H^+(aq) + 2\ e^-] \times 2$
(oxidation half reaction)

Combine the two half reactions.

$4\ H^+(aq) + ClO_4^-(aq) + 2\ H_2O_2(aq) \rightarrow$
$ClO_2^-(aq) + 2\ H_2O(l) + 2\ O_2(g) + 4\ H^+(aq)$

$ClO_4^-(aq) + 2\ H_2O_2(aq) \rightarrow ClO_2^-(aq) + 2\ H_2O(l) + 2\ O_2(g)$

4.67 (a) $S_2O_3^{2-}(aq) \rightarrow S_4O_6^{2-}(aq)$

$2\ S_2O_3^{2-}(aq) \rightarrow S_4O_6^{2-}(aq)$

$2\ S_2O_3^{2-}(aq) \rightarrow S_4O_6^{2-}(aq) + 2\ e^-$ (oxidation half reaction)

$I_2(aq) \rightarrow I^-(aq)$

$I_2(aq) \rightarrow 2\ I^-(aq)$

$I_2(aq) + 2\ e^- \rightarrow 2\ I^-(aq)$ (reduction half reaction)

Combine the two half reactions.

$2\ S_2O_3^{2-}(aq) + I_2(aq) \rightarrow S_4O_6^{2-}(aq) + 2\ I^-(aq)$

(b) $Mn^{2+}(aq) \rightarrow MnO_2(s)$

$Mn^{2+}(aq) + 2\ H_2O(l) \rightarrow MnO_2(s)$

$Mn^{2+}(aq) + 2\ H_2O(l) \rightarrow MnO_2(s) + 4\ H^+(aq)$

$Mn^{2+}(aq) + 2\ H_2O(l) \rightarrow MnO_2(s) + 4\ H^+(aq) + 2\ e^-$
(oxidation half reaction)

$H_2O_2(aq) \rightarrow 2\ H_2O(l)$

$2\ H^+(aq) + H_2O_2(aq) \rightarrow 2\ H_2O(l)$

$2\ H^+(aq) + H_2O_2(aq) + 2\ e^- \rightarrow 2\ H_2O(l)$ (reduction half reaction)

Combine the two half reactions.

$Mn^{2+}(aq) + 2\ H_2O(l) + 2\ H^+(aq) + H_2O_2(aq) \rightarrow$
$MnO_2(s) + 4\ H^+(aq) + 2\ H_2O(l)$

$Mn^{2+}(aq) + H_2O_2(aq) \rightarrow MnO_2(s) + 2\ H^+(aq)$

$Mn^{2+}(aq) + H_2O_2(aq) + 2\ OH^-(aq) \rightarrow MnO_2(s) + 2\ H^+(aq) + 2\ OH^-(aq)$

$Mn^{2+}(aq) + H_2O_2(aq) + 2\ OH^-(aq) \rightarrow MnO_2(s) + 2\ H_2O(l)$

(c) $Zn(s) \rightarrow Zn(OH)_4^{2-}(aq)$

$4\ H_2O(l) + Zn(s) \rightarrow Zn(OH)_4^{2-}(aq)$

$4\ H_2O(l) + Zn(s) \rightarrow Zn(OH)_4^{2-}(aq) + 4\ H^+(aq)$

$[4\ H_2O(l) + Zn(s) \rightarrow Zn(OH)_4^{2-}(aq) + 4\ H^+(aq) + 2\ e^-] \times 4$
(oxidation half reaction)

$NO_3^-(aq) \rightarrow NH_3(aq)$

$NO_3^-(aq) \rightarrow NH_3(aq) + 3\ H_2O(l)$

$9\ H^+(aq) + NO_3^-(aq) \rightarrow NH_3(aq) + 3\ H_2O(l)$

$9\ H^+(aq) + NO_3^-(aq) + 8\ e^- \rightarrow NH_3(aq) + 3\ H_2O(l)$
(reduction half reaction)

Combine the two half reactions.

$16\ H_2O(l) + 4\ Zn(s) + 9\ H^+(aq) + NO_3^-(aq) \rightarrow$
$4\ Zn(OH)_4^{2-}(aq) + 16\ H^+(aq) + NH_3(aq) + 3\ H_2O(l)$

$13\ H_2O(l) + 4\ Zn(s) + NO_3^-(aq) \rightarrow 4\ Zn(OH)_4^{2-}(aq) + 7\ H^+(aq) + NH_3(aq)$

$13\ H_2O(l) + 4\ Zn(s) + NO_3^-(aq) + 7\ OH^-(aq) \rightarrow$
$4\ Zn(OH)_4^{2-}(aq) + 7\ H^+(aq) + 7\ OH^-(aq) + NH_3(aq)$

$13\ H_2O(l) + 4\ Zn(s) + NO_3^-(aq) + 7\ OH^-(aq) \rightarrow$
$4\ Zn(OH)_4^{2-}(aq) + 7\ H_2O(l) + NH_3(aq)$

$6\ H_2O(l) + 4\ Zn(s) + NO_3^-(aq) + 7\ OH^-(aq) \rightarrow 4\ Zn(OH)_4^{2-}(aq) + NH_3(aq)$

(d) $Bi(OH)_3(s) \rightarrow Bi(s)$

$Bi(OH)_3(s) \rightarrow Bi(s) + 3\ H_2O(l)$

$3\ H^+(aq) + Bi(OH)_3(s) \rightarrow Bi(s) + 3\ H_2O(l)$

$[3\ H^+(aq) + Bi(OH)_3(s) + 3\ e^- \rightarrow Bi(s) + 3\ H_2O(l)] \times 2$
(reduction half reaction)

$Sn(OH)_3^-(aq) \rightarrow Sn(OH)_6^{2-}(aq)$

$Sn(OH)_3^-(aq) + 3\ H_2O(l) \rightarrow Sn(OH)_6^{2-}(aq)$

$Sn(OH)_3^-(aq) + 3\ H_2O(l) \rightarrow Sn(OH)_6^{2-}(aq) + 3\ H^+(aq)$

$[Sn(OH)_3^-(aq) + 3\ H_2O(l) \rightarrow Sn(OH)_6^{2-}(aq) + 3\ H^+(aq) + 2\ e^-] \times 3$

Combine the two half reactions.

$6\ H^+(aq) + 2\ Bi(OH)_3(s) + 3\ Sn(OH)_3^-(aq) + 9\ H_2O(l) \rightarrow$
$2\ Bi(s) + 6\ H_2O(l) + 3\ Sn(OH)_6^{2-}(aq) + 9\ H^+(aq)$

$2\ Bi(OH)_3(s) + 3\ Sn(OH)_3^-(aq) + 3\ H_2O(l) \rightarrow$
$2\ Bi(s) + 3\ Sn(OH)_6^{2-}(aq) + 3\ H^+(aq)$

$2\ Bi(OH)_3(s) + 3\ Sn(OH)_3^-(aq) + 3\ H_2O(l) + 3\ OH^-(aq) \rightarrow$
$2\ Bi(s) + 3\ Sn(OH)_6^{2-}(aq) + 3\ H^+(aq) + 3\ OH^-(aq)$

$2\ Bi(OH)_3(s) + 3\ Sn(OH)_3^-(aq) + 3\ H_2O(l) + 3\ OH^-(aq) \rightarrow$
$2\ Bi(s) + 3\ Sn(OH)_6^{2-}(aq) + 3\ H_2O(l)$

$2\ Bi(OH)_3(s) + 3\ Sn(OH)_3^-(aq) + 3\ OH^-(aq) \rightarrow 2\ Bi(s) + 3\ Sn(OH)_6^{2-}(aq)$

4.68 (a) $Zn(s) \rightarrow Zn^{2+}(aq)$

$Zn(s) \rightarrow Zn^{2+}(aq) + 2\ e^-$ (oxidation half reaction)

$VO^{2+}(aq) \rightarrow V^{3+}(aq)$

$VO^{2+}(aq) \rightarrow V^{3+}(aq) + H_2O(l)$

$2\ H^+(aq) + VO^{2+}(aq) \rightarrow V^{3+}(aq) + H_2O(l)$

$[2\ H^+(aq) + VO^{2+}(aq) + e^- \rightarrow V^{3+}(aq) + H_2O(l)] \times 2$
(reduction half reaction)

Combine the two half reactions.

$Zn(s) + 2\ VO^{2+}(aq) + 4\ H^+(aq) \rightarrow Zn^{2+}(aq) + 2\ V^{3+}(aq) + 2\ H_2O(l)$

(b) $Ag(s) \rightarrow Ag^+(aq)$

$Ag(s) \rightarrow Ag^+(aq) + e^-$ (oxidation half reaction)

$NO_3^-(aq) \rightarrow NO_2(g)$

$NO_3^-(aq) \rightarrow NO_2(g) + H_2O(l)$

$2\ H^+(aq) + NO_3^-(aq) \rightarrow NO_2(g) + H_2O(l)$

$2\ H^+(aq) + NO_3^-(aq) + e^- \rightarrow NO_2(g) + H_2O(l)$ (reduction half reaction)

Combine the two half reactions.

$2\ H^+(aq) + Ag(s) + NO_3^-(aq) \rightarrow Ag^+(aq) + NO_2(g) + H_2O(l)$

(c) $Mg(s) \rightarrow Mg^{2+}(aq)$

$[Mg(s) \rightarrow Mg^{2+}(aq) + 2\ e^-] \times 3$ (oxidation half reaction)

$VO_4^{3-}(aq) \rightarrow V^{2+}(aq)$

$VO_4^{3-}(aq) \rightarrow V^{2+}(aq) + 4\ H_2O(l)$

$8\ H^+(aq) + VO_4^{3-}(aq) \rightarrow V^{2+}(aq) + 4\ H_2O(l)$

$[8\ H^+(aq) + VO_4^{3-}(aq) + 3\ e^- \rightarrow V^{2+}(aq) + 4\ H_2O(l)] \times 2$
(reduction half reaction)

Combine the two half reactions.

$3\ Mg(s) + 16\ H^+(aq) + 2\ VO_4^{3-}(aq) \rightarrow 3\ Mg^{2+}(aq) + 2\ V^{2+}(aq) + 8\ H_2O(l)$

(d) $I^-(aq) \rightarrow I_3^-(aq)$

$3\ I^-(aq) \rightarrow I_3^-(aq)$

$[3\ I^-(aq) \rightarrow I_3^-(aq) + 2\ e^-] \times 8$ (oxidation half reaction)

$IO_3^-(aq) \rightarrow I_3^-(aq)$

$3\ IO_3^-(aq) \rightarrow I_3^-(aq)$

$3\ IO_3^-(aq) \rightarrow I_3^-(aq) + 9\ H_2O(l)$

$18\ H^+(aq) + 3\ IO_3^-(aq) \rightarrow I_3^-(aq) + 9\ H_2O(l)$

$18\ H^+(aq) + 3\ IO_3^-(aq) + 16\ e^- \rightarrow I_3^-(aq) + 9\ H_2O(l)$
(reduction half reaction)

Combine the two half reactions.

$18\ H^+(aq) + 3\ IO_3^-(aq) + 24\ I^-(aq) \rightarrow 9\ I_3^-(aq) + 9\ H_2O(l)$

Divide each coefficient by 3.

$6\ H^+(aq) + IO_3^-(aq) + 8\ I^-(aq) \rightarrow 3\ I_3^-(aq) + 3\ H_2O(l)$

4.69 (a) $MnO_4^-(aq) \rightarrow Mn^{2+}(aq)$

$MnO_4^-(aq) \rightarrow Mn^{2+}(aq) + 4\ H_2O(l)$

$8\ H^+(aq) + MnO_4^-(aq) \rightarrow Mn^{2+}(aq) + 4\ H_2O(l)$

$[8\ H^+(aq) + MnO_4^-(aq) + 5\ e^- \rightarrow Mn^{2+}(aq) + 4\ H_2O(l)] \times 4$
(reduction half reaction)

$C_2H_5OH(aq) \rightarrow CH_3COOH(aq)$

$C_2H_5OH(aq) + H_2O(l) \rightarrow CH_3COOH(aq)$

$C_2H_5OH(aq) + H_2O(l) \rightarrow CH_3COOH(aq) + 4\ H^+(aq)$

$[C_2H_5OH(aq) + H_2O(l) \rightarrow CH_3COOH(aq) + 4\ H^+(aq) + 4\ e^-] \times 5$
(oxidation half reaction)

Combine the two half reactions.

$32\ H^+(aq) + 4\ MnO_4^-(aq) + 5\ C_2H_5OH(aq) + 5\ H_2O(l) \rightarrow$
$4\ Mn^{2+}(aq) + 16\ H_2O(l) + 5\ CH_3COOH(aq) + 20\ H^+(aq)$

$12\ H^+(aq) + 4\ MnO_4^-(aq) + 5\ C_2H_5OH(aq) \rightarrow$
$4\ Mn^{2+}(aq) + 11\ H_2O(l) + 5\ CH_3COOH(aq)$

(b) $Cr_2O_7^{2-}(aq) \rightarrow Cr^{3+}(aq)$

$Cr_2O_7^{2-}(aq) \rightarrow 2\ Cr^{3+}(aq)$

$Cr_2O_7^{2-}(aq) \rightarrow 2\ Cr^{3+}(aq) + 7\ H_2O(l)$

$14\ H^+(aq) + Cr_2O_7^{2-}(aq) \rightarrow 2\ Cr^{3+}(aq) + 7\ H_2O(l)$

$14\ H^+(aq) + Cr_2O_7^{2-}(aq) + 6\ e^- \rightarrow 2\ Cr^{3+}(aq) + 7\ H_2O(l)$
(reduction half reaction)

$H_2O_2(aq) \rightarrow O_2(g)$

$H_2O_2(aq) \rightarrow O_2(g) + 2\ H^+(aq)$

$[H_2O_2(aq) \rightarrow O_2(g) + 2\ H^+(aq) + 2\ e^-] \times 3$
(oxidation half reaction)

Combine the two half reactions.

$14\ H^+(aq) + Cr_2O_7^{2-}(aq) + 3\ H_2O_2(aq) \rightarrow 2\ Cr^{3+}(aq) + 7\ H_2O(l) + 3\ O_2(g) + 6\ H^+(aq)$

$8\ H^+(aq) + Cr_2O_7^{2-}(aq) + 3\ H_2O_2(aq) \rightarrow 2\ Cr^{3+}(aq) + 7\ H_2O(l) + 3\ O_2(g)$

(c) $Sn^{2+}(aq) \rightarrow Sn^{4+}(aq)$

$[Sn^{2+}(aq) \rightarrow Sn^{4+}(aq) + 2\ e^-] \times 4$ (oxidation half reaction)

$IO_4^-(aq) \rightarrow I^-(aq)$

$IO_4^-(aq) \rightarrow I^-(aq) + 4\ H_2O(l)$

$8\ H^+(aq) + IO_4^-(aq) \rightarrow I^-(aq) + 4\ H_2O(l)$

$8\ H^+(aq) + IO_4^-(aq) + 8\ e^- \rightarrow I^-(aq) + 4\ H_2O(l)$
(reduction half reaction)

Combine the two half reactions.

$4\ Sn^{2+}(aq) + 8\ H^+(aq) + IO_4^-(aq) \rightarrow 4\ Sn^{4+}(aq) + I^-(aq) + 4\ H_2O(l)$

(d) $PbO_2(s) + Cl^-(aq) \rightarrow PbCl_2(s)$

$PbO_2(s) + 2\ Cl^-(aq) \rightarrow PbCl_2(s)$

$PbO_2(s) + 2\ Cl^-(aq) \rightarrow PbCl_2(s) + 2\ H_2O(l)$

$PbO_2(s) + 4\ H^+(aq) + 2\ Cl^-(aq) \rightarrow PbCl_2(s) + 2\ H_2O(l)$

$[PbO_2(s) + 4\ H^+(aq) + 2\ Cl^-(aq) + 2\ e^- \rightarrow PbCl_2(s) + 2\ H_2O(l)] \times 2$
(reduction half reaction)

$H_2O(l) \rightarrow O_2(g)$

$2\ H_2O(l) \rightarrow O_2(g)$

$2\ H_2O(l) \rightarrow O_2(g) + 4\ H^+(aq)$

$2\ H_2O(l) \rightarrow O_2(g) + 4\ H^+(aq) + 4\ e^-$ (oxidation half reaction)

Combine the two half reactions.

$2\ PbO_2(s) + 8\ H^+(aq) + 4\ Cl^-(aq) + 2\ H_2O(l) \rightarrow$
$2\ PbCl_2(s) + 4\ H_2O(l) + O_2(g) + 4\ H^+(aq)$

$2\ PbO_2(s) + 4\ H^+(aq) + 4\ Cl^-(aq) \rightarrow 2\ PbCl_2(s) + 2\ H_2O(l) + O_2(g)$

Redox Titrations

4.70 $I_2(aq) + 2\ S_2O_3^{2-}(aq) \rightarrow S_4O_6^{2-}(aq) + 2\ I^-(aq)$

35.20 mL = 0.032 50 L

$$0.035\ 20\ L \times \frac{0.150\ mol\ S_2O_3^{2-}}{L} \times \frac{1\ mol\ I_2}{2\ mol\ S_2O_3^{2-}} \times \frac{253.8\ g\ I_2}{1\ mol\ I_2} = 0.670\ g\ I_2$$

4.71 $$2.486\ g\ I_2 \times \frac{1\ mol\ I_2}{253.8\ g\ I_2} \times \frac{2\ mol\ S_2O_3^{2-}}{1\ mol\ I_2} = 1.959 \times 10^{-2}\ mol\ S_2O_3^{2-}$$

$$1.959 \times 10^{-2}\text{ mol} \times \frac{1\text{ L}}{0.250\text{ mol}} = 0.0784\text{ L};\quad 0.0784\text{ L} = 78.4\text{ mL}$$

4.72 $3\ H_3AsO_3(aq) + BrO_3^-(aq) \rightarrow Br^-(aq) + 3\ H_3AsO_4(aq)$

22.35 mL = 0.022 35 L and 50.00 mL = 0.050 00 L

$$0.022\ 35\text{ L} \times \frac{0.100\text{ mol } BrO_3^-}{\text{L}} \times \frac{3\text{ mol } H_3AsO_3}{1\text{ mol } BrO_3^-} = 6.70 \times 10^{-3}\text{ mol } H_3AsO_3$$

$$\text{Molarity} = \frac{6.70 \times 10^{-3}\text{ mol}}{0.050\ 00\text{ L}} = 0.134\text{ M As(III)}$$

4.73 As_2O_3, 197.84 amu; 28.55 mL = 0.028 55 L

$$1.550\text{ g } As_2O_3 \times \frac{1\text{ mol } As_2O_3}{197.84\text{ g } As_2O_3} \times \frac{2\text{ mol } H_3AsO_3}{1\text{ mol } As_2O_3} \times \frac{1\text{ mol } BrO_3^-}{3\text{ mol } H_3AsO_3}$$

$$= 5.223 \times 10^{-3}\text{ mol } BrO_3^-;\quad KBrO_3\text{ molarity} = \frac{5.223 \times 10^{-3}\text{ mol}}{0.028\ 55\text{ L}} = 0.1829\text{ M}$$

4.74 $2\ Fe^{3+}(aq) + Sn^{2+}(aq) \rightarrow 2\ Fe^{2+}(aq) + Sn^{4+}(aq)$

13.28 mL = 0.013 28 L

$$0.013\ 28\text{ L} \times \frac{0.1015\text{ mol } Sn^{2+}}{\text{L}} \times \frac{2\text{ mol } Fe^{3+}}{1\text{ mol } Sn^{2+}} \times \frac{55.847\text{ g } Fe^{3+}}{1\text{ mol } Fe^{3+}} = 0.1506\text{ g } Fe^{3+}$$

$$\text{mass \% Fe} = \frac{0.1506\text{ g}}{0.1875\text{ g}} \times 100\% = 80.32\%$$

4.75 Fe_2O_3, 159.69 amu; 23.84 mL = 0.023 84 L

$$1.4855\text{ g } Fe_2O_3 \times \frac{1\text{ mol } Fe_2O_3}{159.69\text{ g } Fe_2O_3} \times \frac{2\text{ mol } Fe^{3+}}{1\text{ mol } Fe_2O_3} \times \frac{1\text{ mol } Sn^{2+}}{2\text{ mol } Fe^{3+}} = 0.009\ 302\text{ mol } Sn^{2+}$$

Sn^{2+} molarity = $\dfrac{0.009\ 302\ \text{mol}}{0.023\ 84\ \text{L}}$ = 0.3902 M

4.76 $C_2H_5OH(aq) + 2\ Cr_2O_7^{2-}(aq) + 16\ H^+(aq) \rightarrow 2\ CO_2(g) + 4\ Cr^{3+}(aq) + 11\ H_2O(l)$

C_2H_5OH, 46.07 amu; 8.76 mL = 0.008 76 L

$$0.008\ 76\ \text{L} \times \frac{0.049\ 88\ \text{mol}\ Cr_2O_7^{2-}}{\text{L}} \times \frac{1\ \text{mol}\ C_2H_5OH}{2\ \text{mol}\ Cr_2O_7^{2-}} \times \frac{46.07\ \text{g}\ C_2H_5OH}{1\ \text{mol}\ C_2H_5OH}$$

= 0.010 07 g C_2H_5OH

mass % C_2H_5OH = $\dfrac{0.010\ 07\ \text{g}}{10.002\ \text{g}} \times 100\%$ = 0.101%

4.77 21.08 mL = 0.021 08 L

$$0.021\ 08\ \text{L} \times \frac{9.88 \times 10^{-4}\ \text{mol}\ MnO_4^-}{\text{L}} \times \frac{5\ \text{mol}\ H_2C_2O_4}{2\ \text{mol}\ MnO_4^-} \times \frac{1\ \text{mol}\ Ca^{2+}}{1\ \text{mol}\ H_2C_2O_4} \times \frac{40.08\ \text{g}\ Ca^{2+}}{1\ \text{mol}\ Ca^{2+}}$$

= 0.002 09 g = 2.09 mg

General Problems

4.78 (a) $[Fe(CN)_6]^{3-}(aq) \rightarrow Fe(CN)_6]^{4-}(aq)$

$([Fe(CN)_6]^{3-}(aq) + e^- \rightarrow [Fe(CN)_6]^{4-}(aq)) \times 4$ (reduction half reaction)

$N_2H_4(aq) \rightarrow N_2(g)$

$N_2H_4(aq) \rightarrow N_2(g) + 4\ H^+(aq)$

$N_2H_4(aq) \rightarrow N_2(g) + 4\ H^+(aq) + 4\ e^-$

$N_2H_4(aq) + 4\ OH^-(aq) \rightarrow N_2(g) + 4\ H^+(aq) + 4\ OH^-(aq) + 4\ e^-$

$N_2H_4(aq) + 4\ OH^-(aq) \rightarrow N_2(g) + 4\ H_2O(l) + 4\ e^-$
(oxidation half reaction)

Combine the two half reactions.

$4\ [Fe(CN)_6]^{3-}(aq) + N_2H_4(aq) + 4\ OH^-(aq) \rightarrow$
$4\ [Fe(CN)_6]^{4-}(aq) + N_2(g) + 4\ H_2O(l)$

(b) $Cl_2(g) \rightarrow Cl^-(aq)$

$Cl_2(g) \rightarrow 2\ Cl^-(aq)$

$Cl_2(g) + 2\ e^- \rightarrow 2\ Cl^-(aq)$ (reduction half reaction)

$SeO_3^{2-}(aq) \rightarrow SeO_4^{2-}(aq)$

$SeO_3^{2-}(aq) + H_2O(l) \rightarrow SeO_4^{2-}(aq)$

$SeO_3^{2-}(aq) + H_2O(l) \rightarrow SeO_4^{2-}(aq) + 2\ H^+(aq)$

$SeO_3^{2-}(aq) + H_2O(l) \rightarrow SeO_4^{2-}(aq) + 2\ H^+(aq) + 2\ e^-$

$SeO_3^{2-}(aq) + H_2O(l) + 2\ OH^-(aq) \rightarrow$
$SeO_4^{2-}(aq) + 2\ H^+(aq) + 2\ OH^-(aq) + 2\ e^-$

$SeO_3^{2-}(aq) + H_2O(l) + 2\ OH^-(aq) \rightarrow SeO_4^{2-}(aq) + 2\ H_2O(l) + 2\ e^-$

$SeO_3^{2-}(aq) + 2\ OH^-(aq) \rightarrow SeO_4^{2-}(aq) + H_2O(l) + 2\ e^-$
(oxidation half reaction)

Combine the two half reactions.

$SeO_3^{2-}(aq) + Cl_2(g) + 2\ OH^-(aq) \rightarrow SeO_4^{2-}(aq) + 2\ Cl^-(aq) + H_2O(l)$

(c) $CoCl_2(aq) \rightarrow Co(OH)_3(s) + Cl^-(aq)$

$CoCl_2(aq) \rightarrow Co(OH)_3(s) + 2\ Cl^-(aq)$

$CoCl_2(aq) + 3\ H_2O(l) \rightarrow Co(OH)_3(s) + 2\ Cl^-(aq)$

$CoCl_2(aq) + 3\ H_2O(l) \rightarrow Co(OH)_3(s) + 2\ Cl^-(aq) + 3\ H^+(aq)$

$[CoCl_2(aq) + 3\ H_2O(l) \rightarrow Co(OH)_3(s) + 2\ Cl^-(aq) + 3\ H^+(aq) + e^-] \times 2$
(oxidation half reaction)

$HO_2^-(aq) \rightarrow H_2O(l)$

$HO_2^-(aq) \rightarrow 2\ H_2O(l)$

$3\ H^+(aq) + HO_2^-(aq) \rightarrow 2\ H_2O(l)$

$3\ H^+(aq) + HO_2^-(aq) + 2\ e^- \rightarrow 2\ H_2O(l)$ (reduction half reaction)

Combine the two half reactions.

$2\ CoCl_2(aq) + 6\ H_2O(l) + 3\ H^+(aq) + HO_2^-(aq) \rightarrow$
$2\ Co(OH)_3(s) + 4\ Cl^-(aq) + 6\ H^+(aq) + 2\ H_2O(l)$

$2\ CoCl_2(aq) + 4\ H_2O(l) + HO_2^-(aq) \rightarrow$
$2\ Co(OH)_3(s) + 4\ Cl^-(aq) + 3\ H^+(aq)$

$2\ CoCl_2(aq) + 4\ H_2O(l) + HO_2^-(aq) + 3\ OH^-(aq) \rightarrow$
$2\ Co(OH)_3(s) + 4\ Cl^-(aq) + 3\ H^+(aq) + 3\ OH^-(aq)$

$2\ CoCl_2(aq) + 4\ H_2O(l) + HO_2^-(aq) + 3\ OH^-(aq) \rightarrow$
$2\ Co(OH)_3(s) + 4\ Cl^-(aq) + 3\ H_2O(l)$

$2\ CoCl_2(aq) + H_2O(l) + HO_2^-(aq) + 3\ OH^-(aq) \rightarrow$
$2\ Co(OH)_3(s) + 4\ Cl^-(aq)$

4.79 $2\ H_3C_6H_5O_7(aq) + 3\ Ca(OH)_2(aq) \rightarrow Ca_3(C_6H_5O_7)_2(aq) + 6\ H_2O(l)$

$2\ H_3C_6H_5O_7(aq) + 3\ Ca^{2+}(aq) + 6\ OH^-(aq) \rightarrow$
$3\ Ca^{2+}(aq) + 2\ C_6H_5O_7^{3-}(aq) + 6\ H_2O(l)$

Net ionic equation: $H_3C_6H_5O_7(aq) + 3\ OH^-(aq) \rightarrow C_6H_5O_7^{3-}(aq) + 3\ H_2O(l)$

4.80 (a) TiO_2, insoluble (b) $SnCl_4$, soluble

(c) Ag_2S, insoluble (d) $Pd(NO_3)_2$, soluble

4.81 57.91 mL = 0.057 91 L

$$0.057\ 91\ \text{L} \times \frac{0.1018\ \text{mol Ce}^{4+}}{\text{L}} \times \frac{1\ \text{mol Fe}^{2+}}{1\ \text{mol Ce}^{4+}} \times \frac{55.85\ \text{g Fe}^{2+}}{1\ \text{mol Fe}^{2+}} = 0.3292\ \text{g Fe}^{2+}$$

$$\text{mass \% Fe} = \frac{0.3292\text{ g}}{1.2284\text{ g}} \times 100\% = 26.80\%$$

4.82 (a) C_2H_6 H +1, C –3

(b) $Na_2B_4O_7$ O –2, Na +1, B +3

(c) Mg_2SiO_4 O –2, Mg +2, Si +4

4.83 (a) $PbO_2(s) \rightarrow Pb^{2+}(aq)$

$PbO_2(s) \rightarrow Pb^{2+}(aq) + 2\ H_2O(l)$

$4\ H^+(aq) + PbO_2(s) \rightarrow Pb^{2+}(aq) + 2\ H_2O(l)$

$[4\ H^+(aq) + PbO_2(s) + 2\ e^- \rightarrow Pb^{2+}(aq) + 2\ H_2O(l)] \times 5$
(reduction half reaction)

$Mn^{2+}(aq) \rightarrow MnO_4^-(aq)$

$4\ H_2O(l) + Mn^{2+}(aq) \rightarrow MnO_4^-(aq)$

$4\ H_2O(l) + Mn^{2+}(aq) \rightarrow MnO_4^-(aq) + 8\ H^+(aq)$

$[4\ H_2O(l) + Mn^{2+}(aq) \rightarrow MnO_4^-(aq) + 8\ H^+(aq) + 5\ e^-] \times 2$
(oxidation half reaction)

Combine the two half reactions.

$20\ H^+(aq) + 5\ PbO_2(s) + 8\ H_2O(l) + 2\ Mn^{2+}(aq) \rightarrow$
$5\ Pb^{2+}(aq) + 10\ H_2O(l) + 2\ MnO_4^-(aq) + 16\ H^+(aq)$

$4\ H^+(aq) + 5\ PbO_2(s) + 2\ Mn^{2+}(aq) \rightarrow$
$5\ Pb^{2+}(aq) + 2\ H_2O(l) + 2\ MnO_4^-(aq)$

(b) $As_2O_3(s) \rightarrow H_3AsO_4(aq)$

$As_2O_3(s) \rightarrow 2\ H_3AsO_4(aq)$

$5\ H_2O(l) + As_2O_3(s) \rightarrow 2\ H_3AsO_4(aq)$

$5\ H_2O(l) + As_2O_3(s) \rightarrow 2\ H_3AsO_4(aq) + 4\ H^+(aq)$

$5\ H_2O(l)\ +\ As_2O_3(s)\ \rightarrow\ 2\ H_3AsO_4(aq)\ +\ 4\ H^+(aq)\ +\ 4\ e^-$
(oxidation half reaction)

$NO_3^-(aq)\ \rightarrow\ HNO_2(aq)$

$NO_3^-(aq)\ \rightarrow\ HNO_2(aq)\ +\ H_2O(l)$

$3\ H^+(aq)\ +\ NO_3^-(aq)\ \rightarrow\ HNO_2(aq)\ +\ H_2O(l)$

$[3\ H^+(aq)\ +\ NO_3^-(aq)\ +\ 2\ e^-\ \rightarrow\ HNO_2(aq)\ +\ H_2O(l)]\ x\ 2$
(reduction half reaction)

Combine the two half reactions.

$5\ H_2O(l) + As_2O_3(s) + 6\ H^+(aq) + 2\ NO_3^-(aq) \rightarrow$
$2\ H_3AsO_4(aq) + 4\ H^+(aq) + 2\ HNO_2(aq) + 2\ H_2O(l)$

$3\ H_2O(l) + As_2O_3(s) + 2\ H^+(aq) + 2\ NO_3^-(aq) \rightarrow$
$2\ H_3AsO_4(aq) + 2\ HNO_2(aq)$

(c) $Br_2(aq)\ \rightarrow\ Br^-(aq)$

$Br_2(aq)\ \rightarrow\ 2\ Br^-(aq)$

$Br_2(aq)\ +\ 2\ e^-\ \rightarrow\ 2\ Br^-(aq)$ (reduction half reaction)

$SO_2(g)\ \rightarrow\ HSO_4^-(aq)$

$2\ H_2O(l)\ +\ SO_2(g)\ \rightarrow\ HSO_4^-(aq)$

$2\ H_2O(l)\ +\ SO_2(g)\ \rightarrow\ HSO_4^-(aq)\ +\ 3\ H^+(aq)$

$2\ H_2O(l)\ +\ SO_2(g)\ \rightarrow\ HSO_4^-(aq)\ +\ 3\ H^+(aq)\ +\ 2\ e^-$
(oxidation half reaction)

Combine the two half reactions.

$2\ H_2O(l) + Br_2(aq) + SO_2(g) \rightarrow 2\ Br^-(aq) + HSO_4^-(aq) + 3\ H^+(aq)$

(d) $I^-(aq) \rightarrow I_2(s)$

$2\ I^-(aq) \rightarrow I_2(s)$

$2\ I^-(aq) \rightarrow I_2(s) + 2\ e^-$ (oxidation half reaction)

$NO_2^-(aq) \rightarrow NO(g)$

$NO_2^-(aq) \rightarrow NO(g) + H_2O(l)$

$2\ H^+(aq) + NO_2^-(aq) \rightarrow NO(g) + H_2O(l)$

$[2\ H^+(aq) + NO_2^-(aq) + e^- \rightarrow NO(g) + H_2O(l)] \times 2$
(reduction half reaction)

Combine the two half reactions.

$4\ H^+(aq) + 2\ NO_2^-(aq) + 2\ I^-(aq) \rightarrow 2\ NO(g) + I_2(s) + 2\ H_2O(l)$

4.84 Ions below Fe in the activity series (Table 4.3) can be reduced to their elemental forms by Fe. Of Ni^{2+}, Au^{3+}, Zn^{2+}, and Ba^{2+}, only Ni^{2+} and Au^{3+} can be reduced to their elemental forms by Fe.

5.1 Gamma ray $\nu = \dfrac{c}{\lambda} = \dfrac{3.00 \times 10^8 \text{ m/s}}{3.56 \times 10^{-11} \text{ m}} = 8.43 \times 10^{18} \text{ s}^{-1} = 8.43 \times 10^{18} \text{ Hz}$

Radar wave $\nu = \dfrac{c}{\lambda} = \dfrac{3.00 \times 10^8 \text{ m/s}}{10.3 \times 10^{-2} \text{ m}} = 2.91 \times 10^9 \text{ s}^{-1} = 2.91 \times 10^9 \text{ Hz}$

5.2 $\nu = 102.5 \text{ MHz} = 102.5 \times 10^6 \text{ Hz} = 102.5 \times 10^6 \text{ s}^{-1}$

$$\lambda = \frac{c}{\nu} = \frac{3.00 \times 10^8 \text{ m/s}}{102.5 \times 10^6 \text{ s}^{-1}} = 2.93 \text{ m}$$

$\nu = 9.55 \times 10^{17} \text{ Hz} = 9.55 \times 10^{17} \text{ s}^{-1}$

$$\lambda = \frac{c}{\nu} = \frac{3.00 \times 10^8 \text{ m/s}}{9.55 \times 10^{17} \text{ s}^{-1}} = 3.14 \times 10^{-10} \text{ m}$$

5.3 Violet has the highest frequency. Red has the longest wavelength.

5.4 Balmer series: $m = 2$; $R = 1.097 \times 10^{-2} \text{ nm}^{-1}$

$$\frac{1}{\lambda} = R\left[\frac{1}{m^2} - \frac{1}{n^2}\right]; \qquad \frac{1}{\lambda} = R\left[\frac{1}{2^2} - \frac{1}{7^2}\right]$$

$$\frac{1}{\lambda} = 2.519 \times 10^{-3} \text{ nm}^{-1}; \quad \lambda = 397.0 \text{ nm}$$

5.5 Paschen series: $m = 3$; $R = 1.097 \times 10^{-2} \text{ nm}^{-1}$

$$\frac{1}{\lambda} = R\left[\frac{1}{m^2} - \frac{1}{n^2}\right]; \qquad \frac{1}{\lambda} = R\left[\frac{1}{3^2} - \frac{1}{4^2}\right]$$

$$\frac{1}{\lambda} = 5.333 \times 10^{-4} \text{ nm}^{-1}; \quad \lambda = 1875 \text{ nm}$$

5.6 Paschen series: m = 3; $R = 1.097 \times 10^{-2}\ nm^{-1}$

$$\frac{1}{\lambda} = R\left[\frac{1}{m^2} - \frac{1}{n^2}\right]; \qquad \frac{1}{\lambda} = R\left[\frac{1}{3^2} - \frac{1}{\infty^2}\right]$$

$$\frac{1}{\lambda} = 1.219 \times 10^{-3}\ nm^{-1}; \quad \lambda = 820.4\ nm$$

5.7 $\lambda = 91.2\ nm = 91.2 \times 10^{-9}\ m$

$$\nu = \frac{c}{\lambda} = \frac{3.00 \times 10^8\ m/s}{91.2 \times 10^{-9}\ m} = 3.29 \times 10^{15}\ s^{-1}$$

$$E = h\nu = (6.626 \times 10^{-34}\ J{\cdot}s)(3.29 \times 10^{15}\ s^{-1}) = 2.18 \times 10^{-18}\ J/photon$$

$$E = (2.18 \times 10^{-18}\ J/photon)(6.022 \times 10^{23}\ photons/mol)$$

$$E = 1.31 \times 10^6\ J/mol = 1310\ kJ/mol$$

5.8 IR, $\lambda = 1.55 \times 10^{-6}\ m$

$$E = h\frac{c}{\lambda} = (6.626 \times 10^{-34}\ J{\cdot}s)\left(\frac{3.00 \times 10^8\ m/s}{1.55 \times 10^{-6}\ m}\right)(6.022 \times 10^{23}/mol)$$

$$E = 7.72 \times 10^4\ J/mol = 77.2\ kJ/mol$$

UV, $\lambda = 250\ nm = 250 \times 10^{-9}\ m$

$$E = h\frac{c}{\lambda} = (6.626 \times 10^{-34}\ J{\cdot}s)\left(\frac{3.00 \times 10^8\ m/s}{250 \times 10^{-9}\ m}\right)(6.022 \times 10^{23}/mol)$$

$$E = 4.79 \times 10^5\ J/mol = 479\ kJ/mol$$

X ray, $\lambda = 5.49\ nm = 5.49 \times 10^{-9}\ m$

$$E = h\frac{c}{\lambda} = (6.626 \times 10^{-34}\ J{\cdot}s)\left(\frac{3.00 \times 10^8\ m/s}{5.49 \times 10^{-9}\ m}\right)(6.022 \times 10^{23}/mol)$$

$$E = 2.18 \times 10^7\ J/mol = 2.18 \times 10^4\ kJ/mol$$

5.9 $\lambda = \dfrac{h}{mv} = \dfrac{6.626 \times 10^{-34}\ kg\ m^2\ s^{-1}}{(1150\ kg)(24.6\ m/s)} = 2.34 \times 10^{-38}\ m$

5.10 $(\Delta x)(\Delta mv) \geq \dfrac{h}{4\pi}$

uncertainty in velocity = (45 m/s)(0.02) = 0.9 m/s

$$\Delta x \geq \frac{h}{4\pi(\Delta mv)} = \frac{6.626 \times 10^{-34}\ kg\ m^2\ s^{-1}}{4\pi(0.120\ kg)(0.9\ m/s)} = 5 \times 10^{-34}\ m$$

5.11

n	l	m_l	Orbital
5	0	0	5s
	1	−1, 0, +1	5p
	2	−2, −1, 0, +1, +2	5d
	3	−3, −2, −1, 0, +1, +2, +3	5f
	4	−4, −3, −2, −1, 0, +1, +2, +3, +4	5g

There are 25 possible orbitals in the fifth shell.

5.12 (a) For n = 2, l = 0 or 1

(b) For l = 0, m_l = 0

(c) The l quantum number is a positive integer or zero.

5.13 (a) 2p (b) 4f (c) 3d

5.14 (a) 3s orbital: n = 3, l = 0, m_l = 0

(b) 2p orbital: n = 2, l = 1, m_l = −1, 0, +1

(c) 4d orbital: n = 4, l = 2, m_l = −2, −1, 0, +1, +2

5.15 m = 1, n = ∞; R = 1.097×10^{-2} nm^{-1}

$$\frac{1}{\lambda} = R\left[\frac{1}{m^2} - \frac{1}{n^2}\right]; \qquad \frac{1}{\lambda} = R\left[\frac{1}{1^2} - \frac{1}{\infty^2}\right]; \qquad \frac{1}{\lambda} = R\left[\frac{1}{1}\right]$$

$$\frac{1}{\lambda} = 1.097 \times 10^{-2}\ \text{nm}^{-1};\ \lambda = 91.2\ \text{nm}$$

$$E = (6.626 \times 10^{-34}\ \text{J}\cdot\text{s})\left(\frac{3.00 \times 10^{8}\ \text{m/s}}{91.2 \times 10^{-9}\ \text{m}}\right)(6.022 \times 10^{23}/\text{mol})$$

$$E = 1.31 \times 10^{6}\ \text{J/mol} = 1.31 \times 10^{3}\ \text{kJ/mol}$$

5.16 The g orbitals have four nodal planes.

5.17 (a) Ti, $1s^2 2s^2 2p^6 3s^2 3p^6 4s^2 3d^2$ or [Ar] $4s^2 3d^2$

[Ar] ↑↓ (4s) ↑ ↑ __ __ __ (3d)

(b) Zn, $1s^2 2s^2 2p^6 3s^2 3p^6 4s^2 3d^{10}$ or [Ar] $4s^2 3d^{10}$

[Ar] ↑↓ (4s) ↑↓ ↑↓ ↑↓ ↑↓ ↑↓ (3d)

(c) Sn, $1s^2 2s^2 2p^6 3s^2 3p^6 4s^2 3d^{10} 4p^6 5s^2 4d^{10} 5p^2$ or [Kr] $5s^2 4d^{10} 5p^2$

[Kr] ↑↓ (5s) ↑↓ ↑↓ ↑↓ ↑↓ ↑↓ (4d) ↑ ↑ __ (5p)

(d) Pb, [Xe] $6s^2 4f^{14} 5d^{10} 6p^2$

5.18 For Na^+, $1s^2 2s^2 2p^6$

5.19 Cr, Cu, Nb, Mo, Ru, Rh, Pd, Ag, Gd, Pt, Au, Th, Pa, U, Np, Cm

5.20 (a) Ra^{2+} [Rn] (b) La^{3+} [Xe] (d) Ti^{4+} [Ar] (d) N^{3-} [Ne]

Each ion has the ground–state electron configuration of the noble gas closest to it in the periodic table.

5.21 (a) Ba; atoms get larger as you go down a group.

(b) W; atoms get smaller as you go across a period.

(c) Sn; atoms get larger as you go down a group.

(d) Ce; atoms get smaller as you go across a period.

5.22 (a) O^{2-}; decrease in effective nuclear charge and an increase in electron–electron repulsions lead to the larger anion.

(b) S; atoms get larger as you go down a group.

(c) Fe; in Fe^{3+} electrons are removed from a larger valence shell and there is an increase in effective nuclear charge leading to the smaller cation.

(d) H^-; decrease in effective nuclear charge and an increase in electron–electron repulsions lead to the larger anion.

Understanding Key Concepts

1. 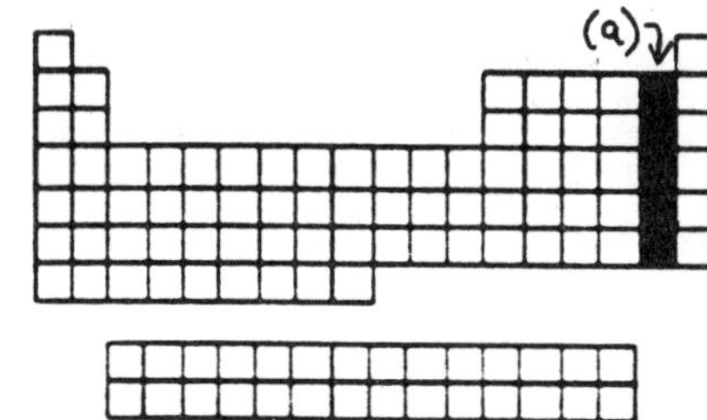

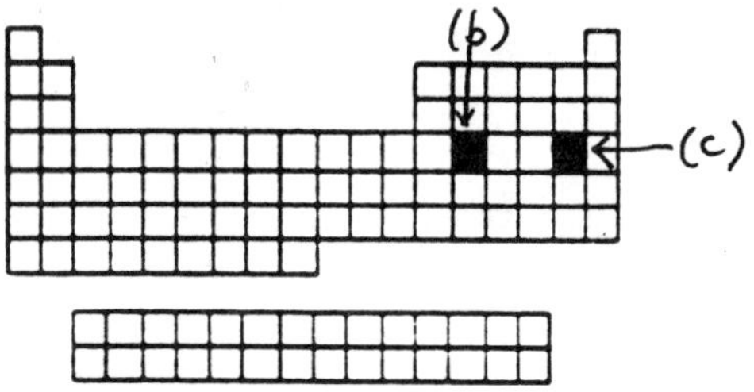

(b)

(c)

2. [Ar] $4s^2 3d^{10} 4p^1$ is Ga.

3. (a) K, [Ar] $\underset{4s}{\uparrow}$

(b) Mo, [Kr] ⇅ (5s) ↑ ↑ ↑ ↑ __ (4d)

(c) Sn, [Kr] ⇅ (5s) ⇅ ⇅ ⇅ ⇅ ⇅ (4d) ↑ ↑ __ (5p)

(d) Ir, [Xe] ⇅ (6s) ⇅ ⇅ ⇅ ⇅ ⇅ ⇅ ⇅ (4f) ⇅ ⇅ ↑ ↑ ↑ (5d)

4. K^+, r = 133 pm; Cl^-, r = 184 pm; K, r = 227 pm

5. Z = 118 [Rn] $7s^2 5f^{14} 6d^{10} 7p^6$

6. Order of orbital filling:

1s→2s→2p→3s→3p→4s→3d→4p→5s→4d→5p→6s→4f→5d→6p→7s→

5f→6d→7p 8s→5g

Z = 121

7. (a) $3p_y$ n = 3, l = 1 (b) $4d_{z^2}$ n = 4, l = 2

Additional Problems

Electromagnetic Radiation

5.23 Violet has the higher frequency and energy. Red has the higher wavelength.

5.24 Ultraviolet light has the higher frequency and the higher energy. Infrared light has the higher wavelength.

5.25 $\lambda = \frac{c}{\nu} = \frac{3.00 \times 10^8 \text{ m/s}}{5.5 \times 10^{15} \text{ s}^{-1}} = 5.5 \times 10^{-8} \text{ m}$

5.26 $\lambda = \frac{c}{\nu} = \frac{3.00 \times 10^8 \text{ m/s}}{4.33 \times 10^{-3} \text{ m}} = 6.93 \times 10^{10} \text{ s}^{-1} = 6.93 \times 10^{10} \text{ Hz}$

5.27 $E = h\nu = (6.626 \times 10^{-34} \text{ J·s})(2.4 \times 10^{11} \text{ s}^{-1}) = 1.6 \times 10^{-22}$ J/photon

$E = (1.6 \times 10^{-22} \text{ J/photon})(6.02 \times 10^{23} \text{ photons/mol}) = 96 \text{ J/mol} = 0.096 \text{ kJ/mol}$

5.28 (a) $\nu = 99.5 \text{ MHz} = 99.5 \times 10^6 \text{ s}^{-1}$

$E = h\nu = (6.626 \times 10^{-34} \text{ J·s})(99.5 \times 10^6 \text{ s}^{-1})(6.022 \times 10^{23} \text{ /mol})$

$E = 3.97 \times 10^{-2} \text{ J/mol} = 3.97 \times 10^{-5} \text{ kJ/mol}$

$\nu = 115.0 \text{ KHz} = 115.0 \times 10^3 \text{ s}^{-1}$

$E = h\nu = (6.626 \times 10^{-34} \text{ J·s})(115.0 \times 10^3 \text{ s}^{-1})(6.022 \times 10^{23} \text{ /mol})$

$E = 4.589 \times 10^{-5} \text{ J/mol} = 4.589 \times 10^{-8} \text{ kJ/mol}$

The FM radio wave (99.5 MHz) has the higher energy.

(b) $\lambda = 3.44 \times 10^{-9}$ m

$$E = h\frac{c}{\lambda} = (6.626 \times 10^{-34} \text{ J·s})\left(\frac{3.00 \times 10^8 \text{ m/s}}{3.44 \times 10^{-9} \text{ m}}\right)(6.022 \times 10^{23}/\text{mol})$$

$E = 3.48 \times 10^7 \text{ J/mol} = 3.48 \times 10^4 \text{ kJ/mol}$

$\lambda = 6.71 \times 10^{-2}$ m

$$E = h\frac{c}{\lambda} = (6.626 \times 10^{-34} \text{ J·s})\left(\frac{3.00 \times 10^8 \text{ m/s}}{6.71 \times 10^{-2} \text{ m}}\right)(6.022 \times 10^{23}/\text{mol})$$

$E = 1.78 \text{ J/mol} = 1.78 \times 10^{-3} \text{ kJ/mol}$

The X ray ($\lambda = 3.44 \times 10^{-9}$ m) has the higher energy.

5.29 $\nu = 400 \text{ MHz} = 400 \times 10^6 \text{ s}^{-1}$

$E = (6.626 \times 10^{-34} \text{ J·s})(400 \times 10^6 \text{ s}^{-1})(6.02 \times 10^{23}/\text{mol})$

$E = 0.160 \text{ J/mol} = 1.60 \times 10^{-4} \text{ kJ/mol}$

5.30 $4.5 \times 10^9 \text{ km} = 4.5 \times 10^{12} \text{ m}$

$d = v \times t; \qquad t = \frac{d}{v} = \frac{4.5 \times 10^{12} \text{ m}}{3.0 \times 10^8 \text{ m/s}} = 15{,}000 \text{ s}$

5.31 (a) $E = 90.5 \text{ kJ/mol} \times \frac{1000 \text{ J}}{1 \text{ kJ}} \times \frac{1 \text{ mol}}{6.02 \times 10^{23}} = 1.50 \times 10^{-19} \text{ J}$

$\nu = \frac{E}{h} = \frac{1.50 \times 10^{-19} \text{ J}}{6.626 \times 10^{-34} \text{ J·s}} = 2.27 \times 10^{14} \text{ s}^{-1}$

$\lambda = \frac{c}{\nu} = \frac{3.00 \times 10^8 \text{ m/s}}{2.27 \times 10^{14} \text{ s}^{-1}} = 1.32 \times 10^{-6} \text{ m} = 1320 \times 10^{-9} \text{ m} = 1320 \text{ nm}$, NEAR IR

(b) $E = 8.05 \times 10^{-4} \text{ kJ/mol} \times \frac{1000 \text{ J}}{1 \text{ kJ}} \times \frac{1 \text{ mol}}{6.02 \times 10^{23}} = 1.34 \times 10^{-24} \text{ J}$

$\nu = \frac{E}{h} = \frac{1.34 \times 10^{-24} \text{ J}}{6.626 \times 10^{-34} \text{ J·s}} = 2.02 \times 10^9 \text{ s}^{-1}$

$\lambda = \frac{c}{\nu} = \frac{3.00 \times 10^8 \text{ m/s}}{2.02 \times 10^9 \text{ s}^{-1}} = 0.149 \text{ m}$, RADIOWAVE

(c) $E = 1.83 \times 10^3 \text{ kJ/mol} \times \frac{1000 \text{ J}}{1 \text{ kJ}} \times \frac{1 \text{ mol}}{6.02 \times 10^{23}} = 3.04 \times 10^{-18} \text{ J}$

$\nu = \frac{E}{h} = \frac{3.04 \times 10^{-18} \text{ J}}{6.626 \times 10^{-34} \text{ J·s}} = 4.59 \times 10^{15} \text{ s}^{-1}$

$\lambda = \frac{c}{\nu} = \frac{3.00 \times 10^8 \text{ m/s}}{4.59 \times 10^{15} \text{ s}^{-1}} = 6.54 \times 10^{-8} \text{ m} = 65.4 \times 10^{-9} \text{ m} = 65.4 \text{ nm}$, UV

5.32 (a) $E = h\nu = (6.626 \times 10^{-34}\ J{\cdot}s)(5.97 \times 10^{19}\ s^{-1})\left(\frac{1\ kJ}{1000\ J}\right)(6.022 \times 10^{23}/mol)$

$E = 2.38 \times 10^{7}$ kJ/mol

(b) $E = h\nu = (6.626 \times 10^{-34}\ J{\cdot}s)(1.26 \times 10^{6}\ s^{-1})\left(\frac{1\ kJ}{1000\ J}\right)(6.022 \times 10^{23}/mol)$

$E = 5.03 \times 10^{-7}$ kJ/mol

(c) $E = h\nu = (6.626 \times 10^{-34}\ J{\cdot}s)\left(\frac{3.00 \times 10^{8}\ m/s}{2.57 \times 10^{2}\ m}\right)\left(\frac{1\ kJ}{1000\ J}\right)(6.022 \times 10^{23}/mol)$

$E = 4.66 \times 10^{-7}$ kJ/mol

5.33 $\lambda = \frac{h}{mv} = \frac{6.626 \times 10^{-34}\ kg\ m^2\ s^{-1}}{(1.673 \times 10^{-27}\ kg)(0.25 \times 3.00 \times 10^{8}\ m/s)} = 5.28 \times 10^{-15}\ m$

5.34 156 km/h = 156×10^{3} m/3600 s = 43.3 m/s; 145 g = 0.145 kg

$$\lambda = \frac{h}{mv} = \frac{6.626 \times 10^{-34}\ kg\ m^2\ s^{-1}}{(0.145\ kg)(43.3\ m/s)} = 1.06 \times 10^{-34}\ m$$

5.35 1.55 mg = 1.55×10^{-3} g = 1.55×10^{-6} kg

$$\lambda = \frac{h}{mv} = \frac{6.626 \times 10^{-34}\ kg\ m^2\ s^{-1}}{(1.55 \times 10^{-6}\ kg)(1.38\ m/s)} = 3.10 \times 10^{-28}\ m$$

5.36 145 g = 0.145 kg; 0.500 nm = 0.500×10^{-9} m

$$\lambda = \frac{h}{mv} = \frac{6.626 \times 10^{-34}\ kg\ m^2\ s^{-1}}{(0.145\ kg)(0.500 \times 10^{-9}\ m)} = 9.14 \times 10^{-24}\ m/s$$

5.37 Different atoms produce very different and distinctive atomic line spectra. These distinctive atomic spectra can be used for element identification.

5.38 For n = 3; λ = 656.3 nm = 656.3×10^{-9} m

$$E = h\frac{c}{\lambda} = (6.626 \times 10^{-34}\ J \cdot s)\left(\frac{2.998 \times 10^{8}\ m/s}{656.3 \times 10^{-9}\ m}\right)\left(\frac{1\ kJ}{1000\ J}\right)(6.022 \times 10^{23}/mol)$$

E = 182.3 kJ/mol

For n = 4; λ = 486.1 nm = 486.1×10^{-9} m

$$E = h\frac{c}{\lambda} = (6.626 \times 10^{-34}\ J \cdot s)\left(\frac{2.998 \times 10^{8}\ m/s}{486.1 \times 10^{-9}\ m}\right)\left(\frac{1\ kJ}{1000\ J}\right)(6.022 \times 10^{23}/mol)$$

E = 246.1 kJ/mol

For n = 5; λ = 434.0 nm = 434.0×10^{-9} m

$$E = h\frac{c}{\lambda} = (6.626 \times 10^{-34}\ J \cdot s)\left(\frac{2.998 \times 10^{8}\ m/s}{434.0 \times 10^{-9}\ m}\right)\left(\frac{1\ kJ}{1000\ J}\right)(6.022 \times 10^{23}/mol)$$

E = 275.6 kJ/mol

5.39 m = 2, n = ∞; R = 1.097×10^{-2} nm^{-1}

$$\frac{1}{\lambda} = R\left[\frac{1}{m^2} - \frac{1}{n^2}\right]; \qquad \frac{1}{\lambda} = R\left[\frac{1}{2^2} - \frac{1}{\infty^2}\right]; \quad \frac{1}{\lambda} = R\left[\frac{1}{4}\right]$$

$$\frac{1}{\lambda} = 2.74 \times 10^{-3}\ nm^{-1};\ \lambda = 364.6\ nm$$

5.40 From problem 5.39, for n = ∞, λ = 364.6 nm = 364.6×10^{-9} m

$$E = h\frac{c}{\lambda} = (6.626 \times 10^{-34}\ J \cdot s)\left(\frac{2.998 \times 10^{8}\ m/s}{364.6 \times 10^{-9}\ m}\right)\left(\frac{1\ kJ}{1000\ J}\right)(6.022 \times 10^{23}/mol)$$

E = 328.1 kJ/mol

5.41 Brackett series: m = 4, n = 5; R = 1.097×10^{-2} nm^{-1}

$$\frac{1}{\lambda} = R\left[\frac{1}{m^2} - \frac{1}{n^2}\right]; \qquad \frac{1}{\lambda} = R\left[\frac{1}{4^2} - \frac{1}{5^2}\right]$$

$$\frac{1}{\lambda} = 2.468 \times 10^{-4}\ nm^{-1};\ \lambda = 4051\ nm$$

$$E = h\frac{c}{\lambda} = (6.626 \times 10^{-34}\ J\cdot s)\left(\frac{2.998 \times 10^{8}\ m/s}{4051 \times 10^{-9}\ m}\right)\left(\frac{1\ kJ}{1000\ J}\right)(6.022 \times 10^{23}/mol)$$

E = 29.55 kJ/mol, IR

Brackett series: m = 4, n = 6; R = 1.097×10^{-2} nm^{-1}

$$\frac{1}{\lambda} = R\left[\frac{1}{m^2} - \frac{1}{n^2}\right]; \qquad \frac{1}{\lambda} = R\left[\frac{1}{4^2} - \frac{1}{6^2}\right]$$

$$\frac{1}{\lambda} = 3.809 \times 10^{-4}\ nm^{-1};\ \lambda = 2625\ nm$$

$$E = h\frac{c}{\lambda} = (6.626 \times 10^{-34}\ J\cdot s)\left(\frac{2.998 \times 10^{8}\ m/s}{2625 \times 10^{-9}\ m}\right)\left(\frac{1\ kJ}{1000\ J}\right)(6.022 \times 10^{23}/mol)$$

E = 45.60 kJ/mol, IR

5.42 $\lambda = 330\ nm = 330 \times 10^{-9}\ m$

$$E = h\frac{c}{\lambda} = (6.626 \times 10^{-34}\ J\cdot s)\left(\frac{3.00 \times 10^{8}\ m/s}{330 \times 10^{-9}\ m}\right)\left(\frac{1\ kJ}{1000\ J}\right)(6.022 \times 10^{23}/mol)$$

E = 363 kJ/mol

Orbitals and Quantum Numbers

5.43 n is the principal quantum number. The size and energy level of an orbital depends on n.

l is the angular–momentum quantum number. l defines the three–dimensional shape of an orbital.

m_l is the magnetic quantum number. m_l defines the spatial orientation of an orbital.

m_s is the spin quantum number. m_s indicates the spin of the electron and can have either of two values, +½ or –½.

5.44 The Heisenberg uncertainty principle states that one can never know both the position and the velocity of an electron beyond a certain level of precision. This means we cannot think of electrons circling the nucleus in specific orbital paths, but we can think of electrons as being found in certain three–dimensional regions of space around the nucleus, called orbitals.

5.45 The probability of finding the electron drops off rapidly as distance from the nucleus increases, although it never drops to zero, even at large distances. As a result, there is no definite boundary or size for an orbital. However, we usually imagine the boundary surface of an orbital enclosing the volume where an electron spends 95% of its time.

5.46

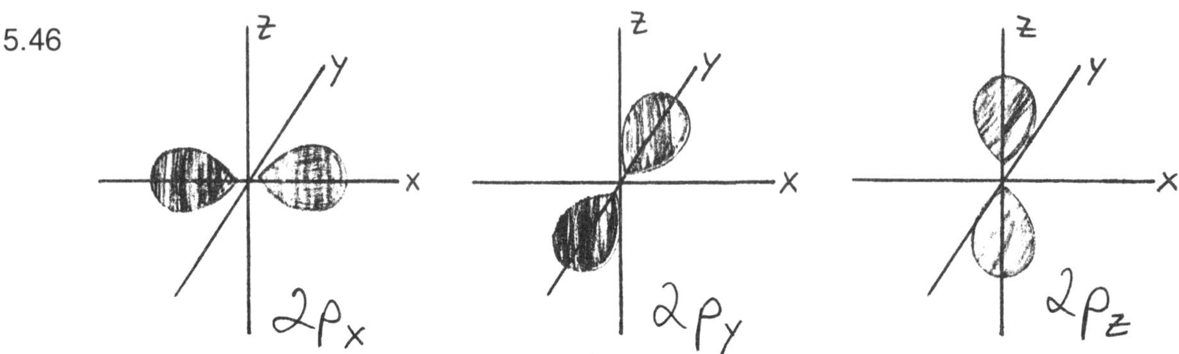

They only differ in their orientation in space.

5.47 Part of the electron–nucleus attraction is canceled by the electron–electron repulsion, an effect we describe by saying that the electrons are shielded from the nucleus by the other electrons. The net nuclear charge actually felt by an electron is called the effective nuclear charge, Z_{eff}, and is often substantially lower than the actual nuclear charge, Z_{actual}.

$$Z_{eff} = Z_{actual} - \text{electron shielding}$$

5.48 Electron shielding gives rise to energy differences among 3s, 3p, and 3d orbitals in multielectron atoms because of the differences in orbital shape. For example, the 3s orbital is spherical and has a large probability density near the nucleus, while the 3p orbital is dumbbell shaped with a node at the nucleus. An electron in a 3s orbital can penetrate closer to the nucleus than an electron in a 3p orbital can and feels less of a shielding effect from other electrons. Generally, for any given value of the principal quantum number n, a lower value of l corresponds to a higher value of Z_{eff} and to a lower energy for the orbital.

5.49 l = 2 for d subshells; l = 3 for f subshells

5.50 A 4s orbital has three nodal surfaces.

nodes are white

regions of maximum electron probability are black

4s orbital

5.51 (a) 4s $n = 4; l = 0; m_l = 0; m_s = \pm\frac{1}{2}$

(b) 3p $n = 3; l = 1; m_l = -1, 0, +1; m_s = \pm\frac{1}{2}$

(c) 5f $n = 5; l = 3; m_l = -3, -2, -1, 0, +1, +2, +3; m_s = \pm\frac{1}{2}$

(d) 5d $n = 5; l = 2; m_l = -2, -1, 0, +1, +2; m_s = \pm\frac{1}{2}$

5.52 (a) 3s (b) 2p (c) 4f (d) 4d

5.53 (a) is not allowed because for $l = 0$, $m_l = 0$ only.

(b) is allowed

(c) is not allowed because for $n = 4$, $l = 0, 1, 2,$ or 3 only.

5.54 For n = 5, the maximum number of electrons will occur when the 5g orbital is filled:

$$[Rn]\ 7s^2 5f^{14} 6d^{10} 7p^6 8s^2 5g^{18} = 138 \text{ electrons}$$

5.55 $n = 4$, $l = 0$ is a 4s orbital. The electron configuration is $1s^2 2s^2 2p^6 3s^2 3p^6 4s^2$. The number of electrons is 20.

5.56 $0.68\ g = 0.68 \times 10^{-3}\ kg$

$$(\Delta x)(\Delta mv) \geq \frac{h}{4\pi}$$

$$\Delta x \geq \frac{h}{4\pi(\Delta mv)} = \frac{6.626 \times 10^{-34}\ kg\ m^2\ s^{-1}}{4\pi(0.68 \times 10^{-3}\ kg)(0.1\ m/s)} = 8 \times 10^{-31}\ m$$

Electron Configurations

5.57 The number of elements in successive periods of the periodic table increases by the progression 2, 8, 18, 32 because the principal quantum number n, increases by 1 as we go from one period to the next. As the principal quantum number increases, the number of orbitals in a shell increases. The progression of elements parallels the number of electrons in a particular shell.

5.58 (a) The aufbau principle says that each successive electron added to an atom occupies the lowest–energy orbital available.

(b) The Pauli exclusion principle says that no two electrons in an atom have the same four quantum numbers.

(c) Hund's rule says that if two or more orbitals with the same energy are available, put one electron (with the same spin quantum number) in each until all are half full.

5.59

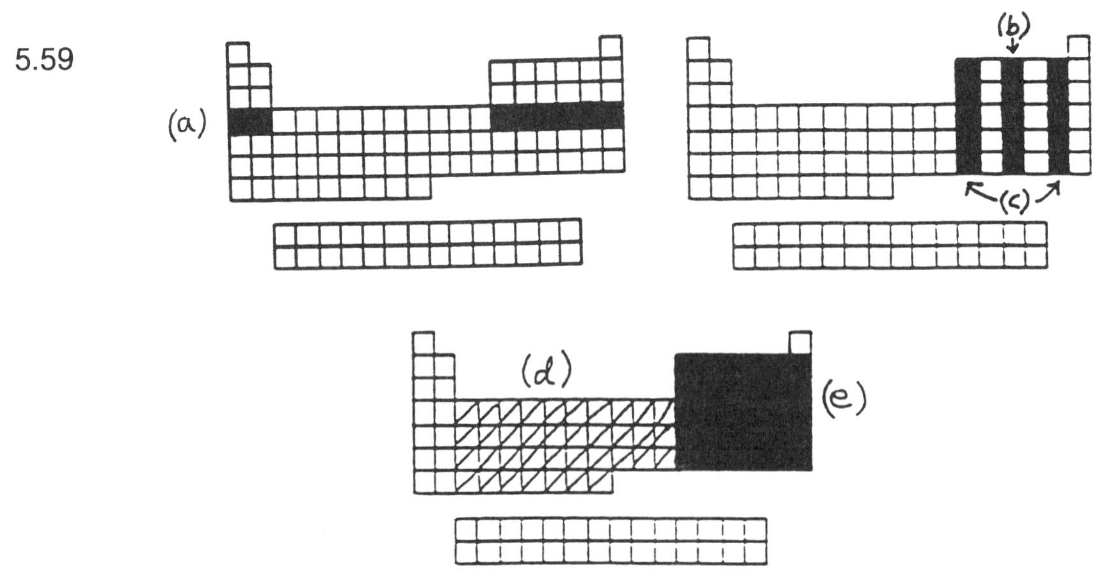

5.60 The n and l quantum numbers determine the energy level of an orbital in a multielectron atom.

5.61 (a) 2p < 3p < 5s < 4d (b) 2s < 4s < 3d < 4p (c) 3d < 4p < 5p < 6s

5.62 (a) 3d after 4s (b) 4p after 3d (c) 6d after 5f (d) 6s after 5p

5.63 (a) 3s before 3p (b) 3d before 4p (c) 6s before 4f (d) 4f before 5d

5.64 (a) Ti, Z = 22 $1s^22s^22p^63s^23p^64s^23d^2$

(b) Ru, Z = 44 $1s^22s^22p^63s^23p^64s^23d^{10}4p^65s^24d^6$

(c) Sn, Z = 50 $1s^22s^22p^63s^23p^64s^23d^{10}4p^65s^24d^{10}5p^2$

(d) Sr, Z = 38 $1s^22s^22p^63s^23p^64s^23d^{10}4p^65s^2$

(e) Se, Z = 34 $1s^22s^22p^63s^23p^64s^23d^{10}4p^4$

5.65 (a) Z = 55, Cs [Kr] $5s^24d^{10}5p^66s^1$ (b) Z = 41, Nb [Kr] $5s^24d^3$

(c) Z = 80, Hg [Xe] $6s^24f^{14}5d^{10}$ (d) Z = 62, Sm [Xe] $6s^24f^6$

5.66 (a) Rb, Z = 37 [Kr] ↑ (5s)

(b) W, Z = 74 [Xe] ↑↓ (6s) ↑↓ ↑↓ ↑↓ ↑↓ ↑↓ ↑↓ ↑↓ (4f) ↑ ↑ ↑ ↑ __ (5d)

(c) Ge, Z = 32 [Ar] ↑↓ (4s) ↑↓ ↑↓ ↑↓ ↑↓ ↑↓ (3d) ↑ ↑ __ (4p)

(d) Zr, Z = 40 [Kr] ↑↓ (5s) ↑ ↑ __ __ __ (4d)

5.67 (a) Z = 25, Mn [Ar] ↑↓ (4s) ↑ ↑ ↑ ↑ ↑ (3d)

(b) Z = 56, Ba [Xe] ↑↓ (6s)

(c) Z = 28, Ni [Ar] $\underline{\uparrow\downarrow}$ (4s) $\underline{\uparrow\downarrow}$ $\underline{\uparrow\downarrow}$ $\underline{\uparrow\downarrow}$ $\underline{\uparrow}$ $\underline{\uparrow}$ (3d)

(d) Z = 47, Ag [Kr] $\underline{\uparrow}$ (5s) $\underline{\uparrow\downarrow}$ $\underline{\uparrow\downarrow}$ $\underline{\uparrow\downarrow}$ $\underline{\uparrow\downarrow}$ $\underline{\uparrow\downarrow}$ (4d)

5.68 4s > 4d > 4f

5.69 Z = 118 [Rn] $7s^2 5f^{14} 6d^{10} 7p^6$

5.70 Z = 119 [Rn] $7s^2 5f^{14} 6d^{10} 7p^6 8s^1$

5.71 Z = 115 [Rn] $7s^2 5f^{14} 6d^{10} 7p^3$

5.72 (a) O $1s^2 2s^2 2p^4$ $\underline{\uparrow\downarrow}$ $\underline{\uparrow}$ $\underline{\uparrow}$ (2p) 2 unpaired e^-

(b) Si $1s^2 2s^2 2p^6 3s^2 3p^2$ $\underline{\uparrow}$ $\underline{\uparrow}$ $\underline{\ \ }$ (3p) 2 unpaired e^-

(c) K [Ar] $4s^1$ 1 unpaired e^-

(d) As [Ar] $4s^2 3d^{10} 4p^3$ $\underline{\uparrow}$ $\underline{\uparrow}$ $\underline{\uparrow}$ (4p) 3 unpaired e^-

5.73 (a) Z = 31, Ga (b) Z = 46, Pd

5.74 (a) La^{3+} [Xe]

(b) Ag^+ [Kr] $4d^{10}$

(c) Sn^{2+} [Kr] $5s^2 4d^{10}$

5.75 Z = 30, Zn

5.76 Cr^{2+} [Ar] $3d^4$ ↑ ↑ ↑ ↑ __ (3d)

Fe^{2+} [Ar] $3d^6$ ↑↓ ↑ ↑ ↑ ↑ (3d)

Atomic Radii and Periodic Properties

5.77 Atomic radii increase down a group because as you go down a group the electron shells are farther away from the nucleus.

5.78 As you go across a period, the effective nuclear charge increases, causing a decrease in atomic radii.

5.79 F < O < S

5.80 (a) K, lower in group 1A (b) Ta, lower in group 5B

(c) V, farther to the left in same period

(d) Ba, four periods lower and only one group to the right

5.81 Mg has a higher ionization energy than Na because Mg has a higher Z_{eff} and a smaller size.

5.82 F has a higher electron affinity than C because of a higher effective nuclear charge and room in the valence shell for the additional electron. In addition, F^- achieves a noble–gas electron configuration.

5.83 Group 2A, right under Ra.

5.84 Cu^{2+} has fewer electrons and a larger effective nuclear charge; therefore it has the smaller ionic radius.

5.85 $S^{2-} > Ca^{2+} > Sc^{3+} > Ti^{4+}$, Z_{eff} increases on going from S^{2-} to Ti^{4+}.

General Problems

5.86 Balmer series: m = 2; R = $1.097 \times 10^{-2}\ nm^{-1}$

$$\frac{1}{\lambda} = R\left[\frac{1}{m^2} - \frac{1}{n^2}\right]; \qquad \frac{1}{\lambda} = R\left[\frac{1}{2^2} - \frac{1}{6^2}\right]$$

$$\frac{1}{\lambda} = 2.438 \times 10^{-3}\ nm^{-1}; \quad \lambda = 410.2\ nm = 410.2 \times 10^{-9}\ m$$

$$E = h\frac{c}{\lambda} = (6.626 \times 10^{-34}\ J\cdot s)\left(\frac{2.998 \times 10^{8}\ m/s}{410.2 \times 10^{-9}\ m}\right)\left(\frac{1\ kJ}{1000\ J}\right)(6.022 \times 10^{23}/mol)$$

$$E = 291.6\ kJ/mol$$

5.87 Lyman series: m = 1, n = 4, 5, 6, and 7; R = $1.097 \times 10^{-2}\ nm^{-1}$

$$\frac{1}{\lambda} = R\left[\frac{1}{m^2} - \frac{1}{n^2}\right]$$

$$n = 4, \quad \frac{1}{\lambda} = R\left[\frac{1}{1^2} - \frac{1}{4^2}\right] = 1.028 \times 10^{-2}\ nm^{-1}; \quad \lambda = 97.2\ nm$$

$$n = 5, \quad \frac{1}{\lambda} = R\left[\frac{1}{1^2} - \frac{1}{5^2}\right] = 1.053 \times 10^{-2}\ nm^{-1}; \quad \lambda = 95.0\ nm$$

$$n = 6, \quad \frac{1}{\lambda} = R\left[\frac{1}{1^2} - \frac{1}{6^2}\right] = 1.067 \times 10^{-2}\ nm^{-1}; \quad \lambda = 93.8\ nm$$

$$n = 7, \quad \frac{1}{\lambda} = R\left[\frac{1}{1^2} - \frac{1}{7^2}\right] = 1.075 \times 10^{-2}\ nm^{-1}; \quad \lambda = 93.1\ nm$$

5.88 Pfund series: m = 5; R = $1.097 \times 10^{-2}\ nm^{-1}$

$$\frac{1}{\lambda} = R\left[\frac{1}{m^2} - \frac{1}{n^2}\right]$$

$n = 6,\ \frac{1}{\lambda} = R\left[\frac{1}{5^2} - \frac{1}{6^2}\right] = 1.341 \times 10^{-4}\ nm^{-1};\quad \lambda = 7458\ nm = 7458\ \times 10^{-9}\ m$

$E = h\frac{c}{\lambda} = (6.626 \times 10^{-34}\ J{\cdot}s)\left(\frac{2.998 \times 10^8\ m/s}{7458 \times 10^{-9}\ m}\right)\left(\frac{1\ kJ}{1000\ J}\right)(6.022 \times 10^{23}/mol)$
$E = 16.04\ kJ/mol$

$n = 7,\ \frac{1}{\lambda} = R\left[\frac{1}{5^2} - \frac{1}{7^2}\right] = 2.149 \times 10^{-4}\ nm^{-1};\quad \lambda = 4653\ nm = 4653 \times 10^{-9}\ m$

$E = h\frac{c}{\lambda} = (6.626 \times 10^{-34}\ J{\cdot}s)\left(\frac{2.998 \times 10^8\ m/s}{4653 \times 10^{-9}\ m}\right)\left(\frac{1\ kJ}{1000\ J}\right)(6.022 \times 10^{23}/mol)$
$E = 25.71\ kJ/mol$

These lines in the Pfund series are in the infrared region of the electromagnetic spectrum.

5.89 Pfund series: $m = 5$, $n = \infty$; $R = 1.097 \times 10^{-2}\ nm^{-1}$

$\frac{1}{\lambda} = R\left[\frac{1}{5^2} - \frac{1}{\infty^2}\right] = R\left[\frac{1}{25}\right] = 4.388 \times 10^{-4}\ nm^{-1};\quad \lambda = 2279\ nm$

5.90 (a) $E = \left(142\frac{kJ}{mol}\right)\left(\frac{1000\ J}{1\ kJ}\right)\left(\frac{1\ mol}{6.02 \times 10^{23}}\right) = 2.36 \times 10^{-19}\ J$

$E = h\frac{c}{\lambda},\quad \lambda = \frac{hc}{E} = \frac{(6.626 \times 10^{-34}\ J{\cdot}s)(3.00 \times 10^8\ m/s)}{2.36 \times 10^{-19}\ J}$

$\lambda = 8.42 \times 10^{-7}\ m$ (infrared)

(b) $E = \left(4.55 \times 10^{-2}\frac{kJ}{mol}\right)\left(\frac{1000\ J}{1\ kJ}\right)\left(\frac{1\ mol}{6.02 \times 10^{23}}\right) = 7.56 \times 10^{-23}\ J$

$E = h\frac{c}{\lambda},\quad \lambda = \frac{hc}{E} = \frac{(6.626 \times 10^{-34}\ J{\cdot}s)(3.00 \times 10^8\ m/s)}{7.56 \times 10^{-23}\ J}$

$\lambda = 2.63 \times 10^{-3}\ m$ (microwaves)

(c) $E = \left(4.81 \times 10^4 \frac{kJ}{mol}\right)\left(\frac{1000\ J}{1\ kJ}\right)\left(\frac{1\ mol}{6.02 \times 10^{23}}\right) = 7.99 \times 10^{-17}\ J$

$E = h\frac{c}{\lambda}, \quad \lambda = \frac{hc}{E} = \frac{(6.626 \times 10^{-34}\ J\cdot s)(3.00 \times 10^8\ m/s)}{7.99 \times 10^{-17}\ J}$

$\lambda = 2.49 \times 10^{-9}\ m$ (X rays)

5.91 (a) $E = h\nu = (6.626 \times 10^{-34}\ J\cdot s)(3.79 \times 10^{11}\ s^{-1})\left(\frac{1\ kJ}{1000\ J}\right)(6.022 \times 10^{23}/mol)$

$E = 0.151\ kJ/mol$

(b) $E = h\nu = (6.626 \times 10^{-34}\ J\cdot s)(5.45 \times 10^4\ s^{-1})\left(\frac{1\ kJ}{1000\ J}\right)(6.022 \times 10^{23}/mol)$

$E = 2.17 \times 10^{-8}\ kJ/mol$

(c) $E = h\nu = (6.626 \times 10^{-34}\ J\cdot s)\left(\frac{3.00 \times 10^8\ m/s}{4.11 \times 10^{-5}\ m}\right)\left(\frac{1\ kJ}{1000\ J}\right)(6.022 \times 10^{23}/mol)$

$E = 2.91\ kJ/mol$

5.92 $\lambda = \frac{h}{mv} = \frac{6.626 \times 10^{-34}\ kg\,m^2 s^{-1}}{(1.673 \times 10^{-27}\ kg)(0.053)(2.998 \times 10^8\ m/s)} = 2.5 \times 10^{-14}\ m$

5.93 $\lambda = \frac{h}{mv} = \frac{6.626 \times 10^{-34}\ kg\ m^2\ s^{-1}}{(0.0126\ kg)(12.6\ m/s)} = 4.17 \times 10^{-33}\ m$

5.94 780 nm is at the red end of the visible region of the electromagnetic spectrum.

$780\ nm = 780 \times 10^{-9}\ m$

$E = h\frac{c}{\lambda} = (6.626 \times 10^{-34}\ J\cdot s)\left(\frac{3.00 \times 10^8\ m/s}{780 \times 10^{-9}\ m}\right)\left(\frac{1\ kJ}{1000\ J}\right)(6.022 \times 10^{23}/mol)$

$E = 153\ kJ/mol$

5.95 795 nm = 795 x 10^{-9} m

$$E = h\frac{c}{\lambda} = (6.626 \times 10^{-34}\ J\cdot s)\left(\frac{3.00 \times 10^{8}\ m/s}{795 \times 10^{-9}\ m}\right)\left(\frac{1\ kJ}{1000\ J}\right)(6.022 \times 10^{23}/mol)$$

E = 151 kJ/mol

5.96 (a) Sr, Z = 38 [Kr] $\underline{\uparrow\downarrow}$ (5s)

(b) Cd, Z = 48 [Kr] $\underline{\uparrow\downarrow}$ (5s) $\underline{\uparrow\downarrow}\ \underline{\uparrow\downarrow}\ \underline{\uparrow\downarrow}\ \underline{\uparrow\downarrow}\ \underline{\uparrow\downarrow}$ (4d)

(c) Z = 22, Ti [Ar] $\underline{\uparrow\downarrow}$ (4s) $\underline{\uparrow}\ \underline{\uparrow}\ \underline{\ \ }\ \underline{\ \ }\ \underline{\ \ }$ (3d)

(d) Z = 34, Se [Ar] $\underline{\uparrow\downarrow}$ (4s) $\underline{\uparrow\downarrow}\ \underline{\uparrow\downarrow}\ \underline{\uparrow\downarrow}\ \underline{\uparrow\downarrow}\ \underline{\uparrow\downarrow}$ (3d) $\underline{\uparrow\downarrow}\ \underline{\uparrow}\ \underline{\uparrow}$ (4p)

5.97 For K, $Z_{eff} = \sqrt{\frac{(418.8\ kJ/mol)(4^2)}{1312\ kJ/mol}} = 2.26$

For Kr, $Z_{eff} = \sqrt{\frac{(1350.7\ kJ/mol)(4^2)}{1312\ kJ/mol}} = 4.06$

5.98 75 W = 75 J/s; 550 nm = 550 x 10^{-9} m

(0.05)(75 J/s) = 3.75 J/s

$$E = h\frac{c}{\lambda} = (6.626 \times 10^{-34}\ J\cdot s)\left(\frac{3.00 \times 10^{8}\ m/s}{550 \times 10^{-9}\ m}\right) = 3.61 \times 10^{-19}\ J/photon$$

$$\text{number of photons} = \frac{3.75\ J/s}{3.61 \times 10^{-19}\ J/photon} = 1 \times 10^{19}\ photons/s$$

5.99 $q = (350\text{ g})(4.184\text{ J/g}\cdot{}^\circ\text{C})(95^\circ\text{C} - 20^\circ\text{C}) = 109{,}830\text{ J}$

$\lambda = 15.0\text{ cm} = 15.0 \times 10^{-2}\text{ m}$

$$E = (6.626 \times 10^{-34}\text{ J}\cdot\text{s})\left(\frac{3.00 \times 10^{8}\text{ m/s}}{15.0 \times 10^{-2}\text{ m}}\right) = 1.33 \times 10^{-24}\text{ J/photon}$$

$$\text{number of photons} = \frac{109{,}830\text{ J}}{1.33 \times 10^{-24}\text{ J/photon}} = 8.3 \times 10^{28}\text{ photons}$$

5.100 $$E = \left(310\,\frac{\text{kJ}}{\text{mol}}\right)\left(\frac{1000\text{ J}}{1\text{ kJ}}\right)\left(\frac{1\text{ mol}}{6.02 \times 10^{23}}\right) = 5.15 \times 10^{-19}\text{ J}$$

$$E = h\frac{c}{\lambda}, \quad \lambda = \frac{hc}{E} = \frac{(6.626 \times 10^{-34}\text{ J}\cdot\text{s})(3.00 \times 10^{8}\text{ m/s})}{5.15 \times 10^{-19}\text{ J}}$$

$\lambda = 3.86 \times 10^{-7}\text{ m} = 386\text{ nm}$

6.1 (a) Br (b) S (c) Se (d) Ne

6.2 (a) Be $1s^2 2s^2$ N $1s^2 2s^2 2p^3$

Be would have the larger third ionization energy because this electron would come from the 1s orbital.

(b) Ga [Ar] $4s^2 3d^{10} 4p^1$ Ge [Ar] $4s^2 3d^{10} 4p^2$

Ga would have the larger fourth ionization energy because this electron would come from the 3d orbitals.

6.3 (b) Cl has the highest E_{i1} and smallest E_{i4}.

6.4 Cr [Ar] $4s^1 3d^5$
Mn [Ar] $4s^2 3d^5$
Fe [Ar] $4s^2 3d^6$

Cr can accept an electron into a 4s orbital. The 4s orbital is lower in energy than a 3d orbital. Both Mn and Fe accept the added electron into a 3d orbital that contains an electron, but Mn has a lower value of Z_{eff}. Therefore, Mn has a less negative E_{ea} than either Cr or Fe.

6.5

$K \rightarrow K^+ + e^-$	+ 418.8 kJ/mol	
$F + e^- \rightarrow F^-$	– 328 kJ/mol	
	+ 91 kJ/mol	(unfavorable)

6.6

$K(s) \rightarrow K(g)$	+89.2 kJ/mol
$K(g) \rightarrow K^+(g) + e^-$	+418.8 kJ/mol
$\frac{1}{2}[F_2(g) \rightarrow 2\ F(g)]$	+79 kJ/mol
$F(g) + e^- \rightarrow F^-(g)$	–328 kJ/mol
$K^+(g) + F^-(g) \rightarrow KF(s)$	–821 kJ/mol
Sum =	–562 kJ/mol

6.7 (a) KCl has the higher lattice energy because of the smaller K^+.

(b) CaF_2 has the higher lattice energy because of the smaller Ca^{2+}.

(c) CaO has the higher lattice energy because of the higher charge on both the cation and anion.

6.8 (a) Li_2O, O –2 (b) K_2O_2, O –1 (c) CsO_2, O –½

6.9 (a) $2\ Cs(s) + 2\ H_2O(l) \rightarrow 2\ Cs^+(aq) + 2\ OH^-(aq) + H_2(g)$

(b) $Na(s) + N_2(g) \rightarrow$ N. R.

(c) $Rb(s) + O_2(g) \rightarrow RbO_2(s)$

(d) $2\ K(s) + 2\ NH_3(g) \rightarrow 2\ KNH_2(s) + H_2(g)$

(e) $2\ Rb(s) + H_2(g) \rightarrow 2\ RbH(s)$

6.10 (a) $Be(s) + Br_2(l) \rightarrow BeBr_2(s)$

(b) $Sr(s) + 2\ H_2O(l) \rightarrow Sr(OH)_2(aq) + H_2(g)$

(c) $2\ Mg(s) + O_2(g) \rightarrow 2\ MgO(s)$

6.11 $BeCl_2 + 2\ K \rightarrow Be + 2\ KCl$

6.12 $Mg(s) + S(s) \rightarrow MgS(s)$

In MgS, the oxidation number of S is – 2.

6.13 $2\ Al(s) + 6\ H^+(aq) \rightarrow 2\ Al^{3+}(aq) + 3\ H_2(g)$

H^+ gains electrons and is the oxidizing agent. Al loses electrons and is the reducing agent.

6.14 $2\ Al(s) + 3\ S(s) \rightarrow Al_2S_3(s)$

6.15 (a) $Br_2(l) + Cl_2(g) \rightarrow 2\ BrCl(g)$

(b) $2\ Al(s) + 3\ F_2(g) \rightarrow 2\ AlF_3(s)$

(c) $H_2(g) + I_2(s) \rightarrow 2\ HI(g)$

6.16 $Br_2(l) + 2\ NaI(s) \rightarrow 2\ NaBr(s) + I_2(s)$

Br_2 gains electrons and is the oxidizing agent. I^- (from NaI) loses electrons and is the reducing agent.

6.17 (a) XeF_2 F −1, Xe +2

(b) XeF_4 F −1, Xe +4

(c) $XeOF_4$ F −1, O −2, Xe +6

6.18 (a) Rb would lose one electron and adopt the Kr noble-gas configuration.

(b) Ba would lose two electrons and adopt the Xe noble-gas configuration.

(c) Ga would lose three electrons and adopt an Ar-like noble-gas configuration (note that Ga^{3+} has ten 3d electrons in addition to the two 3s and six 3p electrons).

(d) F would gain one electron and adopt the Ne noble-gas configuration.

6.19 Group 6A elements will gain 2 electrons.

Understanding Key Concepts

1. (a) At is in Group 7A. The trend going down the group is gas → liquid → solid. At, being at the bottom of the group, should be a solid.

(b) At would likely be dark, like I_2, maybe with a metallic sheen.

(c) At is likely to react with Na just like the other halogens, yielding NaAt.

2. 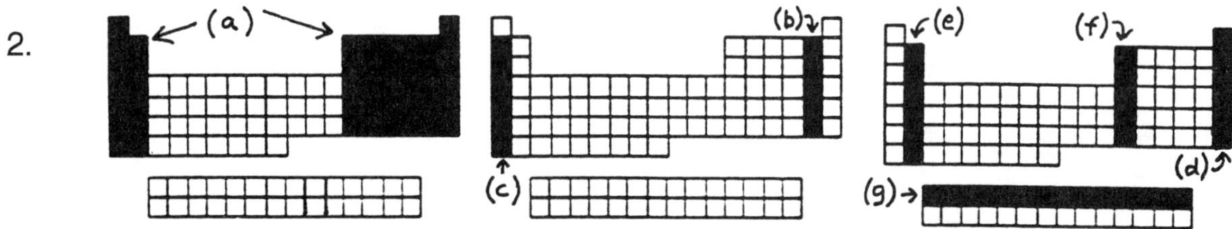

3. (a) shows an extended array, which represents an ionic compound.

   (b) shows discrete units, which represent a covalent compound.

4.

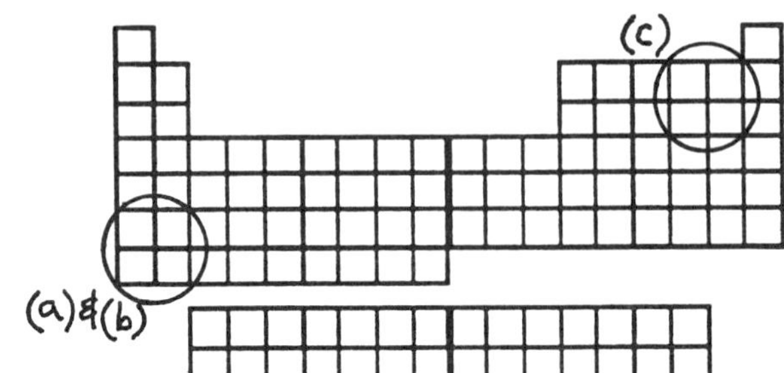

5

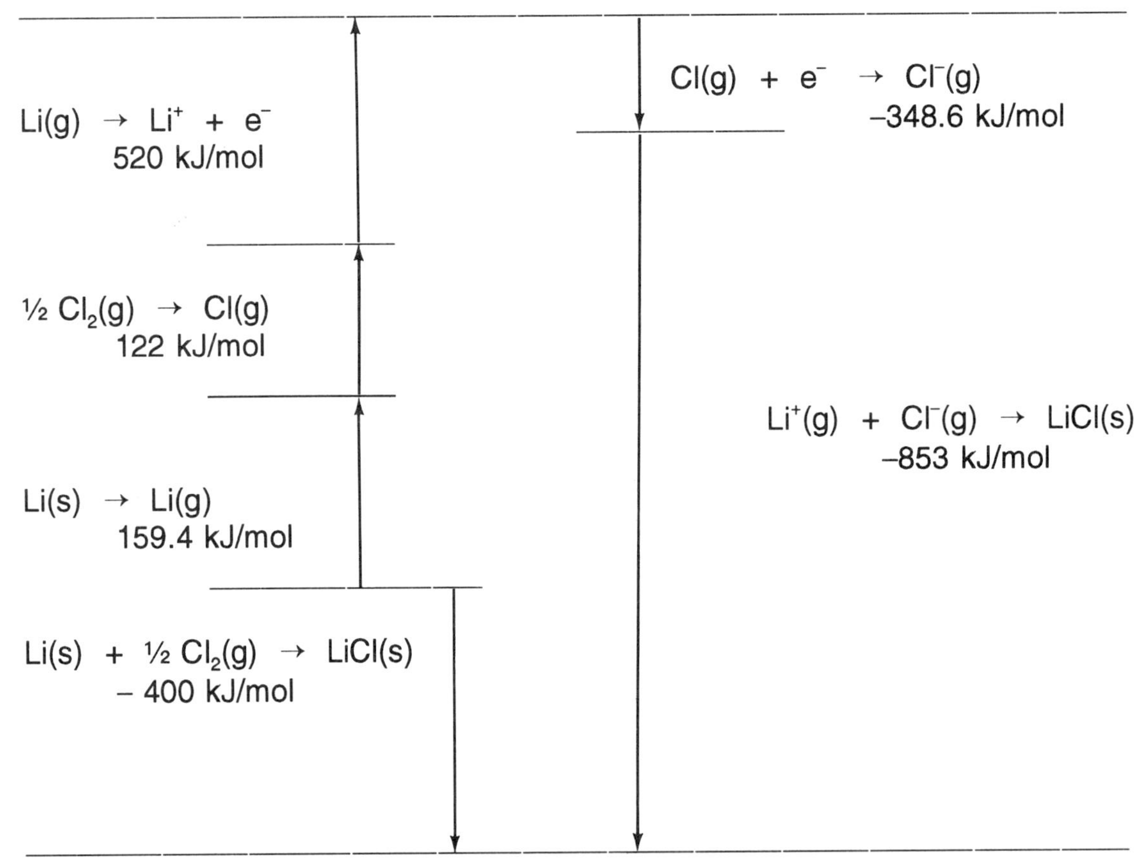

6. Predicted for Fr:

   melting point ≈ 23 °C boiling point ≈ 650 °C

   density ≈ 2 g/cm^3 atomic radius ≈ 275 pm

Chapter 6 – Ionic Bonds and Some Main-Group Chemistry

Additional Problems

Ionization Energy and Electron Affinity

6.20 Ionization energies have a postitve sign because energy is required to remove an electron from an atom of any element.

6.21 The largest E_{i1} are found in Group 8A because of the largest values of Z_{eff}.

The smallest E_{i1} are found in Group 1A because of the smallest values of Z_{eff}.

6.22 Fr would have the smallest ionization energy, and He would have the largest.

6.23 (a) 3p (b) 4p (c) 4s

6.24 (a) K [Ar] $4s^1$ Ca [Ar] $4s^2$

Ca has the smaller second ionization energy because it is easier to remove the second 4s valence electron in Ca than it is to remove the second electron in K from the filled 3p orbitals.

(b) Ca [Ar] $4s^2$ Ga [Ar] $4s^2 3d^{10} 4p^1$

Ca has the larger third ionization energy because it is more difficult to remove the third electron in Ca from the filled 3p orbitals than it is to remove the third electron (second 4s valence electron) from Ga.

6.25 Sn has a smaller fourth ionization energy than Sb because of a smaller Z_{eff}.

Br has a larger sixth ionization energy than Se because of a larger Z_{eff}.

6.26 (a) $1s^2 2s^2 2p^6 3s^2 3p^3$ is P

(b) $1s^2 2s^2 2p^6 3s^2 3p^6$ is Ar

(c) $1s^2 2s^2 2p^6 3s^2 3p^6 4s^2$ is Ca

Ar has the highest E_{i2}. Ar has a higher Z_{eff} than P. The 4s electrons in Ca are easier to remove than any 3p electrons.

Ar has the lowest E_{i7}. It is difficult to remove 3p electrons from Ca, and it is difficult to remove 2p electrons from P.

6.27 (a) Sr, [Kr] $5s^2$; A 5s electron would be lost.

(b) Br, [Ar] $4s^2 3d^{10} 4p^5$; A 4p electron would be lost.

(c) La, [Xe] $6s^2 5d^1$; A 6s electron would be lost.

6.28 (a) Sr, [Kr] $5s^2$; The second 5s electron would be lost.

(b) Br, [Ar] $4s^2 3d^{10} 4p^5$; A second 4p electron would be lost.

(c) La, [Xe] $6s^2 5d^1$; The second 6s electron would be lost.

6.29

	Lowest E_{i1}	Highest E_{i1}
(a)	K	Li
(b)	B	Cl
(c)	Ca	Cl

6.30 (a) Group 2A (b) Group 6A

6.31 (a) Na 496 kJ/mol

(b) Mg (738 + 1451) = 2189 kJ/mol

(c) Al (578 + 1817 + 2745) = 5140 kJ/mol

(d) Cl (1251 + 2297 + 3822 + 5158 + 6540 + 9458 + 11,020) = 39,546 kJ/mol

6.32 (a) F; nonmetals have more negative electron affinities than metals.

(b) Na; Ne (noble gas) has a positive electron affinity.

(c) Br; nonmetals have more negative electron affinities than metals.

6.33 The relationship between the electron affinity of a univalent cation and the ionization energy of the neutral atom is that they have the same magnitude but opposite sign.

6.34 Na^+ has a more negative electron affinity than either Na or Cl because of its positive charge.

6.35 Br would have a more negative electron affinity than Br^- because Br^- has no room in its valence shell for an additional electron.

6.36 Energy is usually released when an electron is added to a neutral atom but absorbed when an electron is removed from a neutral atom because of the positive Z_{eff}.

6.37 E_{i1} increase steadily across the periodic table from group 1A to group 8A because electrons are being removed from the same shell and Z_{eff} is increasing. The electron affinity increases irregularly from 1A to 7A and then falls dramatically for group 8A because the additional electron goes into the next higher shell.

Lattice Energy and Ionic Bonds

6.38 $MgCl_2$ > LiCl > KCl > KBr

6.39 $AlBr_3$ > CaO > $MgBr_2$ > LiBr

6.40 (i) sublimation energy for Li
(ii) bond dissociation energy for Br_2
(iii) E_{i1} for Li
(iv) electron affinity for Br
(v) lattice energy for LiBr

6.41

$Li \rightarrow Li^+ + e^-$	+520 kJ/mol
$Br + e^- \rightarrow Br^-$	<u>–325 kJ/mol</u>
	+195 kJ/mol

6.42 $Li(s) \rightarrow Li(g)$ +159.4 kJ/mol
$Li(s) \rightarrow Li(g) + e^-$ +520 kJ/mol
$\frac{1}{2} [Br_2(l) \rightarrow Br_2(g)]$ +15.4 kJ/mol
$\frac{1}{2} [Br_2(l) \rightarrow 2\ Br(g)]$ +112 kJ/mol
$Br(g) + e^- \rightarrow Br^-(g)$ −325 kJ/mol
$Li^+ (g) + Br^-(g) \rightarrow LiBr(s)$ −807 kJ/mol
Sum = −325 kJ/mol

6.43 (a) $Li(s) \rightarrow Li(g)$ +159.4 kJ/mol
$Li(g) \rightarrow Li^+(g)$ +520 kJ/mol
$\frac{1}{2}[F_2(g) \rightarrow 2\ F(g)]$ +79 kJ/mol
$F(g) + e^- \rightarrow F^-(g)$ −328 kJ/mol
$Li^+(g) + F^-(g) \rightarrow LiF(s)$ −1036 kJ/mol
Sum = −606 kJ/mol

(b) $Ca(s) \rightarrow Ca(g)$ +178.2 kJ/mol
$Ca(g) \rightarrow Ca^+(g)$ +589.5 kJ/mol
$Ca^+(g) \rightarrow Ca^{2+}(g)$ +1145 kJ/mol
$F_2(g) \rightarrow 2\ F(g)$ +158 kJ/mol
$2[F(g) + e^- \rightarrow F^-(g)]$ 2(−328) kJ/mol
$Ca^{2+}(g) + 2\ F^- \rightarrow CaF_2(s)$ −2630 kJ/mol
Sum = −1215 kJ/mol

6.44 $Na(s) \rightarrow Na(g)$ +107.3 kJ/mol
$Na(g) \rightarrow Na^+(g) + e^-$ +495.8 kJ/mol
$\frac{1}{2} [H_2(g) \rightarrow 2\ H(g)]$ +218.0 kJ/mol
$H(g) + e^- \rightarrow H^-(g)$ −72.8 kJ/mol
$Na^+(g) + H^-(g) \rightarrow NaH(s)$ −U
Sum = −60 kJ/mol

$$-U = -60 - 107.3 - 495.8 - \frac{435.9}{2} + 72.8 = -808 \text{ kJ/mol}$$

U = 808 kJ/mol

6.45 $Ca(s) \rightarrow Ca(g)$ +178.2 kJ/mol
$Ca(g) \rightarrow Ca^+(g)$ +589.5 kJ/mol
$Ca^+(g) \rightarrow Ca^{2+}(g)$ +1145 kJ/mol
$H_2(g) \rightarrow 2\ H(g)$ +435.9 kJ/mol
$2[H(g) + e^- \rightarrow H^-]$ 2(−72.8) kJ/mol
$Ca^{2+}(g) + 2\ H^-(g) \rightarrow CaH_2(s)$ −U
Sum = −186.2 kJ/mol

$-U = -186.2 - 178.2 - 589.8 - 1145 - 435.9 + 2(72.8) = -2390$ kJ/mol

$U = 2390$ kJ/mol

6.46

$Cs(s) \rightarrow Cs(g)$	+76.1 kJ/mol
$Cs(g) \rightarrow Cs^+ + e^-$	+375.7 kJ/mol
$\frac{1}{2}[F_2(g) \rightarrow 2\ F(g)]$	+79 kJ/mol
$F(g) + e^- \rightarrow H^-(g)$	−328 kJ/mol
$Cs^+(g) + F^- \rightarrow CsF(g)$	−740 kJ/mol
Sum =	−537 kJ/mol

6.47

$Cs(s) \rightarrow Cs(g)$	+76.1 kJ/mol
$Cs(g) \rightarrow Cs^+(g) + e^-$	+375.7 kJ/mol
$Cs^+(g) \rightarrow Cs^{2+}(g) + e^-$	+2422 kJ/mol
$F_2(g) \rightarrow 2\ F(g)$	+158 kJ/mol
$2[F(g) + e^- \rightarrow F^-(g)]$	2(−328) kJ/mol
$Cs^{2+}(g) + 2\ F^-(g) \rightarrow CsF_2(s)$	−2347 kJ/mol
Sum =	+29 kJ/mol

The overall reaction absorbs 29 kJ/mol.

6.48 In the reaction of cesium with fluorine, CsF will form because the overall energy for the formation of CsF is negative, whereas it is positive for CsF_2.

6.49

$Ca(s) \rightarrow Ca(g)$	+178.2 kJ/mol
$Ca(g) \rightarrow Ca^+(g)$	+589.8 kJ/mol
$\frac{1}{2}[Cl_2(g) \rightarrow 2\ Cl(g)]$	+121.5 kJ/mol
$Cl(g) + e^- \rightarrow Cl^-(g)$	−348.6 kJ/mol
$Ca^+(g) + Cl^-(g) \rightarrow CaCl(s)$	−717 kJ/mol
Sum =	−176 kJ/mol

$Ca(s) \rightarrow Ca(g)$	+178.2 kJ/mol
$Ca(g) \rightarrow Ca^+(g)$	+589.8 kJ/mol
$Ca^+(g) \rightarrow Ca^{2+}(g)$	+1145 kJ/mol
$Cl_2(g) \rightarrow 2\ Cl(g)$	+243 kJ/mol
$2[Cl(g) + e^- \rightarrow Cl^-(g)]$	2(−348.6) kJ/mol
$Ca^{2+}(g) + 2\ Cl^-(g) \rightarrow CaCl_2(s)$	−2258 kJ/mol
Sum =	−799 kJ/mol

6.50 In the reaction of calcium with chlorine, $CaCl_2$ will form because the overall energy for the formation of $CaCl_2$ is much more negative than for the formation of CaCl.

6.51

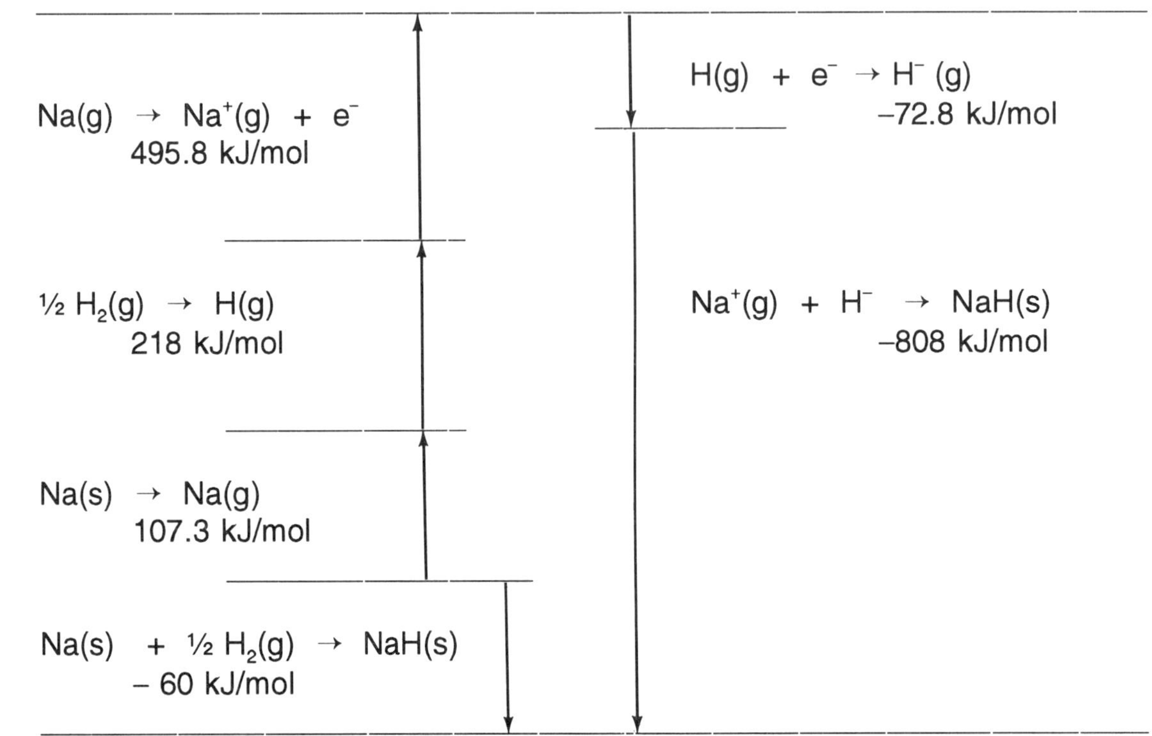

6.52

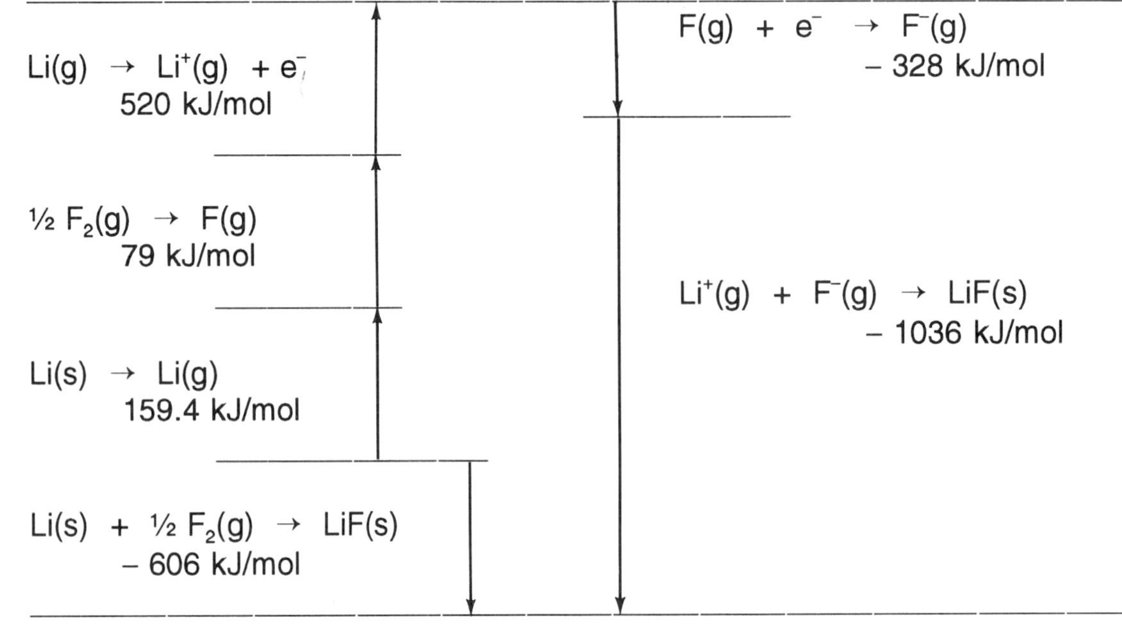

Main Group Chemistry

6.53

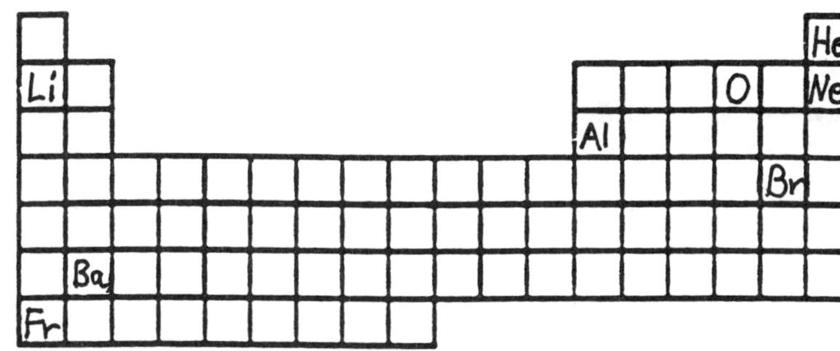

6.54 Solids: I

Liquids: Br

Gases: F, Cl, He, Ne, Ar, Kr, Xe

6.55 (a) Li is used in automotive grease. Li_2CO_3 is a manic depressive drug.

(b) K salts are used in plant fertilizers

(c) $SrCO_3$ is used in color TV picture tubes. Sr salts are used for red fireworks.

(d) Liquid He (bp = 4.2 K) is used for low temperature studies and superconducting magnets.

6.56 (a) $2\ NaCl \xrightarrow[580^\circ C]{\text{electrolysis in } CaCl_2} 2\ Na(l) + Cl_2(g)$

(b) $2\ Al_2O_3 \xrightarrow[980^\circ C]{\text{electrolysis in } Na_3AlF_6} 4\ Al(l) + 3\ O_2(g)$

(c) Ar is obtained from the distillation of liquid air.

(d) $2\ Br^-(aq) + Cl_2(g) \rightarrow Br_2(l) + 2\ Cl^-(aq)$

6.57 Group 1A metals react by losing an electron. As you go down group 1A, the valence electron is more easily removed. This trend parallels chemical reactivity.

Group 7A nonmetals react by gaining an electron. The electron affinity generally increases as you go up the group. This trend parallels chemical reactivity.

6.58 Main–group elements tend to undergo reactions that leave them with eight valence electrons. That is, main–group elements react so that they attain a noble–gas electron configuration with filled s and p sublevels in their valence electron shell.

The octet rule works for valence–shell electrons because taking electrons away from a filled octet is difficult because they are tightly held by a high Z_{eff}; adding more electrons to a filled octet is difficult because, with s and p sublevels full, there is no low–energy orbital available.

6.59 Main group nonmetals in the third period and below occasionally break the octet rule.

6.60 (a) $2\ K(s) + H_2(g) \rightarrow 2\ KH(s)$

(b) $2\ K(s) + 2\ H_2O(l) \rightarrow 2\ K^+(aq) + 2\ OH^-(aq) + H_2(g)$

(c) $2\ K(s) + 2\ NH_3(g) \rightarrow 2\ KNH_2(s) + H_2(g)$

(d) $2\ K(s) + Br_2(l) \rightarrow 2\ KBr(s)$

(e) $K(s) + N_2(g) \rightarrow$ N. R.

(f) $K(s) + O_2(g) \rightarrow KO_2(s)$

6.61 (a) $Ca(s) + H_2(g) \rightarrow CaH_2(s)$

(b) $Ca(s) + 2\ H_2O(l) \rightarrow Ca^{2+}(aq) + 2\ OH^-(aq) + H_2(g)$

(c) $Ca(s) + He(g) \rightarrow$ N. R.

(d) $Ca(s) + Br_2(l) \rightarrow CaBr_2(s)$

(e) $2\ Ca(s) + O_2(g) \rightarrow 2\ CaO(s)$

6.62 (a) $Cl_2(g) + H_2(g) \rightarrow 2\ HCl(g)$

(b) $Cl_2(g) + Ar(g) \rightarrow$ N. R.

(c) $Cl_2(g) + Br_2(l) \rightarrow 2\ BrCl(g)$

(d) $Cl_2(g) + N_2(g) \rightarrow$ N. R.

6.63 $AlCl_3 + 3\ Na \rightarrow Al + 3\ NaCl$

Al^{3+} (from $AlCl_3$) gains electrons and is reduced. Na loses electrons and is oxidized.

6.64 $2\ Mg(s) + O_2(g) \rightarrow 2\ MgO(s)$

$MgO(s) + H_2O(l) \rightarrow Mg(OH)_2(aq)$

6.65 $CaIO_3$, 215.0 amu; 1.00 kg = 1000 g

$$\%\ I = \frac{126.9\ g}{215.0\ g} \times 100\% = 59.02\%;\quad (0.5902)(1000\ g) = 590\ g\ I_2$$

6.66 $2\ Li(s) + 2\ H_2O(l) \rightarrow 2\ LIOH(aq) + H_2(g)$

455 mL = 0.455 L

$$\text{mass of } H_2 = (0.0893\ \frac{g}{L})\ (0.455\ L) = 0.0406\ g\ H_2$$

$$0.0406\ g\ H_2 \times \frac{1\ mol\ H_2}{2.016\ g\ H_2} \times \frac{2\ mol\ Li}{1\ mol\ H_2} \times \frac{6.94\ g\ Li}{1\ mol\ Li} = 0.280\ g\ Li$$

6.67 $SrBr_2$, 247.4 amu

$$5.65\ g \times \frac{1\ mol\ Sr}{87.6\ g\ Sr} \times \frac{1\ mol\ SrBr_2}{1\ mol\ Sr} \times \frac{247.4\ g\ SrBr_2}{1\ mol\ SrBr_2} = 16.0\ g\ SrBr_2$$

6.68 (a) $2\ Cl^-(aq) + F_2(g) \rightarrow 2\ F^-(aq) + Cl_2(g)$

F_2 gains electrons and is the oxidizing agent.

Cl^- loses electrons and is the reducing agent.

(b) $2\ Br^-(aq) + I_2(s) \rightarrow$ N. R.

(c) $2\ I^-(aq) + Br_2(aq) \rightarrow 2\ Br^-(aq) + I_2(aq)$

Br_2 gains electrons and is the oxidizing agent.

I^- loses electrons and is the reducing agent.

6.69 (a) $Mg(s) + 2\ H^+(aq) \rightarrow Mg^{2+}(aq) + H_2(g)$

H^+ gains electrons and is the oxidizing agent. Mg loses electrons and is the reducing agent.

(b) $Kr(g) + F_2(g) \rightarrow KrF_2(s)$

F_2 gains electrons and is the oxidizing agent. Kr loses electrons and is the reducing agent.

(c) $I_2(s) + 3\ Cl_2(g) \rightarrow 2\ ICl_3(l)$

Cl_2 gains electrons and is the oxidizing agent. I_2 loses electrons and is the reducing agent.

6.70 (a) $2\ XeF_2(s) + 2\ H_2O(l) \rightarrow 2\ Xe(g) + 4\ HF(aq) + O_2(g)$

Xe in XeF_2 gains electrons and is the oxidizing agent.

O in H_2O loses electrons and is the reducing agent.

(b) $NaH(s) + H_2O(l) \rightarrow Na^+(aq) + OH^-(aq) + H_2(g)$

H in H_2O gains electrons and is the oxidizing agent.

H in NaH loses electrons and is the reducing agent.

(c) $2\ TiCl_4(l) + H_2(g) \rightarrow 2\ TiCl_3(s) + 2\ HCl(g)$

Ti in $TiCl_4$ gains electrons and is the oxidizing agent.

H_2 loses electrons and is the reducing agent.

General Problems

6.71 58.4 nm = 58.4×10^{-9} m

$$E(\text{photon}) = 6.626 \times 10^{-34}\ J{\cdot}s \times \frac{3.00 \times 10^8\ m/s}{58.4 \times 10^{-9}\ m} \times \frac{1\ kJ}{1000\ J} \times \frac{6.022 \times 10^{23}}{mol} = 2049\ kJ/mol$$

$$E_k = E(\text{electron}) = \tfrac{1}{2}(9.109 \times 10^{-31}\ kg)(2.450 \times 10^6\ m/s)^2\left(\frac{1\ kJ}{1000\ J}\right)\left(\frac{6.022 \times 10^{23}}{mol}\right)$$

$E_k = 1646$ kJ/mol

$E(\text{photon}) = E_i + E_k$; $E_i = E(\text{photon}) - E_k = 2049 - 1646 = 403$ kJ/mol

6.72 142 nm = 142×10^{-9} m

$$E(\text{photon}) = 6.626 \times 10^{-34}\ J{\cdot}s \times \frac{3.00 \times 10^8\ m/s}{142 \times 10^{-9}\ m} \times \frac{1\ kJ}{1000\ J} \times \frac{6.022 \times 10^{23}}{mol} = 843\ kJ/mol$$

$$E_k = E(\text{electron}) = \tfrac{1}{2}(9.109 \times 10^{-31}\ kg)(1.240 \times 10^6\ m/s)^2\left(\frac{1\ kJ}{1000\ J}\right)\left(\frac{6.022 \times 10^{23}}{mol}\right)$$

$E_k = 422$ kJ/mol

$E(\text{photon}) = E_i + E_k$; $E_i = E(\text{photon}) - E_k = 843 - 422 = 421$ kJ/mol

6.73

$Mg(s) \rightarrow Mg(g)$	+147.7	kJ/mol
$Mg(g) \rightarrow Mg^+(g) + e^-$	+737.7	kJ/mol
$\tfrac{1}{2}[F_2(g) \rightarrow 2\ F(g)]$	+79	kJ/mol
$F(g) + e^- \rightarrow F^-(g)$	–328	kJ/mol
$Mg^+(g) + F^-(g) \rightarrow MgF(s)$	–930	kJ/mol
Sum =	–294	kJ/mol

$Mg(s) \rightarrow Mg(g)$	+147.7	kJ/mol
$Mg(g) \rightarrow Mg^{+}(g) + e^{-}$	+737.7	kJ/mo
$Mg^{+}(g) \rightarrow Mg^{2+}(g) + e^{-}$	+1450.7	kJ/mol
$F_2(g) \rightarrow 2\ F(g)$	+158	kJ/mol
$2[F(g) + e^{-} \rightarrow F^{-}(g)]$	2(–328)	kJ/mol
$Mg^{2+}(g) + 2\ F^{-}(g) \rightarrow MgF_2(s)$	–2952	kJ/mol
Sum =	–1114	kJ/mol

6.74 In the reaction of magnesium with fluorine, MgF_2 will form because the overall energy for the formation of MgF_2 is much more negative than for the formation of MgF.

6.75 (a) Na is used in table salt (NaCl), glass, rubber, and pharmaceutical agents.

(b) Mg is used as a structural material when alloyed with Al.

(c) F is used in the manufacture of Teflon, $(C_2F_4)_n$, and in toothpaste as SnF_2.

6.76 (a) $2\ HF(l) \xrightarrow[100^\circ C]{\text{electrolysis}} H_2(g) + F_2(g)$

(b) $3\ CaO(l) + 2\ Al(l) \xrightarrow{\text{high temperature}} 3\ Ca(l) + Al_2O_3(s)$

(c) $2\ NaCl(l) \xrightarrow[580^\circ C]{\text{electrolysis in } CaCl_2} 2Na(l) + Cl_2(g)$

6.77 (a) $2\ Li(s) + H_2(g) \rightarrow 2\ LiH(s)$

(b) $2\ Li(s) + 2\ H_2O(l) \rightarrow 2\ Li^{+}(aq) + 2\ OH^{-}(aq) + H_2(g)$

(c) $2\ Li(s) + 2\ NH_3(g) \rightarrow 2\ LiNH_2(s) + H_2(g)$

(d) $2\ Li(s) + Br_2(l) \rightarrow 2\ LiBr(s)$

(e) $6\ Li(s) + N_2(g) \rightarrow 2\ Li_3N(s)$

(f) $4\ Li(s) + O_2(g) \rightarrow 2\ Li_2O(s)$

6.78 (a) F_2(g) + H_2(g) → 2 HF(g)

(b) F_2(g) + 2 Na(s) → 2 NaF(s)

(c) F_2(g) + Br_2(l) → 2 BrF(g)

(d) F_2(g) + 2 NaBr(s) → 2 NaF(g) + Br_2(l)

6.79 When moving diagonally down and right on the periodic table, the increase in atomic radius caused by going to a larger shell is offset by a decrease caused by a higher Z_{eff}. Thus, there is little net change.

7.1 (a)

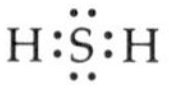

(b)

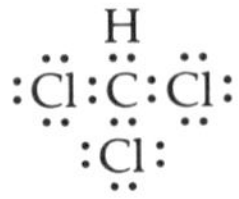

7.2

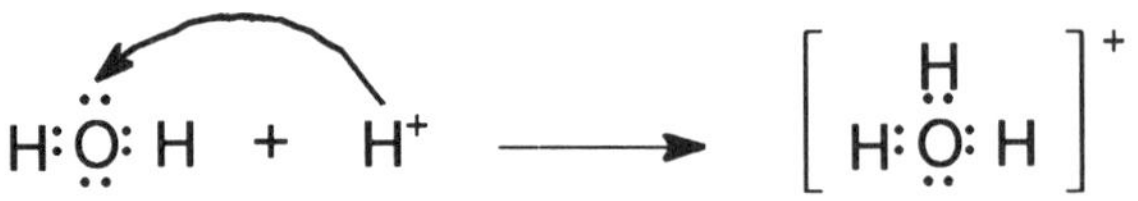

hydronium ion

7.3 (a)

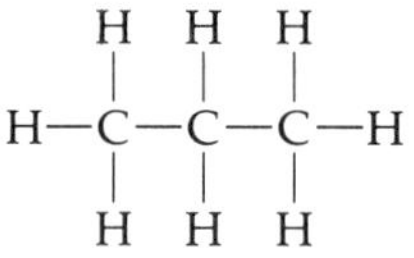

(b)

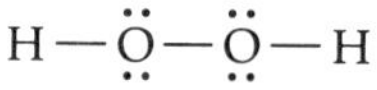

(c)

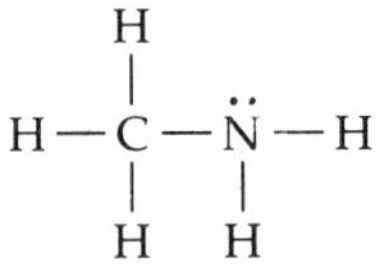

7.4 (a)

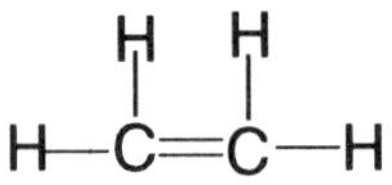

(b)

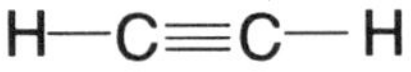

(c)

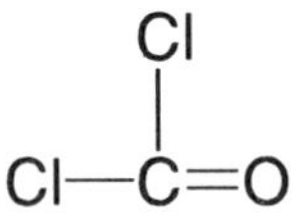

7.5

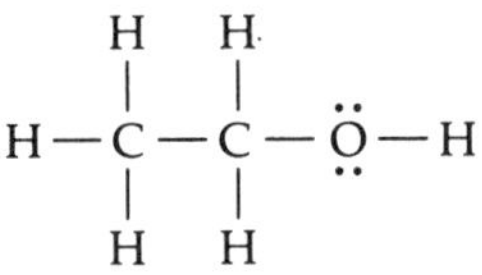

and

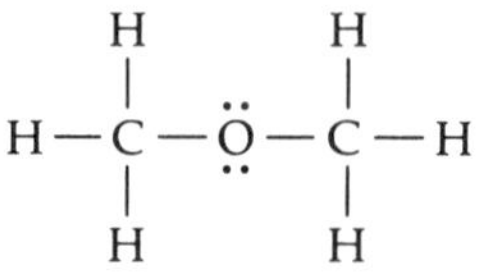

7.6 :C≡O:

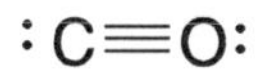

7.7 (a) 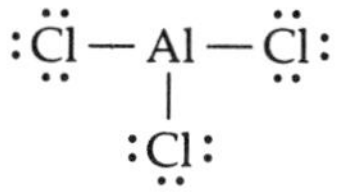 :C̈l—Al—C̈l: with :C̈l: bonded below Al

(b) :C̈l—Ï—C̈l: with :C̈l: bonded below I

(c) :Ö: above Xe; :F̈: and :F̈: upper left and right; :F̈: and :F̈: lower left and right; Xe with lone pair

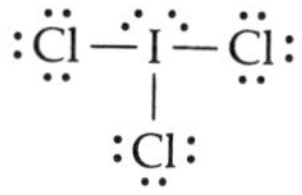

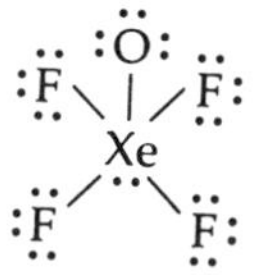

(d) 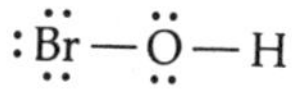 :B̈r—Ö—H

7.8 (a) 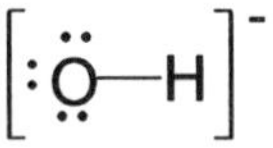 [:Ö—H]⁻

(b) 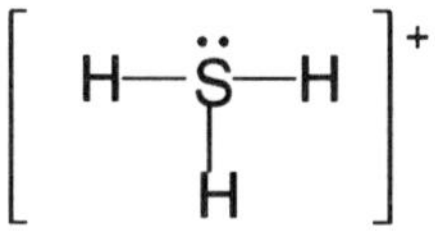 [H—S̈—H, with H bonded below S]⁺

(c) [:Ö: bonded to C—Ö—H, with :Ö: double-bonded below C]⁻

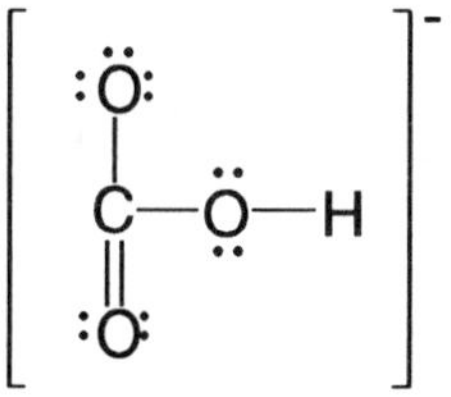

7.9 :N̈=N=Ö: ⟷ :N≡N—Ö:

7.10 (a) 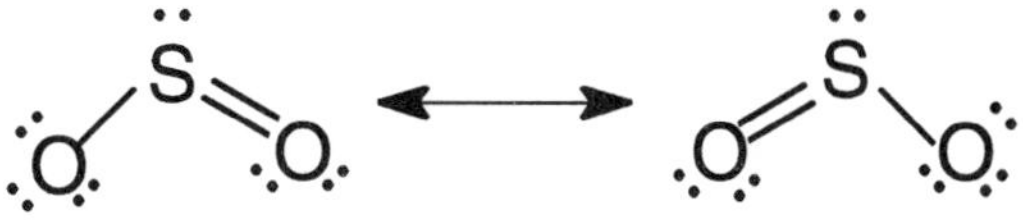 :Ö—S̈=Ö ⟷ O=S̈—Ö:

(b) 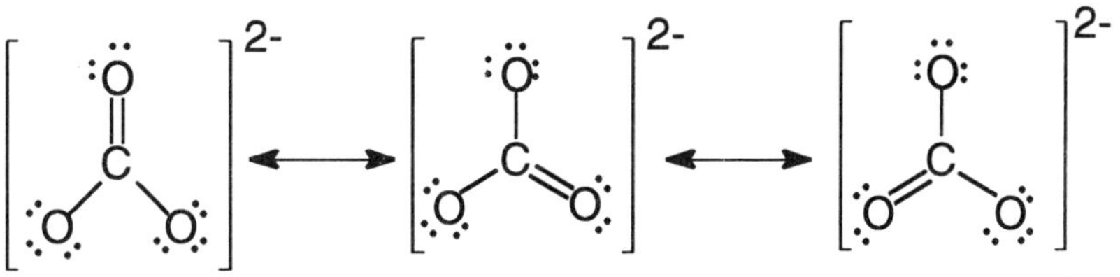 [C with =O above, O– lower left, O– lower right]²⁻ ⟷ [C with O– above, O– lower left, =O lower right]²⁻ ⟷ [C with O– above, =O lower left, O– lower right]²⁻

(c)

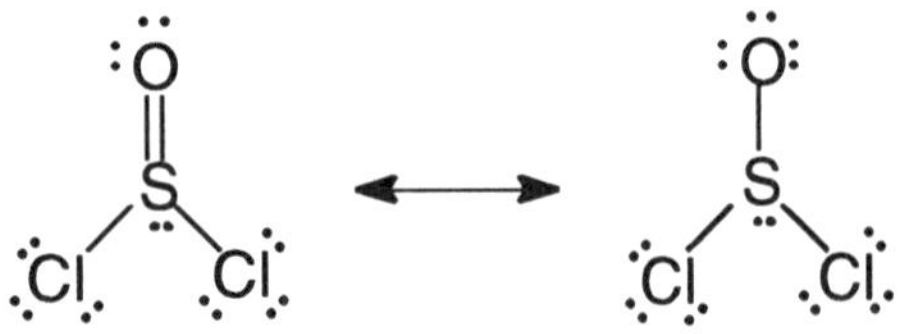

7.11 (a) $SiCl_4$ chlorine EN = 3.0
silicon EN = 1.8
ΔEN = 1.2

The Si–Cl bond is polar covalent.

(b) CsBr bromine EN = 2.8
cesium EN = 0.7
ΔEN = 2.1

The $Cs^{+}Br^{-}$ bond is ionic.

(c) $FeBr_3$ bromine EN = 2.8
iron EN = 1.8
ΔEN = 1.0

The Fe–Br bond is polar covalent.

(d) CH_4 carbon EN = 2.5
hydrogen EN = 2.1
ΔEN = 0.4

The C–H bond is polar covalent.

7.12 (a) CCl_4 chlorine EN = 3.0
carbon EN = 2.5
ΔEN = 0.5

(b) $MgCl_2$ chlorine EN = 3.0
magnesium EN = 1.2
ΔEN = 1.8

(c) $TiCl_3$

chlorine	EN = 3.0
titanium	EN = 1.5
	ΔEN = 1.5

(d) Cl_2O

oxygen	EN = 3.5
chlorine	EN = 3.0
	ΔEN = 0.5

Increasing ionic character: CCl_4 ~ ClO_2 < $TiCl_3$ < $MgCl_2$

7.13 For nitrogen:

Isolated nitrogen valence electrons	5
Bound nitrogen bonding electrons	8
Bound nitrogen nonbonding electrons	0

Formal charge = 5 – ½(8) – 0 = +1

For singly bound oxygen:

Isolated oxygen valence electrons	6
Bound oxygen bonding electrons	2
Bound oxygen nonbonding electrons	6

Formal charge = 6 – ½(2) – 6 = –1

For doubly bound oxygen:

Isolated oxygen valence electrons	6
Bound oxygen bonding electrons	4
Bound oxygen nonbonding electrons	4

Formal charge = 6 – ½(4) – 4 = 0

7.14 (a) $\left[:\ddot{N}{=}C{=}\ddot{O}:\right]^-$

For nitrogen:

Isolated nitrogen valence electrons	5
Bound nitrogen bonding electrons	4
Bound nitrogen nonbonding electrons	4

Formal charge = 5 – ½(4) – 4 = –1

For carbon:

Isolated carbon valence electrons	4
Bound carbon bonding electrons	8
Bound carbon nonbonding electrons	0

Formal charge = 4 – ½ 8) – 0 = 0

For oxygen:	Isolated oxygen valence electrons	6
	Bound oxygen bonding electrons	4
	Bound oxygen nonbonding electrons	4

Formal charge = 6 – ½(4) – 4 = 0

(b) $:\ddot{\underset{..}{O}}-\ddot{O}=\ddot{O}:$

For left oxygen:	Isolated oxygen valence electrons	6
	Bound oxygen bonding electrons	2
	Bound oxygen nonbonding electrons	6

Formal charge = 6 – ½(2) – 6 = –1

For central oxygen:	Isolated oxygen valence electrons	6
	Bound oxygen bonding electrons	6
	Bound oxygen nonbonding electrons	2

Formal charge = 6 – ½(6) – 2 = +1

For right oxygen:	Isolated oxygen valence electrons	6
	Bound oxygen bonding electrons	4
	Bound oxygen nonbonding electrons	4

Formal charge = 6 – ½(4) – 4 = 0

7.15

	Number of Bonded Atoms	Number of Lone Pairs	Shape
(a) O_3	2	1	bent
(b) H_3O^+	3	1	trigonal pyramidal
(c) XeF_2	2	3	linear
(d) PF_6^-	6	0	octahedral
(e) $XeOF_4$	5	1	square pyramidal
(f) AlH_4^-	4	0	tetrahedral
(g) BF_4^-	4	0	tetrahedral

7.15 (cont.)

	Number of Bonded Atoms	Number of Lone Pairs	Shape
(h) $SiCl_4$	4	0	tetrahedral
(i) ICl_4^-	4	2	square planar
(j) $AlCl_3$	3	0	trigonal planar

7.16

```
    H
    |   ..
H—C—O—H
    |   ..
    H
```

bent about O; tetrahedral about C

7.17

```
    H  :O:
    |   ||    ..
H — C — C — O — H
    |         ..
    H
```

```
 H          .O.
   \\       //
H — C — C
   /        \\
 H          .O.— H
```

7.18

```
  H H
  | |
H-C-C-H
  | |
  H H
```

Each C is sp^3 hybridized. The C–C bond is formed by the overlap of one singly occupied sp^3 hybrid orbital from each C. The C–H bonds are formed by the overlap of one singly occupied sp^3 orbital on C with a singly occupied H 1s orbital.

7.19

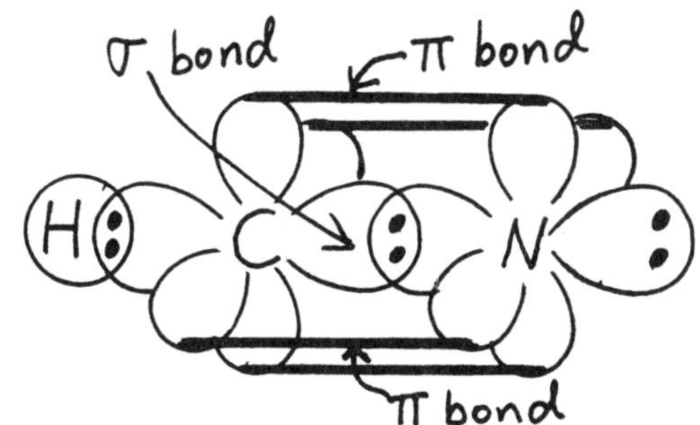

In HCN the carbon and nitrogen are both sp hybridized. The triple bond between carbon and nitrogen is made up of a σ bond formed by head-on overlap of the two sp hybrid orbitals and two mutually perpendicular π bonds formed by sideways overlap of p orbitals. A σ bond joins the carbon and hydrogen. The nitrogen has a lone pair of electrons.

7.20

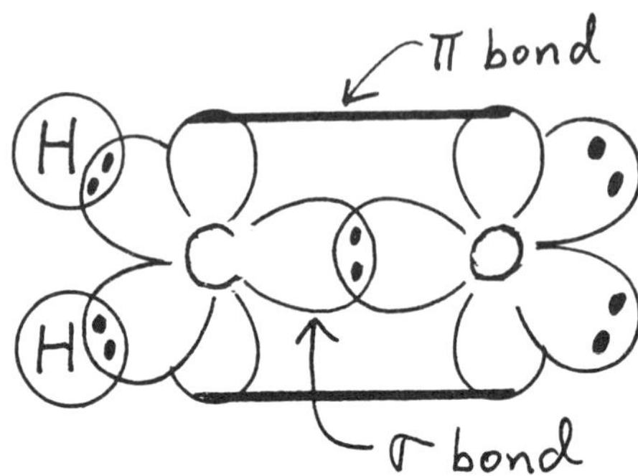

The carbon in formaldehyde is sp^2 hybridized.

7.21 The central I in I_3^- has two single bonds and three lone pairs of electrons. The hybridization of the central I is sp^3d. A sketch of the ion showing the orbitals involved in bonding is shown below.

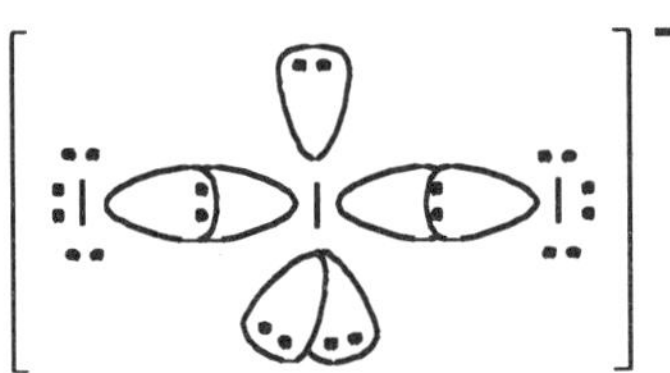

7.22

	Single Bonds	Lone Pairs	S Hybridization
SF_2	2	2	sp^3
SF_4	4	1	sp^3d
SF_6	6	0	sp^3d^2

7.23

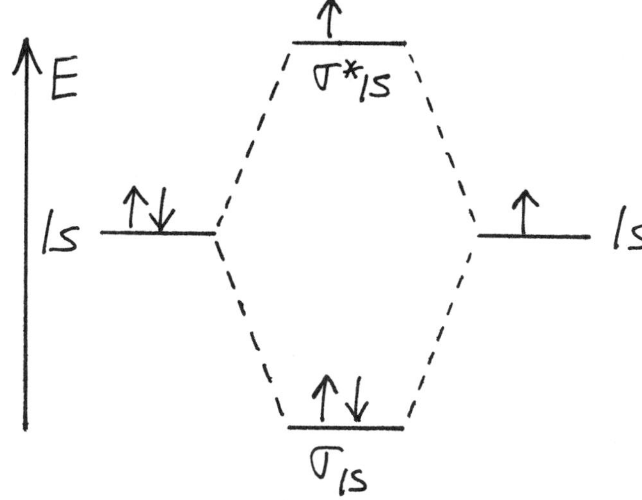

He_2^+ would be stable with a bond order of ½.

7.24 For B_2

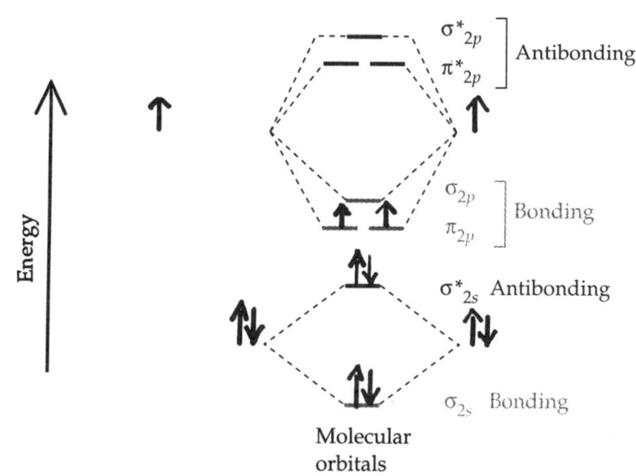

$$B_2 \text{ Bond order } = \frac{\begin{pmatrix}\text{number of}\\ \text{bonding electrons}\end{pmatrix} - \begin{pmatrix}\text{number of}\\ \text{antibonding electrons}\end{pmatrix}}{2} = \frac{4-2}{2} = 1$$

B_2 is paramagnetic because it has two unpaired electrons in the π_{2p} molecular orbitals.

For C_2

Energy

σ^*_{2p} π^*_{2p} Antibonding

σ_{2p} π_{2p} Bonding

σ^*_{2s} Antibonding

σ_{2s} Bonding

Molecular orbitals

$$C_2 \text{ Bond order} = \frac{6 - 2}{2} = 2$$

C_2 is diamagnetic because all electrons are paired.

7.25

$[HCO_2]^- \longleftrightarrow [HCO_2]^-$

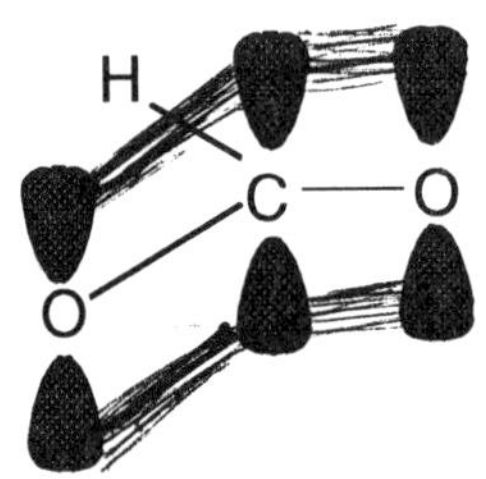

Understanding Key Concepts

1. (a) square pyramidal (b) trigonal pyramidal

   (c) square planar (d) trigonal planar

2. (a) trigonal bipyramidal (b) tetrahedral

   (c) square pyramidal (4 ligands in the horizontal plane, including one hidden)

3. (a)

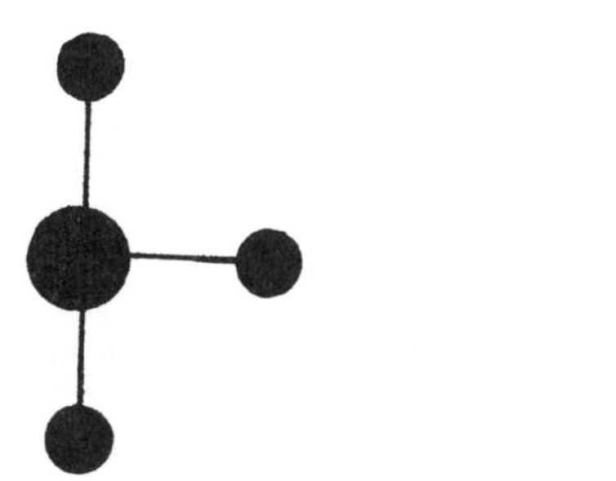

   (b)

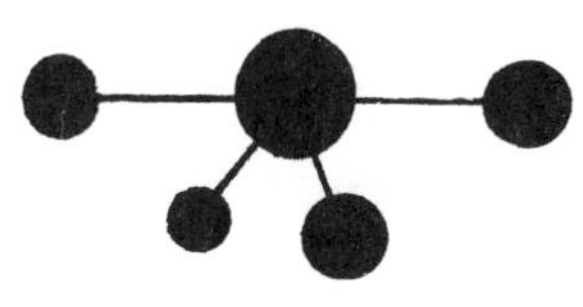

   (c)

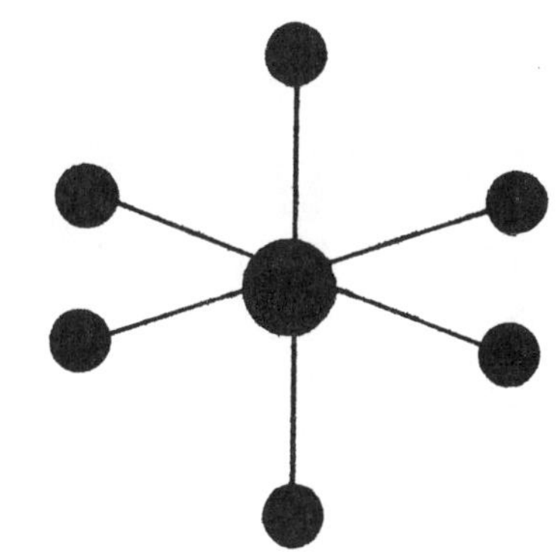

4. (a) sp^2 (b) sp^3d^2 (c) sp^3

5. (a)

   (b)

6. Every carbon is sp^2 hybridized. There are 18 σ bonds and 5 π bonds.

Additional Problems

Lewis Structures

7.26 (a) $AlCl_3$ Al has only 6 electrons around it.

(b) PCl_5 P has 10 electrons around it.

7.27 (a)

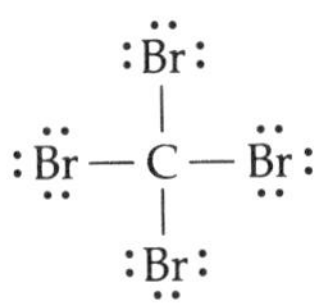

(b)

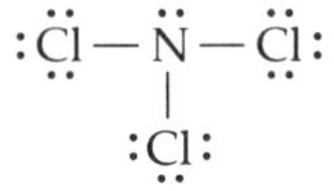

(c)

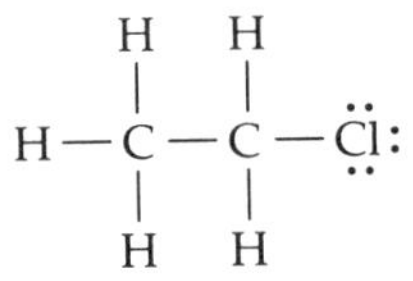

(d)

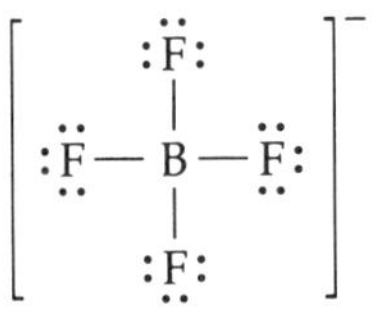

(e)

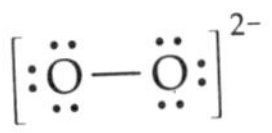

(f)

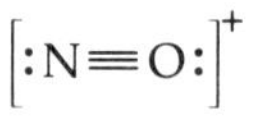

7.28 (a)

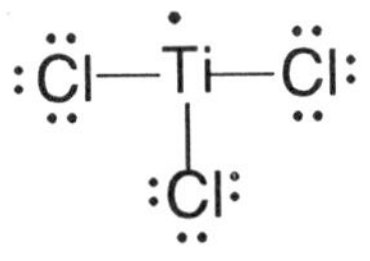

(b)

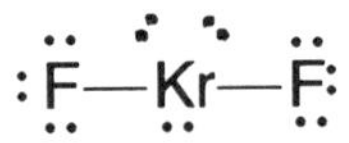

(c)

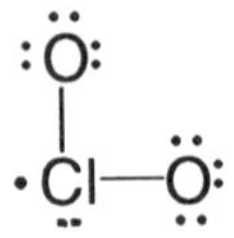

(d)

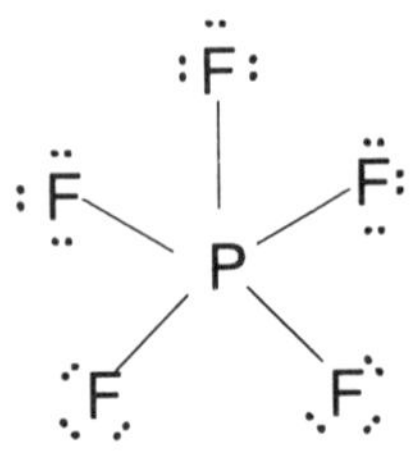

(e)

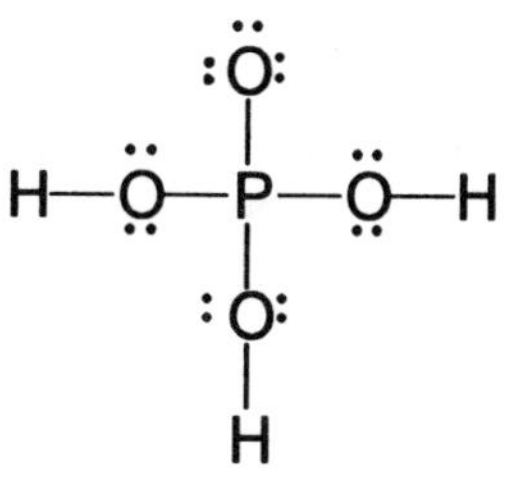

:Ö:
|
H—Ö—P—Ö—H
|
:Ö:
|
H

7.29 (a) 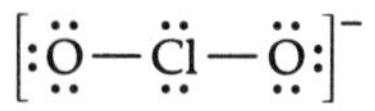 [:Ö—Cl—Ö:]⁻

(b) 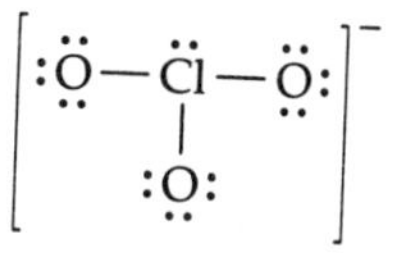 [:Ö—Cl—Ö: | :Ö:]⁻

(c) 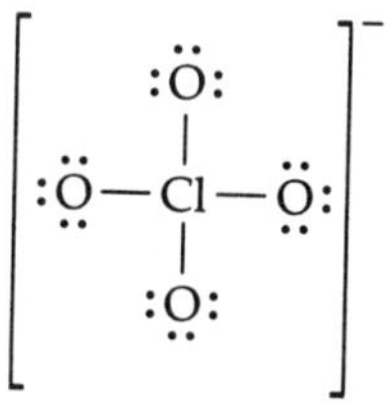 [:Ö: | :Ö—Cl—Ö: | :Ö:]⁻

There are also structures with one Cl=O double bond.

7.30 (a) 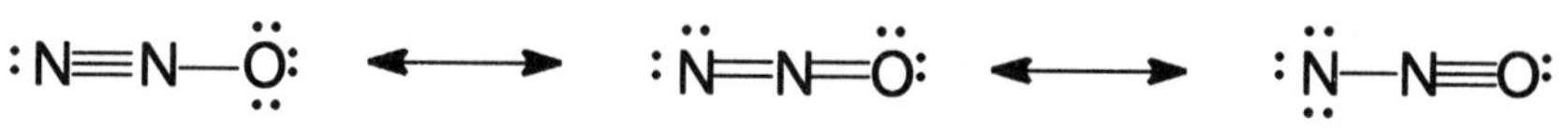 :N≡N—Ö: ⟷ :N=N=Ö: ⟷ :N—N≡O:

(b) 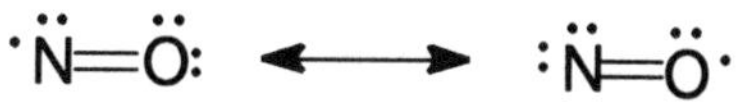 ·N=Ö: ⟷ :N=Ö·

(c) 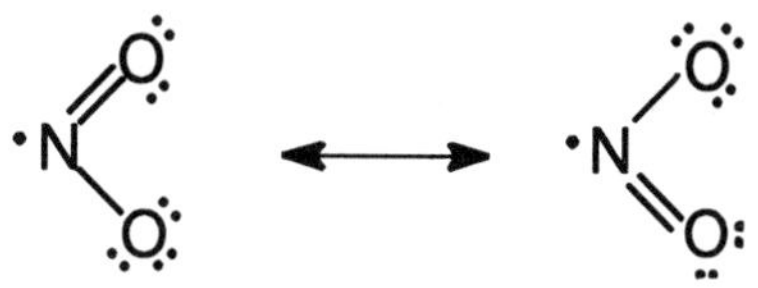 ·N(=O)(—O) ⟷ ·N(—O)(=O)

(d) $:\ddot{O}{=}\ddot{N}{-}N(=\ddot{O}:)(-\ddot{\underset{..}{O}}:) \longleftrightarrow :\ddot{O}{=}\ddot{N}{-}N(-\ddot{\underset{..}{O}}:)(=\ddot{O}:) \longleftrightarrow :\ddot{O}{=}N{=}N(-\ddot{\underset{..}{O}}:)(-\ddot{\underset{..}{O}}:)$

7.31 (a) $COCl_2$ no (b) $SOCl_2$ no

(c) SO_2Cl_2 no (d) H_2CN_2 yes

7.32

$$H{-}\ddot{\underset{..}{O}}{-}\underset{}{\overset{\overset{\ddot{O}:}{\|}}{C}}{-}\overset{\overset{\ddot{O}:}{\|}}{C}{-}\ddot{\underset{..}{O}}{-}H$$

7.33 $\ddot{\underset{..}{O}}{=}C{=}\ddot{\underset{..}{O}}$ CO_2 has two double bonds.

7.34 (a) yes (b) yes (c) yes (d) yes

7.35

$$H{-}\underset{H}{\overset{H}{\underset{|}{\overset{|}{C}}}}{-}\ddot{N}{=}C{=}\ddot{O}: \longleftrightarrow H{-}\underset{H}{\overset{H}{\underset{|}{\overset{|}{C}}}}{-}N{\equiv}C{-}\ddot{\underset{..}{O}}:$$

7.36 (a) The anion has 32 valence electrons. Each Cl has seven valence electrons (28 total). The minus one charge on the anion accounts for one valence electron. This leaves three valence electrons for X. X is Al.

(b) The cation has eight valence electrons. Each H has one valence (4 total). X is left with four valence electrons. Since this is a cation, one valence electron was removed from X. X has five valence electrons. X is P.

7.37 (a) This fourth-row element has six valence electrons. It is Se.

(b) This fourth-row element has eight valence electrons. It is Kr.

7.38 (a) :Cl–C(=O:)–O–C(H)(H)–H

(b) H–C(H)(H)–C≡C–H

7.39 (a) H–C(=:O:)–N(H)–H

(b) H–C(H)(H)–C≡N–O:

Electronegativity and Polar Covalent Bonds

7.40 K < Li < Mg < Pb < C < Br

7.41 Electronegativity increases from left to right across a period, and it decreases down a group.

7.42 (a) HF

fluorine	EN = 4.0
hydrogen	EN = 2.1
	ΔEN = 1.9

HF is polar covalent

(b) HI

iodine	EN = 2.5
hydrogen	EN = 2.1
	ΔEN = 0.4

HI is polar covalent.

(c) $PdCl_2$

chlorine	EN = 3.0
palladium	EN = 2.2
	ΔEN = 0.8

$PdCl_2$ is polar covalent.

(d) BBr_3

bromine	EN = 2.8
boron	EN = 2.0
	ΔEN = 0.8

BBr_3 is polar covalent.

(e) NaOH $Na^+ - OH^-$ is ionic

OH^- oxygen EN = 3.5
hydrogen EN = 2.1
ΔEN = 1.4

OH^- is polar covalent.

7.43 The electronegativity for each element is shown in parentheses.

(a) C (2.5), H (2.1), Cl (3.0): The C–Cl bond is more polar than the C–H bond because of the larger electronegativity difference between the bonded atoms.

(b) Si (1.8), Li (1.0), Cl (3.0): The Si–Cl bond is more polar than the Si–Li bond because of the larger electronegativity difference between the bonded atoms.

(c) N (3.0), Cl (3.0), Mg (1.2): The N–Mg bond is more polar than the N–Cl bond because of the larger electronegativity difference between the bonded atoms.

7.44 (a) $\overset{\delta-}{C} - \overset{\delta+}{H}$ $\overset{\delta+}{C} - \overset{\delta-}{Cl}$

(b) $\overset{\delta-}{Si} - \overset{\delta+}{Li}$ $\overset{\delta+}{Si} - \overset{\delta-}{Cl}$

(c) N – Cl $\overset{\delta-}{N} - \overset{\delta+}{Mg}$

7.45 :C≡O:

For carbon:

Isolated carbon valence electrons	4
Bound carbon bonding electrons	6
Bound carbon nonbonding electrons	2

Formal charge = 4 – ½(6) – 2 = –1

For oxygen:

Isolated oxygen valence electrons	6
Bound oxygen bonding electrons	6
Bound oxygen nonbonding electrons	2

Formal charge = 6 – ½(6) – 2 = +1

7.46 $:N\equiv N-\ddot{\underset{..}{O}}:$

For left nitrogen:

Isolated nitrogen valence electrons	5
Bound nitrogen bonding electrons	6
Bound nitrogen nonbonding electrons	2

Formal charge = 5 – ½(6) – 2 = 0

For central nitrogen:

Isolated nitrogen valence electrons	5
Bound nitrogen bonding electrons	8
Bound nitrogen nonbonding electrons	0

Formal charge = 5 – ½(8) – 0 = +1

For oxygen:

Isolated oxygen valence electrons	6
Bound oxygen bonding electrons	2
Bound oxygen nonbonding electrons	6

Formal charge = 6 – ½(2) – 6 = –1

$:\ddot{N}=N=\ddot{O}:$

For left nitrogen:

Isolated nitrogen valence electrons	5
Bound nitrogen bonding electrons	4
Bound nitrogen nonbonding electrons	4

Formal charge = 5 – ½(4) – 4 = – 1

For central nitrogen:

Isolated nitrogen valence electrons	5
Bound nitrogen bonding electrons	8
Bound nitrogen nonbonding electrons	0

Formal charge = 5 – ½(8) – 0 = +1

For oxygen: Isolated oxygen valence electrons 6
Bound oxygen bonding electrons 4
Bound oxygen nonbonding electrons 4

Formal charge = 6 – ½(4) – 4 = 0

:N≡N—Ö: (0, +1, -1) is the more significant contributor to the resonance hybrid because of the –1 formal charge on the electronegative oxygen.

7.47 (a) H—N(H)—O—H (N with one lone pair, O with two lone pairs)

For hydrogen: Isolated hydrogen valence electrons 1
Bound hydrogen bonding electrons 2
Bound hydrogen nonbonding electrons 0

Formal charge = 1 – ½(2) – 0 = 0

For nitrogen: Isolated nitrogen valence electrons 5
Bound nitrogen bonding electrons 6
Bound nitrogen nonbonding electrons 2

Formal charge = 5 – ½(6) – 2 = 0

For oxygen: Isolated oxygen valence electrons 6
Bound oxygen bonding electrons 4
Bound oxygen nonbonding electrons 4

Formal charge = 6 – ½(4) – 4 = 0

There are no formal charges.

(b) $[H—\ddot{N}—CH_2—H]^-$ (H—N—C(H)(H)—H, N with two lone pairs)

For hydrogen: Isolated hydrogen valence electrons 1
Bound hydrogen bonding electrons 2
Bound hydrogen nonbonding electrons 0

Formal charge = 1 – ½(2) – 0 = 0

For nitrogen:

Isolated nitrogen valence electrons	5
Bound nitrogen bonding electrons	4
Bound nitrogen nonbonding electrons	4

Formal charge = 5 – ½(4) – 4 = –1

For carbon:

Isolated carbon valence electrons	4
Bound carbon bonding electrons	8
Bound carbon nonbonding electrons	0

Formal charge = 4 – ½(8) – 0 = 0

(c)

```
      ..
     :O:
      |
 ..   |   ..
:Cl—P—Cl:
 ..   |   ..
      |
     :Cl:
      ..
```

For chlorine:

Isolated chlorine valence electrons	7
Bound chlorine bonding electrons	2
Bound chlorine nonbonding electrons	6

Formal charge = 7 – ½(2) – 6 = 0

For oxygen:

Isolated oxygen valence electrons	6
Bound oxygen bonding electrons	2
Bound oxygen nonbonding electrons	6

Formal charge = 6 – ½(2) – 6 = –1

For phosphorus:

Isolated phosphorus valence electrons	5
Bound phosphorus bonding electrons	8
Bound phosphorus nonbonding electrons	0

Formal charge = 5 – ½(8) – 0 = +1

7.48 (a)

```
       :O:
 ..     ||    ..
HO—S—OH
 ..     ..    ..
```

For sulfur:

Isolated sulfur valence electrons	6
Bound sulfur bonding electrons	8
Bound sulfur nonbonding electrons	2

Formal charge = 6 – ½(8) – 2 = 0

For doubly bound oxygen:	Isolated oxygen valence electrons	6
	Bound oxygen bonding electrons	4
	Bound oxygen nonbonding electrons	4

Formal charge = 6 – ½(4) – 4 = 0

For oxygen bound to hydrogen:	Isolated oxygen valence electrons	6
	Bound oxygen bonding electrons	4
	Bound oxygen nonbonding electrons	4

Formal charge = 6 – ½(4) – 4 = 0

For hydrogen:	Isolated hydrogen valence electrons	1
	Bound hydrogen bonding electrons	2
	Bound hydrogen nonbonding electrons	0

Formal charge = 1 – ½(2) – 0 = 0

(b)

```
        ..
       :O:
  ..    |    ..
  HO—S—OH
  ..    ..    ..
```

For sulfur:	Isolated sulfur valence electrons	6
	Bound sulfur bonding electrons	6
	Bound sulfur nonbonding electrons	2

Formal charge = 6 – ½(6) – 2 = +1

For oxygen not bound to hydrogen:	Isolated oxygen valence electrons	6
	Bound oxygen bonding electrons	2
	Bound oxygen nonbonding electrons	6

Formal charge = 6 – ½(2) – 6 = – 1

For oxygen bound to hydrogen:	Isolated oxygen valence electrons	6
	Bound oxygen bonding electrons	4
	Bound oxygen nonbonding electrons	4

Formal charge = 6 – ½(4) – 4 = 0

For hydrogen:	Isolated hydrogen valence electrons	1
	Bound hydrogen bonding electrons	2
	Bound hydrogen nonbonding electrons	0

Formal charge = 1 – ½(2) – 0 = 0

(a) is the more important contributor to the resonance hybrid because all formal charges are zero.

7.49 (a)

```
H\
  C=N=N̈:
H/
```

For hydrogen:

Isolated hydrogen valence electrons 1
Bound hydrogen bonding electrons 2
Bound hydrogen nonbonding electrons 0

Formal charge = 1 – ½(2) – 0 = 0

For nitrogen: (central)

Isolated nitrogen valence electrons 5
Bound nitrogen bonding electrons 8
Bound nitrogen nonbonding electrons 0

Formal charge = 5 – ½(8) – 0 = +1

For nitrogen: (terminal)

Isolated nitrogen valence electrons 5
Bound nitrogen bonding electrons 4
Bound nitrogen nonbonding electrons 4

Formal charge = 5 – ½(4) – 4 = –1

For carbon:

Isolated carbon valence electrons 4
Bound carbon bonding electrons 8
Bound carbon nonbonding electrons 0

Formal charge = 4 – ½(8) – 0 = 0

(b)

```
H\
  C–N̈=N̈:
H/
```

For hydrogen:

Isolated hydrogen valence electrons 1
Bound hydrogen bonding electrons 2
Bound hydrogen nonbonding electrons 0

Formal charge = 1 – ½(2) – 0 = 0

For nitrogen: (central)	Isolated nitrogen valence electrons	5
	Bound nitrogen bonding electrons	6
	Bound nitrogen nonbonding electrons	2

Formal charge = $5 - \frac{1}{2}(6) - 2 = 0$

For nitrogen: (terminal)	Isolated nitrogen valence electrons	5
	Bound nitrogen bonding electrons	4
	Bound nitrogen nonbonding electrons	4

Formal charge = $5 - \frac{1}{2}(4) - 4 = -1$

For carbon:	Isolated carbon valence electrons	4
	Bound carbon bonding electrons	6
	Bound carbon nonbonding electrons	0

Formal charge = $4 - \frac{1}{2}(6) - 0 = +1$

Structure (a) is more important.

The VSEPR Model

7.50 From data in Table 7.4:

(a) trigonal planar (b) trigonal bipyramidal

(c) linear (d) octahedral

7.51 From data in Table 7.4:

(a) tetrahedral, 4 (b) octahedral, 6 (c) bent, 3 or 4

(d) linear, 2 or 5 (e) square pyramidal, 6 (f) trigonal pyramidal, 4

7.52 (a) $BeCl_2$ or CO_2 (b) CH_4 or $SiCl_4$ (c) H_2O or SCl_2

7.53 From data in Table 7.4:

(a) seesaw, 5 (b) square planar, 6

(c) trigonal bipyramidal, 5 (d) T shaped, 5

(e) trigonal planar, 3 (f) linear, 2 or 5

7.54

	Number of Bonded Atoms	Number of Lone Pairs	Shape
(a) H_2Se	2	2	bent
(b) $SiCl_4$	4	0	tetrahedral
(c) O_3	2	1	bent
(d) GaH_3	3	0	trigonal planar

7.55

	Number of Bonded Atoms	Number of Lone Pairs	Shape
(a) XeO_4	4	0	tetrahedral
(b) SO_2Cl_2	4	0	tetrahedral
(c) OsO_4	4	0	tetrahedral
(d) SeO_2	2	1	bent

7.56

	Number of Bonded Atoms	Number of Lone Pairs	Shape
(a) SbF_5	5	0	trigonal bipyramidal
(b) IF_4^+	4	1	see saw
(c) SeO_3^{2-}	3	1	trigonal pyramidal
(d) CrO_4^{2-}	4	0	tetrahederal

7.57

	Number of Bonded Atoms	Number of Lone Pairs	Shape
(a) NO_3^-	3	0	trigonal planar
(b) NO_2^+	2	0	linear
(c) NO_2^-	2	1	bent

7.58

	Number of Bonded Atoms	Number of Lone Pairs	Shape
(a) PO_4^{3-}	4	0	tetrahedral
(b) MnO_4^-	4	0	tetrahedral
(c) SO_4^{2-}	4	0	tetrahedral
(d) SO_3^{2-}	3	1	trigonal pyramidal
(e) ClO_4^-	4	0	tetrahedral

7.59

	Number of Bonded Atoms	Number of Lone Pairs	Shape
(a) XeF_3^+	3	2	T shaped
(b) SF_3^+	3	1	trigonal pyramidal
(c) ClF_2^+	2	2	bent
(d) CH_3^+	3	0	trigonal planar

7.60 (a) In SF_2 the sulfur is bound to two fluorines and contains two lone pairs of electrons. SF_2 is bent and the F–S–F bond angle is approximately 109°.

(b) In N_2H_2 each nitrogen is bound to the other nitrogen and one hydrogen. Each nitrogen has one lone pair of electrons. The H–N–N bond angle is approximately 120°.

(c) In KrF_4 the krypton is bound to four fluorines and contains two lone pairs of electrons. KrF_4 is square planar, and the F-Kr-F bound angle is 90°.

(d) In NOCl the nitrogen is bound to one oxygen and one chlorine and contains one lone pair of electrons. NOCl is bent, and the Cl-N-O bond angle is approximately 120°.

7.61

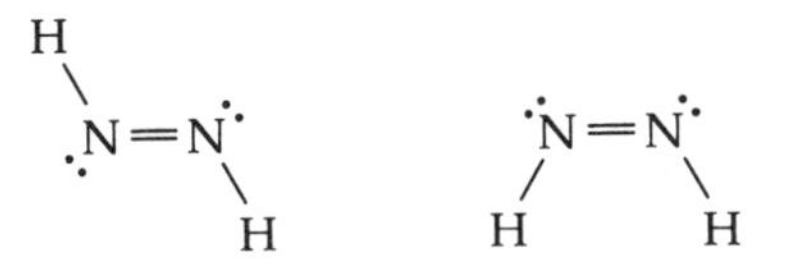

They are geometric isomers not resonance forms. In resonance forms the atoms have the same geometrical arrangement.

7.62

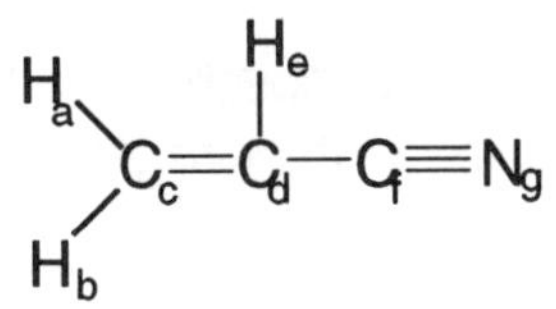

$H_a - C_c - H_b$	~ 120°
$H_a - C_c - C_d$	~ 120°
$H_b - C_c - C_d$	~ 120°
$C_c - C_d - H_e$	~ 120°
$C_c - C_d - C_f$	~ 120°
$H_e - C_d - C_f$	~ 120°
$C_d - C_f - N_g$	~ 180°

7.63

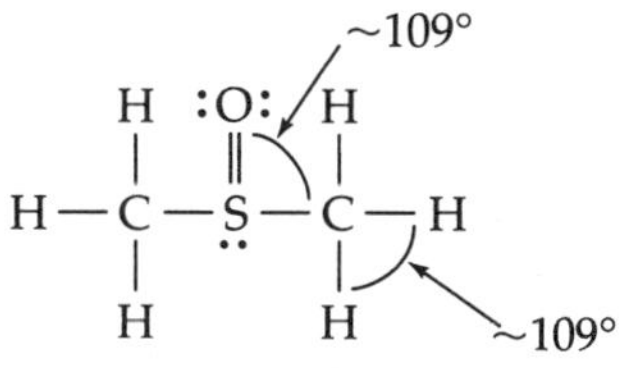

7.64 All six carbons in cyclohexane are bonded to two other carbons and two hydrogens (i.e. four charge clouds). The geometry about each carbon is tetrahedral with a C–C–C bond angle of approximately 109°. Because the geometry about each carbon is tetrahedral, the cyclohexane ring cannot be flat.

7.65 All six carbon atoms are sp^2 hybridized and the bond angle is ~120°. The geometry about each carbon is trigonal planar.

Hybrid Orbitals and Molecular Orbital Theory

7.66 In a π bond, the shared electrons occupy a region above and below a line connecting the two nuclei. A σ bond has its shared electrons located along the axis between the two nuclei.

7.67 Electrons in a bonding molecular orbital spend most of their time in the region between the two nuclei, helping to bond the atoms together. Electrons in an antibonding molecular orbital cannot occupy the central region between the nuclei and cannot contribute to bonding.

7.68 See Table 7.5.

(a) sp (b) sp^3d (c) sp^3d^2 (d) sp^3

7.69 (a) sp^3 (b) sp^3d^2 (c) sp^2 or sp^3 (d) sp or sp^3d (e) sp^3d^2

7.70 See Tables 7.4 and 7.5.

(a) seesaw, 5 charge clouds, sp^3d

(b) square planar, 6 charge clouds, sp^3d^2

(c) trigonal bipyramidal, 5 charge clouds, sp^3d

(d) T shaped, 5 charge clouds, sp^3d

(e) trigonal planar, 3 charge clouds, sp^2

7.71 (a) sp^2 (b) sp^3 (c) sp^3d^2 (d) sp^2

7.72 (a) sp^3 (b) sp^2 (c) sp^2 (d) sp^3

7.73 The C is sp^2 hybridized and the N is sp^3 hybridized.

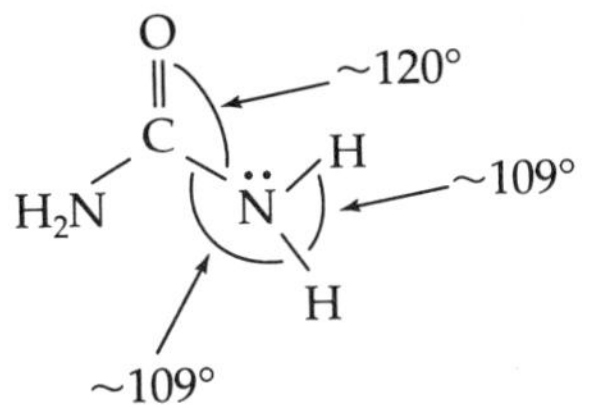

O
‖
C
H_2N N̈ H
H
~120°
~109°
~109°

7.74

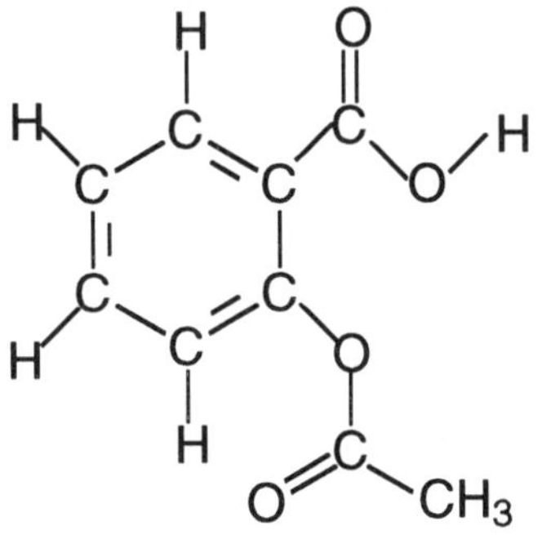

21 sigma bonds

5 pi bonds

7.75

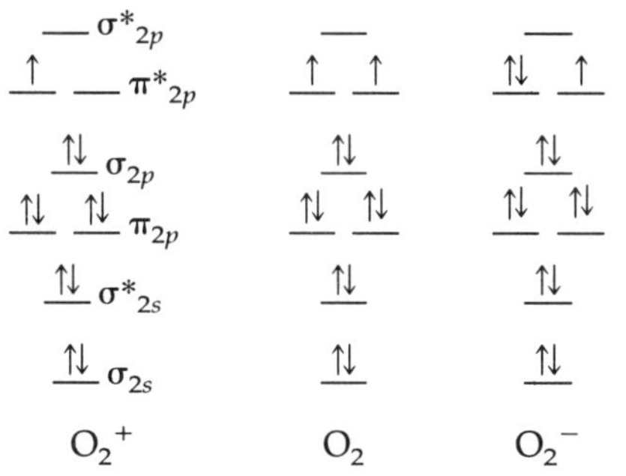

	O_2^+	O_2	O_2^-
σ^*_{2p}	__	__	__
π^*_{2p}	↑ __	↑ ↑	⇅ ↑
σ_{2p}	⇅	⇅	⇅
π_{2p}	⇅ ⇅	⇅ ⇅	⇅ ⇅
σ^*_{2s}	⇅	⇅	⇅
σ_{2s}	⇅	⇅	⇅

	Bond Order
O_2^+	2.5
O_2	2
O_2^-	1.5

All are stable. All have unpaired electrons.

7.76

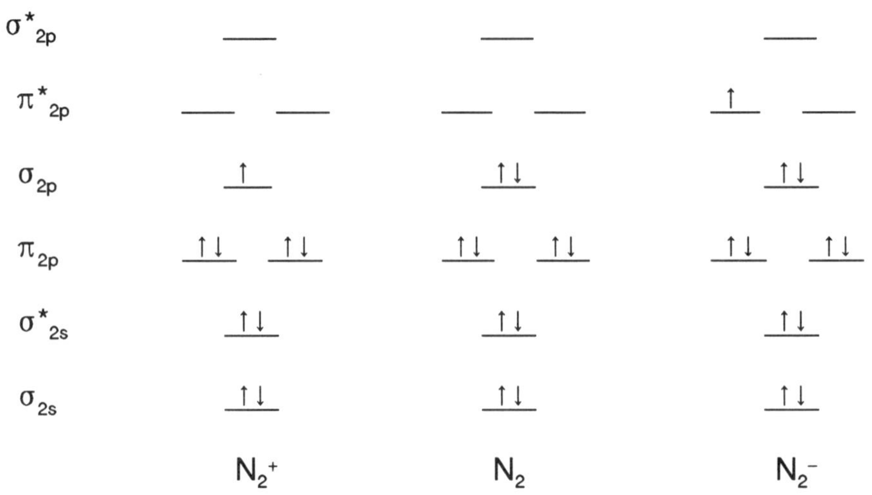

	N_2^+	N_2	N_2^-
σ^*_{2p}	___	___	___
π^*_{2p}	___ ___	___ ___	↑ ___
σ_{2p}	↑	↑↓	↑↓
π_{2p}	↑↓ ↑↓	↑↓ ↑↓	↑↓ ↑↓
σ^*_{2s}	↑↓	↑↓	↑↓
σ_{2s}	↑↓	↑↓	↑↓

$$\text{Bond order} = \frac{\begin{pmatrix}\text{number of}\\ \text{bonding electrons}\end{pmatrix} - \begin{pmatrix}\text{number of}\\ \text{antibonding electrons}\end{pmatrix}}{2}$$

$$N_2^+ \text{ bond order} = \frac{7-2}{2} = 2.5$$

$$N_2 \text{ bond order} = \frac{8-2}{2} = 3$$

$$N_2^- \text{ bond order} = \frac{8-3}{2} = 2.5$$

All are stable with bond orders of either 3 or 2.5.

N_2^+ and N_2^- contain unpaired electrons.

7.77

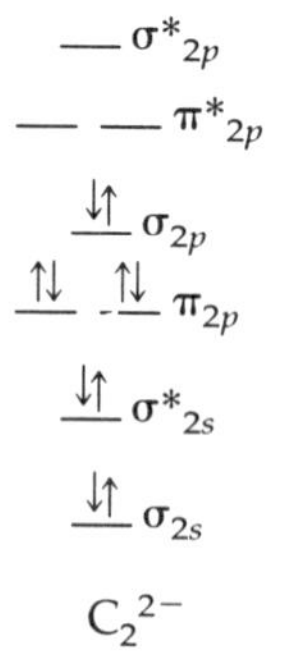

C_2^{2-}	
___ σ^*_{2p}	
___ ___ π^*_{2p}	Bond order = 3
↓↑ σ_{2p}	
↑↓ ↑↓ π_{2p}	$[:C{\equiv}C:]^{2-}$
↓↑ σ^*_{2s}	
↓↑ σ_{2s}	

7.78

p orbitals in allyl cation

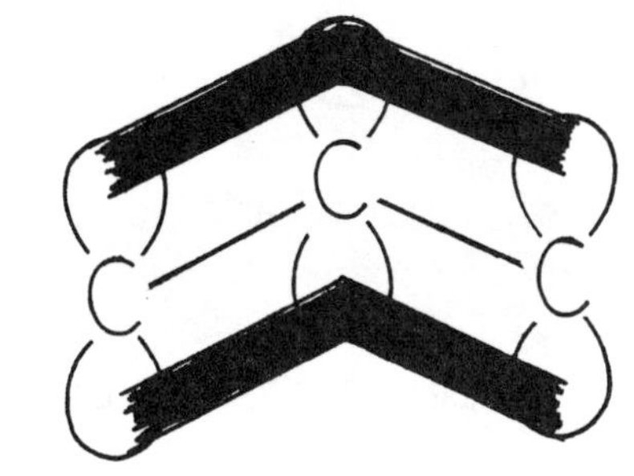

allyl cation showing only the σ bonds (each C is sp^2 hybridized)

delocalized MO model for π bonding in the allyl cation

General Problems

7.79

all sp^2

sp^2

sp^3

7.80 (a)

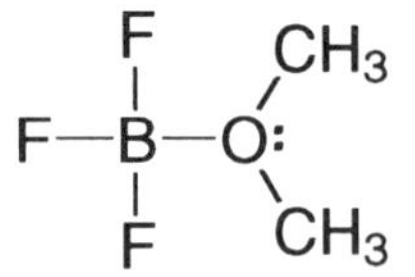

```
     F    CH3
     |   /
  F—B—O:
     |   \
     F    CH3
```

For boron:

For boron:	Isolated boron valence electrons	3
	Bound boron bonding electrons	8
	Bound boron nonbonding electrons	0

Formal charge $= 3 - \frac{1}{2}(8) - 0 = -1$

For oxygen:	Isolated oxygen valence electrons	6
	Bound oxygen bonding electrons	6
	Bound oxygen nonbonding electrons	2

Formal charge $= 6 - \frac{1}{2}(6) - 2 = +1$

(b) In BF_3 the B has three bonding pairs of electrons and no lone pairs. The B is sp^2 hybridized and BF_3 is trigonal planar.

$CH_3\text{-}\ddot{\underset{\cdot\cdot}{O}}\text{—}CH_3$ is bent about the oxygen because of two bonding pairs and two lone pairs of electrons. The O is sp^3 hybridized.

In the product, B is sp^3 hybridized (with four bonding pairs of electrons), and the geometry about it is tetrahedral. The O is also sp^3 hybridized (with three bonding pairs and one lone pair of electrons), and the geometry about it is trigonal pyramidal.

7.81 Both the B and N are sp^2 hybridized. All bond angles are ~120°. The overall geometry of the molecule is planar.

7.82

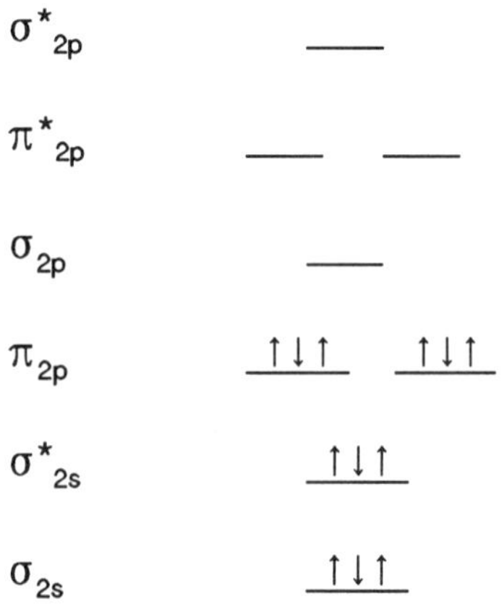

σ^*_{2p}	___
π^*_{2p}	___ ___
σ_{2p}	___
π_{2p}	↑↓↑ ↑↓↑
σ^*_{2s}	↑↓↑
σ_{2s}	↑↓↑

$$O_2 \text{ Bond order} = \frac{\left(\begin{array}{c}\text{number of}\\ \text{bonding electrons}\end{array}\right) - \left(\begin{array}{c}\text{number of}\\ \text{antibonding electrons}\end{array}\right)}{2} = \frac{9 - 3}{2} = 3$$

7.83 The triply bonded carbon atoms are sp hybridized. The theoretical bond angle for C–C≡C is 180°. Benzyne is so reactive because the C–C≡C bond angle is closer to 120° and is very strained.

7.84

```
        ..         ..      2-
       :O:        :O:
  ..    |    ..    |    ..
[ :O—Cr—O—Cr—O: ]
  ..    |    ..    |    ..
       :O:        :O:
        ..         ..
```

The geometry about each Cr is tetrahedral.

7.85 (a) H—C≡C—H

(b)

```
   ..  ..
H—N=N—H
```

(c)

```
         ..
        :O:
  ..     |     ..
:Cl—S—Cl:
  ..     ..    ..
```

7.86 (a) H—C≡C—H

Each carbon is sp hybridized, and the geometry about each carbon is linear.

(b)

```
   ..  ..
H—N=N—H
```

Each nitrogen is sp^2 hybridized, and the geometry about each nitrogen is bent.

(c) :Ö: above :C̈l–S̈–C̈l: (Lewis structure of $SOCl_2$)

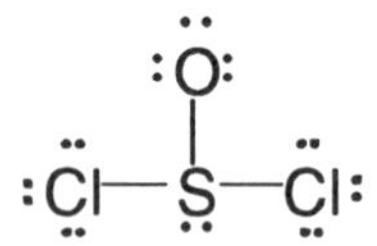

The sulfur is sp^3 hybridized, and the geometry about the sulfur is trigonal pyramidal.

7.87

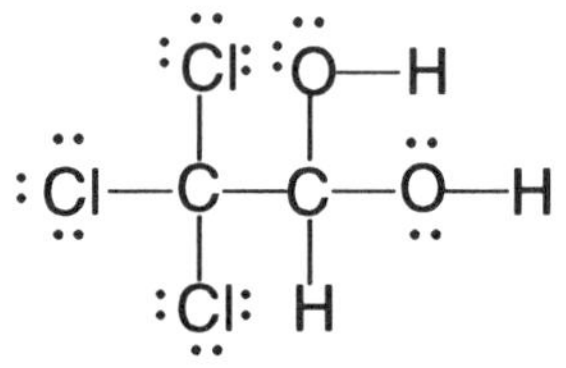

(Lewis structure: CCl_3–C(H)(OH)–OH)

7.88 For simplicity, the lone pairs on O atoms have been omitted.

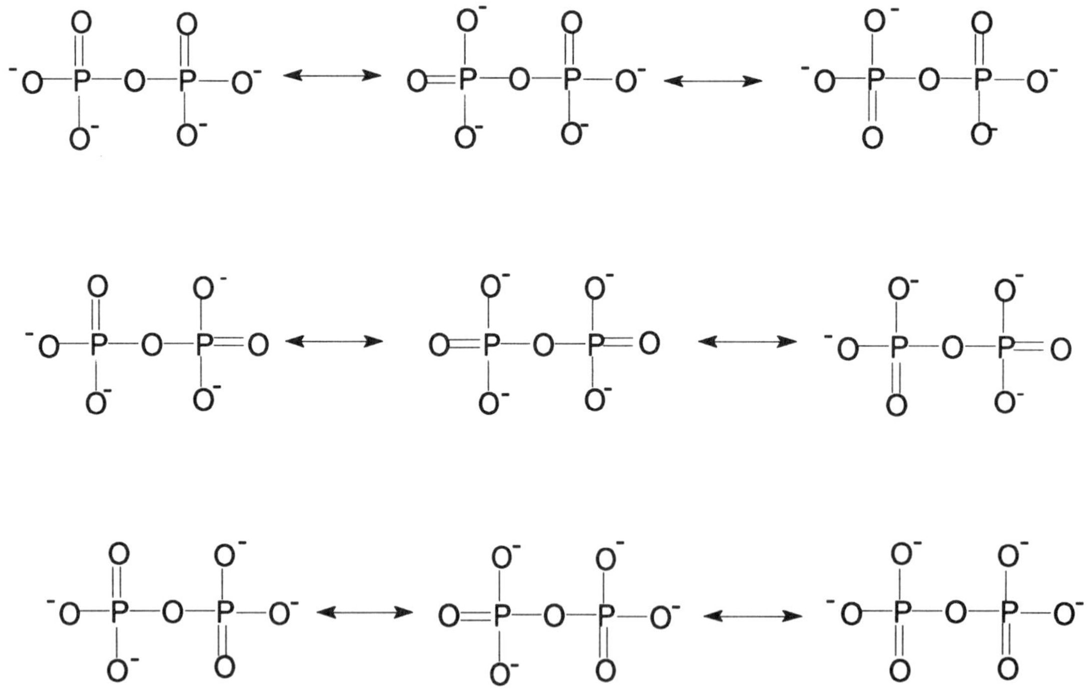

(Resonance structures of $P_2O_7^{4-}$)

7.89 Li_2

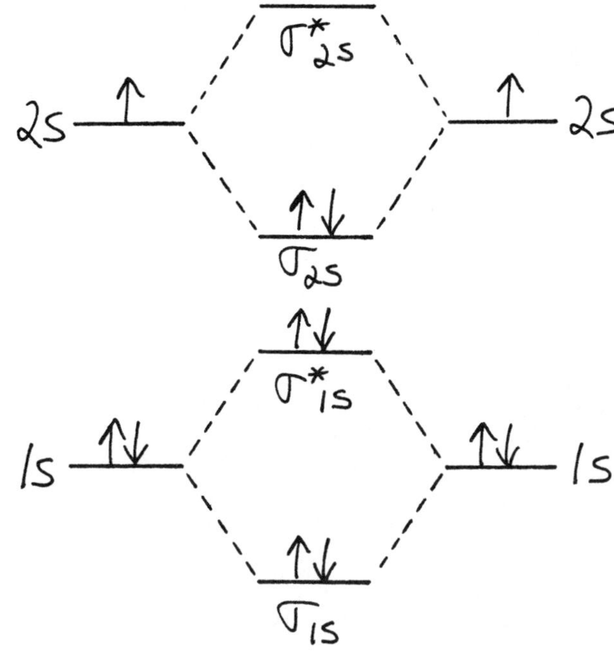

$$Li_2 \text{ Bond order} = \frac{\begin{pmatrix}\text{number of}\\ \text{bonding electrons}\end{pmatrix} - \begin{pmatrix}\text{number of}\\ \text{antibonding electrons}\end{pmatrix}}{2} = \frac{4 - 2}{2} = 1$$

The bond order for Li_2 is 1, and the molecule is likely to be stable.

8.1 $E = \frac{1}{2}mv^2 = \frac{1}{2}(2\ kg)(4\ m/s)^2 = 16\ kg \cdot m^2/s^2 = 16\ J$

8.2 Convert lb to kg.

$$2300\ lb \times \frac{453.59\ g}{1\ lb} \times \frac{1\ kg}{1000\ g} = 1043\ kg$$

Convert mi/h to m/s.

$$55\ \frac{mi}{h} \times \frac{1\ km}{0.62137\ mi} \times \frac{1000\ m}{1 km} \times \frac{1\ h}{3600\ s} = 24.6\ m/s$$

$1\ kg \cdot m^2/s^2 = 1\ J$

$$E = \frac{1}{2}mv^2 = \frac{1}{2}(1043\ kg)(24.6\ m/s)^2 = 3.2 \times 10^5\ J$$

$$E = 3.2 \times 10^5\ J \times \frac{1\ kJ}{1000\ J} = 3.2 \times 10^2\ kJ$$

8.3 (a) and (b) are state functions; (c) is not.

8.4 $\Delta V = (4.3\ L - 8.6\ L) = -4.3\ L$

$$w = -P\Delta V = -(44\ atm)(-4.3\ L) = +189.2\ L \cdot atm$$

$$w = (189.2\ L \cdot atm)(101\ \frac{J}{L \cdot atm}) = +1.9 \times 10^4\ J$$

The positive sign for the work indicates that the surrounding does work on the system. Energy flows into the system.

8.5 $\Delta H^\circ = -484\ \frac{kJ}{2\ mol\ H_2}$

$$P\Delta V = (1.00\ atm)(-5.6\ L) = -5.6\ L \cdot atm$$

$$P\Delta V = (-5.6\ L \cdot atm)(101\ \frac{J}{L \cdot atm}) = -565.6\ J = -570\ J = -0.57\ kJ$$

$w = -P\Delta V = 570\ J = 0.57\ kJ$

$$\Delta H = -121\ \frac{kJ}{0.50\ mol\ H_2}$$

$\Delta E = \Delta H - P\Delta V = -121\ kJ - (-0.57\ kJ) = -120.43\ kJ = -120\ kJ$

8.6 $\Delta V = 448\ L$ and assume $P = 1.00\ atm$

$w = -P\Delta V = -(1.00\ atm)(448\ L) = -448\ L \cdot atm$

$$w = -(448\ L \cdot atm)(101\ \frac{J}{L \cdot atm}) = -4.52 \times 10^4\ J$$

$$w = -4.52 \times 10^4\ J \times \frac{1\ kJ}{1000\ J} = -45.2\ kJ$$

8.7 (a) C_3H_8, 44.10 amu

$\Delta H° = -2217\ kJ/mol\ C_3H_8$

$$15.5\ g \times \frac{1\ mol\ C_3H_8}{44.10\ g\ C_3H_8} \times \frac{-2217\ kJ}{1\ mol\ C_3H_8} = -779\ kJ$$

779 kJ of heat is evolved.

(b) $Ba(OH)_2 \cdot 8\ H_2O$, 315.5 amu

$\Delta H° = +80.3\ kJ/mol\ Ba(OH)_2 \cdot 8\ H_2O$

$$4.88\ g \times \frac{1\ mol\ Ba(OH)_2 \cdot 8\ H_2O}{315.5\ g\ Ba(OH)_2 \cdot 8\ H_2O} \times \frac{80.3\ kJ}{1\ mol\ Ba(OH)_2 \cdot 8\ H_2O} = +1.24\ kJ$$

1.24 kJ of heat is absorbed.

8.8 $q = (\text{specific heat}) \times m \times \Delta T$

$$q = (4.18\ \frac{J}{g \cdot °C})(350\ g)(3°C - 25°C) = -3.2 \times 10^4\ J$$

$$q = -3.2 \times 10^4 \times \frac{1\ kJ}{1000\ J} = -32\ kJ$$

8.9 $q = (\text{specific heat}) \times m \times \Delta T$

$$\text{specific heat} = \frac{q}{m \times \Delta T} = \frac{96\ J}{(75\ g)(10°C)} = 0.13\ J/(g \cdot °C)$$

8.10 $q = (\text{specific heat}) \times m \times \Delta T$

$$m = (25\ mL + 50\ mL)(1.00\ \frac{g}{mL}) = 75\ g$$

$$q = (4.18\ \frac{J}{g \cdot °C})(75\ g)(33.9°C - 25.0°C) = 2790\ J$$

$$\text{mol } H_2SO_4 = 0.025\ L \times 1.00\ \frac{mol}{L}\ H_2SO_4 = 0.025\ mol\ H_2SO_4$$

Heat evolved per mole of H_2SO_4

$$= \frac{2.79 \times 10^3\ J}{0.025\ mol\ H_2SO_4} = 1.1 \times 10^5\ J/mol\ H_2SO_4$$

Since the reaction evolves heat, the sign for ΔH is negative.

$$\Delta H = -1.1 \times 10^5\ J \times \frac{1\ kJ}{1000\ J} = -1.1 \times 10^2\ kJ$$

8.11

$$CH_4(g) + Cl_2(g) \rightarrow CH_3Cl(g) + HCl(g) \qquad \Delta H_1° = -98.3\ kJ$$

$$CH_3Cl(g) + Cl_2(g) \rightarrow CH_2Cl_2(g) + HCl(g) \qquad \Delta H_2° = -104\ kJ$$

Sum $CH_4(g) + 2\ Cl_2(g) \rightarrow CH_2Cl_2(g) + 2\ HCl(g)$

$$\Delta H° = \Delta H_1° + \Delta H_2° = -202\ kJ$$

8.12 $CH_4(g) + Cl_2(g) \rightarrow CH_3Cl(g) + HCl(g)$ $\Delta H_1° = -98.3$ kJ

$CH_3Cl(g) + Cl_2(g) \rightarrow CH_2Cl_2(g) + HCl(g)$ $\Delta H_2° = -104$ kJ

$CH_2Cl_2(g) \rightarrow CH_2Cl(l)$ $\Delta H_3° = -H_{vap} = -29.0$ kJ

Sum $CH_4(g) + 2\ Cl_2(g) \rightarrow CH_2Cl_2(l) + 2\ HCl(g)$

$\Delta H° = \Delta H_1° + \Delta H_2° + \Delta H_3° = -231$ kJ

8.13 Reactants $CH_4 + 2\ Cl_2$

$\Delta H° = -98.3$ kJ

$\Delta H° = -202$ kJ

$CH_3Cl + HCl + Cl_2$

$\Delta H° = -104$ kJ

Products $CH_2Cl_2 + 2\ HCl$

8.14 $4\ NH_3(g) + 5\ O_2(g) \rightarrow 4\ NO(g) + 6\ H_2O(g)$

$\Delta H°_{rxn} = [4\ \Delta H°_f\ (NO) + 6\ \Delta H°_f\ (H_2O)] - [4\ \Delta H°_f\ (NH_3)]$

$\Delta H°_{rxn} = [(4\ mol)(90.2\ kJ/mol) + (6\ mol)(-\ 241.8\ kJ/mol)]$

$-\ [(4\ mol)(-\ 46.1\ kJ/mol)]$

$\Delta H°_{rxn} = -905.6$ kJ

8.15 $6\ CO_2(g) + 6\ H_2O(l) \rightarrow C_6H_{12}O_6(s) + 6\ O_2(g)$

$\Delta H°_{rxn} = \Delta H_f°(C_6H_{12}O_6) - [6\ \Delta H_f°(CO_2) + 6\ \Delta H_f°(H_2O(l))]$

$\Delta H°_{rxn} = [(1\ mol)(-1260\ kJ/mol)] - [(6\ mol)(-393.5\ kJ/mol) + (6\ mol)(-285.8\ kJ/mol)]$

$\Delta H°_{rxn} = +2815.8\ kJ = +2816\ kJ$

8.16 $\Delta H^\circ_{rxn} = \underline{D}$ (bonds broken) – $\underline{D}$ (bonds formed)

$H_2C{=}CH_2(g) + H_2O(g) \rightarrow C_2H_5OH(g)$

$\Delta H^\circ_{rxn} = [\underline{D}(C{=}C) + \underline{D}(O{-}H)] - [\underline{D}(C{-}C) + \underline{D}(C{-}O) + \underline{D}(C{-}H)]$

$\Delta H^\circ_{rxn} = [(1\ mol)(720\ kJ/mol) + (1\ mol)(460\ kJ/mol)]$

$- [(1\ mol)(350\ kJ/mol) + (1\ mol)(350\ kJ/mol) + (1\ mol)(410\ kJ/mol)]$

$\Delta H^\circ_{rxn} = 70\ kJ$

8.17 $2\ NH_3(g) + Cl_2(g) \rightarrow N_2H_4(g) + 2\ HCl(g)$

$\Delta H^\circ_{rxn} = \underline{D}$ (bonds broken) – $\underline{D}$ (bonds formed)

$\Delta H^\circ_{rxn} = [2 \times \underline{D}(N{-}H) + \underline{D}(Cl{-}Cl)] - [\underline{D}(N{-}N) + 2 \times \underline{D}(H{-}Cl)]$

$\Delta H^\circ_{rxn} = [(2\ mol)(390\ kJ/mol) + (1\ mol)(243\ kJ/mol]$

$- [(1\ mol)(240\ kJ/mol) + (2\ mol)(432\ kJ/mol)]$

$\Delta H^\circ_{rxn} = -81\ kJ$

8.18 $C_4H_{10}(l) + \frac{13}{2} O_2(g) \rightarrow 4\ CO_2(g) + 5\ H_2O(g)$

$\Delta H^\circ_{rxn} = [4\ \Delta H^\circ_f(CO_2) + 5\ \Delta H^\circ_f(H_2O)] - \Delta H^\circ_f(C_4H_{10})$

$\Delta H^\circ_{rxn} = [(4\ mol)(-395.5\ kJ/mol) + (5\ mol)(-241.8\ kJ/mol)]$

$- [(1\ mol)(-147.5\ kJ/mol)]$

$\Delta H^\circ_{rxn} = -2643.5\ kJ$

$\Delta H^\circ_C = -2643.5\ kJ/mol$

C_4H_{10}, 58.12 amu

$$\Delta H^\circ_C = \left(-2643.5\ \frac{kJ}{mol}\right)\left(\frac{1\ mol}{58.12\ g}\right) = -45.48\ kJ/g$$

$$\Delta H^\circ_c = \left(-45.48\ \frac{kJ}{g}\right)\left(0.579\ \frac{g}{mL}\right) = -26.3\ kJ/mL$$

8.19 $\Delta S^\circ < 0$ because the reaction decreases the number of moles of gaseous molecules.

8.20 (a) Since ΔG° is negative, the reaction is spontaneous.

(b) Since ΔG° is positive, the reaction is nonspontaneous.

8.21 $\Delta G^\circ = \Delta H^\circ - T\Delta S^\circ$

$\Delta G^\circ = (-92.2\ kJ) - (298\ K)(-0.199\ kJ/K) = -32.9\ kJ$

Because ΔG° is negative, the reaction is spontaneous.

Set $\Delta G^\circ = 0$ and solve for T.

$\Delta G^\circ = 0 = \Delta H^\circ - T\Delta S^\circ$

$$T = \frac{\Delta H^\circ}{\Delta S^\circ} = \frac{-92.2\ kJ}{-0.199\ kJ/K} = 463\ K = 190^\circ C$$

Understanding Key Concepts

1. (a) $w = -P\Delta V$, $\Delta V > 0$, therefore $w < 0$ and the system is doing work on the surroundings.

   (b) Since the temperature has increased there has been an enthalpy change. The system evolved heat, the reaction is exothermic, and $\Delta H < 0$.

2. (a)

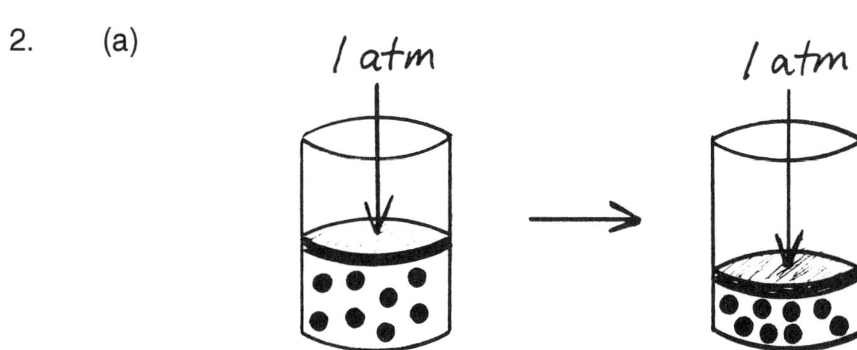

(b)

1 atm
1 atm

3.

reactants
step 1
net reaction
step 2
H°
products

4.

Reactants CH_3CH_2OH + O_2
ΔH° = –200.4 kJ
ΔH° = –492.6 kJ
$CH_3CHO + H_2O + ½ O_2$
ΔH° = –292.2 kJ
Products $CH_3COOH + H_2O$

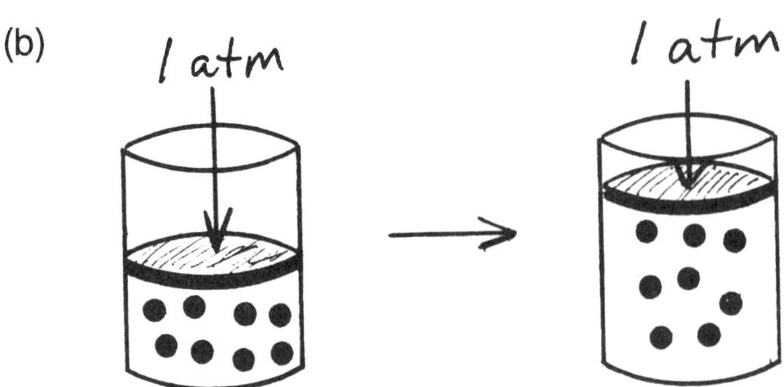

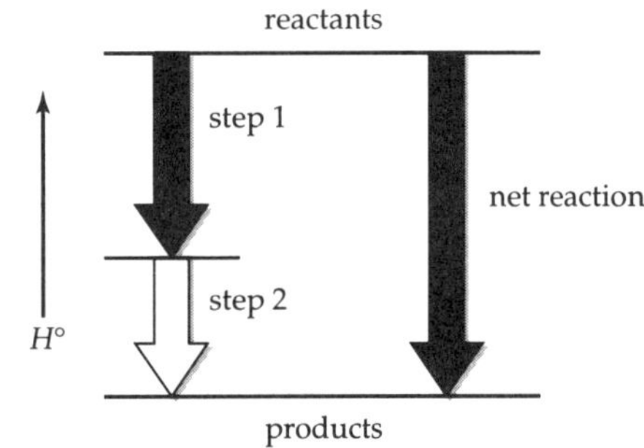

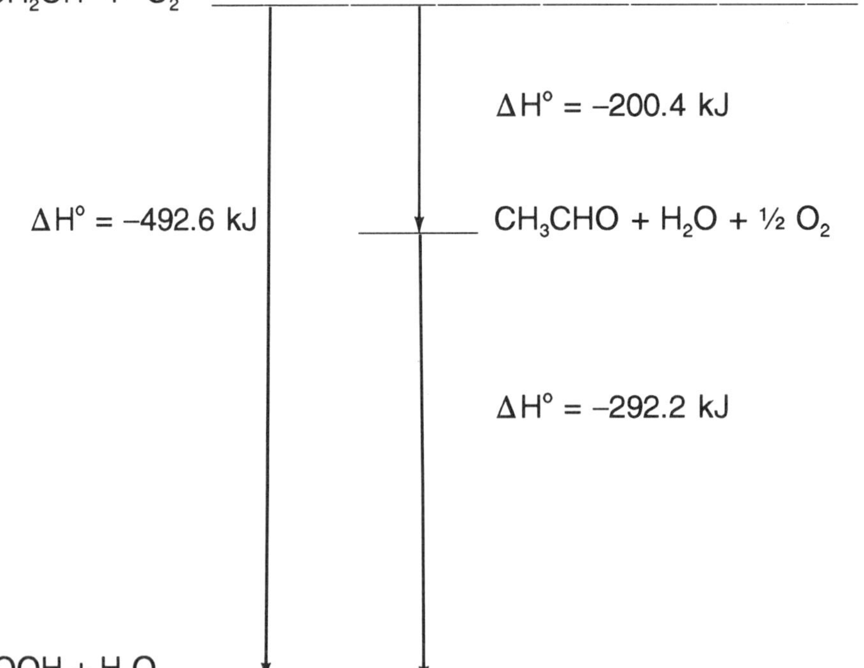

5. (a) The entropy change is positive. (b) The entropy change is negative.

6. Since the process is spontaneous, $\Delta G < 0$. The system is more disordered, $\Delta S > 0$. $\Delta H \approx 0$ for the mixing of gaseous molecules.

Additional Problems

Heat, Work, and Energy

8.22 Heat is the energy transferred from one object to another as the result of a temperature difference between them. Temperature is a measure of the kinetic energy of molecular motion.

Energy is the capacity to do work or supply heat. Work is defined as the distance moved times the force that opposes the motion (w = d x f).

Kinetic energy is the energy of motion. Potential energy is stored energy.

8.23 Car: $E_k = \frac{1}{2}(1400\ \text{kg})\left(\frac{115 \times 10^3\ \text{m}}{3600\ \text{s}}\right)^2 = 7.1 \times 10^5\ \text{J}$

Truck: $E_k = \frac{1}{2}(12{,}000\ \text{kg})\left(\frac{38 \times 10^3\ \text{m}}{3600\ \text{s}}\right)^2 = 6.7 \times 10^5\ \text{J}$

The car has more kinetic energy.

8.24 Heat = q = 7.1×10^5 J (from Problem 8.23)

q = (specific heat) x m x ΔT

$$m = \frac{q}{(\text{specific heat}) \times \Delta T} = \frac{7.1 \times 10^5\ \text{J}}{\left(4.18\ \frac{\text{J}}{\text{g}\cdot{}^\circ\text{C}}\right)(50\ {}^\circ\text{C} - 20\ {}^\circ\text{C})} = 5.7 \times 10^3\ \text{g of water}$$

8.25 $w = -P\Delta V = -(3.6 \text{ atm})(3.4 \text{ L} - 3.2 \text{ L}) = -0.72 \text{ L} \cdot \text{atm}$

$$w = (-0.72 \text{ L} \cdot \text{atm})\left(\frac{101 \text{ J}}{1 \text{ L} \cdot \text{atm}}\right) = -72.7 \text{ J} = -70 \text{ J}$$

The energy change is negative.

8.26 $V_{initial} = 100.0 \text{ mL} = 0.1000 \text{ L}$

$V_{final} = 50.0 \text{ mL} = 0.0500 \text{ L}$

$\Delta V = V_{final} - V_{initial} = (0.0500 \text{ L} - 0.1000 \text{ L}) = -0.0500 \text{ L}$

$w = -P\Delta V = -(1.5 \text{ atm})(-0.0500 \text{ L}) = +0.075 \text{ L} \cdot \text{atm}$

$$w = (+0.075 \text{ L} \cdot \text{atm})\left(101 \frac{\text{J}}{1 \text{ L} \cdot \text{atm}}\right) = +7.6 \text{ J}$$

The positive sign for the work indicates that the surroundings does work on the system. Energy flows into the system.

Energy and Enthalpy

8.27 $\Delta E = q_v$ is the heat change associated with a reaction at constant volume. Since $\Delta V = 0$, no PV work is done.

$\Delta H = q_p$ is the heat change associated with a reaction at constant pressure. Since $\Delta V \neq 0$, PV work can also be done.

8.28 ΔH is negative for an exothermic reaction.

ΔH is positive for an endothermic reaction.

8.29 $\Delta H = \Delta E + P\Delta V$

ΔH and ΔE are nearly equal when there are no gases involved in a chemical reaction, or, if gases are involved, $\Delta V = 0$ (that is, there are the same number of reactant and product gas molecules).

8.30 $P\Delta V = -7.6$ J (from Problem 8.26)

$\Delta H = \Delta E + P\Delta V$

$\Delta E = \Delta H - P\Delta V = -0.31 \text{ kJ} - (-7.6 \times 10^{-3} \text{ kJ})$

$\Delta E = -0.30$ kJ

8.31 $\Delta H = -244$ kJ and $w = -P\Delta V = 35$ kJ, therefore $P\Delta V = -35$ kJ

$\Delta E = \Delta H - P\Delta V = -244 \text{ kJ} - (-35 \text{ kJ}) = -209 \text{ kJ}$

For the system: $\Delta H = -244$ kJ and $\Delta E = -209$ kJ

ΔH and ΔE for the surroundings are just the opposite of what they are for the system.

For the surroundings: $\Delta H = 244$ kJ and $\Delta E = 209$ kJ

8.32 Heat is lost on going from $H_2O(g) \rightarrow H_2O(l) \rightarrow H_2O(s)$.

$H_2O(g)$ has the highest enthalpy content.

$H_2O(s)$ has the lowest enthalpy content.

8.33 $\Delta H = -1255.5$ kJ/mol C_2H_2; C_2H_2, 26.04 amu

$w = -P\Delta V = -(1.00 \text{ atm})(-2.80 \text{ L}) = 2.80 \text{ L}\cdot\text{atm}$

$$w = (2.80 \text{ L}\cdot\text{atm})\left(\frac{101 \text{ J}}{1 \text{ L}\cdot\text{atm}}\right) = 283 \text{ J} = 0.283 \text{ kJ}$$

$$6.50 \text{ g} \times \frac{1 \text{ mol } C_2H_2}{26.04 \text{ g } C_2H_2} = 0.250 \text{ mol } C_2H_2$$

$q = (-1255.5 \text{ kJ/mol})(0.250 \text{ mol}) = -314 \text{ kJ}$

$\Delta E = \Delta H - P\Delta V = -314 \text{ kJ} - (-0.283 \text{ kJ}) = -314 \text{ kJ}$

8.34 mass of $C_4H_{10}O = \left(0.7138\ \frac{g}{mL}\right)(100\ mL) = 71.38\ g$

$C_4H_{10}O$, 74.12 amu

$$\text{mol } C_4H_{10}O = 71.38\ g \times \frac{1\ mol}{74.12\ g} = 0.9626\ mol$$

$$q = n \times \Delta H_{vap} = 0.9626\ mol \times 26.5\ kJ/mol = 25.5\ kJ$$

8.35 Assume 100 mL of H_2O = 100 g H_2O, 18.02 amu

$$100\ g \times \frac{1\ mol\ H_2O}{18.02\ g\ H_2O} \times \frac{40.7\ kJ}{1\ mol\ H_2O} = 226\ kJ$$

The heat to vaporize 100 mL of H_2O is much greater than that to vaporize 100 mL of diethyl ether.

8.36 Al, 26.98 amu

$$\text{mol Al} = 5.00\ g \times \frac{1\ mol}{26.98\ g} = 0.1853\ mol$$

$$q = n \times \Delta H^\circ$$

$$q = 0.1853\ mol\ Al \times \frac{-1408.4\ kJ}{2\ mol\ Al} = -131\ kJ$$

8.37 Na, 22.99 amu; $\Delta H^\circ = -368.4\ kJ/2\ mol\ Na = \Delta H^\circ = -184.2\ kJ/mol\ Na$

$$1.00\ g\ Na \times \frac{1\ mol\ Na}{22.99\ g\ Na} \times \frac{-184.2\ kJ}{1\ mol\ Na} = -8.01\ kJ$$

8.01 kJ of heat is evolved. The reaction is exothermic.

8.38 (a) Fe_2O_3, 159.7 amu

$$\text{mol } Fe_2O_3 = 2.50\ g \times \frac{1\ mol}{159.7\ g} = 0.015\ 65\ mol$$

$q = n \times \Delta H°$

$$q = 0.015\ 65 \text{ mol } Fe_2O_3 \times \frac{-24.8 \text{ kJ}}{1 \text{ mol } Fe_2O_3} = -0.388 \text{ kJ}$$

Because ΔH is negative, the reaction is exothermic.

(b) CaO, 56.08 amu

$$\text{mol CaO} = 233.0 \text{ g} \times \frac{1 \text{ mol}}{56.08 \text{ g}} = 4.155 \text{ mol}$$

$q = n \times \Delta H°$

$$q = 4.155 \text{ mol CaO} \times \frac{464.8 \text{ kJ}}{1 \text{ mol CaO}} = 1931 \text{ kJ}$$

Because ΔH is positive, the reaction is endothermic.

Calorimetry and Heat Capacity

8.39 Heat capacity is the amount of heat required to raise the temperature of a substance a given amount. Specific heat is the amount of heat necessary to raise the temperature of exactly 1g of a substance by exactly 1°C.

8.40 Na, 22.99 amu

$$\text{specific heat} = 28.2 \frac{\text{J}}{\text{mol}\cdot°\text{C}} \times \frac{1 \text{ mol}}{22.99 \text{ g}} = 1.23 \text{ J/(g}\cdot°\text{C)}$$

8.41 $q = (\text{specific heat}) \times m \times \Delta T$

$$\text{specific heat} = \frac{q}{m \times \Delta T} = \frac{89.7 \text{ J}}{(33.0 \text{ g})(5.20°\text{C})} = 0.523 \text{ J/(g}\cdot°\text{C)}$$

$$C_m = (0.523 \text{ J/(g}\cdot°\text{C)})(47.88 \text{ g/mol}) = 25.0 \text{ J/(mol}\cdot°\text{C)}$$

8.42 Assume a negligible volume change when solid CaO is added to 50.0 mL of H_2O.

$$q = (\text{specific heat}) \times m \times \Delta T$$

$$m = 50.0 \text{ mL} \times 1.00 \text{ g/mL} = 50.0 \text{ g}$$

$$q = \left(4.18 \frac{\text{J}}{\text{g}\cdot{}^\circ\text{C}}\right)(50.0 \text{ g})(32.3^\circ\text{C} - 25.0^\circ\text{C}) = 1.53 \times 10^3 \text{ J} = 1.53 \text{ kJ}$$

CaO, 56.08 amu

$$\text{mol CaO} = 1.045 \text{ g} \times \frac{1 \text{ mol}}{56.08 \text{ g}} = 0.018\ 63 \text{ mol}$$

Heat evolved per mole of CaO

$$= \frac{1.53 \text{ kJ}}{0.018\ 63 \text{ mol}} = 82.1 \text{ kJ/mol CaO}$$

Because the reaction evolves heat, the sign for ΔH is negative.

$$\Delta H = -82.1 \text{ kJ}$$

Hess's Law and Heats of Formation

8.43 The standard state of an element is its most stable form at 1 atm and 25°C.

8.44 A compound's standard heat of formation is the amount of heat associated with the formation of 1 mole of a compound from its elements (in their standard states).

8.45 Hess's Law — the overall enthalpy change for a reaction is equal to the sum of the enthalpy changes for the individual steps in the reaction.

Hess's Law works because of the law of conservation of energy.

8.46 $Cu(s) + \frac{1}{2} O_2(g) \rightarrow CuO(s)$ $\Delta H_f° = -157$ kJ/mol CuO

For $2\ CuO(s) \rightarrow 2\ Cu(s) + O_2(g)$,

$\Delta H = (2\ mol)(-\Delta H_f°) = 2\ mol \times 157\ kJ/mol = +314\ kJ$

8.47 $S(s) + O_2(g) \rightarrow SO_2(g)$ $\Delta H_1° = -296.8$ kJ

$SO_2 + \frac{1}{2} O_2(g) \rightarrow SO_3(g)$ $\Delta H_2° = -98.9$ kJ

Sum $S(s) + 3/2\ O_2(g) \rightarrow SO_3(g)$ $\Delta H_3° = \Delta H_1° + \Delta H_2°$

$H_f° = \Delta H_3° = -296.8\ kJ + (-98.9\ kJ) = -395.7$ kJ/mol

8.48 $SO_3(g) + H_2O(l) \rightarrow H_2SO_4(aq)$ $\Delta H_1° = -227.8$ kJ

$H_2(g) + \frac{1}{2} O_2(g) \rightarrow H_2O(l)$ $\Delta H_2° = \Delta H_f° = -285.8$ kJ

$S(s) + 3/2\ O_2(g) \rightarrow SO_3(g)$ $\Delta H_3° = \Delta H_f° = -395.7$ kJ

Sum $S(s) + H_2(g) + 2\ O_2(g) \rightarrow H_2SO_4(aq)$ $\Delta H_f°\ (H_2SO_4) = ?$

$\Delta H_f°\ (H_2SO_4) = \Delta H_1° + \Delta H_2° + \Delta H_3° = -909.3$ kJ

8.49

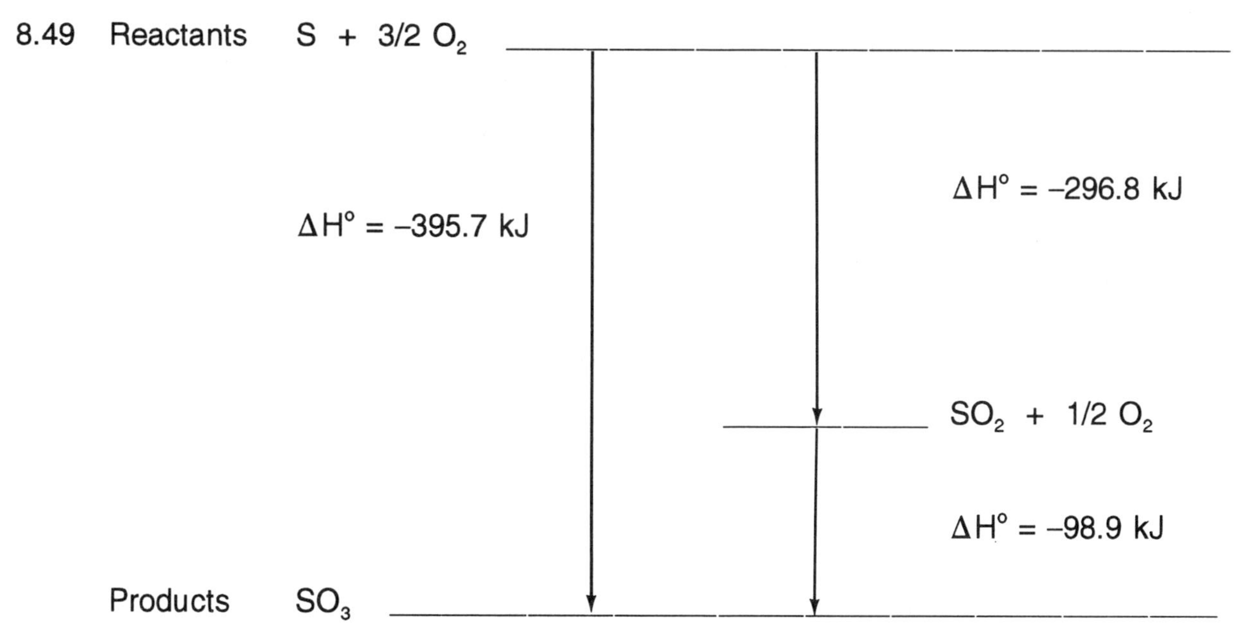

8.50 $\Delta H^\circ_{rxn} = [\Delta H_f^\circ(CH_3COOH) + \Delta H_f^\circ(H_2O)] - \Delta H_f^\circ(CH_3CH_2OH)$

$\Delta H^\circ_{rxn} = [(1\ mol)(-484.5\ kJ/mol) + (1 mol)(-285.8\ kJ/mol)]$

$- [(1\ mol)(-277.7\ kJ/mol)]$

$\Delta H^\circ_{rxn} = -492.6\ kJ$

8.51 $\Delta H^\circ_{rxn} = [12\ \Delta H_f^\circ(CO_2) + 6\ \Delta H_f^\circ(H_2O)] - [2\ \Delta H_f^\circ(C_6H_6)]$

$-6534\ kJ = [(12\ mol)(-393.5\ kJ/mol) + (6\ mol)(-285.8\ kJ/mol)]$

$- [(2\ mol)(\Delta H_f^\circ(C_6H_6))]$

Solve for $\Delta H_f^\circ(C_6H_6)$.

$-6534\ kJ = -6436.8\ kJ - [(2\ mol)(\Delta H_f^\circ(C_6H_6))]$

$97.2\ kJ = (2\ mol)(\Delta H_f^\circ(C_6H_6))$

$\Delta H_f^\circ(C_6H_6) = +48.6\ kJ/mol$

8.52 $C_8H_8(l) + 10\ O_2(g) \rightarrow 8\ CO_2(g) + 4\ H_2O(l)$

$\Delta H^\circ_{rxn} = \Delta H_C^\circ = -\ 4395.2\ kJ$

$\Delta H^\circ_{rxn} = [8\ \Delta H_f^\circ(CO_2) + 4\ \Delta H_f^\circ(H_2O)] - \Delta H_f^\circ(C_8H_8)$

$-\ 4395.2\ kJ = [(8\ mol)(-393.5\ kJ/mol) + (4\ mol)(-285.8\ kJ/mol)]$

$- [(1\ mol)(\Delta H_f^\circ(C_8H_8))]$

Solve for $\Delta H_f^\circ(C_8H_8)$

$-\ 4395.2\ kJ = -4291.2\ kJ - (1\ mol)(\Delta H_f^\circ(C_8H_8))$

$-104.0\ kJ = -(1\ mol)(\Delta H_f^\circ(C_8H_8))$

$$\Delta H_f^\circ(C_8H_8) = \frac{-104.0\ kJ}{-1\ mol} = +104.0\ kJ/mol$$

8.53 $C_3H_6 + 9/2\ O_2 \rightarrow 3\ CO_2 + 3\ H_2O$

$\Delta H^\circ_{rxn} = [3\ \Delta H_f^\circ(CO_2) + 3\ \Delta H_f^\circ(H_2O)] - \Delta H_f^\circ(C_6H_6)$

According to the above reaction, ΔH°_{rxn} will be more negative for cyclopropane because of its more positive ΔH_f°.

8.54 $C_5H_{12}O(l) + 15/2\ O_2(g) \rightarrow 5\ CO_2(g) + 6\ H_2O(l)$

$\Delta H^\circ_{rxn} = [5\ \Delta H_f^\circ(CO_2) + 6\ \Delta H_f^\circ(H_2O)] - \Delta H_f^\circ(C_5H_{12}O)$

$\Delta H^\circ_{rxn} = [(5\ \text{mol})(-393.5\ \text{kJ/mol}) + (6\ \text{mol})(-285.8\ \text{kJ/mol})]$

$- [(1\ \text{mol})(-313.6\ \text{kJ/mol})]$

$\Delta H^\circ_{rxn} = -3369\ \text{kJ}$

8.55 $\Delta H^\circ_{rxn} = \Delta H_f^\circ(\text{MTBE}) - [\Delta H_f^\circ(\text{2-Methylpropene}) + \Delta H_f^\circ(CH_3OH)]$

$-57.8\ \text{kJ} = -313.6\ \text{kJ} - [(1\ \text{mol})(\Delta H_f^\circ(\text{2-Methylpropene})) + (-238.7\ \text{kJ})]$

Solve for ΔH_f°(2-Methylpropene).

$-17.1\ \text{kJ} = (1\ \text{mol})(\Delta H_f^\circ(\text{2-Methylpropene}))$

$\Delta H_f^\circ(\text{2-Methylpropene}) = -17.1\ \text{kJ/mol}$

8.56 $C_{51}H_{88}O_6(l) + 70\ O_2(g) \rightarrow 51\ CO_2(g) + 44\ H_2O(l)$

$\Delta H^\circ_{rxn} = [51\ \Delta H_f^\circ(CO_2) + 44\ \Delta H_f^\circ(H_2O)] - \Delta H_f^\circ(C_{51}H_{88}O_6)$

$\Delta H^\circ_{rxn} = [(51\ \text{mol})(-393.5\ \text{kJ/mol}) + (44\ \text{mol})(-285.8\ \text{KJ/mol})]$

$- [(1\ \text{mol})(-1310\ \text{kJ/mol})]$

$\Delta H^\circ_{rxn} = -3.133 \times 10^4\ \text{kJ/mol}\ C_{51}H_{88}O_6$

$C_{51}H_{88}O_6$, 797.25 amu

$$q = -3.133 \times 10^4\ \frac{\text{kJ}}{\text{mol}} \times \frac{1\ \text{mol}}{797.25\ \text{g}} \times 0.94\ \frac{\text{g}}{\text{mL}} = -37\ \text{kJ/mL}$$

Bond Dissociation Energies

8.57 $H_2C{=}CH_2(g) + H_2(g) \rightarrow CH_3CH_3(g)$

$\Delta H^\circ_{rxn} = \underline{D}$ (bonds broken) – $\underline{D}$ (bonds formed)

$\Delta H^\circ_{rxn} = [\underline{D}(C{=}C) + \underline{D}(H_2)] - [2 \times \underline{D}(C{-}H) + \underline{D}(C{-}C)]$

$\Delta H^\circ_{rxn} = [(1\ mol)(720\ kJ/mol) + (1\ mol)(436\ kJ/mol)]$

$- [(2\ mol)(410\ kJ/mol) + (1\ mol)(350\ kJ/mol)] = -14\ kJ$

8.58 $CH_3CH{=}CH_2 + H_2O \rightarrow CH_3CH(OH)CH_3$

$\Delta H^\circ_{rxn} = \underline{D}$ (bonds broken) – $\underline{D}$ (bonds formed)

$\Delta H^\circ_{rxn} = [\underline{D}(C{=}C) + \underline{D}(O{-}H)] - [\underline{D}(C{-}C) + \underline{D}(C{-}H) + \underline{D}(C{-}O)]$

$\Delta H^\circ_{rxn} = [(1\ mol)(720\ kJ/mol) + (1\ mol)(460\ kJ/mol)]$

$- [(1\ mol)(350\ kJ/mol) + (1\ mol)(410\ kJ/mol) + (1\ mol)(350\ kJ/mol)]$

$\Delta H^\circ_{rxn} = +70\ kJ$

8.59 $C_4H_{10} + 13/2\ O_2 \rightarrow 4\ CO_2 + 5\ H_2O$

$\Delta H^\circ_{rxn} = \underline{D}$ (bonds broken) – $\underline{D}$ (bonds formed)

$\Delta H^\circ_{rxn} = [3 \times \underline{D}(C{-}C) + 10 \times \underline{D}(C{-}H) + 13/2 \times \underline{D}(O_2)]$

$- [8 \times \underline{D}(C{=}O) + 10 \times \underline{D}(O{-}H)]$

$\Delta H^\circ_{rxn} = [(3\ mol)(350\ kJ/mol) + (10\ mol)(410\ kJ/mol) + (13/2\ mol)(498\ kJ/mol)]$

$- [(8\ mol)(804\ kJ/mol) + (10\ mol)(460\ kJ/mol)] = -2645\ kJ$

8.60 $CH_3COOH + CH_3CH_2OH \rightarrow CH_3COOCH_2CH_3 + H_2O$

$\Delta H^\circ_{rxn} = \underline{D}$ (bonds broken) – $\underline{D}$ (bonds formed)

$\Delta H^\circ_{rxn} = [\underline{D}(C{-}O) + \underline{D}(O{-}H)] - [\underline{D}(C{-}O) + \underline{D}(O{-}H)]$

ΔH°_{rxn} = [(1 mol)(350 kJ/mol) + (1 mol)(460 kJ/mol)]

– [(1 mol)(350 kJ/mol) + (1 mol)(460 kJ/mol)]

ΔH°_{rxn} = 0 kJ

Free Energy and Entropy

8.61 Entropy is a measure of molecular disorder.

8.62 $\Delta G = \Delta H - T\Delta S$

ΔH is usually more important because it is usually much larger than $T\Delta S$.

8.63 A reaction can be spontaneous yet endothermic if ΔS is positive (more disorder) and the $T\Delta S$ term is larger than ΔH.

8.64 A reaction can be nonspontaneous yet exothermic if ΔS is negative (more order) and the temperature is high enough so that the $T\Delta S$ term is more negative than ΔH.

8.65 (a) positive (more disorder) (b) negative (more order)

(c) positive (more disorder) (d) positive (more disorder)

8.66 $\Delta S > 0$. The reaction increases the total number of molecules.

8.67 $\Delta S < 0$. The reaction decreases the number of gas molecules.

8.68 $\Delta G = \Delta H - T\Delta S$

(a) $\Delta G = -48\text{ kJ} - (400\text{ K})(135 \times 10^{-3}\text{ kJ/K}) = -102\text{ kJ}$

Because ΔG is negative, the reaction is spontaneous.

Because ΔH is negative, the reaction is exothermic.

(b) $\Delta G = -48\ \text{kJ} - (400\ \text{K})(-135 \times 10^{-3}\ \text{kJ/K}) = +6\ \text{kJ}$

Because ΔG is positive, the reaction is nonspontaneous.

Because ΔH is negative, the reaction is exothermic.

(c) $\Delta G = +48\ \text{kJ} - (400\ \text{K})(135 \times 10^{-3}\ \text{kJ/K}) = -6\ \text{kJ}$

Because ΔG is negative, the reaction is spontaneous.

Because ΔH is positive, the reaction is endothermic.

(d) $\Delta G = +48\ \text{kJ} - (400\ \text{K})(-135 \times 10^{-3}\ \text{kJ/K}) = +102\ \text{kJ}$

Because ΔG is positive, the reaction is nonspontaneous.

Because ΔH is positive, the reaction is endothermic.

8.69 $\Delta G = \Delta H - T\Delta S$

(a) $\Delta G = -128\ \text{kJ} - (500\ \text{K})(35 \times 10^{-3}\ \text{kJ/K}) = -146\ \text{kJ}$

$\Delta G < 0$, spontaneous; $\Delta H < 0$, exothermic

(b) $\Delta G = +67\ \text{kJ} - (250\ \text{K})(-140 \times 10^{-3}\ \text{kJ/K}) = +102\ \text{kJ}$

$\Delta G > 0$, nonspontaneous; $\Delta H > 0$, endothermic

(c) $\Delta G = +75\ \text{kJ} - (800\ \text{K})(95 \times 10^{-3}\ \text{kJ/K}) = -1\ \text{kJ}$

$\Delta G < 0$, spontaneous; $\Delta H > 0$, endothermic

8.70 (a) $\Delta H < 0$ and $\Delta S > 0$; reaction is spontaneous at all temperatures.

(b) $\Delta H < 0$ and $\Delta S < 0$; reaction has a crossover temperature.

(c) $\Delta H > 0$ and $\Delta S > 0$; reaction has a crossover temperature.

(d) $\Delta H > 0$ and $\Delta S < 0$; reaction is nonspontaneous at all temperatures.

8.71 $\Delta G = \Delta H - T\Delta S$

Set $\Delta G = 0$ and solve for T (the crossover temperature).

$$T = \frac{\Delta H}{\Delta S} = \frac{-33 \text{ kJ}}{-0.058 \text{ kJ/K}} = 570 \text{ K}$$

8.72 Because $\Delta H > 0$ and $\Delta S < 0$, the reaction is nonspontaneous at all temperatures. There is no crossover temperature.

8.73 (a) $\Delta H < 0$ and $\Delta S < 0$. The reaction is favored by enthalpy but not by entropy.

$$\Delta G° = \Delta H° - T\Delta S° = -217.5 \text{ kJ/mol} - (298 \text{ K})[-233.9 \times 10^{-3} \text{ kJ/(K} \cdot \text{mol)}]$$

$$\Delta G° = -147.8 \text{ kJ}$$

(b) The reaction has a crossover temperature. Set $\Delta G = 0$ and solve for T (the crossover temperature).

$$\Delta G° = 0 = \Delta H° - T\Delta S°$$

$$T = \frac{\Delta H°}{\Delta S°} = \frac{217.5 \text{ kJ/mol}}{233.9 \times 10^{-3} \text{ kJ/(K}\cdot\text{mol)}} = 929.9 \text{ K}$$

General Problems

8.74 $Mg(s) + 2\ HCl(aq) \rightarrow MgCl_2(aq) + H_2(g)$

$$\text{mol Mg} = 1.50 \text{ g} \times \frac{1 \text{ mol}}{24.3 \text{ g}} = 0.0617 \text{ mol Mg}$$

$$\text{mol HCl} = 0.200 \text{ L} \times 6.00 \frac{\text{mol}}{\text{L}} = 1.20 \text{ mol HCl}$$

There is an excess of HCl. Mg is the limiting reactant.

$$q = \left(4.18 \frac{\text{J}}{\text{g}\cdot°\text{C}}\right)(200 \text{ g})(42.9\ °\text{C} - 25.0\ °\text{C}) + \left(776 \frac{\text{J}}{°\text{C}}\right)(42.9\ °\text{C} - 25.0\ °\text{C})$$

$q = 2.89 \times 10^4\ J$

$q = 2.89 \times 10^4\ J \times \frac{1\ kJ}{1000\ J} = 28.9\ kJ$

Heat evolved per mole of Mg $= \frac{28.9\ kJ}{0.0617\ mol} = 468\ kJ/mol$

Because the reaction evolves heat, the sign for ΔH is negative.

$\Delta H = -468\ kJ$

8.75 (a) $C(s) + CO_2(g) \rightarrow 2\ CO(g)$

$\Delta H^\circ_{rxn} = [2\ \Delta H_f^\circ(CO)] - \Delta H_f^\circ(CO_2)$

$\Delta H^\circ_{rxn} = [(2\ mol)(-110.5\ kJ/mol)] - [(1\ mol)(-393.5\ kJ/mol)] = +172.5\ kJ$

(b) $2\ H_2O_2(l) \rightarrow 2\ H_2O(l) + O_2(g)$

$\Delta H^\circ_{rxn} = [2\ \Delta H_f^\circ(H_2O)] - [2\ \Delta H_f^\circ(H_2O_2)]$

$\Delta H^\circ_{rxn} = [(2\ mol)(-285.8\ kJ/mol)] - [(2\ mol)(-187.8\ kJ/mol)] = -196.0\ kJ$

(c) $Fe_2O_3(s) + 3\ CO(g) \rightarrow 2\ Fe(s) + 3\ CO_2(g)$

$\Delta H^\circ_{rxn} = [3\ \Delta H_f^\circ(CO_2)] - [\Delta H_f^\circ(Fe_2O_3) + 3\ \Delta H_f^\circ(CO)]$

$\Delta H^\circ_{rxn} = [(3\ mol)(-393.5\ kJ/mol)]$

$- [(1\ mol)(-824.2\ kJ/mol) + (3\ mol)(-110.5\ kJ/mol)] = -24.8\ kJ$

8.76

	$2\ NO(g) + O_2(g) \rightarrow 2\ NO_2(g)$	$\Delta H_1^\circ = 2(-57.0\ kJ)$
	$2\ NO_2(g) \rightarrow N_2O_4(g)$	$\Delta H_2^\circ = -57.20\ kJ$
Sum	$2\ NO(g) + O_2(g) \rightarrow N_2O_4(g)$	

$\Delta H^\circ = \Delta H_1^\circ + \Delta H_2^\circ = -171.2\ kJ$

8.77 $\Delta G = \Delta H - T\Delta S$; at equilibrium $\Delta G = 0$. Set $\Delta G = 0$ and solve for T.

$$\Delta G = 0 = \Delta H - T\Delta S$$

$$T = \frac{\Delta H}{\Delta S} = \frac{30.91\ \text{kJ/mol}}{93.2 \times 10^{-3}\ \text{kJ/(K}\cdot\text{mol)}} = 332\ \text{K} = 59°\text{C}$$

8.78 $\Delta G_{fus} = \Delta H_{fus} - T\Delta S_{fus}$; at the melting point $\Delta G = 0$. Set $\Delta G = 0$ and solve for T (the melting point).

$$\Delta G = 0 = \Delta H_{fus} - T\Delta S_{fus}$$

$$T = \frac{\Delta H_{fus}}{\Delta S_{fus}} = \frac{9.95\ \text{kJ}}{0.0357\ \text{kJ/K}} = 279\ \text{K}$$

8.79 $HgS(s) + O_2(g) \rightarrow Hg(l) + SO_2(g)$

(a) $\Delta H°_{rxn} = \Delta H_f°(SO_2) - \Delta H_f°(HgS)$

$\Delta H°_{rxn}$ = [(1 mol)(–296.8 kJ/mol)] – [(1 mol)(–58.2 kJ/mol)] = –238.6 kJ

(b) and (c) Because $\Delta H < 0$ and $\Delta S > 0$, the reaction is spontaneous at <u>all</u> temperatures.

8.80 $\Delta H°_{rxn}$ = $\underline{D}$ (bonds broken) – $\underline{D}$ (bonds formed)

(a) $2\ CH_4(g) \rightarrow C_2H_6(g) + H_2(g)$

$\Delta H°_{rxn}$ = [2 x $\underline{D}$(C–H)] – [$\underline{D}$(C–C) + $\underline{D}$(H–H)]

$\Delta H°_{rxn}$ = [(2 mol)(410 kJ/mol)] – [(1 mol)(350 kJ/mol) + (1 mol)(436 kJ/mol)]

$\Delta H°_{rxn}$ = +34 kJ

(b) $C_2H_6(g) + F_2(g) \rightarrow C_2H_5F(g) + HF(g)$

$\Delta H°_{rxn}$ = [$\underline{D}$(C–H) + $\underline{D}$(F–F)] – [$\underline{D}$(C–F) + $\underline{D}$(H–F)]

ΔH°_{rxn} = [(1 mol)(410 kJ/mol) + (1 mol)(158 kJ/mol)]

– [(1 mol)(450 kJ/mol) + (1 mol)(570 kJ/mol)]

$\Delta H^\circ_{rxn} = -452$ kJ

(c) $N_2(g) + 3\ H_2(g) \rightarrow 2\ NH_3(g)$

The bond dissociation energy for N_2 is 945 kJ/mol.

$\Delta H^\circ_{rxn} = [\underline{D}(N{\equiv}N) + 3 \times \underline{D}(H–H)] - [6 \times \underline{D}(N–H)]$

ΔH°_{rxn} = [(1 mol)(945 kJ/mol) + (3 mol)(436 kJ/mol)] – [(6 mol)(390 kJ/mol)]

$\Delta H^\circ_{rxn} = -87$ kJ

8.81 (a) $\Delta H^\circ_{rxn} = \Delta H_f^\circ(CH_3OH) - \Delta H_f^\circ(CO)$

ΔH°_{rxn} = [(1 mol)(–238.7 kJ/mol)] – [(1 mol)(–110.5 kJ/mol)] = –128.2 kJ

(b) $\Delta G^\circ = \Delta H^\circ - T\Delta S^\circ = -128.2\text{ kJ} - (298\text{ K})(-332 \times 10^{-3}\text{ kJ/K}) = -29.3\text{ kJ}$

(c) Step 1 is spontaneous since $\Delta G^\circ < 0$.

(d) ΔH°, because it is larger than $T\Delta S$.

(e) Set $\Delta G = 0$ and solve for T.

$\Delta G = 0 = \Delta H - T\Delta S$

$$T = \frac{\Delta H}{\Delta S} = \frac{128.2\text{ kJ}}{332 \times 10^{-3}\text{ kJ/K}} = 386\text{ K}$$

The reaction is spontaneous below 386 K.

(f) $\Delta H^\circ_{rxn} = \Delta H_f^\circ(CH_4) - \Delta H_f^\circ(CH_3OH)$

ΔH°_{rxn} = [(1 mol)(–74.8 kJ/mol)] – [(1 mol)(–238.7 kJ/mol)] = +163.9 kJ

(g) $\Delta G^\circ = \Delta H^\circ - T\Delta S^\circ = +163.9\text{ kJ} - (298\text{ K})(162 \times 10^{-3}\text{ kJ/K}) = +115.6\text{ kJ}$

(h) Step 2 is nonspontaneous since $\Delta G^\circ > 0$.

(i) $\Delta H°$, because it is larger than $T\Delta S$.

(j) Set $\Delta G = 0$ and solve for T.

$\Delta G = 0 = \Delta H - T\Delta S$

$$T = \frac{\Delta H}{\Delta S} = \frac{163.9\ \text{kJ}}{162 \times 10^{-3}\ \text{kJ/K}} = 1012\ \text{K}$$

The reaction is spontaneous above 1012 K.

(k) $\Delta G°_{overall} = \Delta G_1° + \Delta G_2° = -29.3\ \text{kJ} + 115.6\ \text{kJ} = +86.3\ \text{kJ}$

$\Delta H°_{overall} = \Delta H_1° + \Delta H_2° = -128.2\ \text{kJ} + 163.9\ \text{kJ} = +35.7\ \text{kJ}$

$\Delta S°_{overall} = \Delta S_1° + \Delta S_2° = -332\ \text{J/K} + 162\ \text{J/K} = -170\ \text{J/K}$

(l) The overall reaction is nonspontaneous since $\Delta G°_{overall} > 0$.

(m) The two reactions should be run separately. Run step 1 below 386 K and run step 2 above 1012 K.

8.82 (a) $2\ C_8H_{18}(l) + 25\ O_2(g) \rightarrow 16\ CO_2(g) + 18\ H_2O(g)$

(b) $C_8H_{18}(l) + 25/2\ O_2(g) \rightarrow 8\ CO_2(g) + 9\ H_2O(g)$

$\Delta H°_{rxn} = \Delta H_C° = -5456.6\ \text{kJ}$

$\Delta H°_{rxn} = [8\ \Delta H_f°(CO_2) + 9\ \Delta H_f°(H_2O)] - \Delta H_f°(C_8H_{18})$

$-5456.6\ \text{kJ} = [(8\ \text{mol})(-393.5\ \text{kJ/mol}) + (9\ \text{mol})(-241.8\ \text{kJ/mol})]$

$- [(1\ \text{mol})(\Delta H_f°(C_8H_{18}))]$

Solve for $\Delta H_f°(C_8H_{18})$.

$-5456.6\ \text{kJ} = -5324\ \text{kJ} - [(1\ \text{mol})(\Delta H_f°(C_8H_{18}))]$

$-132.4\ \text{kJ} = -(1\ \text{mol})(\Delta H_f°(C_8H_{18}))$

$\Delta H_f°(C_8H_{18}) = +132.4\ \text{kJ/mol}$

8.83 Assume 1.00 kg of H_2O.

E_p = (1.00 kg)(9.81 m/s^2)(739 m) = 7250 kg·m^2/s^2 = 7250 J

q = specific heat x m x ΔT

$$\Delta T = \frac{q}{m \times \text{specific heat}} = \frac{7250\ \text{J}}{(1000\ \text{g})(4.18\ \text{J/(g}\cdot{}^\circ\text{C)}} = 1.73^\circ\text{C (temperature rise)}$$

9.1 1 atm = 14.7 psi

$$1.0 \text{ mm Hg} \times \frac{1 \text{ atm}}{760 \text{ mm Hg}} \times \frac{14.7 \text{ psi}}{1 \text{ atm}} = 1.93 \times 10^{-2} \text{ psi}$$

9.2 1.00 atmosphere pressure can support a column of Hg 0.760 m high. Since the density of H_20 is 1.00 g/mL and that of Hg is 13.6 g/mL, 1.00 atmosphere pressure can support a column of H_20 13.6 times higher than that of Hg. The column of H_2O supported by 1.00 atmosphere will be (0.760 m)(13.6) = 10.3 m.

9.3 The pressure in the flask is less than 0.975 atm. The 24.7 cm of Hg is the difference between the two pressures.

$$\text{Pressure difference} = 24.7 \text{ cm Hg} \times \frac{1.00 \text{ atm}}{76.0 \text{ cm Hg}} = 0.325 \text{ atm}$$

Pressure in flask = 0.975 atm – 0.325 atm = 0.650 atm

9.4 $$n = \frac{PV}{RT} = \frac{(1.00 \text{ atm})(100{,}000 \text{ L})}{\left(0.082\ 06 \frac{\text{L}\cdot\text{atm}}{\text{mol}\cdot\text{K}}\right)(273 \text{ K})} = 4.46 \times 10^3 \text{ g } CH_4$$

CH_4, 16.04 amu

$$\text{mass } CH_4 = (4.46 \times 10^3 \text{ mol})\left(\frac{16.04 \text{ g}}{1 \text{ mol}}\right) = 7.15 \times 10^4 \text{ g } CH_4$$

9.5 C_3H_8, 44.10 amu; $P = \frac{nRT}{V}$

V = 350 mL = 0.350 L; T = 20°C = 293 K

$$n = 3.2 \text{ g} \times \frac{1 \text{ mol } C_3H_8}{44.10 \text{ g } C_3H_8} = 0.073 \text{ mol } C_3H_8$$

$$P = \frac{(0.073 \text{ mol})\left(0.082\ 06 \frac{\text{L}\cdot\text{atm}}{\text{mol}\cdot\text{K}}\right)(293 \text{ K})}{0.350 \text{ L}} = 5.0 \text{ atm}$$

9.6 $P = 1.51 \times 10^4 \text{ kPa} \times \frac{1 \text{ atm}}{101.325 \text{ kPa}} = 149 \text{ atm}$

$T = 25.0°C = 298 \text{ K}$

$$n = \frac{PV}{RT} = \frac{(149 \text{ atm})(43.8 \text{ L})}{\left(0.082\ 06 \frac{\text{L}\cdot\text{atm}}{\text{mol}\cdot\text{K}}\right)(298 \text{ K})} = 267 \text{ mol He}$$

9.7 The volume and number of moles of gas remain constant.

$$\frac{nR}{V} = \frac{P_i}{T_i} = \frac{P_f}{T_f}; \quad T_f = \frac{P_f T_i}{P_i} = \frac{(2.37 \text{ atm})(273 \text{ K})}{2.15 \text{ atm}} = 301 \text{ K} = 28°C$$

9.8 $CaCO_3(s) + 2\ HCl(aq) \rightarrow CaCl_2(aq) + CO_2(g) + H_2O(l)$

$CaCO_3$, 100.1 amu; CO_2, 44.01 amu

$$\text{mole } CO_2 = 33.7 \text{ g } CaCO_3 \times \frac{1 \text{ mol } CaCO_3}{100.1 \text{ g } CaCO_3} \times \frac{1 \text{ mol } CO_2}{1 \text{ mol } CaCO_3} = 0.337 \text{ mol } CO_2$$

$$\text{mass } CO_2 = 0.337 \text{ mol } CO_2 \times \frac{44.01 \text{ g } CO_2}{1 \text{ mol } CO_2} = 14.8 \text{ g } CO_2$$

$$V = \frac{nRT}{P} = \frac{(0.337 \text{ mol})\left(0.082\ 06 \frac{\text{L}\cdot\text{atm}}{\text{mol}\cdot\text{K}}\right)(273 \text{ K})}{1.00 \text{ atm}} = 7.55 \text{ L}$$

9.9 $C_3H_8(g) + 5\ O_2(g) \rightarrow 3\ CO_2(g) + 4\ H_2O(l)$

$$n_{propane} = \frac{PV}{RT} = \frac{(4.5 \text{ atm})(15.0 \text{ L})}{\left(0.082\ 06 \frac{\text{L}\cdot\text{atm}}{\text{mol}\cdot\text{K}}\right)(298 \text{ K})} = 2.76 \text{ mol } C_3H_8$$

$$2.76 \text{ mol } C_3H_8 \times \frac{3 \text{ mol } CO_2}{1 \text{ mol } C_3H_8} = 8.28 \text{ mol } CO_2$$

$$V = \frac{nRT}{P} = \frac{(8.28\ \text{mol})\left(0.082\ 06\ \frac{\text{L}\cdot\text{atm}}{\text{mol}\cdot\text{K}}\right)(273\ \text{K})}{1\ \text{atm}} = 186\ \text{L} = 190\ \text{L}$$

9.10 $n = \frac{PV}{RT} = \frac{(1.00\ \text{atm})(1.00\ \text{L})}{\left(0.082\ 06\ \frac{\text{L}\cdot\text{atm}}{\text{mol}\cdot\text{K}}\right)(273\ \text{K})} = 0.0446\ \text{mol}$

$$\text{molar mass} = \frac{1.52\ \text{g}}{0.0446\ \text{moles}} = 34.1\ \text{g/mol}$$

$Na_2S(aq) + 2\ HCl(aq) \rightarrow H_2S(g) + 2\ NaCl(aq)$

The foul-smelling gas is H_2S, hydrogen sulfide.

9.11 $12.45\ \text{g}\ H_2 \times \frac{1\ \text{mol}\ H_2}{2.016\ \text{g}\ H_2} = 6.176\ \text{mol}\ H_2$

$$60.67\ \text{g}\ N_2 \times \frac{1\ \text{mol}\ N_2}{28.01\ \text{g}\ N_2} = 2.166\ \text{mol}\ N_2$$

$$2.38\ \text{g}\ NH_3 \times \frac{1\ \text{mol}\ NH_3}{17.03\ \text{g}\ NH_3} = 0.140\ \text{mol}\ NH_3$$

$$n_{\text{total}} = n_{H_2} + n_{N_2} + n_{NH_3} = 6.176\ \text{mol} + 2.166\ \text{mol} + 0.140\ \text{mol} = 8.482\ \text{mol}$$

$$X_{H_2} = \frac{6.176\ \text{mol}}{8.482\ \text{mol}} = 0.7281$$

$$X_{N_2} = \frac{2.166\ \text{mol}}{8.482\ \text{mol}} = 0.2554$$

$$X_{NH_3} = \frac{0.140\ \text{mol}}{8.482\ \text{mol}} = 0.0165$$

9.12 n_{total} = 8.482 mol (from Problem 9.11).

T = 90°C = 363 K

$$P_{total} = \frac{n_{total}RT}{V} = \frac{(8.482\text{ mol})\left(0.082\ 06\ \frac{\text{L}\cdot\text{atm}}{\text{mol}\cdot\text{K}}\right)(363\text{ K})}{10.00\text{ L}} = 25.27\text{ atm}$$

$$P_{H_2} = X_{H_2}\cdot P_{total} = (0.7281)(25.27\text{ atm}) = 18.4\text{ atm}$$

$$P_{N_2} = X_{N_2}\cdot P_{total} = (0.2554)(25.27\text{ atm}) = 6.45\text{ atm}$$

$$P_{NH_3} = X_{NH_3}\cdot P_{total} = (0.0165)(25.27\text{ atm}) = 0.417\text{ atm}$$

9.13 $P_{H_2O} = X_{H_2O}\cdot P_{Total} = (0.0287)(0.977\text{ atm}) = 0.0280\text{ atm}$

9.14 $\mu = \sqrt{\frac{3RT}{M}}$

M = molar mass

R = 8.314 J/(K · mol)

1 J = 1 kg · m^2/s^2

at 37°C = 310 K

$$\mu = \sqrt{\frac{3 \times 8.314\text{ kg m}^2/(\text{s}^2\text{ K mol}) \times 310\text{ K}}{28.01 \times 10^{-3}\text{ kg/mol}}} = 525\text{ m/s}$$

at –25°C = 248 K

$$\mu = \sqrt{\frac{3 \times 8.314\text{ kg m}^2/(\text{s}^2\text{ K mol}) \times 248\text{ K}}{28.01 \times 10^{-3}\text{ kg/mol}}} = 470\text{ m/s}$$

9.15 (a) $\frac{\text{rate }O_2}{\text{rate Kr}} = \frac{\sqrt{M_{Kr}}}{\sqrt{M_{O_2}}} = \frac{\sqrt{83.8}}{\sqrt{32.0}}$; $\frac{\text{rate }O_2}{\text{rate Kr}} = 1.62$

O_2 diffuses 1.62 times faster than Kr.

(b) $\dfrac{\text{rate } C_2H_2}{\text{rate } N_2} = \dfrac{\sqrt{M_{N_2}}}{\sqrt{M_{C_2H_2}}} = \dfrac{\sqrt{28.0}}{\sqrt{26.0}}$; $\dfrac{\text{rate } C_2H_2}{\text{rate } N_2} = 1.04$

C_2H_2 diffuses 1.04 times faster than N_2.

9.16 $\dfrac{\text{rate } ^{20}\text{Ne}}{\text{rate } ^{22}\text{Ne}} = \dfrac{\sqrt{M\ ^{22}\text{Ne}}}{\sqrt{M\ ^{20}\text{Ne}}} = \dfrac{\sqrt{22}}{\sqrt{20}} = 1.05$

$$\frac{\text{rate } ^{21}\text{Ne}}{\text{rate } ^{22}\text{Ne}} = \frac{\sqrt{M\ ^{22}\text{Ne}}}{\sqrt{M\ ^{21}\text{Ne}}} = \frac{\sqrt{22}}{\sqrt{21}} = 1.02$$

Thus, the relative rates of diffusion are ^{20}Ne(1.05) > ^{21}Ne(1.02) > ^{22}Ne(1.00).

9.17 The amount of ozone is assumed to be constant.

Therefore $nR = \dfrac{P_iV_i}{T_i} = \dfrac{P_fV_f}{T_f}$

Because $V \propto h$, then $\dfrac{P_ih_i}{T_i} = \dfrac{P_fh_f}{T_f}$ where h is the thickness of the O_3 layer.

$$h_f = \frac{P_i}{P_f} \times \frac{T_f}{T_i} \times h_i = \left(\frac{1.6 \times 10^{-9}\text{ atm}}{1\text{ atm}}\right)\left(\frac{273\text{ K}}{230\text{ K}}\right)(20 \times 10^3\text{ m}) = 3.8 \times 10^{-5}\text{ m}$$

(Actually, $V = 4\pi r^2h$, where r = the radius of the earth. When you go out ~30 km to get to the ozone layer, the change in r^2 is less than 1%. Therefore you can neglect the change in r^2 and assume that V is proportional to h.)

Understanding Key Concepts

1. (a) 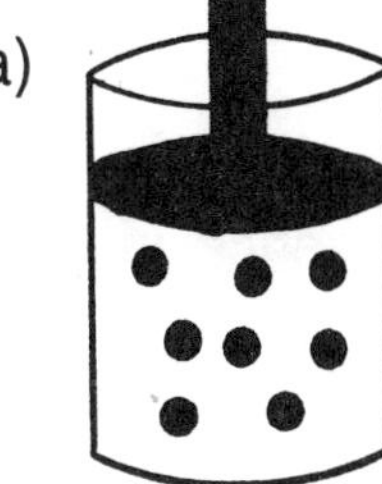

The volume of a gas is proportional to the kelvin temperature at constant pressure. As the temperature increases, the volume will increase as well.

(b) 

The volume of a gas is inversely proportional to pressure at constant temperature. As the pressure increases, the volume will decrease.

(c) 

PV = nRT

The amount of gas (n) is constant.

Therefore $nR = \frac{P_iV_i}{T_i} = \frac{P_fV_f}{T_f}$.

Assume V_i = 1 L and solve for V_f.

$$\frac{P_iV_iT_f}{T_iP_f} = \frac{(3\text{ atm})(1\text{ L})(200\text{ K})}{(300\text{ K})(2\text{ atm})} = V_f = 1\text{ L}$$

There is no change in volume.

2. If the sample remains a gas at 150 K, then drawing (c) represents the gas at this temperature. The gas molecules still fill the container.

3. The two gases should mix randomly and homogeneously. This is represented by drawing (c).

4. The two gases, on average, will be equally distributed among the three flasks.

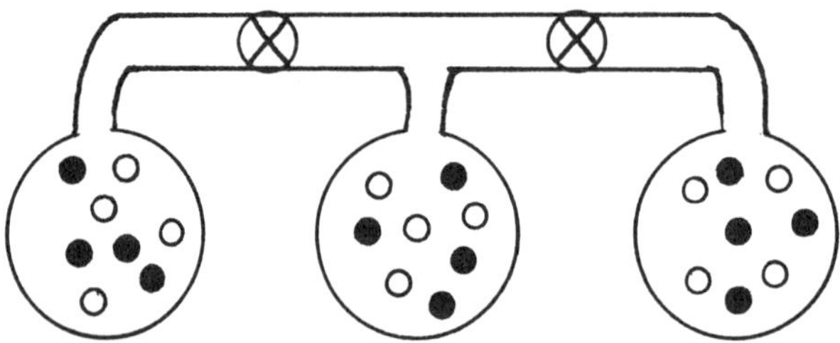

5. The gas pressure in the bulb in mm Hg is equal to the difference in the height of the Hg in the two arms of the manometer.

Additional Problems

Gases and Gas Pressure

9.18 Collisions of gas molecules with the walls of a container exert a force per unit area that is measured as gas pressure.

9.19 Temperature is a measure of the kinetic energy content of a chemical system.

9.20 Gases are much more compressible than solids or liquids because there is a large amount of empty space between individual gas molecules.

9.21 Atmospheric pressure is less at the top of a mountain than at sea level because at higher altitudes there aren't as many gas molecules exerting pressure as there are at sea level.

9.22 $P = 480 \text{ mm Hg} \times \dfrac{1.00 \text{ atm}}{760 \text{ mm Hg}} = 0.632 \text{ atm}$

$P = 480 \text{ mm Hg} \times \dfrac{101{,}325 \text{ Pa}}{760 \text{ mm Hg}} = 6.40 \times 10^4 \text{ Pa}$

9.23 (a) $4.81 \text{ atm} \times \dfrac{101{,}325 \text{ Pa}}{1.00 \text{ atm}} = 4.87 \times 10^5 \text{ Pa}$

(b) $1023 \text{ mm Hg} \times \dfrac{1.00 \text{ atm}}{760 \text{ mm Hg}} = 1.35 \text{ atm}$

(c) $0.0023 \text{ atm} \times \dfrac{101{,}325 \text{ Pa}}{1.00 \text{ atm}} = 230 \text{ Pa}$

9.24 $P = 352 \text{ torr} \times \dfrac{101{,}325 \text{ Pa}}{760 \text{ torr}} \times \dfrac{1 \text{ kPa}}{1000 \text{ Pa}} = 46.9 \text{ kPa}$

$$P = 0.255 \text{ atm} \times \frac{760 \text{ mm Hg}}{1.00 \text{ atm}} = 194 \text{ mm Hg}$$

$$P = 0.0382 \text{ mm Hg} \times \frac{101{,}325 \text{ Pa}}{760 \text{ mm Hg}} = 5.09 \text{ Pa}$$

9.25 $P_{flask} > 754.3$ mm Hg; $P_{flask} = 754.3 \text{ mm Hg} + 176 \text{ mm Hg} = 930 \text{ mm Hg}$

9.26 $P_{flask} > 1.021$ atm (see Figure 9.4).

$$P_{difference} = 28.3 \text{ cm Hg} \times \frac{1.00 \text{ atm}}{76.0 \text{ cm Hg}} = 0.372 \text{ atm}$$

$$P_{flask} = 1.021 \text{ atm} - P_{difference} = 1.021 \text{ atm} - 0.372 \text{ atm} = 0.649 \text{ atm}$$

9.27 $P_{flask} > 752.3$ mm Hg

If the pressure in the flask can support a column of ethyl alcohol (d = 0.7893 g/mL) 55.1 cm high, then it can only support a column of Hg that is much shorter because of the higher density of Hg.

$$55.1 \text{ cm} \times \frac{0.7893 \text{ g/mL}}{13.546 \text{ g/mL}} = 3.21 \text{ cm Hg} = 32.1 \text{ mm Hg}$$

$$P_{flask} = 752.3 \text{ mm Hg} + 32.1 \text{ mm Hg} = 784.4 \text{ mm Hg}$$

$$P_{flask} = 784.4 \text{ mm Hg} \times \frac{101{,}325 \text{ Pa}}{760 \text{ mm Hg}} = 1.046 \times 10^5 \text{ Pa}$$

9.28

	% Volume
N_2	78.08
O_2	20.95
Ar	0.93
CO_2	0.034

The % volume for a particular gas is proportional to the number of molecules of that gas in a mixture of gases.

Average molecular weight of air

$= (0.7808)(\text{mol. wt. } N_2) + (0.2095)(\text{mol. wt. } O_2)$

$+ (0.0093)(\text{at. wt. Ar}) + (0.000\ 34)(\text{mol. wt. } CO_2)$

$= (0.7808)(28.01 \text{ amu}) + (0.2095)(32.00 \text{ amu})$

$+ (0.0093)(39.95 \text{ amu}) + (0.000\ 34)(44.01 \text{ amu}) = 28.96 \text{ amu}$

The Gas Laws

9.29 Standard temperature = 0°C = 273 K

Standard pressure = 1.0 atm

9.30 (a) $\frac{nR}{V} = \frac{P_i}{T_i} = \frac{P_f}{T_f}$

$$\frac{P_i T_f}{T_i} = P_f$$

Let P_i = 1 atm, T_i = 100 K, T_f = 300 K

$$\frac{P_i T_f}{T_i} = \frac{(1 \text{ atm})(300 \text{ K})}{(100 \text{ K})} = P_f = 3 \text{ atm}$$

The pressure would triple.

(b) $\frac{RT}{V} = \frac{P_i}{n_i} = \frac{P_f}{n_f}$

$$\frac{P_i n_f}{n_i} = P_f$$

Let P_i = 1 atm, n_i = 3 mol, n_f = 1 mol

$$\frac{P_i n_f}{n_i} = \frac{(1\ \text{atm})(1\ \text{mol})}{(3\ \text{mol})} = P_f = \frac{1}{3}\ \text{atm}$$

The pressure would be $\frac{1}{3}$ the initial pressure.

(c) $nRT = P_iV_i = P_fV_f$

$$\frac{P_iV_i}{V_f} = P_f$$

Let $P_i = 1$ atm, $V_i = 1$ L, $V_f = 1 - 0.45$ L $= 0.55$ L

$$\frac{P_iV_i}{V_f} = \frac{(1\ \text{atm})(1\ \text{L})}{(0.55\ \text{L})} = P_f = 1.8\ \text{atm}$$

The pressure would increase by 1.8 times.

(d) $nR = \frac{P_iV_i}{T_i} = \frac{P_fV_f}{T_f}$

$$\frac{P_iV_iT_f}{T_iV_f} = P_f$$

Let $P_i = 1$ atm, $V_i = 1$ L, $T_i = 200$ K, $V_f = 3$ L, $T_i = 100$ K

$$\frac{P_iV_iT_f}{T_iV_f} = \frac{(1\ \text{atm})(1\ \text{L})(100\ \text{K})}{(200\ \text{K})(3\ \text{L})} = P_f = 0.17\ \text{atm}$$

The pressure would be 0.17 times the initial pressure.

9.31 They all contain the same number of gas molecules.

9.32 For air, T = 50°C = 323 K.

$$n = \frac{PV}{RT} = \frac{\left(750 \text{ mm Hg} \times \dfrac{1 \text{ atm}}{760 \text{ mm Hg}}\right)(2.50 \text{ L})}{\left(0.082\ 06 \dfrac{\text{L} \cdot \text{atm}}{\text{mol} \cdot \text{K}}\right)(323 \text{ K})} = 0.0931 \text{ mol air}$$

For CO_2, T = –10°C = 263 K

$$n = \frac{PV}{RT} = \frac{\left(765 \text{ mm Hg} \times \dfrac{1 \text{ atm}}{760 \text{ mm Hg}}\right)(2.16 \text{ L})}{\left(0.082\ 06 \dfrac{\text{L} \cdot \text{atm}}{\text{mol} \cdot \text{K}}\right)(263 \text{ K})} = 0.101 \text{ mol } CO_2$$

Since the number of moles of CO_2 is larger than the number of moles of air, the CO_2 sample contains more molecules.

9.33 n and T are constant; therefore nRT = $P_iV_i = P_fV_f$

$$V_f = \frac{P_iV_i}{P_f} = \frac{(150 \text{ atm})(49.0 \text{ L})}{(1.02 \text{ atm})} = 7210 \text{ L}$$

n and P are constant; therefore $\dfrac{nR}{P} = \dfrac{V_i}{T_i} = \dfrac{V_f}{T_f}$

$$V_f = \frac{V_iT_f}{T_i} = \frac{(49.0 \text{ L})(308 \text{ K})}{(293 \text{ K})} = 51.5 \text{ L}$$

9.34 $nR = \dfrac{P_iV_i}{T_i} = \dfrac{P_fV_f}{T_f}$

T_i = 20°C = 293 K

$$V_f = \frac{P_iV_iT_f}{T_iP_f} = \frac{(140 \text{ atm})(8.0 \text{ L})(273 \text{ K})}{(293 \text{ K})(1.00 \text{ atm})} = 1.0 \times 10^3 \text{ L}$$

9.35 $15.0\text{ g } CO_2 \times \dfrac{1\text{ mol } CO_2}{44.0\text{ g } CO_2} = 0.341\text{ mol } CO_2$

$$P = \frac{nRT}{V} = \frac{(0.341\text{ mol})\left(0.082\ 06\ \frac{L\cdot atm}{mol\cdot K}\right)(300\text{ K})}{(0.30\text{ L})} = 27.98\text{ atm}$$

$$27.98\text{ atm} \times \frac{760\text{ mm Hg}}{1\text{ atm}} = 2.1 \times 10^4\text{ mm Hg}$$

9.36 $T = \dfrac{PV}{nR} = \dfrac{(6.0\text{ atm})(0.40\text{ L})}{\left(20.0\text{ g} \times \frac{1\text{ mol}}{28.01\text{ g}}\right)\left(0.082\ 06\ \frac{L\cdot atm}{mol\cdot K}\right)} = 41\text{ K}$

9.37

$$V = \frac{nRT}{P} = \frac{(1\text{ mol})\left(0.082\ 06\ \frac{L\cdot atm}{mol\cdot K}\right)(1\text{ K})}{(1 \times 10^{-14}\text{ mm Hg})\left(\frac{1\text{ atm}}{760\text{ mm Hg}}\right)} = 6 \times 10^{15}\text{ L}$$

$$d = \frac{6.02 \times 10^{23}\text{ molecules}}{6.24 \times 10^{15}\text{ L}} = 9.6 \times 10^7\text{ molecules/L} = 1 \times 10^8\text{ molecules/L}$$

9.38 $n = \dfrac{PV}{RT} = \dfrac{\left(17{,}180\text{ kPa} \times \frac{1000\text{ Pa}}{1\text{ k Pa}} \times \frac{1\text{ atm}}{101{,}325\text{ Pa}}\right)(43.8\text{ L})}{\left(0.082\ 06\ \frac{L\cdot atm}{mol\cdot K}\right)(293K)} = 308.9\text{ mol}$

$$\text{mass Ar} = 308.9\text{ mol} \times \frac{39.948\text{ g}}{1\text{ mol}} = 12340\text{ g} = 1.23 \times 10^4\text{ g}$$

9.39 $V = \dfrac{nRT}{P} = \dfrac{(308.9\text{ mol})\left(0.082\ 06\ \frac{L\cdot atm}{mol\cdot K}\right)(273\text{ K})}{1\text{ atm}} = 6920\text{ L}$

9.40 CH_4, 16.04 amu; 5.54 kg = 5.54×10^3 g

T = 20°C = 293 K

$$P = \frac{nRT}{V} = \frac{\left(5.54 \times 10^3\ g \times \frac{1\ mol}{16.04\ g}\right)\left(0.082\ 06\ \frac{L \cdot atm}{mol \cdot K}\right)(293\ K)}{(43.8\ L)} = 189.6\ atm$$

$$P = 189.6\ atm \times \frac{101{,}325\ Pa}{1\ atm} \times \frac{1\ k\ Pa}{1000\ Pa} = 1.92 \times 10^4\ kPa$$

9.41 $$P = 13{,}800\ kPa \times \frac{1000\ Pa}{1\ kPa} \times \frac{1\ atm}{101{,}325\ Pa} = 136.2\ atm$$

n and T are constant; therefore $nRT = P_iV_i = P_fV_f$

$$V_f = \frac{P_iV_i}{P_f} = \frac{(136.2\ atm)(2.30\ L)}{(1.25\ atm)} = 250.6\ L$$

$$250.6\ L \times \frac{1\ balloon}{1.5\ L} = 167\ balloons$$

Gas Stoichiometry

9.42 For steam, T = 123.0°C = 396 K

$$n = \frac{PV}{RT} = \frac{(0.93\ atm)(15.0\ L)}{\left(0.082\ 06\ \frac{L \cdot atm}{mol \cdot K}\right)(396\ K)} = 0.43\ mol\ steam$$

For ice, H_2O, 18.02 amu

$$n = 10.5\ g \times \frac{1\ mol}{18.02\ g} = 0.583\ mol\ ice$$

Since the number of moles of ice is larger than the number of moles of steam, the ice contains more H_2O molecules.

9.43 The containers are identical. Both containers contain the same number of gas molecules. Weigh the containers. Since the molecular weight for O_2 is greater than the molecular weight for H_2, the heavier container contains O_2.

9.44 room volume = 4.0 m x 5.0 m x 2.5 m = 50 m^3

$$\text{volume} = 50\ m^3 \times \frac{1\ L}{10^{-3}\ m^3} = 5.0 \times 10^4\ L$$

$$n_{total} = \frac{PV}{RT} = \frac{(1.0\ atm)(5.0 \times 10^4\ L)}{\left(0.082\ 06\ \frac{L \cdot atm}{mol \cdot K}\right)(273\ K)} = 2.23 \times 10^3\ mol$$

$$n_{O_2} = (0.2095)n_{total} = (0.2095)(2.23 \times 10^3\ mol) = 467\ mol\ O_2$$

$$\text{mass } O_2 = 467\ mol \times \frac{32.0\ g}{1\ mol} = 1.5 \times 10^4\ g\ O_2$$

9.45 $0.25\ g\ O_2 \times \dfrac{1\ mol\ O_2}{32.0\ g\ O_2} = 7.8 \times 10^{-3}\ mol\ O_2$

$$V = \frac{nRT}{P} = \frac{(7.8 \times 10^{-3}\ mol)\left(0.082\ 06\ \frac{L \cdot atm}{mol \cdot K}\right)(310\ K)}{1.0\ atm}$$

$$V = 0.198\ L = 0.200\ L = 200\ mL\ O_2$$

9.46 (a) CH_4, 16.04 amu

$$d = \frac{16.04\ g}{22.4\ L} = 0.716\ g/L$$

(b) CO_2, 44.01 amu

$$d = \frac{44.01\ g}{22.4\ L} = 1.96\ g/L$$

(c) O_2, 32.00 amu

$$d = \frac{32.00\text{ g}}{22.4\text{ L}} = 1.43\text{ g/L}$$

(d) UF_6, 352.0 amu

$$d = \frac{352.0\text{ g}}{22.4\text{ L}} = 15.7\text{ g/L}$$

9.47 $$n = \frac{PV}{RT} = \frac{\left(356\text{ mm Hg} \times \dfrac{1.00\text{ atm}}{760\text{ mm Hg}}\right)(1.500\text{ L})}{\left(0.082\ 06\ \dfrac{\text{L}\cdot\text{atm}}{\text{mol}\cdot\text{K}}\right)(295.5\text{ K})} = 0.0290\text{ mol}$$

$$\text{molar mass} = \frac{0.9847\text{ g}}{0.0290\text{ mol}} = 34.0\text{ g/mol};\quad \text{molecular weight} = 34.0\text{ amu}$$

9.48 (a) assume 1.000 L gas sample

$$n = \frac{PV}{RT} = \frac{(1.00\text{ atm})(1.000\text{ L})}{\left(0.082\ 06\ \dfrac{\text{L}\cdot\text{atm}}{\text{mol}\cdot\text{K}}\right)(273\text{ K})} = 0.0446\text{ mol}$$

$$\text{molar mass} = \frac{1.342\text{ g}}{0.0446\text{ mol}} = 30.1\text{ g/mol};\quad \text{molecular weight} = 30.1\text{ amu}$$

(b) assume 1.000 L gas sample

$$n = \frac{PV}{RT} = \frac{\left(752\text{ mm Hg} \times \dfrac{1\text{ atm}}{760\text{ mm Hg}}\right)(1.000\text{ L})}{\left(0.082\ 06\ \dfrac{\text{L}\cdot\text{atm}}{\text{mol}\cdot\text{K}}\right)(298\text{ K})} = 0.0405\text{ mol}$$

$$\text{molar mass} = \frac{1.053\text{ g}}{0.0405\text{ mol}} = 26.0\text{ g/mol};\quad \text{molecular weight} = 26.0\text{ amu}$$

9.49 Assume a 1.000 L gas sample.

$$n = \frac{PV}{RT} = \frac{(1.000\ \text{atm})(1.000\ \text{L})}{\left(0.082\ 06\ \frac{\text{L}\cdot\text{atm}}{\text{mol}\cdot\text{K}}\right)(273.15\ \text{K})} = 0.044\ 61\ \text{mol}$$

$$\text{average molar mass} = \frac{1.413\ \text{g}}{0.044\ 61\ \text{mol}} = 31.67\ \text{g/mol}$$

$$31.67 = x \cdot M_{Ar} + (1 - x) \cdot M_{N_2}$$

$$31.67 = (x)(39.948) + (1 - x)(28.013)$$

Solve for x: $x = 0.3064$, $1 - x = 0.6936$

The mixture contains 30.64% Ar and 69.36% N_2

Assume 100 moles of gas.

$$X_{Ar} = \frac{30.64\ \text{mol}}{100\ \text{mol}} = 0.3064 \qquad X_{N_2} = \frac{69.36\ \text{mol}}{100\ \text{mol}} = 0.6936$$

9.50 Average molar mass = (0.270)(molar mass F_2) + (0.730)(molar mass He)

= (0.270)(38.00 g/mol) + (0.730)(4.003 g/mol) = 13.18 g/mol

Assume 1.00 mole of the gas mixture. T = 27.5°C = 300.6 K

$$V = \frac{nRT}{P} = \frac{(1.00\ \text{mol})\left(0.082\ 06\ \frac{\text{L}\cdot\text{atm}}{\text{mol}\cdot\text{K}}\right)(300.6\ \text{K})}{\left(714\ \text{mm Hg} \times \frac{1.00\ \text{atm}}{760\ \text{mm Hg}}\right)} = 26.3\ \text{L}$$

$$d = \frac{13.18\ \text{g}}{26.3\ \text{L}} = 0.501\ \text{g/L}$$

9.51 HgO, 216.59 amu

$$10.57 \text{ g HgO} \times \frac{1 \text{ mol HgO}}{216.59 \text{ g HgO}} \times \frac{1 \text{ mol } O_2}{2 \text{ mol HgO}} = 0.024\ 40 \text{ mol } O_2$$

$$V = \frac{nRT}{P} = \frac{(0.024\ 40 \text{ mol})\left(0.082\ 06\ \frac{\text{L}\cdot\text{atm}}{\text{mol}\cdot\text{K}}\right)(273.15 \text{ K})}{1.000 \text{ atm}} = 0.5469 \text{ L}$$

9.52 HgO, 216.59 amu

$$\text{mass HgO} = 0.0155 \text{ mol } O_2 \times \frac{2 \text{ mol HgO}}{1 \text{ mol } O_2} \times \frac{216.59 \text{ g HgO}}{1 \text{ mol HgO}} = 6.71 \text{ g HgO}$$

9.53 (a) $25.5 \text{ g Zn} \times \frac{1 \text{ mol Zn}}{65.39 \text{ g Zn}} \times \frac{1 \text{ mol } H_2}{1 \text{ mol Zn}} = 0.390 \text{ mol } H_2$

$$V = \frac{nRT}{P} = \frac{(0.390 \text{ mol})\left(0.082\ 06\ \frac{\text{L}\cdot\text{atm}}{\text{mol}\cdot\text{K}}\right)(288 \text{ K})}{\left(742 \text{ mm Hg} \times \frac{1 \text{ atm}}{760 \text{ mm Hg}}\right)} = 9.44 \text{ L}$$

(b) $$n = \frac{PV}{RT} = \frac{\left(350 \text{ mm Hg} \times \frac{1 \text{ atm}}{760 \text{ mm Hg}}\right)(5.00 \text{ L})}{\left(0.082\ 06\ \frac{\text{L}\cdot\text{atm}}{\text{mol}\cdot\text{K}}\right)(303.15 \text{ K})} = 0.092\ 56 \text{ mol } H_2$$

$$0.092\ 56 \text{ mol } H_2 \times \frac{1 \text{ mol Zn}}{1 \text{ mol } H_2} \times \frac{65.39 \text{ g Zn}}{1 \text{ mol Zn}} = 6.05 \text{ g Zn}$$

9.54 $2\ NH_4NO_3(s) \rightarrow 2\ N_2(g) + 4\ H_2O(g) + O_2(g)$

NH_4NO_3, 80.04 amu

$$\text{Total moles of gas} = 450 \text{ g } NH_4NO_3 \times \frac{1 \text{ mol } NH_4NO_3}{80.04 \text{ g } NH_4NO_3} \times \frac{7 \text{ mol gas}}{2 \text{ mol } NH_4NO_3}$$

Total moles of gas = 19.68 mol gas

T = 450°C = 723 K

$$V = \frac{nRT}{P} = \frac{(19.68 \text{ mol})\left(0.082\ 06 \frac{L \cdot atm}{mol \cdot K}\right)(723 \text{ K})}{(1.00 \text{ atm})} = 1.17 \times 10^3 \text{ L}$$

9.55 (a) $V_{24h} = (4.50 \text{ L/min})(60 \text{ min/h})(24 \text{ h/day}) = 6480 \text{ L}$

$V_{CO_2} = (0.034)V_{24h} = (0.034)(6480 \text{ L}) = 220 \text{ L}$

$$n = \frac{PV}{RT} = \frac{\left(735 \text{ mm Hg} \times \frac{1 \text{ atm}}{760 \text{ mm Hg}}\right)(220 \text{ L})}{\left(0.082\ 06 \frac{L \cdot atm}{mol \cdot K}\right)(298 \text{ K})} = 8.70 \text{ mol } CO_2$$

$$8.70 \text{ mol } CO_2 \times \frac{44.01 \text{ g } CO_2}{1 \text{ mol } CO_2} = 383 \text{ g} = 380 \text{ g } CO_2$$

(b) Na_2O_2, 77.98 amu

$$3650 \text{ g } Na_2O_2 \times \frac{1 \text{ mol } Na_2O_2}{77.98 \text{ g } Na_2O_2} \times \frac{2 \text{ mol } CO_2}{2 \text{ mol } Na_2O_2} \times \frac{1 \text{ day}}{8.70 \text{ mol } CO_2} = 5.4 \text{ days}$$

Dalton's Law

9.56 Because of Avogadro's Law ($V \propto n$), the % volumes are also % moles.

	% mole
N_2	78.08
O_2	20.95
Ar	0.93
CO_2	0.034

In decimal form, % mole = mole fraction

$P_{N_2} = X_{N_2} \cdot P_{total} = (0.7808)(1.000 \text{ atm}) = 0.7808 \text{ atm}$

$P_{O_2} = X_{O_2} \cdot P_{total} = (0.2095)(1.000 \text{ atm}) = 0.2095 \text{ atm}$

$P_{Ar} = X_{Ar} \cdot P_{total} = (0.0093)(1.000 \text{ atm}) = 0.0093 \text{ atm}$

$P_{CO_2} = X_{CO_2} \cdot P_{total} = (0.000\ 34)(1.000 \text{ atm}) = 0.000\ 34 \text{ atm}$

Pressures of the rest are negligible.

9.57 $X_{CH_4} = \dfrac{94 \text{ mol}}{100 \text{ mol}} = 0.94;\quad X_{C_2H_6} = \dfrac{4 \text{ mol}}{100 \text{ mol}} = 0.040;$

$X_{C_3H_8} = \dfrac{1.5 \text{ mol}}{100 \text{ mol}} = 0.015;\quad X_{C_4H_{10}} = \dfrac{0.5 \text{ mol}}{100 \text{ mol}} = 0.0050$

$P_{CH_4} = X_{CH_4} \cdot P_{total} = (0.94)(1.48 \text{ atm}) = 1.4 \text{ atm}$

$P_{C_2H_6} = X_{C_2H_6} \cdot P_{total} = (0.040)(1.48 \text{ atm}) = 0.059 \text{ atm}$

$P_{C_3H_8} = X_{C_3H_8} \cdot P_{total} = (0.015)(1.48 \text{ atm}) = 0.022 \text{ atm}$

$P_{C_4H_{10}} = X_{C_4H_{10}} \cdot P_{total} = (0.0050)(1.48 \text{ atm}) = 0.0074 \text{ atm}$

9.58 Assume a 100.0 g sample.

g CO_2 = 1.00 g

g O_2 = 99.0 g

mole $CO_2 = 1.00 \text{ g} \times \dfrac{1 \text{ mol}}{44.01 \text{ g}} = 0.0227 \text{ mole } CO_2$

mole $O_2 = 99.0 \text{ g} \times \dfrac{1 \text{ mol}}{32.00 \text{ g}} = 3.09 \text{ mol } O_2$

$n_{total} = 3.09 \text{ mol} + 0.0227 \text{ mol} = 3.11 \text{ mol}$

$$X_{O_2} = \frac{3.09 \text{ mol}}{3.11 \text{ mol}} = 0.994$$

$$X_{CO_2} = \frac{0.0227 \text{ mol}}{3.11 \text{ mol}} = 0.00730$$

$$P_{O_2} = X_{O_2} \cdot P_{total} = (0.994)(0.977 \text{ atm}) = 0.971 \text{ atm}$$

$$P_{CO_2} = X_{CO_2} \cdot P_{total} = (0.007\ 30)(0.977 \text{ atm}) = 0.007\ 13 \text{ atm}$$

9.59 Assume a 100.0 g sample.

$$\text{g HCl} = (0.0500)(100.0 \text{ g}) = 5.00 \text{ g}; \quad 5.00 \text{ g HCl} \times \frac{1 \text{ mol HCl}}{36.5 \text{ g HCl}} = 0.137 \text{ mol HCl}$$

$$\text{g } H_2 = (0.0100)(100.0 \text{ g}) = 1.00 \text{ g}; \quad 1.00 \text{ g } H_2 \times \frac{1 \text{ mol } H_2}{2.016 \text{ g } H_2} = 0.496 \text{ mol } H_2$$

$$\text{g Ne} = (0.94)(100.0 \text{ g}) = 94 \text{ g}; \quad 94 \text{ g Ne} \times \frac{1 \text{ mol Ne}}{20.18 \text{ g Ne}} = 4.66 \text{ mol Ne}$$

$$n_{total} = 0.137 + 0.496 + 4.66 = 5.3 \text{ mol}$$

$$X_{HCl} = \frac{0.137 \text{ mol}}{5.3 \text{ mol}} = 0.026$$

$$X_{H_2} = \frac{0.496 \text{ mol}}{5.3 \text{ mol}} = 0.094$$

$$X_{Ne} = \frac{4.66 \text{ mol}}{5.3 \text{ mol}} = 0.88$$

9.60 From Problem 9.59: $X_{HCl} = 0.026$, $X_{H_2} = 0.094$, $X_{Ne} = 0.88$

$$P_{HCl} = X_{HCl} \cdot P_{total} = (0.026)(13{,}800 \text{ kPa}) = 3.6 \times 10^2 \text{ kPa}$$

$$P_{H_2} = X_{H_2} \cdot P_{total} = (0.094)(13{,}800 \text{ kPa}) = 1.3 \times 10^3 \text{ kPa}$$

$$P_{Ne} = X_{Ne} \cdot P_{total} = (0.88)(13{,}800) \text{ kPa}) = 1.2 \times 10^4 \text{ kPa}$$

9.61 $P_{total} = P_{H_2} + P_{H_2O}$

$P_{H_2} = P_{total} - P_{H_2O}$ = 747 mm Hg – 23.8 mm Hg = 723 mm Hg

$$n = \frac{PV}{RT} = \frac{\left(723 \text{ mm Hg} \times \frac{1 \text{ atm}}{760 \text{ mm Hg}}\right)(3.557 \text{ L})}{\left(0.082\ 06 \frac{\text{L} \cdot \text{atm}}{\text{mol} \cdot \text{K}}\right)(298 \text{ K})} = 0.1384 \text{ mol } H_2$$

$$0.1384 \text{ mol } H_2 \times \frac{1 \text{ mol Mg}}{1 \text{ mol } H_2} \times \frac{24.3 \text{ g Mg}}{1 \text{ mol Mg}} = 3.36 \text{ g Mg}$$

Kinetic Molecular Theory and Graham's Law

9.62 The kinetic-molecular theory is based on the following assumptions:

1. A gas consists of tiny particles, either atoms or molecules, moving about at random.

2. The volume of the particles themselves is negligible compared with the total volume of the gas; most of the volume of a gas is empty space.

3. The gas particles act independently; there are no attractive or repulsive forces between particles.

4. Collisions of the gas particles, either with other particles or with the walls of the container, are elastic; that is, the total kinetic energy of the gas particles is constant at constant T.

5. The average kinetic energy of the gas particles is proportional to the Kelvin temperature of the sample.

9.63 Diffusion – The mixing of different gases by random molecular motion and with frequent collisions.

Effusion – The process in which gas molecules escape through a tiny hole in a membrane without collisions.

9.64 Heat is the energy transferred from one object to another as the result of a temperature difference between them.

Temperature is a measure of the kinetic energy of molecular motion.

9.65 The atomic weight of He is much less than the molecular weights of the major components of air (N_2 and O_2). The rate of effusion of He through the balloon skin is much faster.

9.66 $$\mu = \sqrt{\frac{3RT}{M}} = \sqrt{\frac{3 \times 8.314 \text{ kg m}^2/(\text{s}^2 \text{ K mol}) \times 220 \text{ K}}{28.0 \times 10^{-3} \text{ kg/mol}}} = 443 \text{ m/s}$$

9.67 For Br_2: $$\mu = \sqrt{\frac{3RT}{M}} = \sqrt{\frac{3 \times 8.314 \text{ kg m}^2/(\text{s}^2 \text{ K mol}) \times 293 \text{ K}}{159.8 \times 10^{-3} \text{ kg/mol}}} = 214 \text{ m/s}$$

For Xe: $$\mu = 214 \text{ m/s} = \sqrt{\frac{3 \times 8.314 \text{ kg m}^2/(\text{s}^2 \text{ K mol}) \times T}{131.3 \times 10^{-3} \text{ kg/mol}}}$$

Square both sides of the equation and solve for T.

$$45796 \text{ m}^2/\text{s}^2 = \frac{3 \times 8.314 \text{ kg m}^2/(\text{s}^2 \text{ K mol}) \times T}{131.3 \times 10^{-3} \text{ kg/mol}}$$

T = 241 K = –32°C

9.68 For H_2, $$\mu = \sqrt{\frac{3RT}{M}} = \sqrt{\frac{3 \times 8.314 \text{ kg m}^2/(\text{s}^2 \text{ K mol}) \times 150 \text{ K}}{2.02 \times 10^{-3} \text{ kg/mol}}} = 1360 \text{ m/s}$$

For He, $$\mu = \sqrt{\frac{3 \times 8.314 \text{ kg m}^2/(\text{s}^2 \text{ K mol}) \times 648 \text{ K}}{4.00 \times 10^{-3} \text{ kg/mol}}} = 2010 \text{ m/s}$$

He at 375°C has the higher average speed.

9.69 $\frac{\text{rate}_{H_2}}{\text{rate}_X} = \frac{\sqrt{M_X}}{\sqrt{M_{H_2}}}$; $\frac{2.92}{1} = \frac{\sqrt{M_X}}{\sqrt{2.02}}$; $2.92\sqrt{2.02} = \sqrt{M_X}$

Solve for M_X: $M_X = 17.2$ g/mol; molecular weight = 17.2 amu

9.70 HCl, 36.5 amu; F_2, 38.0 amu; Ar, 39.9 amu

$$\frac{\text{rate HCl}}{\text{rate Ar}} = \frac{\sqrt{M_{Ar}}}{\sqrt{M_{HCl}}} = \frac{\sqrt{39.9}}{\sqrt{36.5}} = 1.05$$

$$\frac{\text{rate } F_2}{\text{rate Ar}} = \frac{\sqrt{M_{Ar}}}{\sqrt{M_{F_2}}} = \frac{\sqrt{39.9}}{\sqrt{38.0}} = 1.02$$

The relative rates of diffusion are HCl(1.05) > F_2(1.02) > Ar(1.00).

9.71 Since CO and N_2 have the same mass, they will have the same diffusion rates.

9.72 $$\mu = 45 \text{ m/s} = \sqrt{\frac{3 \times 8.314 \text{ kg m}^2/(\text{s}^2 \text{ K mol}) \times T}{4.00 \times 10^{-3} \text{ kg/mol}}}$$

Square both sides of the equation and solve for T.

$$2025 \text{ m}^2/\text{s}^2 = \frac{3 \times 8.314 \text{ kg m}^2/(\text{s}^2 \text{ K mol}) \times T}{4.00 \times 10^{-3} \text{ kg/mol}}$$

T = 0.325 K = –272.83°C (near absolute zero)

9.73 $$230 \text{ km/h} \times \frac{1000 \text{ m}}{1 \text{ km}} \times \frac{1 \text{ h}}{3600 \text{ s}} = 63.9 \text{ m/s}$$

$$\mu = 63.9 \text{ m/s} = \sqrt{\frac{3 \times 8.314 \text{ kg m}^2/(\text{s}^2 \text{ K mol}) \times T}{32.0 \times 10^{-3} \text{ kg/mol}}}$$

Square both sides of the equation and solve for T.

$$4083\ m^2/s^2 = \frac{3 \times 8.314\ kg\ m^2/(s^2\ K\ mol) \times T}{32.0 \times 10^{-3}\ kg/mol}$$

$T = 5.24\ K = -268°C$

General Problems

9.74 Average molecular weight of air = 28.96 amu

CO_2, 44.01 amu

$$P = 760\ mm\ Hg \times \frac{44.01\ g/mol}{28.96\ g/mol} = 1155\ mm\ Hg$$

9.75 $$n = \frac{PV}{RT} = \frac{(2.15\ atm)(7.35\ L)}{\left(0.082\ 06\ \frac{L \cdot atm}{mol \cdot K}\right)(293\ K)} = 0.657\ mol\ Ar$$

$$0.657\ mol\ Ar \times \frac{39.948\ g\ Ar}{1\ mol\ Ar} = 26.2\ g\ Ar$$

$m_{total} = 478.1\ g + 26.2\ g = 504.3\ g$

9.76 This is initially a Boyle's Law problem, since only P and V are changing while n and T remain fixed. The initial volume for each gas is the volume of their individual bulbs. The final volume for each gas is the total volume of the three bulbs.

$nRT = P_iV_i = P_fV_f$; $V_f = 1.50 + 1.00 + 2.00 = 4.50\ L$

For CO_2: $$P_f = \frac{P_iV_i}{V_f} = \frac{(2.13\ atm)(1.50\ L)}{(4.50\ L)} = 0.710\ atm$$

For H_2: $$P_f = \frac{P_iV_i}{V_f} = \frac{(0.861\ atm)(1.00\ L)}{(4.50\ L)} = 0.191\ atm$$

For Ar: $$P_f = \frac{P_iV_i}{V_f} = \frac{(1.15\ atm)(2.00\ L)}{(4.50\ L)} = 0.511\ atm$$

From Dalton's Law, $P_{total} = P_{CO_2} + P_{H_2} + P_{Ar}$

$P_{total} = 0.710 \text{ atm} + 0.191 \text{ atm} + 0.511 \text{ atm} = 1.412 \text{ atm}$

9.77 (a) Bulb A contains $CO_2(g)$ and $N_2(g)$

Bulb B contains $CO_2(g)$, $N_2(g)$, and $H_2O(s)$

(b) Initial moles of gas = n = $\frac{PV}{RT} = \frac{\left(564 \text{ mm Hg} \times \frac{1 \text{ atm}}{760 \text{ mm Hg}}\right)(1.000 \text{ L})}{\left(0.082\ 06 \frac{\text{L}\cdot\text{atm}}{\text{mol}\cdot\text{K}}\right)(298 \text{ K})}$

Initial moles of gas = 0.030 35 mol

$$\text{mol gas in Bulb A} = n = \frac{PV}{RT} = \frac{\left(219 \text{ mm Hg} \times \frac{1 \text{ atm}}{760 \text{ mm Hg}}\right)(1.000 \text{ L})}{\left(0.082\ 06 \frac{\text{L}\cdot\text{atm}}{\text{mol}\cdot\text{K}}\right)(298 \text{ K})} = 0.011\ 78 \text{ mol}$$

$$\text{mol gas in Bulb B} = n = \frac{PV}{RT} = \frac{\left(219 \text{ mm Hg} \times \frac{1 \text{ atm}}{760 \text{ mm Hg}}\right)(1.000 \text{ L})}{\left(0.082\ 06 \frac{\text{L}\cdot\text{atm}}{\text{mol}\cdot\text{K}}\right)(203 \text{ K})} = 0.017\ 29 \text{ mol}$$

$n_{H_2O} = n_{initial} - n_A - n_B = 0.030\ 35 - 0.011\ 78 - 0.017\ 29 = 0.001\ 28 \text{ mol}$

$n_{H_2O} = 0.0013 \text{ mol } H_2O$

(c) Bulb A contains $N_2(g)$

Bulb B contains $N_2(g)$ and $H_2O(s)$

Bulb C contains $N_2(g)$ and $CO_2(s)$

(d) $$n_A = \frac{PV}{RT} = \frac{\left(33.5 \text{ mm Hg x } \dfrac{1 \text{ atm}}{760 \text{ mm Hg}}\right)(1.000 \text{ L})}{\left(0.082\ 06 \dfrac{\text{L}\cdot\text{atm}}{\text{mol}\cdot\text{K}}\right)(298 \text{ K})} = 0.001\ 803 \text{ mol}$$

$$n_B = \frac{PV}{RT} = \frac{\left(33.5 \text{ mm Hg x } \dfrac{1 \text{ atm}}{760 \text{ mm Hg}}\right)(1.000 \text{ L})}{\left(0.082\ 06 \dfrac{\text{L}\cdot\text{atm}}{\text{mol}\cdot\text{K}}\right)(203 \text{ K})} = 0.002\ 646 \text{ mol}$$

$$n_C = \frac{PV}{RT} = \frac{\left(33.5 \text{ mm Hg x } \dfrac{1 \text{ atm}}{760 \text{ mm Hg}}\right)(1.000 \text{ L})}{\left(0.082\ 06 \dfrac{\text{L}\cdot\text{atm}}{\text{mol}\cdot\text{K}}\right)(83 \text{ K})} = 0.006\ 472 \text{ mol}$$

$n_{N_2} = n_A + n_B + n_C = 0.001\ 803 + 0.002\ 646 + 0.006\ 472 = 0.010\ 92 \text{ mol } N_2$

(e) $n_{CO_2} = n_{initial} - n_{H_2O} - n_{N_2} = 0.030\ 35 - 0.0013 - 0.010\ 92 = 0.0181 \text{ mol } CO_2$

9.78 $C_3H_5N_3O_9$, 227.1 amu

(a) moles $C_3H_5N_3O_9 = 1.00 \text{ g x } \dfrac{1 \text{ mol}}{227.1 \text{ g}} = 0.004\ 40 \text{ mol}$

$$n_{air} = \frac{PV}{RT} = \frac{(1.00 \text{ atm})(0.500 \text{ L})}{\left(0.082\ 06 \dfrac{\text{L}\cdot\text{atm}}{\text{mol}\cdot\text{K}}\right)(293 \text{ K})} = 0.0208 \text{ mol air}$$

(b) moles gas from $C_3H_5N_3O_9 = 0.004\ 40 \text{ mol x } \dfrac{29 \text{ mol gas}}{4 \text{ mol nitro}}$

moles gas from $C_3H_5N_3O_9$ = 0.0319 mol gas from $C_3H_5N_3O_9$

$n_{total} = 0.0319 \text{ mol} + 0.0208 \text{ mol} = 0.0527 \text{ mol}$

(c) $$P = \frac{nRT}{V} = \frac{(0.0527\ \text{mol})\left(0.082\ 06\ \frac{\text{L}\cdot\text{atm}}{\text{mol}\cdot\text{K}}\right)(698\ \text{K})}{(0.500\ \text{L})} = 6.04\ \text{atm}$$

9.79 $$P = \frac{nRT}{V} = \frac{(0.60\ \text{mol})\left(0.082\ 06\ \frac{\text{L}\cdot\text{atm}}{\text{mol}\cdot\text{K}}\right)(293\ \text{K})}{(0.200\ \text{L})} = 72\ \text{atm}$$

$$P = \frac{nRT}{(V - nb)} - \frac{an^2}{V^2}$$

$$P = \frac{(0.60\ \text{mol})\left(0.082\ 06\ \frac{\text{L}\cdot\text{atm}}{\text{mol}\cdot\text{K}}\right)(293\ \text{K})}{[(0.200\ \text{L}) - (0.60\ \text{mol})(0.0391\ \text{L/mol})]} - \frac{\left(1.39\ \frac{\text{L}^2\cdot\text{atm}}{\text{mol}^2}\right)(0.60\ \text{mol})^2}{(0.200\ \text{L})^2}$$

$P = 81.716\ \text{atm} - 12.51\ \text{atm} = 69\ \text{atm}$

9.80 NH_3, 17.03 amu

$$\text{mol } NH_3 = 45.0\ \text{g} \times \frac{1\ \text{mol}}{17.03\ \text{g}} = 2.64\ \text{mol}$$

$$P = \frac{nRT}{(V - nb)} - \frac{an^2}{V^2}$$

(a) At T = 0°C = 273 K

$$P = \frac{(2.64\ \text{mol})\left(0.082\ 06\ \frac{\text{L}\cdot\text{atm}}{\text{mol}\cdot\text{K}}\right)(273\ \text{K})}{[(0.250\ \text{L}) - (2.64\ \text{mol})(0.0371\ \text{L/mol})]} - \frac{\left(4.17\ \frac{\text{L}^2\cdot\text{atm}}{\text{mol}^2}\right)(2.64\ \text{mol})^2}{(0.250\ \text{L})^2}$$

$P = 389\ \text{atm} - 465\ \text{atm} = -76\ \text{atm}$

(b) At T = 50°C = 323 K

$$P = \frac{(2.64\text{ mol})\left(0.082\ 06\ \frac{\text{L}\cdot\text{atm}}{\text{mol}\cdot\text{K}}\right)(323\text{ K})}{[(0.250\text{ L}) - (2.64\text{ mol})(0.0371\text{ L/mol})]} - \frac{\left(4.17\ \frac{\text{L}^2\cdot\text{atm}}{\text{mol}^2}\right)(2.64\text{ mol})^2}{(0.250\text{ L})^2}$$

P = 461 atm – 465 atm = –4 atm

(c) At T = 100°C = 373 K

$$P = \frac{(2.64\text{ mol})\left(0.082\ 06\ \frac{\text{L}\cdot\text{atm}}{\text{mol}\cdot\text{K}}\right)(373\text{ K})}{[(0.250\text{ L}) - (2.64\text{ mol})(0.0371\text{ L/mol})]} - \frac{\left(4.17\ \frac{\text{L}^2\cdot\text{atm}}{\text{mol}^2}\right)(2.64\text{ mol})^2}{(0.250\text{ L})^2}$$

P = 531 atm – 465 atm = 66 atm

The negative pressures at 0°C and 50°C indicate that the van der Waals equation does not apply under these conditions. At the high ammonia concentration (n/V = 10.56 M) and at temperatures only moderately above the normal boiling point of –33°C, ammonia liquefies.

9.81 (a) $2\ C_8H_{18}(l)\ +\ 25\ O_2(g)\ \rightarrow\ 16\ CO_2(g)\ +\ 18\ H_2O(g)$

(b) $4.6 \times 10^{10}\text{ L } C_8H_{18} \times \frac{1000\text{ mL}}{1\text{ L}} \times \frac{0.792\text{ g}}{1\text{ mL}} = 3.64 \times 10^{13}\text{ g } C_8H_{18}$

$$3.64 \times 10^{13}\text{ g } C_8H_{18} \times \frac{1\text{ mol } C_8H_{18}}{114.2\text{ g } C_8H_{18}} \times \frac{16\text{ mol } CO_2}{2\text{ mol } C_8H_{18}} = 2.55 \times 10^{12}\text{ mol } CO_2$$

$$2.55 \times 10^{12}\text{ mol } CO_2 \times \frac{44.0\text{ g } CO_2}{1\text{ mol } CO_2} \times \frac{1\text{ kg}}{1000\text{ g}} = 1.1 \times 10^{11}\text{ kg } CO_2$$

(c) $V = \frac{nRT}{P} = \frac{(2.55 \times 10^{12}\text{ mol})\left(0.082\ 06\ \frac{\text{L}\cdot\text{atm}}{\text{mol}\cdot\text{K}}\right)(273\text{ K})}{(1.00\text{ atm})} = 5.7 \times 10^{13}\text{ L of } CO_2$

9.82 12.5 moles of O_2 are needed (from Problem 9.81).

$12.5 \text{ mol } O_2 = (0.210)(n_{air})$

$$n_{air} = \frac{12.5 \text{ mol}}{0.210} = 59.5 \text{ mol air}$$

$$V = \frac{nRT}{P} = \frac{(59.5 \text{ mol})\left(0.082\ 06 \frac{L \cdot atm}{mol \cdot K}\right)(273K)}{(1.00 \text{ atm})} = 1.33 \times 10^3 \text{ L}$$

9.83 Freezing point of H_2O on the Rankine scale is (9/5)(273.15) = 492°R.

$$R = \frac{PV}{nT} = \frac{(1.00 \text{ atm})(22.414 \text{ L})}{(1.00 \text{ mol})(492\,^\circ\text{R})} = 0.0456 \frac{L \cdot atm}{mol \cdot {}^\circ R}$$

9.84 (a) $$n_{total} = \frac{PV}{RT} = \frac{\left(258 \text{ mm Hg} \times \frac{1.00 \text{ atm}}{760 \text{ mm Hg}}\right)(0.500 \text{ L})}{\left(0.082\ 06 \frac{L \cdot atm}{mol \cdot K}\right)(293 \text{ K})} = 0.007\ 06 \text{ mol}$$

(b) $$n_B = \frac{PV}{RT} = \frac{\left(344 \text{ mm Hg} \times \frac{1 \text{ atm}}{760 \text{ mm Hg}}\right)(0.250 \text{ L})}{\left(0.082\ 06 \frac{L \cdot atm}{mol \cdot K}\right)(293 \text{ K})} = 0.004\ 71 \text{ moles}$$

(c) $$d = \frac{0.218 \text{ g}}{0.250 \text{ L}} = 0.872 \text{ g/L}$$

(d) $$\text{molar mass} = \frac{0.218 \text{ g}}{0.004\ 71 \text{ mol}} = 46.3 \text{ g/mol}, NO_2; \quad \text{mol. wt.} = 46.3 \text{ amu}$$

(e) $Hg_2CO_3(s) + 6\ HNO_3(aq) \rightarrow$

$2\ Hg(NO_3)_2(aq) + 3\ H_2O(l) + CO_2(g) + 2\ NO_2(g)$

9.85 $n = \frac{PV}{RT} = \frac{(1\text{ atm})(1323\text{ L})}{\left(0.082\ 06\ \frac{\text{L}\cdot\text{atm}}{\text{mol}\cdot\text{K}}\right)(2223\text{ K})} = 7.25\text{ mol of all gases}$

(a) $0.004\ 00\text{ mol "nitro"} \times \frac{7.25\text{ mol gases}}{1\text{ mol "nitro"}} = 0.0290\text{ mol hot gases}$

(b) $n = \frac{PV}{RT} = \frac{\left(623\text{ mm Hg} \times \frac{1\text{ atm}}{760\text{ mm Hg}}\right)(0.500\text{ L})}{\left(0.082\ 06\ \frac{\text{L}\cdot\text{atm}}{\text{mol}\cdot\text{K}}\right)(263\text{ K})} = 0.0190\text{ mol B + C + D}$

$n_A = n_{total} - n_{(B+C+D)} = 0.0290 - 0.0190 = 0.0100$ mol A; A = H_2O

(c) $n = \frac{PV}{RT} = \frac{\left(260\text{ mm Hg} \times \frac{1\text{ atm}}{760\text{ mm Hg}}\right)(0.500\text{ L})}{\left(0.082\ 06\ \frac{\text{L}\cdot\text{atm}}{\text{mol}\cdot\text{K}}\right)(298\text{ K})} = 0.007\ 00\text{ mol C + D}$

$n_B = n_{(B+C+D)} - n_{(C+D)} = 0.0190 - 0.007\ 00 = 0.0120$ mol B; B = CO_2

(d) $n = \frac{PV}{RT} = \frac{\left(223\text{ mm Hg} \times \frac{1\text{ atm}}{760\text{ mm Hg}}\right)(0.500\text{ L})}{\left(0.082\ 06\ \frac{\text{L}\cdot\text{atm}}{\text{mol}\cdot\text{K}}\right)(298\text{ K})} = 0.006\ 00\text{ mol D}$

$n_C = n_{(C+D)} - n_D = 0.007\ 00 - 0.006\ 00 = 0.001\ 00$ mol C; C = O_2

molar mass D = $\frac{0.168\text{ g}}{0.006\ 00\text{ mol}} = 28.0$ g/mol; D = N_2

(e) $0.004\ C_3H_5N_3O_9(l) \rightarrow 0.0100\ H_2O(g) + 0.012\ CO_2(g) + 0.001\ O_2(g) + 0.006\ N_2(g)$

Multiply each coefficient by 1000 to obtain integers.

$4\ C_3H_5N_3O_9(l) \rightarrow 10\ H_2O(g) + 12\ CO_2(g) + O_2(g) + 6\ N_2(g)$

9.86 $$\frac{\text{rate } ^{35}Cl_2}{\text{rate } ^{37}Cl_2} = \frac{\sqrt{M\ ^{37}Cl_2}}{\sqrt{M\ ^{35}Cl_2}} = \frac{\sqrt{74.0}}{\sqrt{70.0}} = 1.03$$

$$\frac{\text{rate } ^{35}Cl\,^{37}Cl}{\text{rate } ^{37}Cl_2} = \frac{\sqrt{M\ ^{37}Cl_2}}{\sqrt{M\ ^{35}Cl^{37}Cl}} = \frac{\sqrt{74.0}}{\sqrt{72.0}} = 1.01$$

The relative rates of diffusion are $^{35}Cl_2(1.03) > {}^{35}Cl^{37}Cl(1.01) > {}^{37}Cl_2(1.00)$.

9.87 $$V = \frac{nRT}{P} = \frac{(1.00\text{ mol})\left(0.082\ 06\ \frac{L\cdot atm}{mol\cdot K}\right)(1050\text{ K})}{(75\text{ atm})} = 1.1\text{ L}$$

9.88 (a) $CaC_2(s) + 2\ H_2O(l) \rightarrow C_2H_2(g) + Ca(OH)_2(aq)$

(b) CaC_2, 64.10 amu

$$\text{mol } C_2H_2 = 25.0\text{ g } CaC_2 \times \frac{1\text{ mol } CaC_2}{64.10\text{ g } CaC_2} \times \frac{1\text{ mol } C_2H_2}{1\text{ mol } CaC_2} = 0.390\text{ mol } C_2H_2$$

$$V = \frac{nRT}{P} = \frac{(0.390\text{ mol})\left(0.082\ 06\ \frac{L\cdot atm}{mol\cdot K}\right)(300\text{ K})}{\left(755\text{ mm Hg} \times \frac{1\text{ atm}}{760\text{ mm Hg}}\right)} = 9.66\text{ L}$$

10.1 $\mu = Q \times r = (1.60 \times 10^{-19}\text{ C})(92 \times 10^{-12}\text{ m})\left(\frac{1\text{ D}}{3.336 \times 10^{-30}\text{ C}\cdot\text{m}}\right) = 4.41\text{ D}$

% ionic character for HF = $\frac{1.82\text{ D}}{4.41\text{ D}} \times 100\% = 41\%$

HF has more ionic character than HCl. HCl has only 17% ionic character.

10.2

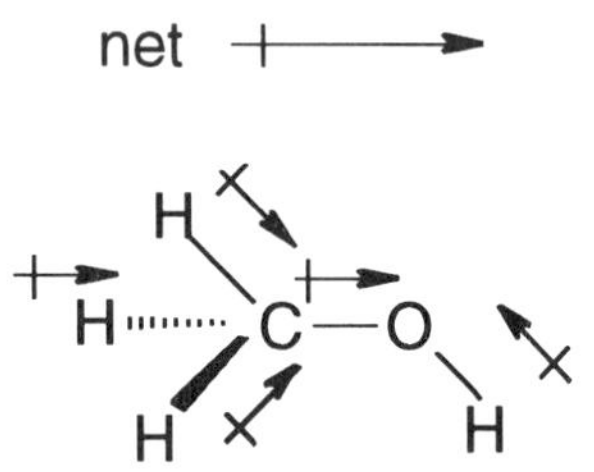

net

10.3 (a) SF_6 has polar covalent bonds but the molecule is symmetrical (octahedral). The individual bond polarities cancel, and the molecule has no dipole moment.

(b) $H_2C{=}CH_2$ can be assumed to have nonpolar C–H bonds. In addition the molecule is symmetrical. The molecule has no dipole moment.

(c) The C–Cl bonds in $CHCl_3$ are polar covalent bonds, and the molecule is polar.

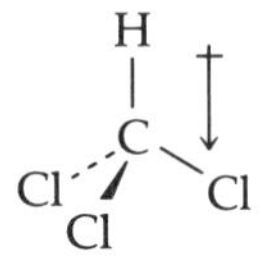

(d) The C–Cl bonds in CH_2Cl_2 are polar covalent bonds, and the molecule is polar.

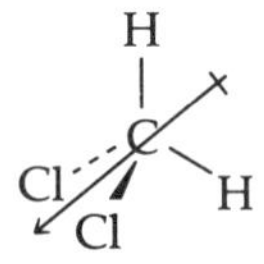

10.4 (a) Of the four substances, only HNO_3 has a net dipole moment.

(b) Only HNO_3 can hydrogen-bond.

(c) Ar has fewer electrons than Cl_2 and CCl_4, and has the smallest dispersion forces.

10.5 H_2S dipole–dipole, dispersion
CH_3OH hydrogen bonding, dipole–dipole, dispersion
CBr_4 dispersion
Ne dispersion

$Ne < H_2S < CH_3OH < CBr_4$

Even though CBr_4 experiences only dispersion forces, it has the highest boiling point of the group because of its large number of electrons.

10.6 (a) $CO_2(s) \rightarrow CO_2(g)$, ΔS is positive.

(b) $H_2O(g) \rightarrow H_2O(l)$, ΔS is negative.

(c) ΔS is positive (more disorder).

10.7 $\Delta G = \Delta H - T\Delta S$; at the boiling point (phase change), $\Delta G = 0$.

$$\Delta H = T\Delta S;\ T = \frac{\Delta H_{vap}}{\Delta S_{vap}} = \frac{29.2\ \text{kJ/mol}}{87.5 \times 10^{-3}\ \text{kJ/(K}\cdot\text{mol)}} = 334\ \text{K}$$

10.8 The boiling point is the temperature where the vapor pressure of a liquid equals the external pressure.

$P_1 = 760$ mm Hg; $P_2 = 260$ mm Hg; $T_1 = 80.1°C$

$\Delta H_{vap} = 30.8$ kJ/mol

$$\log P_2 = \log P_1 + \frac{\Delta H_{vap}}{2.303\,R}\left(\frac{1}{T_1} - \frac{1}{T_2}\right)$$

$$(\log P_2 - \log P_1)\left(\frac{2.303\,R}{\Delta H_{vap}}\right) = \frac{1}{T_1} - \frac{1}{T_2}$$

Solve for T_2 (the boiling point for benzene at 260 mm Hg).

$$\frac{1}{T_1} - (\log P_2 - \log P_1)\left(\frac{2.303\,R}{\Delta H_{vap}}\right) = \frac{1}{T_2}$$

$$\frac{1}{353.2\ \text{K}} - [\log(260) - \log(760)]\left(\frac{2.303\left(8.3145\ \frac{\text{J}}{\text{K}\cdot\text{mol}}\right)}{30{,}800\ \text{J/mol}}\right) = \frac{1}{T_2}$$

$$\frac{1}{T_2} = 0.003\ 121$$

$T_2 = 320\ \text{K} = 47°\text{C}$ (boiling point is lower at lower pressure)

10.9 $$\Delta H_{vap} = \frac{(\log P_2 - \log P_1)(2.303\ R)}{\left(\frac{1}{T_1} - \frac{1}{T_2}\right)}$$

$P_1 = 400$ mm Hg; $T_1 = 41.0\ °\text{C} = 314.2$ K

$P_2 = 760$ mm Hg; $T_2 = 331.9$ K

$$\Delta H_{vap} = \frac{[\log(760) - \log(400)][2.303\ (8.3145\ \frac{\text{J}}{\text{K}\cdot\text{mol}})]}{\left(\frac{1}{314.2\ \text{K}} - \frac{1}{331.9\ \text{K}}\right)} = 31{,}448\ \text{J/mol} = 31.4\ \text{kJ/mol}$$

10.10 Two cubes share a common face, and four cubes have a common edge.

10.11 (a) 1/8 atom at 8 corners and 1 atom at body center = 2 atoms

(b) 1/8 atom at 8 corners and 1/2 atom at 6 faces = 4 atoms

10.12 For a simple cube, d = 2r

$$r = \frac{d}{2} = \frac{334\ \text{pm}}{2} = 167\ \text{pm}$$

10.13 For a simple cube, there is one atom per unit cell.

$$\text{mass of one Po atom} = 209\ \text{g/mol} \times \frac{1\ \text{mol}}{6.022 \times 10^{23}\ \text{atoms}}$$

$$= 3.4706 \times 10^{-22}\ \text{g/atom}$$

unit cell edge = d = 334 pm = 334×10^{-12} m = 3.34×10^{-8} cm

unit cell volume = $d^3 = (3.34 \times 10^{-8}\ \text{cm})^3 = 3.7260 \times 10^{-23}\ \text{cm}^3$

$$\text{density} = \frac{\text{mass}}{\text{volume}} = \frac{3.4706 \times 10^{-22}\ \text{g}}{3.7260 \times 10^{-23}\ \text{cm}^3} = 9.31\ \text{g/cm}^3$$

10.14 Because Ar crystallizes in a face–centered cubic unit cell, there are four Ar atoms in the unit cell.

$$\text{mass of one Ar atom} = 39.95\ \text{g/mol} \times \frac{1\ \text{mol}}{6.022 \times 10^{23}\ \text{atom}}$$

$$= 6.634 \times 10^{-23}\ \text{g/atom}$$

unit cell mass = 4 atoms x mass of one Ar atom

= 4 atoms x 6.634×10^{-23} g/atom = 2.654×10^{-22} g

$$\text{density} = \frac{\text{mass}}{\text{volume}}$$

$$\text{unit cell volume} = \frac{\text{unit cell mass}}{\text{density}} = \frac{2.654 \times 10^{-22}\ \text{g}}{1.623\ \text{g/cm}^3} = 1.635 \times 10^{-22}\ \text{cm}^3$$

$$\text{unit cell edge} = d = \sqrt[3]{1.635 \times 10^{-22}\ \text{cm}^3} = 5.468 \times 10^{-8}\ \text{cm}$$

$$d = 5.468 \times 10^{-8}\ \text{cm} \times \frac{1\text{m}}{100\ \text{cm}} = 5.468 \times 10^{-10}\ \text{m} = 546.8 \times 10^{-12}\ \text{m}$$

d = 546.8 pm

$$r = \sqrt{\frac{d^2}{8}} = \frac{\sqrt{(546.8\ \text{pm})^2}}{8} = 193.3\ \text{pm}$$

10.15 For CuCl:

1/8 Cl^- at 8 corners and 1/2 Cl^- at 6 faces = 4 Cl^- (4 minuses)

4 Cu^+ inside (4 pluses)

For $BaCl_2$:

1/8 Ba^{2+} at 8 corners and 1/2 Ba^{2+} at 6 faces = 4 Ba^{2+} (8 pluses)

8 Cl^- inside (8 minuses)

10.16 (a) $CO_2(s) \rightarrow CO_2(g)$

(b) $CO_2(l) \rightarrow CO_2(g)$

(c) $CO_2(g) \rightarrow CO_2(l) \rightarrow$ supercritical CO_2

Understanding Key Concepts

1. (a) toluene, $C_7H_8(l)$; dispersion forces

   (b) mercury, Hg(l); dispersion forces

   (c) $NH_3(l)$; hydrogen bonding

2.

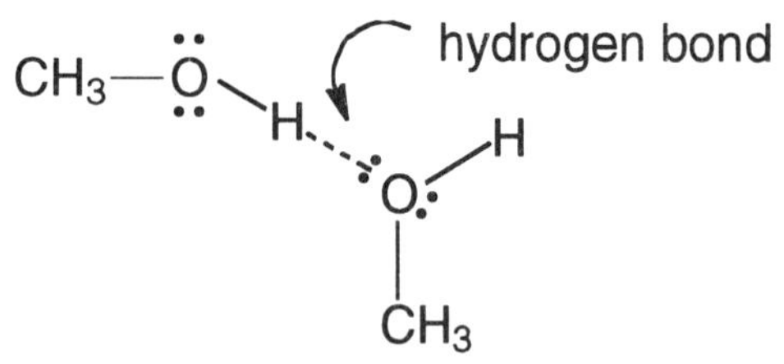

CH_3—O—H···O(H)—CH_3

hydrogen bond

3. (a) cubic closest–packed

   (b) simple cubic

   (c) hexagonal closest–packed

   (d) body–centered cubic

4. primitive–cubic unit cell

face–centered cubic unit cell

body–centered cubic unit cell

5. (a) cubic closest–packed

(b) 1/8 S^{2-} at 8 corners and 1/2 S^{2-} at 6 faces = 4 S^{2-}

4 Zn^{2+} inside

6. (a) normal boiling point ≈ 300 K; normal melting point ≈ 180 K

(b) (i) solid (ii) gas (iii) supercritical fluid

Additional Problems

Dipole Moments and Intermolecular Forces

10.17 The measure of net molecular polarity is called the dipole moment, μ. $\mu = \underline{Q} \times \underline{r}$, where $\underline{Q}$ is the magnitude of the charge at either end of the dipole, and $\underline{r}$ is the distance between the charges.

10.18 If a molecule has polar covalent bonds, the molecular shape (and location of lone pairs of electrons) determines whether a molecule has a dipole moment or not. The molecular shape will determine whether the bond dipoles cancel or not.

10.19 ion–dipole forces
dipole–dipole forces
dispersion forces
hydrogen bonding

10.20 The attraction of opposite charges is the fundamental interaction that gives rise to all intermolecular forces.

10.21 Dipole–dipole forces arise between molecules that have permanent dipole moments. London dispersion forces arise between molecules as a result of induced temporary dipoles.

10.22 (a) $CHCl_3$ has a permanent dipole moment. Dipole–dipole intermolecular forces are important. London dispersion forces are also present.

(b) O_2 has no dipole moment. London dispersion intermolecular forces are important.

(c) polyethylene, C_nH_{2n+2}. London dispersion intermolecular forces are important.

(d) CH_3OH has a permanent dipole moment. Dipole–dipole intermolecular forces and hydrogen bonding are important. London dispersion forces are also present.

10.23 For CH_3OH and CH_4, dispersion forces are small. CH_3OH can hydrogen bond; CH_4 cannot. This accounts for the large difference in boiling points.

For 1–decanol and decane, dispersion forces are comparable and relatively large along the C–H chain. 1–decanol can hydrogen bond; decane cannot. This accounts for the 55°C higher boiling point for 1–decanol.

10.24 (a)

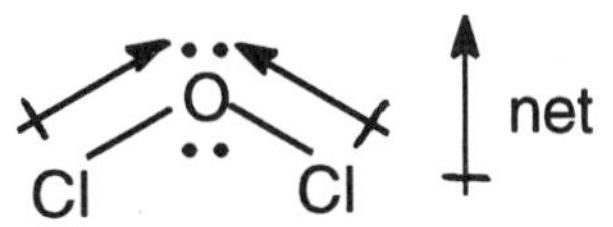

Cl O Cl net

(b)

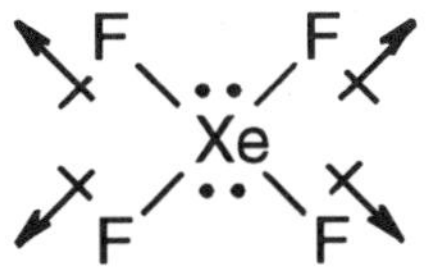

F F Xe F F net dipole moment =0

(c)

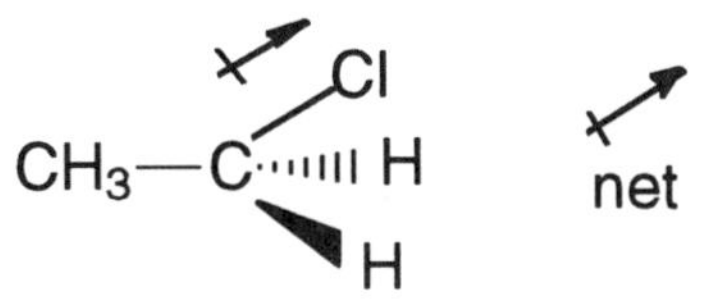

CH_3—C Cl H H net

(d)

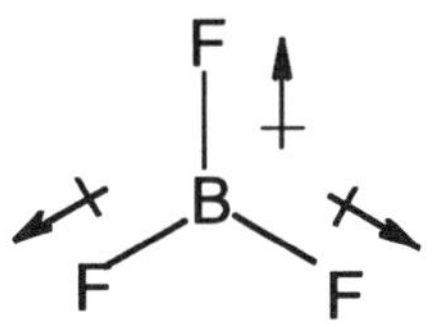

F B F F net dipole moment =0

10.25 (a) F N F F (b) CH_3 N H H

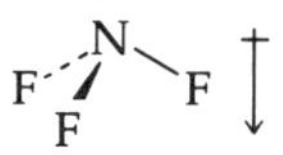

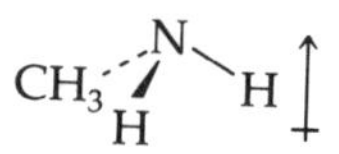

(c) XeF_2 is linear with zero dipole moment.

(d) PCl_5 is trigonal bipyramidal with zero dipole moment.

10.26

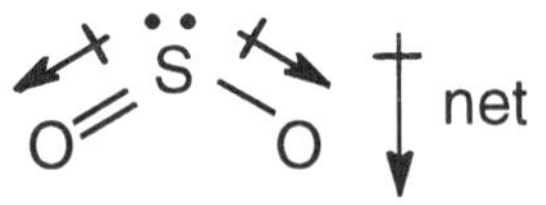

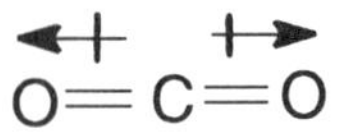

SO_2 (bent; bond dipoles add; net dipole shown) — net

$O=C=O$ — net dipole moment =0

SO_2 is bent and the individual bond dipole moments add to give the molecule a net dipole moment.

CO_2 is linear and the individual bond dipole moments point in opposite directions to cancel each other out. CO_2 has no net dipole moment.

10.27 (a) F_2 and Kr have no dipole-dipole forces

(b) HF has the largest hydrogen bond forces

(c) Kr has the largest dispersion forces

10.28 (a) C_8H_{18} has the larger dispersion forces because of its longer hydrocarbon chain.

(b) HI has the larger dispersion forces because of the larger, more polarizable iodine.

(c) H_2Se has the larger dispersion forces because of the more polarizable and less electronegative Se.

10.29

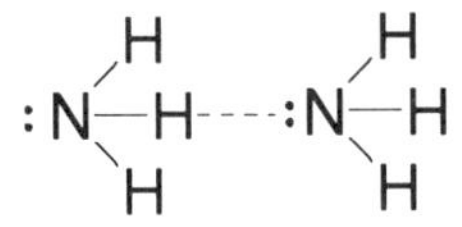

$:NH_2\text{—}H\cdots:NH_3$ (H–N–H···:N–H, hydrogen bond between two NH_3 molecules)

Vapor Pressure and Changes of State

10.30 ΔH_{vap} is the heat necessary to convert 1 mole of a liquid to 1 mole of its vapor (gas) at the normal boiling point.

ΔH_{fusion} is the heat necessary to convert 1 mole of a solid to 1 mol of the liquid at the normal melting point.

Both ΔH_{vap} and ΔH_{fusion} are positive (endothermic).

10.31 ΔH_{vap} is usually larger than ΔH_{fusion} because ΔH_{vap} is the heat required to overcome all intermolecular forces.

10.32 ΔH_{vap} and ΔH_{fusion} are both usually larger than the molar heat capacity because the molar heat capacity does not involve any changes of state and no intermolecular forces need to be overcome. ΔH_{vap} and ΔH_{fusion} involve changes of state and the overcoming of intermolecular forces.

10.33 (a) Hg(l) → Hg(g)

(b) no change of state, Hg remains a liquid

(c) Hg(g) → Hg(l) → Hg(s)

10.34 As the pressure over the liquid H_2O is lowered, H_2O vapor is removed by the pump. As H_2O vapor is removed, more of the liquid H_2O is converted to H_2O vapor. This conversion is an endothermic process and the temperature decreases. The combination of both a decrease in pressure and temperature takes you across the liquid/solid boundary in the phase diagram so the H_2O that remains turns to ice.

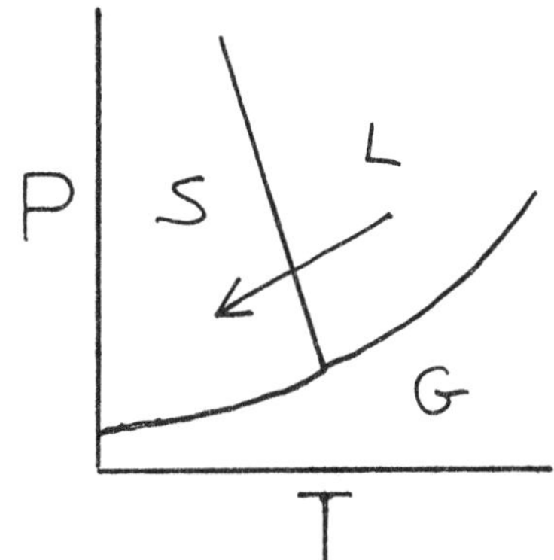

10.35 The normal boiling point for ether is relatively low (34.6°C). As the pressure is reduced by the pump, the relatively high vapor pressure of the ether equals the external pressure produced by the pump and the liquid boils.

10.36 H_2O, 18.02 amu

$$5.00 \text{ g } H_2O \times \frac{1 \text{ mol } H_2O}{18.02 \text{ g } H_2O} = 0.2775 \text{ mol } H_2O$$

q_1 = (0.2775 mol)[75.3 x 10^{-3} kJ/(K · mol)](273 K – 263 K) = 0.2090 kJ

q_2 = (0.2775 mol)(6.01 kJ/mol) = 1.668 kJ

q_3 = (0.2775 mol)(75.3 x 10^{-3} kJ/(K · mol)](303 K – 273K) = 0.6269 kJ

$q_{total} = q_1 + q_2 + q_3$ = 2.50 kJ

2.50 kJ of heat is required.

10.37 H_2O, 18.02 amu

$$7.55 \text{ g } H_2O \times \frac{1 \text{ mol } H_2O}{18.02 \text{ g } H_2O} = 0.4190 \text{ mol } H_2O$$

q_1 = (0.4190 mol)[75.3 x 10^{-3} kJ/(K · mol)](–33.5 K) = –1.057 kJ

q_2 = –(0.4190 mol)(6.01 kJ/mol) = –2.518 kJ

q_3 = (0.4190 mol)[75.3 x 10^{-3} kJ/(K · mol)](–10.0 K) = –0.3155 kJ

$q_{total} = q_1 + q_2 + q_3$ = –3.89 kJ

3.89 kJ of heat is released

10.38 H_2O, 18.02 amu

$$15.3 \text{ g } H_2O \times \frac{1 \text{ mol } H_2O}{18.02 \text{ g } H_2O} = 0.8491 \text{ mol } H_2O$$

q_1 = (0.8491 mol)[75.3 x 10^{-3} kJ/(K · mol)](373K – 388K) = –0.9591 kJ

q_2 = –(0.8491 mol)(40.67 kJ/mol) = –34.53 kJ

q_3 = (0.8491 mol)[75.3 x 10^{-3} kJ/(K · mol)](348 K – 373 K) = –1.598 kJ

$q_{total} = q_1 + q_2 + q_3$ = –37.1 kJ

37.1 kJ of heat is released.

10.39

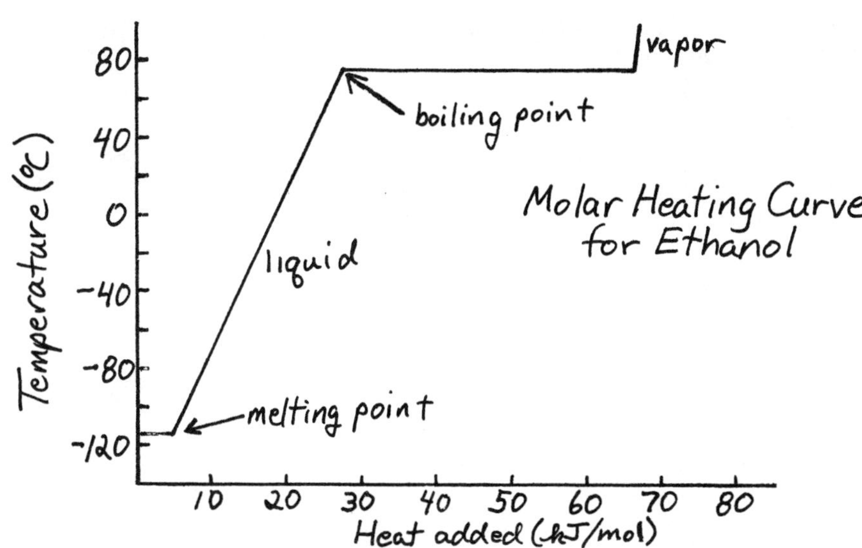

10.40 (a) Air is released into a larger volume; entropy (disorder) increases; ΔS is positive

(b) The process involves solid → gas; entropy (disorder) increases; ΔS is positive.

(c) In the chemical reaction, the number of moles of liquid decreases. The entropy (disorder) decreases; ΔS is negative.

10.41 $\Delta G = \Delta H - T\Delta S$; at the boiling point (phase change), $\Delta G = 0$.

$$\Delta H = T\Delta S;\ \ T = \frac{\Delta H_{vap}}{\Delta S_{vap}} = \frac{30.8\ \text{kJ/mol}}{87.2 \times 10^{-3}\ \text{kJ/(K·mol)}} = 353.2\ \text{K} = 80.1^{\circ}\text{C}$$

10.42 boiling point = 218°C = 491 K

$\Delta G = \Delta H_{vap} - T\Delta S_{vap}$

At the boiling point (phase change), $\Delta G = 0$

$\Delta H_{vap} = T\Delta S_{vap}$

$$\Delta S_{vap} = \frac{\Delta H_{vap}}{T} = \frac{43.3 \text{ kJ/mol}}{491 \text{ K}} = 0.0882 \text{ kJ/(K}\cdot\text{mol)}$$

$$\Delta S_{vap} = 88.2 \text{ J/(K}\cdot\text{mol)}$$

10.43 $$\Delta S_{fus} = \frac{\Delta H_{fus}}{T} = \frac{2.64 \text{ kJ/mol}}{371 \text{ K}} = 0.007\ 12 \text{ kJ/(K}\cdot\text{mol)}$$

$$\Delta S_{fus} = 7.12 \text{ J/(K}\cdot\text{mol)}$$

10.44 $$\Delta H_{vap} = \frac{(\log P_2 - \log P_1)(2.303)(R)}{\left(\frac{1}{T_1} - \frac{1}{T_2}\right)}$$

$T_1 = -5.1°C = 268.0$ K; $P_1 = 100$ mm Hg

$T_2 = 46.5°C = 319.6$ K; $P_2 = 760$ mm Hg

$$\Delta H_{vap} = \frac{[\log(760) - \log(100)](2.303)(8.3145 \times 10^{-3} \text{ kJ/K}\cdot\text{mol})}{\left(\frac{1}{268.0 \text{ K}} - \frac{1}{319.6 \text{ K}}\right)} = 28.0 \text{ kJ/mol}$$

10.45 $$\log P_2 = \log P_1 + \frac{\Delta H_{vap}}{2.303 R}\left(\frac{1}{T_1} - \frac{1}{T_2}\right)$$

$\Delta H_{vap} = 28.0$ kJ/mol

$P_1 = 100$ mm Hg; $T_1 = -5.1°C = 268.0$ K; $T_2 = 20.0°C = 293.2$ K

Solve for P_2.

$$\log P_2 = \log(100) + \frac{28.0 \text{ kJ/mol}}{(2.303)(8.3145 \times 10^{-3} \text{ kJ/K}\cdot\text{mol})}\left(\frac{1}{268.0 \text{ K}} - \frac{1}{293.2 \text{ K}}\right)$$

$\log P_2 = 2.46895$; $P_2 = 294.4$ mm Hg $= 294$ mm Hg

10.46 $$\Delta H_{vap} = \frac{(\log P_2 - \log P_1)(2.303)(R)}{\left(\frac{1}{T_1} - \frac{1}{T_2}\right)}$$

P_1 = 100 mm Hg; T_1 = 5.4°C = 278.6 K

P_2 = 760 mm Hg; T_2 = 56.8°C = 330.0 K

$$\Delta H_{vap} = \frac{[\log(760) - \log(100)](2.303)(8.3145 \times 10^{-3}\ kJ/K\cdot mol)}{\left(\frac{1}{278.6\ K} - \frac{1}{330.0\ K}\right)} = 30.2\ kJ/mol$$

10.47 $$\log P_2 = \log P_1 + \frac{\Delta H_{vap}}{2.303\ R}\left(\frac{1}{T_1} - \frac{1}{T_2}\right)$$

ΔH_{vap} = 30.2 kJ/mol

P_1 = 100 mm Hg; T_1 = 5.4°C = 278.6 K; T_2 = 30.0°C = 303.2 K

Solve for P_2.

$$\log P_2 = \log (100) + \frac{30.2\ kJ/mol}{(2.303)(8.3145 \times 10^{-3}\ kJ/K\cdot mol)}\left(\frac{1}{278.6\ K} - \frac{1}{303.2\ K}\right)$$

$\log P_2$ = 2.4593; P_2 = 287.9 mm Hg = 288 mm Hg

10.48

T(K)	P_{vap}(mm Hg)	$\log P_{vap}$	1/T
263	80.1	1.904	0.003 802
273	133.6	2.1258	0.003 663
283	213.3	2.3290	0.003 534
293	329.6	2.5180	0.003 413
303	495.4	2.6950	0.003 300
313	724.4	2.8600	0.003 195

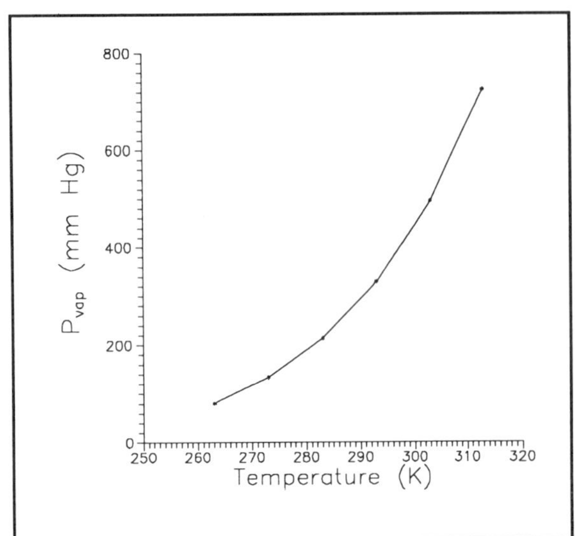

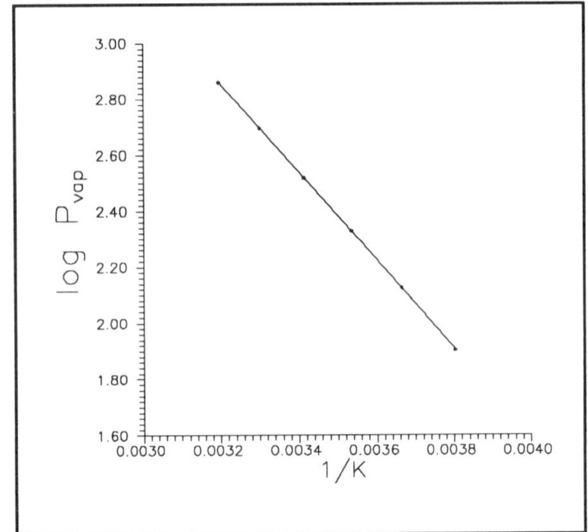

$$\log P_{vap} = \left(-\frac{\Delta H_{vap}}{2.303\ R}\right)\frac{1}{T} + C$$

$$\text{slope} = -1573/\text{K} = -\frac{\Delta H_{vap}}{2.303\ R}$$

$$\Delta H_{vap} = (1573/\text{K})(2.303)(R)$$

$$\Delta H_{vap} = (1573/\text{K})(2.303)[8.3145 \times 10^{-3}\ \text{kJ/(K}\cdot\text{mol)}]$$

$$\Delta H_{vap} = 30.1\ \text{kJ/mol}; \qquad C = 7.89$$

10.49 $\Delta H_{vap} = 30.1$ kJ/mol, $C = 7.89$

10.50 $$\Delta H_{vap} = \frac{(\log P_2 - \log P_1)(2.303)(R)}{\left(\frac{1}{T_1} - \frac{1}{T_2}\right)}$$

$P_1 = 80.1$ mm Hg; $\qquad T_1 = 263$ K

$P_2 = 724.4$ mm Hg; $\qquad T_2 = 313$ K

$$\Delta H_{vap} = \frac{[\log(724.4) - \log(80.1)](2.303)(8.3145 \times 10^{-3}\ \text{kJ/K}\cdot\text{mol})}{\left(\frac{1}{263\ \text{K}} - \frac{1}{313\ \text{K}}\right)} = 30.1\ \text{kJ/mol}$$

The calculated ΔH_{vap} and that obtained from the plot in Problem 10.49 are the same

Structures of Solids

10.51 molecular solid, CO_2, I_2

metallic solid, any metallic element

covalent network solid, diamond

ionic solid, NaCl

10.52 molecular solid, covalent molecules

metallic solid, metal atoms

covalent network solid, nonmetal atoms

ionic solid, cations and anions

10.53 The unit cell is the smallest repeating unit in a crystal.

10.54 From Table 10.9.

Hexagonal and cubic closest packing are the most efficient because 74% of the available space is used.

Simple cubic packing is the least efficient because only 52% of the available space is used.

10.55 Cu is face–centered cubic. d = 362 pm

$$r = \sqrt{\frac{d^2}{8}} = \sqrt{\frac{(362\text{ pm})^2}{8}} = 128\text{ pm}$$

10.56 $d = 362\text{ pm} = 362 \times 10^{-12}\text{ m} = 3.62 \times 10^{-8}\text{ cm}$

unit cell volume = $(3.62 \times 10^{-8}\text{ cm})^3 = 4.74 \times 10^{-23}\text{ cm}^3$

mass of one Cu atom = $63.55\text{ g/mol} \times \dfrac{1\text{ mol}}{6.022 \times 10^{23}\text{ atom}}$

$= 1.055 \times 10^{-22}$ g/atom

Cu is face–centered cubic; there are therefore four Cu atoms in the unit cell.

unit cell mass = $(4\text{ atoms})(1.055 \times 10^{-22}\text{ g/atom}) = 4.22 \times 10^{-22}\text{ g}$

$$\text{density} = \frac{\text{mass}}{\text{volume}} = \frac{4.22 \times 10^{-22}\text{g}}{4.74 \times 10^{-23}\text{ cm}^3} = 8.90\text{ g/cm}^3$$

$$\text{molar volume} = \frac{63.55\text{ g/mol}}{8.90\text{ g/cm}^3} = 7.14\text{ cm}^3\text{/mol}$$

10.57 Pb is face–centered cubic. $d = 495\text{ pm} = 4.95 \times 10^{-8}\text{ cm}$

$$r = \sqrt{\frac{d^2}{8}} = \sqrt{\frac{(495\text{ pm})^2}{8}} = 175\text{ pm}$$

unit cell volume = $(4.95 \times 10^{-8}\text{ cm})^3 = 1.2129 \times 10^{-22}\text{ cm}^3$

mass of one Pb atom = $207.2\text{ g/mol} \times \dfrac{1\text{ mol}}{6.022 \times 10^{23}\text{ atoms}}$

$= 3.4407 \times 10^{-22}$ g/atom

Pb is face–centered cubic; there are therefore four Pb atoms in the unit cell.

$$\text{density} = \frac{\text{mass}}{\text{volume}} = \frac{4(3.4407 \times 10^{-22}\text{ g})}{1.2129 \times 10^{-22}\text{ cm}^3} = 11.3\text{ g/cm}^3$$

10.58 mass of one Al atom = $26.98\text{ g/mol} \times \dfrac{1\text{ mol}}{6.022 \times 10^{23}\text{ atom}}$

$= 4.480 \times 10^{-23}$ g/atom

Al is face–centered cubic; there are therefore four Al atoms in the unit cell.

unit cell mass = (4 atoms)(4.480×10^{-23} g/atom) = 1.792×10^{-22} g

$$\text{density} = \frac{\text{mass}}{\text{volume}}$$

$$\text{unit cell volume} = \frac{\text{unit cell mass}}{\text{density}} = \frac{1.792 \times 10^{-22}\ \text{g}}{2.699\ \text{g/cm}^3} = 6.640 \times 10^{-23}\ \text{cm}^3$$

$$\text{unit cell edge} = d = \sqrt[3]{6.640 \times 10^{-23}\ \text{cm}^3} = 4.049 \times 10^{-8}\ \text{cm}$$

$$d = 4.049 \times 10^{-8}\ \text{cm} \times \frac{1\ \text{m}}{100\ \text{cm}} = 4.049 \times 10^{-10}\ \text{m}$$

$$d = 404.9 \times 10^{-12}\ \text{m} = 404.9\ \text{pm}$$

10.59 W is body–centered cubic. d = 317 pm

a = edge = d; b = face diagonal; c = body diagonal

$$b^2 = 2a^2$$

$$c^2 = a^2 + b^2$$

$$c^2 = a^2 + 2a^2 = 3a^2$$

$$c = \sqrt{3}\,a$$

$$\text{unit cell body diagonal} = \sqrt{3}\,d = \sqrt{3}\,(317\ \text{pm}) = 549\ \text{pm}$$

10.60 unit cell body diagonal = 4r = 549 pm

$$\text{For W, } r = \frac{549\ \text{pm}}{4} = 137\ \text{pm}$$

10.61 mass of one Na atom = $23.0\ \text{g/mol} \times \dfrac{1\ \text{mol}}{6.022 \times 10^{23}\ \text{atoms}} = 3.82 \times 10^{-23}$ g/atom

Because Na is body–centered cubic; there are two Na atoms in the unit cell.

unit cell mass = 2(3.82×10^{-23} g) = 7.64×10^{-23} g

$$\text{unit cell volume} = \frac{\text{unit cell mass}}{\text{density}} = \frac{7.64 \times 10^{-23}\ \text{g}}{0.971\ \text{g/cm}^3} = 7.87 \times 10^{-23}\ \text{cm}^3$$

$$\text{unit cell edge} = d = \sqrt[3]{7.87 \times 10^{-23}\ \text{cm}^3} = 4.29 \times 10^{-8}\ \text{cm} = 429\ \text{pm}$$

$$4R = \sqrt{3}\,d;\quad R = \frac{\sqrt{3}\,d}{4} = \frac{\sqrt{3}\,(429\ \text{pm})}{4} = 186\ \text{pm}$$

10.62 mass of one Ti atom = $47.88\ \text{g/mol} \times \dfrac{1\ \text{mol}}{6.022 \times 10^{23}\ \text{atoms}}$

$= 7.951 \times 10^{-23}\ \text{g/atom}$

$r = 144.8\ \text{pm} = 144.8 \times 10^{-12}\ \text{m}$

$$r = 144.8 \times 10^{-12}\ \text{m} \times \frac{100\ \text{cm}}{1\ \text{m}} = 1.448 \times 10^{-8}\ \text{cm}$$

Calculate the volume and then the density for Ti assuming it is simple cubic, body-centered cubic, and face-centered cubic. Compare the calculated density with the actual density to identify the unit cell.

For simple cubic:

$$d = 2r;\ \text{volume} = d^3 = [2(1.448 \times 10^{-8}\ \text{cm})]^3 = 2.429 \times 10^{-23}\ \text{cm}^3$$

$$\text{density} = \frac{\text{unit cell mass}}{\text{volume}} = \frac{7.951 \times 10^{-23}\ \text{g}}{2.429 \times 10^{-23}\ \text{cm}^3} = 3.273\ \text{g/cm}^3$$

For face-centered cubic:

$$d = 2\sqrt{2}\,r;\ \text{volume} = d^3 = [2\sqrt{2}(1.448 \times 10^{-8}\ \text{cm})]^3 = 6.870 \times 10^{-23}\ \text{cm}^3$$

$$\text{density} = \frac{4(7.951 \times 10^{-23}\ \text{g})}{6.870 \times 10^{-23}\ \text{cm}^3} = 4.630\ \text{g/cm}^3$$

For body-centered cubic:

From Problems 10.59 and 10.60,

$$d = \frac{4r}{\sqrt{3}}; \text{ volume} = d^3 = \left[\frac{4(1.448 \times 10^{-8}\text{ cm})}{\sqrt{3}}\right]^3 = 3.739 \times 10^{-23}\text{ cm}^3$$

$$\text{density} = \frac{2(7.951 \times 10^{-23}\text{ g})}{3.739 \times 10^{-23}\text{ cm}^3} = 4.253\text{ g/cm}^3$$

The calculated density for a face–centered cube (4.630 g/cm^3) is closest to the actual density of 4.54 g/cm^3. Ti crystallizes in the face–centered cubic unit cell.

10.63 mass of one Ca = 40.08 g/mol x $\frac{1\text{ mol}}{6.022 \times 10^{23}\text{ atom}}$ = 6.656 x 10^{-23} g/atom

unit cell edge = d = 558.2 pm = 5.582 x 10^{-8} cm

unit cell volume = d^3 = (5.582 x 10^{-8} cm)^3 = 1.739 x 10^{-22} cm^3

unit cell mass = (1.739 x 10^{-22} cm^3)(1.55 g/cm^3) = 2.695 x 10^{-22} g

(a) number of Ca atoms in unit cell = $\frac{\text{unit cell mass}}{\text{mass of one Ca atom}}$

$$= \frac{2.695 \times 10^{-22}\text{ g}}{6.656 \times 10^{-23}\text{ g/atom}} = 4.05 = 4\text{ Ca atoms}$$

(b) Because the unit cell contains 4 Ca atoms, the unit cell is face–centered cubic.

10.64 Six Na^+ ions touch each H^- ion and six H^- ions touch each Na^+ ion.

10.65 Na^+ H^- Na^+

← 488 pm →

unit cell edge = d = 488 pm; Na–H bond = d/2 = 244 pm

10.66 For CsCl:

(1/8 x 8 corners), so 1 Cl^- and 1 minus per unit cell

1 Cs^+ inside, so 1 plus per unit cell

10.67 See problem 10.59

body diagonal = $\sqrt{3}\,d = \sqrt{3}\,(412.3\ \text{pm}) = 714.12\ \text{pm}$

Cs–Cl bond = body diagonal/2 = (714.12 pm)/2 = 357.1 pm

10.68 From Problem 10.67, Cs–Cl bond length = 357.1 pm.

Cs–Cl bond length = $r_{Cs^+} + r_{Cl^-}$

$357.1\ \text{pm} = r_{Cs^+} + r_{Cl^-}$

$357.1\ \text{pm} = r_{Cs^+} + 181\ \text{pm}$

$r_{Cs^+} = 357.1\ \text{pm} - 181\ \text{pm} = 176\ \text{pm}$

10.69

```
           \   /
            N—B
    \      /   \     /
     N—B       N—B
    /   \     /   \
 —B      N—B       N—
    \   /   \     /
     N—B       N—B
    /   \     /   \
 —B      N—B       N—
    \   /   \     /
     N—B       N—B
    /   \     /   \
            N—B
           /   \
```

Phase Diagrams

10.70 Pressure on the y–axis and temperature on the x–axis.

10.71 (a) gas (b) liquid (c) solid

10.72 (a) $H_2O(l) \rightarrow H_2O(s)$

(b) 380°C is above the critical temperature; therefore, the water cannot be liquefied. At the higher pressure, it will behave as a supercritical fluid.

10.73

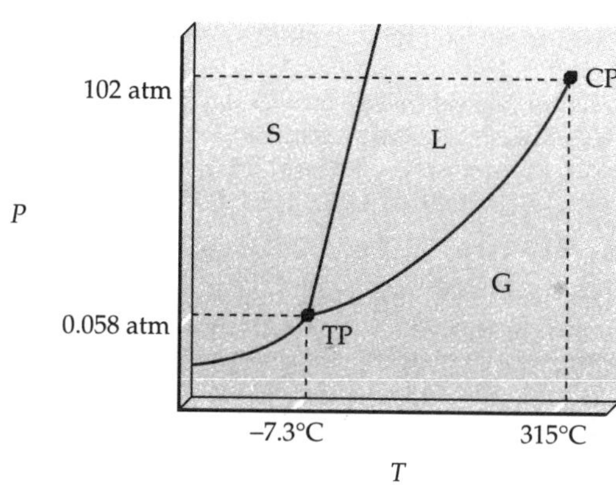

10.74 (a) $Br_2(s)$ (b) $Br_2(l)$

10.75

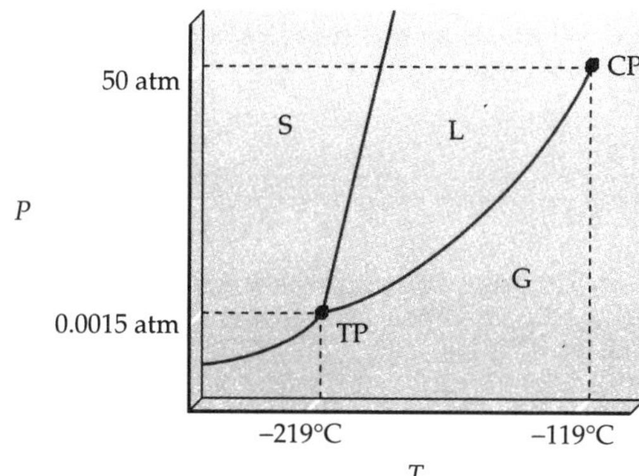

10.76 Solid O_2 does not melt when pressure is applied because the solid is denser than the liquid and the solid/liquid boundary in the phase diagram slopes to the right.

10.77

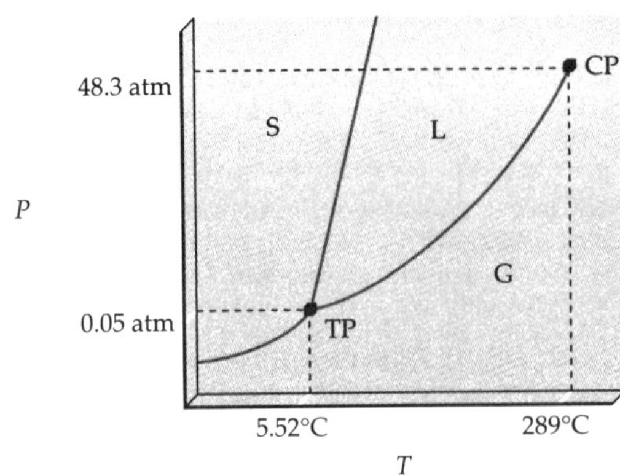

10.78

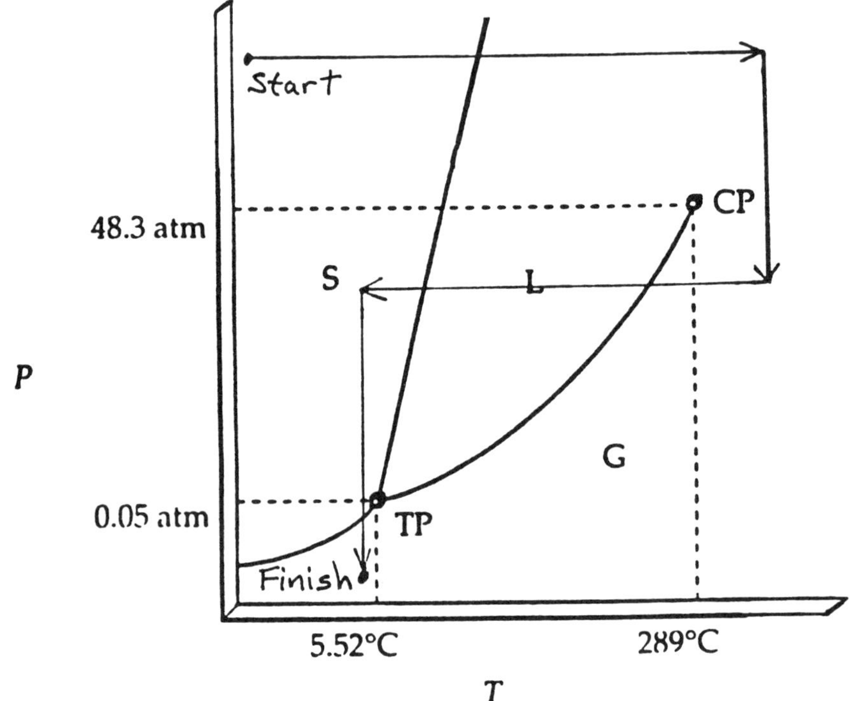

The starting phase is benzene as a solid, and the final phase is benzene as a gas.

10.79 solid → liquid → supercritical fluid → liquid → solid → gas

10.80 Ammonia can be liquefied at 25°C because this temperature is below T_c (132.5°C).

Methane cannot be liquefied at 25°C because this temperature is above T_c (–82.1°C).

Sulfur dioxide can be liquefied at 25°C because this temperature is below T_c (157.8°C).

General Problems

10.81 (a) C_8H_{18}, dispersion

(b) $C_2H_5NH_2$, dipole–dipole, hydrogen bonding, dispersion

(c) $C_6H_6(OH)_6$, dipole–dipole, hydrogen bonding, dispersion

(d) aqueous NaOH, ion–dipole, hydrogen bonding, dispersion

10.82 $7.50 \text{ g} \times \dfrac{1 \text{ mol}}{200.6 \text{ g}} = 0.037\ 39 \text{ mol Hg}$

$q_1 = (0.037\ 39 \text{ mol})[28.3 \times 10^{-3} \text{ kJ/(K} \cdot \text{mol)}](234.2 \text{ K} - 223.2 \text{ K}) = 0.011\ 64 \text{ kJ}$

$q_2 = (0.037\ 39 \text{ mol})(2.33 \text{ kJ/mol}) = 0.087\ 12 \text{ kJ}$

$q_3 = (0.037\ 39 \text{ mol})[28.3 \times 10^{-3} \text{ kJ/(K} \cdot \text{mol)}](323.2 \text{ K} - 234.2 \text{ K}) = 0.094\ 17 \text{ kJ}$

$q_{total} = q_1 + q_2 + q_3 = 0.193 \text{ kJ}$

0.193 kJ of heat is required.

10.83

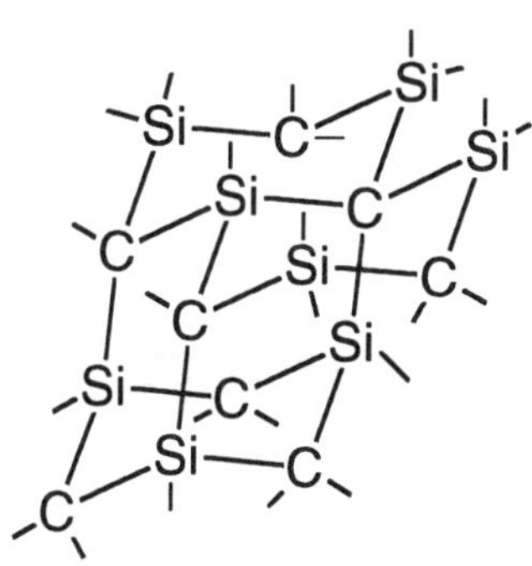

10.84 $\log P_2 = \log P_1 + \frac{\Delta H_{vap}}{2.303\,R}\left(\frac{1}{T_1} - \frac{1}{T_2}\right)$

ΔH_{vap} = 40.67 kJ/mol

At 1 atm, H_2O boils at 100°C; therefore set

T_1 = 100°C = 373 K, and P_1 = 1.00 atm.

Let T_2 = 95°C = 368 K, and solve for P_2.

(P_2 is the atmospheric pressure in Denver.)

$$\log P_2 = \log(1) + \frac{40.67\text{ kJ/mol}}{(2.303)(8.3145 \times 10^{-3}\text{ kJ/K}\cdot\text{mol})}\left(\frac{1}{373\text{ K}} - \frac{1}{368\text{ K}}\right)$$

$\log P_2 = -0.077\,37$; $\quad P_2$ = 0.837 atm

10.85

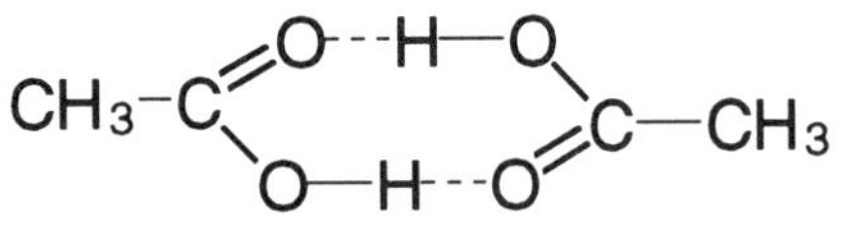

$CH_3-C(=O\cdots H-O)(-O-H\cdots O=)C-CH_3$

10.86 C_2H_5OH, 46.07 amu

$15.0\text{ g} \times \frac{1\text{ mol}}{46.07\text{ g}} = 0.326\text{ mol } C_2H_5OH$

q_1 = (0.326 mol)[0.0657 kJ/(K · mol)](351.6 K – 363.2 K) = –0.2485 kJ

q_2 = –(0.326 mol)(38.6 kJ/mol) = –12.58 kJ

q_3 = (0.326 mol)[0.113 kJ/(K · mol)](298.2 K – 351.6 K) = –1.967 kJ

$q_{total} = q_1 + q_2 + q_3$ = –14.8 kJ

14.8 kJ of heat is released.

10.87 $\Delta G = \Delta H - T\Delta S$; at the melting point (phase change), $\Delta G = 0$.

$$\Delta H = T\Delta S;\ \ T = \frac{\Delta H_{vap}}{\Delta S_{vap}} = \frac{9.037\ \text{kJ/mol}}{9.79 \times 10^{-3}\ \text{kJ/(K}\cdot\text{mol)}} = 923\ \text{K} = 650^\circ\text{C}$$

10.88 melting point = –23.2°C = 250.0 K

$$\Delta G = \Delta H_{fusion} - T\Delta S_{fusion}$$

At the melting point (phase change), $\Delta G = 0$

$$\Delta H_{fusion} = T\Delta S_{fusion}$$

$$\Delta S_{fusion} = \frac{\Delta H_{fusion}}{T} = \frac{9.37\ \text{kJ/mol}}{250.0\ \text{K}} = 0.0375\ \text{kJ/(K}\cdot\text{mol)}$$

$$\Delta S_{fusion} = 37.5\ \text{J/(K}\cdot\text{mol)}$$

10.89 $$\Delta H_{vap} = \frac{(\log P_2 - \log P_1)(2.303\ R)}{\left(\frac{1}{T_1} - \frac{1}{T_2}\right)}$$

$P_1 = 40.0$ mm Hg; $T_1 = 191.6$ K

$P_2 = 400$ mm Hg; $T_2 = 229.2$ K

$$\Delta H_{vap} = \frac{[\log(400) - \log(40.0)]\left[(2.303)\left(8.3145 \times 10^{-3}\frac{\text{kJ}}{\text{K}\cdot\text{mol}}\right)\right]}{\left(\frac{1}{191.6\ \text{K}} - \frac{1}{229.2\ \text{K}}\right)} = 22.36\ \text{kJ/mol}$$

Using $\Delta H_{vap} = 22.36$ kJ/mol

$$\log P_2 = \log P_1 + \frac{\Delta H_{vap}}{2.303\ R}\left(\frac{1}{T_1} - \frac{1}{T_2}\right)$$

$$(\log P_2 - \log P_1)\left(\frac{2.303\,R}{\Delta H_{vap}}\right) = \frac{1}{T_1} - \frac{1}{T_2}$$

$$\frac{1}{T_1} - (\log P_2 - \log P_1)\left(\frac{2.303\,R}{\Delta H_{vap}}\right) = \frac{1}{T_2}$$

P_1 = 40.0 mm Hg; T_1 = 191.6 K

P_2 = 760 mm Hg

Solve for T_2 (the normal boiling point).

$$\frac{1}{194.6\text{ K}} - [\log(760) - \log(40.0)]\left(\frac{2.303\left(8.3145 \times 10^{-3}\frac{\text{kJ}}{\text{K}\cdot\text{mol}}\right)}{22.36\text{ kJ/mol}}\right) = \frac{1}{T_2}$$

$\frac{1}{T_2}$ = 0.004 124 33; T_2 = 242.46 K = –30.7°C

10.90 (a) $$\log P_2 = \log P_1 + \frac{\Delta H_{vap}}{2.303\,R}\left(\frac{1}{T_1} - \frac{1}{T_2}\right)$$

$$(\log P_2 - \log P_1)\left(\frac{2.303\,R}{\Delta H_{vap}}\right) = \frac{1}{T_1} - \frac{1}{T_2}$$

$$\frac{1}{T_1} - (\log P_2 - \log P_1)\left(\frac{2.303\,R}{\Delta H_{vap}}\right) = \frac{1}{T_2}$$

P_1 = 100.0 mm Hg; T_1 = –23°C = 250 K

P_2 = 760.0 mm Hg

Solve for T_2, the normal boiling point for CCl_3F.

$$\frac{1}{250\text{ K}} - [\log(760.0) - \log(100.0)]\left(\frac{2.303\left(8.3145 \times 10^{-3}\frac{\text{kJ}}{\text{K}\cdot\text{mol}}\right)}{24.77\text{ kJ/mol}}\right) = \frac{1}{T_2}$$

$$\frac{1}{T_2} = 0.003\ 319; \qquad T_2 = 301.3\ \text{K} = 28.1°\text{C}$$

(b) $$\Delta S_{vap} = \frac{\Delta H_{vap}}{T} = \frac{24.77\ \text{kJ/mol}}{301.3\ \text{K}} = 0.082\ 21\ \text{kJ/(K} \cdot \text{mol)}$$

$$\Delta S_{vap} = 82.2\ \text{J/(K} \cdot \text{mol)}$$

10.91 $$\Delta H_{vap} = \frac{(\log P_2 - \log P_1)(2.303\ R)}{\left(\frac{1}{T_1} - \frac{1}{T_2}\right)}$$

$P_1 = 100$ mm Hg; $T_1 = 162.85$ K

$P_2 = 760$ mm Hg; $T_2 = 184.65$ K

$$\Delta H_{vap} = \frac{(\log(760) - \log(100))\left[(2.303)\left(8.3145 \times 10^{-3} \frac{\text{kJ}}{\text{K}\cdot\text{mol}}\right)\right]}{\left(\frac{1}{162.85\ \text{K}} - \frac{1}{184.65\ \text{K}}\right)} = 23.3\ \text{kJ/mol}$$

10.92 $$\log P_2 = \log P_1 + \frac{\Delta H_{vap}}{2.303\ R}\left(\frac{1}{T_1} - \frac{1}{T_2}\right)$$

$$(\log P_2 - \log P_1)\left(\frac{2.303\ R}{\Delta H_{vap}}\right) = \frac{1}{T_1} - \frac{1}{T_2}$$

$$\frac{1}{T_1} - (\log P_2 - \log P_1)\left(\frac{2.303\ R}{\Delta H_{vap}}\right) = \frac{1}{T_2}$$

$P_1 = 760$ mm Hg; $T_1 = 56.2°\text{C} = 329.4$ K

$P_2 = 105$ mm Hg

Solve for T_2.

$$\frac{1}{329.4\ K} - [\log(105) - \log(760)]\left(\frac{2.303\left(8.3145 \times 10^{-3}\frac{kJ}{K \cdot mol}\right)}{29.1.77\ kJ/mol}\right) = \frac{1}{T_2}$$

$$\frac{1}{T_2} = 0.003\ 601; \qquad T_2 = 277.7\ K = 4.5°C$$

10.93

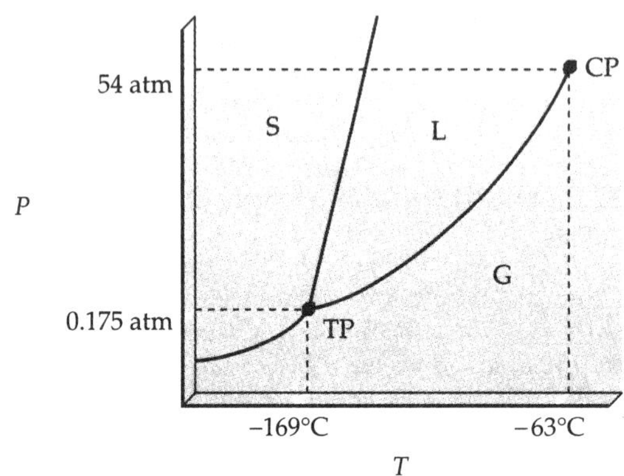

Kr cannot be liquified at room temperature because room temperature is above T_c (–63°C).

10.94 (a) Kr(l) (b) supercritical Kr

10.95 For a body–centered cube

$4r = \sqrt{3}$ edge; $\qquad$ edge $= \dfrac{4r}{\sqrt{3}}$

volume of sphere $= \dfrac{4}{3}\pi r^3$

volume of unit cell $= \left(\dfrac{4r}{\sqrt{3}}\right)^3 = \dfrac{64\ r^3}{3\sqrt{3}}$

volume of 2 spheres $= 2\left(\dfrac{4}{3}\pi r^3\right) = \dfrac{8}{3}\pi r^3$

$$\% \text{ volume occupied} = \frac{\left(\frac{8}{3}\pi r^3\right)}{\left(\frac{64 r^3}{3\sqrt{3}}\right)} \times 100\% = 68\%$$

10.96 From Problem 10.59, $4r = \sqrt{3}d$

$$r = \frac{\sqrt{3}\,d}{4} = \frac{\sqrt{3}(287 \text{ pm})}{4} = 124 \text{ pm}$$

10.97 unit cell edge = d = 287 pm = 287×10^{-12} m = 2.87×10^{-8} cm

unit cell volume = $d^3 = (2.87 \times 10^{-8} \text{ cm})^3 = 2.364 \times 10^{-23} \text{ cm}^3$

unit cell mass = $(2.364 \times 10^{-23} \text{ cm}^3)(7.86 \text{ g/cm}^3) = 1.858 \times 10^{-22}$ g

Fe is body-centered cubic; therefore there are two Fe atoms per unit cell.

$$\text{mass of one Fe atom} = \frac{1.858 \times 10^{-22} \text{ g}}{2 \text{ Fe atoms}} = 9.290 \times 10^{-23} \text{ g/atom}$$

$$\text{Avogadro's number} = 55.85 \text{ g/mol} \times \frac{1 \text{ atom}}{9.290 \times 10^{-23} \text{ g}} = 6.01 \times 10^{23} \text{ atoms/mol}$$

10.98 unit cell edge = d = 408 pm = 408×10^{-12} m = 4.08×10^{-8} cm

unit cell volume = $(4.08 \times 10^{-8} \text{ cm})^3 = 6.792 \times 10^{-23} \text{ cm}^3$

unit cell mass = $(10.50 \text{ g/cm}^3)(6.792 \times 10^{-23} \text{ cm}^3) = 7.132 \times 10^{-22}$ g

Ag is face-centered cubic; therefore there are four Ag atoms in the unit cell.

$$\text{mass of one Ag atom} = \frac{7.132 \times 10^{-22} \text{g}}{4 \text{ Ag atoms}} = 1.783 \times 10^{-22} \text{ g/atom}$$

$$\text{Avogadro's number} = 107.9 \text{ g/mol} \times \frac{1 \text{ atom}}{1.783 \times 10^{-22} \text{ g}} = 6.05 \times 10^{23} \text{ atoms/mol}$$

10.99 unit cell edge = $2r_{Cl^-} + 2r_{Na^+}$ = 2(181 pm) + 2(97 pm) = 556 pm

10.100 unit cell edge = d = 556 pm = 556×10^{-12} m = 5.56×10^{-8} cm

unit cell volume = $(5.56 \times 10^{-8}\ \text{cm})^3 = 1.719 \times 10^{-22}\ \text{cm}^3$

The unit cell contains 4 Na^+ ions and 4 Cl^- ions.

$$\text{mass of one Na}^+ \text{ ion} = 22.99\ \text{g/mol} \times \frac{1\ \text{mol}}{6.022 \times 10^{23}\ \text{ions}} = 3.818 \times 10^{-23}\ \text{g/Na}^+$$

$$\text{mass of one Cl}^- \text{ ion} = 35.45\ \text{g/mol} \times \frac{1\ \text{mol}}{6.022 \times 10^{23}\ \text{ions}} = 5.887 \times 10^{-23}\ \text{g/Cl}^-$$

unit cell mass = $4(3.818 \times 10^{-23}\ \text{g}) + 4(5.887 \times 10^{-23}\ \text{g}) = 3.882 \times 10^{-22}$ g

$$\text{density} = \frac{\text{unit cell mass}}{\text{unit cell volume}} = \frac{3.882 \times 10^{-22}\ \text{g}}{1.719 \times 10^{-22}\ \text{cm}^3} = 2.26\ \text{g/cm}^3$$

11.1 Toluene is nonpolar and is insoluble in water.

Br_2 is nonpolar but because of its size is polarizable and is soluble in water.

KBr is an ionic compound and is very soluble in water.

toluene < Br_2 < KBr (solubility in H_2O)

11.2 Na^+ has the larger (more negative) hydration energy because the Na^+ ion is smaller than the Cs^+ ion and water molecules can approach more closely and bind more tightly to the Na^+ ion.

Ba^{2+} has the larger (more negative) hydration energy because of its higher charge.

11.3 NaCl, 58.44 amu; 1.00 mol NaCl = 58.44 g

1.00 L H_2O = 1000 mL = 1000 g (assuming a density of 1.00 g/mL)

$$\text{weight \% NaCl} = \frac{58.44 \text{ g}}{1000 \text{ g} + 58.44 \text{ g}} \times 100\% = 5.52 \text{ wt \%}$$

11.4 $$\text{ppm} = \frac{\text{mass of } CO_2}{\text{total mass of solution}} \times 10^6 \text{ ppm}$$

total mass of solution = density x volume = (1.3 g/L)(1.0 L) = 1.3 g

$$35 \text{ ppm} = \frac{\text{mass of } CO_2}{1.3 \text{ g}} \times 10^6 \text{ ppm}$$

$$\text{mass of } CO_2 = \frac{(35 \text{ ppm})(1.3 \text{ g})}{10^6 \text{ ppm}} = 4.6 \times 10^{-5} \text{ g } CO_2$$

11.5 Assume 1.00 L of sea water.

mass of 1.00 L = (1000 mL)(1.025 g/mL) = 1025 g

$$\frac{\text{mass NaCl}}{1025 \text{ g}} \times 100\% = 3.50 \text{ wt \%}$$

$$\text{mass NaCl} = \frac{1025\text{ g} \times 3.50}{100} = 35.88\text{ g}$$

There are 35.88 g NaCl per 1.00 L of solution.

$$M = \frac{(35.88\text{ g NaCl})\left(\dfrac{1\text{ mol NaCl}}{58.44\text{ g NaCl}}\right)}{1.00\text{ L}} = 0.614\text{ M}$$

11.6 $C_{27}H_{46}O$, 386.7 amu; $CHCl_3$, 119.4 amu

$$40.0\text{ g} \times \frac{1\text{ kg}}{1000\text{ g}} = 0.0400\text{ kg}$$

$$\text{molality} = \frac{\text{mol } C_{27}H_{46}O}{\text{kg } CHCl_3} = \frac{\left(0.385\text{ g} \times \dfrac{1\text{ mol}}{386.7\text{ g}}\right)}{0.0400\text{ kg}} = 0.0249\ m$$

$$X_{C_{27}H_{46}O} = \frac{\text{mol } C_{27}H_{46}O}{\text{mol } C_{27}H_{46}O + \text{mol } CHCl_3}$$

$$X_{C_{27}H_{46}O} = \frac{\left(0.385\text{ g} \times \dfrac{1\text{ mol}}{386.7\text{ g}}\right)}{\left[\left(0.385\text{ g} \times \dfrac{1\text{ mol}}{386.7\text{ g}}\right) + \left(40.0\text{ g} \times \dfrac{1\text{ mol}}{119.4\text{ g}}\right)\right]} = 2.96 \times 10^{-3}$$

11.7 CH_3CO_2Na, 82.03 amu

$$\text{kg } H_2O = (0.150\text{ mol } CH_3CO_2Na)\left(\frac{1\text{ kg } H_2O}{0.500\text{ mol } CH_3CO_2Na}\right) = 0.300\text{ kg } H_2O$$

$$(0.150\text{ mol } CH_3CO_2Na)\left(\frac{82.03\text{ g } CH_3CO_2Na}{1\text{ mol } CH_3CO_2Na}\right) = 12.3\text{ g } CH_3CO_2Na$$

mass of solution needed = 300 g + 12.3 g = 312 g

11.8 Assume you have a solution with 1.000 kg (1000 g) of H_2O. If this solution is 0.258 *m*, then it must also contain 0.258 mol glucose.

mass of glucose = $0.258 \text{ mol} \times \frac{180.2 \text{ g}}{1 \text{ mol}} = 46.5$ g glucose

mass of solution = 1000 g + 46.5 g = 1046.5 g

density = 1.0173 g/mL

volume of solution = $1046.5 \text{ g} \times \frac{1 \text{ mL}}{1.0173 \text{ g}} = 1028.7$ mL

volume = $1028.7 \text{ mL} \times \frac{1 \text{ L}}{1000 \text{ mL}} = 1.029$ L

molarity = $\frac{0.258 \text{ mol}}{1.029 \text{ L}} = 0.251$ M

11.9 Assume 1.00 L of solution.

mass of 1.00 L = (1.0042 g/mL)(1000 mL) = 1004.2 g of solution

$0.500 \text{ mol } CH_3COOH \times \frac{60.05 \text{ g } CH_3COOH}{1 \text{ mol } CH_3COOH} = 30.02 \text{ g } CH_3COOH$

1004.2 g – 30.02 g = 974.2 g = 0.9742 kg of H_2O

molality = $\frac{0.500 \text{ mol}}{0.9742 \text{ kg}} = 0.513\ m$

11.10 Assume you have 100.0 g of seawater.

mass NaCl = (0.0350)(100.0 g) = 3.50 g NaCl

mass H_2O = 100.0 g – 3.50 g = 96.5 g H_2O

NaCl, 58.44 amu

mol NaCl = $3.50 \text{ g} \times \frac{1 \text{ mol}}{58.44 \text{ g}} = 0.0599$ mol NaCl

mass H_2O = $96.5 \text{ g} \times \frac{1 \text{ kg}}{1000 \text{ g}} = 0.0965 \text{ kg } H_2O$

$$\text{molality} = \frac{0.0599 \text{ mol}}{0.0965 \text{ kg}} = 0.621\ m$$

11.11 $M = k \cdot P;\ k = \frac{M}{P} = \frac{3.2 \times 10^{-2} \text{ M}}{1.0 \text{ atm}} = 3.2 \times 10^{-2} \text{ mol/(L} \cdot \text{atm)}$

11.12 (a) $M = k \cdot P = [3.2 \times 10^{-2} \text{ mol/(L} \cdot \text{atm)}](2.5 \text{ atm}) = 0.080 \text{ M}$

(b) $M = k \cdot P = [3.2 \times 10^{-2} \text{ mol/(L} \cdot \text{atm)}](4.0 \times 10^{-4} \text{ atm}) = 1.3 \times 10^{-5} \text{ M}$

11.13 $C_7H_6O_2$, 122.1 amu; C_2H_6O, 46.07 amu

$P_{soln} = P_{solv} \cdot X_{solv}$

$$X_{solv} = \frac{\text{mol } C_2H_6O}{\text{mol } C_2H_6O + \text{mol } C_7H_6O_2} = \frac{\left(100 \text{ g} \times \frac{1 \text{ mol}}{46.07 \text{ g}}\right)}{\left(100 \text{ g} \times \frac{1 \text{ mol}}{46.07 \text{ g}}\right) + \left(5.00 \text{ g} \times \frac{1 \text{ mol}}{122.1 \text{ g}}\right)} = 0.981$$

$P_{soln} = (100.5 \text{ mm Hg})(0.981) = 98.6 \text{ mm Hg}$

11.14 $P_{soln} = P_{solv} \cdot X_{solv}$

$$X_{solv} = \frac{P_{soln}}{P_{solv}}$$

$$X_{solv} = \frac{(55.3 - 1.30) \text{ mm Hg}}{55.3 \text{ mm Hg}} = 0.976$$

NaBr dissociates into two ions in aqueous solution.

$$X_{solv} = \frac{\text{mol } H_2O}{\text{mol } H_2O + \text{mol } Na^+ + \text{mol } Br^-}$$

$$X_{solv} = 0.976 = \frac{\left(250 \text{ g} \times \frac{1 \text{ mol}}{18.02 \text{ g}}\right)}{\left(250 \text{ g} \times \frac{1 \text{ mol}}{18.02 \text{ g}}\right) + x \text{ mol } Na^+ + x \text{ mol } Br^-}$$

$$0.976 = \frac{13.9 \text{ mol}}{13.9 \text{ mol} + 2x \text{ mol}}; \qquad \text{solve for x.}$$

x = 0.171 mol Na^+ = 0.171 mol Br^- = 0.171 mol NaBr

NaBr, 102.9 amu

$$\text{mass NaBr} = 0.171 \text{ mol} \times \frac{102.9 \text{ g}}{1 \text{ mol}} = 17.6 \text{ g NaBr}$$

11.15 C_2H_6O, 46.07 amu; H_2O, 18.02 amu

Part 1:

$$25.0 \text{ g } C_2H_6O \times \frac{1 \text{ mol } C_2H_6O}{46.07 \text{ g } C_2H_6O} = 0.5426 \text{ mol } C_2H_6O$$

$$100.0 \text{ g } H_2O \times \frac{1 \text{ mol } H_2O}{18.02 \text{ g } H_2O} = 5.549 \text{ mol } H_2O$$

$$X_{C_2H_6O} = \frac{0.5426 \text{ mol}}{0.5426 \text{ mol} + 5.549 \text{ mol}} = 0.08907$$

$$X_{H_2O} = \frac{5.549 \text{ mol}}{0.5426 \text{ mol} + 5.549 \text{ mol}} = 0.9109$$

$$P_{soln} = X_{C_2H_6O} P^{o}_{C_2H_6O} + X_{H_2O} P^{o}_{H_2O}$$

$$P_{soln} = (0.08907)(61.2 \text{ mm Hg}) + (0.9109)(23.8 \text{ mm Hg}) = 27.1 \text{ mm Hg}$$

Part 2:

$$100 \text{ g } C_2H_6O \times \frac{1 \text{ mol } C_2H_6O}{46.07 \text{ g } C_2H_6O} = 2.171 \text{ mol } C_2H_6O$$

$$25.0 \text{ g } H_2O \times \frac{1 \text{ mol } H_2O}{18.02 \text{ g } H_2O} = 1.387 \text{ mol } H_2O$$

$$X_{C_2H_6O} = \frac{2.171 \text{ mol}}{2.171 \text{ mol} + 1.387 \text{ mol}} = 0.6102$$

$$X_{H_2O} = \frac{1.387 \text{ mol}}{2.171 \text{ mol} + 1.387 \text{ mol}} = 0.3898$$

$$P_{soln} = X_{C_2H_6O} P^{o}_{C_2H_6O} + X_{H_2O} P^{o}_{H_2O}$$

$$P_{soln} = (0.6102)(61.2 \text{ mm Hg}) + (0.3898)(23.8 \text{ mm Hg}) = 46.6 \text{ mm Hg}$$

11.16 $C_9H_8O_4$, 180.2 amu

$CHCl_3$ is the solvent. For $CHCl_3$, $K_b = 3.63 \frac{^oC \cdot kg}{mol}$

$$75.00 \text{ g} \times \frac{1 \text{ kg}}{1000 \text{ g}} = 0.075\ 00 \text{ kg}$$

$$\Delta T_b = K_b \cdot m = \left(3.63 \frac{^oC \cdot kg}{mol}\right) \frac{\left(1.50 \text{ g} \times \frac{1 \text{ mol}}{180.2 \text{ g}}\right)}{(0.07500 \text{ kg})} = 0.40^oC$$

solution boiling point = 61.7°C + ΔT_b = 61.7°C + 0.40°C = 62.1°C

11.17 K_2SO_4, 174.3 amu; there are 3 ions (solute particles)/K_2SO_4

$$\Delta T_f = K_f \cdot (3 \cdot m) = \left(1.86 \frac{^oC \cdot kg}{mol}\right)(3) \frac{\left(7.40 \text{ g} \times \frac{1 \text{ mol}}{174.3 \text{ g}}\right)}{0.110 \text{ kg}} = 2.15^oC$$

Solution freezing point = 0.00°C – ΔT_f = 0.00°C – 2.15°C = –2.15°C

11.18 There are 2 ions/KBr.

$$\Delta T_f = K_f \cdot 2 \cdot m$$

freezing point = –2.95°C = 0.00°C – ΔT_f; ΔT_f = 2.95°C

$$m = \frac{\Delta T_f}{K_f \cdot 2} = \frac{2.95^oC}{\left(1.86 \frac{^oC \cdot kg}{mol}\right)(2)} = 0.793\ m$$

11.19 For $CaCl_2$ there are 3 ions (solute particles)/$CaCl_2$

$\Pi = MRT$; For $CaCl_2$, $\Pi = 3MRT$

$$\Pi = (3)(0.125 \text{ mol/L})\left(0.082\ 06 \frac{L \cdot atm}{mol \cdot K}\right)(310 \text{ K}) = 9.54 \text{ atm}$$

11.20 $\Pi = MRT$

$$M = \frac{\Pi}{RT} = \frac{(3.85 \text{ atm})}{\left(0.082\ 06 \frac{L \cdot atm}{mol \cdot K}\right)(300 \text{ K})} = 0.156 \text{ M}$$

11.21 $\Delta T_f = K_f \cdot m$

$$m = \frac{\Delta T_f}{K_f} = \frac{2.10°C}{37.7 \frac{°C \cdot kg}{mol}} = 0.0557\ m$$

$$0.0557 \frac{mol}{kg} = \frac{\left(0.250 \text{ g} \times \frac{1}{\text{molar mass}}\right)}{0.035\ 00 \text{ kg}};$$ solve for the molar mass.

molar mass = 128 g/mol

11.22 $\Pi = MRT$

$$M = \frac{\Pi}{RT} = \frac{\left(149 \text{ mm Hg} \times \frac{1 \text{ atm}}{760 \text{ mm Hg}}\right)}{\left(0.08206 \frac{L \cdot atm}{mol \cdot K}\right)(298 \text{ K})} = 8.02 \times 10^{-3} \frac{mol}{L}$$

$$\text{molarity} = 8.02 \times 10^{-3} \frac{mol}{L} = \frac{\left(0.822 \text{ g} \times \frac{1}{\text{molar mass}}\right)}{(0.300 \text{ L})}$$

Solve for the molar mass.

molar mass = 342 g/mol

11.23 (a) bp ≈ 107°C

(b) $X_{toluene}$ ≈ 0.64 and $X_{benzene}$ ≈ 0.36

Understanding Key Concepts

1. (a) < (b) < (c)

2. The surface area of a solid plays an important role in determining how rapidly a solid dissolves. The larger the surface area, the more solid–solvent interactions, and the more rapidly the solid will dissolve. Powdered NaCl has a much larger surface area than a large block of NaCl, and it will dissolve more rapidly.

3. Assume that only the blue (open) spheres (solvent) can pass through the semipermeable membrane. There will be a net transfer of solvent from the right compartment (pure solvent) to the left compartment (solution) to achieve equilibrium.

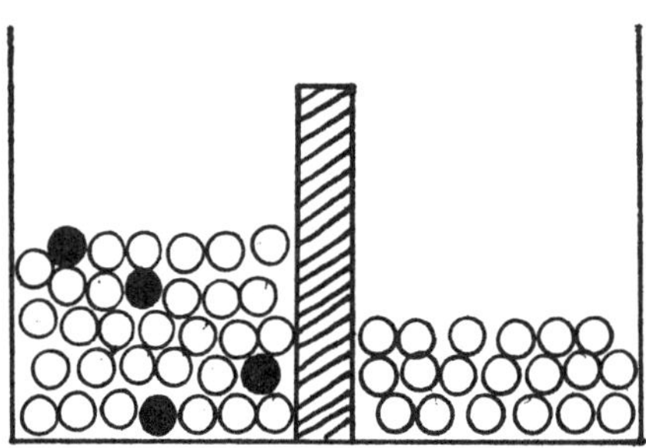

4. NaCl is a nonvolatile solute. Methyl alcohol is a volatile solute. When NaCl is added to water, the vapor pressure of the solution is decreased, which means that the boiling point of the solution will increase. When methyl alcohol is added to water, the vapor pressure of the solution is increased which means that the boiling point of the solution will decrease.

5. K_f for snow (H_2O) is 1.86 $\frac{°C \cdot kg}{mol}$. Reasonable amounts of salt are capable of lowering the freezing point (ΔT_f) of the snow below an air temperature of –2°C. Reasonable amounts of salt, however, are not capable of causing a ΔT_f of more than 30°C which would be required if it is to melt snow when the air temperature is –30°C.

6. (b) ~95°C

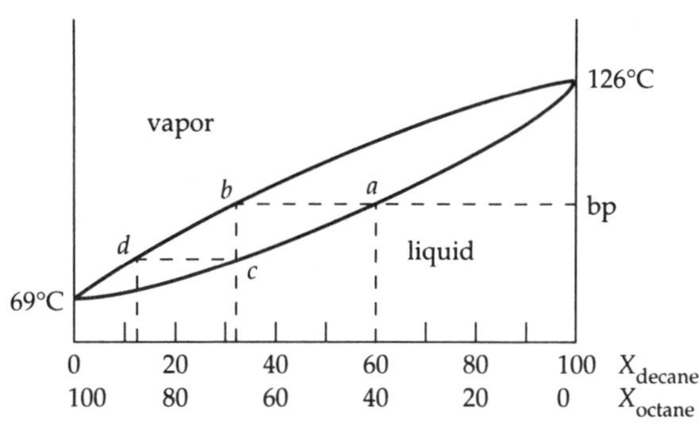

Additional Problems

Solutions and Energy Changes

11.24 Homogeneous mixtures can be classified according to the size of their constituent particles.

Colloids contain particles with diameters in the range 2 - 1000 nm.

Suspensions contain particles that are greater than about 1000 nm in diameter.

Solutions contain particles the size of a typical ion or covalent molecule, 0.2 - 2 nm in diameter.

11.25 (a) a gas in a liquid – carbonated soft drink

(b) a solid in a solid – metal alloys (14–karat gold)

(c) a liquid in a solid – dental amalgam (Hg in Ag)

11.26 Substances tend to dissolve when the solute and solvent have the same type and magnitude of intermolecular forces; thus the rule of thumb "like dissolves like."

11.27 Energy is required to overcome intermolecular forces holding solute particles together in the crystal. For an ionic solid, this is the lattice energy. Substances with higher lattice energies tend to be less soluble than substances with lower lattice energies.

11.28 SO_4^{2-} has the larger hydration energy because of its higher charge. Both SO_4^{2-} and ClO_4^- are comparable in size, so size is not a factor.

11.29 Both Br_2 and CCl_4 are nonpolar, and intermolecular forces for both are dispersion forces. H_2O is a polar molecule with dipole–dipole forces and hydrogen bonding. Therefore, Br_2 is more soluble in CCl_4.

11.30 Dissolve the solid mixture in H_2O. Only the KBr will dissolve. Filter the solution to separate the solid, undissolved cholesterol. Evaporate the H_2O to recover the KBr as a solid.

11.31 Ethyl alcohol and water are both polar with small dispersion forces. They both can hydrogen bond, and are miscible.

Pentyl alcohol is slightly polar and can hydrogen bond. It has, however, a relatively large dispersion force because of its size, which limits its water solubility.

11.32 The intermolecular forces associated with octane are dispersion forces. Both pentyl alcohol and methyl alcohol can hydrogen bond. Pentyl alcohol has relatively large dispersion forces because of its size. Methyl alcohol does not. Pentyl alcohol is soluble in octane; methyl alcohol is not.

11.33 ΔH_{soln} for HBr is exothermic. The HBr solution will be warm to touch.

ΔH_{soln} for $AgNO_3$ is endothermic. The $AgNO_3$ solution will be cool to touch.

11.34 $CaCl_2$, 110.98 amu

For a 1.00 *m* solution:

heat released = 81,300 J

mass of solution = 1000 g H_2O + 110.98 g $CaCl_2$ = 1110.98 g

$$\Delta T = \frac{q}{(\text{specific heat})(\text{mass of solution})}$$

$$\Delta T = \frac{81{,}300\ \text{J}}{[4.18\ \text{J/(K}\cdot\text{g)}](1110.98\ \text{g})} = 17.5\ \text{K} = 17.5\,^{\circ}\text{C}$$

Final temperature = 25.0°C + 17.5°C = 42.5°C

11.35 NH_4ClO_4, 117.48 amu

For a 1.00 *m* solution:

heat absorbed = 33,500 J

mass of solution = 1000 g H_2O + 117.48 g NH_4ClO_4 = 1117.48 g

$$\Delta T = \frac{q}{(\text{specific heat})(\text{mass of solution})}$$

$$\Delta T = \frac{-33{,}500\ \text{J}}{[4.18\ \text{J/(K}\cdot\text{g)}](1117.48\ \text{g})} = -7.2\ \text{K} = -7.2^\circ\text{C}$$

Final temperature = 25.0°C – 7.2°C = 17.8°C

11.36 Entropies of solution are usually positive because molecular randomness usually increases during dissolution.

Units of Concentration

11.37 $\text{molarity} = \dfrac{\text{moles of solute}}{\text{liters of solution}}$; $\text{molality} = \dfrac{\text{moles of solute}}{\text{kg of solvent}}$

11.38 A saturated solution contains enough solute so that there is an equilibrium between dissolved solute and undissolved solid.

A supersaturated solution contains a greater-than-equilibrium amount of solute.

11.39 (a) Dissolve 1.50 moles of glucose in enough water to make 1.00 L of solution.

(b) Dissolve 1.135 moles of KBr in 1.00 kg of H_2O.

(c) Mix together 0.15 moles of CH_3OH with 0.85 moles of H_2O.

11.40 $C_7H_6O_2$, 122.12 amu, 165 mL = 0.165 L

$$\text{mol } C_7H_6O_2 = 0.165 \text{ L} \times \frac{0.0268 \text{ mol}}{1.00 \text{ L}} = 0.004\ 42 \text{ mol}$$

$$\text{mass } C_7H_6O_2 = 0.004\ 42 \text{ mol} \times \frac{122.12 \text{ g}}{1 \text{ mol}} = 0.540 \text{ g}$$

Dissolve 4.42×10^{-3} mol (0.540 g) of $C_7H_6O_2$ in enough $CHCl_3$ to make 165 mL of solution.

11.41 $C_7H_6O_2$, 122.12 amu

$$0.0268 \text{ mol } C_7H_6O_2 \times \frac{122.12 \text{ g } C_7H_6O_2}{1 \text{ mol } C_7H_6O_2} = 3.27 \text{ g } C_7H_6O_2$$

Dissolve 3.27 g of $C_7H_6O_2$ in 1.000 kg of $CHCl_3$, and take 165 mL of the solution.

11.42 (a) KCl, 74.6 amu

A 0.500 M KCl solution contains 37.3 g of KCl per 1.00 L of solution.

A 0.500 wt % KCl solution contains 5.00 g of KCl per 995 g of water.

The 0.500 M KCl solution is more concentrated (that is, it contains more solute per amount of solvent).

(b) Both solutions contain the same amount of solute. The 1.75 M solution contains less solvent than the 1.75 m solution. The 1.75 M solution is more concentrated.

11.43 (a) $C_6H_8O_7$, 192.12 amu

$$0.655 \text{ mol } C_6H_8O_7 \times \frac{192.12 \text{ g } C_6H_8O_7}{1 \text{ mol } C_6H_8O_7} = 126 \text{ g } C_6H_8O_7$$

$$\text{weight \% } C_6H_8O_7 = \frac{126 \text{ g}}{126 \text{ g} + 1000 \text{ g}} \times 100\% = 11.2 \text{ wt \%}$$

(b) 0.135 mg = 0.135×10^{-3} g

(5.00 mL H_2O)(1.00 g/mL) = 5.00 g H_2O

$$\text{weight \% KBr} = \frac{0.135 \times 10^{-3}\ \text{g}}{(0.135 \times 10^{-3}\ \text{g}) + 5.00\ \text{g}} \times 100\% = 0.002\ 70\ \text{wt \% KBr}$$

(c) $$\text{weight \% aspirin} = \frac{5.50\ \text{g}}{5.50\ \text{g} + 145\ \text{g}} \times 100\% = 3.65\ \text{wt \% aspirin}$$

11.44 (a) $$\text{molality} = \frac{0.655\ \text{mol}}{1.00\ \text{kg}} = 0.655\ m$$

(b) KBr, 119.00 amu; 5.00 g = 0.005 00 kg

$$\text{molality} = \frac{\left(0.135 \times 10^{-3}\ \text{g} \times \dfrac{1\ \text{mol}}{119.00\ \text{g}}\right)}{0.005\ 00\ \text{kg}} = 2.27 \times 10^{-4}\ m$$

(c) $C_9H_8O_4$, 180.16 amu; 145 g = 0.145 kg

$$\text{molality} = \frac{\left(5.50\ \text{g} \times \dfrac{1\ \text{mol}}{180.16\ \text{g}}\right)}{0.145\ \text{kg}} = 0.211\ m$$

11.45 $$\text{volume} = (3\ \text{m} \times 5\ \text{m} \times 2.5\ \text{m})\left(\frac{1\ \text{L}}{10^{-3}\ \text{m}^3}\right) = 37{,}500\ \text{L}$$

mass of air = (37,500 L)(1.3 g/L) = 4.9×10^4 g = 5×10^4 g

$$0.1\ \text{ppm}\ O_3 = \frac{\text{mass}\ O_3}{5 \times 10^4\ \text{g}} \times 10^6\ \text{ppm}$$

mass O_3 = 0.005 g = 5 mg

11.46 $P_{O_3} = P_{total} \cdot X_{O_3}$

$$X_{O_3} = \frac{P_{O_3}}{P_{total}} = \frac{1.6 \times 10^{-9}\text{ atm}}{1.3 \times 10^{-2}\text{ atm}} = 1.2 \times 10^{-7}$$

Assume one mole of air (29 g/mol)

$$\text{mol } O_3 = n_{air} \cdot X_{O_3} = (1\text{ mol})(1.2 \times 10^{-7}) = 1.2 \times 10^{-7}\text{ mol } O_3$$

O_3, 48.00 amu

$$\text{mass } O_3 = 1.2 \times 10^{-7}\text{ mol} \times \frac{48.0\text{ g}}{1\text{ mol}} = 5.8 \times 10^{-6}\text{ g } O_3$$

$$\text{ppm } O_3 = \frac{5.8 \times 10^{-6}\text{ g}}{29\text{ g}} \times 10^6 = 0.20\text{ ppm}$$

11.47 (a) H_2SO_4, 98.08 amu

$$\text{molality} = \frac{\left(25.0\text{ g} \times \frac{1\text{ mol}}{98.08\text{ g}}\right)}{1.30\text{ kg}} = 0.196\ m$$

(b) $C_{10}H_{14}N_2$, 162.23 amu; CH_2Cl_2, 84.93 amu

$$2.25\text{ g } C_{10}H_{14}N_2 \times \frac{1\text{ mol } C_{10}H_{14}N_2}{162.23\text{ g } C_{10}H_{14}N_2} = 0.0139\text{ mol } C_{10}H_{14}N_2$$

$$80.0\text{ g } CH_2Cl_2 \times \frac{1\text{ mol } CH_2Cl_2}{84.93\text{ g } CH_2Cl_2} = 0.942\text{ mol } CH_2Cl_2$$

$$X_{C_{10}H_{14}N_2} = \frac{0.0139\text{ mol}}{0.942\text{ mol} + 0.0139\text{ mol}} = 0.0145$$

$$X_{CH_2Cl_2} = \frac{0.942\text{ mol}}{0.942\text{ mol} + 0.0139\text{ mol}} = 0.985$$

11.48 $C_{12}H_{22}O_{11}$, 342.30 amu

$$32.5\text{ g } C_{12}H_{22}O_{11} \times \frac{1\text{ mol } C_{12}H_{22}O_{11}}{342.30\text{ g } C_{12}H_{22}O_{11}} = 0.0949\text{ mol } C_{12}H_{22}O_{11}$$

$$0.85\ m = 0.85\frac{\text{mol}}{\text{kg}} = \frac{0.0949\text{ mol}}{\text{kg of } H_2O}$$

$$\text{kg of } H_2O = \frac{0.0949\text{ mol}}{0.85\frac{\text{mol}}{\text{kg}}} = 0.112\text{ kg}$$

$$\text{mass of } H_2O = 0.112\text{ kg} \times \frac{1000\text{ g}}{1\text{ kg}} = 112\text{ g } H_2O$$

11.49 NaOCl, 74.44 amu

A 5.0 wt % aqueous solution of NaOCl contains 5.0 g NaOCl and 95 g H_2O.

$$\text{molality} = \frac{\left(5.0\text{ g} \times \frac{1\text{ mol}}{74.44\text{ g}}\right)}{0.095\text{ kg}} = 0.71\ m$$

$$5.0\text{ g NaOCl} \times \frac{1\text{ mol NaOCl}}{74.44\text{ g NaOCl}} = 0.0672\text{ mol NaOCl}$$

$$95\text{ g } H_2O \times \frac{1\text{ mol } H_2O}{18.02\text{ g } H_2O} = 5.27\text{ mol } H_2O$$

$$X_{NaOCl} = \frac{0.0672\text{ mol}}{5.27\text{ mol} + 0.0672\text{ mol}} = 0.013$$

11.50 $$16.0\text{ wt \%} = \frac{16.0\text{ g } H_2SO_4}{16.0\text{ g } H_2SO_4 + 84.0\text{ g } H_2O}$$

H_2SO_4, 98.08 amu

density = 1.1094 g/mL

$$\text{volume of solution} = 100.0\text{ g} \times \frac{1\text{ mL}}{1.1094\text{ g}} = 90.14\text{ mL} = 0.090\ 14\text{ L}$$

$$\text{molarity} = \frac{\left(16.0\text{ g} \times \frac{1\text{ mol}}{98.08\text{ g}}\right)}{0.090\ 14\text{ L}} = 1.81\text{ M}$$

11.51 $C_2H_6O_2$, 62.07 amu

A 40.0 wt % aqueous solution of $C_2H_6O_2$ contains 40.0 g $C_2H_6O_2$ and 60.0 g H_2O.

density = 1.0514 g/mL

$$\text{volume of solution} = 100.0\text{ g} \times \frac{1\text{ mL}}{1.0514\text{ g}} = 95.1\text{ mL} = 0.0951\text{ L}$$

$$\text{molarity} = \frac{\left(40.0\text{ g} \times \frac{1\text{ mol}}{62.07\text{ g}}\right)}{0.0951\text{ L}} = 6.78\text{ M}$$

11.52 $$\text{molality} = \frac{\left(40.0\text{ g} \times \frac{1\text{ mol}}{62.07\text{ g}}\right)}{0.0600\text{ kg}} = 10.7\ m$$

11.53 $C_{19}H_{21}NO_3$, 311.34 amu; 1.5 mg = 1.5×10^{-3} g

$$1.3 \times 10^{-3}\ \frac{\text{mol}}{\text{kg}} = \frac{\left(1.5 \times 10^{-3}\text{ g} \times \frac{1\text{ mol}}{311.34\text{ g}}\right)}{\text{kg of solvent}};\quad \text{solve for kg of solvent.}$$

kg of solvent = 0.0037 kg

Because the solution is very dilute, kg of solvent ≈ kg of solution.

$$\text{g of solution} = (0.0037\text{ kg})\left(\frac{1000\text{ g}}{1\text{ kg}}\right) = 3.7\text{ g}$$

11.54 $C_6H_{12}O_6$, 180.16 amu; H_2O, 18.02 amu

Assume 1.00 L of solution.

mass of solution = (1000 mL)(1.0624 g/mL) = 1062.4 g

$$\text{mass of solute} = 0.944 \text{ mol} \times \frac{180.16 \text{ g}}{1 \text{ mol}} = 170.1 \text{ g } C_6H_{12}O_6$$

mass of H_2O = 1062.4 g – 170.1 g = 892.3 g H_2O

mol $C_6H_{12}O_6$ = 0.944 mol

$$\text{mol } H_2O = 892.3 \text{ g} \times \frac{1 \text{ mol}}{18.02 \text{ g}} = 49.5 \text{ mol}$$

(a) $$X_{C_6H_{12}O_6} = \frac{\text{mol } C_6H_{12}O_6}{\text{mol } C_6H_{12}O_6 + \text{mol } H_2O} = \frac{0.944 \text{ mol}}{0.944 \text{ mol} + 49.5 \text{ mol}} = 0.0187$$

(b) $$\text{wt \%} = \frac{\text{mass } C_6H_{12}O_6}{\text{total mass of solution}} \times 100\% = \frac{170.1 \text{ g}}{1062.4 \text{ g}} \times 100\% = 16.0\%$$

(c) $$\text{molality} = \frac{\text{mol } C_6H_{12}O_6}{\text{kg } H_2O} = \frac{0.944 \text{ mol}}{0.8923 \text{ kg}} = 1.06\ m$$

11.55 $C_{12}H_{22}O_{11}$, 342.30 amu

Assume 1.00 L of solution.

mass of solution = (1000 mL)(1.0432 g/mL) = 1043.2 g

$$\text{mass of solute} = 0.335 \text{ mol } C_{12}H_{22}O_{11} \times \frac{342.30 \text{ g } C_{12}H_{22}O_{11}}{1 \text{ mol } C_{12}H_{22}O_{11}} = 114.7 \text{ g } C_{12}H_{22}O_{11}$$

mass of H_2O = 1043.2 g – 114.7 g = 928.5 g H_2O

$$\text{mol } C_{12}H_{22}O_{11} = 0.335 \text{ mol}; \quad 928.5 \text{ g } H_2O \times \frac{1 \text{ mol } H_2O}{18.02 \text{ g } H_2O} = 51.53 \text{ mol } H_2O$$

$$X_{C_{12}H_{22}O_{11}} = \frac{0.335 \text{ mol}}{51.53 \text{ mol} + 0.335 \text{ mol}} = 0.006\ 46$$

$$\text{weight \% } C_{12}H_{22}O_{11} = \frac{114.7 \text{ g}}{1043.2 \text{ g}} \times 100\% = 11.0 \text{ wt \% } C_{12}H_{22}O_{11}$$

$$\text{molality} = \frac{0.335 \text{ mol}}{0.9285 \text{ kg}} = 0.361\ m$$

Solubility and Henry's Law

11.56 $M = k \cdot P = (0.091 \frac{\text{mol}}{\text{L} \cdot \text{atm}})(0.75 \text{ atm}) = 0.068 \text{ M}$

11.57 $M = k \cdot P;\ k = \frac{M}{P} = \frac{0.195 \text{ M}}{1.00 \text{ atm}} = 0.195 \text{ mol/(L} \cdot \text{atm)}$

11.58 $P = 25.5 \text{ mm Hg} \times \frac{1.00 \text{ atm}}{760 \text{ mm Hg}} = 0.0336 \text{ atm}$

$M = k \cdot P = (0.195 \frac{\text{mol}}{\text{L} \cdot \text{atm}})(0.0336 \text{ atm}) = 6.55 \times 10^{-3} \text{ M}$

11.59 [Xe] = 10 mmol/L = 0.010 M at STP

$M = k \cdot P;\ k = \frac{M}{P} = \frac{0.010 \text{ M}}{1.0 \text{ atm}} = 0.010 \text{ mol/(L} \cdot \text{atm)}$

11.60 $M = k \cdot P$

Calculate k: $k = \frac{M}{P} = \frac{2.21 \times 10^{-3} \text{ mol/L}}{1.00 \text{ atm}} = 2.21 \times 10^{-3} \frac{\text{mol}}{\text{L} \cdot \text{atm}}$

Convert 4 mg/L to mol/L:

$4 \text{ mg} = 4 \times 10^{-3} \text{ g}$

$$O_2 \text{ molarity} = \frac{\left(4 \times 10^{-3}\,\text{g} \times \dfrac{1\,\text{mol}}{32.00\,\text{g}}\right)}{1.00\,\text{L}} = 1.25 \times 10^{-4}\,\text{M}$$

$$P_{O_2} = \frac{M}{k} = \frac{1.25 \times 10^{-4}\,\dfrac{\text{mol}}{\text{L}}}{2.21 \times 10^{-3}\,\dfrac{\text{mol}}{\text{L}\cdot\text{atm}}} = 0.06\,\text{atm}$$

11.61 Assuming H_2O as the solvent, NH_3 does not obey Henry's law because NH_3 can both hydrogen bond and react with H_2O.

Colligative Properties

11.62 The difference in entropy between a solution and a pure solvent is responsible for colligative properties.

11.63 (a) vapor pressure lowering (b) boiling point elevation

(c) freezing point depression (d) osmosis

11.64 Osmotic pressure is the amount of pressure that needs to be applied to cause osmosis to stop.

11.65 solid < liquid < solution < gas

11.66

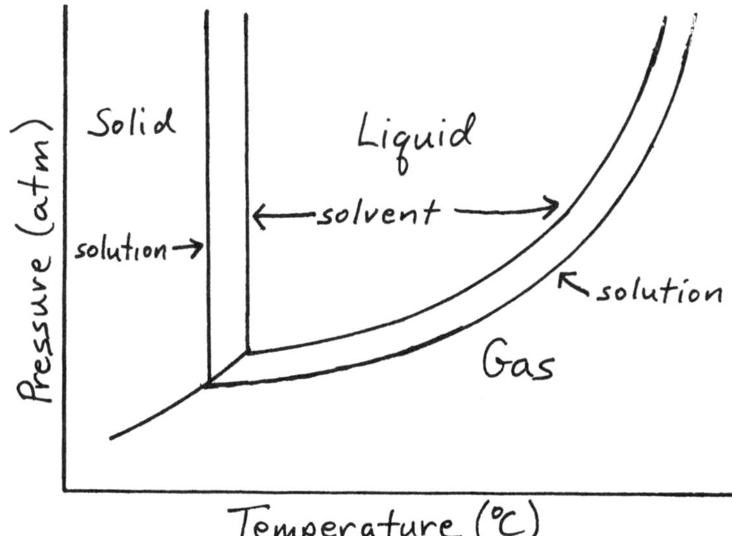

11.67 Molality is a temperature independent concentration unit. For freezing point depression and boiling point elevation, molality is used so that the solute concentration is independent of temperature changes. Molarity is temperature dependent. Molarity can be used for osmotic pressure because osmotic pressure is measured at a fixed temperature.

11.68 (a) CH_4N_2O, 60.06 amu; H_2O, 18.02 amu

$$10.0 \text{ g } CH_4N_2O \times \frac{1 \text{ mol } CH_4N_2O}{60.06 \text{ g } CH_4N_2O} = 0.167 \text{ mol } CH_4N_2O$$

$$150.0 \text{ g } H_2O \times \frac{1 \text{ mol } H_2O}{18.02 \text{ g } H_2O} = 8.32 \text{ mol } H_2O$$

$$X_{H_2O} = \frac{8.32 \text{ mol}}{8.32 \text{ mol} + 0.167 \text{ mol}} = 0.980$$

$$P_{soln} = X_{H_2O}P^{o}_{H_2O} = (0.980)(71.93 \text{ mm Hg}) = 70.5 \text{ mm Hg}$$

(b) LiCl, 42.39 amu

$$10.0 \text{ g LiCl} \times \frac{1 \text{ mol LiCl}}{42.39 \text{ g LiCl}} = 0.236 \text{ mol LiCl}$$

LiCl dissociates into $Li^+(aq)$ and $Cl^-(aq)$ in H_2O.

mol Li^+ = mol Cl^- = mol LiCl = 0.236 mol

$$150.0 \text{ g } H_2O \times \frac{1 \text{ mol } H_2O}{18.02 \text{ g } H_2O} = 8.32 \text{ mol } H_2O$$

$$X_{H_2O} = \frac{8.32 \text{ mol}}{8.32 \text{ mol} + 0.236 \text{ mol} + 0.236 \text{ mol}} = 0.946$$

$$P_{soln} = X_{H_2O}P^{o}_{H_2O} = (0.946)(71.93 \text{ mm Hg}) = 68.0 \text{ mm Hg}$$

11.69 (a) $\Delta T_b = K_b \cdot m = \left(0.51\ \frac{°C \cdot kg}{mol}\right)\left(\frac{0.167\ mol}{0.150\ kg}\right) = 0.57°C$

solution boiling point = 100.00°C + ΔT_b = 100.00°C + 0.57°C = 100.57°C

(b) $\Delta T_b = K_b \cdot m = \left(0.51\ \frac{°C \cdot kg}{mol}\right)\left(\frac{2(0.236\ mol)}{0.150\ kg}\right) = 1.6°C$

solution boiling point = 100.00°C + ΔT_b = 100.00°C + 1.6°C = 101.6°C

11.70 For H_2O, $K_f = 1.86\ \frac{°C \cdot kg}{mol}$; 150.0 g = 0.1500 kg

(a) $\Delta T_f = K_f \cdot m = \left(1.86\ \frac{°C \cdot kg}{mol}\right)\left(\frac{0.167\ mol}{0.1500\ kg}\right) = 2.07°C$

solution freezing point = 0.00°C − ΔT_f = 0.00°C − 2.07°C = −2.07°C

(b) $\Delta T_f = K_f \cdot m = \left(1.86\ \frac{°C \cdot kg}{mol}\right)\left(\frac{2(0.236\ mol)}{0.1500\ kg}\right) = 5.85°C$

solution freezing point = 0.00°C − ΔT_f = 0.00°C − 5.85°C = −5.85°C

11.71 Br_2, 159.81 amu; CCl_4, 153.82 amu

$$P^{o}_{Br_2} = 30.5\ kPa \times \frac{760\ mm\ Hg}{101.325\ kPa} = 228.8\ mm\ Hg$$

$$P^{o}_{CCl_4} = 16.5\ kPa \times \frac{760\ mm\ Hg}{101.325\ kPa} = 123.8\ mm\ Hg$$

$$1.50\ g\ Br_2 \times \frac{1\ mol\ Br_2}{159.81\ g\ Br_2} = 9.39 \times 10^{-3}\ mol\ Br_2$$

$$145.0\ g\ CCl_4 \times \frac{1\ mol\ CCl_4}{153.82\ g\ CCl_4} = 0.943\ mol\ CCl_4$$

$$X_{Br_2} = \frac{9.39 \times 10^{-3}\text{ mol}}{(0.943\text{ mol}) + (9.39 \times 10^{-3}\text{ mol})} = 0.009\ 86$$

$$X_{CCl_4} = \frac{0.943\text{ mol}}{(0.943\text{ mol}) + (9.39 \times 10^{-3}\text{ mol})} = 0.990$$

$$P_{soln} = X_{Br_2}P^{o}_{Br_2} + X_{CCl_4}P^{o}_{CCl_4}$$

$$P_{soln} = (0.009\ 86)(228.8\text{ mm Hg}) + (0.990)(123.8\text{ mm Hg}) = 125\text{ mm Hg}$$

11.72 Acetone, C_3H_6O, 58.08 amu, $P^{o}_{C_3H_6O}$ = 285 mm Hg

Ethyl acetate, $C_4H_8O_2$, 88.11 amu, $P^{o}_{C_4H_8O_2}$ = 118 mm Hg

$$25.0\text{ g } C_3H_6O \times \frac{1\text{ mol } C_3H_6O}{58.08\text{ g } C_3H_6O} = 0.430\text{ mol } C_3H_6O$$

$$25.0\text{ g } C_4H_8O_2 \times \frac{1\text{ mol } C_4H_8O_2}{88.11\text{ g } C_4H_8O_2} = 0.284\text{ mol } C_4H_8O_2$$

$$X_{C_3H_6O} = \frac{0.430\text{ mol}}{0.430\text{ mol} + 0.284\text{ mol}} = 0.602$$

$$X_{C_4H_8O_2} = \frac{0.284\text{ mol}}{0.430\text{ mol} + 0.284\text{ mol}} = 0.398$$

$$P_{soln} = X_{C_3H_6O}P^{o}_{C_3H_6O} + X_{C_4H_8O_2}P^{o}_{C_4H_8O_2}$$

$$P_{soln} = (0.602)(285\text{ mm Hg}) + (0.398)(118\text{ mm Hg}) = 219\text{ mm Hg}$$

11.73 In the liquid, $X_{acetone} = 0.602$ and $X_{ethyl\ acetate} = 0.398$

In the vapor, P_{Total} = 219 mm Hg

$$P_{acetone} = X_{acetone}\ P^{o}_{acetone} = (0.602)(285\text{ mm Hg}) = 172\text{ mm Hg}$$

$P_{\text{ethyl acetate}} = X_{\text{ethyl acetate}}\, P^{o}_{\text{ethyl acetate}} = (0.398)(118 \text{ mm Hg}) = 47 \text{ mm Hg}$

$$X_{\text{acetone}} = \frac{P_{\text{acetone}}}{P_{\text{total}}} = \frac{172 \text{ mm Hg}}{219 \text{ mm Hg}} = 0.785$$

$$X_{\text{ethyl acetate}} = \frac{P_{\text{ethyl acetate}}}{P_{\text{total}}} = \frac{47 \text{ mm Hg}}{219 \text{ mm Hg}} = 0.215$$

11.74 $CHCl_3$, 119.38 amu, $P^{o}_{CHCl_3} = 205$ mm Hg

CH_2Cl_2, 84.93 amu, $P^{o}_{CH_2Cl_2} = 415$ mm Hg

$$15.0 \text{ g } CHCl_3 \times \frac{1 \text{ mol } CHCl_3}{119.38 \text{ g } CHCl_3} = 0.126 \text{ mol } CHCl_3$$

$$37.5 \text{ g } CH_2Cl_2 \times \frac{1 \text{ mol } CH_2Cl_2}{84.93 \text{ g } CH_2Cl_2} = 0.442 \text{ mol } CH_2Cl_2$$

$$X_{CHCl_3} = \frac{0.126 \text{ mol}}{0.126 \text{ mol} + 0.442 \text{ mol}} = 0.222$$

$$X_{CH_2Cl_2} = \frac{0.442 \text{ mol}}{0.126 \text{ mol} + 0.442 \text{ mol}} = 0.778$$

$P_{\text{soln}} = X_{CHCl_3} P^{o}_{CHCl_3} + X_{CH_2Cl_2} P^{o}_{CH_2Cl_2}$

$P_{\text{soln}} = (0.222)(205 \text{ mm Hg}) + (0.778)(415 \text{ mm Hg}) = 368 \text{ mm Hg}$

11.75 In the liquid, $X_{CHCl_3} = 0.222$ and $X_{CH_2Cl_2} = 0.778$

In the vapor, $P_{\text{total}} = 368$ mm Hg

$P_{CHCl_3} = X_{CHCl_3}\, P^{o}_{CHCl_3} = (0.222)(205 \text{ mm Hg}) = 45.5 \text{ mm Hg}$

$$P_{CH_2Cl_2} = X_{CH_2Cl_2} P^{o}_{CH_2Cl_2} = (0.778)(415 \text{ mm Hg}) = 323 \text{ mm Hg}$$

$$X_{CHCl_3} = \frac{P_{CHCl_3}}{P_{total}} = \frac{45.5 \text{ mm Hg}}{368 \text{ mm Hg}} = 0.124$$

$$X_{CH_2Cl_2} = \frac{P_{CH_2Cl_2}}{P_{total}} = \frac{323 \text{ mm Hg}}{368 \text{ mm Hg}} = 0.876$$

11.76 $C_6H_{12}O_6$, 180.16 amu

For ethyl alcohol, $K_b = 1.22 \ \frac{°C \cdot kg}{mol}$; 285 g = 0.285 kg

$$\Delta T_b = K_b \cdot m = \left(1.22 \ \frac{°C \cdot kg}{mol}\right)\left(\frac{26.0 \text{ g} \times \frac{1 \text{ mol}}{180.16 \text{ g}}}{0.285 \text{ kg}}\right) = 0.618°C$$

solution boiling point = normal boiling point + ΔT_b = 79.1°C

normal boiling point = 79.1°C – ΔT_b = 79.1°C – 0.618°C = 78.5°C

11.77 $C_9H_8O_4$, 180.16 amu; 215 g = 0.215 kg

$$\Delta T_b = K_b \cdot m = 0.47 \text{ °C}$$

$$K_b = \frac{\Delta T_b}{m} = \frac{0.47°C}{\left(\frac{5.00 \text{ g} \times \frac{1 \text{ mol}}{180.16 \text{ g}}}{0.215 \text{ kg}}\right)} = 3.6 \ \frac{°C \cdot kg}{mol}$$

11.78 $C_6H_8O_6$, 176.13 amu; 50.0 g = 0.0500 kg

$$\Delta T_f = K_f \cdot m = 1.33°C$$

$$K_f = \frac{\Delta T_f}{m} = \frac{1.33 \text{ °C}}{\left(\frac{3.00 \text{ g} \times \frac{1 \text{ mol}}{176.13 \text{ g}}}{0.0500 \text{ kg}}\right)} = 3.90 \ \frac{°C \cdot kg}{mol}$$

11.79 $\Delta T_b = K_b \cdot m = 1.76\ ^\circ C$

$$m = \frac{\Delta T_b}{K_b} = \frac{1.76\ ^\circ C}{3.07\ \frac{^\circ C \cdot kg}{mol}} = 0.573\ m$$

11.80 $\Pi = MRT$

(a) NaCl 58.44 amu; 350.0 mL = 0.3500 L

There are 2 moles of ions/mole of NaCl

$$\Pi = (2)\left(\frac{5.00\ g \times \frac{1 mol}{58.44\ g}}{0.3500\ L}\right)\left(0.082\ 06\ \frac{L \cdot atm}{mol \cdot K}\right)(323\ K) = 13.0\ atm$$

(b) CH_3CO_2Na, 82.03 amu; 55.0 mL = 0.0550 L

There are 2 moles of ions/mole of CH_3CO_2Na

$$\Pi = (2)\left(\frac{6.33\ g \times \frac{1\ mol}{82.03\ g}}{0.0550\ L}\right)\left(0.082\ 06\ \frac{L \cdot atm}{mol \cdot K}\right)(283K) = 65.2\ atm$$

11.81 $\Pi = MRT$

$$\Pi = \left(\frac{11.5 \times 10^{-3}\ g \times \frac{1\ mol}{5990\ g}}{0.006\ 60\ L}\right)\left(0.082\ 06\ \frac{L \cdot atm}{mol \cdot K}\right)(298\ K) = 0.007\ 11\ atm$$

$$\Pi = 0.007\ 11\ atm \times \frac{760\ mm\ Hg}{1\ atm} = 5.41\ mm\ Hg$$

11.82 From Problem 11.81, Π = 5.41 mm Hg

$$\text{height of } H_2O \text{ column} = 5.41\ mm\ Hg \times \frac{13.534\ mm\ H_2O}{1.00\ mm\ Hg} = 73.2\ mm$$

$$\text{height of } H_2O \text{ column} = 73.2 \text{ mm} \times \frac{1 \text{ m}}{1000 \text{ mm}} = 0.0732 \text{ m}$$

11.83 $$M = \frac{\Pi}{RT} = \frac{4.85 \text{ atm}}{\left(0.082\ 06 \frac{\text{L}\cdot\text{atm}}{\text{mol}\cdot\text{K}}\right)(300 \text{ K})} = 0.197 \text{ M}$$

11.84 $\Pi = MRT$

$$M = \frac{\Pi}{RT} = \frac{7.7 \text{ atm}}{\left(0.082\ 06 \frac{\text{L}\cdot\text{atm}}{\text{mol}\cdot\text{K}}\right)(310 \text{ K})} = 0.30 \text{ M}$$

Uses of Colligative Properties

11.85 Osmotic pressure is most often used for the determination of molecular weight because, of the four colligative properties, osmotic pressure gives the largest colligative property change per mole of solute.

11.86 $C_6H_{12}O_6$ does not dissociate in aqueous solution. LiCl and NaCl both dissociate into two solute particles per formula unit in aqueous solution. $CaCl_2$ dissociates into three solute particles per formula unit in aqueous solution. Assume that you have 1.00 g of each substance. Calculate the number of moles of solute particles in 1.00 g of each substance.

$C_6H_{12}O_6$, 180.2 amu

$$\text{moles solute particles} = 1.00 \text{ g} \times \frac{1 \text{ mol}}{180.2 \text{ g}} = 0.005\ 55 \text{ moles}$$

LiCl, 42.4 amu

$$\text{moles solute particles} = 2\left(1.00 \text{ g} \times \frac{1 \text{ mol}}{42.4 \text{ g}}\right) = 0.0472 \text{ moles}$$

NaCl, 58.4 amu

$$\text{moles solute particles} = 2\left(1.00 \text{ g} \times \frac{1 \text{ mol}}{58.4 \text{ g}}\right) = 0.0342 \text{ moles}$$

$CaCl_2$, 111.0 amu

$$\text{moles solute particles} = 3\left(1.00\text{ g} \times \frac{1\text{ mol}}{111.0\text{ g}}\right) = 0.0270\text{ moles}$$

LiCl produces more solute particles/gram than any of the other three substances. LiCl would be the most efficient per unit mass.

11.87 $\Pi = 407.2\text{ mm Hg} \times \dfrac{1\text{ atm}}{760\text{ mm Hg}} = 0.5358\text{ atm}$

$$M = \frac{\Pi}{RT} = \frac{0.5358\text{ atm}}{\left(0.082\ 06\,\frac{\text{L}\cdot\text{atm}}{\text{mol}\cdot\text{K}}\right)(298.15\text{ K})} = 0.021\ 90\text{ M} = 0.021\ 90\text{ mol/L}$$

$$0.021\ 90\text{ mol/L} = \frac{\left(\frac{1.500\text{ g}}{\text{molar mass}}\right)}{0.2000\text{ L}}$$

molar mass of cellobiose = 342.5 g/mol; molecular weight = 342.5 amu

11.88 height of Hg column = $32.9\text{ cm }H_2O \times \dfrac{1.00\text{ cm Hg}}{13.534\text{ cm }H_2O} = 2.43\text{ cm Hg}$

$$\Pi = 2.43\text{ cm Hg} \times \frac{1.00\text{ atm}}{76.0\text{ cm Hg}} = 0.0320\text{ atm}$$

$$\Pi = MRT; \quad M = \frac{\Pi}{RT} = \frac{0.0320\text{ atm}}{\left(0.082\ 06\,\frac{\text{L}\cdot\text{atm}}{\text{mol}\cdot\text{K}}\right)(298\text{ K})} = 0.001\ 31\,\frac{\text{mol}}{\text{L}}$$

$$M = 0.001\ 31\,\frac{\text{mol}}{\text{L}} = \frac{\left(\frac{15.0 \times 10^{-3}\text{ g}}{\text{molar mass}}\right)}{0.0200\text{ L}};$$ solve for the molar mass.

molar mass of met–enkephalin = 573 g/mol

11.89 HCl is a strong electrolyte in H_2O and completely dissociates into two solute particles per each HCl.

HF is a weak electrolyte in H_2O. Only a few percent of the HF molecules dissociates into ions.

11.90 First, determine the empirical formula:

Assume 100.0 g of β-carotene.

10.51% H $\quad 10.51 \text{ g H} \times \frac{1 \text{ mol H}}{1.008 \text{ g H}} = 10.43 \text{ mol H}$

89.49% C $\quad 89.49 \text{ g C} \times \frac{1 \text{ mol C}}{12.01 \text{ g C}} = 7.45 \text{ mol C}$

$C_{7.45}H_{10.43}$

Divide each subscript by the smaller, 7.45.

$C_{7.45\,/\,7.45}H_{10.43\,/\,7.45}$

$CH_{1.4}$

Multiply each subscript by 5 to obtain integers.

Empirical formula is C_5H_7, 67.1 amu.

Second, calculate the molecular weight:

$\Delta T_f = K_f \cdot m$

$$m = \frac{\Delta T_f}{K_f} = \frac{1.17°\text{C}}{37.7 \frac{°\text{C} \cdot \text{kg}}{\text{mol}}} = 0.0310 \frac{\text{mol}}{\text{kg}}$$

$$m = 0.0310 \frac{\text{mol}}{\text{kg}} = \frac{\left(\frac{0.0250 \text{ g}}{\text{molar mass}}\right)}{0.001\ 50 \text{ kg}};$$ solve for the molar mass.

molar mass of β–carotene = 538 g/mol; molecular weight = 538 amu

Finally, determine the molecular formula:

Divide the molecular weight by the empirical formula weight.

$\frac{538 \text{ amu}}{67.1 \text{ amu}} = 8$; molecular formula is $C_{(8 \times 5)}H_{(8 \times 7)}$, or $C_{40}H_{56}$

11.91 First, determine the empirical formula:

Assume a 100.0 g sample of lysine.

49.29% C $49.29 \text{ g C} \times \frac{1 \text{ mol C}}{12.011 \text{ g C}} = 4.10 \text{ mol C}$

9.65% H $9.65 \text{ g H} \times \frac{1 \text{ mol H}}{1.008 \text{ g H}} = 9.57 \text{ mol H}$

19.16% N $19.16 \text{ g N} \times \frac{1 \text{ mol N}}{14.007 \text{ g N}} = 1.37 \text{ mol N}$

21.89% O $21.89 \text{ g O} \times \frac{1 \text{ mol O}}{15.999 \text{ g O}} = 1.37 \text{ mol O}$

$C_{4.10}H_{9.57}N_{1.37}O_{1.37}$

Divide each subscript by the smallest, 1.37.

$C_{4.10 / 1.37}H_{9.57 / 1.37}N_{1.37 / 1.37}O_{1.37 / 1.37}$

Empirical formula is C_3H_7NO, 73.09 amu

Second, calculate the molecular weight:

$\Delta T_f = K_f \cdot m = 1.37 \text{ °C}$

$$m = \frac{\Delta T_f}{K_f} = \frac{1.37 \text{ °C}}{8.00 \frac{\text{°C} \cdot \text{kg}}{\text{mol}}} = 0.171\ m = 0.171 \text{ mol/kg}$$

$$0.171 \text{ mol/kg} = \frac{\left(\dfrac{30.0 \times 10^{-3}\text{ g}}{\text{molar mass}}\right)}{0.001\ 200 \text{ kg}};$$ solve for the molar mass.

molar mass of lysine = 146 g/mol; molecular weight = 146 amu

Finally, determine the molecular formula:

Divide the molecular weight by the empirical formula weight.

$\frac{146 \text{ amu}}{73.09 \text{ amu}} = 2$; molecular formula is $C_{(2 \times 3)}H_{(2 \times 7)}N_{(2 \times 1)}O_{(2 \times 1)}$, or $C_6H_{14}N_2O_2$

11.92 The mixture will begin to boil at ~ 55°C.

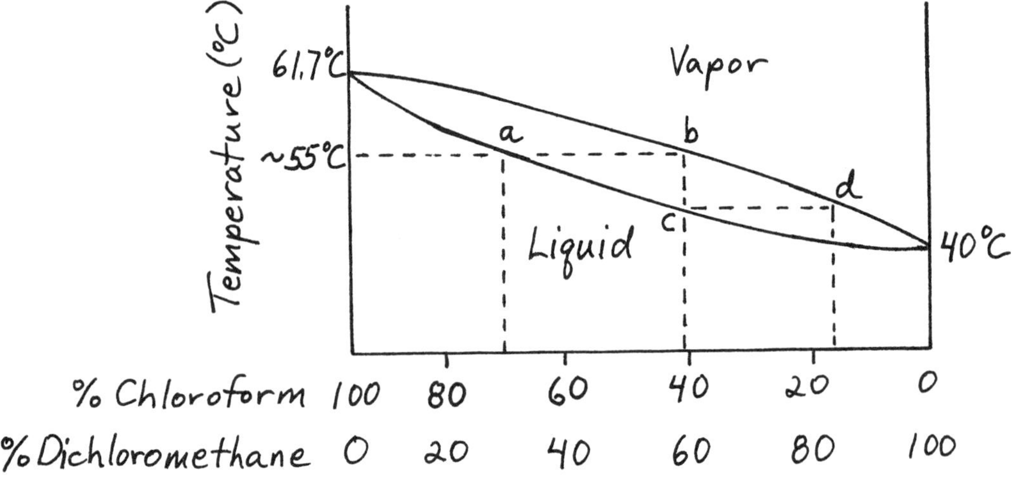

General Problems

11.93 KBr, 119.00 amu; there are 2 ions/KBr; 125 g = 0.125 kg

$\Delta T_b = 103.2\ °C - 100.0\ °C = 3.2\ °C$

$\Delta T_b = K_b \cdot 2 \cdot m$

$$m = \frac{\Delta T_b}{2K_b} = \frac{3.2\ ^\circ C}{2\left(0.51\ \frac{^\circ C\cdot kg}{mol}\right)} = 3.137\ mol/kg$$

$$3.137\ mol/kg = \frac{\left(\text{mass of KBr x } \frac{1\ mol}{119.00\ g}\right)}{0.125\ kg};\quad \text{solve for the mass of KBr.}$$

mass of KBr = 47 g

11.94 $C_2H_6O_2$, 62.07 amu

$\Delta T_f = 22.0^\circ C$

$\Delta T_f = K_f \cdot m$

$$m = \frac{\Delta T_f}{K_f} = \frac{22.0^\circ C}{1.86\ \frac{^\circ C\cdot kg}{mol}} = 11.8\ m$$

$$m = 11.8\ \frac{mol}{kg} = \frac{\left((\text{mass of ethylene glycol}) \text{ x } \frac{1\ mol}{62.07\ g}\right)}{3.55\ kg}$$

Solve for the mass of ethylene glycol.

mass of ethylene glycol = 2600 g = 2.60×10^3 g

11.95 The KCl solution has the higher boiling point because it is a strong electrolyte, yielding two ions per formula unit.

11.96 For KBr, the solute particle molality is 2 x 0.665 m = 1.33 m

For $CaCl_2$, the solute particle molality is 3 x 0.400 m = 1.20 m

The $CaCl_2$ solution has the smaller solute particle molality, and its freezing point depression (ΔT_f) will not be as large. The $CaCl_2$ solution will have the higher freezing point.

11.97 The vapor pressure of toluene is lower than the vapor pressure of benzene at the same temperature. When 1 mL of toluene is added to 100 mL of benzene, the vapor pressure of the solution decreases, which means that the boiling point of the solution will increase. When 1 mL of benzene is added to 100 mL of toluene, the vapor pressure of the solution increases, which means that the boiling point of the solution will decrease.

11.98 When solid $CaCl_2$ is added to liquid water, the temperature rises because ΔH_{soln} for $CaCl_2$ is exothermic.

When solid $CaCl_2$ is added to ice at 0°C, some of the ice will melt and the temperature will fall because the $CaCl_2$ lowers the freezing piont of an ice/water mixture.

11.99 AgCl, 143.32 amu; there are 2 ions/AgCl

$\Pi = 2MRT$

$$\Pi = 2\left(\frac{0.007 \text{ x } 10^{-3}\text{ g x } \dfrac{1\text{ mol}}{143.32\text{ g}}}{0.001\text{ L}}\right)\left(0.082\ 06\ \frac{\text{L}\cdot\text{atm}}{\text{mol}\cdot\text{K}}\right)(278\text{ K}) = 0.002\text{ atm}$$

11.100 $C_{10}H_8$, 128.17 amu; 150.0 g = 0.1500 kg

$\Delta T_f = 0.35°C$

$\Delta T_f = K_f \cdot m$

$$m = \frac{\Delta T_f}{K_f} = \frac{0.35°\text{ C}}{5.12\ \dfrac{°\text{C}\cdot\text{kg}}{\text{mol}}} = 0.0684\ \frac{\text{mol}}{\text{kg}}$$

$$m = 0.0684\ \frac{\text{mol}}{\text{kg}} = \frac{\left((\text{mass of } C_{10}H_8) \text{ x } \dfrac{1\text{ mol}}{128.17\text{ g}}\right)}{0.1500\text{ kg}}$$

Solve for the mass of $C_{10}H_8$.

mass of $C_{10}H_8$ = 1.3 g

11.101 NaCl, 58.44 amu; there are 2 ions/NaCl

A 3.5 wt % aqueous solution of NaCl contains 3.5 g NaCl and 96.5 g H_2O.

$$\text{molality} = \frac{\left(3.5\text{ g} \times \dfrac{1\text{ mol}}{58.44\text{ g}}\right)}{0.0965\text{ kg}} = 0.62\ m$$

$$\Delta T_f = K_f \cdot 2 \cdot m = \left(1.86\ \frac{°\text{C}\cdot\text{kg}}{\text{mol}}\right)(2)(0.62\text{ mol/kg}) = 2.3°\text{C}$$

freezing point = 0.0°C − ΔT_f = 0.0°C − 2.3°C = −2.3°C

$$\Delta T_b = K_b \cdot 2 \cdot m = \left(0.51\ \frac{°\text{C}\cdot\text{kg}}{\text{mol}}\right)(2)(0.62\text{ mol/kg}) = 0.63°\text{C}$$

boiling point = 100.00°C + ΔT_b = 100.00°C + 0.63°C = 100.63°C

11.102 (a) Assume a total mass of solution of 1000.0 g.

$$\text{ppm} = \frac{\text{mass of solute ion}}{\text{total mass of solution}} \times 10^6$$

For each ion:

$$\text{mass of solute ion} = \frac{(\text{ppm})(1000.0\text{ g})}{10^6}$$

Ion	Mass	Moles
Cl^-	19.0 g	0.536 mol
Na^+	10.5 g	0.457 mol
SO_4^{2-}	2.65 g	0.0276 mol
Mg^{2+}	1.35 g	0.0555 mol
Ca^{2+}	0.400 g	0.009 98 mol
K^+	0.380 g	0.009 72 mol
HCO_3^-	0.140 g	0.002 29 mol
Br^-	0.065 g	0.000 81 mol
Total	34.5 g	1.099 mol

Mass of H_2O = 1000.0 g − 34.5 g = 965.5 g H_2O = 0.9655 kg H_2O

$$\text{molality} = \frac{1.099\text{ mol}}{0.9655\text{ kg}} = 1.138\ m$$

(b) Assume M = *m* for a dilute solution.

$$\Pi = \text{MRT} = (1.138\ \frac{\text{mol}}{\text{L}})(0.082\ 06\ \frac{\text{L}\cdot\text{atm}}{\text{mol}\cdot\text{K}})(300\text{ K}) = 28.0\text{ atm}$$

11.103 90 wt % isopropyl alcohol = $\dfrac{10.5\text{ g}}{10.5\text{ g} + \text{mass of } H_2O} \times 100\%$

Solve for the mass of H_2O.

$$\text{mass of } H_2O = \left(10.5\text{ g} \times \frac{100}{90}\right) - 10.5\text{ g} = 1.2\text{ g}$$

mass of solution = 10.5 g + 1.2 g = 11.7 g

11.7 g of rubbing alcohol contains 10.5 g of isopropyl alcohol.

11.04 C_3H_8O, 60.10 amu

mass C_3H_8O = (0.90)(50.0 g) = 45 g

$$45\text{ g } C_3H_8O \times \frac{1\text{ mol } C_3H_8O}{60.10\text{ g } C_3H_8O} = 0.75\text{ mol } C_3H_8O$$

11.105 $B(OH)_3$, 61.83 amu

A 0.50 wt % aqueous solution of $B(OH)_3$ contains 0.50 g $B(OH)_3$ and 99.5 g H_2O.

$$\text{molality} = \frac{\left(0.50\text{ g} \times \frac{1\text{ mol}}{61.83\text{ g}}\right)}{0.0995\text{ kg}} = 0.081\ m$$

11.106 $C_6H_{12}O_6$, 180.16 amu; 50.0 mL = 0.0500 L; 17.5 mg = 17.5 x 10^{-3} g

$$\Pi = MRT$$

$$T = \frac{\Pi}{MR} = \frac{\left(37.8\text{ mm Hg x } \dfrac{1\text{ atm}}{760\text{ mm Hg}}\right)}{\left[\dfrac{\left(17.5 \text{ x } 10^{-3}\text{g x } \dfrac{1\text{ mol}}{180.16\text{ g}}\right)}{0.0500\text{ L}}\right]\left(0.082\ 06\ \dfrac{\text{L}\cdot\text{atm}}{\text{mol}\cdot\text{K}}\right)} = 312\text{ K}$$

12.1 $2\ N_2O_5(g) \rightarrow 4\ NO_2(g) + O_2(g)$

time	$[N_2O_5]$	$[O_2]$
200 s	0.0142 M	0.0029 M
300 s	0.0120 M	0.0040 M

$$\text{rate of decomposition of } N_2O_5 = -\frac{\Delta[N_2O_5]}{\Delta t} = -\frac{0.0120\text{ M} - 0.0142\text{ M}}{300\text{ s} - 200\text{ s}}$$

$$= 2.2 \times 10^{-5}\text{ M/s}$$

$$\text{rate of formation of } O_2 = \frac{\Delta[O_2]}{\Delta t} = \frac{0.0040\text{ M} - 0.0029\text{ M}}{300\text{ s} - 200\text{ s}} = 1.1 \times 10^{-5}\text{ M/s}$$

12.2 $3\ I^-(aq) + H_3AsO_4(aq) + 2\ H^+(aq) \rightarrow I_3^-(aq) + H_3AsO_3(aq) + H_2O(l)$

(a) $-\dfrac{\Delta[I^-]}{\Delta t} = 4.8 \times 10^{-4}\text{ M/s}$

$$\frac{\Delta[I_3^-]}{\Delta t} = \frac{1}{3}\left(-\frac{\Delta[I^-]}{\Delta t}\right) = \left(\frac{1}{3}\right)(4.8 \times 10^{-4}\text{ M/s}) = 1.6 \times 10^{-4}\text{ M/s}$$

(b) $-\dfrac{\Delta[H^+]}{\Delta t} = 2\left(\dfrac{\Delta[I_3^-]}{\Delta t}\right) = (2)(1.6 \times 10^{-4}\text{ M/s}) = 3.2 \times 10^{-4}\text{ M/s}$

12.3 rate = $k[BrO_3^-][Br^-][H^+]^2$
1st order in BrO_3^-
1st order in Br^-
2nd order in H^+
4th order overall

rate = $k[H_2][I_2]$
1st order in H_2
1st order in I_2
2nd order overall

rate = $k[CH_3CHO]^{3/2}$
3/2 order in CH_3CHO
3/2 order overall

12.4 $H_2O_2(aq) + 3\ I^-(aq) + 2\ H^+(aq) \rightarrow I_3^-(aq) + 2\ H_2O(l)$

$$\text{rate} = \frac{\Delta[I_3^-]}{\Delta t} = k[H_2O_2]^m[I^-]^n$$

(a) $$\frac{\text{rate}_3}{\text{rate}_1} = \frac{2.30 \times 10^{-4}\ M/s}{1.15 \times 10^{-4}\ M/s} = 2$$

$$\frac{[H_2O_2]_3}{[H_2O_2]_1} = \frac{0.200\ M}{0.100\ M} = 2$$

Since both ratios are the same, m = 1.

$$\frac{\text{rate}_2}{\text{rate}_1} = \frac{2.30 \times 10^{-4}\ M/s}{1.15 \times 10^{-4}\ M/s} = 2$$

$$\frac{[I^-]_2}{[I^-]_1} = \frac{0.200\ M}{0.100\ M} = 2$$

Since both ratios are the same, n = 1.

The rate law is: rate = $k[H_2O_2][I^-]$

(b) $$k = \frac{\text{rate}}{[H_2O_2][I^-]}$$

Using data from Experiment 1:

$$k = \frac{1.15 \times 10^{-4}\ M/s}{(0.100\ M)(0.100\ M)} = 1.15 \times 10^{-2}\ /(M \cdot s)$$

(c) rate = $k[H_2O_2][I^-] = [1.15 \times 10^{-2}/(M \cdot s)](0.300\ M)(0.400\ M) = 1.38 \times 10^{-3}\ M/s$

12.5 Rate Law | Units of k

Rate Law	Units of k
Rate = $k[(CH_3)_3CBr]$	1/s
Rate = $k[Br_2]$	1/s
Rate = $k[BrO_3^-][Br^-][H^+]^2$	$1/(M^3 \cdot s)$
Rate = $k[H_2][I_2]$	$1/(M \cdot s)$
Rate = $[CH_3CHO]^{3/2}$	$1/(M^{1/2} \cdot s)$

12.6 (a) $\log \dfrac{[Co(NH_3)_5Br^{2+}]_t}{[Co(NH_3)_5Br^{2+}]_0} = \dfrac{-kt}{2.303}$

$k = 6.3 \times 10^{-6}/s$

$t = 10\ h \times \dfrac{3600\ s}{1\ h} = 36{,}000\ s$

$\log[Co(NH_3)_5Br^{2+}]_t = \dfrac{-kt}{2.303} + \log[Co(NH_3)_5Br^{2+}]_0$

$\log[Co(NH_3)_5Br^{2+}]_t = \dfrac{-(6.3 \times 10^{-6}/s)(36{,}000\ s)}{2.303} + \log(0.100)$

$\log[Co(NH_3)_5Br^{2+}]_t = -1.0985$

After 10 h, $[Co(NH_3)_5Br^{2+}] = 10^{-1.0985} = 0.080$ M

(b) $[Co(NH_3)_5Br^{2+}]_0 = 0.100$ M

If 75% of the $Co(NH_3)_5Br^{2+}$ reacts then 25% remains.

$[Co(NH_3)_5Br^{2+}]_t = (0.25)(0.100\ M) = 0.025$ M

$\log \dfrac{[Co(NH_3)_5Br^{2+}]_t}{[Co(NH_3)_5Br^{2+}]_0} = \dfrac{-kt}{2.303}$

$$t = \frac{(2.303)\log\dfrac{[Co(NH_3)_5Br^{2+}]_t}{[Co(NH_3)_5Br^{2+}]_0}}{-k}$$

$$t = \frac{(2.303)\log\left(\dfrac{0.025}{0.100}\right)}{-(6.3 \times 10^{-6}/s)} = 2.2 \times 10^5 \text{ s}$$

$$t = 2.2 \times 10^5 \text{ s} \times \frac{1 \text{ h}}{3600 \text{ s}} = 61 \text{ h}$$

12.7

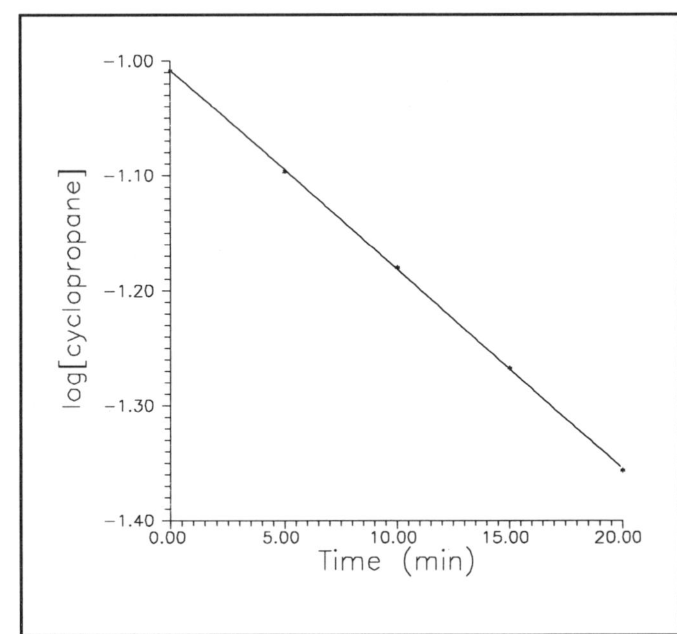

Slope = -0.0173/min = -2.9×10^{-4}/s and k = -2.303(slope)

A plot of log[cyclopropane] versus time is linear, indicating that the data fit the equation for a first order reaction. k = 6.7×10^{-4}/s (0.040/min)

12.8 (a) k = 1.8×10^{-5}/s

$$t_{1/2} = \frac{0.693}{k} = \frac{0.693}{1.8 \times 10^{-5}/s} = 38{,}500 \text{ s}$$

$$t_{1/2} = 38{,}500 \text{ s} \times \frac{1 \text{ h}}{3600 \text{ s}} = 11 \text{ h}$$

(b) $0.30\text{ M} \xrightarrow{t_{1/2}} 0.15\text{ M} \xrightarrow{t_{1/2}} 0.075\text{ M} \xrightarrow{t_{1/2}} 0.0375\text{ M} \xrightarrow{t_{1/2}} 0.019\text{ M}$

(c) Since 25% of the initial concentration corresponds to 1/4 or $(1/2)^2$ of the initial concentration, the time required is two half-lives:

$t = 2t_{1/2} = 2(11\text{ h}) = 22\text{ h}$

12.9

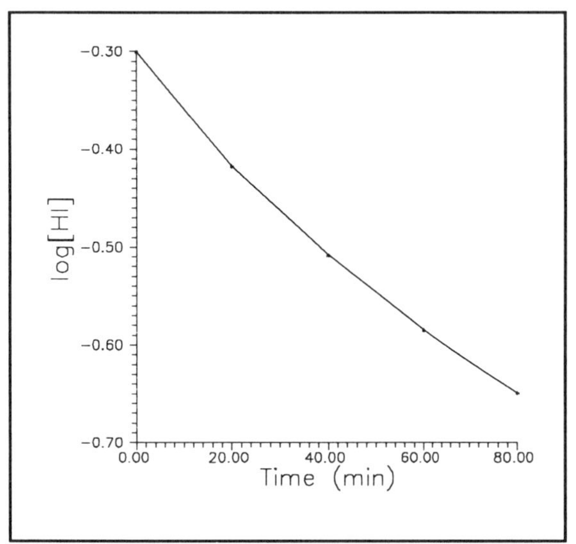

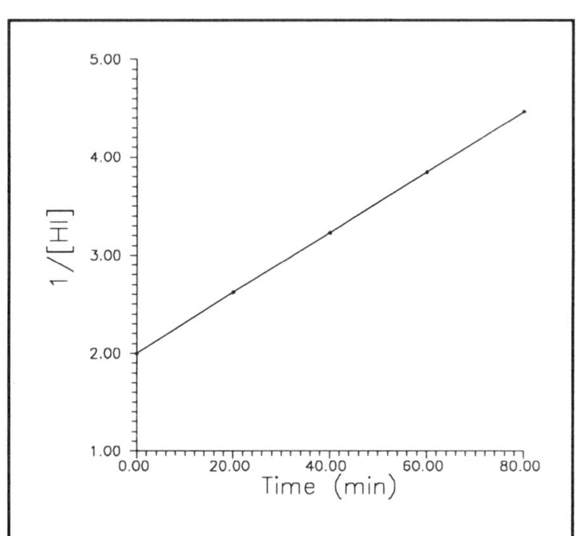

(a) A plot of 1/[HI] versus time is linear. The reaction is second order.

(b) $k = \text{slope} = 0.0308/(\text{M} \cdot \text{min})$

(c) $$t = \frac{1}{k}\left[\frac{1}{[\text{HI}]_t} - \frac{1}{[\text{HI}]_o}\right] = \frac{1}{0.0308/(\text{M}\cdot\text{min})}\left[\frac{1}{0.100\text{ M}} - \frac{1}{0.500\text{ M}}\right] = 260\text{ min}$$

(d) It requires one half-life ($t_{1/2}$) for the [HI] to drop from 0.400 M to 0.200 M.

$$t_{1/2} = \frac{1}{k[\text{HI}]_o} = \frac{1}{[0.0308/(\text{M}\cdot\text{min})](0.400\text{ M})} = 81.2\text{ min}$$

12.10 (a)

$$NO_2(g) + F_2(g) \rightarrow NO_2F(g) + F(g)$$
$$F(g) + NO_2(g) \rightarrow NO_2F(g)$$

Overall reaction $$2\ NO_2(g) + F_2(g) \rightarrow 2\ NO_2F(g)$$

Because F(g) is produced in the first reaction and consumed in the second, it is a reaction intermediate.

(b) In each reaction there are two reactants, so each elementary reaction is bimolecular.

12.11 (a) rate = $k[O_3][O]$

(b) rate = $k[Br]^2[Ar]$

(c) rate = $k[Co(CN)_5(H_2O)^{2-}]$

12.12

$$Co(CN)_5(H_2O)^{2-}(aq) \rightarrow Co(CN)_5^{2-}(aq) + H_2O(l) \quad \text{(slow)}$$

$$Co(CN)_5^{2-}(aq) + I^-(aq) \rightarrow Co(CN)_5I^{3-}(aq) \quad \text{(fast)}$$

Overall reaction

$$Co(CN)_5(H_2O)^{2-}(aq) + I^-(aq) \rightarrow Co(CN)_5I^{3-}(aq) + H_2O(l)$$

The overall reaction is the stoichiometric reaction.

The rate law for the first (slow) elementary reaction is:

rate = $k[Co(CN)_5(H_2O)^{2-}]$

12.13 (a) $\log\left(\frac{k_2}{k_1}\right) = \left(\frac{-E_a}{2.303\,R}\right)\left(\frac{1}{T_2} - \frac{1}{T_1}\right)$

$k_1 = 3.7 \times 10^{-5}$/s, $T_1 = 25°C = 298$ K

$k_2 = 1.7 \times 10^{-3}$/s, $T_2 = 55°C = 328$ K

$$E_a = -\frac{[\log k_2 - \log k_1](2.303)(R)}{\left(\frac{1}{T_2} - \frac{1}{T_1}\right)}$$

$$E_a = -\frac{[\log(1.7 \times 10^{-3}) - \log(3.7 \times 10^{-5})](2.303)[8.314 \times 10^{-3}\ \text{kJ/(K·mol)}]}{\left(\frac{1}{328\ \text{K}} - \frac{1}{298\ \text{K}}\right)}$$

$E_a = 104$ kJ/mol

(b) $k_1 = 3.7 \times 10^{-5}$/s, $T_1 = 25°C = 298$ K

solve for k_2, $T_2 = 35°C = 308$ K

$$\log k_2 = \left(\frac{-E_a}{2.303\,R}\right)\left(\frac{1}{T_2} - \frac{1}{T_1}\right) + \log k_1$$

$$\log k_2 = \left(\frac{-104\ \text{kJ/mol}}{(2.303)[8.314 \times 10^{-3}\ \text{kJ/(K·mol)}]}\right)\left(\frac{1}{308\ \text{K}} - \frac{1}{298\ \text{K}}\right) + \log(3.7 \times 10^{-5})$$

$\log k_2 = -3.84$ $\qquad k_2 = 10^{-3.84} = 1.4 \times 10^{-4}$/s

Understanding Key Concepts

1. Because rate = k[A][B], the rate is proportional to the product of the number of A molecules and the number of B molecules.

   The relative rates of the reaction in vessels (a) – (d) are 2 : 1 : 4 : 2

2. Because the same reaction takes place in each vessel, the ks are all the same.

3. (a) For the first-order reaction, half of the A molecules are converted to B each minute.

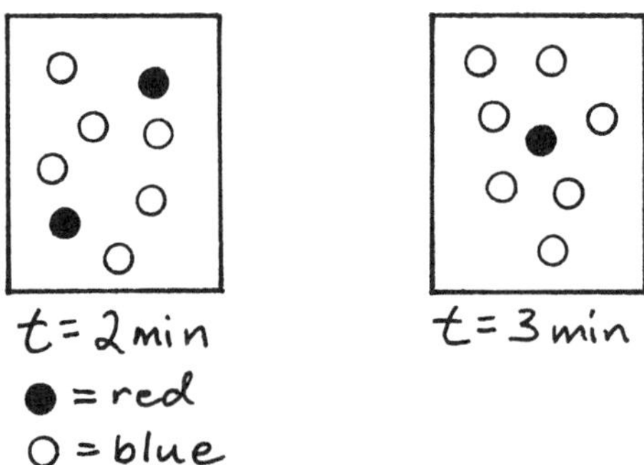

(b) Because half of the A molecules are converted to B molecules in 1 min, the half-life is 1 min.

4. (a) Because two A molecules combine to form A_2, the reaction is second order in A.

(b) rate = $k[A]^2$

(c) The second box represents the passing of one half-life, and the third box represents the passing of a second half-life for a second–order reaction.

A relative value of k can be calculated.

$$k = \frac{0.693}{t_{1/2}[A]} = \frac{0.693}{(1)(16)} = 0.0433$$

$t_{1/2}$ in going from box 3 to box 4 is: $t_{1/2} = \dfrac{0.693}{k[A]} = \dfrac{0.693}{(0.0433)(4)} = 4 \text{ min}$

(For fourth box, t = 7 min)

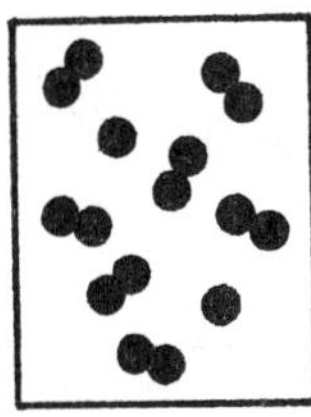

t = 3 min + 4 min = 7 min

5. (a) bimolecular (b) unimolecular (c) termolecular

6. (a) rate = $k[B_2][C]$

(b) $B_2 + C \rightarrow CB + B$ (slow)

$CB + A \rightarrow AB + C$ (fast)

(c) C is a catalyst. C does not appear in the chemical equation because it is consumed in the first step and regenerated in the second step.

Additional Problems

Reaction Rates

12.14 M/s or $\frac{\text{mol}}{\text{L} \cdot \text{s}}$

12.15 molecules/($cm^3 \cdot s$)

12.16 (a) $\text{rate} = \frac{-\Delta[\text{cyclopropane}]}{\Delta t} = -\frac{0.080\text{ M} - 0.098\text{ M}}{5.0\text{ min} - 0.0\text{ min}} = 3.6 \times 10^{-3}\text{ M/min}$

$$\text{rate} = 3.6 \times 10^{-3}\ \frac{\text{M}}{\text{min}} \times \frac{1\text{ min}}{60\text{ s}} = 6.0 \times 10^{-5}\text{ M/s}$$

(b) $\text{rate} = \frac{-\Delta[\text{cyclopropane}]}{\Delta t} = -\frac{0.044\text{ M} - 0.054\text{ M}}{20.0\text{ min} - 15.0\text{ min}} = 2.0 \times 10^{-3}\text{ M/min}$

$$\text{rate} = 2.0 \times 10^{-3}\ \frac{\text{M}}{\text{min}} \times \frac{1\text{ min}}{60\text{ s}} = 3.3 \times 10^{-5}\text{ M/s}$$

12.17 (a) $\text{rate} = -\frac{\Delta[NO_2]}{\Delta t} = -\frac{(5.59 \times 10^{-3}\text{ M}) - (6.58 \times 10^{-3}\text{ M})}{100\text{ s} - 50\text{ s}} = 2.0 \times 10^{-5}\text{ M/s}$

(b) $\text{rate} = -\frac{\Delta[NO_2]}{\Delta t} = -\frac{(4.85 \times 10^{-3}\text{ M}) - (5.59 \times 10^{-3}\text{ M})}{150\text{ s} - 100\text{ s}} = 1.5 \times 10^{-5}\text{ M/s}$

12.18

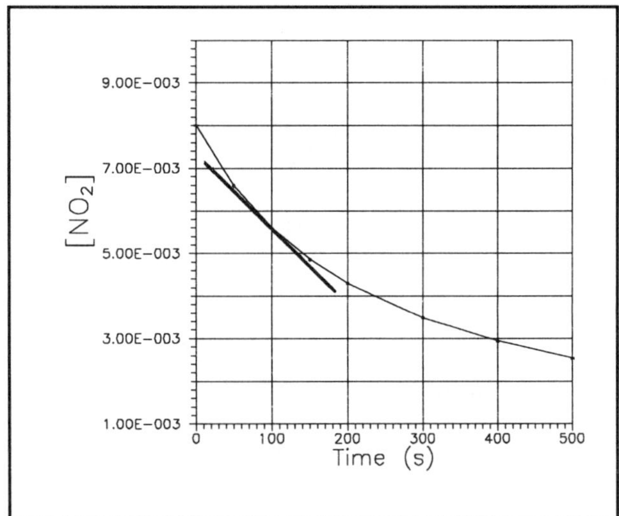

(a) The instantaneous rate of decomposition of NO_2 at t = 100 s is determined from the slope of the curve at t = 100 s.

$$\text{rate} = -\frac{\Delta[NO_2]}{\Delta t} = -\text{slope} = -\frac{(4.00 \times 10^{-3}\ M) - (7.00 \times 10^{-3}\ M)}{190\ s - 20\ s} = 1.8 \times 10^{-5}\ M/s$$

(b) The initial rate of decomposition of NO_2 is determined from the slope of the curve at t = 0 s. This is equivalent to the slope of the curve from 0 s to 50 s because in this time interval the curve is almost linear.

$$\text{initial rate} = -\frac{\Delta[NO_2]}{\Delta t} = -\text{slope} = -\frac{(6.58 \times 10^{-3}\ M) - (8.00 \times 10^{-3}\ M)}{50\ s - 0\ s}$$

initial rate = 2.8×10^{-5} M/s

12.19 (a) $-\frac{\Delta[H_2]}{\Delta t} = -3\frac{\Delta[N_2]}{\Delta t}$

The rate of consumption of H_2 is 3 times faster.

(b) $\frac{\Delta[NH_3]}{\Delta t} = -2\frac{\Delta[N_2]}{\Delta t}$

The rate of formation of NH_3 is 2 times faster.

12.20 $N_2(g) + 3\ H_2(g) \rightarrow 2\ NH_3(g)$

$$-\frac{\Delta[N_2]}{\Delta t} = -\frac{1}{3}\frac{\Delta[H_2]}{\Delta t} = \frac{1}{2}\frac{\Delta[NH_3]}{\Delta t}$$

12.21 (a) $-\frac{\Delta[I^-]}{\Delta t} = -3\frac{\Delta[S_2O_8^{2-}]}{\Delta t} = 3(1.5 \times 10^{-3}\ M/s) = 4.5 \times 10^{-3}\ M/s$

(b) $\frac{\Delta[SO_4^{2-}]}{\Delta t} = -2\frac{\Delta[S_2O_8^{2-}]}{\Delta t} = 2(1.5 \times 10^{-3}\ M/s) = 3.0 \times 10^{-3}\ M/s$

Rate Laws

12.22 rate = $k[NO]^2[Br_2]$

2nd order in NO
1st order in Br_2
3rd order overall

12.23 rate = $k[CHCl_3][Cl_2]^{1/2}$

1st order in $CHCl_3$
1/2 order in Cl_2
3/2 order overall

12.24 rate = $k[H_2][ICl]$

units for k are $\frac{L}{mol \cdot s}$ or $1/(M \cdot s)$

12.25 rate = $k[NO]^2[H_2]$, units for k are $1/(M^2 \cdot s)$

12.26 (a) rate = $k[CH_3COCH_3]^m$

$$m = \frac{\log\left(\frac{\text{rate}_2}{\text{rate}_1}\right)}{\log\left(\frac{[CH_3COCH_3]_2}{[CH_3COCH_3]_1}\right)} = \frac{\log\left(\frac{7.8 \times 10^{-5}}{5.2 \times 10^{-5}}\right)}{\log\left(\frac{9.0 \times 10^{-3}}{6.0 \times 10^{-3}}\right)} = 1$$

rate = $k[CH_3COCH_3]$

(b) From experiment #1,

$$k = \frac{\text{rate}}{[CH_3COCH_3]} = \frac{5.2 \times 10^{-5}\ \text{M/s}}{6.0 \times 10^{-3}\ \text{M}} = 8.7 \times 10^{-3}/\text{s}$$

(c) rate = $k[CH_3COCH_3] = (8.7 \times 10^{-3}/\text{s})(1.8 \times 10^{-3}\text{M}) = 1.6 \times 10^{-5}$ M/s

12.27 (a) rate = $k[CH_3NNCH_3]^m$

$$m = \frac{\log\left(\frac{\text{rate}_2}{\text{rate}_1}\right)}{\log\left(\frac{[CH_3NNCH_3]_2}{[CH_3NNCH_3]_1}\right)} = \frac{\log\left(\frac{2.0 \times 10^{-6}}{6.0 \times 10^{-6}}\right)}{\log\left(\frac{8.0 \times 10^{-3}}{2.4 \times 10^{-2}}\right)} = 1$$

rate = $k[CH_3NNCH_3]$

(b) From Experiment #1, $k = \frac{\text{rate}}{[CH_3NNCH_3]} = \frac{6.0 \times 10^{-6}\ \text{M/s}}{2.4 \times 10^{-2}\ \text{M}} = 2.5 \times 10^{-4}/\text{s}$

(c) rate = $k[CH_3NNCH_3] = (2.5 \times 10^{-4}/\text{s})(0.020\ \text{M}) = 5.0 \times 10^{-6}$ M/s

12.28 (a) rate = $k[NH_4^+]^m[NO_2^-]^n$

$$m = \frac{\log\left(\frac{\text{rate}_2}{\text{rate}_1}\right)}{\log\left(\frac{[NH_4^+]_2}{[NH_4^+]_1}\right)} = \frac{\log\left(\frac{3.6 \times 10^{-6}}{7.2 \times 10^{-6}}\right)}{\log\left(\frac{0.12}{0.24}\right)} = 1$$

$$n = \frac{\log\left(\frac{\text{rate}_3}{\text{rate}_2}\right)}{\log\left(\frac{[NO_2^-]_3}{[NO_2^-]_2}\right)} = \frac{\log\left(\frac{5.4 \times 10^{-6}}{3.6 \times 10^{-6}}\right)}{\log\left(\frac{0.15}{0.10}\right)} = 1$$

rate = $k[NH_4^+][NO_2^-]$

(b) From Experiment #1,

$$k = \frac{\text{rate}}{[NH_4^+][NO_2^-]} = \frac{7.2 \times 10^{-6}\ \text{M/s}}{(0.24\ \text{M})(0.10\ \text{M})} = 3.0 \times 10^{-4}/(\text{M} \cdot \text{s})$$

(c) rate = $k[NH_4^+][NO_2^-]$ = $[3.0 \times 10^{-4}/(\text{M} \cdot \text{s})](0.39\ \text{M})(0.052\ \text{M}) = 6.1 \times 10^{-6}$ M/s

12.29 (a) rate = $k[NO]^m[Cl_2]^n$

$$m = \frac{\log\left(\frac{\text{rate}_2}{\text{rate}_1}\right)}{\log\left(\frac{[NO]_2}{[NO]_1}\right)} = \frac{\log\left(\frac{4.0 \times 10^{-2}}{1.0 \times 10^{-2}}\right)}{\log\left(\frac{0.26}{0.13}\right)} = 2$$

$$n = \frac{\log\left(\frac{\text{rate}_3}{\text{rate}_1}\right)}{\log\left(\frac{[Cl_2]_3}{[Cl_2]_1}\right)} = \frac{\log\left(\frac{5.0 \times 10^{-3}}{1.0 \times 10^{-2}}\right)}{\log\left(\frac{0.10}{0.20}\right)} = 1$$

rate = $k[NO]^2[Cl_2]$

(b) From Experiment #1, $k = \frac{\text{rate}}{[NO]^2[Cl_2]} = \frac{1.0 \times 10^{-2}\text{ M/s}}{(0.13\text{ M})^2(0.20\text{ M})} = 3.0/(M^2 \cdot s)$

(c) rate = $k[NO]^2[Cl_2] = [3.0/(M^2 \cdot s)](0.12\text{ M})^2(0.12\text{ M}) = 5.2 \times 10^{-3}$ M/s

Integrated Rate Law; Half-Life

12.30 $\log\frac{[C_3H_6]_t}{[C_3H_6]_o} = \frac{-kt}{2.303}$, $k = 6.7 \times 10^{-4}/s$

(a) $t = 30\text{ min} \times \frac{60\text{ s}}{1\text{ min}} = 1800\text{ s}$

$$\log[C_3H_6]_t = \frac{-kt}{2.303} + \log[C_3H_6]_o$$

$$\log[C_3H_6]_t = \frac{-(6.7 \times 10^{-4}/s)(1800\text{ s})}{2.303} + \log(0.0500)$$

$\log[C_3H_6]_t = -1.825$ $\qquad$ $[C_3H_6]_t = 10^{-1.825} = 0.015$ M

(b) $$t = \frac{(2.303)\log\frac{[C_3H_6]_t}{[C_3H_6]_o}}{-k}$$

$$t = \frac{(2.303)\log\left(\frac{0.0100}{0.0500}\right)}{-(6.7 \times 10^{-4}/s)} = 2403 \text{ s}$$

$$t = 2403 \text{ s} \times \frac{1 \text{ min}}{60 \text{ s}} = 40 \text{ min}$$

(c) $[C_3H_6]_0 = 0.0500$ M

If 25% of the C_3H_6 reacts then 75% remains.

$[C_3H_6]_t = (0.75)(0.0500 \text{ M}) = 0.0375$ M.

$$t = \frac{(2.303)\log\frac{[C_3H_6]_t}{[C_3H_6]_0}}{-k}$$

$$t = \frac{(2.303)\log\left(\frac{0.0375}{0.0500}\right)}{-(6.7 \times 10^{-4}/s)} = 429 \text{ s}$$

$$t = 429 \text{ s} \times \frac{1 \text{ min}}{60 \text{ s}} = 7.2 \text{ min}$$

12.31 $t_{1/2} = \frac{0.693}{k} = \frac{0.693}{6.7 \times 10^{-4}/s} = 1034 \text{ s} = 17 \text{ min}$

$$t = \left(\frac{-2.303}{k}\right)\log\frac{[C_3H_6]_t}{[C_3H_6]_0} = \left(\frac{-2.303}{6.7 \times 10^{-4}/s}\right)\log\frac{(0.0625)(0.0500)}{(0.0500)} = 4140 \text{ s}$$

$$t = 4140 \text{ s} \times \frac{1 \text{ min}}{60 \text{ s}} = 69 \text{ min}$$

This is also 4 half-lives. 100 → 50 → 25 → 12.5 → 6.25

12.32 $kt = \frac{1}{[C_4H_6]_t} - \frac{1}{[C_4H_6]_o}$, $k = 4.0 \times 10^{-2}/(M \cdot s)$

(a) $t = 1.00 \text{ h} \times \frac{3600 \text{ s}}{1 \text{ h}} = 3600 \text{ s}$

$$\frac{1}{[C_4H_6]_t} = kt + \frac{1}{[C_4H_6]_o}$$

$$\frac{1}{[C_4H_6]_t} = (4.0 \times 10^{-2}/(M \cdot s))(3600 \text{ s}) + \frac{1}{0.0200 \text{ M}}$$

$$\frac{1}{[C_4H_6]_t} = 194/M \text{ and } [C_4H_6] = 5.2 \times 10^{-3} \text{ M}$$

(b) $t = \frac{1}{k}\left[\frac{1}{[C_4H_6]_t} - \frac{1}{[C_4H_6]_o}\right]$

$$t = \frac{1}{4.0 \times 10^{-2}/(M \cdot s)}\left[\frac{1}{(0.0020 \text{ M})} - \frac{1}{(0.0200 \text{ M})}\right] = 11{,}250 \text{ s}$$

$$t = 11{,}250 \text{ s} \times \frac{1 \text{ h}}{3600 \text{ s}} = 3.1 \text{ h}$$

12.33 $t_{1/2} = \frac{1}{k[C_4H_6]_o} = \frac{1}{[4.0 \times 10^{-2}/(M \cdot s)](0.0200 \text{ M})} = 1250 \text{ s} = 21 \text{ min}$

$$t = t_{1/2} = \frac{1}{k[C_4H_6]_o} = \frac{1}{[4.0 \times 10^{-2}/(M \cdot s)](0.0100 \text{ M})} = 2500 \text{ s} = 42 \text{ min}$$

12.34

time (min)	$[N_2O]$	$\log[N_2O]$	$1/[N_2O]$
0	0.250	–0.602	4.00
60	0.218	–0.662	4.59
90	0.204	–0.690	4.90
120	0.190	–0.721	5.26
180	0.166	–0.780	6.02

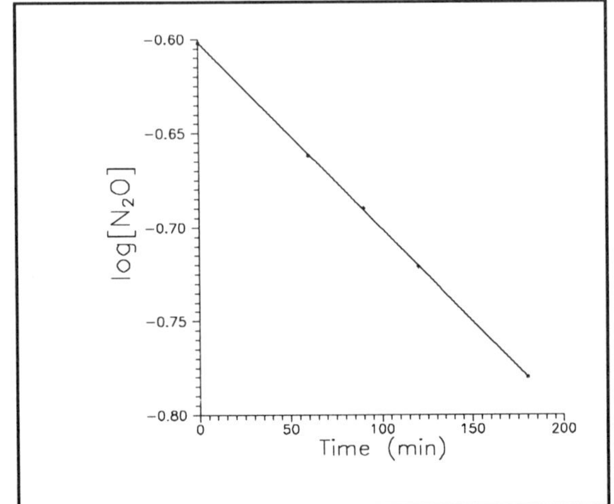

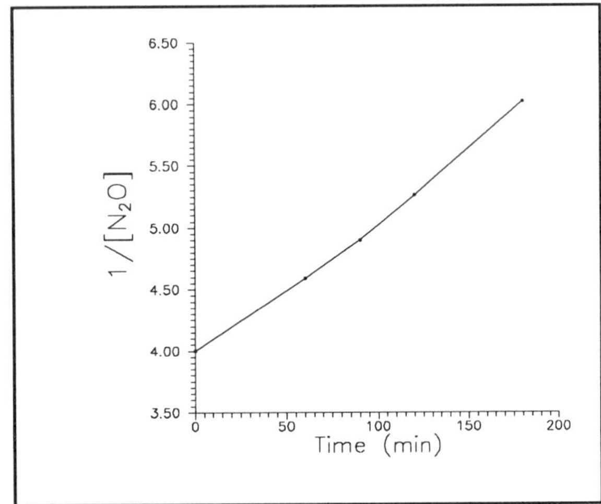

A plot of $\log[N_2O]$ versus time is linear. The reaction is first order in N_2O.

$$k = -2.303(\text{slope}) = -2.303(-9.88 \times 10^{-4}/\text{min}) = 2.28 \times 10^{-3}/\text{min}$$

$$k = 2.28 \times 10^{-3}/\text{min} \times \frac{1 \text{ min}}{60 \text{ s}} = 3.79 \times 10^{-5}/\text{s}$$

12.35

time (s)	[NOBr]	log[NOBr]	1/[NOBr]
0	0.0400	–1.398	25.0
10	0.0303	–1.519	33.0
20	0.0244	–1.613	41.0
30	0.0204	–1.690	49.0
40	0.0175	–1.757	57.1

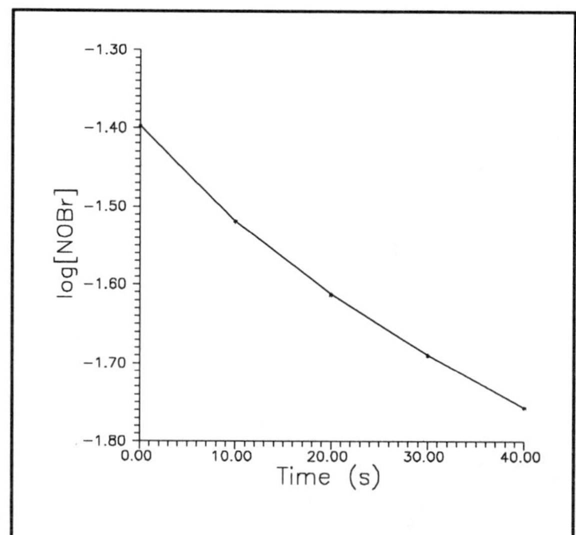

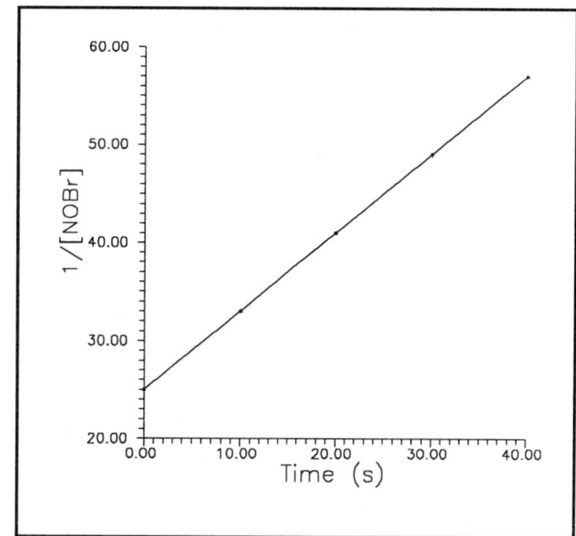

A plot of 1/[NOBr] versus time is linear. The reaction is second order in NOBr.

k = slope = 0.80/(M · s)

12.36 Since 0.060 M is half of 0.120 M, 5.2 h is the half-life.

For a first–order reaction, the half-life is independent of initial concentration. Since 0.015 M is half of 0.030 M, it will take one half-life, 5.2 h.

$$k = \frac{0.693}{t_{1/2}} = \frac{0.693}{5.2\ \text{h}} = 0.133/\text{h}$$

$$\log\frac{[N_2O_5]_t}{[N_2O_5]_o} = \frac{-kt}{2.303}$$

$$t = \frac{(2.303)\log\dfrac{[N_2O_5]_t}{[N_2O_5]_o}}{-k}$$

$$t = \frac{(2.303)\log\left(\dfrac{0.015}{0.480}\right)}{-(0.133/\text{h})} = 26\ \text{h} \qquad \text{(Note that t is five half-lives)}$$

12.37 $k = \dfrac{0.693}{t_{1/2}} = \dfrac{0.693}{248\ \text{s}} = 2.79 \times 10^{-3}/\text{s}$

Reaction Mechanisms

12.38 (a) The reaction mechanism is a sequence of elementary reactions that defines the pathway from reactants to products.

(b) An elementary reaction is a single step in a reaction mechanism.

(c) Molecularity is the number of molecules (or atoms) on the reactant side of a chemical equation for an elementary reaction.

(d) A reaction intermediate is a species formed in one step of a reaction mechanism and consumed in a subsequent step.

12.39 The rate determining step is the slowest step in a multistep reaction.

The rate law shows the molecularity of the rate determining step. The observed rate law is the rate law for the rate determining step.

12.40 (a)

$$H_2(g) + ICl(g) \rightarrow HI(g) + HCl(g)$$
$$HI(g) + ICl(g) \rightarrow I_2(g) + HCl(g)$$

Overall reaction $$H_2(g) + 2\ ICl(g) \rightarrow I_2(g) + 2\ HCl(g)$$

(b) Because HI(g) is produced in the first step and consumed in the second step, it is a reaction intermediate.

(c) In each reaction there are two reactant molecules, so each elementary reaction is bimolecular.

12.41 (a)

$$NO(g) + Cl_2(g) \rightarrow NOCl_2(g)$$
$$NOCl_2(g) + NO(g) \rightarrow 2\ NOCl(g)$$

Overall reaction $$2\ NO(g) + Cl_2(g) \rightarrow 2\ NOCl(g)$$

(b) Because $NOCl_2$ is produced in the first step and consumed in the second step, $NOCl_2$ is a reaction intermediate.

(c) Each elementary step is bimolecular.

12.42 (a) bimolecular, rate = $k[O_3][Cl]$

(b) unimolecular, rate = $k[NO_2]$

(c) bimolecular, rate = $k[ClO][O]$

(d) termolecular, rate = $k[Cl]^2[N_2]$

12.43 (a) unimolecular, rate = $k[I_2]$

(b) termolecular, rate = $k[NO]^2[Br_2]$

(c) bimolecular, rate = $k[CH_3Br][OH^-]$

(d) unimolecular, rate = $k[N_2O_5]$

12.44 (a)

$$NO_2Cl(g) \rightarrow NO_2(g) + Cl(g)$$
$$Cl(g) + NO_2Cl(g) \rightarrow NO_2(g) + Cl_2(g)$$

Overall reaction $2\ NO_2Cl(g) \rightarrow 2\ NO_2(g) + Cl_2(g)$

(b) rate = $k_1[NO_2Cl]$

12.45 $H_2(g) + ICl(g) \rightarrow HI(g) + HCl(g)$ (slow)

$HI(g) + ICl(g) \rightarrow I_2(g) + HCl(g)$ (fast)

12.46 $A + B \rightarrow AB$ (slow)

$AB + B \rightarrow AB_2$ (fast)

The Arrhenius Equation

12.47 Plot log k versus 1/T to determine the activation energy, E_a.

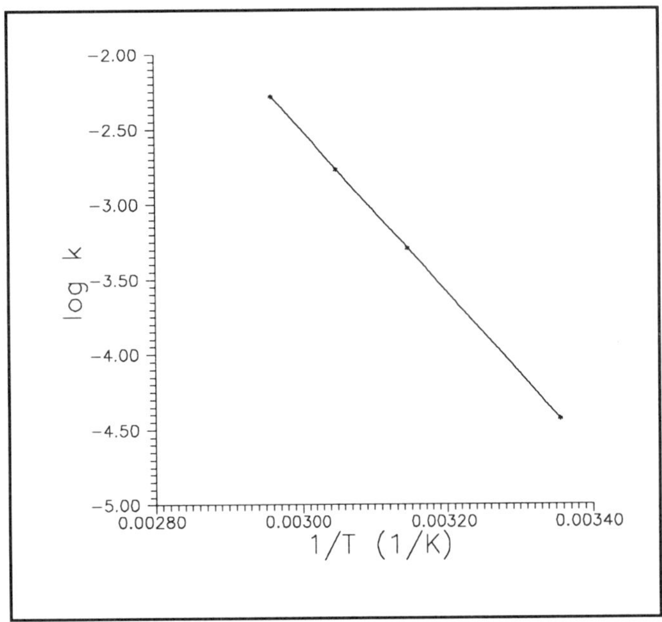

Slope = –5411 K

$E_a = -2.303R(\text{slope}) = -(2.303)(8.314 \times 10^{-3}\ \text{kJ/(K} \cdot \text{mol))}(-5411\ \text{K}) = 104\ \text{kJ/mol}$

12.48 Plot log k versus 1/T to determine the activation energy, E_a.

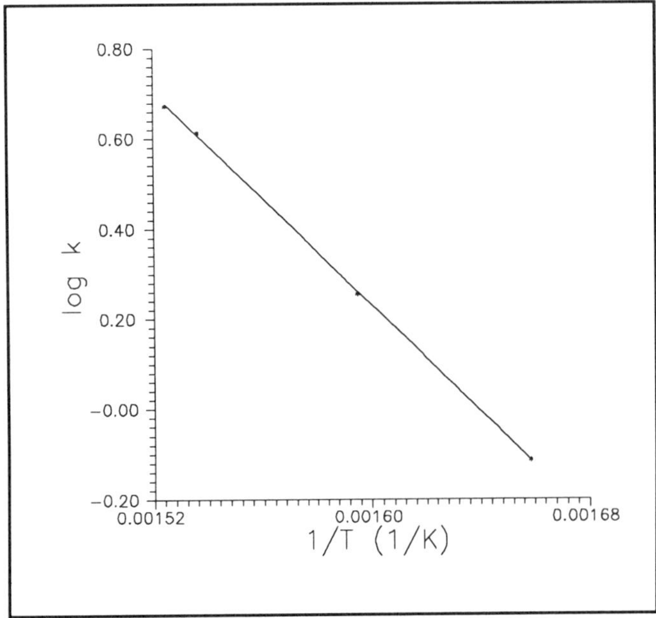

Slope = –5903 K

$E_a = -2.303\ R(\text{slope}) = -(2.303)(8.314 \times 10^{-3}\ \text{kJ/(K} \cdot \text{mol)})(-5903\text{K}) = 113\ \text{kJ/mol}$

12.49 (a) $$\log\left(\frac{k_2}{k_1}\right) = \left(\frac{-E_a}{2.303\ R}\right)\left(\frac{1}{T_2} - \frac{1}{T_1}\right)$$

$k_1 = 1.3/(\text{M} \cdot \text{s})$, $T_1 = 700$ K

$k_2 = 23.0/(\text{M} \cdot \text{s})$, $T_2 = 800$ K

$$E_a = -\frac{[\log k_2 - \log k_1](2.303)(R)}{\left(\frac{1}{T_2} - \frac{1}{T_1}\right)}$$

$$E_a = -\frac{[\log(23.0) - \log(1.3)](2.303)[8.314 \times 10^{-3}\ \text{kJ/(K}\cdot\text{mol)}]}{\left(\frac{1}{800\ \text{K}} - \frac{1}{700\ \text{K}}\right)} = 134\ \text{kJ/mol}$$

(b) $k_1 = 1.3/(\text{M} \cdot \text{s})$, $T_1 = 700$ K

solve for k_2, $T_2 = 750$ K

$$\log k_2 = \left(\frac{-E_a}{2.303\ R}\right)\left(\frac{1}{T_2} - \frac{1}{T_1}\right) + \log k_1$$

$$\log k_2 = \left(\frac{-133.8\ \text{kJ/mol}}{(2.303)[8.314 \times 10^{-3}\ \text{kJ/(K}\cdot\text{mol)}]}\right)\left(\frac{1}{750\ \text{K}} - \frac{1}{700\ \text{K}}\right) + \log(1.3)$$

$\log k_2 = 0.779$ $\quad k_2 = 10^{0.779} = 6.0/(\text{M} \cdot \text{s})$

12.50 $\log\left(\frac{k_2}{k_1}\right) = \left(\frac{-E_a}{2.303\,R}\right)\left[\frac{1}{T_2} - \frac{1}{T_1}\right]$

(a) Since rate doubles, $k_2 = 2k_1$

$k_1 = 1.0 \times 10^{-3}/s$, $T_1 = 25°C = 298\ K$

$k_2 = 2.0 \times 10^{-3}/s$, $T_2 = 35°C = 308\ K$

$$E_a = -\frac{[\log k_2 - \log k_1](2.303)(R)}{\left(\frac{1}{T_2} - \frac{1}{T_1}\right)}$$

$$E_a = -\frac{[\log(2.0 \times 10^{-3}) - \log(1.0 \times 10^{-3})](2.303)[8.314 \times 10^{-3}\ kJ/(K\cdot mol)]}{\left(\frac{1}{308\ K} - \frac{1}{298\ K}\right)}$$

$E_a = 53\ kJ/mol$

(b) Since rate triples, $k_2 = 3k_1$

$k_1 = 1.0 \times 10^{-3}/s$, $T_1 = 25°C = 298\ K$

$k_2 = 3.0 \times 10^{-3}/s$, $T_2 = 35°C = 308\ K$

$$E_a = -\frac{[\log(3.0 \times 10^{-3}) - \log(1.0 \times 10^{-3})](2.303)[8.314 \times 10^{-3}\ kJ/(K\cdot mol)]}{\left(\frac{1}{308\ K} - \frac{1}{298\ K}\right)}$$

$E_a = 84\ kJ/mol$

12.51 The reactant molecules may not have sufficient kinetic energy or correct orientation to react.

12.52 (a) The frequency factor is the pre–exponential term in the Arrhenius equation, related both to the probability that reactants are favorably oriented and to the collision frequency.

(b) The steric factor represents the fraction of collisions having proper orientation for conversion of reactants to products.

(c) The activation energy (E_a) is the height of the energy barrier between reactants and products.

(d) The transition state is the configuration of atoms at the maximum in the potential energy profile for a reaction.

12.53 (a) (b)

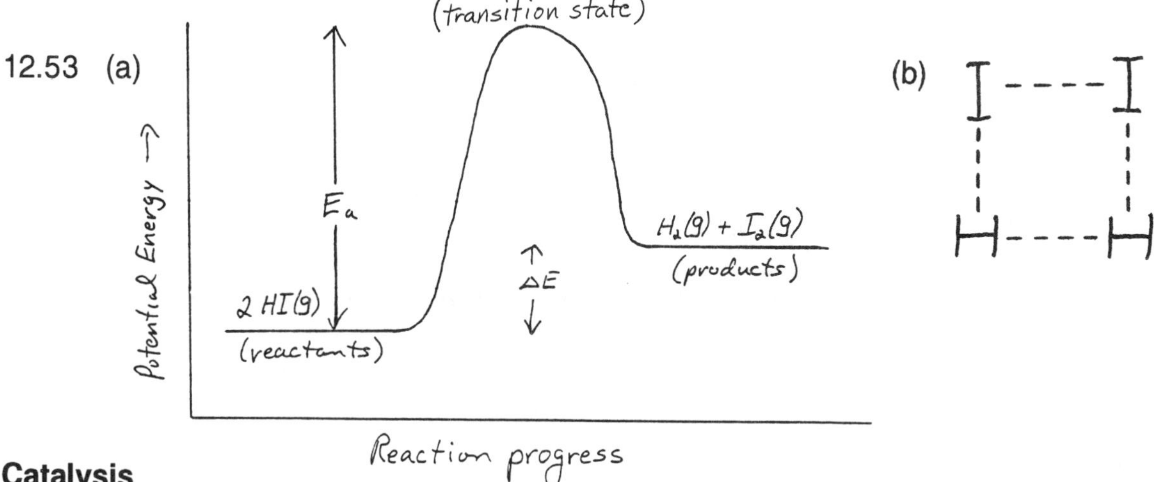

Catalysis

12.54 A catalyst increases the rate of a reaction by changing the reaction mechanism and lowering the activation energy.

12.55 A catalyst doesn't appear in the chemical equation for a reaction because a catalyst is consumed in one reaction step and regenerated in a later step.

12.56 A homogeneous catalyst is one that exists in the same phase as the reactants.

example: NO(g) acts as a homogeneous catalyst for the conversion of $O_2(g)$ to $O_3(g)$.

A heterogeneous catalyst is one that exists in a different phase from the reactants.

example: solid Ni, Pd, or Pt for catalytic hydrogenation, $C_2H_4(g) + H_2(g) \rightarrow C_2H_6(g)$.

12.57 (a) $O_3(g) + O(g) \rightarrow 2\ O_2(g)$

(b) Cl acts as a catalyst.

(c) ClO is a reaction intermediate.

(d) A catalyst is consumed in one step and regenerated in a subsequent step. A reaction intermediate is produced in one step and consumed in another.

12.58 (a)

$$2\ SO_2(g) + 2\ NO_2(g) \rightarrow 2\ SO_3(g) + 2\ NO(g)$$
$$2\ NO(g) + O_2(g) \rightarrow 2\ NO_2(g)$$

Overall reaction $2\ SO_2(g) + O_2(g) \rightarrow 2\ SO_3(g)$

(b) $NO_2(g)$ acts as a catalyst because it is used in the first step and regenerated in the second.

NO(g) is a reaction intermediate because it is produced in the first step and consumed in the second.

General Problems

12.59 Reactant concentration, temperature, and a catalyst.

Reactants and a catalyst may show up in the rate law. If they do, changes in their concentrations will affect the rate. The rate constant changes with temperature; it increases with an increase in temperature and decreases with a decrease in temperature.

12.60 (a) The reaction rate will increase with an increase in temperature at constant volume.

(b) The reaction rate will decrease with an increase in volume at constant temperature because reactant concentrations will decrease.

(c) The reaction rate will increase with the addition of a catalyst.

(d) Addition of an inert gas at constant volume will not affect the reaction rate.

12.61 (a) From the data in the table for Experiment 1, we see that 0.20 mol of A reacts with 0.10 mol of B to produce 0.10 mol of D. The balanced equation for the reaction is: $2\ A + B \rightarrow D$

(b) From the data in the table, initial rates $= -\dfrac{\Delta A}{\Delta t}$ have been calculated.

For example, from Experiment 1:

$$\text{Initial Rate} = -\frac{\Delta A}{\Delta t} = -\frac{(4.80\text{ M} - 5.00\text{ M})}{60\text{ s}} = 3.33 \times 10^{-3}\text{ M/s}$$

Initial concentrations and initial rate data have been collected in the table below.

EXPT	$[A]_o$ (M)	$[B]_o$ (M)	$[C]_o$ (M)	Initial Rate (M/s)
1	5.00	2.00	1.00	3.33×10^{-3}
2	10.00	2.00	1.00	6.66×10^{-3}
3	5.00	4.00	1.00	3.33×10^{-3}
4	5.00	2.00	2.00	6.66×10^{-3}

Rate = $k[A]^m[B]^n[C]^p$

From Expts 1 and 2, [A] doubles and the initial rate doubles; therefore m = 1.

From Expts 1 and 3, [B] doubles but the initial rate does not change; therefore n = 0

From Expts 1 and 4, [C] doubles and the initial rate doubles; therefore p = 1.

The reaction is: first order in A; zero order in B; first order in C; second order overall

(c) rate = k[A][C]

(d) C is a catalyst. C appears in the rate law, but it is not consumed in the reaction.

(e) $A + C \rightarrow AC$ (slow)

$AC + B \rightarrow AB + C$ (fast)

$A + AB \rightarrow D$ (fast)

(f) From data in Expt 1:

$$k = \frac{\text{rate}}{[A][C]} = \frac{\Delta D/\Delta t}{[A][C]} = \frac{0.10\text{ M}/60\text{ s}}{(5.00\text{ M})(1.00\text{ M})} = 3.4 \times 10^{-4}/(\text{M} \cdot \text{s})$$

12.62 As the temperature of a gas is raised by 10°C, even though the collision frequency increases by only ~2%, the reaction rate increases by 100% or more because there is an exponential increase in the fraction of the collisions that leads to products.

12.63 For $E_a = 50$ kJ/mol

$$f = e^{-E_a/RT} = \exp\left\{\frac{-50 \text{ kJ/mol}}{[8.314 \times 10^{-3} \text{ kJ/(K}\cdot\text{mol)}](300 \text{ K})}\right\} = 2.0 \times 10^{-9}$$

For $E_a = 100$ kJ/mol

$$f = e^{-E_a/RT} = \exp\left\{\frac{-100 \text{ kJ/mol}}{[8.314 \times 10^{-3} \text{ kJ/(K}\cdot\text{mol)}](300 \text{ K})}\right\} = 3.9 \times 10^{-18}$$

12.64 The higher the height of the barrier in the potential energy profile, the slower the rate of the reaction because the rate is proportional to $e^{-E_a/RT}$.

12.65 (a) 1 → 1/2 → 1/4 → 1/8

After three half-lives, 1/8 of the strontium-90 will remain.

(b) $k = \frac{0.693}{t_{1/2}} = \frac{0.693}{29 \text{ y}} = 0.0239/\text{y} = 0.024/\text{y}$

(c) $t = \left(\frac{-2.303}{k}\right)\log\frac{(\text{Sr-90})_t}{(\text{Sr-90})_o} = \left(\frac{-2.303}{0.0239/\text{y}}\right)\log\frac{(0.01)}{(1)} = 193 \text{ y}$

12.66 $\log\left(\frac{k_2}{k_1}\right) = \left(\frac{-E_a}{2.303\,R}\right)\left(\frac{1}{T_2} - \frac{1}{T_1}\right)$

$k_2 = 2.5k_1$

$k_1 = 1.0, \quad T_1 = 20°\text{C} = 293 \text{ K}$

$k_2 = 2.5, \quad T_2 = 30°C = 303\ K$

$$E_a = -\frac{[\log k_2 - \log k_1](2.303)(R)}{\left(\frac{1}{T_2} - \frac{1}{T_1}\right)}$$

$$E_a = -\frac{[\log(2.5) - \log(1.0)](2.303)[8.314 \times 10^{-3}\ kJ/(K \cdot mol)]}{\left(\frac{1}{303\ K} - \frac{1}{293\ K}\right)}$$

$E_a = 68\ kJ/mol$

$k_1 = 1.0, \quad T_1 = 120°C = 393\ K$

$k_2 = ?, \quad T_2 = 130°C = 403\ K$

Solve for k_2.

$$\log k_2 = \frac{-E_a}{2.303\ R}\left(\frac{1}{T_2} - \frac{1}{T_1}\right) + \log k_1$$

$$\log k_2 = \frac{-68\ kJ/mol}{(2.303)[8.314 \times 10^{-3}\ kJ/(K \cdot mol)]}\left(\frac{1}{403\ K} - \frac{1}{393\ K}\right) + \log(1.0)$$

$\log k_2 = 0.224 \qquad k_2 = 10^{0.224} = 1.7$

The rate is increased by a factor of 1.7.

12.67 (a) $2\ NO(g) + Br_2(g) \rightarrow 2\ NOBr(g)$

(b) Since $NOBr_2$ is generated in the first step and consumed in the second step, $NOBr_2$ is a reaction intermediate.

(c) rate = $k[NO][Br_2]$

(d) It can't be the first step. It must be the second step.

13.1 $\dfrac{[NO_2]^2}{[N_2O_4]} = 4.64 \times 10^{-3}$ M

$$[N_2O_4] = \frac{[NO_2]^2}{4.64 \times 10^{-3}\ M} = \frac{(0.0200\ M)^2}{(4.64 \times 10^{-3}\ M)} = 0.0862\ M$$

13.2 (a) $K_c = \dfrac{[SO_3]^2}{[SO_2]^2[O_2]}$ (b) $K_c = \dfrac{[SO_2]^2[O_2]}{[SO_3]^2}$

13.3 (a) $K_c = \dfrac{[SO_3]^2}{[SO_2]^2[O_2]} = \dfrac{(5.0 \times 10^{-2})^2}{(3.0 \times 10^{-3})^2(3.5 \times 10^{-3})} = 7.9 \times 10^4$

(b) $K_c = \dfrac{[SO_2]^2[O_2]}{[SO_3]^2} = \dfrac{(3.0 \times 10^{-3})^2(3.5 \times 10^{-3})}{(5.0 \times 10^{-2})^2} = 1.3 \times 10^{-5}$

13.4 $K_p = \dfrac{(P_{CO_2})(P_{H_2})}{(P_{CO})(P_{H_2O})} = \dfrac{(6.12)(20.3)}{(1.31)(10.0)} = 9.48$

13.5 $2\ NO(g) + O_2 \rightleftarrows 2\ NO_2(g)$

$\Delta n = 2 - 3 = -1$

$K_p = K_c(RT)^{\Delta n}$, $K_c = K_p(1/RT)^{\Delta n}$

at 500 K: $K_p = (6.9 \times 10^5)[(0.0821)(500)]^{-1} = 1.7 \times 10^4$

at 1000 K: $K_c = (1.3 \times 10^{-2})\left(\dfrac{1}{(0.0821)(1000)}\right)^{-1} = 1.1$

13.6 (a) $K_c = \dfrac{[H_2]^3}{[H_2O]^3}$ $K_p = \dfrac{(P_{H_2})^3}{(P_{H_2O})^3}$

(b) $K_c = [H_2]^2[O_2]$ $K_p = (P_{H_2})^2(P_{O_2})$

13.7 $K_c = 1.2 \times 10^{-42}$. Since K_c is very small, the equilibrium mixture contains mostly H_2 molecules.

13.8 The container volume of 5.0 L must be included to calculate molar concentrations.

(a) $$Q_c = \frac{[NO_2]_t^2}{[NO]_t^2[O_2]_t} = \frac{(0.80\ \text{mol}/5.0\ \text{L})^2}{(0.060\ \text{mol}/5.0\ \text{L})^2(1.0\ \text{mol}/5.0\ \text{L})} = 890$$

Since $Q_c < K_c$, the reaction is not at equilibrium. The reaction will proceed to the right to reach equilibrium.

(b) $$Q_c = \frac{[NO_2]_t^2}{[NO]_t^2[O_2]_t} = \frac{(4.0\ \text{mol}/5.0\ \text{L})^2}{(5.0 \times 10^{-3}\ \text{mol}/5.0\ \text{L})^2(0.20\ \text{mol}/5.0\ \text{L})} = 1.6 \times 10^7$$

Since $Q_c > K_c$, the reaction is not at equilibrium. The reaction will proceed to the left to reach equilibrium.

13.9 $$K_c = \frac{[H]^2}{[H_2]} = 1.2 \times 10^{-42}$$

(a) $[H] = \sqrt{K_c[H_2]} = \sqrt{(1.2 \times 10^{-42})(0.10)} = 3.5 \times 10^{-22}$ M

(b) H atoms = $(3.5 \times 10^{-22}\ \text{mol/L})(1.0\ \text{L})(6.022 \times 10^{23}\ \text{atoms/mol})$

= 210 H atoms

H_2 molecules = $(0.10\ \text{mol/L})(1.0\ \text{L})(6.022 \times 10^{23}\ \text{molecules/mol})$

= 6.0×10^{22} H_2 molecules

13.10

	$CO(g)$	+	$H_2O(g)$	⇄	$CO_2(g)$	+	$H_2(g)$
initial (M)	0.150		0.150		0		0
change (M)	–x		–x		+x		+x
equil (M)	0.150 – x		0.150 – x		x		x

$$K_c = 4.24 = \frac{[CO_2][H_2]}{[CO][H_2O]} = \frac{x^2}{(0.150 - x)^2}$$

Take the square root of both sides and solve for x.

$$\sqrt{4.24} = \sqrt{\frac{x^2}{(0.150 - x)^2}}$$

$$2.06 = \frac{x}{0.150 - x}$$

$x = 0.101$

At equilibrium, $[CO_2] = [H_2] = x = 0.101$ M

$[CO] = [H_2O] = 0.150 - x = 0.150 - 0.101 = 0.049$ M

13.11

	$N_2O_4(g)$	⇄	$2\ NO_2(g)$
initial (M)	0.0500M		0
change (M)	–x		+2x
equil (M)	0.0500 – x		2x

$$K_c = 4.64 \times 10^{-3} = \frac{[NO_2]^2}{[N_2O_4]} = \frac{(2x)^2}{(0.0500 - x)}$$

$4x^2 + (4.64 \times 10^{-3})x - (2.32 \times 10^{-4}) = 0$

Use the quadratic formula to solve for x.

$$x = \frac{(-4.64 \times 10^{-3}) \pm \sqrt{(4.64 \times 10^{-3})^2 + 4(4)(2.32 \times 10^{-4})}}{2(4)} = \frac{-0.00464 \pm 0.06110}{8}$$

$x = -0.008\ 22$ and $0.007\ 06$

Discard the solution that uses the negative square root (–0.008 22) because it will lead to negative concentrations and that is impossible.

$[N_2O_4] = 0.0500 - x = 0.0500 - 0.007\ 06 = 0.0429$ M

$[NO_2] = 2x = 2(0.007\ 06) = 0.0141$ M

13.12 (a) CO(reactant) added, H_2 concentration increases.

(b) CO_2 (product) added, H_2 concentration decreases.

(c) H_2O (reactant) removed, H_2 concentration decreases.

(d) CO_2 (product) removed, H_2 concentration increases.

At equilibrium $Q_c = K_c = \frac{[CO_2][H_2]}{[CO][H_2O]}$, if some CO_2 is removed from the equilibrium mixture, the numerator in Q_c is decreased, which means that $Q_c < K_c$ and the reaction will shift to the right increasing the H_2 concentration.

13.13 (a) Because there are 2 mol of gas on both sides of the balanced equation, the composition of the equilibrium mixture is unaffected by a change in pressure. The number of moles of reaction products remains the same.

(b) Because there are 2 mol of gas on the left side and 1 mol of gas on the right side of the balanced equation, the stress of an increase in pressure is relieved by a shift in the reaction to the side with fewer molels of gas (in this case to products). The number of moles of reaction products increases.

(c) Because there is 1 mol of gas on the left side and 2 mol of gas on the right side of the balanced equation, the stress of an increase in pressure is relieved by a shift in the reaction to the side with fewer moles of gas (in this case, to reactants). The number of moles of reaction product decreases.

13.14 The reaction is exothermic. As the temperature is increased the reaction shifts from right to left. The amount of ethyl acetate decreases.

$$K_c = \frac{[CH_3COOC_2H_5][H_2O]}{[CH_3COOH][C_2H_5OH]}$$

As the temperature is decreased, the reaction shifts from left to right. The product concentrations increase, and the reactant concentrations decrease. This corresponds to an increase in K_c

13.15 (a) A catalyst does not affect the equilibrium composition. The amount of CO remains the same.

(b) The reaction is exothermic. An increase in temperature shifts the reaction toward reactants. The amount of CO increases.

(c) Because there are 3 mol of gas on the left side and 2 mol of gas on the right side of the balanced equation, the stress of an increase in pressure is relieved by a shift in the reaction to the side with fewer moles of gas (in this case, to products). The amount of CO decreases.

(d) An increase in pressure as a result of the addition of an inert gas (with no volume change) does not affect the equilibrium composition. The amount of CO remains the same.

(e) Adding O_2 increases the O_2 concentration and shifts the reaction toward products. The amount of CO decreases.

3.16 (a) Since $k_f >> k_r$, K_c will be large and the equilibrium mixture will have a larger concentration of products than reactants.

(b) $K_c = \dfrac{k_f}{k_r} = \dfrac{1.32 \times 10^{-4} M^{-1}s^{-1}}{1.22 \times 10^{-6} M^{-1}s^{-1}} = 108$

Understanding Key Concepts

1. (a) (1) and (3) since the number of A and B's are the same in the third and fourth box.

(b) $K_c = \dfrac{[B]}{[A]} = \dfrac{6}{4} = 1.5$

(c) Because the same number of molecules appear on both sides of the equation, the volume terms in K_c all cancel. Therefore, we can calculate K_c without including the volume.

2. (a) $A_2 + C_2 \rightleftarrows 2\ AC$ (most product molecules)

(b) $A_2 + B_2 \rightleftarrows 2\ AB$ (fewest product molecules)

3. (a) (2) (b) (1), reverse; (3), forward

4. (a) $A_2 + 2\ B \rightleftarrows 2\ AB$

(b) The number of AB molecules will increase, because as the volume is decreased at constant temperature, the pressure will increase and the reaction will shift to the side of fewer molecules to reduce the pressure.

5. When the stopcock is opened, the reaction will go in the reverse direction because there will be initially an excess of AB molecules.

6. As the temperature is raised, the reaction proceeds in the reverse direction. This is consistent with an exothermic reaction where "heat" can be considered as a product.

Additional Problems

Equilibrium Expressions and Equilibrium Constants

13.17 (a) $K_c = \dfrac{[PCl_3][Cl_2]}{[PCl_5]}$ (b) $K_c = \dfrac{[NOCl]^2}{[NO]^2[Cl_2]}$ (c) $K_c = \dfrac{[CS_2][H_2]^4}{[CH_4][H_2S]^2}$

13.18 (a) $K_p = \dfrac{(P_{PCl_3})(P_{Cl_2})}{(P_{PCl_5})}$, $\Delta n = 1$ and $K_c = K_p\left(\dfrac{1}{RT}\right)$

(b) $K_p = \dfrac{(P_{NOCl})^2}{(P_{NO})^2(P_{Cl_2})}$, $\Delta n = -1$ and $K_c = K_p\left(\dfrac{1}{RT}\right)^{-1}$

(c) $K_p = \dfrac{(P_{CS_2})(P_{H_2})^4}{(P_{CH_4})(P_{H_2S})^2}$, $\Delta n = 2$ and $K_c = K_p\left(\dfrac{1}{RT}\right)^2$

13.19 $K_c = \dfrac{[C_2H_5OC_2H_5][H_2O]}{[C_2H_5OH]^2}$

13.20 $K_c = \dfrac{[HOCH_2CH_2OH]}{[C_2H_4O][H_2O]}$

13.21 (a) $K_c = \dfrac{[\text{malic acid}]}{[\text{fumaric acid}]}$ (b) $K_c = \dfrac{[\text{citric acid}]}{[\text{acetic acid}][\text{oxaloacetic acid}]}$

13.22 The two reactions are the reverse of each other.

$$K_c(\text{reverse}) = \frac{1}{K_c(\text{forward})} = \frac{1}{7.5 \times 10^{-9}} = 1.3 \times 10^{8}$$

13.23 $K_c = \dfrac{[PCl_3][Cl_2]}{[PCl_5]} = \dfrac{(1.5 \times 10^{-2})(3.2 \times 10^{-2})}{(8.3 \times 10^{-3})} = 0.058$

13.24 $K_p = \dfrac{(P_{NOCl})^2}{(P_{NO})^2(P_{Cl_2})} = \dfrac{(1.35)^2}{(0.240)^2(0.608)} = 52.0$

13.25 The container volume of 2.00 L must be included to calculate molar concentrations.

Initial [HI] = 9.30×10^{-3} mol/2.00 L = 4.65×10^{-3} M = 0.004 65 M

	$H_2(g)$	+ $I_2(g)$	$\rightleftarrows$ 2 HI(g)
initial (M)	0	0	0.004 65
change (M)	+x	+x	–2x
equil (M)	x	x	0.004 65 – 2x

x = $[H_2] = [I_2] = 6.29 \times 10^{-4}$ M = 0.000 629 M

[HI] = 0.004 65 – 2x = 0.004 65 – 2(0.000 629) = 0.003 39 M

$$K_c = \frac{[HI]^2}{[H_2][I_2]} = \frac{(0.003\ 39)^2}{(0.000\ 629)^2} = 29.0$$

13.26 (a) $K_c = \dfrac{[H^+][CH_3CO_2^-]}{[CH_3COOH]}$

(b)

	$CH_3COOH(aq)$	⇌	$H^+(aq)$	+	$CH_3CO_2^-(aq)$
initial (M)	1.0		0		0
change (M)	–0.0042		+0.0042		+0.0042
equil (M)	1.0 – 0.0042		0.0042		0.0042

$$K_c = \frac{(0.0042)(0.0042)}{(1.0 - 0.0042)} = 1.8 \times 10^{-5}$$

13.27 (a) $K_c = \dfrac{[CH_3COOC_2H_5][H_2O]}{[CH_3COOH][C_2H_5OH]}$

(b)

	$CH_3COOH(l)$	+	$C_2H_5OH(l)$	⇌	$CH_3COOC_2H_5(l)$	+	$H_2O(l)$
initial (mol)	1.00		1.00		0		0
change (mol)	–x		–x		+x		+x
equil (mol)	1.00 – x		1.00 – x		x		x

x = 0.65 mol; 1.00 – x = 0.35 mol

$$K_c = \frac{(0.65)^2}{(0.35)^2} = 3.4$$

Because there are the same number of molecules on both sides of the equation, the volume terms in K_c cancel. Therefore, we can calculate K_c without including the volume.

13.28 $CH_3COOC_2H_5(l) + H_2O(l) \rightleftharpoons CH_3COOH(l) + C_2H_5OH(l)$

$$K_c(\text{hydrolysis}) = \frac{1}{K_c(\text{forward})} = \frac{1}{3.4} = 0.29$$

13.29 Because $\Delta n = 0$, $K_p = K_c = 4.24$

13.30 $2\ SO_2(g) + O_2(g) \rightleftarrows 2\ SO_3(g)$

$\Delta n = 2 - (2 + 1) = -1$ and $K_p = 3.30$

$$K_c = K_p\left(\frac{1}{RT}\right)^{\Delta n} = (3.30)\left(\frac{1}{(0.0821)(1000)}\right)^{-1} = 271$$

13.31 $K_p = P_{H_2O} = 0.0313$ atm; $\Delta n = 1$

$$K_c = K_p\left(\frac{1}{RT}\right) = (0.0313)\left(\frac{1}{(0.0821)(298)}\right) = 1.28 \times 10^{-3}$$

13.32 $P_{C_{10}H_8} = 0.10 \text{ mm Hg} \times \dfrac{1 \text{ atm}}{760 \text{ mm Hg}} = 1.3 \times 10^{-4}$ atm

$K_p = P_{C_{10}H_8} = 1.3 \times 10^{-4}$

$\Delta n = 1 - 0 = 1$, $T = 27°C = 300$ K

$$K_c = K_p\left(\frac{1}{RT}\right)^{\Delta n} = (1.3 \times 10^{-4})\left(\frac{1}{(0.0821)(300)}\right) = 5.3 \times 10^{-6}$$

13.33 (a) $K_c = \dfrac{[CO_2]^3}{[CO]^3}$ $\qquad K_p = \dfrac{(P_{CO_2})^3}{(P_{CO})^3}$

(b) $K_c = \dfrac{1}{[O_2]^3}$ $\qquad K_p = \dfrac{1}{(P_{O_2})^3}$

(c) $K_c = [SO_3]$ $\qquad K_p = P_{SO_3}$

(d) $K_c = [Ba^{2+}][SO_4^{2-}]$

Using the Equilibrium Constant

13.34 (a) Because K_c is very large, the equilibrium mixture contains mostly product.

(b) Because K_c is very small, the equilibrium mixture contains mostly reactant.

13.35 (a) proceeds hardly at all toward completion

(b) goes almost all the way to completion

13.36 (a) Because K_c is very small, the equilibrium mixture contains mostly reactant.

(b) Because K_c is very large, the equilibrium mixture contains mostly product.

(c) Because K_c = 1.8 (close to 1), the equilibrium mixture contains an appreciable concentration of both reactants and products.

13.37 $K_c = 1.2 \times 10^{82}$ is very large. When equilibrium is reached, very little if any ethanol will remain because the reaction goes to completion.

13.38 Because K_c is very small, pure air will contain very little O_3 (ozone) at equilibrium.

$$3\ O_2(g) \rightleftarrows 2\ O_3(g)$$

$$K_c = \frac{[O_3]^2}{[O_2]^3} = 1.7 \times 10^{-56}$$

$$[O_2] = 8 \times 10^{-3}\ M$$

$$[O_3] = \sqrt{[O_2]^3 \times K_c} = \sqrt{(8 \times 10^{-3})^3(1.7 \times 10^{-56})} = 9 \times 10^{-32}\ M$$

13.39 Q_c has the same form as K_c but contains arbitrary (not necessarily equilibrium concentrations), so it can have any value.

13.40 The container volume of 10 L must be included to calculate molar concentrations.

$$Q_c = \frac{[CS_2]_t[H_2]_t^4}{[CH_4]_t[H_2S]_t^2} = \frac{(3.0\ mol/10\ L)(3.0\ mol/10\ L)^4}{(2.0\ mol/10\ L)(4.0\ mol/10\ L)^2} = 7.6 \times 10^{-2}$$

$$K_c = 2.5 \times 10^{-3}$$

The reaction is not at equilibrium because $Q_c > K_c$. The reaction will proceed from right to left to reach equilibrium.

13.41 $Q_c = \dfrac{[CO][H_2]^3}{[H_2O][CH_4]} = \dfrac{(0.15)(0.20)^3}{(0.035)(0.050)} = 0.69$; $K_c = 4.7$

The reaction is not at equilibrium because $Q_c < K_c$. The reaction will proceed from left to right to reach equilibrium.

13.42 $K_c = \dfrac{[NH_3]^2}{[N_2][H_2]^3} = 1.7 \times 10^2$

At equilibrium, $[N_2] = 0.020$ M and $[H_2] = 0.18$ M

$[NH_3] = \sqrt{[N_2] \times [H_2]^3 \times K_c} = \sqrt{(0.020)(0.18)^3(1.7 \times 10^2)} = 0.14$ M

13.43 $K_c = 2.7 \times 10^2 = \dfrac{[SO_3]^2}{[SO_2]^2[O_2]}$

Because $[SO_3] = [SO_2]$, then $2.7 \times 10^2 = \dfrac{1}{[O_2]}$

$[O_2] = 3.7 \times 10^{-3}$ M

13.44

	$N_2(g)$	+ $O_2(g)$	$\rightleftarrows$ 2 NO(g)
initial (M)	1.40	1.40	0
change (M)	–x	–x	+2x
equil (M)	1.40 – x	1.40 – x	2x

$K_c = 1.7 \times 10^{-3} = \dfrac{[NO]^2}{[N_2][O_2]} = \dfrac{(2x)^2}{(1.40 - x)^2}$

Take the square root of both sides and solve for x.

$\sqrt{1.7 \times 10^{-3}} = \sqrt{\dfrac{(2x)^2}{(1.40 - x)^2}}$

$4.1 \times 10^{-2} = \dfrac{2x}{1.40 - x}$

$x = 2.8 \times 10^{-2}$

At equilibrium, [NO] = 2x = $2(2.8 \times 10^{-2})$ = 0.056 M

$[N_2] = [O_2] = 1.40 - x = 1.40 - (2.8 \times 10^{-2}) = 1.37$ M

13.45 $N_2(g) + O_2(g) \rightleftharpoons 2\ NO(g)$

	$N_2(g)$	$O_2(g)$	2 NO(g)
initial (M)	2.24	0.56	0
change (M)	–x	–x	+2x
equil (M)	2.24 – x	0.56 – x	2x

$$K_c = \frac{[NO]^2}{[N_2][O_2]} = 1.7 \times 10^{-3} = \frac{(2x)^2}{(2.24 - x)(0.56 - x)}$$

$4x^2 + (4.8 \times 10^{-3})x - (2.1 \times 10^{-3}) = 0$

Use the quadratic formula to solve for x.

$$x = \frac{(-4.8 \times 10^{-3}) \pm \sqrt{(4.8 \times 10^{-3})^2 + 4(4)(2.1 \times 10^{-3})}}{2(4)} = \frac{-0.0048 \pm 0.1834}{8}$$

x = –0.0235 and 0.0223

Discard the solution that uses the negative square root (x = –0.0235) because it gives a negative NO concentration and that is impossible.

$[N_2] = 2.24 - x = 2.24 - 0.0223 = 2.22$ M

$[O_2] = 0.56 - x = 0.56 - 0.0223 = 0.54$ M

[NO] = 2x = 2(0.0223) = 0.045 M

13.46 $PCl_5(g) \rightleftharpoons PCl_3(g) + Cl_2(g)$

	$PCl_5(g)$	$PCl_3(g)$	$Cl_2(g)$
initial (M)	0.160	0	0
change (M)	–x	+x	+x
equil (M)	0.160 – x	x	x

$$K_c = \frac{[PCl_3][Cl_2]}{[PCl_5]} = 5.8 \times 10^{-2} = \frac{x^2}{0.160 - x}$$

$x^2 + (5.8 \times 10^{-2})x - 0.00928 = 0$

Use the quadratic formula to solve for x.

$$x = \frac{(-5.8 \times 10^{-2}) \pm \sqrt{(5.8 \times 10^{-2})^2 + 4(0.00928)}}{2(1)} = \frac{(-5.8 \times 10^{-2}) \pm 0.20}{2}$$

x = 0.071 and –0.129

Discard the solution that uses the negative square root (x = –0.129) because it gives negative concentrations of PCl_3 and Cl_2 and that is impossible.

$[PCl_3] = [Cl_2] = x = 0.071$ M

$[PCl_5] = 0.160 - x = 0.160 - 0.071 = 0.089$ M

13.47 (a) $$K_c = \frac{[CH_3COOC_2H_5][H_2O]}{[CH_3COOH][C_2H_5OH]} = 3.4 = \frac{(x)(12.0)}{(4.0)(6.0)}$$

x = 6.8 moles $CH_3COOC_2H_5$

Note that the volume cancels because the same number of molecules appear on both sides of the chemical equation.

(b) $CH_3COOH(l) + C_2H_5OH(l) \rightleftarrows CH_3COOC_2H_5(l) + H_2O(l)$

	$CH_3COOH(l)$	$C_2H_5OH(l)$	$CH_3COOC_2H_5(l)$	$H_2O(l)$
initial (mol)	1.00	10.00	0	0
change (mol)	–x	–x	+x	+x
equil (mol)	1.00 – x	10.00 – x	x	x

$$K_c = 3.4 = \frac{x^2}{(1.00 - x)(10.00 - x)}$$

$2.4x^2 - 37.4x + 34 = 0$

Use the quadratic formula to solve for x.

$$x = \frac{(37.4) \pm \sqrt{(-37.4)^2 - 4(2.4)(34)}}{2(2.4)} = \frac{37.4 \pm 32.75}{4.8}$$

x = 0.969 and 14.6

Discard the solution that uses the positive square root (x = 14.6) because it leads to negative concentrations and that is impossible.

mol CH_3COOH = 1.00 – x = 1.00 – 0.969 = 0.03 mol

mol C_2H_5OH = 10.00 – x = 10.00 – 0.969 = 9.03 mol

mol $CH_3COOC_2H_5$ = mol H_2O = x = 0.97 mol

13.48 When equal volumes of two solutions are mixed together, their concentrations are cut in half.

	CH_3Cl	+	OH^-	⇄	CH_3OH	+	Cl^-
initial (M)	0.05		0.1		0		0
assume complete reaction (M)	0		0.05		0.05		0.05
assume small back reaction (M)	+x		+x		–x		–x
equil (M)	x		0.05 + x		0.05 – x		0.05 – x

$$K_c = \frac{[CH_3OH][Cl^-]}{[CH_3Cl][OH^-]} = 10^{16} = \frac{(0.05-x)^2}{x(0.05+x)}$$

Because K_c is very large, x << 0.05.

$$10^{16} \approx \frac{(0.05)^2}{x(0.05)}$$

$x = 5 \times 10^{-18}$

$[CH_3Cl] = x = 5 \times 10^{-18}$ M

$[OH^-] = [CH_3OH] = [Cl^-] \approx 0.05$ M

Le Châtelier's Principle

13.49 (a) Cl^- (reactant) added, AgCl(s) increases

(b) Ag^+ (reactant) added, AgCl(s) increases

(c) Cl^- (reactant) removed, AgCl(s) decreases

(d) Ag^+ (reactant) removed, AgCl(s) decreases

13.50 (a) $ClNO_2$ (reactant) added, NO_2 concentration increases

(b) ClNO (product) added, NO_2 concentration decreases

(c) NO (reactant) added, NO_2 concentration increases

(d) NO (reactant) removed, NO_2 concentration decreases

At equilibrium, $Q_c = K_c = \dfrac{[ClNO][NO_2]}{[ClNO_2][NO]}$. If some $ClNO_2$ is added to the equilibrium mixture, the denominator in Q_c increases and $Q_c < K_c$. The reaction shifts to the right, increasing the NO_2 concentration.

13.51 (a) Because there are 2 mol of gas on the left side and 3 mol of gas on the right side of the balanced equation, the stress of an increase in pressure is relieved by a shift in the reaction to the side with fewer moles of gas (in this case, to reactants). The number of moles of reaction products decreases.

(b) Because there are 2 mol of gas on both sides of the balanced equation, the composition of the equilibrium mixture is unaffected by a change in pressure. The number of moles of reaction product remains the same.

(c) Because there are 2 mol of gas on the left side and 1 mol of gas on the right side of the balanced equation, the stress of an increase in pressure is relieved by a shift in the reaction to the side with fewer moles of gas (in this case, to products). The number of moles of reaction products increases.

13.52 As the volume increases, the pressure decreases at constant temperature.

(a) Because there is 1 mol of gas on the left side and 2 mol of gas on the right side of the balanced equation, the stress of an increase in volume (decrease in pressure) is relieved by a shift in the reaction to the side with the larger number of moles of gas (in this case, to products).

(b) Because there are 3 mol of gas on the left side and 2 mol of gas on the right side of the balanced equation, the stress of an increase in volume (decrease in pressure) is relieved by a shift in the reaction to the side with the larger number of moles of gas (in this case, to reactants).

(c) Because there are 3 mol of gas on both sides of the balanced equation, the composition of the equilibrium mixture is unaffected by an increase in volume (decrease in pressure). There is no net reaction in either direction.

13.53 $CO(g) + H_2O(g) \rightleftharpoons CO_2(g) + H_2(g)$ $\Delta H° = -41.2$ kJ

The reaction is exothermic. $[H_2]$ decreases when the temperature is increased.

As the temperature is decreased, the reaction shifts to the right. $[CO_2]$ and $[H_2]$ increase, [CO] and $[H_2O]$ decrease, and K_c increases.

13.54 Since $\Delta H°$ is positive, the reaction is endothermic.

heat + $3\ O_2(g) \rightleftharpoons 2\ O_3(g)$

$$K_c = \frac{[O_3]^2}{[O_2]^3}$$

As the temperature increases, heat is added to the reaction, causing a shift to the right. The $[O_3]$ increases, and the $[O_2]$ decreases. This results in an increase in K_c.

13.55 [HI] (product) decreases when the temperature increases because the reaction is exothermic.

13.56 (a) $Fe(NO_3)_3$ is a source of Fe^{3+}. Fe^{3+} (reactant) added; the $FeCl^{2+}$ concentration increases.

(b) Cl^- (reactant) removed; the $FeCl^{2+}$ concentration decreases.

(c) An endothermic reaction shifts to the right as the temperature increases; the $FeCl^{2+}$ concentration increases.

(d) A catalyst does not affect the composition of the equilibrium mixture; no change in $FeCl^{2+}$ concentration.

13.57 (a) The reaction is exothermic. The amount of CH_3OH (product) decreases as the temperature increases.

(b) When the volume decreases, the reaction shifts to the side with fewer gas molecules. The amount of CH_3OH increases.

(c) Addition of an inert gas (He) does not affect the equilibrium composition. There is no change.

(d) Addition of CO (reactant) shifts the reaction toward product. The amount of CH_3OH increases.

(e) Addition or removal of a catalyst does not affect the equilibrium composition. There is no change.

13.58 (a) An endothermic reaction shifts to the right as the temperature increases. The amount of acetone increases.

(b) Because there is 1 mol of gas on the left side and 2 mol of gas on the right side of the balanced equation, the stress of an increase in volume (decrease in pressure) is relieved by a shift in the reaction to the side with the larger number of moles of gas (in this case, to products). The amount of acetone increases.

(c) The addition of Ar (an inert gas) with no volume change does not affect the composition of the equilibrium mixture. The amount of acetone does not change.

(d) H_2 (product) added; the amount of acetone decreases.

(e) A catalyst does not affect the composition of the equilibrium mixture. The amount of acetone does not change.

Chemical Equilibrium and Chemical Kinetics

13.59 $A + B \rightleftarrows C$

$rate_f = k_f[A][B]$ and $rate_r = k_r[C]$

at equilibrium, $rate_f = rate_r$

$k_f[A][B] = k_r[C]$

$$\frac{k_f}{k_r} = \frac{[C]}{[A][B]} = K_c$$

13.60 An equilibrium mixture that contains large amounts of reactants and small amounts of products has a small K_c. A small K_c has $k_f < k_r$ (c).

13.61 $K_c = \frac{k_f}{k_r} = \frac{0.13}{6.2 \times 10^{-4}} = 210$

13.62 $K_c = \frac{k_f}{k_r}$; $k_r = \frac{k_f}{K_c} = \frac{6 \times 10^{-6}/(M \cdot s)}{10^{16}} = 6 \times 10^{-22}/(M \cdot s)$

General Problems

13.63 (i) change concentrations (ii) change volume (iii) change temperature

A change in temperature will cause a change in the value of the equilibrium constant.

13.64 A catalyst increases k_f and k_r by the same factor, so the equilibrium composition is unchanged.

13.65 The activation energy (E_a) is positive, and for an exothermic reaction, E_a(reverse) > E_a(forward).

$k_f = A_f\, e^{-E_a(f)/RT}$, $k_r = A_r\, e^{-E_a(r)/RT}$

$$K_c = \frac{k_f}{k_r} = \frac{A_f e^{-E_a(f)/RT}}{A_r e^{-E_a(r)/RT}} = \frac{A_f}{A_r} e^{[E_a(r) - E_a(f)]/RT}$$

$[E_a(r) - E_a(f)]$ is positive, so the exponent is always positive. As the temperature increases, the exponent $([E_a(r) - E_a(f)]/RT)$ decreases and the value for K_c decreases as well.

13.66 $Hb + O_2 \rightleftarrows Hb(O_2)$

If CO binds to Hb, Hb is removed from the reaction and the reaction will shift to the left resulting in O_2 being released from $Hb(O_2)$. This will decrease the effectiveness of Hb for carrying O_2.

13.67

	C(s)	+	$CO_2(g)$	⇄	2 CO(g)
initial (M)	excess		1.50 mol/20.0 L		0
change (M)			–x		+2x
equil (M)			0.0750 – x		2x

$[CO] = 2x = 7.00 \times 10^{-2}$ M

x = 0.0350 M

(a) $[CO_2] = 0.0750 - x = 0.0750 - 0.0350 = 0.0400$ M

(b) $K_c = \dfrac{[CO]^2}{[CO_2]} = \dfrac{(7.00 \times 10^{-2})^2}{(0.0400)} = 0.122$

13.68 $[N_2O_4] = \dfrac{0.500 \text{ mol}}{4.00 \text{ L}} = 0.125$ M

	$N_2O_4(g)$	⇄	$2\ NO_2(g)$
initial (M)	0.125		0
change (M)	–(0.793)(0 .125)		+(2)(0.793)(0.125)
equil (M)	0.125 – (0.793)(0.125)		(2)(0.793)(0.125)

At equilibrium, $[N_2O_4] = 0.125 - (0.793)(0.125) = 0.0259$ M

$[NO_2] = (2)(0.793)(0.125) = 0.198$ M

$K_c = \dfrac{[NO_2]^2}{[N_2O_4]} = \dfrac{(0.198)^2}{(0.0259)} = 1.51$

$\Delta n = 2 - 1 = 1$ and $K_p = K_c(RT)^{\Delta n}$

$K_p = K_c(RT) = (1.51)(0.0821)(400) = 49.6$

13.69 K_c is very large. The reaction goes to completion.

$K_c = \dfrac{1}{[CO_2]}$; $[CO_2] = \dfrac{1}{K_c} = \dfrac{1}{1.6 \times 10^{24}} = 6.2 \times 10^{-25}$ M

13.70 $K_c = \dfrac{[NH_3]^2}{[N_2][H_2]^3} = 0.291$

At equilibrium, $[N_2] = 1.0 \times 10^{-3}$ M and $[H_2] = 2.0 \times 10^{-3}$ M

$[NH_3] = \sqrt{[N_2] \times [H_2]^3 \times K_c} = \sqrt{(1.0 \times 10^{-3})(2.0 \times 10^{-3})^3(0.291)} = 1.5 \times 10^{-6}$ M

13.71 $K_p = \dfrac{(P_F)^2}{P_{F_2}} = 7.83$

$P_F = \sqrt{K_p P_{F_2}} = \sqrt{(7.83)(0.200)} = 1.25$ atm

	$F_2(g)$	⇄	2 F(g)
initial (atm)	x		0
change (atm)	–y		+2y
equil (atm)	x – y		2y

2y = 1.25 so y = 0.625

x – y = 0.200; x = 0.200 + y = 0.200 + 0.625 = 0.825

$f = \dfrac{0.625}{0.825} = 0.758$

13.72 $2\ HI(g) \rightleftarrows H_2(g) + I_2(g)$

Calculate K_c.

$K_c = \dfrac{[H_2][I_2]}{[HI]^2} = \dfrac{(0.13)(0.70)}{(2.1)^2} = 0.0206$

$[HI] = \dfrac{0.20\ \text{mol}}{0.500\ \text{L}} = 0.40$ M

	$2\ HI(g)$	⇄	$H_2(g)$	+	$I_2(g)$
initial (M)	0.40		0		0
change (M)	–2x		+x		+x
equil (M)	0.40 – 2x		x		x

$$K_c = 0.0206 = \frac{[H_2][I_2]}{[HI]^2} = \frac{x^2}{(0.40 - 2x)^2}$$

Take the square root of both sides, and solve for x.

$$\sqrt{0.0206} = \sqrt{\frac{x^2}{(0.40 - 2x)^2}}$$

$$0.144 = \frac{x}{0.40 - 2x}$$

$x = 0.045$

At equilibrium, $[H_2] = [I_2] = x = 0.045$ M

$[HI] = 0.40 - 2x = 0.40 - 2(0.045) = 0.31$ M

13.73 Note the container volume is 5.00 L

$[H_2] = [I_2] = 1.00$ mol/5.00 L = 0.200 M

$[HI] = 2.50$ mol/5.00 L = 0.500 M

	$H_2(g)$	+	$I_2(g)$	⇄	$2\ HI(g)$
initial (M)	0.200		0.200		0.500
change (M)	–x		–x		+2x
equil (M)	0.200 – x		0.200 – x		0.500 + 2x

$$K_c = \frac{[HI]^2}{[H_2][I_2]} = 129 = \frac{(0.500 + 2x)^2}{(0.200 - x)^2}$$

Take the square root of both sides, and solve for x.

$$\sqrt{129} = \sqrt{\frac{(0.500 + 2x)^2}{(0.200 - x)^2}}$$

$$11.4 = \frac{0.500 + 2x}{0.200 - x}$$

$x = 0.133$

$[H_2] = [I_2] = 0.200 - x = 0.200 - 0.133 = 0.067$ M

$[HI] = 0.500 + 2x = 0.500 + 2(0.133) = 0.766$ M

13.74 $[H_2O] = \dfrac{6.00 \text{ mol}}{5.00 \text{ L}} = 1.20$ M

	$C(s)$	+	$H_2O(g)$	$\rightleftarrows$	$CO_2(g)$	+	$H_2(g)$
initial (M)			1.20		0		0
change (M)			–x		+x		+x
equil (M)			1.20 – x		x		x

$$K_c = \frac{[CO_2][H_2]}{[H_2O]} = 3.0 \times 10^{-2} = \frac{x^2}{1.20 - x}$$

$x^2 + (3.0 \times 10^{-2})x - 0.036 = 0$

Use the quadratic formula to solve for x.

$$x = \frac{(-0.030) \pm \sqrt{(0.030)^2 + 4(0.036)}}{2(1)} = \frac{-0.030 \pm 0.381}{2}$$

$x = 0.176$ and -0.206

Discard the solution that uses the negative square root ($x = -0.206$) because it leads to negative concentrations and that is impossible.

$[CO_2] = [H_2] = x = 0.18$ M

$[H_2O] = 1.20 - x = 1.20 - 0.18 = 1.02$ M

13.75 (a) Because K_c decreases as the temperature increases, the reaction is exothermic.

(b) There is no change for all three.

13.76 A decrease in volume (a), an increase in temperature (b), and the addition of reactants (c) will affect the value of Q_c. (a), (b), and (c) all cause the reaction to shift in one direction or the other. This causes a change in reactant and product concentrations, which changes the value of Q_c.

Only a change in temperature (b) affects the value of K_c.

Addition of a catalyst (d) or the addition of an inert gas (e) does not affect the equilibrium composition. K_c and Q_c are not affected by (d) or (e).

13.77 (a) Addition of a solid does not affect the equilibrium composition. There is no change in the number of moles of CO_2.

(b) Adding a product causes the reaction to shift toward reactants. The number of moles of CO_2 decreases.

(c) Decreasing the volume causes the reaction to shift toward the side with fewer mol of gas (reactant side). The number of moles of CO_2 decreases.

(d) The reaction is endothermic. An increase in temperature shifts the reaction toward products. The number of moles of CO_2 increases.

13.78 2 monomer $\rightleftarrows$ dimer

(a) In benzene, $K_c = 1.51 \times 10^2$

	2 monomer	$\rightleftarrows$	dimer
initial (M)	0.100		0
change (M)	–2x		+x
equil (M)	0.100 – 2x		x

$$K_c = \frac{[\text{dimer}]}{[\text{monomer}]^2} = 1.51 \times 10^2 = \frac{x}{(0.100 - 2x)^2}$$

$$604x^2 - 61.4x + 1.51 = 0$$

Use the quadratic formula to solve for x.

$$x = \frac{61.4 \pm \sqrt{(-61.4)^2 - (4)(604)(1.51)}}{2(604)} = \frac{61.4 \pm 11.04}{1208}$$

x = 0.0600 and 0.0417

Discard the solution that uses the positive square root (x = 0.0600) because it gives a negative concentration of the monomer and that is impossible.

[monomer] = 0.100 – 2x = 0.100 – 2(0.0417) = 0.017 M

[dimer] = x = 0.0417 M

$$\frac{[\text{dimer}]}{[\text{monomer}]} = \frac{0.0417\ \text{M}}{0.017\ \text{M}} = 2.5$$

(b) In H_2O, $K_c = 3.7 \times 10^{-2}$

	2 monomer	⇄	dimer
initial (M)	0.100		0
change (M)	–2x		+x
equil (M)	0.100 – 2x		x

$$K_c = \frac{[\text{dimer}]}{[\text{monomer}]^2} = 3.7 \times 10^{-2} = \frac{x}{(0.100 - 2x)^2}$$

$0.148x^2 - 1.0148x + 0.000\,37 = 0$

Use the quadratic formula to solve for x.

$$x = \frac{1.0148 \pm \sqrt{(-1.0148)^2 - (4)(0.148)(0.00037)}}{2(0.148)} = \frac{1.0148 \pm 1.0147}{0.296}$$

x = 6.86 and 3.7×10^{-4}

Discard the solution that uses the positive square root (x = 6.86) because it gives a negative concentration of the monomer and that is impossible.

[monomer] = 0.100 – 2x = 0.100 – 2(3.7×10^{-4}) = 0.099 M

[dimer] = x = 3.7×10^{-4} M

$$\frac{[\text{dimer}]}{[\text{monomer}]} = \frac{3.7 \times 10^{-4}\text{ M}}{0.099\text{ M}} = 0.0038$$

(c) K_c for the water solution is so much smaller than K_c for the benzene solution because H_2O can hydrogen bond with acetic acid, thus preventing acetic acid dimer formation. Benzene cannot hydrogen bond with acetic acid.

13.79 (a) $K_c = \frac{[C_2H_6][C_2H_4]}{[C_4H_{10}]}$ $\qquad K_p = \frac{(P_{C_2H_6})(P_{C_2H_4})}{P_{C_4H_{10}}}$

(b) $K_p = 12$; $\Delta n = 1$

$$K_c = K_p\left(\frac{1}{RT}\right) = (12)\left(\frac{1}{(0.0821)(773)}\right) = 0.19$$

(c)

	$C_4H_{10}(g)$	⇄	$C_2H_6(g)$	+	$C_2H_4(g)$
initial (atm)	50		0		0
change (atm)	–x		+x		+x
equil (atm)	50 – x		x		x

$$K_p = 12 = \frac{x^2}{50 - x}$$

$x^2 + 12x - 600 = 0$

Use the quadratic formula to solve for x.

$$x = \frac{(-12) \pm \sqrt{(12)^2 + 4(1)(600)}}{2(1)} = \frac{-12 \pm 50.44}{2}$$

x = –31.22 and 19.22

Discard the solution that uses the negative square root (x = –31.22) because it leads to negative concentrations and that is impossible.

$$\%\ C_4H_{10}\text{ converted} = \frac{19.22}{50} \times 100\% = 38\%$$

$$P_{total} = P_{C_4H_{10}} + P_{C_2H_6} + P_{C_2H_4} = (50 - x) + x + x = (50 - 19) + 19 + 19 = 69\text{ atm}$$

(d) A decrease in volume would decrease the % conversion of C_4H_{10}.

13.80 (a) Because $\Delta n = 0$, $K_p = K_c = 1.0 \times 10^5$

(b) $K_p = 1.0 \times 10^5 = \dfrac{P_{propene}}{P_{cyclopropane}}$

$$P_{cyclopropane} = \frac{P_{propene}}{1.0 \times 10^5} = \frac{5.0 \text{ atm}}{1.0 \times 10^5} = 5.0 \times 10^{-5} \text{ atm}$$

(c) The ratio of the two concentrations is equal to K_c. The ratio (K_c) cannot be changed by adding cyclopropane. The individual concentrations can change but the ratio of concentrations can't.

Because there is one mole of gas on each side of the balanced equation, the composition of the equilibrium mixture is unaffected by a decrease in volume. The ratio of the two concentrations will not change.

(d) Because K_c is large, $k_f > k_r$.

13.81 (a) $K_p = 3.45$; $\Delta n = 1$

$$K_c = K_p\left(\frac{1}{RT}\right) = (3.45)\left(\frac{1}{(0.0821)(500)}\right) = 0.0840$$

(b) $[(CH_3)_3CCl] = 1.00 \text{ mol}/5.00 \text{ L} = 0.200 \text{ M}$

	$(CH_3)_3CCl(g)$	$\rightleftarrows$	$(CH_3)_2C{=}CH_2(g)$	+	$HCl(g)$
initial (M)	0.200		0		0
change (M)	–x		+x		+x
equil (M)	0.200 – x		x		x

$$K_c = 0.0840 = \frac{x^2}{0.200 - x}$$

$x^2 + 0.0840x - 0.0168 = 0$

Use the quadratic formula to solve for x.

$$x = \frac{(-0.0840) \pm \sqrt{(0.0840)^2 + 4(1)(0.0168)}}{2(1)} = \frac{-0.0840 \pm 0.272}{2}$$

x = –0.178 and 0.094

Discard the solution that uses the negative square root (x = –0.178) because it leads to negative concentrations and that is impossible.

$[(CH_3)_2C{=}CCH_2] = [HCl] = x = 0.094$ M

$[(CH_3)_3CCl] = 0.200 - x = 0.200 - 0.094 = 0.106$ M

(c) $K_p = 3.45$

	$(CH_3)_3CCl(g)$	⇄	$(CH_3)_2C{=}CH_2(g)$	+	$HCl(g)$
initial (atm)	0		0.400		0.600
change (atm)	+x		–x		–x
equil (atm)	x		0.400 – x		0.600 – x

$$K_p = 3.45 = \frac{(0.400 - x)(0.600 - x)}{x}$$

$x^2 - 4.45x + 0.240 = 0$

Use the quadratic formula to solve for x.

$$x = \frac{(4.45) \pm \sqrt{(-4.45)^2 - 4(1)(0.240)}}{2(1)} = \frac{4.45 \pm 4.34}{2}$$

x = 0.055 and 4.40

Discard the solution that uses the positive square root (x = 4.40) because it leads to a negative partial pressures and that is impossible.

$P_{t\text{-butyl chloride}} = x = 0.055$ atm

$P_{isobutylene} = 0.400 - x = 0.400 - 0.055 = 0.345$ atm

$P_{HCl} = 0.600 - x = 0.600 - 0.055 = 0.545$ atm

13.82 a A + bB $\rightleftarrows$ a C + d D

$$K_p = \frac{(P_C)^c(P_D)^d}{(P_A)^a(P_B)^b}$$

From the ideal gas law:

P_A = [A]RT; P_B = [B]RT; P_C = [C]RT; P_D = [D]RT

Substitute for the partial pressures in K_p.

$$K_p = \frac{([C]RT)^c([D]RT)^d}{([A]RT)^a([B]RT)^b}$$

$$K_p = \frac{[C]^c(RT)^c[D]^d(RT)^d}{[A]^a(RT)^a[B]^b(RT)^b}$$

$$K_p = \frac{[C]^c[D]^d}{[A]^a[B]^b} \times \frac{(RT)^c(RT)^d}{(RT)^a(RT)^b}$$

$$K_p = \frac{[C]^c[D]^d}{[A]^a[B]^b} \times \frac{(RT)^{(c+d)}}{(RT)^{(a+b)}}$$

$$K_c = \frac{[C]^c[D]^d}{[A]^a[B]^b}$$

$K_p = K_c(RT)^{[(c + d) - (a + b)]}$

$\Delta n = [(c + d) - (a + b)]$

$K_p = K_c(RT)^{\Delta n}$

14.1 PV = nRT

$$PV = \frac{g}{\text{molar mass}}RT$$

$$d_{H_2} = \frac{g}{V} = \frac{P(\text{molar mass})}{RT} = \frac{(1.00\text{ atm})(2.016\text{ g/mol})}{\left(0.08206\ \frac{L\cdot atm}{mol\cdot K}\right)(298\text{ K})} = 0.0824\text{ g/L}$$

1 L = 1000 mL = 1000 cm^3

d_{H_2} = 0.0824 g/1000 cm^3 = 8.24 x 10^{-5} g/cm^3

$$\frac{d_{air}}{d_{H_2}} = \frac{1.185\times10^{-3}\text{ g/cm}^3}{8.24\times10^{-5}\text{ g/cm}^3} = 14.4;$$ Air is 14 times more dense than H_2.

14.2 For every 100.0 g, there are:

61.4 g O, 22.9 g C, 10.0 g H, 2.6 g N, and 3.1 g other

$$22.9\text{ g C} \times \frac{1\text{ mol C}}{12.011\text{ g C}} = 1.907\text{ mol C}$$

$$10.0\text{ g H} \times \frac{1\text{ mol H}}{1.008\text{ g H}} = 9.921\text{ mol H}$$

Assume the sample contains 1.907 mol ^{13}C and 9.921 mol D.

$$\text{mass }^{13}C = 1.907\text{ mol }^{13}C \times \frac{13.0034\text{ g }^{13}C}{1\text{ mol }^{13}C} = 24.8\text{ g }^{13}C$$

$$\text{mass D} = 9.921\text{ mol D} \times \frac{2.0141\text{ g D}}{1\text{ mol D}} = 20.0\text{ g D}$$

(a) Total mass if all H is D is:

61.4 g O + 22.9 C + 20.0 g D + 2.6 g N + 3.1 g other = 110.0 g

$$\text{mass \% D} = \frac{20.0\text{ g D}}{110.0\text{ g}} = 18.2\%\text{ D}$$

(b) Total mass if all C is ^{13}C is:

61.4 g O + 24.8 g ^{13}C + 10.0 g H + 2.6 g N + 3.1 g other = 101.9 g

$$\text{mass \% } ^{13}C = \frac{24.8 \text{ g } ^{13}C}{101.9 \text{ g}} \times 100\% = 24.3\% \ ^{13}C$$

(c) The isotope effect for H is larger than that for C because D is two times the mass of ^{1}H while ^{13}C is only about 8% heavier than ^{12}C.

14.3 $2\ Ga(s) + 6\ H^+(aq) \rightarrow 3\ H_2(g) + 2\ Ga^{3+}(aq)$

14.4 $CaH_2(s) + 2\ H_2O(l) \rightarrow 2\ H_2(g) + Ca^{2+}(aq) + 2\ OH^-(aq)$

PV = nRT and 25°C = 298 K

$$n_{H_2} = \frac{PV}{RT} = \frac{(1.00 \text{ atm})(2.0 \times 10^5 \text{ L})}{\left(0.082\ 06 \frac{\text{L}\cdot\text{atm}}{\text{mol}\cdot\text{K}}\right)(298 \text{ K})} = 8.18 \times 10^3 \text{ mol } H_2$$

CaH_2, 42.09 amu

$$\text{mass } CaH_2 = 8.18 \times 10^3 \text{ mol } H_2 \times \frac{1 \text{ mol } CaH_2}{2 \text{ mol } H_2} \times \frac{42.09 \text{ g } CaH_2}{1 \text{ mol } CaH_2} \times \frac{1 \text{ kg}}{1000 \text{ g}}$$

mass CaH_2 = 1.7×10^2 kg CaH_2

14.5 Assume 12.0 g of Pd with a volume of 1.0 cm^3.

V_{H_2} = 935 cm^3 = 935 mL = 0.935 L

PV = nRT

$$n_{H_2} = \frac{PV}{RT} = \frac{(1.00 \text{ atm})(0.935 \text{ L})}{\left(0.082\ 06 \frac{\text{L}\cdot\text{atm}}{\text{mol}\cdot\text{K}}\right)(273 \text{ K})} = 0.0417 \text{ mol } H_2$$

$n_H = 2n_{H_2}$ = 0.0834 mol H

$$12.0 \text{ g Pd} \times \frac{1 \text{ mol Pd}}{106.42 \text{ g Pd}} = 0.113 \text{ mol Pd}$$

$Pd_{0.113}H_{0.0834}$

$Pd_{0.113\,/\,0.113}H_{0.0834\,/\,0.113}$

$PdH_{0.74}$

g H = (0.0834 mol H)(1.008 g/mol) = 0.0841 g H

$d_H = 0.0841 \text{ g/cm}^3$

$$M_H = \frac{0.0834 \text{ mol}}{0.001 \text{ L}} = 83.4 \text{ M}$$

14.6 $2\ KMnO_4(s) \rightarrow K_2MnO_4(s) + MnO_2(s) + O_2(g)$

$KMnO_4$, 158.03 amu

$$\text{mol } O_2 = 0.20 \text{ g } KMnO_4 \times \frac{1 \text{ mol } KMnO_4}{158.03 \text{ g } KMnO_4} \times \frac{1 \text{ mol } O_2}{2 \text{ mol } KMnO_4}$$

$\text{mol } O_2 = 6.33 \times 10^{-4} \text{ mol } O_2$

PV = nRT and 25°C = 298 K

$$V = \frac{nRT}{P} = \frac{(6.33 \times 10^{-4} \text{ mol})\left(0.082\ 06 \frac{\text{L} \cdot \text{atm}}{\text{mol} \cdot \text{K}}\right)(298 \text{ K})}{1.00 \text{ atm}} = 0.0155 \text{ L}$$

$$V = 0.0155 \text{ L} \times \frac{1000 \text{ mL}}{1 \text{ L}} = 16 \text{ mL } O_2$$

14.7 (a) $Li_2O(s) + H_2O(l) \rightarrow 2\ Li^+(aq) + 2\ OH^-(aq)$

(b) $SO_3(l) + H_2O(l) \rightarrow H^+(aq) + HSO_4^-(aq)$

(c) $Cr_2O_3(s) + 6\ H^+(aq) \rightarrow 2\ Cr^{3+}(aq) + 3\ H_2O(l)$

(d) $Cr_2O_3(s) + 2\ OH^-(aq) + 3\ H_2O(l) \rightarrow 2\ Cr(OH)_4^-(aq)$

14.8 (a) Rb_2O_2 Rb +1, O –1, peroxide

(b) CaO Ca +2, O –2, oxide

(c) CsO_2 Cs +1, O –1/2, superoxide

(d) SrO_2 Sr +2, O –1, peroxide

(e) CO_2 C +4, O –2, oxide

14.9 (a) $Rb_2O_2(s) + H_2O(l) \rightarrow 2\ Rb^+(aq) + HO_2^-(aq) + OH^-(aq)$

(b) $CaO(s) + H_2O(l) \rightarrow Ca^{2+}(aq) + 2\ OH^-(aq)$

(c) $2\ CsO_2(s) + H_2O(l) \rightarrow O_2(g) + 2\ Cs^+(aq) + HO_2^-(aq) + OH^-(aq)$

(d) $SrO_2(s) + H_2O(l) \rightarrow Sr^{2+}(aq) + HO_2^-(aq) + OH^-(aq)$

(e) $CO_2(g) + H_2O(l) \rightarrow H^+(aq) + HCO_3^-(aq)$

14.10 $H-\ddot{\underset{..}{O}}-\ddot{\underset{..}{O}}-H$

The Lewis electron dot structure indicates a single bond (see text Table 14.2) which is consistent with an O–O bond length of 148 pm.

14.11 $PbS(s) + 4\ H_2O_2(aq) \rightarrow PbSO_4(s) + 4\ H_2O(l)$

14.12 (a) $2\ Li(s) + 2\ H_2O(l) \rightarrow H_2(g) + 2\ Li^+(aq) + 2\ OH^-(aq)$

(b) $Sr(s) + 2\ H_2O(l) \rightarrow H_2(g) + Sr^{2+}(aq) + 2\ OH^-(aq)$

(c) $Br_2(l) + H_2O(l) \rightleftarrows HOBr(aq) + H^+(aq) + Br^-(aq)$

14.13 mass of H_2O = 5.62 g – 3.10 g = 2.52 g H_2O

$$2.52\text{ g } H_2O \times \frac{1\text{ mol } H_2O}{18.02\text{ g } H_2O} = 0.140\text{ mol } H_2O$$

$$3.10 \text{ g } NiSO_4 \times \frac{1 \text{ mol } NiSO_4}{154.8 \text{ g } NiSO_4} = 0.0200 \text{ mol } NiSO_4$$

$$\text{number of } H_2O\text{'s in hydrate} = \frac{n_{H_2O}}{n_{NiSO_4}} = \frac{0.140 \text{ mol}}{0.0200 \text{ mol}} = 7$$

Hydrate formula is $NiSO_4 \cdot 7H_2O$

Understanding Key Concepts

1. (a) (1) covalent (2) ionic (3) covalent (4) interstitial

   (b) (1) H, +1; other element, –3
   (2) H, –1; other element, +1
   (3) H, +1; other element, –2

2. (a) A, NaH; B, PdH_x; C, H_2S; D, HI

   (b) NaH (ionic); PdH_x (interstitial); H_2S and HI (covalent)

   (c) H_2S and HI (molecular); NaH and PdH_x (3–dimensional crystal)

   (d) NaH: Na +1, H –1
   H_2S: S –2, H +1
   HI: I –1, H +1

3. React H_2O with a reducing agent to produce H_2. Ca or Al could be used.

4. (a) 6

   $^{16}O_2$, $^{17}O_2$, $^{18}O_2$, $^{16}O^{17}O$, $^{16}O^{18}O$, $^{17}O^{18}O$

   (b) 18

$^{16}O_3$	$^{16}O_2{}^{17}O$	$^{18}O_3$
$^{17}O_2{}^{16}O$	$^{17}O_3$	$^{16}O_2{}^{18}O$
$^{18}O_2{}^{16}O$	$^{18}O_2{}^{17}O$	$^{17}O_2{}^{18}O$

$^{16}O^{17}O^{16}O$	$^{17}O^{18}O^{17}O$	$^{16}O^{17}O^{18}O$
$^{16}O^{18}O^{16}O$	$^{17}O^{16}O^{17}O$	$^{18}O^{16}O^{18}O$
$^{17}O^{18}O^{16}O$	$^{18}O^{16}O^{17}O$	$^{18}O^{17}O^{18}O$

5. (a) A, CaO; B, Al_2O_3; C, SO_3; D, SeO_3

   (b) CaO (basic); Al_2O_3 (amphoteric); SO_3 and SeO_3 (acidic)

   (c) CaO (most ionic); SO_3 (most covalent)

   (d) CaO and Al_2O_3 (3–dimensional crystal); SO_3 and SeO_3 (molecular)

   (e) CaO (highest melting point); SO_3 (lowest melting point)

6. Water does not undergo a disproportionation reaction because H is already in its highest oxidation state and O is already in its lowest oxidation state.

7. K is oxidized by water. F_2 is reduced by water.

   Cl_2 and Br_2 disproportionate when treated with water.

Additional Problems

Isotopes of Hydrogen

14.14

Isotope	Nucleus Composition	Atom %
protium	1 proton	99.9844%
deuterium	1 proton, 1 neutron	0.0156%
tritium	1 proton, 2 neutrons	$\sim 10^{-16}$%

14.15 Deuterium can be separated from protium by passing an electric current through a solution of an inert electrolyte in ordinary water. H_2O is preferentially electrolyzed. Over time, the solution becomes enriched in D_2O.

14.16 Quantitative differences in properties that arise from the differences in the masses of the isotopes are known as isotope effects.

H_2 and D_2 have different melting and boiling points.

H_2O and D_2O have different dissociation constants.

14.17

	H_2O	D_2O
melting point	0.00°C	3.81°C
boiling point	100.00°C	101.42°C
density at 25°C	0.997 g/mL	1.104 g/mL

14.18 $$\frac{\text{mass }^2\text{D} - \text{mass }^1\text{H}}{\text{mass }^1\text{H}} \times 100\% = \frac{2.0141\text{ amu} - 1.0078\text{ amu}}{1.0078\text{ amu}} \times 100\% = 98.85\%$$

$$\frac{\text{mass }^3\text{H} - \text{mass }^2\text{H}}{\text{mass }^2\text{H}} \times 100\% = \frac{3.0160\text{ amu} - 2.0141\text{ amu}}{2.0141\text{ amu}} \times 100\% = 49.74\%$$

The differences in properties will be larger for H_2O and D_2O rather than for D_2O and T_2O because of the larger relative difference in mass for H and D versus D and T. This is supported by the data in Table 14.1.

14.19 (a) ocean volume = $(1.35 \times 10^{18}\text{ m}^3)(100\text{ cm/m})^3 = 1.35 \times 10^{24}\text{ cm}^3$

mass of H_2O = $(1.35 \times 10^{24}\text{ cm}^3)(1.00\text{ g/cm}^3) = 1.35 \times 10^{24}$ g

$$1.35 \times 10^{24}\text{ g H}_2\text{O} \times \frac{1\text{ mol H}_2\text{O}}{18.02\text{ g H}_2\text{O}} = 7.49 \times 10^{22}\text{ mol H}_2\text{O}$$

$$\text{mol H} = 7.49 \times 10^{22}\text{ mol H}_2\text{O} \times \frac{2\text{ mol H}}{1\text{ mol H}_2\text{O}} = 1.50 \times 10^{23}\text{ mol H}$$

mol D = $(1.50 \times 10^{23}$ mol H)(0.000 156 atom fraction D) = 2.34×10^{19} mol D

mass of D = $(2.34 \times 10^{19}$ mol D)(2.0141 g/mol)(1 kg/1000 g) = 4.71×10^{16} kg D

(b) mol T = $(1.50 \times 10^{23}$ mol H$)(0.01 \times 10^{-16}$ atom fraction T$) = 1.50 \times 10^{5}$ mol T

mass of T = $(1.50 \times 10^{5}$ mol T)(3.0160 g/mol)(1 kg/1000 g) = 5×10^{2} kg T

14.20 There are 18 kinds of H_2O

$H_2{}^{16}O$	$H_2{}^{17}O$	$H_2{}^{18}O$
$D_2{}^{16}O$	$D_2{}^{17}O$	$D_2{}^{18}O$
$T_2{}^{16}O$	$T_2{}^{17}O$	$T_2{}^{18}O$
$HD^{16}O$	$HD^{17}O$	$HD^{18}O$
$HT^{16}O$	$HT^{17}O$	$HT^{18}O$
$DT^{16}O$	$DT^{17}O$	$DT^{18}O$

14.21 There are 10 kinds of PH_3.

PH_3	PH_2D	PH_2T
PD_3	PD_2H	PD_2T
PT_3	PT_2H	PT_2D
PHDT		

14.22 (a) $Zn(s) + 2\ H^{+}(aq) \rightarrow Zn^{2+}(aq) + H_2(g)$

(b) $C(s) + H_2O(g) \xrightarrow{1000^{\circ}C} CO(g) + H_2(g)$

(c) $CH_4(g) + H_2O(g) \xrightarrow[\text{Ni catalyst}]{1100^{\circ}C} CO(g) + 3\ H_2(g)$

(d) $2\ H_2O(l) \xrightarrow{\text{electrolysis}} 2\ H_2(g) + O_2(g)$

14.23 (a) $Fe(s) + 2\ H^+(aq) \rightarrow Fe^{2+}(aq) + H_2(g)$

(b) $Ca(s) + 2\ H_2O(l) \rightarrow H_2(g) + Ca^{2+}(aq) + 2\ OH^-(aq)$

(c) $2\ Al(s) + 6\ H^+(aq) \rightarrow 2\ Al^{3+}(aq) + 3\ H_2(g)$

(d) $C_2H_6(g) + 2\ H_2O(g) \rightarrow 2\ CO(g) + 5\ H_2(g)$

14.24 The steam–hydrocarbon reforming process is the most important industirial preparation of hydrogen.

$$CH_4(g) + H_2O(g) \xrightarrow[\text{Ni catalyst}]{1100^\circ C} CO(g) + 3\ H_2(g)$$

$$CO(g) + H_2O(g) \xrightarrow[\text{catalyst}]{400^\circ C} CO_2(g) + H_2(g)$$

$$CO_2(g) + 2\ OH^-(aq) \rightarrow CO_3^{2-}(aq) + H_2O(l)$$

14.25 (a) $3\ Fe(s) + 4\ H_2O(g) \rightarrow Fe_3O_4(s) + 4\ H_2(g)$

(b) $C_3H_8(g) + 3\ H_2O(g) \rightarrow 3\ CO(g) + 7\ H_2(g)$

(c) $CO(g) + H_2O(g) \rightarrow CO_2(g) + H_2(g)$

14.26 (a) $LiH(s) + H_2O(l) \rightarrow H_2(g) + Li^+(aq) + OH^-(aq)$

$CaH_2(s) + 2\ H_2O(l) \rightarrow 2\ H_2(g) + Ca^{2+}(aq) + 2\ OH^-(aq)$

LiH, 7.95 amu; CaH_2, 42.09 amu

Obtain $\dfrac{1\ \text{mol}\ H_2}{7.95\ \text{g LiH}} = 0.126\ \text{mol}\ H_2/\text{g LiH}$

and $\dfrac{2\ \text{mol}\ H_2}{42.09\ \text{g}\ CaH_2} = 0.0475\ \text{mol}\ H_2/\text{g}\ CaH_2$

Therefore, LiH gives more H_2.

(b) PV = nRT and 25°C = 298 K

$$n_{H_2} = \frac{PV}{RT} = \frac{(150\text{ atm})(100\text{ L})}{\left(0.082\ 06\ \frac{L\cdot atm}{mol\cdot K}\right)(298K)} = 613.4\text{ mol } H_2$$

$$\text{mass } CaH_2 = 613.4\text{ mol } H_2 \times \frac{1\text{ mol } CaH_2}{2\text{ mol } H_2} \times \frac{42.09\text{ g } CaH_2}{1\text{ mol } CaH_2} \times \frac{1\text{ kg}}{1000\text{ g}}$$

mass CaH_2 = 12.9 kg CaH_2

14.27 PV = nRT

$$n_{H_2} = \frac{PV}{RT} = \frac{\left(740\text{ mm Hg} \times \frac{1\text{ atm}}{760\text{ mm Hg}}\right)(1.99 \times 10^8\text{ L})}{\left(0.082\ 06\ \frac{L\cdot atm}{mol\cdot K}\right)(293\text{ K})} = 8.059 \times 10^6\text{ mol } H_2$$

$n_C = n_{H_2} = 8.059 \times 10^6$ mol C

mass C = $(8.059 \times 10^6$ mol C)(12.011 g C/mol)(1 kg/1000 g) = 9.68×10^4 kg C

Reactivity of Hydrogen

14.28 Hydrogen is located in group 1A of the periodic table because it has one valence electron. Hydrogen's chemical behavior is so different from that of the alkali metals because its ionization potential is so high it prefers to form covalent compounds.

14.29 Hydrogen could be located in group 7A of the periodic table because it only needs one electron to fill its valence shell. Hydrogen behaves like other group 7A elements by forming covalent compounds with single covalent bonds.

14.30 (a) HCl(g) covalently bound H; HCl(aq) H^+

(b) CaH_2 H^-

(c) SiH_4 covalently bound H

(d) RbH H^-

14.31 (a) KH, H^- (b) PH_3, covalent (c) $H_2S(g)$, covalent (d) BaH_2, H^-

Binary Hydrides

14.32

Ionic Hydrides	Covalent Hydrides	Metallic Hydrides
NaH	NH_3	$TiH_{1.7}$
CaH_2	H_2O	$ZrH_{1.9}$
SrH_2	CH_4	UH_3

14.33 (a) H_2S, covalent (b) NaH, ionic (c) PdH_x, interstitial

(d) SrH_2, ionic (e) CH_4, covalent

14.34 (a) HF F –1, H +1

(b) LiH Li +1, H –1

(c) SnH_4 Sn +4, H –1

(d) MgH_2 Mg +2, H –1

14.35 (a) SrH_2, H –1 (b) H_2Se, H +1 (c) HBr, H +1 (d) NaH, H –1

14.36 H_2S Covalent hydride
gas
weak acid in H_2O

NaH ionic hydride
solid (salt like)
reacts with H_2O to produce H_2

PdH_x metallic (interstitial) hydride
solid
stores hydrogen

14.37 (a) KH, ionic bonding (b) PH_3, covalent bonding

14.38 (a) H–Se–H (with two lone pairs on Se) bent

(b) H–As–H with H below As (one lone pair on As) trigonal pyramidal

(c) H–Si–H with H above and below Si tetrahedral

14.39 (a) GeH_4, tetrahedral (b) H_2S, bent (c) NH_3, trigonal pyramidal

40.40 (a) $2\ K(l) + H_2(g) \rightarrow 2\ KH(s)$

(b) $KH(s) + H_2O(l) \rightarrow H_2(g) + K^+(aq) + OH^-(aq)$

(c) $CaH_2(s) + 2\ H^+(aq) \rightarrow Ca^{2+}(aq) + 2\ H_2(g)$

(d) $Sr(l) + H_2(g) \rightarrow SrH_2(s)$

14.41 (a) $MgH_2(s) + 2\ H^+(aq) \rightarrow Mg^{2+}(aq) + 2\ H_2(g)$

(b) $NaH(s) + H_2O(l) \rightarrow H_2(g) + Na^+(aq) + OH^-(aq)$

(c) $2\ Li(l) + H_2(g) \rightarrow 2\ LiH(s)$

(d) $BaH_2(s) + H^+(aq) + HSO_4^-(aq) \rightarrow 2\ H_2(g) + BaSO_4(s)$

14.42 A nonstoichiometric compound is a compound whose atomic composition cannot be expressed as a ratio of small whole numbers. An example is PdH_x. The lack of stoichiometry results from the hydrogen occupying holes in the solid state structure.

14.43 TiH_2, 49.90 amu

Assume 1.0 cm^3 of TiH_2 which has a mass of 3.9 g.

$$3.9 \text{ g } TiH_2 \times \frac{1 \text{ mol } TiH_2}{49.90 \text{ g } TiH_2} = 0.078 \text{ mol } TiH_2$$

$$0.078 \text{ mol } TiH_2 \times \frac{2 \text{ mol H}}{1 \text{ mol } TiH_2} = 0.156 \text{ mol H}$$

$$0.156 \text{ mol H} \times \frac{1.008 \text{ g H}}{1 \text{ mol H}} = 0.157 \text{ g H}$$

$d_H = 0.16$ g/cm^3; the density of H in TiH_2 is about 2.25 times the density of liquid H_2.

$PV = nRT$

$$V = \frac{nRT}{P} = \frac{\left(0.16 \text{ g} \times \frac{1 \text{ mol}}{2.016 \text{ g}}\right)\left(0.082\ 06 \frac{\text{L}\cdot\text{atm}}{\text{mol}\cdot\text{K}}\right)(273 \text{ K})}{1.00 \text{ atm}} = 1.8 \text{ L}$$

$1.8 \text{ L} = 1.8 \times 10^3 \text{ mL} = 1.8 \times 10^3 \text{ cm}^3$

Preparation and Uses of Oxygen

14.44 (a) O_2 is obtained in industry by the fractional distillation of liquid air.

(b) In the laboratory, O_2 is prepared by the thermal decomposition of $KClO_3(s)$.

$$2\ KClO_3(s) \xrightarrow[MnO_2]{heat} 2\ KCl(s) + 3\ O_2(g)$$

14.45 Some uses for O_2 are:

(1) steel making
(2) sewage treatment
(3) human respiration

14.46 $2\ H_2O_2(aq) \xrightarrow{\text{catalyst}} 2\ H_2O(l) + O_2(g)$

H_2O_2, 34.01 amu

$$\text{mol } O_2 = 20.4 \text{ g } H_2O_2 \times \frac{1 \text{ mol } H_2O_2}{34.01 \text{ g } H_2O_2} \times \frac{1 \text{ mol } O_2}{2 \text{ mol } H_2O_2} = 0.300 \text{ mol } O_2$$

PV = nRT and 25°C = 298 K

$$V = \frac{nRT}{P} = \frac{(0.300 \text{ mol})\left(0.082\ 06 \frac{\text{L}\cdot\text{atm}}{\text{mol}\cdot\text{K}}\right)(298 \text{ K})}{1.00 \text{ atm}} = 7.34 \text{ L } O_2$$

14.47 $2\ C_2H_2(g) + 5\ O_2(g) \rightarrow 4\ CO_2(g) + 2\ H_2O(g) \qquad \Delta H° = -2511 \text{ kJ}$

C_2H_2, 26.04 amu

$$\text{mass } C_2H_2 = 1000 \text{ kJ} \times \frac{2 \text{ mol } C_2H_2}{2511 \text{ kJ}} \times \frac{26.04 \text{ g } C_2H_2}{1 \text{ mol } C_2H_2} = 20.74 \text{ g} = 20.7 \text{ g } C_2H_2$$

$$V = \frac{nRT}{P} = \frac{\left(\frac{5}{2} \times 20.74 \text{ g} \times \frac{1 \text{ mol}}{26.04 \text{ g}}\right)\left(0.082\ 06 \frac{\text{L}\cdot\text{atm}}{\text{mol}\cdot\text{K}}\right)(273 \text{ K})}{1.00 \text{ atm}} = 44.6 \text{ L}$$

Properties of Oxygen

14.48 Some physical properties of O_2 are:

(1) melting point = –291°C
(2) boiling point = –183°C
(3) paramagnetic
(4) pale blue color as liquid or solid

14.49 Some chemical properties of O_2 are:

(1) O_2 forms ionic oxides

$$2\ M(s) + O_2(g) \rightarrow 2\ MO(s)$$

(2) O_2 forms covalent oxides

$X(s) + O_2(g) \rightarrow XO_2(g)$

(3) O_2 supports combustion

$C_3H_8(g) + 5\ O_2(g) \rightarrow 3\ CO_2(g) + 4\ H_2O(g)$

14.50 $:\ddot{O}::\ddot{O}:$

The Lewis electron dot structure shows an O=O double bond. It also shows all electrons paired. This is not consistent with the fact that O_2 is paramagnetic.

14.51 The highest occupied molecular orbital in O_2 is the doubly degenerate π^*_{2p} orbital which contains 2 unpaired electrons. The bond order is 2 because O_2 has four π_{2p}, two σ_{2p} bonding electrons, and two π^*_{2p} antibonding electrons. (See text, Figure 7.20.)

14.52 (a) $P_4(s) + 5\ O_2(g) \rightarrow P_4O_{10}(s)$

(b) $4\ Li(s) + O_2(g) \rightarrow 2\ Li_2O(s)$

(c) $4\ Al(s) + 3\ O_2(g) \rightarrow 2\ Al_2O_3(s)$

(d) $Si(s) + O_2(g) \rightarrow SiO_2(s)$

14.53 (a) $2\ Ca(s) + O_2(g) \rightarrow 2\ CaO(s)$

(b) $C(s) + O_2(g) \rightarrow CO_2(g)$

(c) $4\ As(s) + 5\ O_2(g) \rightarrow 2\ As_2O_5(s)$

(d) $4\ B(s) + 3\ O_2(g) \rightarrow 2\ B_2O_3(s)$

Oxides

14.54 Acidic oxides are covalent and are formed by nonmetals. Examples are CO_2 and Cl_2O_7.

Basic oxides are ionic and are formed by the active metals on the left side of the periodic table. Examples are CaO and K_2O.

Amphoteric oxides exhibit both acidic and basic properties. Examples are Al_2O_3 and Ga_2O_3.

14.55 $Li_2O < BeO < B_2O_3 < CO_2 < N_2O_5$ (see Figure 14.6)

14.56 $P_4O_{10} < SiO_2 < GeO_2 < Ga_2O_3 < K_2O$ (see Figure 14.6).

14.57 $Cl_2O_7 < Al_2O_3 < Na_2O < Cs_2O$ (see Figure 14.6)

14.58 $BaO < SnO_2 < CO_2 < N_2O_5$ (see Figure 14.6)

14.59 (a) CrO_3 (higher Cr oxidation state)

(b) N_2O_5 (higher N oxidation state)

(c) SO_3 (higher S oxidation)

14.60 (a) CrO (lower Cr oxidation state)

(b) SnO (lower Sn oxidation state)

(c) As_2O_3 (lower As oxidation state)

14.61 (a) $Cl_2O_7(l) + H_2O(l) \rightarrow 2\ H^+(aq) + 2\ ClO_4^-(aq)$

(b) $K_2O(s) + H_2O(l) \rightarrow 2\ K^+(aq) + 2\ OH^-(aq)$

(c) $SO_3(l) + H_2O(l) \rightarrow H^+(aq) + HSO_4^-(aq)$

14.62 (a) $BaO(s) + H_2O(l) \rightarrow Ba^{2+}(aq) + 2\ OH^-(aq)$

(b) $Cs_2O(s) + H_2O(l) \rightarrow 2\ Cs^+(aq) + 2\ OH^-(aq)$

(c) $N_2O_5(s) + H_2O(l) \rightarrow 2\ H^+(aq) + 2\ NO_3^-(aq)$

14.63 (a) $ZnO(s) + 2\ H^+(aq) \rightarrow Zn^{2+}(aq) + H_2O(l)$

(b) $ZnO(s) + 2\ OH^-(aq) + H_2O(l) \rightarrow Zn(OH)_4^{2-}(aq)$

14.64 (a) $Ga_2O_3(s) + 3\ H^+(aq) + 3\ HSO_4^-(aq)$

$\rightarrow 2\ Ga^{3+}(aq) + 3\ SO_4^{2-}(aq) + 3\ H_2O(l)$

(b) $Ga_2O_3(s) + 2\ OH^-(aq) + 2\ H_2O(l) \rightarrow 2\ Ga(OH)_4^-(aq)$

Peroxides and Superoxides; Hydrogen Peroxide

14.65 Na_2O_2 is a peroxide. It is a diamagnetic, ionic solid. It forms basic solutions on reaction with water.

$Na_2O_2(s) + H_2O(l) \rightarrow 2\ Na^+(aq) + HO_2^-(aq) + OH^-(aq)$

KO_2 is a superoxide. It is a paramagnetic, ionic solid. It produces $O_2(g)$ on reaction with water.

$2\ KO_2(s) + H_2O(l) \rightarrow O_2(g) + 2\ K^+(aq) + HO_2^-(aq) + OH^-(aq)$

14.66 Some properties of H_2O_2 are:

(1) pure H_2O_2 has a melting point of –0.4°C
(2) pure H_2O_2 explodes when heated
(3) H_2O_2 behaves as a weak acid in H_2O

Some uses of H_2O_2 are:

(1) antiseptic
(2) bleach

14.67 (a) BaO_2 (b) CaO (c) CsO_2 (d) Li_2O (e) Na_2O_2

14.68 (a) $BaO_2(s) + H_2O(l) \rightarrow Ba^{2+}(aq) + HO_2^-(aq) + OH^-(aq)$

(b) $2\ RbO_2(s) + H_2O(l) \rightarrow O_2(g) + 2\ Rb^+(aq) + HO_2^-(aq) + OH^-(aq)$

14.69

	O_2	O_2^-	O_2^{2-}
σ^*_{2p}	—	—	—
π^*_{2p}	↑ ↑	↑↓ ↑	↑↓ ↑↓
π_{2p}	↑↓ ↑↓	↑↓ ↑↓	↑↓ ↑↓
σ_{2p}	↑↓	↑↓	↑↓
	Bond order = 2	Bond order = 1.5	Bond order = 1

(a) The O–O bond length increases because the bond order decreases. The bond order decreases because of the increased occupancy of antibonding orbitals.

(b) O_2^- has 1 unpaired electron and is paramagnetic. O_2^{2-} has no unpaired electrons and is diamagnetic.

14.70

	O_2^+
σ^*_{2p}	—
π^*_{2p}	↑ —
π_{2p}	↑↓ ↑↓
σ_{2p}	↑↓

(a) O_2^+ bond order = 2 1/2, so the O–O bond should be shorter than that in O_2.

(b) O_2^+ is be paramagnetic, with one unpaired electron.

14.71 A reaction in which one substance is both oxidized and reduced is called a disproportionation reaction.

$2\ KO_2(s)\ +\ H_2O(l)\ \rightarrow\ O_2(g)\ +\ 2\ K^+(aq)\ +\ HO_2^-(aq)\ +\ OH^-(aq)$

The oxygen in KO_2 undergoes disproportionation.

14.72 $2\ H_2O_2(l)\ \xrightarrow{\text{catalyst}}\ 2\ H_2O(l)\ +\ O_2(g)$

14.73 (a) $H_2O_2(aq)\ +\ 2\ H^+(aq)\ +\ 2\ I^-(aq)\ \rightarrow\ I_2(aq)\ +\ 2\ H_2O(l)$

(b) $3\ H_2O_2(aq) + 8\ H^+(aq) + Cr_2O_7^{2-}(aq) \rightarrow 2\ Cr^{3+}(aq) + 3\ O_2(g) + 7\ H_2O(l)$

14.74 (a) $2\ Fe^{2+}(aq)\ +\ 2\ H^+(aq)\ +\ H_2O_2(aq)\ \rightarrow\ 2\ Fe^{3+}(aq)\ +\ 2\ H_2O(l)$

(b) $IO_4^-(aq)\ +\ H_2O_2(aq)\ \rightarrow\ IO_3^-(aq)\ +\ O_2(g)\ +\ H_2O(l)$

Ozone

14.75

dioxygen	ozone
O_2	O_3
:Ö=Ö:	:Ö=Ö—Ö: (with lone pairs)

14.76 (resonance structures of ozone: O=O–O ⟷ O–O=O)

Ozone has two resonance structures which leads to two equivalent O–O bond lengths.

14.77 $3\ O_2(g)\ \xrightarrow{\text{electric discharge}}\ 3\ O_3(g)$

14.78 $3\ O_2(g) \rightarrow 2\ O_3(g) \quad \Delta H^\circ = 285\ kJ$

O_2, 32.00 amu

$$\text{energy required} = 10.0\ g\ O_2 \times \frac{1\ mol\ O_2}{32.00\ g\ O_2} \times \frac{285\ kJ}{3\ mol\ O_2} = 29.7\ kJ$$

14.79 Some physical properties of ozone are:

(1) melting point = −192°C
(2) boiling point = −112°C
(3) pale blue gas
(4) sharp, penetrating odor

Some chemical properties of ozone are:

(1) powerful oxidizing agent
(2) $O_3(l)$ can decompose explosively

14.80 To distinguish O_3 from O_2, bubble the gas into a basic solution containing KI and a starch indicator. If the solution turns blue, the gas is O_3.

$O_3(g) + 2\ I^-(aq) + H_2O(l) \rightarrow O_2(g) + I_2(aq) + 2\ OH^-(aq)$

The I_2 that is produced combines with the starch indicator, giving the blue color.

O_2 does not oxidize I^-.

Water and Hydrates

14.81 (a) $2\ F_2(g) + 2\ H_2O(l) \rightarrow O_2(g) + 4\ HF(aq)$

(b) $I_2(s) + H_2O(l) \rightarrow HOI(aq) + H^+(aq) + I^-(aq)$

(c) $2\ K(s) + 2\ H_2O(l) \rightarrow H_2(g) + 2\ K^+(aq) + 2\ OH^-(aq)$

14.82 (a) $2\ Na(s) + 2\ H_2O(l) \rightarrow H_2(g) + 2\ Na^+(aq) + 2\ OH^-(aq)$

(b) $Ba(s) + 2\ H_2O(l) \rightarrow H_2(g) + Ba^{2+}(aq) + 2\ OH^-(aq)$

(c) $Cl_2(g) + H_2O(l) \rightleftharpoons HOCl(aq) + H^+(aq) + Cl^-(aq)$

14.83 $AlCl_3 \cdot 6\ H_2O$

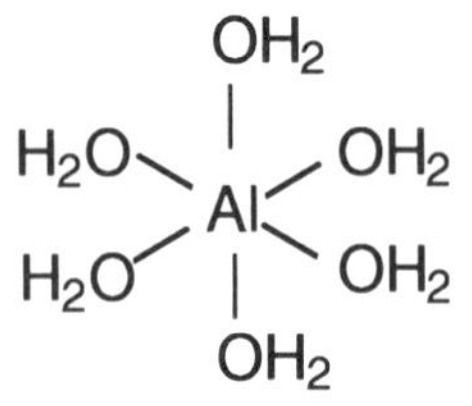

$$[Al(OH_2)_6]$$

(octahedral structure: Al bonded to six OH_2 groups)

14.84 $CaSO_4 \cdot H_2O$, 154.16 amu; H_2O, 18.02 amu

Assume one mole of $CaSO_4 \cdot H_2O$

$$\text{mass \% } H_2O = \frac{\text{mass } H_2O}{\text{mass hydrate}} \times 100\ \% = \frac{18.02\ g}{154.16\ g} \times 100\% = 11.69\%$$

14.85 $CaSO_4 \cdot \frac{1}{2}\ H_2O$, 145.15 amu; H_2O, 18.02 amu

mass of H_2O lost = 3.44 g – 2.90 g = 0.54 g H_2O

$$2.90\ g\ CaSO_4 \cdot \tfrac{1}{2}\ H_2O \times \frac{1\ mol}{145.15\ g} = 0.020\ mol\ CaSO_4 \cdot \tfrac{1}{2}\ H_2O$$

$$0.54\ g\ H_2O \times \frac{1\ mol}{18.02\ g} = 0.030\ mol$$

$$\text{number of } H_2O\text{s lost} = \frac{0.030\ mol}{0.020\ mol} = 1.5\ H_2O \text{ per } CaSO_4 \cdot \tfrac{1}{2}\ H_2O \text{ formed}$$

The mineral gypsum is $CaSO_4 \cdot 2\ H_2O$; x = 2

14.86 Assume that you start with 1.00 g $CoCl_2$ sample. When the $CoCl_2$ is exposed to moist air, there is an 83.0% increase in mass to 1.83 g.

mass of $CoCl_2 \cdot xH_2O$ = 1.83 g

$CoCl_2$, 129.84 amu; H_2O, 18.02 amu

mass $CoCl_2$ = 1.00 g

mass H_2O = 1.83 g – 1.00 g = 0.83 g

$$0.83\text{ g } H_2O \times \frac{1\text{ mol } H_2O}{18.02\text{ g } H_2O} = 0.046\text{ mol } H_2O$$

$$1.00\text{ g } CoCl_2 \times \frac{1\text{ mol } CoCl_2}{129.84\text{ g } CoCl_2} = 0.007\ 70\text{ mol } CoCl_2$$

$$x = \frac{\text{mol } H_2O}{\text{mol } CoCl_2} = \frac{0.046\text{ mol}}{0.007\ 70\text{ mol}} = 6.0$$

The hydrate formula is $CoCl_2 \cdot 6\ H_2O$

Natural Waters

14.87 In drinking water, Cl_2 is used to kill bacteria.

The sedimentation of unspended matter in drinking water is accelerated by the addition of CaO and $Al_2(SO_4)_3$. The CaO makes the water slightly basic, which precipitates Al^{3+} as $Al(OH)_3$. The gelatinous $Al(OH)_3$ precipitate slowly settles, carrying with it suspended solid and colloidal material and most of the bacteria.

14.88 Convert mi^3 to cm^3

$$(1.0\text{ mi}^3)\left(\frac{1609\text{ m}}{1\text{ mi}}\right)^3\left(\frac{100\text{ cm}}{1\text{ m}}\right)^3 = 4.2 \times 10^{15}\text{ cm}^3$$

mass of sea water = volume x density = $(4.2 \times 10^{15}\text{ cm}^3)(1.025\text{ g/cm}^3)$

mass of sea water = 4.3×10^{15} g sea water

$$\text{mass of salts} = (0.035)(4.3 \times 10^{15}\text{ g})\left(\frac{1\text{ kg}}{1000\text{ g}}\right) = 1.5 \times 10^{11}\text{ kg}$$

14.89 volume = $(1.00\text{ m}^3)(100\text{ cm/m})^3 = 1.00 \times 10^6\text{ cm}^3$

mass = $(1.00 \times 10^6\text{ cm}^3)(1.025\text{ g/cm}^3) = 1.025 \times 10^6\text{ g} = 1025\text{ kg}$

kg Mg^{2+} = (1025 kg)(1.35 g Mg^{2+}/kg)(1 kg/1000 g) = 1.38 kg Mg^{2+}

14.90 mass = $(2.0 \times 10^6 \text{ kg bromine})\left(\frac{1000 \text{ g}}{1 \text{ kg}}\right)\left(\frac{1 \text{ kg sea water}}{0.065 \text{ g Br}^-}\right)$

mass = 3.08×10^{10} kg sea water

$$\text{volume} = \frac{\text{mass}}{\text{density}} = \frac{(3.08 \times 10^{10} \text{kg})\left(\frac{1000 \text{ g}}{1 \text{ kg}}\right)}{1.025 \text{ g/cm}^3} = 3.00 \times 10^{13} \text{ cm}^3$$

volume = 3.00×10^{13} mL

$$\text{volume} = 3.00 \times 10^{13} \text{ mL} \times \frac{1 \text{ L}}{1000 \text{ mL}} = 3.0 \times 10^{10} \text{ L}$$

If recovery is only 20%, you need to process five times this volume.

processed volume = $5(3.0 \times 10^{10} \text{ L}) = 1.5 \times 10^{11}$ L

General Problems

14.91 Some physical properties of hydrogen are:

(1) melting point = 13.96 K.
(2) boiling point = 20.39 K.
(3) density = 0.0824 g/L at 25°C and 1.00 atm.

14.92 Some chemical properties of hydrogen are:

(1) forms binary hydrides.
(2) reacts explosively with O_2 to produce H_2O.
(3) reacts with N_2 to produce NH_3.
(4) reacts with CO to produce CH_3OH.

14.93 Two industrial uses of hydrogen are:

(1) Haber process for ammonia synthesis

$$N_2(g) + 3\ H_2(g) \rightleftarrows 2\ NH_3(g)$$

(2) Methanol synthesis

$$CO(g) + 2\ H_2(g) \xrightarrow{\text{Co catalyst}} CH_3OH(l)$$

14.94 Some advantages of hydrogen as a fuel are:

(1) High energy content to mass ratio.
(2) The combustion product is the nonpolluting H_2O.

$$2\ H_2(g) + O_2(g) \rightarrow 2\ H_2O(g)$$

Some disadvantages of hydrogen as a fuel are:

(1) Difficulty with storage and handling.
(2) The reaction above, if not controlled, occurs explosively.

14.95 $N_2(g) + 3\ H_2(g) \rightarrow 2\ NH_3(g)$

H_2, 2.016 amu; NH_3, 17.03 amu

$(17 \times 10^6 \text{ tons } NH_3)(2000 \text{ pounds/ton})(453.5 \text{ g/pound}) = 1.54 \times 10^{13} \text{ g } NH_3$

$$1.54 \times 10^{13} \text{ g } NH_3 \times \frac{1 \text{ mol } NH_3}{17.03 \text{ g } NH_3} = 9.04 \times 10^{11} \text{ mol } NH_3$$

$$9.04 \times 10^{11} \text{ mol } NH_3 \times \frac{3 \text{ mol } H_2}{2 \text{ mol } NH_3} = 1.36 \times 10^{12} \text{ mol } H_2$$

$$1.36 \times 10^{12} \text{ mol } H_2 \times \frac{2.016 \text{ g } H_2}{1 \text{ mol } H_2} \times \frac{1 \text{ pound}}{453.5 \text{ g}} \times \frac{1 \text{ ton}}{2000 \text{ pounds}} = 3.0 \times 10^6 \text{ tons}$$

mass H_2 = 3.0 million tons H_2

14.96 Butadiene, C_4H_6, 54.09 amu; 2.7 kg = 2700 g

$$\text{moles } H_2 = 2700 \text{ g } C_4H_6 \times \frac{1 \text{ mol } C_4H_6}{54.09 \text{ g } C_4H_6} \times \frac{2 \text{ mol } H_2}{1 \text{ mol } C_4H_6} = 99.8 \text{ mol } H_2$$

PV = nRT; at STP, P = 1.00 atm and T = 273 K

$$V = \frac{nRT}{P} = \frac{(99.8\ \text{mol})\left(0.082\ 06\ \frac{\text{L}\cdot\text{atm}}{\text{mol}\cdot\text{K}}\right)(273\ \text{K})}{1.00\ \text{atm}} = 2.2 \times 10^3\ \text{L of } H_2$$

14.97 (a) NH_3, H +1 (b) BaH_2, H −1 (c) $Ca(OH)_2$, H +1

(d) HBr, H +1 (e) H_3PO_4, H +1

14.98 In nature, oxygen is found as $O_2(g)$ in the atmosphere, in the form of H_2O in the hydrosphere, and combined with other elements in the form of silicates, carbonates, oxides, and other oxygen containing minerals in the lithosphere.

14.99

Oxygen Oxidation State	Compound	Name
−2	CaO	calcium oxide
−1	H_2O_2	hydrogen peroxide
−1/2	KO_2	potassium superoxide

14.100
(a) B_2O_3 boron oxide
(b) H_2O_2 hydrogen peroxide
(c) SrH_2 strontium hydride
(d) CsO_2 cesium superoxide
(e) $HClO_4$ perchloric acid
(f) BaO_2 barium peroxide

14.101 (a) $Ca(OH)_2$ (b) Cr_2O_3 (c) RbO_2 (d) Na_2O_2

(e) BaH_2 (f) H_2Se

14.102 (a) $2\ H_2(g) + O_2(g) \rightarrow 2\ H_2O(l)$

(b) $O_3(g) + 2\ I^-(aq) + H_2O(l) \rightarrow O_2(g) + I_2(aq) + 2\ OH^-(aq)$

(c) $H_2O_2(aq) + 2\ H^+(aq) + 2\ Br^-(aq) \rightarrow 2\ H_2O(l) + Br_2(aq)$

(d) $2\ Na(l) + H_2(g) \rightarrow 2\ NaH(s)$

(e) $2\ Na(s) + 2\ H_2O(l) \rightarrow H_2(g) + 2\ Na^+(aq) + 2\ OH^-(aq)$

14.103 (a) $2\ H_2(g) + O_2(g) \rightarrow 2\ H_2O(l)$

(b) $5\ H_2O_2(aq) + 2\ MnO_4^-(aq) + 6\ H^+(aq) \rightarrow 5\ O_2(g) + 2\ Mn^{2+}(aq) + 8\ H_2O(l)$

(c) $2\ F_2(g) + 2\ H_2O(l) \rightarrow O_2(g) + 4\ HF(aq)$

14.104 (a) $CO(g) + 2\ H_2(g) \rightarrow CH_3OH(l)$

$\Delta H° = \Delta H_f°(CH_3OH) - \Delta H_f°(CO)$

$\Delta H° = (1\ mol)(-238.7\ kJ/mol) - (1\ mol)(-110.5\ kJ/mol) = -128.2\ kJ$

(b) $CO(g) + H_2O(g) \rightarrow CO_2(g) + H_2(g)$

$\Delta H° = \Delta H_f°(CO_2) - [\Delta H_f°(CO) + \Delta H_f°(H_2O)]$

$\Delta H° = (1\ mol)(-393.5\ kJ/mol) - [(1\ mol)(-110.5\ kJ/mol)$

$+ (1\ mol)(-241.8\ kJ/mol)] = -41.2\ kJ$

(c) $2\ KClO_3(s) \rightarrow 2\ KCl(s) + 3\ O_2(g)$

$\Delta H° = 2\ \Delta H_f°(KCl) - 2\ \Delta H_f°(KClO_3)$

$\Delta H° = (2\ mol)(-436.7\ kJ/mol) - (2\ mol)(-397.7\ kJ/mol) = -78.0\ kJ$

(d) $6\ CO_2(g) + 6\ H_2O(l) \rightarrow 6\ O_2(g) + C_6H_{12}O_6(s)$

$\Delta H° = \Delta H_f°(C_6H_{12}O_6) - [6\ \Delta H_f°(CO_2) + 6\ \Delta H_f°(H_2O)]$

$\Delta H° = (1\ mol)(-1260\ kJ/mol) - [(6\ mol)(-393.5\ kJ/mol)$

$+ (6\ mol)(-285.8\ kJ/mol)] = 2816\ kJ$

15.1 (a) $H_2SO_4(aq) + H_2O(l) \rightleftarrows H_3O^+(aq) + HSO_4^-(aq)$
conjugate base

(b) $HSO_4^-(aq) + H_2O(l) \rightleftarrows H_3O^+(aq) + SO_4^{2-}(aq)$
conjugate base

(c) $H_3O^+(aq) + H_2O(l) \rightleftarrows H_3O^+(aq) + H_2O(l)$
conjugate base

15.2 (a) $H_2CO_3(aq) + H_2O(l) \rightleftarrows H_3O^+(aq) + HCO_3^-(aq)$
conjugate acid

(b) $HCO_3^-(aq) + H_2O(l) \rightleftarrows H_3O^+(aq) + CO_3^{2-}(aq)$
conjugate acid

(c) $Base(aq) + H_2O(l) \rightleftarrows BaseH^+(aq) + OH^-(aq)$
conjugate acid

15.3 $$[H_3O^+] = \frac{K_w}{[OH^-]} = \frac{1.0 \times 10^{-14}}{5.0 \times 10^{-6}} = 2.0 \times 10^{-9}\text{ M}$$

Because $[OH^-] > [H_3O^+]$, the solution is basic.

15.4 $K_w = [H_3O^+][OH^-]$

In a neutral solution, $[H_3O^+] = [OH^-]$

At 50°C, $[H_3O^+] = [OH^-] = \sqrt{K_w} = \sqrt{5.5 \times 10^{-14}} = 2.3 \times 10^{-7}\text{ M}$

15.5 (a) $$[H_3O^+] = \frac{K_w}{[OH^-]} = \frac{1.0 \times 10^{-14}}{1.58 \times 10^{-6}} = 6.3 \times 10^{-9}\text{ M}$$

$pH = -\log[H_3O^+] = -\log(6.3 \times 10^{-9}) = 8.20$

(b) $pH = -\log[H_3O^+] = -\log(6.0 \times 10^{-5}) = 4.22$

15.6 $[H_3O^+] = 10^{-pH} = 10^{-7.40} = 4.0 \times 10^{-8}$ M

$$[OH^-] = \frac{K_w}{[H_3O^+]} = \frac{1.0 \times 10^{-14}}{4.0 \times 10^{-8}} = 2.5 \times 10^{-7} \text{ M}$$

15.7 (a) Because $HClO_4$ is a strong acid, $[H_3O^+] = 0.050$ M.

pH = $-\log[H_3O^+] = -\log(0.050) = 1.30$

(b) Because HCl is a strong acid, $[H_3O^+] = 1.0$ M.

pH = $-\log[H_3O^+] = -\log(1.0) = 0.00$

(c) pH = $-\log[H_3O^+] = -\log(6.0) = -0.78$

15.8 (a) Because KOH is a strong base, $[OH^-] = 0.020$ M.

$$[H_3O^+] = \frac{K_w}{[OH^-]} = \frac{1.0 \times 10^{-14}}{0.020} = 5.0 \times 10^{-13} \text{ M}$$

pH = $-\log[H_3O^+] = -\log(5.0 \times 10^{-13}) = 12.30$

(b) Because $Ba(OH)_2$ is a strong base, $[OH^-] = 2(0.010 \text{ M}) = 0.020$ M.

$$[H_3O^+] = \frac{K_w}{[OH^-]} = \frac{1.0 \times 10^{-14}}{0.020} = 5.0 \times 10^{-13} \text{ M}$$

pH = $-\log[H_3O^+] = -\log(5.0 \times 10^{-13}) = 12.30$

15.9 $BaO(s) + H_2O(l) \rightarrow Ba(OH)_2(aq)$

BaO, 153.33 amu

$$0.25 \text{ g BaO} \times \frac{1 \text{ mol BaO}}{153.33 \text{ g BaO}} \times \frac{1 \text{ mol Ba(OH)}_2}{1 \text{ mol BaO}} \times \frac{2 \text{ mol OH}^-}{1 \text{ mol Ba(OH)}_2} = 3.26 \times 10^{-3} \text{ mol OH}^-$$

$$[OH^-] = \frac{3.26 \times 10^{-3} \text{ mol } OH^-}{0.500 \text{ L}} = 6.52 \times 10^{-3} \text{ M}$$

$$[H_3O^+] = \frac{K_w}{[OH^-]} = \frac{1.0 \times 10^{-14}}{6.52 \times 10^{-3}} = 1.53 \times 10^{-12} \text{ M}$$

$pH = -\log[H_3O^+] = -\log(1.53 \times 10^{-12}) = 11.81$

15.10 The strongest base that can exist in H_2O is OH^-. Any stronger base will be protonated by H_2O leaving OH^- in solution.

15.11

	$HOCl(aq) + H_2O(l)$	$\rightleftarrows$	$H_3O^+(aq)$	$+ OCl^-(aq)$
initial (M)	0.10		~0	0
change (M)	–x		+x	+x
equil (M)	0.10 – x		x	x

$x = [H_3O^+] = 10^{-pH} = 10^{-4.23} = 5.9 \times 10^{-5}$ M

$[OCl^-] = x = 5.9 \times 10^{-5}$ M

$[HOCl] = 0.10 - x = (0.10 - 5.9 \times 10^{-5})$M

$$K_a = \frac{[H_3O^+][OCl^-]}{[HOCl]} = \frac{(5.9 \times 10^{-5})(5.9 \times 10^{-5})}{(0.10 - 5.9 \times 10^{-5})} = 3.5 \times 10^{-8}$$

This value of K_a agrees with the value in Table 15.2.

15.12 (a)

	$CH_3COOH(aq) + H_2O(l)$	$\rightleftarrows$	$H_3O^+(aq)$	$+ CH_3COO^-(aq)$
initial (M)	1.00		~0	0
change (M)	–x		+x	+x
equil (M)	1.00 – x		x	x

$$K_a = \frac{[H_3O^+][CH_3COO^-]}{[CH_3COOH]} = 1.8 \times 10^{-5} = \frac{x^2}{1.00 - x} \approx \frac{x^2}{1.00}$$

Solve for x. $x = [H_3O^+] = 4.2 \times 10^{-3}$ M

$pH = -\log[H_3O^+] = -\log(4.2 \times 10^{-3}) = 2.38$

$[CH_3COO^-] = x = 4.2 \times 10^{-3}$ M

$[CH_3COOH] = 1.00 - x = 1.00$ M

$$[OH^-] = \frac{K_w}{[H_3O^+]} = \frac{1.0 \times 10^{-14}}{4.2 \times 10^{-3}} = 2.4 \times 10^{-12} \text{ M}$$

(b) $CH_3COOH(aq) + H_2O(l) \rightleftarrows H_3O^+(aq) + CH_3COO^-(aq)$

	CH_3COOH	H_3O^+	CH_3COO^-
initial (M)	0.0100	~0	0
change (M)	–x	+x	+x
equil (M)	0.0100 – x	x	x

$$K_a = \frac{[H_3O^+][CH_3COO^-]}{[CH_3COOH]} = 1.8 \times 10^{-5} = \frac{x^2}{0.0100 - x}$$

$x^2 + (1.8 \times 10^{-5})x - (1.8 \times 10^{-7}) = 0$

Use the quadratic formula to solve for x.

$$x = \frac{(-1.8 \times 10^{-5}) \pm \sqrt{(1.8 \times 10^{-5})^2 + 4(1.8 \times 10^{-7})}}{2(1)} = \frac{(-1.8 \times 10^{-5}) \pm (8.5 \times 10^{-4})}{2}$$

$x = 4.2 \times 10^{-4}$ and -4.3×10^{-4}

Of the two solutions for x, only the positive value of x has physical meaning because x is the $[H_3O^+]$.

$x = [H_3O^+] = 4.2 \times 10^{-4}$ M

$pH = -\log[H_3O^+] = -\log(4.2 \times 10^{-4}) = 3.38$

$[CH_3COO^-] = x = 4.2 \times 10^{-4}$ M

$[CH_3COOH] = 0.0100 - x = 0.0100 - (4.2 \times 10^{-4}) = 0.0096$ M

$$[OH^-] = \frac{K_w}{[H_3O^+]} = \frac{1.0 \times 10^{-14}}{4.2 \times 10^{-4}} = 2.4 \times 10^{-11} \text{ M}$$

15.13 (a) From Example 15.8:

$[H_3O^+] = [HF]_{diss} = 4.0 \times 10^{-3}$ M

$$\% \text{ dissociation} = \frac{[HF]_{diss}}{[HF]_{initial}} \times 100\% = \frac{4.0 \times 10^{-3}\text{ M}}{0.050\text{ M}} \times 100\% = 8.0\% \text{ dissociation}$$

(b)

	$HF(aq)$	+	$H_2O(l)$	⇄	$H_3O^+(aq)$	+	$F^-(aq)$
initial (M)	0.50				~0		0
change (M)	–x				+x		+x
equil (M)	0.50 – x				x		x

$$K_a = \frac{[H_3O^+][F^-]}{[HF]} = 3.5 \times 10^{-4} = \frac{x^2}{0.50 - x}$$

$x^2 + (3.5 \times 10^{-4})x - (1.75 \times 10^{-4}) = 0$

Use the quadratic formula to solve for x.

$$x = \frac{(-3.5 \times 10^{-4}) \pm \sqrt{(3.5 \times 10^{-4})^2 + 4(1)(1.75 \times 10^{-4})}}{2(1)} = \frac{(-3.5 \times 10^{-4}) \pm 0.0265}{2}$$

x = 0.0131 and –0.0134

Of the two solutions for x, only the positive value of x has physical meaning, because x is the $[H_3O^+]$.

$[H_3O^+] = [HF]_{diss} = 0.013$ M

$$\% \text{ dissociation} = \frac{[HF]_{diss}}{[HF]_{initial}} \times 100\% = \frac{0.013\text{ M}}{0.50\text{ M}} \times 100\% = 2.6\% \text{ dissociation}$$

15.14

	$H_2SO_3(aq)$	+	$H_2O(l)$	⇌	$H_3O^+(aq)$	+	$HSO_3^-(aq)$
initial (M)	0.10				~0		0
change (M)	–x				+x		+x
equil (M)	0.10 – x				x		x

$$K_{a1} = \frac{[H_3O^+][HSO_3^-]}{[H_2SO_3]} = 1.5 \times 10^{-2} = \frac{x^2}{0.10 - x}$$

$x^2 + 0.015x - 0.0015 = 0$

Use the quadratic formula to solve for x.

$$x = \frac{(-0.015) \pm \sqrt{(0.015)^2 + (4)(0.0015)}}{2(1)} = \frac{-0.015 \pm 0.079}{2}$$

x = 0.032 and −0.047

Of the two solutions for x, only the positive value of x has physical meaning since x is the $[H_3O^+]$.

$x = [H_3O^+] = [HSO_3^-] = 0.032$ M

$[H_2SO_3] = 0.10 - x = 0.10 - 0.032 = 0.07$ M

The second dissociation of H_2SO_3 produces a negligible amount of H_3O^+ compared with that from the first dissociation.

$HSO_3^-(aq) + H_2O(l) \rightleftarrows H_3O^+(aq) + SO_3^{2-}(aq)$

$$K_{a2} = \frac{[H_3O^+][SO_3^{2-}]}{[HSO_3^-]} = 6.3 \times 10^{-8} = \frac{(0.032)[SO_3^{2-}]}{(0.032)}$$

$[SO_3^{2-}] = K_{a2} = 6.3 \times 10^{-8}$ M

$$[OH^-] = \frac{K_w}{[H_3O^+]} = \frac{1.0 \times 10^{-14}}{0.032} = 3.1 \times 10^{-13} \text{ M}$$

$pH = -\log[H_3O^+] = -\log(0.032) = 1.49$

15.15 From the complete dissociation of the first proton, $[H_3O^+] = [HSO_4^-] = 0.50$ M.

For the dissociation of the second proton, the following equilibrium must be considered:

	$HSO_4^-(aq)$ + $H_2O(l)$	⇄	$H_3O^+(aq)$	+ $SO_4^{2-}(aq)$
initial (M)	0.50		0.50	0
change (M)	–x		+x	+x
equil (M)	0.50 – x		0.50 + x	x

$$K_{a2} = \frac{[H_3O^+][SO_4^{2-}]}{[HSO_4^-]} = 1.2 \times 10^{-2} = \frac{(0.50 + x)(x)}{0.50 - x}$$

$x^2 + 0.512x - 0.0060 = 0$

Use the quadratic formula to solve for x.

$$x = \frac{(-0.512) \pm \sqrt{(0.512)^2 + 4(1)(0.0060)}}{2(1)} = \frac{-0.512 \pm 0.535}{2}$$

x = 0.011 and –0.524

Of the two solutions for x, only the positive value of x has physical meaning, since x is the $[SO_4^{2-}]$.

$[H_2SO_4] = 0$ M

$[HSO_4^-] = 0.50 - x = 0.49$ M

$[SO_4^{2-}] = x = 0.011$ M

$[H_3O^+] = 0.50 + x = 0.51$ M

$pH = -\log[H_3O^+] = -\log(0.51) = 0.29$

$$[OH^-] = \frac{K_w}{[H_3O^+]} = \frac{1.0 \times 10^{-14}}{0.51} = 2.0 \times 10^{-14}\ M$$

15.16

	$NH_3(aq)$ + $H_2O(l)$	⇄	$NH_4^+(aq)$	+ $OH^-(aq)$
initial (M)	0.40		0	~0
change (M)	–x		+x	+x
equil (M)	0.40 – x		x	x

$$K_b = \frac{[NH_4^+][OH^-]}{[NH_3]} = 1.8 \times 10^{-5} = \frac{x^2}{0.40 - x} \approx \frac{x^2}{0.40}$$

Solve for x. $x = [OH^-] = 2.7 \times 10^{-3}$ M

$[NH_4^+] = x = 2.7 \times 10^{-3}$ M

$[NH_3] = 0.40 - x = 0.40$ M

$$[H_3O^+] = \frac{K_w}{[OH^-]} = \frac{1.0 \times 10^{-14}}{2.7 \times 10^{-3}} = 3.7 \times 10^{-12}\ M$$

$pH = -\log[H_3O^+] = -\log(3.7 \times 10^{-12}) = 11.43$

15.17 (a) $K_a = \dfrac{K_w}{K_b \text{ for } C_5H_{11}N} = \dfrac{1.0 \times 10^{-14}}{1.3 \times 10^{-3}} = 7.7 \times 10^{-12}$

(b) $K_b = \dfrac{K_w}{K_a \text{ for HOCl}} = \dfrac{1.0 \times 10^{-14}}{3.5 \times 10^{-8}} = 2.9 \times 10^{-7}$

15.18 (a) 0.25 M NH_4Br

NH_4^+ is an acidic cation. Br^- is a neutral anion. The salt solution is acidic.

For NH_4^+, $K_a = \dfrac{K_w}{K_b \text{ for } NH_3} = \dfrac{1.0 \times 10^{-14}}{1.8 \times 10^{-5}} = 5.6 \times 10^{-10}$

	$NH_4^+(aq)$ + $H_2O(l)$	⇄	$H_3O^+(aq)$	+ $NH_3(aq)$
initial (M)	0.25		~0	0
change (M)	$-x$		$+x$	$+x$
equil (M)	$0.25 - x$		x	x

$$K_a = \frac{[H_3O^+][NH_3]}{[NH_4^+]} = 5.6 \times 10^{-10} = \frac{x^2}{0.25 - x} \approx \frac{x^2}{0.25}$$

Solve for x. $x = [H_3O^+] = 1.2 \times 10^{-5}$ M

pH = –log[H_3O^+] = –log(1.2 x 10^{-5}) = 4.92

(b) 0.40 M $ZnCl_2$

Zn^{2+} is an acidic cation. Cl^- is a neutral anion. The salt solution is acidic.

$$Zn(H_2O)_6^{2+}(aq) + H_2O(l) \rightleftarrows H_3O^+(aq) + Zn(H_2O)_5(OH)^+(aq)$$

	$Zn(H_2O)_6^{2+}$	H_3O^+	$Zn(H_2O)_5(OH)^+$
initial (M)	0.40	~0	0
change (M)	–x	+x	+x
equil(M)	0.40 – x	x	x

$$K_a = \frac{[H_3O^+][Zn(H_2O)_5(OH)^+]}{[Zn(H_2O)_6^{2+}]} = 2.5 \times 10^{-10} = \frac{x^2}{0.40 - x} \approx \frac{x^2}{0.40}$$

Solve for x. x = [H_3O^+] = 1.0 x 10^{-5} M

pH = –log[H_3O^+] = –log(1.0 x 10^{-5}) = 5.00

15.19 For NO_2^-, $K_b = \dfrac{K_w}{K_a \text{ for } HNO_2} = \dfrac{1.0 \times 10^{-14}}{4.6 \times 10^{-4}} = 2.2 \times 10^{-11}$

$$NO_2^-(aq) + H_2O(l) \rightleftarrows HNO_2(aq) + OH^-(aq)$$

	NO_2^-	HNO_2	OH^-
initial (M)	0.20	0	~0
change (M)	–x	+x	+x
equil (M)	0.20 – x	x	x

$$K_b = \frac{[HNO_2][OH^-]}{[NO_2^-]} = 2.2 \times 10^{-11} = \frac{x^2}{0.20 - x} \approx \frac{x^2}{0.20}$$

Solve for x. x = [OH^-] = 2.1 x 10^{-6} M

$$[H_3O^+] = \frac{K_w}{[OH^-]} = \frac{1.0 \times 10^{-14}}{2.1 \times 10^{-6}} = 4.8 \times 10^{-9} \text{ M}$$

pH = –log[H_3O^+] = –log(4.8 x 10^{-9}) = 8.32

5.20 For NH_4^+, $K_a = \dfrac{K_w}{K_b \text{ for } NH_3} = \dfrac{1.0 \times 10^{-14}}{1.8 \times 10^{-5}} = 5.6 \times 10^{-10}$

For CN^-, $K_b = \dfrac{K_w}{K_a \text{ for } HCN} = \dfrac{1.0 \times 10^{-14}}{4.9 \times 10^{-10}} = 2.0 \times 10^{-5}$

Because $K_b > K_a$, the solution is basic.

15.21 (a) KBr: K^+, neutral cation; Br^-, neutral anion; solution is neutral

(b) $NaNO_2$: Na^+, neutral cation; NO_2^-, basic anion; solution is basic

(c) NH_4Br: NH_4^+, acidic cation; Br^-, neutral anion; solution is acidic

(d) $ZnCl_2$: Zn^{2+}, acidic cation; Cl^-, neutral anion; solution is acidic

(e) NH_4F

For NH_4^+, $K_a = \dfrac{K_w}{K_b \text{ for } NH_3} = \dfrac{1.0 \times 10^{-14}}{1.8 \times 10^{-5}} = 5.6 \times 10^{-10}$

For F^-, $K_b = \dfrac{K_w}{K_a \text{ for } HF} = \dfrac{1.0 \times 10^{-14}}{3.5 \times 10^{-4}} = 2.9 \times 10^{-11}$

Because $K_a > K_b$, the solution is acidic.

15.22 (a) H_2Se is a stronger acid than H_2S because Se is below S in the 6A group and the H–Se bond is weaker than the H–S bond.

(b) HI is a stronger acid than H_2Te because I is to the right of Te in the same row of the periodic table, I is more electronegative than Te, and the H–I bond is more polar.

(c) HNO_3 is a stronger acid than HNO_2 because acid strength increases with increasing oxidation number of N. The oxidation number for N is +5 in HNO_3 and +3 in HNO_2.

(d) H_2SO_3 is a stronger acid than H_2SeO_3 because acid strength increases with increasing electronegativity of the central atom. S is more electronegative than Se.

15.23 (a) Lewis acid, $AlCl_3$; Lewis base, Cl^-

(b) Lewis acid, Ag^+; Lewis base, NH_3

(c) Lewis acid, SO_2; Lewis base, OH^-

(d) Lewis acid, Cr^{3+}; Lewis base, H_2O

Understanding Key Concepts

1. (a) acids, HCO_3^- and H_3O^+; bases, H_2O and CO_3^{2-}

   (b) acids, HF and H_2CO_3; base HCO_3^- and F^-

2. (a) HX^-, Y^-, Z^- (b) $Z^- < Y^- < HX^-$

3. (a) HX < HZ < HY (b) HY (c) HX (d) (2/10) x 100% = 20%

4. (c) represents a solution of a weak diprotic acid, H_2A. Because K_{a2} is always less than K_{a1}, (a) and (d) represent impossible situations. (b) contains no H_2A.

5. (a) $A^-(aq) + H_2O(l) \rightleftarrows HA(aq) + OH^-(aq)$; basic

   (b) $M(H_2O)_6^{3+}(aq) + H_2O(l) \rightleftarrows H_3O^+(aq) + M(H_2O)_5(OH)^{2+}(aq)$; acidic

   (c) $2\ H_2O(l) \rightleftarrows H_3O^+(aq) + OH^-(aq)$; neutral

   (d) $M(H_2O)_6^{3+}(aq) + A^-(aq) \rightleftarrows HA(aq) + M(H_2O)_5(OH)^{2+}(aq)$; acidic because K_a for $M(H_2O)_6^{3+}$ (10^{-4}) is greater than K_b for A^- (10^{-9}).

6. (a) H_2S, weakest; HBr, strongest. Acid strength for H_nX increases with increasing polarity of the H–X bond and with increasing size of X.

   (b) H_2SeO_3, weakest; $HClO_3$, strongest. Acid strength for H_nYO_3 increases with increasing electronegativity of Y.

7. (a) Brønsted–Lowry acids: NH_4^+, $H_2PO_4^-$

Brønsted–Lowry bases: SO_3^{2-}, OCl^-, $H_2PO_4^-$

(b) Lewis acids: Fe^{3+}, BCl_3

Lewis bases: SO_3^{2-}, OCl^-, $H_2PO_4^-$

Additional Problems

Acid–Base Concepts

15.24 (a) An Arrhenius acid is a substance that dissociates in water to produce hydrogen ions (H^+). (examples: HCl, HNO_3)

(b) An Arrhenius base is a substance that dissociates in water to produce hydroxide ions (OH^-). (examples: NaOH, $Ca(OH)_2$)

(c) A Brønsted–Lowry acid is a proton donor. (examples: HCl, $HClO_4$)

(d) A Brønsted–Lowry base in a proton acceptor. (examples: OH^-, NH_3)

15.25 NH_3, CN^- and NO_2^-

15.26 (a) $H_3O^+(aq) + NO_3^-(aq) + K^+(aq) + OH^-(aq) \rightarrow$

$$K^+(aq) + NO_3^-(aq) + 2\ H_2O(l)$$

(b) $2\ H_3O^+(aq) + 2\ NO_3^-(aq) + Ba^{2+}(aq) + 2\ OH^-(aq) \rightarrow$

$$Ba^{2+}(aq) + 2\ NO_3^-(aq) + 4\ H_2O(l)$$

After removing spectator ions, the net ionic equation for both reactions is:

$$H_3O^+(aq) + OH^-(aq) \rightarrow 2\ H_2O(l)$$

15.27 (a) OCl^- (b) PO_4^{3-} (c) OH^-

(d) CH_3NH_2 (e) HCO_3^- (f) H^-

15.28 (a) HF (b) H_3O^+ (c) $(CH_3)_2NH_2^+$

(d) $H_2PO_4^-$ (e) H_2O (f) HSO_4^-

15.29 (a) $CH_3COOH(aq) + NH_3(aq) \rightleftharpoons NH_4^+(aq) + CH_3COO^-(aq)$
acid base——acid base

(b) $CO_3^{2-}(aq) + H_3O^+(aq) \rightleftharpoons H_2O(l) + HCO_3^-(aq)$
base acid——base acid

(c) $HSO_3^-(aq) + H_2O(l) \rightleftharpoons H_3O^+(aq) + SO_3^{2-}(aq)$
acid base——acid base

(d) $HSO_3^-(aq) + H_2O(l) \rightleftharpoons H_2SO_3(aq) + OH^-(aq)$
base acid——acid base

15.30 A Brønsted–Lowry base is a proton acceptor. To be able to accept a proton the Brønsted–Lowry base must have at least one lone pair of electrons to form a bond with H^+.

15.31 From data in Table 15.1:

Strong acids: HNO_3 and H_2SO_4

Strong bases: H^- and O^{2-}

15.32 The weaker acid of the pair has the stronger conjugate base.

(a) H_2CO_3 is a weak acid. H_2SO_4 is a strong acid. H_2CO_3 has the stronger conjugate base.

(b) HCl is a strong acid. HF is a weak acid. HF has the stronger conjugate base.

(c) NH_4^+ is a weaker acid than HF. NH_4^+ has the stronger conjugate base.

(d) HCN is a weaker acid than HSO_4^-. HCN has the stronger conjugate base.

15.33 (a) left, HCO_3^- is the stronger base

(b) left, F^- is the stronger base

(c) right, NH_3 is the stronger base

(d) right, CN^- is the stronger base

Dissociation of Water; pH

15.34 $H{-}\ddot{\underset{\cdot\cdot}{O}}{-}H$

H_2O as a proton donor (acid):

$B{:}(aq) + H_2O(l) \rightleftarrows BH^+(aq) + OH^-(aq)$

H_2O as a proton acceptor (base):

$HA(aq) + H_2O(l) \rightleftarrows H_3O^+(aq) + A^-(aq)$

15.35 $2\ H_2O(l) \rightleftarrows H_3O^+(aq) + OH^-(aq)$; $K_w = [H_3O^+][OH^-]$

$[H_2O]$ is omitted because it is a constant.

15.36 If $[H_3O^+] > 1.0 \times 10^{-7}$ M, solution is acidic.

If $[H_3O^+] < 1.0 \times 10^{-7}$ M, solution is basic.

If $[H_3O^+] = [OH^-] = 1.0 \times 10^{-7}$ M, solution is neutral.

If $[OH^-] > 1.0 \times 10^{-7}$ M, solution is basic

If $[OH^-] < 1.0 \times 10^{-7}$ M, solution is acidic.

(a) basic (b) basic (c) acidic (d) neutral (e) acidic

15.37 (a) $[OH^-] = \dfrac{K_w}{[H_3O^+]} = \dfrac{1.0 \times 10^{-14}}{2.5 \times 10^{-4}} = 4.0 \times 10^{-11}$ M

(b) $[OH^-] = \dfrac{K_w}{[H_3O^+]} = \dfrac{1.0 \times 10^{-14}}{2.0} = 5.0 \times 10^{-15}$ M

(c) $[H_3O^+] = \dfrac{K_w}{[OH^-]} = \dfrac{1.0 \times 10^{-14}}{5.6 \times 10^{-9}} = 1.8 \times 10^{-6}$ M

(d) $[H_3O^+] = \dfrac{K_w}{[OH^-]} = \dfrac{1.0 \times 10^{-14}}{1.5 \times 10^{-3}} = 6.7 \times 10^{-12}$ M

(e) $[H_3O^+] = \dfrac{K_w}{[OH^-]} = \dfrac{1.0 \times 10^{-14}}{1.0 \times 10^{-7}} = 1.0 \times 10^{-7}$ M

15.38 (a) $pH = -\log[H_3O^+] = -\log(2.0 \times 10^{-5}) = 4.70$

(b) $[H_3O^+] = \dfrac{K_w}{[OH^-]} = \dfrac{1.0 \times 10^{-14}}{4 \times 10^{-3}} = 2.5 \times 10^{-12}$ M

$pH = -\log[H_3O^+] = -\log(2.5 \times 10^{-12}) = 11.6$

(c) $pH = -\log[H_3O^+] = -\log(10^{-3}) = 3$

(d) $[H_3O^+] = \dfrac{K_w}{[OH^-]} = \dfrac{1.0 \times 10^{-14}}{1.0 \times 10^{-7}} = 1.0 \times 10^{-7}$

$pH = -\log[H_3O^+] = -\log(1.0 \times 10^{-7}) = 7.00$

(e) $pH = -\log[H_3O^+] = -\log(5.0) = -0.70$

(f) $[H_3O^+] = \dfrac{K_w}{[OH^-]} = \dfrac{1.0 \times 10^{-14}}{12} = 8.3 \times 10^{-16}$ M

$pH = -\log[H_3O^+] = -\log(8.3 \times 10^{-16}) = 15.08$

15.39 $[H_3O^+] = 10^{-pH}$

(a) 8×10^{-5} M (b) 1.5×10^{-11} M (c) 1.0 M

(d) 5.6×10^{-15} M (e) 10 M

15.40 $\Delta pH = \log(\Delta[H_3O^+])$

(a) $\Delta pH = \log(1000) = 3$

(b) $\Delta pH = \log(1.0 \times 10^5) = 5.00$

(c) $\Delta pH = \log(2.0) = 0.30$

15.41 (a) $10^1 = 10$ (b) 1×10^{10} (c) $10^{0.10} = 1.3$

15.42 (a) $pH = -\log[H_3O^+] = -\log(10^{-2}) = 2$

(b) $pH = -\log[H_2O^+] = -\log(4 \times 10^{-8}) = 7.4$

(c) $[H_3O^+] = \dfrac{K_w}{[OH^-]} = \dfrac{1.0 \times 10^{-14}}{8 \times 10^{-8}} = 1.25 \times 10^{-7}$ M

$pH = -\log[H_3O^+] = -\log(1.25 \times 10^{-7}) = 6.9$

(d) $[H_3O^+] = \dfrac{K_w}{[OH^-]} = \dfrac{1.0 \times 10^{-14}}{6 \times 10^{-10}} = 1.7 \times 10^{-5}$ M

$pH = -\log[H_3O^+] = -\log(1.7 \times 10^{-5}) = 4.8$

$[H_3O^+] = \dfrac{K_w}{[OH^-]} = \dfrac{1.0 \times 10^{-14}}{2 \times 10^{-6}} = 5 \times 10^{-9}$ M

$pH = -\log[H_3O^+] = -\log(5 \times 10^{-9}) = 8.3$

pH = 4.8 to 8.3

Strong Acids and Strong Bases

15.43 (a) $[H_3O^+] = 0.20$ M; pH = $-\log[H_3O^+] = -\log(0.20) = 0.70$

(b) $[OH^-] = 6.3 \times 10^{-3}$ M

$$[H_3O^+] = \frac{K_w}{[OH^-]} = \frac{1.0 \times 10^{-14}}{6.3 \times 10^{-3}} = 1.6 \times 10^{-12} \text{ M}$$

pH = $-\log[H_3O^+] = -\log(1.6 \times 10^{-12}) = 11.80$

(c) $[OH^-] = 2(4.0 \times 10^{-3}$ M$) = 8.0 \times 10^{-3}$ M

$$[H_3O^+] = \frac{K_w}{[OH^-]} = \frac{1.0 \times 10^{-14}}{8.0 \times 10^{-3}} = 1.25 \times 10^{-12} \text{ M}$$

pH = $-\log[H_3O^+] = -\log(1.25 \times 10^{-12}) = 11.90$

(d) $[H_3O^+] = 1.5$ M; pH = $-\log[H_3O^+] = -\log(1.5) = -0.18$

(e) $[OH^-] = 1.5$ M

$$[H_3O^+] = \frac{K_w}{[OH^-]} = \frac{1.0 \times 10^{-14}}{1.5} = 6.7 \times 10^{-15} \text{ M}$$

pH = $-\log[H_3O^+] = -\log(6.7 \times 10^{-15}) = 14.18$

15.44 (a) LiOH, 23.95 amu; 250 mL = 0.250 L

$$\text{molarity of LiOH(aq)} = \frac{\left(4.8 \text{ g} \times \frac{1 \text{ mol}}{23.95 \text{ g}}\right)}{0.250 \text{ L}} = 0.80 \text{ M}$$

LiOH is a strong base; therefore $[OH^-] = 0.80$ M.

$$[H_3O^+] = \frac{K_w}{[OH^-]} = \frac{1.0 \times 10^{-14}}{0.80} = 1.25 \times 10^{-14} \text{ M}$$

$pH = -\log[H_3O^+] = -\log(1.25 \times 10^{-14}) = 13.90$

(b) HCl, 36.46 amu

$$\text{molarity of HCl(aq)} = \frac{\left(0.93\ \text{g} \times \frac{1\ \text{mol}}{36.46\ \text{g}}\right)}{0.40\ \text{L}} = 0.064\ \text{M}$$

HCl is a strong acid; therefore $[H_3O^+] = 0.064$ M

$pH = -\log[H_3O^+] = -\log(0.064) = 1.19$

(c) Na_2O, 61.98 amu; 100 mL = 0.100 L

$$\text{moles of } Na_2O = 0.20\ \text{g} \times \frac{1\ \text{mol}}{61.98\ \text{g}} = 0.0032\ \text{mol}$$

$$O^{2-}(aq) + H_2O(l) \xrightarrow{100\%} 2\ OH^-(aq)$$

moles of OH^- = (2)(0.0032 mol) = 0.0064 mol

$$[OH^-] = \frac{0.0064\ \text{mol}}{0.100\ \text{L}} = 0.064\ \text{M}$$

$$[H_3O^+] = \frac{K_w}{[OH^-]} = \frac{1.0 \times 10^{-14}}{0.064} = 1.6 \times 10^{-13}\ \text{M}$$

$pH = -\log[H_3O^+] = -\log(1.6 \times 10^{-13}) = 12.80$

15.45 (a) $M_f \cdot V_f = M_i \cdot V_i$

$$M_f = \frac{M_i \cdot V_i}{V_f} = \frac{(0.10\ \text{M})(50\ \text{mL})}{(1000\ \text{mL})} = 5.0 \times 10^{-3}\ \text{M}$$

$pH = -\log[H_3O^+] = -\log(5.0 \times 10^{-3}) = 2.30$

(b) For HCl, $M_f = \dfrac{M_i \cdot V_i}{V_f} = \dfrac{(2.0 \times 10^{-3}\ \text{M})(100\ \text{mL})}{(500\ \text{mL})} = 4.0 \times 10^{-4}$ M

For $HClO_4$, $M_f = \dfrac{M_i \cdot V_i}{V_f} = \dfrac{(1.0 \times 10^{-3}\ M)(400\ mL)}{(500\ mL)} = 8.0 \times 10^{-4}\ M$

$[H_3O^+] = (4.0 \times 10^{-4}\ M) + (8.0 \times 10^{-4}\ M) = 1.2 \times 10^{-3}\ M$

$pH = -\log[H_3O^+] = -\log(1.2 \times 10^{-3}) = 2.92$

(c) On mixing equal volumes of the two strong acids, both acid concentrations are cut in half.

$[H_3O^+] = 0.10\ M + 0.25\ M = 0.35\ M$

$pH = -\log[H_3O^+] = -\log(0.35) = 0.46$

15.46 (a) NaOH (b) CaO (c) $CaCO_3$ (d) $Ca(OH)_2$

15.47 (a) $HClO_2(aq) + H_2O(l) \rightleftarrows H_3O^+(aq) + ClO_2^-(aq)$; $K_a = \dfrac{[H_3O^+][ClO_2^-]}{[HClO_2]}$

(b) $HOBr(aq) + H_2O(l) \rightleftarrows H_3O^+(aq) + OBr^-(aq)$; $K_a = \dfrac{[H_3O^+][OBr^-]}{[HOBr]}$

(c) $HCOOH(aq) + H_2O(l) \rightleftarrows H_3O^+(aq) + HCOO^-(aq)$; $K_a = \dfrac{[H_3O^+][HCOO^-]}{[HCOOH]}$

15.48 (a) The larger the K_a, the stronger the acid.

$C_6H_5OH < HOCl < CH_3COOH < HNO_3$

(b) The larger the K_a, the larger the percent dissociation for the same concentration.

$HNO_3 > CH_3COOH > HOCl > C_6H_5OH$

1 M HNO_3, $[H_3O^+] = 1\ M$

1 M CH_3COOH, $[H_3O^+] = \sqrt{K_a} = \sqrt{1.8 \times 10^{-5}} = 4 \times 10^{-3}\ M$

1 M HOCl, $[H_3O^+] = \sqrt{K_a} = \sqrt{3.5 \times 10^{-8}} = 2 \times 10^{-4}$ M

1 M C_6H_5OH, $[H_3O^+] = \sqrt{K_a} = \sqrt{1.3 \times 10^{-10}} = 1 \times 10^{-5}$ M

5.49

	HOBr(aq) + $H_2O(l)$ ⇄	$H_3O^+(aq)$ +	$OBr^-(aq)$
initial (M)	0.040	~0	0
change (M)	–x	+x	+x
equil (M)	0.040 – x	x	x

$x = [H_3O^+] = 10^{-pH} = 10^{-5.05} = 8.9 \times 10^{-6}$ M

$$K_a = \frac{[H_3O^+][OBr^-]}{[HOBr]} = \frac{x^2}{0.040 - x} = \frac{(8.9 \times 10^{-6})^2}{0.040 - (8.9 \times 10^{-6})} = 2.0 \times 10^{-9}$$

15.50

	$C_3H_6O_3(aq)$ + $H_2O(l)$ ⇄	$H_3O^+(aq)$ +	$C_3H_5O_3^-(aq)$
initial (M)	0.10	~0	0
change (M)	–x	+x	+x
equil (M)	0.10 – x	x	x

$x = [H_3O^+] = 10^{-pH} = 10^{-2.43} = 3.7 \times 10^{-3}$ M

$[C_3H_5O_3^-] = x = 3.7 \times 10^{-3}$ M

$[C_3H_6O_3] = 0.10 - x = 0.10 - (3.7 \times 10^{-3})$ M $= 0.10$ M

$$K_a = \frac{[H_3O^+][C_3H_5O_3^-]}{[C_3H_6O_3]} = \frac{(3.7 \times 10^{-3})(3.7 \times 10^{-3})}{0.10} = 1.4 \times 10^{-4}$$

15.51

	$C_6H_5OH(aq)$ + $H_2O(l)$ ⇄	$H_3O^+(aq)$ +	$C_6H_5O^-(aq)$
initial (M)	0.10	~0	0
change (M)	–x	+x	+x
equil (M)	0.10 – x	x	x

$$K_a = \frac{[H_3O^+][C_6H_5O^-]}{[C_6H_5OH]} = 1.3 \times 10^{-10} = \frac{x^2}{0.10 - x} \approx \frac{x^2}{0.10}$$

Solve for x. $x = 3.6 \times 10^{-6}$ M = $[H_3O^+] = [C_6H_5O^-]$

$[C_6H_5OH] = 0.10 - x = 0.10$ M

$pH = -\log[H_3O^+] = -\log(3.6 \times 10^{-6}) = 5.44$

$$[OH^-] = \frac{K_w}{[H_3O^+]} = \frac{1.0 \times 10^{-14}}{3.6 \times 10^{-6}} = 2.8 \times 10^{-9} \text{ M}$$

$$\% \text{ dissociation} = \frac{[C_6H_5OH]_{diss}}{[C_6H_5OH]_{initial}} \times 100\% = \frac{3.6 \times 10^{-6} \text{ M}}{0.10 \text{ M}} \times 100\% = 0.0036\%$$

15.52 $HCOOH(aq) + H_2O(l) \rightleftarrows H_3O^+(aq) + HCOO^-(aq)$

	HCOOH		H_3O^+	$HCOO^-$
initial (M)	0.20		~0	0
change (M)	–x		+x	+x
equil (M)	0.20 – x		x	x

$$K_a = \frac{[H_3O^+][HCOO^-]}{[HCOOH]} = 1.8 \times 10^{-4} = \frac{x^2}{0.20 - x}$$

$x^2 + (1.8 \times 10^{-4})x - (3.6 \times 10^{-5}) = 0$

Use the quadratic formula to solve for x.

$$x = \frac{(-1.8 \times 10^{-4}) \pm \sqrt{(1.8 \times 10^{-4})^2 + (4)(3.6 \times 10^{-5})}}{2(1)} = \frac{(-1.8 \times 10^{-4}) \pm 0.012}{2}$$

x = 0.0059 and –0.0061

Of the two solutions for x, only the positive value of x has physical meaning because x is the $[H_3O^+]$.

$[H_3O^+] = [HCOO^-] = x = 0.0059$ M

$[HCOOH] = 0.20 - x = 0.20 - 0.0059 = 0.19$ M

$$[OH^-] = \frac{K_w}{[H_3O^+]} = \frac{1.0 \times 10^{-14}}{0.0059} = 1.7 \times 10^{-12}\ M$$

$pH = -\log[H_3O^+] = -\log(0.0059) = 2.23$

15.53 $$HNO_2(aq) + H_2O(l) \rightleftarrows H_3O^+(aq) + NO_2^-(aq)$$

	HNO_2	H_3O^+	NO_2^-
initial (M)	1.5	~0	0
change (M)	–x	+x	+x
equil (M)	1.5 – x	x	x

$$K_a = \frac{[H_3O^+][NO_2^-]}{[HNO_2]} = 4.5 \times 10^{-4} = \frac{x^2}{1.5 - x} \approx \frac{x^2}{1.5}$$

Solve for x. $x = 0.026\ M = [H_3O^+]$

$pH = -\log[H_3O^+] = -\log(0.026) = 1.59$

$$\%\ \text{dissociation} = \frac{[HNO_2]_{diss}}{[HNO_2]_{initial}} \times 100\% = \frac{0.026\ M}{1.5\ M} \times 100\% = 1.7\%$$

15.54 $C_9H_8O_4$, 180.16 amu; 300 mL = 0.300 L;

324 mg = 0.324 g

$$[C_9H_8O_4] = \frac{\left((2)(0.324\ g) \times \frac{1\ mol}{180.16\ g}\right)}{0.300\ L} = 0.0120\ M$$

$$C_9H_8O_4(aq) + H_2O(l) \rightleftarrows H_3O^+(aq) + C_9H_7O_4^-(aq)$$

	$C_9H_8O_4$	H_3O^+	$C_9H_7O_4^-$
initial (M)	0.0120	~0	0
change (M)	–x	+x	+x
equil (M)	0.0120 – x	x	x

$$K_a = \frac{[H_3O^+][C_9H_7O_4^-]}{[C_9H_8O_4]} = 3.0 \times 10^{-4} = \frac{x^2}{0.0120 - x}$$

$x^2 + (3.0 \times 10^{-4})x - (3.6 \times 10^{-6}) = 0$

Use the quadratic formula to solve for x.

$$x = \frac{(-3.0 \times 10^{-4}) \pm \sqrt{(3.0 \times 10^{-4})^2 + (4)(3.6 \times 10^{-6})}}{2(1)} = \frac{(-3.0 \times 10^{-4}) \pm (3.8 \times 10^{-3})}{2}$$

$x = 1.8 \times 10^{-3}$ and -2.1×10^{-3}

Of the two solutions for x, only the positive value of x has physical meaning because x is the $[H_3O^+]$.

$[H_3O^+] = x = 1.8 \times 10^{-3}$ M

$pH = -\log[H_3O^+] = -\log(1.8 \times 10^{-3}) = 2.74$

$$\% \text{ dissociation} = \frac{[C_9H_8O_4]_{diss}}{[C_9H_8O_4]_{initial}} \times 100\% = \frac{1.8 \times 10^{-3}\text{ M}}{0.0120\text{ M}} \times 100\% = 15\%$$

15.55 (a) $H_3PO_4(aq) + H_2O(l) \rightleftarrows H_3O^+(aq) + H_2PO_4^-(aq)$; $K_{a1} = \dfrac{[H_3O^+][H_2PO_4^-]}{[H_3PO_4]}$

(b) $H_2PO_4^-(aq) + H_2O(l) \rightleftarrows H_3O^+(aq) + HPO_4^{2-}(aq)$; $K_{a2} = \dfrac{[H_3O^+][HPO_4^{2-}]}{[H_2PO_4^-]}$

(c) $HPO_4^{2-}(aq) + H_2O(l) \rightleftarrows H_3O^+(aq) + PO_4^{3-}(aq)$; $K_{a3} = \dfrac{[H_3O^+][PO_4^{3-}]}{[HPO_4^{2-}]}$

15.56

	$H_2CO_3(aq)$	+ $H_2O(l)$	$\rightleftarrows$	$H_3O^+(aq)$	+ $HCO_3^-(aq)$
initial (M)	0.010			~0	0
change (M)	–x			+x	+x
equil (M)	0.010 – x			x	x

$$K_{a1} = \frac{[H_3O^+][HCO_3^-]}{[H_2CO_3]} = 4.3 \times 10^{-7} = \frac{x^2}{0.010 - x} \approx \frac{x^2}{0.010}$$

Solve for x. $x = 6.6 \times 10^{-5}$

$[H_3O^+] = [HCO_3^-] = x = 6.6 \times 10^{-5}$ M

The second dissociation of H_2CO_3 produces a negligible amount of H_3O^+ compared with that from the first dissociation.

$HCO_3^-(aq) + H_2O(l) \rightleftarrows H_3O^+(aq) + CO_3^{2-}(aq)$

$$K_{a2} = \frac{[H_3O^+][CO_3^{2-}]}{[HCO_3^-]} = 5.6 \times 10^{-11} = \frac{(6.6 \times 10^{-5})[CO_3^{2-}]}{(6.6 \times 10^{-5})}$$

$[CO_3^{2-}] = K_{a2} = 5.6 \times 10^{-11}$ M

$$[OH^-] = \frac{K_w}{[H_3O^+]} = \frac{1.0 \times 10^{-14}}{6.6 \times 10^{-5}} = 1.5 \times 10^{-10} \text{ M}$$

$pH = -\log[H_3O^+] = -\log(6.6 \times 10^{-5}) = 4.18$

15.57 For the dissociation of the first proton, the following equilibrium must be considered:

	$H_2C_2O_4(aq)$	+	$H_2O(l)$	$\rightleftarrows$	$H_3O^+(aq)$	+	$HC_2O_4^-(aq)$
initial (M)	0.20				~0		0
change (M)	–x				+x		+x
equil (M)	0.20 – x				x		x

$$K_{a1} = \frac{[H_3O^+][HC_2O_4^-]}{[H_2C_2O_4]} = 5.9 \times 10^{-2} = \frac{x^2}{0.20 - x}$$

$x^2 + 0.059x - 0.0118 = 0$

Use the quadratic formula to solve for x.

$$x = \frac{(-0.059) \pm \sqrt{(0.059)^2 + 4(1)(0.0118)}}{2(1)} = \frac{-0.059 \pm 0.225}{2}$$

x = 0.083 and –0.142

Of the two solutions for x, only the positive value of x has physical meaning, because x is the $[H_3O^+]$.

$[H_3O^+] = [HC_2O_4^-] = 0.083$ M

For the dissociation of the second proton, the following equilibrium must be considered:

	$HC_2O_4^-(aq)$ + $H_2O(l)$	⇄	$H_3O^+(aq)$	+	$C_2O_4^{2-}(aq)$
initial (M)	0.083		0.083		0
change (M)	–x		+x		+x
equil (M)	0.083 – x		0.083 + x		x

$$K_{a2} = \frac{[H_3O^+][C_2O_4^{2-}]}{[HC_2O_4^-]} = 6.4 \times 10^{-5} = \frac{(0.083 + x)(x)}{0.083 - x} \approx \frac{(0.083)(x)}{0.083} = x$$

$[H_3O^+] = 0.083 + x = 0.083$ M

$pH = -\log[H_3O^+] = -\log(0.083) = 1.08$

$[C_2O_4^{2-}] = x = 6.4 \times 10^{-5}$ M

15.58

	$H_2SO_3(aq)$ + $H_2O(l)$	⇄	$H_3O^+(aq)$	+	$HSO_3^-(aq)$
initial (M)	0.025		~0		0
change (M)	–x		+x		+x
equil (M)	0.025 – x		x		x

$$K_{a1} = \frac{[H_3O^+][HSO_3^-]}{[H_2SO_3]} = 1.5 \times 10^{-2} = \frac{x^2}{0.025 - x}$$

$x^2 + 0.015x - (3.75 \times 10^{-4}) = 0$

Use the quadratic formula to solve for x.

$$x = \frac{(-0.015) \pm \sqrt{(0.015)^2 + (4)(3.75 \times 10^{-4})}}{2(1)} = \frac{-0.015 \pm 0.0415}{2}$$

x = 0.013 and –0.028

Of the two solutions for x, only the positive value of x has physical meaning because x is the $[H_3O^+]$.

$x = [H_3O^+] = [HSO_3^-] = 0.013$ M

$[H_2SO_3] = 0.025 - x = 0.025 - 0.013 = 0.012$ M

The second dissociation of H_2SO_3 produces a negligible amount of H_3O^+ compared with that from the first dissociation.

$HSO_3^-(aq) + H_2O(l) \rightleftharpoons H_3O^+(aq) + SO_3^{2-}(aq)$

$$K_{a2} = \frac{[H_3O^+][SO_3^{2-}]}{[HSO_3^-]} = 6.3 \times 10^{-8} = \frac{(0.013)[SO_3^{2-}]}{(0.013)}$$

$[SO_3^{2-}] = K_{a2} = 6.3 \times 10^{-8}$ M

15.59 Mixing equal volumes of the two strong acids cuts the initial acid concentrations in half.

$[H_3O^+] = 0.1$ M (from HCl) + 0.3 M (from H_2SO_4) = 0.4 M

For the dissociation of the second proton from H_2SO_4, the following equilibrium must be considered:

	$HSO_4^-(aq)$	+	$H_2O(l)$	$\rightleftharpoons$	$H_3O^+(aq)$	+	$SO_4^{2-}(aq)$
initial (M)	0.3				0.4		0
change (M)	–x				+x		+x
equil	0.3 – x				0.4 + x		x

$$K_{a2} = \frac{[H_3O^+][SO_4^{2-}]}{[HSO_4^-]} = 1.2 \times 10^{-2} = \frac{(0.4 + x)(x)}{0.3 - x}$$

$x^2 + 0.412x - 0.0036 = 0$

Use the quadratic formula to solve for x.

$$x = \frac{(-0.412) \pm \sqrt{(0.412)^2 + 4(1)(0.0036)}}{2(1)} = \frac{-0.412 \pm 0.429}{2}$$

x = 0.0085 and –0.420

Of the two solutions for x, only the positive value of x has physical meaning, because x is the $[SO_4^{2-}]$.

$[H_3O^+] = 0.4 + x = 0.4$ M; $[SO_4^{2-}] = x = 0.008$ M

Weak Bases; Relation Between K_a and K_b

15.60 (a) $(CH_3)_2NH(aq) + H_2O(l) \rightleftarrows (CH_3)_2NH_2^+(aq) + OH^-(aq)$

$$K_b = \frac{[(CH_3)_2NH_2^+][OH^-]}{[(CH_3)_2NH]}$$

(b) $C_6H_5NH_2(aq) + H_2O(l) \rightleftarrows C_6H_5NH_3^+(aq) + OH^-(aq)$

$$K_b = \frac{[C_6H_5NH_3^+][OH^-]}{[C_6H_5NH_2]}$$

(c) $C_5H_5N(aq) + H_2O(l) \rightleftarrows C_5H_5NH^+(aq) + OH^-(aq)$

$$K_b = \frac{[C_5H_5NH^+][OH^-]}{[C_5H_5N]}$$

(d) $CN^-(aq) + H_2O(l) \rightleftarrows HCN(aq) + OH^-(aq)$

$$K_b = \frac{[HCN][OH^-]}{[CN^-]}$$

15.61 $[H_3O^+] = 10^{-pH} = 10^{-9.5} = 3.16 \times 10^{-10}$ M

$$[OH^-] = \frac{K_w}{[H_3O^+]} = \frac{1.0 \times 10^{-14}}{3.16 \times 10^{-10}} = 3.16 \times 10^{-5}\ M$$

	$C_{17}H_{19}NO_3(aq)$ + $H_2O(l)$ ⇌	$C_{17}H_{20}NO_3^+(aq)$	+ $OH^-(aq)$
initial (M)	7.0×10^{-4}	0	~0
change (M)	$-x$	$+x$	$+x$
equil (M)	$(7.0 \times 10^{-4}) - x$	x	x

$x = [OH^-] = 3.16 \times 10^{-5}$ M

$$K_b = \frac{[C_{17}H_{20}NO_3^+][OH^-]}{[C_{17}H_{19}NO_3]} = \frac{x^2}{(7.0 \times 10^{-4}) - x} = 1.49 \times 10^{-6} = 1 \times 10^{-6}$$

15.62 (a)

	$CH_3NH_2(aq)$ + $H_2O(l)$ ⇌	$CH_3NH_3^+(aq)$	+ $OH^-(aq)$
initial (M)	0.24	0	~0
change (M)	$-x$	$+x$	$+x$
equil (M)	$0.24 - x$	x	x

$$K_b = \frac{[CH_3NH_3^+][OH^-]}{[CH_3NH_2]} = 3.7 \times 10^{-4} = \frac{x^2}{0.24 - x}$$

$x^2 + (3.7 \times 10^{-4})x - (8.9 \times 10^{-5}) = 0$

Use the quadratic formula to solve for x.

$$x = \frac{(-3.7 \times 10^{-4}) \pm \sqrt{(3.7 \times 10^{-4})^2 + (4)(8.9 \times 10^{-5})}}{2(1)} = \frac{(-3.7 \times 10^{-4}) \pm 0.0189}{2}$$

$x = 0.0093$ and -0.0096

Of the two solutions for x, only the positive value of x has physical meaning because x is the $[OH^-]$.

$[OH^-] = x = 0.0093$ M

$$[H_3O^+] = \frac{K_w}{[OH^-]} = \frac{1.0 \times 10^{-14}}{0.0093} = 1.1 \times 10^{-12}\ M$$

$pH = -\log[H_3O^+] = -\log(1.1 \times 10^{-12}) = 11.96$

(b) $C_5H_5N(aq) + H_2O(l) \rightleftarrows C_5H_5NH^+(aq) + OH^-(aq)$

	C_5H_5N	$C_5H_5NH^+$	OH^-
initial (M)	0.040	0	~0
change (M)	–x	+x	+x
equil (M)	0.040 – x	x	x

$$K_b = \frac{[C_5H_5NH^+][OH^-]}{[C_5H_5N]} = 1.8 \times 10^{-9} = \frac{x^2}{0.040 - x} \approx \frac{x^2}{0.040}$$

Solve for x. $x = [OH^-] = 8.5 \times 10^{-6}\ M$

$$[H_3O^+] = \frac{K_w}{[OH^-]} = \frac{1.0 \times 10^{-14}}{8.5 \times 10^{-6}} = 1.2 \times 10^{-9}\ M$$

$pH = -\log[H_3O^+] = -\log(1.2 \times 10^{-9}) = 8.92$

(c) $NH_2OH(aq) + H_2O(l) \rightleftarrows NH_3OH^+(aq) + OH^-(aq)$

	NH_2OH	NH_3OH^+	OH^-
initial (M)	0.075	0	~0
change (M)	–x	+x	+x
equil (M)	0.075 – x	x	x

$$K_b = \frac{[NH_3OH^+][OH^-]}{[NH_2OH]} = 9.1 \times 10^{-9} = \frac{x^2}{0.075 - x} \approx \frac{x^2}{0.075}$$

Solve for x. $x = [OH^-] = 2.6 \times 10^{-5}\ M$

$$[H_3O^+] = \frac{K_w}{[OH^-]} = \frac{1.0 \times 10^{-14}}{2.6 \times 10^{-5}} = 3.8 \times 10^{-10}\ M$$

$pH = -\log[H_3O^+] = -\log(3.8 \times 10^{-10}) = 9.42$

15.63 $C_6H_5NH_2(aq) + H_2O(l) \rightleftarrows C_6H_5NH_3^+(aq) + OH^-(aq)$

	$C_6H_5NH_2(aq)$	$C_6H_5NH_3^+(aq)$	$OH^-(aq)$
initial (M)	0.15	0	~0
change (M)	–x	+x	+x
equil (M)	0.15 – x	x	x

$$K_b = \frac{[C_6H_5NH_3^+][OH^-]}{[C_6H_5NH_2]} = 4.3 \times 10^{-10} = \frac{x^2}{0.15 - x} \approx \frac{x^2}{0.15}$$

Solve for x. $x = 8.0 \times 10^{-6}$ M = $[C_6H_5NH_3^+] = [OH^-]$

$[C_6H_5NH_2] = 0.15 - x = 0.15$ M

$$[H_3O^+] = \frac{K_w}{[OH^-]} = \frac{1.0 \times 10^{-14}}{8.0 \times 10^{-6}} = 1.2 \times 10^{-9}\ M$$

pH = $-\log[H_3O^+] = -\log(1.25 \times 10^{-9}) = 8.90$

15.64 $HCN(aq) + H_2O(l) \rightleftarrows H_3O^+(aq) + CN^-(aq);\quad K_a = \frac{[H_3O^+][CN^-]}{[HCN]}$

$CN^-(aq) + H_2O(l) \rightleftarrows HCN(aq) + OH^-(aq);\quad K_b = \frac{[HCN][OH^-]}{[CN^-]}$

$$K_a \times K_b = \left(\frac{[H_3O^+][CN^-]}{[HCN]}\right)\left(\frac{[HCN][OH^-]}{[CN^-]}\right) = [H_3O^+][OH^-] = K_w$$

15.65 (a) $K_a = \frac{K_w}{K_b \text{ for } C_3H_7NH_2} = \frac{1.0 \times 10^{-14}}{5.1 \times 10^{-4}} = 2.0 \times 10^{-11}$

(b) $K_a = \frac{K_w}{K_b \text{ for } NH_2OH} = \frac{1.0 \times 10^{-14}}{9.1 \times 10^{-9}} = 1.1 \times 10^{-6}$

(c) $K_a = \frac{K_w}{K_b \text{ for } C_6H_5NH_2} = \frac{1.0 \times 10^{-14}}{4.3 \times 10^{-10}} = 2.3 \times 10^{-5}$

(d) $K_a = \dfrac{K_w}{K_b \text{ for } C_5H_5N} = \dfrac{1.0 \times 10^{-14}}{1.8 \times 10^{-9}}$ = 5.6 x 10–6

15.66 (a) $K_b = \dfrac{K_w}{K_a \text{ for HF}} = \dfrac{1.0 \times 10^{-14}}{3.5 \times 10^{-4}} = 2.9 \times 10^{-11}$

(b) $K_b = \dfrac{K_w}{K_a \text{ for HOBr}} = \dfrac{1.0 \times 10^{-14}}{2.0 \times 10^{-9}} = 5.0 \times 10^{-6}$

(c) $K_b = \dfrac{K_w}{K_a \text{ for } H_2S} = \dfrac{1.0 \times 10^{-14}}{1.0 \times 10^{-7}} = 1.0 \times 10^{-7}$

(d) $K_b = \dfrac{K_w}{K_a \text{ for } HS^-} = \dfrac{1.0 \times 10^{-14}}{1 \times 10^{-19}} = 1 \times 10^{5}$

Acid–Base Properties of Salts

15.67 (a) $CH_3NH_3^+(aq) + H_2O(l) \rightleftarrows H_3O^+(aq) + CH_3NH_2(aq)$
acid base——acid base

(b) $Cr(H_2O)_6^{3+}(aq) + H_2O(l) \rightleftarrows H_3O^+(aq) + Cr(H_2O)_5(OH)^{2+}(aq)$
acid base——acid base

(c) $CH_3COO^-(aq) + H_2O(l) \rightleftarrows CH_3COOH(aq) + OH^-(aq)$
base acid acid base

(d) $PO_4^{3-}(aq) + H_2O(l) \rightleftarrows HPO_4^{2-}(aq) + OH^-(aq)$
base acid acid base

15.68 (a) F^- (conjugate base of a weak acid), basic solution

(b) Br^- (anion of a strong acid), neutral solution

(c) NH_4^+ (conjugate acid of a weak base), acidic solution

(d) $K(H_2O)_6^+$ (neutral cation), neutral solution

(e) SO_3^{2-} (conjugate base of a weak acid), basic solution

(f) $Cr(H_2O)_6^{3+}$ (acidic cation), acidic solution

15.69 (a) $Fe(NO_3)_3$: Fe^{3+}, acidic cation; NO_3^-, neutral anion; solution is acidic

(b) $Ba(NO_3)_2$: Ba^{2+}, neutral cation; NO_3^-, neutral anion; solution is neutral

(c) NaOCl: Na^+, neutral cation; OCl^-, basic anion; solution is basic

(d) NH_4I: NH_4^+, acidic cation; I^-, neutral anion; solution is acidic

(e) NH_4NO_2

For NH_4^+, $K_a = 5.6 \times 10^{-10}$; for NO_2^-, $K_b = 2.2 \times 10^{-11}$

Because $K_a > K_b$, the solution is acidic.

15.70 (a) Na_2CO_3

Na^+ is a neutral cation. CO_3^{2-} is a basic anion.

$CO_3^{2-}(aq) + H_2O(l) \rightleftarrows HCO_3^-(aq) + OH^-(aq)$

(b) NH_4NO_3

NH_4^+ is an acidic cation. NO_3^- is a neutral anion.

$NH_4^+(aq) + H_2O(l) \rightleftarrows H_3O^+(aq) + NH_3(aq)$

(c) NaCl

Na^+ is a neutral cation. Cl^- is a neutral anion.

$$2\ H_2O(l) \rightleftarrows H_3O^+(aq) + OH^-(aq)$$

15.71 (a) $(C_2H_5NH_3)NO_3$: $C_2H_5NH_3^+$, acidic cation; NO_3^-, neutral anion

$C_2H_5NH_2$, $K_b = 6.4 \times 10^{-4}$

$$C_2H_5NH_3^+,\ K_a = \frac{K_w}{K_b \text{ for } C_2H_5NH_2} = \frac{1.0 \times 10^{-14}}{6.4 \times 10^{-4}} = 1.56 \times 10^{-11}$$

	$C_2H_5NH_3^+(aq)$ + $H_2O(l)$	⇄ $H_3O^+(aq)$	+ $C_2H_5NH_2(aq)$
initial (M)	0.10	~0	0
change (M)	–x	+x	+x
equili (M)	0.10 – x	x	x

$$K_a = \frac{[H_3O^+][C_2H_5NH_2]}{[C_2H_5NH_3^+]} = 1.56 \times 10^{-11} = \frac{x^2}{0.10 - x} \approx \frac{x^2}{0.10}$$

Solve for x. $x = 1.2 \times 10^{-6}\ M = [H_3O^+] = [C_2H_5NH_2]$

$pH = -\log[H_3O^+] = -\log(1.25 \times 10^{-6}) = 5.90$

$[C_2H_5NH_3^+] = 0.10 - x = 0.10\ M$

$[NO_3^-] = 0.10\ M$

$$[OH^-] = \frac{K_w}{[H_3O^+]} = \frac{1.0 \times 10^{-14}}{1.25 \times 10^{-6}\ M} = 8.0 \times 10^{-9}$$

(b) $Na(CH_3CO_2)$: Na^+, neutral cation; $CH_3CO_2^-$, basic anion

CH_3CO_2H, $K_a = 1.8 \times 10^{-5}$

$$CH_3CO_2^-,\ K_b = \frac{K_w}{K_a \text{ for } CH_3CO_2H} = \frac{1.0 \times 10^{-14}}{1.8 \times 10^{-5}} = 5.6 \times 10^{-10}$$

	$CH_3CO_2^-(aq)$ + $H_2O(aq)$	⇄ $CH_3CO_2H(aq)$	+ $OH^-(aq)$
initial (M)	0.10	0	~0
change (M)	–x	+x	+x
equil (M)	0.10 – x	x	x

$$K_b = \frac{[CH_3CO_2H][OH^-]}{[CH_3CO_2^-]} = 5.6 \times 10^{-10} = \frac{x^2}{0.10 - x} \approx \frac{x^2}{0.10}$$

Solve for x. $x = 7.5 \times 10^{-6}$ M = $[CH_3COOH]$ = $[OH^-]$

$[CH_3COO^-] = 0.10 - x = 0.10$ M

$[Na^+] = 0.10$ M

$$[H_3O^+] = \frac{K_w}{[OH^-]} = \frac{1.0 \times 10^{-14}}{7.5 \times 10^{-6}} = 1.3 \times 10^{-9} \text{ M}$$

$pH = -\log[H_3O^+] = -\log(1.3 \times 10^{-9}) = 8.89$

(c) $NaNO_3$: Na^+, neutral cation; NO_3^-, neutral anion

$[Na^+] = [NO_3^-] = 0.10$ M

$[H_3O^+] = [OH^-] = 1.0 \times 10^{-7}$ M; pH = 7.00

15.72 (a) $Fe(H_2O)_6^{2+}(aq) + H_2O(l) \rightleftarrows H_3O^+(aq) + Fe(H_2O)_5(OH)^+(aq)$

	$Fe(H_2O)_6^{2+}(aq)$	$H_3O^+(aq)$	$Fe(H_2O)_5(OH)^+(aq)$
initial (M)	0.020	~0	0
change (M)	–x	+x	+x
equil (M)	0.020 – x	x	x

$$K_a = \frac{[H_3O^+][Fe(H_2O)_5(OH)^+]}{[Fe(H_2O)_6^{2+}]} = 3.2 \times 10^{-10} = \frac{x^2}{0.020 - x} \approx \frac{x^2}{0.020}$$

Solve for x. $x = [H_3O^+] = 2.5 \times 10^{-6}$ M

$pH = -\log[H_3O^+] = -\log(2.5 \times 10^{-6}) = 5.60$

$$\% \text{ dissociation} = \frac{[Fe(H_2O)_6^{2+}]_{diss}}{[Fe(H_2O)_6^{2+}]_{initial}} \times 100\% = \frac{2.5 \times 10^{-6} \text{ M}}{0.020 \text{ M}} \times 100\% = 0.012\%$$

(b) $Fe(H_2O)_6^{3+}(aq) + H_2O(l) \rightleftarrows H_3O^+(aq) + Fe(H_2O)_5(OH)^{2+}(aq)$

	$Fe(H_2O)_6^{3+}$	H_3O^+	$Fe(H_2O)_5(OH)^{2+}$
initial (M)	0.020	~0	0
change (M)	–x	+x	+x
equil (M)	0.020 – x	x	x

$$K_a = \frac{[H_3O^+][Fe(H_2O)_5(OH)^{2+}]}{[Fe(H_2O)_6^{3+}]} = 6.3 \times 10^{-3} = \frac{x^2}{0.020 - x}$$

$x^2 + (6.3 \times 10^{-3})x - (1.26 \times 10^{-4}) = 0$

Use the quadratic formula to solve for x.

$$x = \frac{(-6.3 \times 10^{-3}) \pm \sqrt{(6.3 \times 10^{-3})^2 + (4)(1.26 \times 10^{-4})}}{2(1)} = \frac{(-6.3 \times 10^{-3}) \pm 0.0233}{2}$$

x = 0.0085 and –0.0148

Of the two solutions for x, only the positive value of x has physical meaning because x is the $[H_3O^+]$.

$x = [H_3O^+] = 0.0085$ M

$pH = -\log[H_3O^+] = -\log(0.0085) = 2.07$

$$\text{\% dissociation} = \frac{[Fe(H_2O)_6^{3+}]_{diss}}{[Fe(H_2O)_6^{3+}]_{initial}} \times 100\% = \frac{0.0085\ M}{0.020\ M} \times 100\% = 42\%$$

Factors That Affect Acid Strength

15.73 (a) $PH_3 < H_2S < HCl$; electronegativity increases from P to Cl

(b) $NH_3 < PH_3 < AsH_3$; X–H bond strength decreases from N to As

(c) $HBrO < HBrO_2 < HBrO_3$; acid strength increases with the number of Os

15.74

```
     H   ..
     |   O:
     |   ||    ..
H—C—C—O—H
     |         ..
     H
```

```
     Cl  ..
     |   O:
     |   ||    ..
Cl—C—C—O—H
     |         ..
     Cl
```

Because Cl is more electronegative than H, CCl_3COOH is the stronger acid with the larger K_a.

15.75 (a) H_2Te, weaker X–H bond

(b) H_3PO_4, P has higher electronegativity

(c) $H_2PO_4^-$, lower negative charge

(d) NH_4^+, higher positive charge and N is more electronegative than C.

15.76 (a) ClO_2^- (conjugate base of $HClO_2$) is a stronger base than ClO_3^- (conjugate base of $HClO_3$) because $HClO_2$ is the weaker acid.

(b) $HSeO_4^-$ (conjugate base of H_2SeO_4) is a stronger base than HSO_4^- (conjugate base of H_2SO_4) because H_2SeO_4 is the weaker acid.

(c) OH^- (conjugate base of H_2O) is a stronger base than HS^- (conjugate base of H_2S) because H_2O is the weaker acid.

(d) HS^- is the stronger base because Br^- has no basic properties. HBr is a strong acid.

Lewis Acids and Bases

15.77 A Lewis acid is an electron pair acceptor, and a Lewis base is an electron pair donor.

The Lewis definition is more general than the Brønsted–Lowry definition because not only reactions with H^+ are included but also reactions with other cations and neutral molecules that have vacant valence orbitals.

15.78 Al^{3+}, Cu^{2+}, and BF_3.

15.79 (a) Lewis acid, SiF_4; Lewis base, F^-

(b) Lewis acid, Zn^{2+}; Lewis base, NH_3

(c) Lewis acid, $Al(OH)_3$; Lewis base, OH^-

(d) Lewis acid, AgCl; Lewis base, Cl^-

15.80 (a) CN^-, Lewis base

(b) H^+, Lewis acid

(c) H_2O, Lewis base

(d) Fe^{3+}, Lewis acid

(e) OH^-, Lewis base

(f) CO_2, Lewis acid

(g) $P(CH_3)_3$, Lewis base

(h) $B(CH_3)_3$, Lewis acid

15.81 (a) BF_3, F is more electronegative than H

(b) SO_3, more O atoms draw electron density away from S

(c) Sn^{4+}, higher charge

(d) CH_3^+, electron deficient

General Problems

15.82 $C_6H_8O_6$, 176.13 amu; 250 mg = 0.250 g; 250 mL = 0.250 L

$$[C_6H_8O_6] = \frac{\left(0.250\text{ g} \times \frac{1\text{ mol}}{176.13\text{ g}}\right)}{0.250\text{ L}} = 5.68 \times 10^{-3}\text{ M}$$

$$C_6H_8O_6(aq) + H_2O(l) \rightleftarrows H_3O^+(aq) + C_6H_7O_6^-(aq)$$

	$C_6H_8O_6(aq)$	$H_3O^+(aq)$	$C_6H_7O_6^-(aq)$
initial (M)	5.68×10^{-3}	~0	0
change (M)	–x	+x	+x
equil (M)	$(5.68 \times 10^{-3}) - x$	x	x

$$K_a = \frac{[H_3O^+][C_6H_7O_6^-]}{[C_6H_8O_6]} = 8.0 \times 10^{-5} = \frac{x^2}{(5.68 \times 10^{-3}) - x}$$

$$x^2 + (8.0 \times 10^{-5})x - (4.54 \times 10^{-7}) = 0$$

Use the quadratic formula to solve for x.

$$x = \frac{(-8.0 \times 10^{-5}) \pm \sqrt{(8.0 \times 10^{-5})^2 + (4)(4.54 \times 10^{-7})}}{2(1)} = \frac{(-8.0 \times 10^{-5}) \pm 0.001\ 35}{2}$$

$x = 6.35 \times 10^{-4}$ and -7.15×10^{-4}

Of the two solutions for x, only the positive value of x has physical meaning because x is the $[H_3O^+]$.

$x = [H_3O^+] = 6.35 \times 10^{-4}$ M

$pH = -\log[H_3O^+] = -\log(6.35 \times 10^{-4}) = 3.20$

15.83 $C_{21}H_{22}N_2O_2$, 334.42 amu; 16 mg = 0.016 g

$$\text{molarity} = \frac{\left(0.016\ \text{g} \times \frac{1\ \text{mol}}{334.42\ \text{g}}\right)}{0.100\ \text{L}} = 4.8 \times 10^{-4}\ \text{M}$$

$$C_{21}H_{22}N_2O_2(aq) + H_2O(l) \rightleftarrows C_{21}H_{23}N_2O_2^+(aq) + OH^-(aq)$$

	$C_{21}H_{22}N_2O_2(aq)$	$C_{21}H_{23}N_2O_2^+(aq)$	$OH^-(aq)$
initial (M)	4.8×10^{-4}	0	~0
change (M)	–x	+x	+x
equil (M)	$(4.8 \times 10^{-4}) - x$	x	x

$$K_b = \frac{[C_{21}H_{23}N_2O_2^+][OH^-]}{[C_{21}H_{22}N_2O_2]} = 1.8 \times 10^{-6} = \frac{x^2}{(4.8 \times 10^{-4}) - x}$$

$x^2 + (1.8 \times 10^{-6})x - (8.6 \times 10^{-10}) = 0$

Use the quadratic formula to solve for x.

$$x = \frac{(-1.8 \times 10^{-6}) \pm \sqrt{(1.8 \times 10^{-6})^2 + (4)(8.6 \times 10^{-10})}}{2(1)} = \frac{(-1.8 \times 10^{-6}) \pm (5.87 \times 10^{-5})}{2}$$

$x = 2.84 \times 10^{-5}$ and -3.02×10^{-5}

Of the two solutions for x, only the positive value of x has physical meaning, because x is the $[OH^-]$.

$[OH^-] = 2.84 \times 10^{-5}$ M

$$[H_3O^+] = \frac{K_w}{[OH^-]} = \frac{1.0 \times 10^{-14}}{2.84 \times 10^{-5}} = 3.52 \times 10^{-10} \text{ M}$$

$pH = -\log[H_3O^+] = -\log(3.52 \times 10^{-10}) = 9.45$

15.84 (a) $HBr(aq) + H_2O(l) \rightleftarrows H_3O^+(aq) + Br^-(aq)$

acid (HBr), base (H_2O), acid (H_3O^+), base (Br^-); conjugate pairs: HBr / Br^- and H_2O / H_3O^+

(b) $NH_4^+(aq) + H_2O(l) \rightleftarrows H_3O^+(aq) + NH_3(aq)$

acid (NH_4^+), base (H_2O), acid (H_3O^+), base (NH_3); conjugate pairs: NH_4^+ / NH_3 and H_2O / H_3O^+

(c) $HS^-(aq) + H_2O(l) \rightleftarrows H_2S(aq) + OH^-(aq)$

base (HS^-), acid (H_2O), acid (H_2S), base (OH^-); conjugate pairs: HS^- / H_2S and H_2O / OH^-

15.85 (a) $HCO_3^-(aq) + H_2O(l) \rightleftarrows H_3O^+(aq) + CO_3^{2-}(aq)$

(b) $HCO_3^-(aq) + H_2O(l) \rightleftarrows H_2CO_3(aq) + OH^-(aq)$

15.86 In aqueous solution:

(1) H_2S acts as an acid only.
(2) HS^- can act as both an acid and a base.
(3) S^{2-} can act as a base only.
(4) H_2O can act as both an acid and a base.
(5) H_3O^+ acts as an acid only.
(6) OH^- acts as a base only.

15.87 C_6H_5COONa: Na^+, neutral cation, $C_6H_5COO^-$, basic anion

C_6H_5COOH, $K_a = 6.5 \times 10^{-5}$

$$C_6H_5COO^-,\ K_b = \frac{K_w}{K_a \text{ for } C_6H_6COOH} = \frac{1.0 \times 10^{-14}}{6.5 \times 10^{-5}} = 1.54 \times 10^{-10}$$

$$C_6H_5COO^-(aq) + H_2O(l) \rightleftharpoons C_6H_5COOH(aq) + OH^-(aq)$$

	$C_6H_5COO^-(aq)$	$C_6H_5COOH(aq)$	$OH^-(aq)$
initial (M)	0.050	0	~0
change (M)	−x	+x	+x
equil (M)	0.050 − x	x	x

$$K_b = \frac{[C_6H_5COOH][OH^-]}{[C_6H_5COO^-]} = 1.54 \times 10^{-10} = \frac{x^2}{0.050 - x} \approx \frac{x^2}{0.050}$$

Solve for x. $x = 2.77 \times 10^{-6} = 2.8 \times 10^{-6}$ M = $[C_6H_5COOH] = [OH^-]$

$[Na^+] = 0.050$ M

$[C_6H_5COO^-] = 0.050 - x = 0.050$ M

$$[H_3O^+] = \frac{K_w}{[OH^-]} = \frac{1.0 \times 10^{-14}}{2.77 \times 10^{-6}} = 3.61 \times 10^{-9}\text{ M} = 3.6 \times 10^{-9}\text{ M}$$

$pH = -\log[H_3O^+] = -\log(3.61 \times 10^{-9}) = 8.44$

15.88 $[H_3O^+] = 10^{-pH}$ and $[OH^-] = \dfrac{K_w}{[H_3O^+]}$

(a) $[H_3O^+] = 10^{-3.2} = 6 \times 10^{-4}$ M, $[OH^-] = \dfrac{1.0 \times 10^{-14}}{6 \times 10^{-4}} = 2 \times 10^{-11}$ M

(b) $[H_3O^+] = 10^{-7.8} = 2 \times 10^{-8}$ M, $[OH^-] = \dfrac{1.0 \times 10^{-14}}{2 \times 10^{-8}} = 5 \times 10^{-7}$ M

(c) $[H_3O^+] = 10^{-3.1} = 8 \times 10^{-4}$ M, $[OH^-] = \dfrac{1.0 \times 10^{-14}}{8 \times 10^{-4}} = 1 \times 10^{-11}$ M

(d) $[H_3O^+] = 10^{-6.4} = 4 \times 10^{-7}$ M, $[OH^-] = \dfrac{1.0 \times 10^{-14}}{4 \times 10^{-7}} = 2 \times 10^{-8}$ M

(e) $[H_3O^+] = 10^{-4.2} = 6 \times 10^{-5}$ M, $[OH^-] = \dfrac{1.0 \times 10^{-14}}{6 \times 10^{-5}} = 2 \times 10^{-10}$ M

(f) $[H_3O^+] = 10^{-1.9} = 1 \times 10^{-2}$ M, $[OH^-] = \dfrac{1.0 \times 10^{-14}}{1 \times 10^{-2}} = 1 \times 10^{-12}$ M

15.89

$$\left[\mathrm{H{:}\ddot{O}{:}H}\right]^+ + \mathrm{H{:}\ddot{O}{:}H} \rightarrow \left[\mathrm{H{-}\ddot{O}{-}H\cdots\ddot{O}{-}H}\right]^+$$

(with an H on each O, drawn below)

H_3O^+ can hydrogen bond with additional H_2O molecules.

15.90 (a) HF is a stronger acid than H_2O because F is more electronegative than O. HCl is a stronger acid than HF because of a lower H–X bond strength. HCl is the strongest acid of the three.

(b) $HClO_3$ is a stronger acid than $HClO_2$ because in this case acid strength increases with increasing oxidation number of Cl, which increases with an increasing number of oxygen atoms. $HClO_3$ is a stronger acid than $HBrO_3$ because Cl is more electronegative than Br. $HClO_3$ is the strongest acid of the three.

(c) H_2Se is a stronger acid than H_2S because of a lower H–X bond strength. HBr is a stronger acid than H_2Se because Br is more electronegative than Se. HBr is the strongest acid of the three.

15.91 H_2O, 18.02 amu

$$\text{at } 0\ ^\circ\text{C}, [H_2O] = \frac{\left(0.9987\text{ g x }\frac{1\text{ mol}}{18.02\text{ g}}\right)}{0.001\text{ L}} = 55.42\text{ M}$$

$K_w = [H_3O^+][OH^-]$, for a neutral solution $[H_3O^+] = [OH^-]$

$$[H_3O^+] = \sqrt{K_w} = \sqrt{1.14 \times 10^{-15}} = 3.376 \times 10^{-8}\text{ M}$$

$$pH = -\log[H_3O^+] = -\log(3.376 \times 10^{-8}) = 7.472$$

$$\text{fraction dissociated} = \frac{[H_2O]_{diss}}{[H_2O]_{initial}} = \frac{3.376 \times 10^{-8}\text{ M}}{55.42\text{ M}} = 6.09 \times 10^{-10}$$

$$\%\text{ dissociation} = \frac{[H_2O]_{diss}}{[H_2O]_{initial}} \times 100\% = \frac{3.376 \times 10^{-8}\text{ M}}{55.42\text{ M}} \times 100\% = 6.09 \times 10^{-8}\%$$

15.92 $\Delta[H_3O^+] = 10^{\Delta pH} = 10^{(9.8 - 8.2)} = 40$

The $[H_3O^+]$ decreases by a factor of 40 on going from pH = 8.2 to pH = 9.8.

15.93 For the dissociation of the first proton, the following equilibrium must be considered:

$$H_3PO_4(aq) + H_2O(l) \rightleftarrows H_3O^+(aq) + H_2PO_4^-(aq)$$

	H_3PO_4	H_3O^+	$H_2PO_4^-$
initial (M)	0.10	~0	0
change (M)	–x	+x	+x
equil (M)	0.10 – x	x	x

$$K_{a1} = \frac{[H_3O^+][H_2PO_4^-]}{[H_3PO_4]} = 7.5 \times 10^{-3} = \frac{x^2}{0.10 - x}$$

$$x^2 + (7.5 \times 10^{-3})x - (7.5 \times 10^{-4}) = 0$$

Solve for x using the quadratic formula.

$$x = \frac{(-7.5 \times 10^{-3}) \pm \sqrt{(7.5 \times 10^{-3})^2 + (4)(7.5 \times 10^{-4})}}{2(1)} = \frac{(-7.5 \times 10^{-3}) \pm 0.055}{2}$$

x = 0.024 and –0.031

Of the two solutions for x, only the positive value of x has physical meaning, because x is the $[H_3O^+]$.

$x = 0.024\ M = [H_2PO_4^-] = [H_3O^+]$

For the dissociation of the second proton, the following equilibrium must be considered:

$$H_2PO_4^-(aq) + H_2O(l) \rightleftarrows H_3O^+(aq) + HPO_4^{2-}(aq)$$

	$H_2PO_4^-$	H_3O^+	HPO_4^{2-}
initial (M)	0.024	0.024	0
change (M)	–y	+y	+y
equil (M)	0.024 – y	0.024 + y	y

$$K_{a2} = \frac{[H_3O^+][HPO_4^{2-}]}{[H_2PO_4^-]} = 6.2 \times 10^{-8} = \frac{(0.024 + y)(y)}{0.024 - y} \approx \frac{(0.024)(y)}{0.024} = y$$

$y = 6.2 \times 10^{-8}\ M = [HPO_4^{2-}]$

For the dissociation of the third proton, the following equilibrium must be considered:

$$HPO_4^{2-}(aq) + H_2O(l) \rightleftarrows H_3O^+(aq) + PO_4^{3-}(aq)$$

	HPO_4^{2-}	H_3O^+	PO_4^{3-}
initial (M)	6.2×10^{-8}	0.024	0
change (M)	–z	+z	+z
equil (M)	(6.2×10^{-8}) – z	0.024 + z	z

$$K_{a3} = \frac{[H_3O^+][PO_4^{3-}]}{[HPO_4^{2-}]} = 4.8 \times 10^{-13} = \frac{(0.024 + z)(z)}{(6.2 \times 10^{-8}) - z} \approx \frac{(0.024)(z)}{6.2 \times 10^{-8}}$$

$z = 1.2 \times 10^{-18}\ M = [PO_4^{3-}]$

$[H_3PO_4] = 0.10 - x = 0.076\ M$

$[H_2PO_4^-] = [H_3O^+] = 0.024$ M

$[HPO_4^{2-}] = 6.2 \times 10^{-8}$ M

$[PO_4^{3-}] = 1.2 \times 10^{-18}$ M

$$[OH^-] = \frac{K_w}{[H_3O^+]} = \frac{1.0 \times 10^{-14}}{0.024} = 4.2 \times 10^{-13}\ M$$

$pH = -\log[H_3O^+] = -\log(0.024) = 1.62$

15.94 For $C_{10}H_{14}N_2H^+$, $K_{a1} = \dfrac{K_w}{K_{b1} \text{ for } C_{10}H_{14}N_2} = \dfrac{1.0 \times 10^{-14}}{1.0 \times 10^{-6}} = 1.0 \times 10^{-8}$

For $C_{10}H_{14}N_2H_2^{2+}$, $K_{a2} = \dfrac{K_w}{K_{b2} \text{ for } C_{10}H_{14}N_2H^+} = \dfrac{1.0 \times 10^{-14}}{1.3 \times 10^{-11}} = 7.7 \times 10^{-4}$

15.95 $HCO_3^-(aq) + Al(H_2O)_6^{3+}(aq) \rightarrow H_2O(l) + CO_2(g) + Al(H_2O)_5(OH)^{2+}(aq)$

15.96 (a) $Zn(NO_3)_2$, weakly acidic salt

(b) Na_2O, strong base

(c) NaOCl, weakly basic salt

(d) $NaClO_4$, neutral salt

(e) $HClO_4$, strong acid

$Na_2O < NaOCl < NaClO_4 < Zn(NO_3)_2 < HClO_4$

15.97 (a) NaBr, neutral salt

(b) HBr, strong acid

(c) NH_4Br, weakly acidic salt

(d) $(NH_4)_2CO_3$, weakly basic salt with basic anion and acidic cation

(e) Na_2CO_3, weakly basic salt with basic anion and neutral cation

$HBr < NH_4Br < NaBr < (NH_4)_2CO_3 < Na_2CO_3$

15.98 (a) $HOCl(aq) + H_2O(l) \rightleftarrows H_3O^+(aq) + OCl^-(aq)$

	HOCl		H_3O^+	OCl^-
initial (M)	2.0		~0	0
change (M)	–x		+x	+x
equil (M)	2.0 – x		x	x

$$K_a = \frac{[H_3O^+][OCl^-]}{[HOCl]} = 3.5 \times 10^{-8} = \frac{x^2}{2.0 - x} \approx \frac{x^2}{2.0}$$

Solve for x. $x = 2.6 \times 10^{-4}$ M

$$\% \text{ dissociation} = \frac{[HOCl]_{diss}}{[HOCl]_{initial}} \times 100\% = \frac{2.6 \times 10^{-4}\text{ M}}{2.0\text{ M}} \times 100\% = 0.013\%$$

(b) $HOCl(aq) + H_2O(l) \rightleftarrows H_3O^+(aq) + OCl^-(aq)$

	HOCl		H_3O^+	OCl^-
initial (M)	0.020		~0	0
change (M)	–x		+x	+x
equil (M)	0.020 – x		x	x

$$K_a = \frac{[H_3O^+][OCl^-]}{[HOCl]} = 3.5 \times 10^{-8} = \frac{x^2}{0.020 - x} \approx \frac{x^2}{0.020}$$

Solve for x. $x = 2.6 \times 10^{-5}$ M

$$\% \text{ dissociation} = \frac{[HOCl]_{diss}}{[HOCl]_{initial}} \times 100\% = \frac{2.6 \times 10^{-5}\text{ M}}{0.020\text{ M}} \times 100\% = 0.13\%$$

(c) $HF(aq) + H_2O(l) \rightleftarrows H_3O^+(aq) + F^-(aq)$

	HF		H_3O^+	F^-
initial (M)	2.0		~0	0
change (M)	–x		+x	+x
equil (M)	2.0 – x		x	x

$$K_a = \frac{[H_3O^+][F^-]}{[HF]} = 3.5 \times 10^{-4} = \frac{x^2}{2.0 - x} \approx \frac{x^2}{2.0}$$

Solve for x. x = 0.026 M

$$\% \text{ dissociation} = \frac{[HF]_{diss}}{[HF]_{initial}} \times 100\% = \frac{0.026 \text{ M}}{2.0 \text{ M}} \times 100\% = 1.3\%$$

The percent dissociation increases as the acid concentration decreases for the same acid. The percent dissociation increases as K_a increases for the same concentration of acid.

15.99 $\% \text{ dissociation} = \frac{[HA]_{diss}}{[HA]_{initial}}$

For a weak acid, $[HA]_{diss} = [H_3O^+] = [A^-]$

$$K_a = \frac{[H_3O^+][A^-]}{[HA]} = \frac{[H_3O^+]^2}{[HA]}$$

$$[H_3O^+] = \sqrt{K_a[HA]}$$

$$\% \text{ dissociation} = \frac{[HA]_{diss}}{[HA]} = \frac{[H_3O^+]}{[HA]} = \frac{\sqrt{K_a[HA]}}{[HA]} = \sqrt{\frac{K_a}{[HA]}}$$

15.100 HCl is a strong acid. $[H_3O^+]_{initial} = 0.10$ M

The dissociation of HF produces a negligible amount of H_3O^+ compared with that produced from HCl.

$$HF(aq) + H_2O(l) \rightleftarrows H_3O^+(aq) + F^-(aq)$$

$$K_a = \frac{[H_3O^+][F^-]}{[HF]} = 3.5 \times 10^{-4} = \frac{(0.10)[F^-]}{(0.10)}$$

$[F^-] = K_a = 3.5 \times 10^{-4}$ M

[HF] = 0.10 M

$[H_3O^+] = 0.10$ M

$$[OH^-] = \frac{K_w}{[H_3O^+]} = \frac{1.0 \times 10^{-14}}{0.10} = 1.0 \times 10^{-13} \text{ M}$$

pH = $-\log[H_3O^+] = -\log(0.10) = 1.00$

15.101 Both reactions occur together.

x = $[H_3O^+]$ from CH_3COOH

y = $[H_3O^+]$ from C_6H_5COOH

The following equilibria must be considered.

	$CH_3COOH(aq)$ + $H_2O(l)$	⇄ $H_3O^+(aq)$	+ $CH_3COO^-(aq)$
initial (M)	0.10	y	0
change (M)	–x	+x	+x
equil (M)	0.10 – x	x + y	x

	$C_6H_5COOH(aq)$ + $H_2O(l)$	⇄ $H_3O^+(aq)$	+ $C_6H_5COO^-(aq)$
initial (M)	0.10	x	0
change (M)	–y	+y	+y
equil (M)	0.10 – y	x + y	y

$$K_a(\text{for } CH_3COOH) = \frac{[H_3O^+][CH_3COO^-]}{[CH_3COOH]} = 1.8 \times 10^{-5} = \frac{(x+y)(x)}{0.10-x} \approx \frac{(x+y)(x)}{0.10}$$

$1.8 \times 10^{-6} = (x + y)(x)$

$$K_a(\text{for } C_6H_5COOH) = \frac{[H_3O^+][C_6H_5COO^-]}{[C_6H_5COOH]} = 6.5 \times 10^{-5} = \frac{(x+y)(y)}{0.10-y} \approx \frac{(x+y)(y)}{0.10}$$

$6.5 \times 10^{-6} = (x + y)(y)$

$1.8 \times 10^{-6} = (x + y)(x)$
$6.5 \times 10^{-6} = (x + y)(y)$

These two equations must be solved simultaneously for x and y. Divide the first equation by the second.

$\frac{x}{y} = \frac{1.8 \times 10^{-6}}{6.5 \times 10^{-6}}$; $x = 0.277y$

$6.5 \times 10^{-6} = (x + y)(y)$; substitute $x = 0.277y$ into this equation and solve for y.

$6.5 \times 10^{-6} = (0.277y + y)(y) = 1.277y^2$

$y = 0.002\ 256$

$x = 0.277y = (0.277)(0.002\ 256) = 0.000\ 624\ 9$

$[H_3O^+] = (x + y) = (0.000\ 624\ 9 + 0.002\ 256) = 0.002\ 881$ M

$pH = -\log[H_3O^+] = -\log(0.002\ 881) = 2.54$

15.102 Let $C_7H_5NO_3S$ = HSac

$x = [H_3O^+]$ from HSac

$y = [H_3O^+]$ from H_2O

	HSac(aq) + H_2O(l) ⇄	H_3O^+(aq) +	Sac^-(aq)
initial (M)	0.019	y	0
change (M)	–x	+x	+x
equil (M)	0.019 – x	x + y	x

$$K_a = \frac{[H_3O^+][Sac^-]}{[HSac]} = 2.1 \times 10^{-12} = \frac{(x + y)(x)}{0.019 - x} \approx \frac{(x + y)(x)}{0.019}$$

$4.0 \times 10^{-14} = (x + y)(x)$

$$2\ H_2O(l) \rightleftarrows H_3O^+(aq) + OH^-(aq)$$

	$H_3O^+(aq)$	$OH^-(aq)$
initial (M)	x	~0
change (M)	+y	+y
equil (M)	x + y	y

$K_w = [H_3O^+][OH^-] = 1.0 \times 10^{-14} = (x + y)(y)$

$4.0 \times 10^{-14} = (x + y)(x)$
$1.0 \times 10^{-14} = (x + y)(y)$

Solve these two equations simultaneously for x and y. Divide the first equation by the second.

$$\frac{x}{y} = \frac{4.0 \times 10^{-14}}{1.0 \times 10^{-14}} = 4.0$$

$x = 4.0y$

$1.0 \times 10^{-14} = (x + y)(y) = (4.0y + y)y = 5.0y^2$

$$y = \sqrt{\frac{1.0 \times 10^{-14}}{5.0}} = 4.5 \times 10^{-8}$$

$x = 4.0y = (4.0)(4.5 \times 10^{-8}) = 1.8 \times 10^{-7}$

$[H_3O^+] = (x + y) = (1.8 \times 10^{-7}) + (4.5 \times 10^{-8}) = 2.25 \times 10^{-7}$ M

$pH = -\log[H_3O^+] = -\log(2.25 \times 10^{-7}) = 6.65$

15.103 For 1.0×10^{-10} M HCl, pH = 7.00, and the principal source of H_3O^+ is the dissociation of H_2O.

For 1.0×10^{-7} M HCl,

$$2\ H_2O(l) \rightleftarrows H_3O^+(aq) + OH^-(aq)$$

	$H_3O^+(aq)$	$OH^-(aq)$
initial (M)	1.0×10^{-7}	~0
change (M)	+x	+x
equil (M)	$(1.0 \times 10^{-7}) + x$	x

$K_w = 1.0 \times 10^{-14} = [H_3O^+][OH^-] = [(1.0 \times 10^{-7}) + x](x)$

$x^2 + (1.0 \times 10^{-7})x - (1.0 \times 10^{-14}) = 0$

Solve for x using the quadratic formula.

$$x = \frac{(-1.0 \times 10^{-7}) \pm \sqrt{(1.0 \times 10^{-7})^2 + (4)(1.0 \times 10^{-14})}}{2(1)} = \frac{(-1.0 \times 10^{-7}) \pm (2.236 \times 10^{-7})}{2}$$

$x = 6.18 \times 10^{-8}$ and -1.62×10^{-7}

Of the two solutions for x, only the positive value of x has physical meaning, because x is the $[OH^-]$.

$[H_3O^+] = (1.0 \times 10^{-7}) + x = (1.0 \times 10^{-7}) + (6.18 \times 10^{-8}) = 1.618 \times 10^{-7}$ M

$pH = -\log[H_3O^+] = -\log(1.618 \times 10^{-7}) = 6.79$

15.104 (a) NH_4F

For NH_4^+, $K_a = 5.6 \times 10^{-10}$ and for F^-, $K_b = 2.9 \times 10^{-11}$

Because $K_a > K_b$, the salt solution is acidic.

(b) $NH_4(CH_3CO_2)$

For NH_4^+, $K_a = 5.6 \times 10^{-10}$ and for $CH_3CO_2^-$, $K_b = 5.6 \times 10^{-10}$

Because $K_a = K_b$, the salt solution is neutral.

(c) $(NH_4)_2SO_3$

For NH_4^+, $K_a = 5.6 \times 10^{-10}$ and for SO_3^{2-}, $K_b = 1.6 \times 10^{-7}$

Because $K_b > K_a$, the salt solution is basic.

15.105 $NH_4^+(aq) + H_2O(l) \rightleftarrows H_3O^+(aq) + NH_3(aq)$ $K_1 = K_a$

$F^-(aq) + H_2O(l) \rightleftarrows HF(aq) + OH^-(aq)$ $K_2 = K_b$

$H_3O^+(aq) + OH^-(aq) \rightleftarrows 2\ H_2O(l)$ $K_3 = 1/K_w$

Overall reaction $NH_4^+(aq) + F^-(aq) \rightleftarrows HF(aq) + NH_3(aq)$ $K = K_1K_2K_3$

$$K = K_1K_2K_3 = \frac{K_aK_b}{K_w}$$

The equilibrium constant for the transfer of a proton from the cation of a salt to the anion of the salt is equal to $(K_aK_b)/K_w$.

(a)

	$NH_4^+(aq)$ +	$F^-(aq)$ $\rightleftarrows$	$HF(aq)$ +	$NH_3(aq)$
initial (M)	0.25	0.25	0	0
change (M)	–x	–x	+x	+x
equil (M)	0.25 – x	0.25 – x	x	x

$$K = \frac{K_aK_b}{K_w} = \frac{(5.56 \times 10^{-10})(2.86 \times 10^{-11})}{(1.0 \times 10^{-14})} = \frac{[HF][NH_3]}{[NH_4^+][F^-]} = \frac{x^2}{(0.25 - x)^2}$$

where K_a is K_a for NH_4^+ and K_b is K_b for F^-.

Take the square root of both side and solve for x.

$x = 3.14 \times 10^{-4}$

$[NH_4^+] = [F^-] = 0.25 - x = 0.25$ M

$[NH_3] = [HF] = x = 3.14 \times 10^{-4}$ M

$$K_a(NH_4^+) = \frac{[H_3O^+][NH_3]}{[NH_4^+]} = \frac{[H_3O^+](3.14 \times 10^{-4})}{0.25} = 5.56 \times 10^{-10}$$

$[H_3O^+] = 4.4 \times 10^{-7}$ M

$$[OH^-] = \frac{K_w}{[H_3O^+]} = \frac{1.0 \times 10^{-14}}{4.4 \times 10^{-7}} = 2.3 \times 10^{-8}\ M$$

$pH = -\log[H_3O^+] = -\log(4.4 \times 10^{-7}) = 6.36$

(b) $NH_4^+(aq) + CH_3CO_2^-(aq) \rightleftarrows CH_3CO_2H(aq) + NH_3(aq)$

	$NH_4^+(aq)$	$CH_3CO_2^-(aq)$	$CH_3CO_2H(aq)$	$NH_3(aq)$
initial (M)	0.25	0.25	0	0
change (M)	–x	–x	+x	+x
equil (M)	0.25 – x	0.25 – x	x	x

$$K = \frac{K_aK_b}{K_w} = \frac{(5.56 \times 10^{-10})^2}{(1.0 \times 10^{-14})} = \frac{[CH_3CO_2H][NH_3]}{[NH_4^+][CH_3CO_2^-]} = \frac{x^2}{(0.25 - x)^2}$$

where K_a is K_a for NH_4^+ and K_b is K_b for $CH_3CO_2^-$.

Take the square root of both side and solve for x.

$x = 1.38 \times 10^{-3}$

$[NH_4^+] = [CH_3COO^-] = 0.25 - x = 0.25$ M

$[NH_3] = [CH_3COOH] = x = 1.4 \times 10^{-3}$ M

$$K_a(NH_4^+) = \frac{[H_3O^+][NH_3]}{[NH_4^+]} = \frac{[H_3O^+](1.4 \times 10^{-3})}{0.25} = 5.56 \times 10^{-10}$$

$[H_3O^+] = 1.0 \times 10^{-7}$ M

$$[OH^-] = \frac{K_w}{[H_3O^+]} = \frac{1.0 \times 10^{-14}}{1.0 \times 10^{-7}} = 1.0 \times 10^{-7} \text{ M}$$

$pH = -\log[H_3O^+] = -\log(1.0 \times 10^{-7}) = 7.00$

(c) $NH_4^+(aq) + SO_3^{2-}(aq) \rightleftarrows HSO_3^-(aq) + NH_3(aq)$

	$NH_4^+(aq)$	$SO_3^{2-}(aq)$	$HSO_3^-(aq)$	$NH_3(aq)$
initial (M)	0.50	0.25	0	0
change (M)	–x	–x	+x	+x
equil (M)	0.50 – x	0.25 – x	x	x

$$K = \frac{K_aK_b}{K_w} = \frac{(5.56 \times 10^{-10})(1.59 \times 10^{-7})}{(1.0 \times 10^{-14})} = \frac{[HSO_3^-][NH_3]}{[NH_4^+][SO_3^{2-}]} = \frac{x^2}{(0.50 - x)(0.25 - x)}$$

where K_a is K_a for NH_4^+ and K_b is K_b for SO_3^{2-}.

$0.9912x^2 + 0.006\ 63x - 0.001\ 11 = 0$

Use the quadratic formula to solve for x.

$$x = \frac{(-0.006\ 63) \pm \sqrt{(0.00663)^2 + 4(0.9912)(0.001\ 11)}}{2(0.9912)} = \frac{-0.006\ 63 \pm 0.066\ 67}{2(0.9912)}$$

x = 0.0303 and –0.0370

Of the two solutions for x, only the positive value of x has physical meaning, because x is the $[NH_3]$ and $[HSO_3^-]$.

x = 0.030

$[NH_4^+] = 0.50 - x = 0.47$ M

$[SO_3^{2-}] = 0.25 - x = 0.22$ M

$[NH_3] = [HSO_3^-] = x = 0.030$ M

$$K_b(SO_3^{2-}) = \frac{[HSO_3^-][OH^-]}{[SO_3^{2-}]} = \frac{(0.030)[OH^-]}{0.22} = 1.59 \times 10^{-7}$$

$[OH^-] = 1.166 \times 10^{-6}$ M = 1.2×10^{-6} M

$$[H_3O^+] = \frac{K_w}{[OH^-]} = \frac{1.0 \times 10^{-14}}{1.166 \times 10^{-6}} = 8.6 \times 10^{-9}\ M$$

$pH = -\log[H_3O^+] = -\log(8.6 \times 10^{-9}) = 8.06$

16.1 (a) $HNO_2(aq) + OH^-(aq) \rightleftarrows NO_2^-(aq) + H_2O(l)$; NO_2^- (basic anion), pH > 7.00

(b) $H_3O^+(aq) + NH_3(aq) \rightleftarrows NH_4^+(aq) + H_2O(l)$; NH_4^+ (acidic cation), pH < 7.00

(c) $OH^-(aq) + H_3O^+(aq) \rightleftarrows 2\ H_2O(l)$; pH = 7.00

16.2 (a) $HF(aq) + OH^-(aq) \rightleftarrows H_2O(l) + F^-(aq)$

$$K_n = \frac{K_a}{K_w} = \frac{3.5 \times 10^{-4}}{1.0 \times 10^{-14}} = 3.5 \times 10^{10}$$

(b) $H_3O^+(aq) + OH^-(aq) \rightleftarrows 2\ H_2O(l)$

$$K_n = \frac{1}{K_w} = \frac{1}{1.0 \times 10^{-14}} = 1.0 \times 10^{14}$$

(c) $HF(aq) + NH_3(aq) \rightleftarrows NH_4^+(aq) + F^-(aq)$

$$K_n = \frac{K_aK_b}{K_w} = \frac{(3.5 \times 10^{-4})(1.8 \times 10^{-5})}{1.0 \times 10^{-14}} = 6.3 \times 10^5$$

The tendency to proceed to completion is determined by the magnitude of K_n. The larger the value of K_n, the further does the reaction proceed to completion.

The tendency to proceed to completion is:

reaction (c) < reaction (a) < reaction (b)

16.3 $HCN(aq) + H_2O(l) \rightleftarrows H_3O^+(aq) + CN^-(aq)$

	HCN(aq)	H_3O^+(aq)	CN^-(aq)
initial (M)	0.025	~0	0.010
change (M)	–x	+x	+x
equil (M)	0.025 – x	x	0.010 + x

$$K_a = \frac{[H_3O^+][CN^-]}{[HCN]} = 4.9 \times 10^{-10} = \frac{x(0.010 + x)}{0.025 - x} \approx \frac{x(0.010)}{0.025}$$

Solve for x. $x = 1.2 \times 10^{-9}$ M = $[H_3O^+]$

$pH = -\log[H_3O^+] = -\log(1.23 \times 10^{-9}) = 8.91$

$$[OH^-] = \frac{K_w}{[H_3O^+]} = \frac{1.0 \times 10^{-14}}{1.23 \times 10^{-9}} = 8.2 \times 10^{-6}\ M$$

$[Na^+] = [CN^-] = 0.010\ M$

$[HCN] = 0.025\ M$

$$\%\ \text{dissociation} = \frac{[HCN]_{diss}}{[HCN]_{initial}} \times 100\% = \frac{1.23 \times 10^{-9}\ M}{0.025\ M} \times 100\% = 4.9 \times 10^{-6}\ \%$$

16.4 From $NH_4Cl(s)$, $[NH_4^+]_{initial} = \dfrac{0.10\ mol}{0.500\ L} = 0.20\ M$

	$NH_3(aq)$ + $H_2O(l)$	⇄	$NH_4^+(aq)$	+ $OH^-(aq)$
initial (M)	0.40		0.20	~0
change (M)	–x		+x	+x
equil (M)	0.40 – x		0.20 + x	x

$$K_b = \frac{[NH_4^+][OH^-]}{[NH_3]} = 1.8 \times 10^{-5} = \frac{(0.20 + x)(x)}{(0.40 - x)} \approx \frac{(0.20)(x)}{(0.40)}$$

Solve for x. $x = [OH^-] = 3.6 \times 10^{-5}\ M$

$$[H_3O^+] = \frac{K_w}{[OH^-]} = \frac{1.0 \times 10^{-14}}{3.6 \times 10^{-5}} = 2.8 \times 10^{-10}\ M$$

$pH = -\log[H_3O^+] = -\log(2.8 \times 10^{-10}) = 9.55$

16.5

	$HF(aq)$ + $H_2O(l)$	⇄	$H_3O^+(aq)$	+ $F^-(aq)$
initial (M)	0.25		~0	0.50
change (M)	–x		+x	+x
equil (M)	0.25 – x		x	0.50 + x

$$K_a = \frac{[H_3O^+][F^-]}{[HF]} = 3.5 \times 10^{-4} = \frac{x(0.50 + x)}{0.25 - x} \approx \frac{x(0.50)}{0.25}$$

Solve for x. $x = 1.75 \times 10^{-4}$ M = $[H_3O^+]$

For the buffer, pH = $-\log[H_3O^+] = -\log(1.75 \times 10^{-4}) = 3.76$

(a) mol HF = 0.025 mol; mol F^- = 0.050 mol; vol = 0.100 L

$$F^-(aq) + H_3O^+(aq) \xrightarrow{100\%} HF(aq) + H_2O(l)$$

	$F^-(aq)$	$H_3O^+(aq)$	$HF(aq)$
before (mol)	0.050	0.002	0.025
change (mol)	−0.002	−0.002	+0.002
after (mol)	0.048	0	0.027

$$[H_3O^+] = K_a \frac{[HF]}{[F^-]} = (3.5 \times 10^{-4})\left(\frac{0.27}{0.48}\right) = 1.97 \times 10^{-4} \text{ M}$$

pH = $-\log[H_3O^+] = -\log(1.97 \times 10^{-4}) = 3.71$

(b) mol HF = 0.025 mol; mol F^- = 0.050 mol; vol = 0.100 L

$$HF(aq) + OH^-(aq) \xrightarrow{100\%} F^-(aq) + H_2O(l)$$

	$HF(aq)$	$OH^-(aq)$	$F^-(aq)$
before (mol)	0.025	0.004	0.050
change (mol)	−0.004	−0.004	+0.004
after (mol)	0.021	0	0.054

$$[H_3O^+] = K_a \frac{[HF]}{[F^-]} = (3.5 \times 10^{-4})\left(\frac{0.21}{0.54}\right) = 1.36 \times 10^{-4} \text{ M}$$

pH = $-\log[H_3O^+] = -\log(1.36 \times 10^{-4}) = 3.87$

16.6

$$HF(aq) + H_2O(l) \rightleftharpoons H_3O^+(aq) + F^-(aq)$$

	$HF(aq)$	$H_3O^+(aq)$	$F^-(aq)$
initial (M)	0.050	~0	0.100
change (M)	−x	+x	+x
equil (M)	0.050 − x	x	0.100 + x

$$K_a = \frac{[H_3O^+][F^-]}{[HF]} = 3.5 \times 10^{-4} = \frac{x(0.100 + x)}{0.050 - x} \approx \frac{x(0.100)}{0.050}$$

Solve for x. $x = [H_3O^+] = 1.75 \times 10^{-4}$ M

$pH = -\log[H_3O^+] = -\log(1.75 \times 10^{-4}) = 3.76$

mol HF = 0.050 mol/L x 0.100 L = 0.0050 mol HF

mol F^- = 0.100 mol/L x 0.100 L = 0.0100 mol F^-

mol HNO_3 = mol H_3O^+ = 0.002 mol

Neutralization reaction: $F^-(aq) + H_3O^+(aq) \xrightarrow{100\%} HF(aq) + H_2O(l)$

	$F^-(aq)$	$H_3O^+(aq)$	$HF(aq)$
before reaction (mol)	0.0100	0.002	0.0050
change (mol)	–0.002	–0.002	+0.002
after reaction (mol)	0.008	0	0.007

$$[HF] = \frac{0.007 \text{ mol}}{0.100 \text{ L}} = 0.07 \text{ M}$$

$$[F^-] = \frac{0.008 \text{ mol}}{0.100 \text{ L}} = 0.08 \text{ M}$$

$$[H_3O^+] = K_a \frac{[HF]}{[F^-]} = (3.5 \times 10^{-4})\frac{(0.07)}{(0.08)} = 3 \times 10^{-4} \text{ M}$$

$pH = -\log[H_3O^+] = -\log(3 \times 10^{-4}) = 3.5$

This solution has less buffering capacity than the solution in Problem 16.5 because it contains less HF and F^- per 100 mL. Note that the change in pH is greater than that in Problem 16.5.

16.7 When equal volumes of two solutions are mixed together, the concentration of each solution is cut in half.

$$pH = pK_a + \log\frac{[\text{base}]}{[\text{acid}]} = pK_a + \log\frac{[CO_3^{2-}]}{[HCO_3^-]}$$

For HCO_3^-, $K_a = 5.6 \times 10^{-11}$

$$pH = -\log(5.6 \times 10^{-11}) + \log\left(\frac{0.050}{0.10}\right) = 10.25 - 0.30 = 9.95$$

16.8 $pH = pK_a + \log\frac{[CO_3^{2-}]}{[HCO_3^-]}$

For HCO_3^-, $K_a = 5.6 \times 10^{-11}$, $pK_a = -\log K_a = -\log(5.6 \times 10^{-11}) = 10.25$

$$10.40 = 10.25 + \log\frac{[CO_3^{2-}]}{[HCO_3^-]}$$

$$\log\frac{[CO_3^{2-}]}{[HCO_3^-]} = 10.40 - 10.25 = 0.15$$

$$\frac{[CO_3^{2-}]}{[HCO_3^-]} = 10^{0.15} = 1.4$$

To obtain a buffer solution with pH 10.40, make the Na_2CO_3 concentration 1.4 times the concentration of $NaHCO_3$.

16.9 Look for an acid with $pK_a = pH$

required pH = 7.50

$K_a = 10^{-pH} = 10^{-7.50} = 3.2 \times 10^{-8}$

Suggested buffer system: HOCl ($K_a = 3.5 \times 10^{-8}$) and NaOCl.

16.10 (a) mol HCl = mol H_3O^+ = 0.100 mol/L x 0.0400 L = 0.004 00 mol

mol NaOH = mol OH^- = 0.100 mol/L x 0.0350 L = 0.003 50 mol

Neutralization reaction:	$H_3O^+(aq)$ +	$OH^-(aq)$	→ $2\ H_2O(l)$
before reaction (mol)	0.004 00	0.003 50	
change (mol)	−0.003 50	−0.003 50	
after reaction (mol)	0.000 50	0	

$$[H_3O^+] = \frac{0.000\ 50\ \text{mol}}{(0.0400\ \text{L} + 0.0350\ \text{L})} = 6.7 \times 10^{-3}\ \text{M}$$

$pH = -\log[H_3O^+] = -\log(6.7 \times 10^{-3}) = 2.17$

(b) mol HCl = mol H_3O^+ = 0.100 mol/L x 0.0400 L = 0.004 00 mol

mol NaOH = mol OH^- = 0.100 mol/L x 0.0450 L = 0.004 50 mol

Neutralization reaction:	$H_3O^+(aq)$ +	$OH^-(aq)$ →	$2\ H_2O(l)$
before reaction (mol)	0.004 00	0.004 50	
change (mol)	–0.004 00	–0.004 00	
after reaction (mol)	0	0.000 50	

$$[OH^-] = \frac{0.000\ 50\ \text{mol}}{(0.0400\ \text{L} + 0.0450\ \text{L})} = 5.9 \times 10^{-3}\ \text{M}$$

$$[H_3O^+] = \frac{K_w}{[OH^-]} = \frac{1.0 \times 10^{-14}}{5.9 \times 10^{-3}} = 1.7 \times 10^{-12}\ \text{M}$$

$pH = -\log[H_3O^+] = -\log(1.7 \times 10^{-12}) = 11.77$

The results obtained here are consistent with the pH data in Table 16.1

16.11 (a) mol NaOH = mol OH^- = 0.100 mol/L x 0.0400 L = 0.004 00 mol

mol HCl = mol H_3O^+ = 0.0500 mol/L x 0.0600 L = 0.003 00 mol

Neutralization reaction:	$H_3O^+(aq)$ +	$OH^-(aq)$ →	$2\ H_2O(l)$
before reaction (mol)	0.003 00	0.004 00	
change (mol)	–0.003 00	–0.003 00	
after reaction (mol)	0	0.001 00	

$$[OH^-] = \frac{0.001\ 00\ \text{mol}}{(0.0400\ \text{L} + 0.0600\ \text{L})} = 1.0 \times 10^{-2}\ \text{M}$$

$$[H_3O^+] = \frac{K_w}{[OH^-]} = \frac{1.0 \times 10^{-14}}{1.0 \times 10^{-2}} = 1.0 \times 10^{-12}\ \text{M}$$

$pH = -\log(1.0 \times 10^{-12}) = 12.00$

(b) mol NaOH = mol OH^- = 0.100 mol/L x 0.0400 L = 0.004 00 mol

mol HCl = mol H_3O^+ = 0.0500 mol/L x 0.0802 L = 0.004 01 mol

Neutralization reaction:	$H_3O^+(aq)$ +	$OH^-(aq)$	→ $2\ H_2O(l)$
before reaction (mol)	0.004 01	0.004 00	
change (mol)	–0.004 00	–0.004 00	
after reaction (mol)	0.000 01	0	

$$[H_3O^+] = \frac{0.000\ 01\ \text{mol}}{(0.0400\ \text{L} + 0.0802\ \text{L})} = 8.3 \times 10^{-5}\ \text{M}$$

pH = $-\log(8.3 \times 10^{-5}) = 4.08$

(c) mol NaOH = mol OH^- = 0.100 mol/L x 0.0400 L = 0.004 00 mol

mol HCl = mol H_3O^+ = 0.0500 mol/L x 0.1000 L = 0.005 00 mol

Neutralization reaction:	$H_3O^+(aq)$ +	$OH^-(aq)$	→ $2\ H_2O(l)$
before reaction (mol)	0.005 00	0.004 00	
change (mol)	–0.004 00	–0.004 00	
after reaction (mol)	0.001 00	0	

$$[H_3O^+] = \frac{0.001\ 00\ \text{mol}}{(0.0400\ \text{L} + 0.1000\ \text{L})} = 7.1 \times 10^{-3}\ \text{M}$$

pH = $-\log(7.1 \times 10^{-3}) = 2.15$

16.12 $$\text{mol NaOH required} = \left(\frac{0.016\ \text{mol HOCl}}{\text{L}}\right)(0.100\ \text{L})\left(\frac{1\ \text{mol NaOH}}{1\ \text{mol HOCl}}\right) = 0.0016\ \text{mol}$$

$$\text{vol NaOH required} = (0.0016\ \text{mol})\left(\frac{1\ \text{L}}{0.0400\ \text{mol}}\right) = 0.040\ \text{L} = 40\ \text{mL}$$

40 mL of 0.0400 M NaOH are required to reach the equivalence point.

(a) mmol HOCl = 0.016 mmol/mL x 100.0 mL = 1.6 mmol

mmol NaOH = mmol OH^- = 0.0400 mmol/mL x 10.0 mL = 0.400 mmol

Neutralization reaction:	$HOCl(aq)$	+ $OH^-(aq)$	$\rightarrow$	$OCl^-(aq)$	+ $H_2O(l)$
before reaction (mmol)	1.6	0.400		0	
change (mmol)	–0.400	–0.400		+0.400	
after reaction (mmol)	1.2	0		0.400	

$$[HOCl] = \frac{1.2\text{ mmol}}{(100.0\text{ mL} + 10.0\text{ mL})} = 1.09 \times 10^{-2}\text{ M}$$

$$[OCl^-] = \frac{0.400\text{ mmol}}{(100.0\text{ mL} + 10.0\text{ mL})} = 3.64 \times 10^{-3}\text{ M}$$

	$HOCl(aq)$	+ $H_2O(l)$	$\rightleftarrows$	$H_3O^+(aq)$	+ $OCl^-(aq)$
initial (M)	0.0109			~0	0.003 64
change (M)	–x			+x	+x
equil (M)	0.0109 – x			x	0.003 64 + x

$$K_a = \frac{[H_3O^+][OCl^-]}{[HOCl]} = 3.5 \times 10^{-8} = \frac{x(0.003\ 64 + x)}{0.0109 - x} \approx \frac{x(0.003\ 64)}{0.0109}$$

Solve for x. $x = [H_3O^+] = 1.05 \times 10^{-7}$ M

$pH = -\log[H_3O^+] = -\log(1.05 \times 10^{-7}) = 6.98$

(b) Halfway to the equivalence point, $[OCl^-] = [HOCl]$

$pH = pK_a = -\log K_a = -\log(3.5 \times 10^{-8}) = 7.46$

(c) At the equivalence point the solution contains the salt, NaOCl.

mol NaOCl = initial mol HOCl = 0.0016 mol = 1.6 mmol

$$[OCl^-] = \frac{1.6\text{ mmol}}{(100.0\text{ mL} + 40.0\text{ mL})} = 1.1 \times 10^{-2}\text{ M}$$

$$\text{For } OCl^-,\ K_b = \frac{K_w}{K_a \text{ for HOCl}} = \frac{1.0 \times 10^{-14}}{3.5 \times 10^{-8}} = 2.9 \times 10^{-7}$$

	$OCl^-(aq)$ + $H_2O(l)$	$\rightleftarrows$	$HOCl(aq)$	+ $OH^-(aq)$
initial (M)	0.011		0	~0
change (M)	–x		+x	+x
equil (M)	0.011 – x		x	x

$$K_b = \frac{[HOCl][OH^-]}{[OCl^-]} = 2.9 \times 10^{-7} = \frac{x^2}{0.011 - x} \approx \frac{x^2}{0.011}$$

Solve for x. $x = [OH^-] = 5.65 \times 10^{-5}$ M

$$[H_3O^+] = \frac{K_w}{[OH^-]} = \frac{1.0 \times 10^{-14}}{5.65 \times 10^{-5}} = 1.77 \times 10^{-10} = 1.8 \times 10^{-10}\ M$$

$pH = -\log[H_3O^+] = -\log(1.77 \times 10^{-10}) = 9.75$

16.13 From Problem 16.12, pH = 9.75 at the equivalence point.

Use thymolphthalein (pH 9.4 – 10.6). Bromothymol blue is unacceptable because it changes color halfway to the equivalence point.

16.14 (a) mol NaOH required to reach first equivalence point

$$= \left(\frac{0.0800 \text{ mol } H_2SO_3}{L}\right)(0.0400\ L)\left(\frac{1 \text{ mol NaOH}}{1 \text{ mol } H_2SO_3}\right) = 0.003\ 20 \text{ mol}$$

vol NaOH required to reach first equivalence point

$$= (0.003\ 20 \text{ mol})\left(\frac{1\ L}{0.160 \text{ mol}}\right) = 0.020\ L = 20.0\ mL$$

20.0 mL is enough NaOH solution to reach the first equivalence point for the titration of the diprotic acid, H_2SO_3.

For H_2SO_3,

$K_{a1} = 1.5 \times 10^{-2}$, $pK_{a1} = -\log K_{a1} = -\log(1.5 \times 10^{-2}) = 1.82$

$K_{a2} = 6.3 \times 10^{-8}$, $pK_{a2} = -\log K_{a2} = -\log(6.3 \times 10^{-8}) = 7.20$

At the first equivalence point,

$$pH = \frac{pK_{a1} + pK_{a2}}{2} = \frac{1.82 + 7.20}{2} = 4.51$$

(b) mol NaOH required to reach second equivalence point

$$= \left(\frac{0.0800 \text{ mol } H_2SO_3}{L}\right)(0.0400 \text{ L})\left(\frac{2 \text{ mol NaOH}}{1 \text{ mol } H_2SO_3}\right) = 0.006\ 40 \text{ mol}$$

vol NaOH required to reach second equivalence point

$$= (0.006\ 40 \text{ mol})\left(\frac{1 \text{ L}}{0.160 \text{ mol}}\right) = 0.040 \text{ L} = 40.0 \text{ mL}$$

30.0 mL is enough NaOH solution to reach halfway to the second equivalent point.

Halfway to the second equivalence point

$$pH = pK_{a2} = -\log K_{a2} = -\log(6.3 \times 10^{-8}) = 7.20$$

(c) mmol HSO_3^- = 0.0800 mmol/mL x 40.0 mL = 3.2 mmol

volume NaOH added after first equivalence point

= 35.0 mL – 20.0 mL = 15.0 mL

mmol NaOH = mmol OH^- = 0.160 mmol/L x 15.0 mL = 2.4 mmol

Neutralization reaction:	$HSO_3^-(aq)$ +	$OH^-(aq)$	⇄	$SO_3^{2-}(aq)$ +	$H_2O(l)$
before reaction (mmol)	3.2	2.4		0	
change (mmol)	–2.4	–2.4		+2.4	
after reaction (mmol)	0.8	0		2.4	

$$[HSO_3^-] = \frac{0.8 \text{ mmol}}{(40.0 \text{ mL} + 35.0 \text{ mL})} = 0.0107 \text{ M}$$

$$[SO_3^{2-}] = \frac{2.4\text{ mmol}}{(40.0\text{ mL} + 35.0\text{ mL})} = 0.032\text{ M}$$

	$HSO_3^-(aq)$ + $H_2O(l)$	⇌	$H_3O^+(aq)$	+ $SO_3^{2-}(aq)$
initial (M)	0.0107		~0	0.032
change (M)	–x		+x	+x
equil (M)	0.0107 – x		x	0.032 + x

$$K_a = \frac{[H_3O^+][SO_3^{2-}]}{[HSO_3^-]} = 6.3 \times 10^{-8} = \frac{x(0.032 + x)}{0.0107 - x} \approx \frac{x(0.032)}{0.0107}$$

Solve for x. $x = [H_3O^+] = 2.1 \times 10^{-8}$ M

$pH = -\log[H_3O^+] = -\log(2.1 \times 10^{-8}) = 7.7$

16.15 (a) $K_{sp} = [Ag^+][Cl^-]$ (b) $K_{sp} = [Pb^{2+}][I^-]^2$ (c) $K_{sp} = [Ca^{2+}]^3[PO_4^{3-}]^2$

16.16 $K_{sp} = [Ca^{2+}]^3[PO_4^{3-}]^2 = (2.01 \times 10^{-8})^3(1.6 \times 10^{-5})^2 = 2.1 \times 10^{-33}$

16.17 $[Ba^{2+}] = [SO_4^{2-}] = 1.05 \times 10^{-5}$ M

$K_{sp} = [Ba^{2+}][SO_4^{2-}] = (1.05 \times 10^{-5})^2 = 1.10 \times 10^{-10}$

16.18 (a)

	$AgCl(s)$ ⇌	$Ag^+(aq)$	+ $Cl^-(aq)$
equil (M)		x	x

$K_{sp} = [Ag^+][Cl^-] = 1.8 \times 10^{-10} = (x)(x)$

molar solubility $= x = \sqrt{K_{sp}} = 1.3 \times 10^{-5}$ mol/L

(b)

	$Ag_2SO_4(s)$ ⇌	$2\ Ag^+(aq)$	+ $SO_4^{2-}(aq)$
equil (M)		2x	x

$K_{sp} = [Ag^+]^2[SO_4^{2-}] = 1.2 \times 10^{-5} = (2x)^2(x) = 4x^3$

$$\text{molar solubility} = x = \sqrt[3]{\frac{1.2 \times 10^{-5}}{4}} = 1.44 \times 10^{-2} \text{ mol/L}$$

Ag_2SO_4, 311.80 amu

$$\text{solubility} = \frac{\left(1.44 \times 10^{-2} \text{ mol} \times \frac{311.80}{1 \text{ mol}}\right)}{1 \text{ L}} = 4.5 \text{ g/L}$$

16.19

	$MgF_2(s)$	$\rightleftarrows$	$Mg^{2+}(aq)$	+	$2\ F^-(aq)$
initial (M)			0.10		0
change (M)			+x		+2x
equil (M)			0.10 + x		2x

$K_{sp} = 7.4 \times 10^{-11} = [Mg^{2+}][F^-]^2 = (0.10 + x)(2x)^2 \approx (0.10)(4x^2)$

$x = 1.4 \times 10^{-5}$, molar solubility = $x = 1.4 \times 10^{-5}$ M

16.20 Compounds that contain basic anions are more soluble in acidic solution than in pure water. AgCN, $Al(OH)_3$, and ZnS all contain basic anions.

16.21 $[Cu^{2+}] = (5.0 \times 10^{-3} \text{ mol})/(0.500 \text{ L}) = 0.010$ M

	$Cu^{2+}(aq)$	+	$4\ NH_3(aq)$	$\rightleftarrows$	$Cu(NH_3)_4{}^{2+}(aq)$
before reaction (M)	0.010		0.40		0
assume 100 % reaction (M)	–0.010		–4(0.010)		+0.010
after reaction (M)	0		0.36		0.010
assume small back reaction (M)	+x		+4x		–x
equil (M)	x		0.36 + 4x		0.010 – x

$$K_f = \frac{[Cu(NH_3)_4^{2+}]}{[Cu^{2+}][NH_3]^4} = 1.1 \times 10^{13} = \frac{(0.010 - x)}{(x)(0.36 + 4x)^4} \approx \frac{0.010}{x(0.36)^4}$$

Solve for x. $x = [Cu^{2+}] = 5.4 \times 10^{-14}$ M

16.22 $AgBr(s) \rightleftharpoons Ag^+(aq) + Br^-(aq)$ $K_{sp} = 5.4 \times 10^{-13}$

$Ag^+(aq) + 2\ S_2O_3^{2-} \rightarrow Ag(S_2O_3)_2^{3-}(aq)$ $K_f = 2.9 \times 10^{13}$

dissolution reaction $AgBr(s) + 2\ S_2O_3^{2-}(aq) \rightleftharpoons Ag(S_2O_3)_2^{3-}(aq) + Br^-(aq)$

$K = (K_{sp})(K_f) = (5.4 \times 10^{-13})(2.9 \times 10^{13}) = 16$

	$AgBr(s) + 2\ S_2O_3^{2-}(aq)$	$\rightleftharpoons Ag(S_2O_3)_2^{3-}(aq)$	$+ Br^-(aq)$
initial (M)	0.10	0	0
change (M)	–2x	x	x
equil (M)	0.10 – 2x	x	x

$$K = \frac{[Ag(S_2O_3)_2^{3-}][Br^-]}{[S_2O_3^{2-}]^2} = 16 = \frac{x^2}{(0.10-2x)^2}$$

Take the square root of both sides and solve for x.

$$\sqrt{16} = \sqrt{\frac{x^2}{(0.10-2x)^2}}$$

$$4.0 = \frac{x}{0.10-2x}$$

x = molar solubility = 0.044 mol/L

16.23 On mixing equal volumes of two solutions, the concentrations of both solutions are cut in half.

For $BaCO_3$, $K_{sp} = 2.6 \times 10^{-9}$

(a) $IP = [Ba^{2+}][CO_3^{2-}] = (1.5 \times 10^{-3})(1.0 \times 10^{-3}) = 1.5 \times 10^{-6}$

$IP > K_{sp}$; a precipitate of $BaCO_3$ will form.

(b) $IP = [Ba^{2+}][CO_3^{2-}] = (5.0 \times 10^{-6})(2.0 \times 10^{-5}) = 1.0 \times 10^{-10}$

$IP < K_{sp}$; no precipitate will form.

16.24 $MS(s) + 2\ H_3O^+(aq) \rightleftarrows M^{2+}(aq) + H_2S(aq) + 2\ H_2O(l)$

$$K_{spa} = \frac{[M^{2+}][H_2S]}{[H_3O^+]^2}$$

For ZnS, $K_{spa} = 3 \times 10^{-2}$; for CdS, $K_{spa} = 8 \times 10^{-7}$

$[Cd^{2+}] = [Zn^{2+}] = 0.005$ M

Because the two cation concentrations are equal, Q_c is the same for both.

$$Q_c = \frac{[M^{2+}]_t[H_2S]_t}{[H_3O^+]_t^2} = \frac{(0.005)(0.10)}{(0.3)^2} = 6 \times 10^{-3}$$

$Q_c > K_{spa}$ for CdS; CdS will precipitate.

$Q_c < K_{spa}$ for ZnS; Zn^{2+} will remain in solution.

Understanding Key Concepts

1. (a) (1) and (3). Both pictures show equal concentrations of HA and A^-.

   (b) (3). It contains a higher concentration of HA and A^-.

2. (a) (2) has the highest pH, $[A^-] > [HA]$

   (3) has the lowest pH, $[HA] > [A^-]$

   (b) 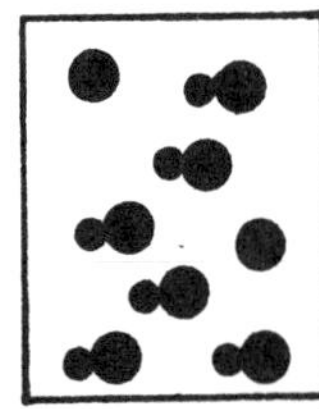 (c) 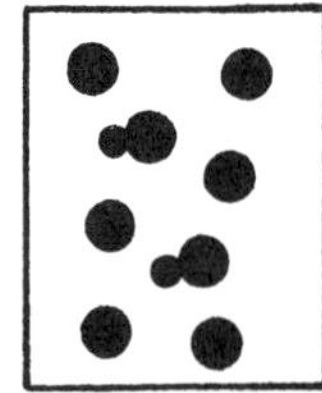

3. (4); only A^- and water should be present

4. (a)

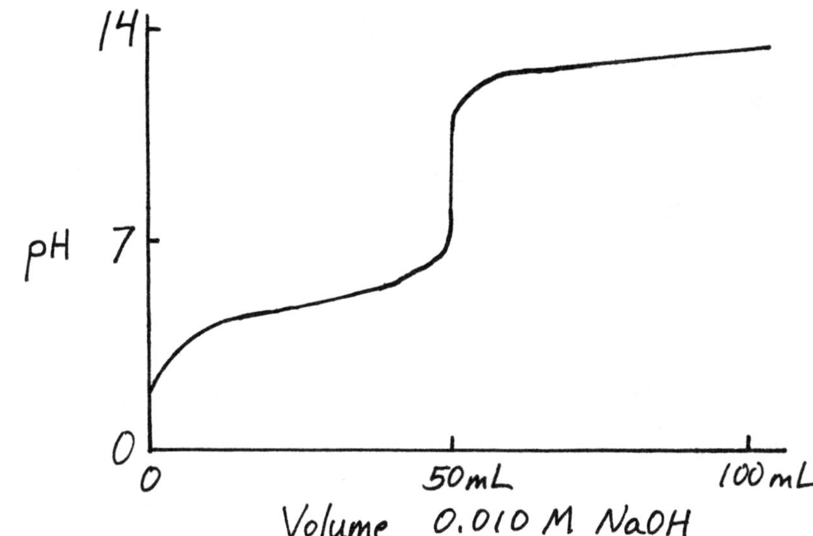

(b) mol NaOH required = $\left(\frac{0.010 \text{ mol HA}}{\text{L}}\right)(0.0500 \text{ L})\left(\frac{1 \text{ mol NaOH}}{1 \text{ mol HA}}\right)$ = 0.000 50 mol

vol NaOH required = $(0.000\ 50 \text{ mol})\left(\frac{1 \text{ L}}{0.010 \text{ mol}}\right)$ = 0.050 L = 50 mL

(c) A basic salt is present at the equivalence point; pH > 7.00

(d) Halfway to the equivalence point, the pH = pK_a = 4.00

5. (a) $AgBr(s) \rightleftarrows Ag^+(aq) + Br^-(aq)$

(a) HBr is a source of Br^- (reaction product). The solubility of AgBr is decreased.

(b) unaffected

(c) $AgNO_3$ is a source of Ag^+ (reaction product). The solubility of AgBr is decreased.

(d) NH_3 forms a complex with Ag^+, removing it from solution. The solubility of AgBr is increased.

(b) $BaCO_3(s) \rightleftarrows Ba^{2+}(aq) + CO_3^{2-}(aq)$

(a) unaffected

(b) HNO_3 reacts with CO_3^{2-}, removing it from the solution. The solubility of $BaCO_3$ is increased.

(c) $Ba(NO_3)_2$ is a source of Ba^{2+} (reaction product). The solubility of $BaCO_3$ is decreased.

(d) $NaCO_3$ is a source of CO_3^{2-} (reaction product). The solubility of $BaCO_3$ is decreased.

Additional Problems

Neutralization Reactions

16.25 The four types of neutralization reactions are:

(1) Strong Acid–Strong Base

$HNO_3(aq) + KOH(aq) \rightarrow H_2O(l) + KNO_3(aq)$

net ionic equation: $H_3O^+(aq) + OH^-(aq) \rightarrow 2\ H_2O(l)$

(2) Weak Acid–Strong Base

$HF(aq) + NaOH(aq) \rightarrow H_2O(l) + NaF(aq)$

net ionic equation: $HF(aq) + OH^-(aq) \rightarrow H_2O(l) + F^-(aq)$

(3) Strong Acid–Weak Base

$HCl(aq) + NH_3(aq) \rightarrow NH_4Cl(aq)$

net ionic equation: $H_3O^+(aq) + NH_3(aq) \rightarrow H_2O(l) + NH_4^+(aq)$

(4) Weak Acid–Weak Base

$CH_3COOH(aq) + NH_3(aq) \rightleftarrows NH_4^+(aq) + CH_3CO_2^-(aq)$

16.26 (a) $HI(aq) + LiOH(aq) \rightarrow H_2O(l) + LiI(aq)$

net ionic equation: $H_3O^+(aq) + OH^-(aq) \rightarrow 2\ H_2O(l)$

The solution at neutralization contains a neutral salt (LiI); pH = 7.00.

(b) $2\ HOCl(aq) + Ba(OH)_2(aq) \rightarrow 2\ H_2O(l) + Ba(OCl)_2(aq)$

net ionic equation: $HOCl(aq) + OH^-(aq) \rightarrow H_2O(l) + OCl^-(aq)$

The solution at neutralization contains a basic anion (OCl^-); pH > 7.00

(c) $HNO_3(aq) + C_6H_5NH_2(aq) \rightarrow C_6H_5NH_3NO_3(aq)$

net ionic equation: $H_3O^+(aq) + C_6H_5NH_2(aq) \rightarrow H_2O(l) + C_6H_5NH_3^+(aq)$

The solution at neutralization contains an acidic cation ($C_6H_5NH_3^+$); pH < 7.00.

(d) $C_6H_5COOH(aq) + KOH(aq) \rightarrow H_2O(l) + C_6H_5COOK(aq)$

net ionic equation: $C_6H_5COOH(aq) + OH^-(aq) \rightarrow H_2O(l) + C_6H_5COO^-(aq)$

The solution at neutralization contains a basic anion ($C_6H_5COO^-$); pH > 7.00.

16.27 (a) Strong acid – strong base reaction

$$K_n = \frac{1}{K_w} = \frac{1}{1.0 \times 10^{-14}} = 1.0 \times 10^{14}$$

(b) Weak acid – strong base reaction

$$K_n = \frac{K_a}{K_w} = \frac{3.5 \times 10^{-8}}{1.0 \times 10^{-14}} = 3.5 \times 10^{6}$$

(c) Strong acid – weak base reaction

$$K_n = \frac{K_b}{K_w} = \frac{4.3 \times 10^{-10}}{1.0 \times 10^{-14}} = 4.3 \times 10^{4}$$

(d) Weak acid – strong base reaction

$$K_n = \frac{K_a}{K_w} = \frac{6.5 \times 10^{-5}}{1.0 \times 10^{-14}} = 6.5 \times 10^9$$

(c) < (b) < (d) < (a)

16.28 Mixture (a) is a weak acid–strong base neutralization reaction. The solution at neutralization contains a basic anion (F^-), so the pH > 7.00.

Mixtures (b) and (c) are strong acid–strong base neutralizations reactions. The solutions at neutralization contain neutral salts, so the pH = 7.00.

Mixture (a) has the highest pH.

16.29 Weak acid – weak base reaction

$$K_n = \frac{K_a K_b}{K_w} = \frac{(1.3 \times 10^{-10})(1.8 \times 10^{-9})}{1.0 \times 10^{-14}} = 2.3 \times 10^{-5}$$

K_n is small so the neutralization reaction does not proceed very far to completion.

The Common–Ion Effect

16.30 The common–ion effect is the shift in the position of an equilibrium on addition of a substance that provides an ion in common with one of the ions already involved in the equilibrium.

Le Châtelier's principle states that if a stress is applied to a reaction mixture at equilibrium, reaction occurs in the direction that relieves the stress. In the common–ion effect, the addition of a substance that provides a common ion is the stress, and the reaction shifts accordingly.

The common–ion effect is a specific example of Le Châtelier's principle.

16.31 (a) $NaNO_2$, source of NO_2^-

(c) HCl, source of H^+ (H_3O^+)

(d) $Ba(NO_2)_2$, source of NO_2^-

16.32 $NH_3(aq) + H_2O(l) \rightleftharpoons NH_4^+(aq) + OH^-(aq)$

(a) KOH is a strong base, and it increases the $[OH^-]$. The pH increases.

(b) NH_4NO_3 is a source of NH_4^+ (reaction product). The equilibrium shifts towards reactants, and the $[OH^-]$ decreases. The pH decreases.

(c) NH_4Br is a source of NH_4^+ (reaction product). The equilibrium shifts towards reactants, and the $[OH^-]$ decreases. The pH decreases.

(d) KBr does not affect the pH of the solution.

16.33 (a) LiF is a source of F^- (reaction product). The equilibrium shifts toward reactants, and the $[H_3O^+]$ decreases. The pH increases.

(b) and (c) pH remains the same.

(d) NH_4Cl is a source of NH_4^+ (reaction product). The equilibrium shifts toward reactants, and the $[OH^-]$ decreases. The pH decreases.

16.34 For 0.25 M HF and 0.10 M NaF

	$HF(aq)$ + $H_2O(l)$	$\rightleftharpoons$ $H_3O^+(aq)$	+ $F^-(aq)$
initial (M)	0.25	~0	0.10
change (M)	–x	+x	+x
equil (M)	0.25 – x	x	0.10 + x

$$K_a = \frac{[H_3O^+][F^-]}{[HF]} = 3.5 \times 10^{-4} = \frac{x(0.10 + x)}{0.25 - x} \approx \frac{x(0.10)}{0.25}$$

Solve for x. $x = [H_3O^+] = 8.8 \times 10^{-4}$ M

$pH = -\log[H_3O^+] = -\log(8.8 \times 10^{-4}) = 3.06$

16.35 For 0.10 M HN_3:

	$HN_3(aq)$ + $H_2O(l)$	⇄	$H_3O^+(aq)$	+ $N_3^-(aq)$
initial (M)	0.10		~0	0
change (M)	–x		+x	+x
equil (M)	0.10 – x		x	x

$$K_a = \frac{[H_3O^+][N_3^-]}{[HN_3]} = 1.9 \times 10^{-5} = \frac{x^2}{0.10 - x} \approx \frac{x^2}{0.10}$$

Solve for x. x = 1.4×10^{-3} M

$$\% \text{ dissociation} = \frac{[HN_3]_{diss}}{[HN_3]_{initial}} \times 100\% = \frac{1.4 \times 10^{-3}\ M}{0.10\ M} \times 100\% = 1.4\%$$

For 0.10 M HN_3 in 0.10 M HCl:

	$HN_3(aq)$ + $H_2O(l)$	⇄	$H_3O^+(aq)$	+ $N_3^-(aq)$
initial (M)	0.10		0.10	0
change (M)	–x		+x	+x
equil (M)	0.10 – x		0.10 + x	x

$$K_a = \frac{[H_3O^+][N_3^-]}{[HN_3]} = 1.9 \times 10^{-5} = \frac{(0.10 + x)(x)}{0.10 - x} \approx \frac{(0.10)(x)}{0.10} = x$$

Solve for x. x = 1.9×10^{-5} M

$$\% \text{ dissociation} = \frac{[HN_3]_{diss}}{[HN_3]_{initial}} \times 100\% = \frac{1.9 \times 10^{-5}\ M}{0.10\ M} \times 100\% = 0.019\%$$

The % dissociation is less because of the common ion (H_3O^+) effect.

16.36 NH_4NO_3, 80.04 amu

$$[NH_4^+] = \text{molarity of } NH_4NO_3 = \frac{\left(4.0\ g \times \frac{1\ mol}{80.04\ g}\right)}{0.100\ L} = 0.50\ M$$

$$NH_3(aq) + H_2O(l) \rightleftharpoons NH_4^+(aq) + OH^-(aq)$$

	$NH_3(aq)$	$NH_4^+(aq)$	$OH^-(aq)$
initial (M)	0.30	0.50	~0
change (M)	–x	+x	+x
equil (M)	0.30 – x	0.50 + x	x

$$K_b = \frac{[NH_4^+][OH^-]}{[NH_3]} = 1.8 \times 10^{-5} = \frac{(0.50 + x)x}{0.30 - x} \approx \frac{(0.50)x}{0.30}$$

Solve for x. $x = [OH^-] = 1.1 \times 10^{-5}$ M

$$[H_3O^+] = \frac{K_w}{[OH^-]} = \frac{1.0 \times 10^{-14}}{1.1 \times 10^{-5}} = 9.1 \times 10^{-10} \text{ M}$$

$pH = -\log[H_3O^+] = -\log(9.1 \times 10^{-10}) = 9.04$

Buffer Solutions

16.37 A buffer solution is a solution of a weak acid and its conjugate base that resists drastic changes in pH.

A solution containing equal concentrations of CH_3COOH and CH_3COONa is a buffer solution.

16.38 (a) A solution containing equal concentrations of CH_3COOH and CH_3COONa is a buffer solution with pH < 7.

(b) A solution containing equal concentrations of NH_4Cl and NH_3 is a buffer solution with pH < 7.

16.39 Solutions (a), (c) and (d) are buffer solutions. Neutralization reactions give solutions with equal concentrations of HF and F^-.

16.40 Both solutions are buffers with the same pH because the $[NO_2^-]/[HNO_2] = 1$ in both cases. Solution (a), however, has a higher concentration of both HNO_2 and NO_2^-, and therefore it has the greater buffer capacity.

16.41 When blood absorbs acid, the equilibrium shifts to the left, decreasing the pH, but not by much because the $[HCO_3^-]/[H_2CO_3]$ ratio remains nearly constant. When blood absorbs base, the equilibrium shifts to the right, increasing the pH, but not by much because the $[HCO_3^-]/[H_2CO_3]$ ratio remains nearly constant.

16.42 $H_2PO_4^-(aq) + H_2O(l) \rightleftarrows H_3O^+(aq) + HPO_4^{2-}(aq)$

For $H_2PO_4^-$, $K_{a2} = 6.2 \times 10^{-8}$, $pK_{a2} = -\log K_{a2} = 7.21$

$$pH = 7.4 = pK_{a2} + \log\frac{[HPO_4^{2-}]}{[H_2PO_4^-]} = 7.21 + \log\frac{[HPO_4^{2-}]}{[H_2PO_4^-]}$$

To maintain pH near 7.4, need $\log\dfrac{[HPO_4^{2-}]}{[H_2PO_4^-]} = 0.19$ and $\dfrac{[HPO_4^{2-}]}{[H_2PO_4^-]} = 10^{0.19} = 1.5$

The principal buffer reactions are:

$H_3O^+(aq) + HPO_4^{2-}(aq) \rightarrow H_2PO_4^-(aq) + H_2O(l)$

$OH^-(aq) + H_2PO_4^-(aq) \rightarrow HPO_4^{2-}(aq) + H_2O(l)$

16.43 $pH = pK_a + \log\dfrac{[\text{base}]}{[\text{acid}]} = pK_a + \log\dfrac{[CN^-]}{[HCN]}$

For HCN, $K_a = 4.9 \times 10^{-10}$

$$pH = -\log(4.9 \times 10^{-10}) + \log\left(\frac{0.12}{0.20}\right) = 9.09$$

16.44 $NaHCO_3$, 84.01 amu; Na_2CO_3, 105.99 amu

$$[HCO_3^-] = \text{molarity of } NaHCO_3 = \frac{\left(4.2\text{ g} \times \dfrac{1\text{ mol}}{84.01\text{ g}}\right)}{0.20\text{ L}} = 0.25\text{ M}$$

$$[CO_3^{2-}] = \text{molarity of } Na_2CO_3 = \frac{\left(5.3\text{ g} \times \dfrac{1\text{ mol}}{105.99\text{ g}}\right)}{0.20\text{ L}} = 0.25\text{ M}$$

$$pH = pK_a + \log\frac{[CO_3^{2-}]}{[HCO_3^-]}$$

For HCO_3^-, $K_{a2} = 5.6 \times 10^{-11}$, $pK_{a2} = -\log K_{a2} = 10.25$

$$pH = 10.25 + \log\frac{[0.25]}{[0.25]} = 10.25$$

The pH of a buffer solution will not change on dilution because the acid and base concentrations will change by the same amount and their ratio will remain the same.

16.45 $pH = pK_a + \log\dfrac{[\text{base}]}{[\text{acid}]} = pK_a + \log\dfrac{[NH_3]}{[NH_4^+]}$

For NH_4^+, $K_a = 5.6 \times 10^{-10}$ and $pK_a = 9.25$

For the buffer: $pH = 9.25 + \log\dfrac{(0.200)}{(0.200)} = 9.25$

(a) add 0.005 mol NaOH, $[OH^-] = 0.005\text{ mol}/0.500\text{ L} = 0.01\text{ M}$

	$NH_4^+(aq)$ +	$OH^-(aq)$ ⇄	$NH_3(aq)$ + $H_2O(l)$
before reaction (M)	0.200	0.01	0.200
change (M)	–0.01	–0.01	+0.01
after reaction (M)	0.200 – 0.01	0	0.200 + 0.01

$$pH = 9.25 + \log\frac{[NH_3]}{[NH_4^+]} = 9.25 + \log\frac{(0.200 + 0.01)}{(0.200 - 0.01)} = 9.29$$

(b) add 0.020 mol HCl, $[H_3O^+]$ = 0.020 mol/0.500 L = 0.040 M

	$NH_3(aq)$ +	$H_3O^+(aq)$ ⇄	$NH_4^+(aq)$ + $H_2O(l)$
before reaction (M)	0.200	0.040	0.200
change (M)	–0.040	–0.040	+0.040
after reaction (M)	0.200 – 0.040	0	0.200 + 0.040

$$pH = 9.25 + \log \frac{[NH_3]}{[NH_4^+]} = 9.25 + \log \frac{(0.200 - 0.040)}{(0.200 + 0.040)} = 9.07$$

16.46

	Acid	K_a	$pK_a = -\log K_a$
(a)	H_3BO_3	5.8×10^{-10}	9.24
(b)	HCOOH	1.8×10^{-4}	3.74
(c)	HOCl	3.5×10^{-8}	7.46

The stronger the acid (the larger the K_a), the smaller is the pK_a.

16.47 (a) $K_a = 10^{-pK_a} = 10^{-5.00} = 1.0 \times 10^{-5}$

(b) $K_a = 10^{-pK_a} = 10^{-8.70} = 2.0 \times 10^{-9}$

(b) is the weaker acid

16.48 $pH = pK_a + \log \frac{[HCOO^-]}{[HCOOH]}$

For HCOOH, $K_a = 1.8 \times 10^{-4}$; $pK_a = -\log K_a = 3.74$

$$pH = 3.74 + \log \frac{(0.50)}{(0.25)} = 4.04$$

16.49 $pH = pK_a + \log\frac{[HCO_3^-]}{[H_2CO_3]}$

For H_2CO_3, $K_a = 4.3 \times 10^{-7}$; $pK_a = -\log K_a = 6.37$

$$7.40 = 6.37 + \log\frac{[HCO_3^-]}{[H_2CO_3]}$$

$$1.03 = \log\frac{[HCO_3^-]}{[H_2CO_3]}$$

$$\frac{[HCO_3^-]}{[H_2CO_3]} = 10^{1.03} = 10.7$$

$$\frac{[H_2CO_3]}{[HCO_3^-]} = 0.093$$

16.50 $pH = pK_a + \log\frac{[NH_3]}{[NH_4^+]}$

For NH_4^+, $K_a = 5.6 \times 10^{-10}$; $pK_a = -\log K_a = 9.25$

$$9.80 = 9.25 + \log\frac{[NH_3]}{[NH_4^+]}$$

$$\log\frac{[NH_3]}{[NH_4^+]} = 9.80 - 9.25 = 0.55$$

$$\frac{[NH_3]}{[NH_4^+]} = 10^{0.55} = 3.5$$

The volume of the 1.0 M NH_3 solution should be 3.5 times the volume of the 1.0 M NH_4Cl solution so that the mixture will buffer at pH 9.80.

16.51 $pH = pK_a + \log\dfrac{[CH_3COO^-]}{[CH_3COOH]}$

For CH_3COOH, $K_a = 1.8 \times 10^{-5}$; $pK_a = -\log K_a = 4.74$

$$4.44 = 4.74 + \log\frac{[CH_3COO^-]}{[CH_3COOH]}$$

$$-0.30 = \log\frac{[CH_3COO^-]}{[CH_3COOH]}$$

$$\frac{[CH_3COO^-]}{[CH_3COOH]} = 10^{-0.30} = 0.50$$

The solution should have 0.50 mol of CH_3COO^- per mole of CH_3COOH. For example, you could dissolve 41g of CH_3COONa in 1.00 L of 1.00 M CH_3COOH.

16.52 H_3PO_4, $K_{a1} = 7.5 \times 10^{-3}$; $pK_{a1} = -\log K_{a1} = 2.12$

$H_2PO_4^-$, $K_{a2} = 6.2 \times 10^{-8}$; $pK_{a2} = -\log K_{a2} = 7.21$

HPO_4^{2-}, $K_{a3} = 4.8 \times 10^{-13}$; $pK_{a3} = -\log K_{a3} = 12.32$

The buffer system of choice for pH 7.00 is (b) $H_2PO_4^-$ – HPO_4^{2-} because the pK_a for $H_2PO_4^-$ (7.21) is closest to 7.00.

pH Titration Curves

16.53 (a) (0.060 L)(0.150 mol/L)(1000 mmol/mol) = 9.00 mmol HNO_3

(b) $\text{vol NaOH} = (9.00 \text{ mmol } HNO_3)\left(\dfrac{1 \text{ mmol NaOH}}{1 \text{ mmol } HNO_3}\right)\left(\dfrac{1 \text{ mL NaOH}}{0.450 \text{ mmol NaOH}}\right)$

$= 20.0$ mL NaOH

(c) At the equivalence point the solution contains the neutral salt $NaNO_3$. The pH is 7.00.

(d)

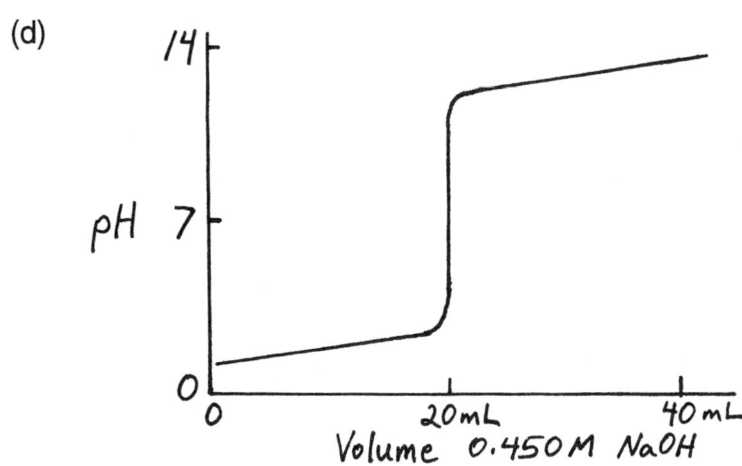

16.54 mmol OH^- = (20.0 mL)(0.150 mmol/mL) = 3.00 mmol

mmol acid present = mmol OH^- added = 3.00 mmol

$$\text{acid concentration} = \frac{3.00\text{ mmol}}{60.0\text{ mL}} = 0.0500\text{ M}$$

16.55 $HBr(aq) + NaOH(aq) \rightarrow Na^+(aq) + Br^-(aq) + H_2O(l)$

(a) $[H_3O^+]$ = 0.120 M; pH = –log (0.120) = 0.92

(b) (50.0 mL)(0.120 mmol/mL) = 6.00 mmol HBr

(20.0 mL)(0.240 mmol/mL) = 4.80 mmol NaOH

6.00 mmol HBr – 4.80 mmol NaOH = 1.20 mmol HBr after neutralization

$$[H_3O^+] = \frac{1.20\text{ mmol}}{(50.0\text{ mL} + 20.0\text{ mL})} = 0.0171\text{ M}$$

pH = –log(0.0171) = 1.77

(c) (24.9 mL)(0.240 mmol/mL) = 5.98 mmol NaOH

6.00 mmol HBr – 5.98 mmol NaOH = 0.02 mmol HBr after neutralization

$$[H_3O^+] = \frac{0.02\text{ mmol}}{(50.0\text{ mL} + 24.9\text{ mL})} = 3 \times 10^{-4}\text{ M}$$

$pH = -\log(3 \times 10^{-4}) = 3.5$

(d) The titration reaches the equivalence point when 25.0 mL of 0.240 M NaOH is added. At the equivalence point the solution contains the neutral salt NaBr. The pH is 7.00.

(e) (25.1 mL)(0.240 mmol/mL) = 6.024 mmol NaOH

6.024 mmol NaOH – 6.00 mmol HBr = 0.024 mmol NaOH after neutralization

$$[OH^-] = \frac{0.024\ \text{mmol}}{(50.0\ \text{mL} + 25.1\ \text{mL})} = 3.2 \times 10^{-4}\ \text{M}$$

$$[H_3O^+] = \frac{K_w}{[OH^-]} = \frac{1.0 \times 10^{-14}}{3.2 \times 10^{-4}} = 3.1 \times 10^{-11}\ \text{M}$$

$pH = -\log(3.1 \times 10^{-11}) = 10.5$

(f) (40.0 mL)(0.240 mmol/mL) = 9.60 mmol NaOH

9.60 mmol NaOH – 6.00 mmol HBr = 3.60 mmol NaOH after neutralization

$$[OH^-] = \frac{3.60\ \text{mmol}}{(50.0\ \text{mL} + 40.0\ \text{mL})} = 0.040\ \text{M}$$

$$[H_3O^+] = \frac{K_w}{[OH^-]} = \frac{1.0 \times 10^{-14}}{0.040} = 2.5 \times 10^{-13}\ \text{M}$$

$pH = -\log(2.5 \times 10^{-13}) = 12.60$

16.56 mmol HF = (40.0 mL)(0.250 mmol/mL) = 10.0 mmol

mmol NaOH required = mmol HF = 10.0 mmol

$$\text{mL NaOH required} = (10.0\ \text{mmol})\left(\frac{1.00\ \text{mL}}{0.200\ \text{mmol}}\right) = 50.0\ \text{mL}$$

50.0 mL of 0.200 M NaOH is required to reach the equivalence point.

For HF, $K_a = 3.5 \times 10^{-4}$; $pK_a = -\log K_a = 3.46$

(a) mmol HF = 10.0 mmol

mmol NaOH = (0.200 mmol/mL)(10.0 mL) = 2.00 mmol

Neutralization reaction:	$HF(aq)$	+ $OH^-(aq)$	→ $F^-(aq)$	+ $H_2O(l)$
before reaction (mmol)	10.0	2.00	0	
change (mmol)	–2.00	–2.00	+2.00	
after reaction (mmol)	8.0	0	2.00	

$$[HF] = \frac{8.0\ \text{mmol}}{(40.0\ \text{mL} + 10.0\ \text{mL})} = 0.16\ \text{M}$$

$$[F^-] = \frac{2.00\ \text{mmol}}{(40.0\ \text{mL} + 10.0\ \text{mL})} = 0.0400\ \text{M}$$

	$HF(aq)$	+ $H_2O(l)$	⇄ $H_3O^+(aq)$	+ $F^-(aq)$
initial (M)	0.16		~0	0.0400
change (M)	–x		+x	+x
equil (M)	0.16 – x		x	0.0400 + x

$$K_a = \frac{[H_3O^+][F^-]}{[HF]} = 3.5 \times 10^{-4} = \frac{x(0.0400 + x)}{0.16 - x} \approx \frac{x(0.0400)}{0.16}$$

Solve for x. $x = [H_3O^+] = 1.4 \times 10^{-3}$ M

$pH = -\log[H_3O^+] = -\log(1.4 \times 10^{-3}) = 2.85$

(b) Halfway to the equivalence point,

$pH = pK_a = -\log K_a = -\log(3.5 \times 10^{-4}) = 3.46$

(c) At the equivalence point only the salt NaF is in solution.

$$[F^-] = \frac{10.0\ \text{mmol}}{(40.0\ \text{mL} + 50.0\ \text{mL})} = 0.111\ \text{M}$$

$$F^-(aq) + H_2O(l) \rightleftarrows HF(aq) + OH^-(aq)$$

	$F^-(aq)$	$HF(aq)$	$OH^-(aq)$
initial (M)	0.111	0	~0
change (M)	–x	+x	+x
equil (M)	0.111 – x	x	x

For F^-, $K_b = \dfrac{K_w}{K_a \text{ for HF}} = \dfrac{1.0 \times 10^{-14}}{3.5 \times 10^{-4}} = 2.9 \times 10^{-11}$

$$K_b = \frac{[HF][OH^-]}{[F^-]} = 2.9 \times 10^{-11} = \frac{x^2}{0.111 - x} \approx \frac{x^2}{0.111}$$

Solve for x. $x = [OH^-] = 1.8 \times 10^{-6}$ M

$$[H_3O^+] = \frac{K_w}{[OH^-]} = \frac{1.0 \times 10^{-14}}{1.8 \times 10^{-6}} = 5.6 \times 10^{-9} \text{ M}$$

$pH = -\log[H_3O^+] = -\log(5.6 \times 10^{-9}) = 8.25$

(d) mmol HF = 10.0 mmol

mol NaOH = (0.200 mmol/mL)(80.0 mL) = 16.0 mmol

Neutralization reaction: $HF(aq) + OH^-(aq) \rightarrow F^-(aq) + H_2O(l)$

	$HF(aq)$	$OH^-(aq)$	$F^-(aq)$
before reaction (mmol)	10.0	16.0	0
change (mmol)	–10.0	–10.0	+10.0
after reaction (mmol)	0	6.0	10.0

After the equivalence point, the pH of the solution is determined by the $[OH^-]$.

$$[OH^-] = \frac{6.0 \text{ mmol}}{(40.0 \text{ mL} + 80.0 \text{ mL})} = 5.0 \times 10^{-2} \text{ M}$$

$$[H_3O^+] = \frac{K_w}{[OH^-]} = \frac{1.0 \times 10^{-14}}{5.0 \times 10^{-2}} = 2.0 \times 10^{-13} \text{ M}$$

$pH = -\log[H_3O^+] = -\log(2.0 \times 10^{-13}) = 12.70$

16.57

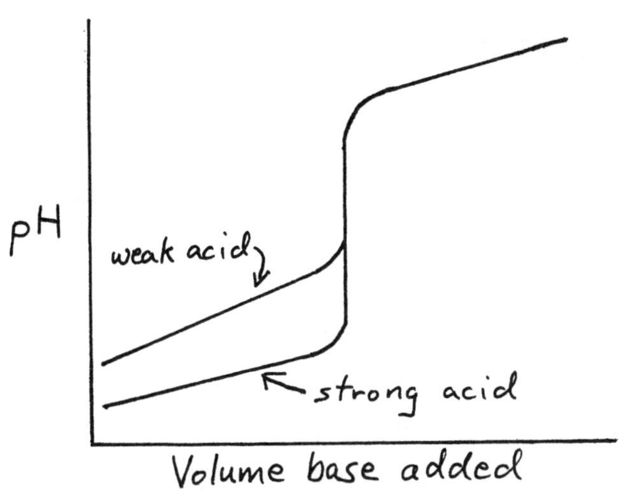

(a) The pH for the weak acid is higher.

(b) Initially, the pH rises more quickly for the weak acid, but then the curve becomes more level in the region halfway to the equivalence point.

(c) The pH is higher at the equivalence point for the weak acid.

(d) Both curves are identical beyond the equivalence point because the pH is determined by the $[OH^-]$.

(e) If the acid concentrations are the same, the volume of base needed to reach the equilavence point is the same.

16.58 mmol CH_3NH_2 = (100.0 mL)(0.100 mmol/mL) = 10.0 mmol

mmol HNO_3 required = mmol CH_3NH_2 = 10.0 mmol

$$\text{vol } HNO_3 \text{ required} = (10.0 \text{ mmol})\left(\frac{1.00 \text{ mL}}{0.250 \text{ mmol}}\right) = 40.0 \text{ mL}$$

40.0 mL of 0.250 M HNO_3 are required to reach the equivalence point.

(a) $CH_3NH_2(aq) + H_2O(l) \rightleftarrows CH_3NH_3^+(aq) + OH^-(aq)$

	$CH_3NH_2(aq)$	$CH_3NH_3^+(aq)$	$OH^-(aq)$
initial (M)	0.100	0	~0
change (M)	–x	+x	+x
equil (M)	0.100 – x	x	x

$$K_b = \frac{[CH_3NH_3^+][OH^-]}{[CH_3NH_2]} = 3.7 \times 10^{-4} = \frac{x^2}{0.100 - x}$$

$x^2 + (3.7 \times 10^{-4})x - (3.7 \times 10^{-5}) = 0$

Use the quadratic formula to solve for x.

$$x = \frac{(-3.7 \times 10^{-4}) \pm \sqrt{(3.7 \times 10^{-4})^2 + (4)(3.7 \times 10^{-5})}}{2(1)} = \frac{-3.7 \times 10^{-4} \pm 0.0122}{2}$$

x = 0.0059 and –0.0063

Of the two solutions for x, only the positive value of x has physical meaning because x is the $[OH^-]$.

$[OH^-] = x = 0.0059$ M

$$[H_3O^+] = \frac{K_w}{[OH^-]} = \frac{1.0 \times 10^{-14}}{5.9 \times 10^{-3}} = 1.7 \times 10^{-12}\ M$$

$pH = -\log[H_3O^+] = -\log(1.7 \times 10^{-12}) = 11.77$

(b) 20.0 mL of HNO_3 is halfway to the equivalence point.

$$\text{For } CH_3NH_3^+,\ K_a = \frac{K_w}{K_b \text{ for } CH_3NH_2} = \frac{1.0 \times 10^{-14}}{3.7 \times 10^{-4}} = 2.7 \times 10^{-11}$$

$pH = pK_a = -\log(2.7 \times 10^{-11}) = 10.57$

(c) At the equivalence point only the salt $CH_3NH_3NO_3$ is in solution.

mmol $CH_3NH_3NO_3$ = (0.100 mmol/mL)(100.0 mL) = 10.0 mmol

$$[CH_3NH_3^+] = \frac{10.0\ \text{mmol}}{(100.0\ \text{mL} + 40.0\ \text{mL})} = 0.0714\ M$$

$$CH_3NH_3^+(aq) + H_2O(l) \rightleftharpoons H_3O^+(aq) + CH_3NH_2(aq)$$

	$CH_3NH_3^+(aq)$	$H_3O^+(aq)$	$CH_3NH_2(aq)$
initial (M)	0.0714	~0	0
change (M)	–x	+x	+x
equil (M)	0.0714 – x	x	x

$$K_a = \frac{[H_3O^+][CH_3NH_2]}{[CH_3NH_3^+]} = 2.7 \times 10^{-11} = \frac{x^2}{0.0714 - x} \approx \frac{x^2}{0.0714}$$

Solve for x. $x = [H_3O^+] = 1.4 \times 10^{-6}$ M

$pH = -\log[H_3O^+] = -\log(1.4 \times 10^{-6}) = 5.85$

(d) mmol CH_3NH_2 = (0.100 mmol/mL)(100.0 mL) = 10.0 mmol

mmol HNO_3 = (0.250 mmol/mL)(60.0 mL) = 15.0 mmol

Neutralization reaction: $CH_3NH_2(aq) + H_3O^+(aq) \rightarrow CH_3NH_3^+(aq) + H_2O(l)$

	$CH_3NH_2(aq)$	$H_3O^+(aq)$	$CH_3NH_3^+(aq)$
before reaction (mmol)	10.0	15.0	0
change (mmol)	–10.0	–10.0	+10.0
after reaction (mmol)	0	5.0	10.0

After the equivalence point the pH of the solution is determined by the $[H_3O^+]$.

$$[H_3O^+] = \frac{5.0\text{ mmol}}{(100.0\text{ mL} + 60.0\text{ mL})} = 3.1 \times 10^{-2}\text{ M}$$

$pH = -\log[H_3O^+] = -\log(3.1 \times 10^{-2}) = 1.51$

16.59 (a) (10.0 mL)(0.100 mmol/mL) = 1.00 mmol NaOH added = 1.00 mmol HA produced.

(50.0 mL)(0.100 mmol/mL) = 5.00 mmol H_2A^+

5.00 mmol H_2A^+ – 1.00 mmol NaOH = 4.00 mmol H_2A^+ after neutralization

$$[H_2A^+] = \frac{4.00\text{ mmol}}{(50.0\text{ mL} + 10.0\text{ mL})} = 6.67 \times 10^{-2}\text{ M}$$

$$[HA] = \frac{1.00 \text{ mmol}}{(50.0 \text{ mL} + 10.0 \text{ mL})} = 1.67 \times 10^{-2} \text{ M}$$

$$pH = pK_{a1} + \log\frac{[HA]}{[H_2A^+]} = -\log(4.6 \times 10^{-3}) + \log\left(\frac{1.67 \times 10^{-2}}{6.67 \times 10^{-2}}\right) = 1.74$$

(b) Halfway to the first equivalence point, pH = pK_{a1} = 2.34

(c) At the first equivalence point, pH = $\frac{pK_{a1} + pK_{a2}}{2}$ = 6.02

(d) Halfway between the first and second equivalence points, pH = pK_{a2} = 9.70

16.60 When equal volumes of acid and base react, all concentrations are cut in half.

(a) At the equivalence point only the salt $NaNO_2$ is in solution.

$[NO_2^-]$ = 0.050 M

For NO_2^-, $K_b = \frac{K_w}{K_a \text{ for } HNO_2} = \frac{1.0 \times 10^{-14}}{4.5 \times 10^{-4}} = 2.2 \times 10^{-11}$

	$NO_2^-(aq)$ + $H_2O(l)$	⇄	$HNO_2(aq)$	+ $OH^-(aq)$
Initial (M)	0.050		0	~0
change (M)	–x		+x	+x
equil (M)	0.050 – x		x	x

$$K_b = \frac{[HNO_2][OH^-]}{[NO_2^-]} = 2.2 \times 10^{-11} = \frac{x^2}{0.050 - x} \approx \frac{x^2}{0.050}$$

Solve for x. x = $[OH^-]$ = 1.1×10^{-6} M

$$[H_3O^+] = \frac{K_w}{[OH^-]} = \frac{1.0 \times 10^{-14}}{1.1 \times 10^{-6}} = 9.1 \times 10^{-9} \text{ M}$$

pH = $-\log[H_3O^+]$ = $-\log(9.1 \times 10^{-9})$ = 8.04

Cresol red, phenol red, or m–nitrophenol would be suitable indicators. (See Text Figure 15.3)

(b) The pH is 7.00 at the equivalence point for the titration of a strong acid (HI) with a strong base (NaOH).

Alizarin, bromothymol blue, or phenol red would be suitable indicators. (Any indicator that changes color in the pH range 4 – 10 is satisfactory for a strong acid – strong base titration.)

(c) At the equivalence point only the salt CH_3NH_3Cl is in solution.

$[CH_3NH_3^+] = 0.050$ M

For $CH_3NH_3^+$, $K_a = \dfrac{K_w}{K_b \text{ for } CH_3NH_2} = \dfrac{1.0 \times 10^{-14}}{3.7 \times 10^{-4}} = 2.7 \times 10^{-11}$

$$CH_3NH_3^+(aq) + H_2O(l) \rightleftharpoons H_3O^+(aq) + CH_3NH_2(aq)$$

	$CH_3NH_3^+$	H_3O^+	CH_3NH_2
initial (M)	0.050	~0	0
change (M)	–x	+x	+x
equil (M)	0.050 – x	x	x

$$K_a = \frac{[H_3O^+][CH_3NH_2]}{[CH_3NH_3^+]} = 2.7 \times 10^{-11} = \frac{x^2}{0.050 - x} \approx \frac{x^2}{0.050}$$

Solve for x. $x = [H_3O^+] = 1.2 \times 10^{-6}$ M

$pH = -\log[H_3O^+] = -\log(1.2 \times 10^{-6}) = 5.92$

Eriochrome black T and bromocresol purple would be suitable indicators.

(d) At the equivalence point only the salt Na_2SO_3 is in the solution.

$[SO_3^{2-}] = 0.050$ M

For SO_3^{2-}, $K_b = \dfrac{K_w}{K_a \text{ for } HSO_3^-} = \dfrac{1.0 \times 10^{-14}}{6.3 \times 10^{-8}} = 1.6 \times 10^{-7}$

$$SO_3^{2-}(aq) + H_2O(l) \rightleftarrows HSO_3^-(aq) + OH^-(aq)$$

	SO_3^{2-}	HSO_3^-	OH^-
initial (M)	0.050	0	~0
change (M)	–x	+x	+x
equil (M)	0.050 – x	x	x

$$K_b = \frac{[HSO_3^-][OH^-]}{[SO_3^{2-}]} = 1.6 \times 10^{-7} = \frac{x^2}{0.050 - x} \approx \frac{x^2}{0.050}$$

Solve for x. $x = [OH^-] = 8.9 \times 10^{-5}$ M

The second dissociation for SO_3^{2-} contributes a negligible amount of additional OH^-.

$$[H_3O^+] = \frac{K_w}{[OH^-]} = \frac{1.0 \times 10^{-14}}{8.9 \times 10^{-5}} = 1.1 \times 10^{-10}\ M$$

$pH = -\log[H_3O^+] = -\log(1.1 \times 10^{-10}) = 9.96$

Thymolphthalein would be a suitable indicator.

Solubility Equilibria

16.61 Solubility refers to the amount of substance that dissolves in a given volume of solvent.

The solubility product is K_{sp}, the equilibrium constant for a dissolution reaction.

16.62 (a) $Ag_2CO_3(s) \rightleftarrows 2\,Ag^+(aq) + CO_3^{2-}(aq)$ $\quad K_{sp} = [Ag^+]^2[CO_3^{2-}]$

(b) $PbCrO_4(s) \rightleftarrows Pb^{2+}(aq) + CrO_4^{2-}(aq)$ $\quad K_{sp} = [Pb^{2+}][CrO_4^{2-}]$

(c) $Al(OH)_3(s) \rightleftarrows Al^{3+}(aq) + 3\,OH^-(aq)$ $\quad K_{sp} = [Al^{3+}][OH^-]^3$

(d) $Hg_2Cl_2(s) \rightleftarrows Hg_2^{2+}(aq) + 2\,Cl^-(aq)$ $\quad K_{sp} = [Hg_2^{2+}][Cl^-]^2$

16.63 (a) $K_{sp} = [Ca^{2+}][OH^-]^2$ (b) $K_{sp} = [Ag^+]^3[PO_4^{3-}]$

(c) $K_{sp} = [Ba^{2+}][CO_3^{2-}]$ (d) $K_{sp} = [Ca^{2+}]^5[PO_4^{3-}]^3[OH^-]$

16.64 (a) $K_{sp} = [Pb^{2+}][I^-]^2 = (5.0 \times 10^{-3})(1.3 \times 10^{-3})^2 = 8.4 \times 10^{-9}$

(b) $$[I^-] = \sqrt{\frac{K_{sp}}{[Pb^{2+}]}} = \sqrt{\frac{(8.4 \times 10^{-9}}{(2.5 \times 10^{-4})}} = 5.8 \times 10^{-3}\ M$$

(c) $$[Pb^{2+}] = \frac{K_{sp}}{[I^-]^2} = \frac{(8.4 \times 10^{-9})}{(2.5 \times 10^{-4})^2} = 0.13\ M$$

16.65

	$Ag_2CO_3(s)$	⇌	$2\ Ag^+(aq)$	+	$CO_3^{2-}(aq)$
equil (M)			2x		x

$[Ag^+] = 2x = 2.56 \times 10^{-4}$ M

$[CO_3^{2-}] = x = (2.56 \times 10^{-4}\ M)/2 = 1.28 \times 10^{-4}$ M

$K_{sp} = [Ag^+]^2[CO_3^{2-}] = (2.56 \times 10^{-4})^2(1.28 \times 10^{-4}) = 8.39 \times 10^{-12}$

16.66 (a) $[Cd^{2+}] = [CO_3^{2-}] = 2.5 \times 10^{-6}$ M

$K_{sp} = [Cd^{2+}][CO_3^{2-}] = (2.5 \times 10^{-6})^2 = 6.2 \times 10^{-12}$

(b) $[Ca^{2+}] = 1.06 \times 10^{-2}$ M

$[OH^-] = 2[Ca^{2+}] = 2(1.06 \times 10^{-2}\ M) = 2.12 \times 10^{-2}$ M

$K_{sp} = [Ca^{2+}][OH^-]^2 = (1.06 \times 10^{-2})(2.12 \times 10^{-2})^2 = 4.76 \times 10^{-6}$

(c) $PbBr_2$, 367.01 amu

$$[Pb^{2+}] = \text{molarity of } PbBr_2 = \frac{\left(4.34\ g \times \frac{1\ mol}{367.01\ g}\right)}{1\ L} = 1.18 \times 10^{-2}\ M$$

$[Br^-] = 2[Pb^{2+}] = 2(1.18 \times 10^{-2}\ M) = 2.36 \times 10^{-2}$ M

$K_{sp} = [Pb^{2+}][Br^-]^2 = (1.18 \times 10^{-2})(2.36 \times 10^{-2})^2 = 6.57 \times 10^{-6}$

(d) $BaCrO_4$, 253.32 amu

$$[Ba^{2+}] = [CrO_4^{2-}] = \text{molarity of } BaCrO_4 = \frac{\left(2.8 \times 10^{-3}\ g \times \dfrac{1\ mol}{253.32\ g}\right)}{1\ L} = 1.1 \times 10^{-5}\ M$$

$$K_{sp} = [Ba^{2+}][CrO_4^{2-}] = (1.1 \times 10^{-5})^2 = 1.2 \times 10^{-10}$$

16.67 (a)

	$CuCO_3(s)$	$\rightleftarrows$	$Cu^{2+}(aq)$	+	$CO_3^{2-}(aq)$
equil (M)			x		x

$$K_{sp} = [Cu^{2+}][CO_3^{2-}] = 2.5 \times 10^{-10} = (x)(x)$$

$$\text{molar solubility} = x = \sqrt{2.5 \times 10^{-10}} = 1.6 \times 10^{-5}\ M$$

(b)

	$Ag_2SO_4(s)$	$\rightleftarrows$	$2\ Ag^{+}(aq)$	+	$SO_4^{2-}(aq)$
equil (M)			2x		x

$$K_{sp} = [Ag^{+}]^2[SO_4^{2-}] = 1.2 \times 10^{-5} = (2x)^2x = 4x^3$$

$$\text{molar solubility} = x = \sqrt[3]{\frac{1.2 \times 10^{-5}}{4}} = 1.4 \times 10^{-2}\ M$$

(c)

	$Cr(OH)_3(s)$	$\rightleftarrows$	$Cr^{3+}(aq)$	+	$3\ OH^{-}(aq)$
equil (M)			x		3x

$$K_{sp} = [Cr^{3+}][OH^{-}]^3 = 6.7 \times 10^{-31} = (x)(3x)^3 = 27x^4$$

$$\text{molar solubility} = x = \sqrt[4]{\frac{6.7 \times 10^{-31}}{27}} = 1.3 \times 10^{-8}\ M$$

16.68 (a)

	$Ag_2CrO_4(s)$	$\rightleftarrows$	$2\ Ag^{+}(aq)$	+	$CrO_4^{2-}(aq)$
equil (M)			2x		x

$K_{sp} = [Ag^+]^2[CrO_4^{2-}] = 1.1 \times 10^{-12} = (2x)^2(x) = 4x^3$

molar solubility = $x = \sqrt[3]{\dfrac{1.1 \times 10^{-12}}{4}} = 6.5 \times 10^{-5}$ M

Ag_2CrO_4, 331.73 amu

solubility = $(6.5 \times 10^{-5}$ mol/L)(331.73 g/mol) = 0.022 g/L

(b) $CuBr(s) \rightleftarrows Cu^+(aq) + Br^-(aq)$

		Cu^+	Br^-
equil (M)		x	x

$K_{sp} = [Cu^+][Br^-] = 6.3 \times 10^{-9} = (x)(x)$

molar solubility = $x = \sqrt{6.3 \times 10^{-9}} = 7.9 \times 10^{-5}$ M

CuBr, 143.45 amu

solubility = $(7.9 \times 10^{-5}$ mol/L)(143.45 g/mol) = 0.011 g/L

(c) $Cu_3(PO_4)_2(s) \rightleftarrows 3\ Cu^{2+}(aq) + 2\ PO_4^{3-}(aq)$

		Cu^{2+}	PO_4^{3-}
equil (M)		3x	2x

$K_{sp} = [Cu^{2+}]^3[PO_4^{3-}]^2 = 1.4 \times 10^{-37} = (3x)^3(2x)^2 = 108x^5$

$x^5 = \dfrac{1.4 \times 10^{-37}}{108} = 1.3 \times 10^{-39}$

$\log(x^5) = 5 \log x = \log(1.3 \times 10^{-39}) = -38.89$

$\log x = \dfrac{-38.89}{5} = -7.78$

molar solubility = $x = 10^{-7.78} = 1.7 \times 10^{-8}$ M

$Cu_3(PO_4)_2$, 380.58 amu

solubility = $(1.7 \times 10^{-8}$ mol/L)(380.58 g/mol) = 6.5×10^{-6} g/L

Factors That Affect Solubility

16.69 $Ag_2CO_3(s) \rightleftarrows 2\ Ag^+(aq) + CO_3^{2-}(aq)$

(a) $AgNO_3$, source of Ag^+; equilibrium shifts left

(b) HNO_3, source of H_3O^+, removes CO_3^{2-}; equilibrium shifts right

(c) Na_2CO_3, source of CO_3^{2-}; equilibrium shifts left

(d) NH_3, forms $Ag(NH_3)_2^+$; removes Ag^+; equilibrium shifts right

16.70 (a)

	AgBr(s) ⇄	$Ag^+(aq)$ +	$Br^-(aq)$
equil (M)		x	x

$K_{sp} = [Ag^+][Br^-] = 5.4 \times 10^{-13} = (x)(x)$

molar solubility = $x = \sqrt{5.4 \times 10^{-13}} = 7.3 \times 10^{-7}$ M

(b) $[Br^-] = 0.050$ M

	AgBr(s) ⇄	$Ag^+(aq)$ +	$Br^-(aq)$
initial (M)		0	0.050
equil (M)		x	0.050 + x

$K_{sp} = [Ag^+][Br^-] = 5.4 \times 10^{-13} = x(0.050 + x) \approx x(0.050)$

molar solubility = $x = \dfrac{5.4 \times 10^{-13}}{0.050} = 1.1 \times 10^{-11}$ M

16.71 (a)

	$PbI_2(s)$ ⇄	$Pb^{2+}(aq)$ +	$2\ I^-(aq)$
initial (M)		0.20	0
equil (M)		0.20 + x	2x

$K_{sp} = [Pb^{2+}][I^-]^2 = 8.5 \times 10^{-9} = (0.20 + x)(2x)^2 \approx (0.20)(2x)^2 = 0.80\ x^2$

$$\text{molar solubility} = x = \sqrt{\frac{8.5 \times 10^{-9}}{0.80}} = 1.0 \times 10^{-4}\ \text{M}$$

(b) $PbI_2(s) \rightleftarrows Pb^{2+}(aq) + 2\ I^-(aq)$

	$Pb^{2+}(aq)$	$2\ I^-(aq)$
initial (M)	0	0.20
equil (M)	x	0.20 + 2x

$K_{sp} = [Pb^{2+}][I^-]^2 = 8.5 \times 10^{-9} = (x)(0.20 + 2x)^2 \approx (x)(0.20)^2 = x(0.040)$

$$\text{molar solubility} = x = \frac{8.5 \times 10^{-9}}{0.040} = 2.1 \times 10^{-7}\ \text{M}$$

16.72 (a) pH = 12.00

$[H_3O^+] = 10^{-pH} = 10^{-12.00} = 1.0 \times 10^{-12}\ \text{M}$

$$[OH^-] = \frac{K_w}{[H_3O^+]} = \frac{1.0 \times 10^{-14}}{1.0 \times 10^{-12}} = 0.010\ \text{M}$$

$Mg(OH)_2(s) \rightleftarrows Mg^{2+}(aq) + 2\ OH^-(aq)$

	$Mg^{2+}(aq)$	$2\ OH^-(aq)$
equil (M)	x	0.010 (fixed by buffer)

$K_{sp} = [Mg^{2+}][OH^-]^2 = 5.6 \times 10^{-12} = x(0.010)^2$

$$\text{molar solubility} = x = \frac{5.6 \times 10^{-12}}{(0.010)^2} = 5.6 \times 10^{-8}\ \text{M}$$

(b) pH = 9.00

$[H_3O^+] = 10^{-pH} = 10^{-9.00} = 1.0 \times 10^{-9}\ \text{M}$

$$[OH^-] = \frac{K_w}{[H_3O^+]} = \frac{1.0 \times 10^{-14}}{1.0 \times 10^{-9}} = 1.0 \times 10^{-5}\ \text{M}$$

	$Mg(OH)_2(s)$ ⇄	$Mg^{2+}(aq)$ +	$2\ OH^-(aq)$
equil (M)		x	1.0×10^{-5} (fixed by buffer)

$K_{sp} = [Mg^{2+}][OH^-]^2 = 5.6 \times 10^{-12} = x(1.0 \times 10^{-5})^2$

$$\text{molar solubility} = x = \frac{5.6 \times 10^{-12}}{(1.0 \times 10^{-5})^2} = 0.056\ M$$

16.73 (b) $CaCO_3$, (c) $Ni(OH)_2$ and (d) $Ca_3(PO_4)_2$ are more soluble in acidic solution because the acid reacts with the aqueous anion from each of these compounds removing it from solution.

16.74 (a) $MnS(s) + H_3O^+(aq) \rightleftarrows Mn^{2+}(aq) + HS^-(aq) + H_2O(l)$

(b) $Fe(OH)_3(s) + 3\ H_3O^+(aq) \rightleftarrows Fe^{3+}(aq) + 6\ H_2O(l)$

(c) $AgCl(s) \rightleftarrows Ag^+(aq) + Cl^-(aq)$

(d) $BaCO_3(s) + H_3O^+(aq) \rightleftarrows Ba^{2+}(aq) + HCO_3^-(aq) + H_2O(l)$

16.75 On mixing equal volumes of two solutions, the concentrations of both solutions are cut in half.

	$Ag^+(aq)$ +	$2\ CN^-(aq)$ ⇄	$Ag(CN)_2^-(aq)$
before reaction (M)	0.0010	0.10	0
assume 100% reaction	–0.0010	–2(0.0010)	0.0010
after reaction (M)	0	0.098	0.0010
assume small back rxn	+x	+2x	–x
equil (M)	x	0.098 + 2x	0.0010 – x

$$K_f = 1 \times 10^{21} = \frac{[Ag(CN)_2^-]}{[Ag^+][CN^-]^2} = \frac{(0.0010 - x)}{x(0.098 + 2x)^2} \approx \frac{0.0010}{x(0.098)^2}$$

Solve for x. $x = [Ag^+] = 1 \times 10^{-22}$ M

16.76 (a)

$$AgI(s) \rightleftarrows Ag^+(aq) + I^-(aq) \qquad K_{sp} = 8.5 \times 10^{-17}$$

$$Ag^+(aq) + 2\ CN^-(aq) \rightarrow Ag(CN)_2^-(aq) \qquad K_f = 1 \times 10^{21}$$

dissolution reaction $AgI(s) + 2\ CN^-(aq) \rightleftarrows Ag(CN)_2^-(aq) + I^-(aq)$

$K = (K_{sp})(K_f) = (8.5 \times 10^{-17})(1 \times 10^{21}) = 8 \times 10^4$

(b)

$$Al(OH)_3(s) \rightleftarrows Al^{3+}(aq) + 3\ OH^-(aq) \qquad K_{sp} = 1.9 \times 10^{-33}$$

$$Al^{3+}(aq) + 4\ OH^-(aq) \rightarrow Al(OH)_4^-(aq) \qquad K_f = 2.1 \times 10^{34}$$

dissolution reaction $Al(OH)_3(s) + OH^-(aq) \rightleftarrows Al(OH)_4^-(aq)$

$K = (K_{sp})(K_f) = (1.9 \times 10^{-33})(2.1 \times 10^{34}) = 40$

(c)

$$Zn(OH)_2(s) \rightleftarrows Zn^{2+}(aq) + 2\ OH^-(aq) \qquad K_{sp} = 4.1 \times 10^{-17}$$

$$Zn^{2+}(aq) + 4\ NH_3(aq) \rightarrow Zn(NH_3)_4^{2+}(aq) \qquad K_f = 2.9 \times 10^9$$

dissolution reaction $Zn(OH)_2(s) + 4\ NH_3(aq) \rightleftarrows Zn(NH_3)_4^{2+} + 2\ OH^-(aq)$

$K = (K_{sp})(K_f) = (4.1 \times 10^{-17})(2.9 \times 10^9) = 1.2 \times 10^{-7}$

16.77 (a)

$$AgI(s) \rightleftarrows Ag^+(aq) + I^-(aq)$$

equil (M) x x

$K_{sp} = [Ag^+][I^-] = 8.5 \times 10^{-17} = (x)(x)$

molar solubility $= x = \sqrt{8.5 \times 10^{-17}} = 9.2 \times 10^{-9}$ M

(b) $AgI(s) + 2\,CN^-(aq) \rightleftarrows Ag(CN)_2^-(aq) + I^-(aq)$

initial (M)	0.10	0	0
change (M)	–2x	+x	+x
equil (M)	0.10 – 2x	x	x

$K = (K_{sp})(K_f) = (8.5 \times 10^{-17})(1 \times 10^{21}) = 8.5 \times 10^4$

$$K = 8.5 \times 10^4 = \frac{[Ag(CN)_2^-][I^-]}{[CN^-]^2} = \frac{x^2}{(0.10 - 2x)^2}$$

Take the square root of both sides and solve for x.

molar solubility = x = 0.050 M

Precipitation; Qualitative Analysis

16.78 For $BaSO_4$, $K_{sp} = 1.1 \times 10^{-10}$

Total volume = 300 mL + 100 mL = 400 mL

$$[Ba^{2+}] = \frac{(4.0 \times 10^{-3}\ M)(100\ mL)}{(400\ mL)} = 1.0 \times 10^{-3}\ M$$

$$[SO_4^{2-}] = \frac{(6.0 \times 10^{-4}\ M)(300\ mL)}{(400\ mL)} = 4.5 \times 10^{-4}\ M$$

$IP = [Ba^{2+}]_t[SO_4^{2-}]_t = (1.0 \times 10^{-3})(4.5 \times 10^{-4}) = 4.5 \times 10^{-7}$

$IP > K_{sp}$; $BaSO_4(s)$ will precipitate.

16.79 On mixing equal volumes of two solutions, the concentrations of both solutions are cut in half.

For $PbCl_2$, $K_{sp} = 1.2 \times 10^{-5} = [Pb^{2+}][Cl^-]^2$

$IP = (0.0050)(0.0050)^2 = 1.2 \times 10^{-7}$

$IP < K_{sp}$; no precipitate will form.

$$[Cl^-] = \sqrt{\frac{K_{sp}}{[Pb^{2+}]}} = \sqrt{\frac{1.2 \times 10^{-5}}{5.0 \times 10^{-3}}} = 0.049 \text{ M}$$

A $[Cl^-]$ just greater than 0.049 M will result in precipitation.

16.80 For $BaSO_4$, $K_{sp} = 1.1 \times 10^{-10}$

For $Fe(OH)_3$, $K_{sp} = 2.6 \times 10^{-39}$

Total volume = 80 mL + 20 mL = 100 mL

$$[Ba^{2+}] = \frac{(1.0 \times 10^{-5} \text{ M})(80 \text{ mL})}{(100 \text{ mL})} = 8.0 \times 10^{-6} \text{ M}$$

$$[OH^-] = 2[Ba^{2+}] = 2(8.0 \times 10^{-6}) = 1.6 \times 10^{-5} \text{ M}$$

$$[Fe^{3+}] = \frac{2(1.0 \times 10^{-5} \text{ M})(20 \text{ mL})}{(100 \text{ mL})} = 4.0 \times 10^{-6} \text{ M}$$

$$[SO_4^{2-}] = \frac{3(1.0 \times 10^{-5} \text{ M})(20 \text{ mL})}{(100 \text{ mL})} = 6.0 \times 10^{-6} \text{ M}$$

For $BaSO_4$, IP = $[Ba^{2+}]_t[SO_4^{2-}]_t = (8.0 \times 10^{-6})(6.0 \times 10^{-6}) = 4.8 \times 10^{-11}$

IP < K_{sp}; $BaSO_4$ will not precipitate.

For $Fe(OH)_3$, IP = $[Fe^{3+}]_t[OH^-]_t^3 = (4.0 \times 10^{-6})(1.6 \times 10^{-5})^3 = 1.6 \times 10^{-20}$

IP > K_{sp}; $Fe(OH)_3(s)$ will precipitate.

16.81 (a) $$[CO_3^{2-}] = \frac{(2.0 \times 10^{-3} \text{ M})(0.10 \text{ mL})}{(250 \text{ mL})} = 8.0 \times 10^{-7} \text{ M}$$

$K_{sp} = 5.0 \times 10^{-9} = [Ca^{2+}][CO_3^{2-}]$

IP = $[Ca^{2+}][CO_3^{2-}] = (8.0 \times 10^{-4})(8.0 \times 10^{-7}) = 6.4 \times 10^{-10}$

IP < K_{sp}; no precipitate will form.

(b) Na_2CO_3, 106 amu; 10 mg = 0.010 g

$$[CO_3^{2-}] = \frac{\left(0.010\text{ g} \times \frac{1\text{ mol}}{106\text{ g}}\right)}{0.250\text{ L}} = 3.8 \times 10^{-4}\text{ M}$$

$IP = [Ca^{2+}][CO_3^{2-}] = (8.0 \times 10^{-4})(3.8 \times 10^{-4}) = 3.0 \times 10^{-7}$

$IP > K_{sp}$; $CaCO_3(s)$ will precipitate.

16.82 pH = 10.80

$[H_3O^+] = 10^{-pH} = 10^{-10.80} = 1.6 \times 10^{-11}$ M

$$[OH^-] = \frac{K_w}{[H_3O^+]} = \frac{1.0 \times 10^{-14}}{1.6 \times 10^{-11}} = 6.2 \times 10^{-4}\text{ M}$$

For $Mg(OH)_2$, $K_{sp} = 5.6 \times 10^{-12}$

$IP = [Mg^{2+}]_t[OH^-]_t^2 = (2.5 \times 10^{-4})(6.2 \times 10^{-4})^2 = 9.6 \times 10^{-11}$

$IP > K_{sp}$; $Mg(OH)_2(s)$ will precipitate

16.83 $$K_{spa} = \frac{[M^{2+}][H_2S]}{[H_3O^+]^2}$$

For FeS, $K_{spa} = 6 \times 10^2$; for SnS, $K_{spa} = 1 \times 10^{-5}$

Fe^{2+} and Sn^{2+} can be separated by bubbling H_2S through an acidic solution containing the two cations because their K_{spa} values are so different.

For FeS and SnS, $Q_c = \frac{(0.01)(0.10)}{(0.3)^2} = 1.1 \times 10^{-2}$

For FeS, $Q_c < K_{spa}$, and no FeS will precipitate.

For SnS, $Q_c > K_{spa}$, and SnS will precipitate.

16.84 For CoS, $K_{spa} = \dfrac{[Co^{2+}][H_2S]}{[H_3O^+]^2} = 3$

(i) In 0.5 M HCl, $[H_3O^+] = 0.5$ M

$$Q_c = \frac{[Co^{2+}]_t[H_2S]_t}{[H_3O^+]_t^2} = \frac{(0.10)(0.10)}{(0.5)^2} = 0.04$$

$Q_c < K_{spa}$; CoS will not precipitate

(ii) At pH = 8, $[H_3O^+] = 10^{-pH} = 10^{-8}$ M

$$Q_c = \frac{[Co^{2+}]_t[H_2S]_t}{[H_3O^+]_t^2} = \frac{(0.10)(0.10)}{(10^{-8})^2} = 1 \times 10^{14}$$

$Q_c > K_{spa}$; CoS(s) will precipitate

16.85 $Mg(OH)_2$, $K_{sp} = 5.6 \times 10^{-12}$; $Al(OH)_3$, $K_{sp} = 1.9 \times 10^{-33}$

At pH = 8, $[H_3O^+] = 10^{-pH} = 10^{-8} = 1 \times 10^{-8}$ M

$$[OH^-] = \frac{K_w}{[H_3O^+]} = \frac{1.0 \times 10^{-14}}{1 \times 10^{-8}} = 1 \times 10^{-6}\ M$$

For $Mg(OH)_2$, IP = $[Mg^{2+}][OH^-]^2 = (0.01)(1 \times 10^{-6})^2 = 1 \times 10^{-14}$

IP < K_{sp}; no Mg(OH) will precipitate.

For $Al(OH)_3$, IP = $[Al^{3+}][OH^-]^3 = (0.01)(1 \times 10^{-6})^3 = 1 \times 10^{-20}$

IP > K_{sp}; $Al(OH)_3$ will precipitate.

16.86 (a) add Cl^- to precipitate AgCl

(b) add CO_3^{2-} to precipitate $CaCO_3$

(c) add H_2S to precipitate MnS

(d) add NH_3 and NH_4Cl to precipitate $Cr(OH)_3$
(Need buffer to control $[OH^-]$; excess OH^- produces the soluble $Cr(OH)_4^-$.)

16.87 (a) add Cl^- to precipitate Hg_2Cl_2

(b) add $(NH_4)_2HPO_4$ to precipitate $MgNH_4PO_4$

(c) add HCl and H_2S to precipitate HgS

(d) add Cl^- to precipitate $PbCl_2$

16.88 Prepare aqueous solutions of the three salts. Add a solution of $(NH_4)_2HPO_4$. If a white precipitate forms, the solution contains Mg^{2+}. Perform flame test on the other two solutions. A yellow flame test indicates Na^+. A violet flame test indicates K^+.

General Problems

16.89 (a), solution contains HCN and CN^-

(c), solution can contain HCN and CN^-

(e), solution can contain HCN and CN^-

16.90 $AgCl(s) \rightleftarrows Ag^+(aq) + Cl^-(aq)$

AgCl is less soluble in an aqueous HCl solution because HCl is a source of Cl^-, the common ion.

The common–ion effect changes the solubility of a substance but does not change the solubility product (K_{sp}). K_{sp} is an equilibrium constant.

16.91 (a) $H_3O^+(aq) + NH_3(aq) \rightleftarrows NH_4^+(aq) + H_2O(l)$

(b) $H_3O^+(aq) + CH_3NH_2(aq) \rightleftarrows CH_3NH_3^+(aq) + H_2O(l)$

(c) $H_3O^+(aq) + OH^-(aq) \rightleftarrows 2\ H_2O(l)$

Reaction (c) proceeds farthest to the right because it is the reaction of a strong acid with a strong base.

16.92 $HOCl(aq) + H_2O(l) \rightleftharpoons H_3O^+(aq) + OCl^-(aq)$

(a) The addition of KCl (a neutral salt) does not affect the equilibrium. The pH does not change.

(b) HCl is a strong acid and a source of H_3O^+. The pH decreases.

(c) NaOCl is a source of OCl^- (reaction product). The equilibrium shifts toward reactants, and the $[H_3O^+]$ decreases. The pH increases.

(d) NaOH is a strong base that reacts with H_3O^+, and so the $[H_3O^+]$ decreases. The pH increases.

16.93 (a), solution contains H_2CO_3 and HCO_3^-

(b), solution contains HCO_3^- and CO_3^{2-}

(d), solution contains HCO_3^- and CO_3^{2-}

16.94

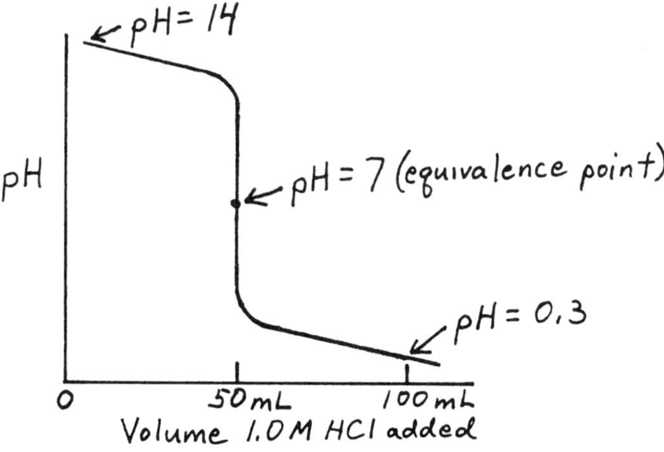

mmol NaOH = (50 mL)(1.0 mmol/mL) = 50 mmol

mmol HCl = mmol NaOH = 50 mmol

$$\text{vol HCl} = (50\ \text{mmol})\left(\frac{1.0\ \text{mL}}{1.0\ \text{mmol}}\right) = 50\ \text{mL}$$

50 mL of 1.0 M HCl is needed to reach the equivalence point.

16.95 A solution that is 0.30 M HF and 0.30 M NaF is a buffer solution. The pH of a buffer solution is not changed on dilution because $[H_3O^+] = K_a\frac{[HA]}{[A^-]}$, and the concentration ratio $[HA]/[A^-]$ is independent of volume.

16.96 For NH_4^+, $K_a = \frac{K_w}{K_b \text{ for } NH_3} = \frac{1.0 \times 10^{-14}}{1.8 \times 10^{-5}} = 5.6 \times 10^{-10}$

$pK_a = -\log K_a = -\log(5.6 \times 10^{-10}) = 9.25$

$$pH = pK_a + \log\frac{[NH_3]}{[NH_4^+]}$$

$$9.40 = 9.25 + \log\frac{[NH_3]}{[NH_4^+]}$$

$$\log\frac{[NH_3]}{[NH_4^+]} = 9.40 - 9.25 = 0.15$$

$$\frac{[NH_3]}{[NH_4^+]} = 10^{0.15} = 1.41$$

Because the volume is the same for both NH_3 and NH_4^+, $\frac{\text{mol } NH_3}{\text{mol } NH_4^+} = 1.41$.

mol NH_3 = (0.20 mol/L)(0.250 L) = 0.050 mol NH_3

$$\text{mol } NH_4^+ = \frac{\text{mol } NH_3}{1.41} = \frac{0.050}{1.41} = 0.035 \text{ mol } NH_4^+$$

$$\text{vol } NH_4^+ = (0.035 \text{ mol})\left(\frac{1 \text{ L}}{3.0 \text{ mol}}\right) = 0.012 \text{ L} = 12 \text{ mL}$$

12 mL of 3.0 M NH_4Cl must be added to 250 mL of 0.20 M NH_3 to obtain a buffer solution having pH 9.40.

16.97 pH = 2.00, $[H_3O^+] = 10^{-pH} = 10^{-2.00} = 1.0 \times 10^{-2}$ M

mol H_3O^+ = $(1.0 \times 10^{-2}$ mol/L)(0.300 L) = 3.0×10^{-3} mol

mol OH^- required = mol H_3O^+ = 3.0×10^{-3} mol

vol OH^- required = $(3.0 \times 10^{-3}\text{ mol})\left(\frac{1\text{ L}}{0.100\text{ mol}}\right) = 3.0 \times 10^{-2}\text{ L} = 30\text{ mL}$

16.98 $$Ca(OH)_2(s) \rightleftarrows Ca^{2+}(aq) + 2\,OH^-(aq)$$

equil (M) x 2x

$K_{sp} = [Ca^{2+}][OH^-]^2 = 4.7 \times 10^{-6} = x(2x)^2 = 4x^3$

$$x = \sqrt[3]{\frac{4.7 \times 10^{-6}}{4}} = 1.06 \times 10^{-2}$$

$[OH^-] = 2x = 2(1.06 \times 10^{-2}) = 0.021$ M

$$[H_3O^+] = \frac{K_w}{[OH^-]} = \frac{1.0 \times 10^{-14}}{0.021} = 4.8 \times 10^{-13}\text{ M}$$

$pH = -\log[H_3O^+] = -\log(4.8 \times 10^{-13}) = 12.32$

16.99 On mixing equal volumes of two solutions, the concentrations of both solutions are cut in half.

$$pH = pK_a + \log\frac{[\text{base}]}{[\text{acid}]} = pK_a + \log\frac{[NH_3]}{[NH_4^+]}$$

For NH_4^+, $K_a = \frac{K_w}{K_b \text{ for } NH_3} = \frac{1.0 \times 10^{-14}}{1.8 \times 10^{-5}} = 5.6 \times 10^{-10}$, $pK_a = -\log K_a = 9.25$

$$pH = 9.25 + \log\left(\frac{0.20}{0.30}\right) = 9.07$$

16.100 pH = 10.35

$[H_3O^+] = 10^{-pH} = 10^{-10.35} = 4.5 \times 10^{-11}$ M

$$[OH^-] = \frac{K_w}{[H_3O^+]} = \frac{1.0 \times 10^{-14}}{4.5 \times 10^{-11}} = 2.2 \times 10^{-4}\ M$$

$$[Mg^{2+}] = \frac{[OH^-]}{2} = \frac{2.2 \times 10^{-4}}{2} = 1.1 \times 10^{-4}\ M$$

$$K_{sp} = [Mg^{2+}][OH^-]^2 = (1.1 \times 10^{-4})(2.2 \times 10^{-4})^2 = 5.3 \times 10^{-12}$$

16.101 $HClO_4(aq) + KOH(aq) \rightarrow K^+(aq) + ClO_4^-(aq) + H_2O(l)$

(0.010 L)(0.44 mol/L) = 0.0044 mol KOH

(0.030 L)(0.20 mol/L) = 0.0060 mol $HClO_4$

(0.0060 – 0.0044) = 0.0016 mol $HClO_4$ after neutralization

$HClO_4$ is a strong acid.

$$[H_3O^+] = \frac{0.0016\ \text{mol}}{0.040\ \text{L}} = 0.040\ M$$

pH = –log(0.040) = 1.40

16.102 mmol Hg_2^{2+} = (0.010 mmol/mL)(1.0 mL) = 0.010 mmol

mmol Cl^- = (6 mmol/mL)(0.05 mL) = 0.3 mmol

Assume complete reaction.

	$Hg_2^{2+}(aq)$	+	$2\ Cl^-(aq)$	$\rightarrow$	$Hg_2Cl_2(s)$
before reaction (mmol)	0.010		0.3		
change (mmol)	–0.010		–2(0.010)		
after reaction (mmol)	0		0.28		

$$[Cl^-] = \frac{0.28\ \text{mmol}}{1.05\ \text{mL}} = 0.27\ M$$

Allow Hg_2Cl_2 to establish a new equilibrium.

$$Hg_2Cl_2(s) \rightleftarrows Hg_2^{2+}(aq) + 2\ Cl^-(aq)$$

	Hg_2^{2+}	Cl^-
initial (M)	0	0.27
equil (M)	x	0.27 + 2x

$K_{sp} = [Hg_2^{2+}][Cl^-]^2 = 1.4 \times 10^{-18} = x(0.27 + 2x)^2 \approx x(0.27)^2$

$$x = [Hg_2^{2+}] = \frac{1.4 \times 10^{-18}}{(0.27)^2} = 2 \times 10^{-17}\ \text{mol/L}$$

Hg_2^{2+}, 401.18 amu

Hg_2^{2+} concentration = $(2 \times 10^{-17}\ \text{mol/L})(401.18\ \text{g/mol}) = 8 \times 10^{-15}\ \text{g/L}$

16.103 $pH = pK_a + \log\frac{[\text{base}]}{[\text{acid}]} = pK_a + \log\frac{[NH_3]}{[NH_4^+]}$

For NH_4^+, $K_a = \dfrac{K_w}{K_b \text{ for } NH_3} = \dfrac{1.0 \times 10^{-14}}{1.8 \times 10^{-5}} = 5.6 \times 10^{-10}$; $pK_a = -\log K_a = 9.25$

Because $[NH_4^+] = [NH_3]$ for the buffer, $pH = pK_a = 9.25$

pH = 9.25, $[H_3O^+] = 10^{-pH} = 10^{-9.25} = 5.6 \times 10^{-10}$ M

$[OH^-] = \dfrac{K_w}{[H_3O^+]} = \dfrac{1.0 \times 10^{-14}}{5.6 \times 10^{-10}} = 1.8 \times 10^{-5}$ M (fixed by buffer)

$$Fe(OH)_3(s) \rightleftarrows Fe^{3+}(aq) + 3\ OH^-(aq)$$

	Fe^{3+}	OH^-
equil (M)	x	1.8×10^{-5} (fixed by buffer)

$K_{sp} = [Fe^{3+}][OH^-]^3 = 2.6 \times 10^{-39} = (x)(1.8 \times 10^{-5})^3$

Solve for x.

molar solubility = x = 4.5×10^{-25} M

16.104 mol HCl = (0.600 mol/L)(0.100 L) = 0.0600 mol

mol NaF = (0.200 mol/L)(0.500 L) = 0.100 mol

Neutralization reaction: $H_3O^+(aq) + F^-(aq) \rightarrow HF(aq) + H_2O(l)$

	H_3O^+	F^-	HF
before reaction (mol)	0.0600	0.100	0
change (mol)	–0.0600	–0.0600	+0.0600
after reaction (mol)	0	0.040	0.0600

After the neutralization reaction, the solution contains HF and F^- and behaves as a buffer.

For HF, $K_a = 3.5 \times 10^{-4}$; $pK_a = -\log K_a = 3.46$

Total volume = 0.100 L + 0.500 L = 0.600 L

$$[HF] = \frac{0.0600 \text{ mol}}{0.600 \text{ L}} = 0.100 \text{ M}; \quad [F^-] = \frac{0.040 \text{ mol}}{0.600 \text{ L}} = 0.067 \text{ M}$$

$$pH = pK_a + \log\frac{[F^-]}{[HF]} = 3.46 + \log\frac{(0.067)}{(0.100)} = 3.29$$

16.105 $Cr^{3+}(aq) + 4\ OH^-(aq) \rightleftarrows Cr(OH)_4^-(aq)$

	Cr^{3+}	OH^-	$Cr(OH)_4^-$
before reaction (M)	0.0050	1.0	0
assume 100% reaction	–0.0050	–(4)(0.0050)	+0.0050
after reaction(M)	0	0.98	0.0050
assume small back rxn	+x	+4x	–x
equil (M)	x	0.98 + 4x	0.0050 – x

$$K_f = \frac{[Cr(OH)_4^-]}{[Cr^{3+}][OH^-]^4} = 8 \times 10^{29} = \frac{(0.0050 - x)}{(x)(0.98 + 4x)^4} \approx \frac{(0.0050)}{(x)(0.98)^4}$$

Solve for x. $x = [Cr^{3+}] = 6.8 \times 10^{-33}$ M $= 7 \times 10^{-33}$ M

$$\text{fraction uncomplexed } Cr^{3+} = \frac{[Cr^{3+}]}{[Cr(OH)_4^-]} = \frac{7 \times 10^{-33} \text{ M}}{0.0050 \text{ M}} = 1.4 \times 10^{-30} = 1 \times 10^{-30}$$

16.106 (a) AgCl, $K_{sp} = [Ag^+][Cl^-] = 1.8 \times 10^{-10}$

$$[Cl^-] = \frac{K_{sp}}{[Ag^+]} = \frac{1.8 \times 10^{-10}}{0.030} = 6.0 \times 10^{-9} \text{ M}$$

(b) Hg_2Cl_2, $K_{sp} = [Hg_2^{2+}][Cl^-]^2 = 1.4 \times 10^{-18}$

$$[Cl^-] = \sqrt{\frac{K_{sp}}{[Hg_2^{2+}]}} = \sqrt{\frac{1.4 \times 10^{-18}}{0.030}} = 6.8 \times 10^{-9} \text{ M}$$

(c) $PbCl_2$, $K_{sp} = [Pb^{2+}][Cl^-]^2 = 1.2 \times 10^{-5}$

$$[Cl^-] = \sqrt{\frac{K_{sp}}{[Pb^{2+}]}} = \sqrt{\frac{1.2 \times 10^{-5}}{0.030}} = 0.020 \text{ M}$$

AgCl(s) will begin to precipitate when the $[Cl^-]$ just exceeds 6.0×10^{-9} M. At this Cl^- concentration, IP < K_{sp} for $PbCl_2$ so all of the Pb^{2+} will remain in solution.

16.107 NaOH, 40.0 amu

$$20 \text{ g} \times \frac{1 \text{ mol}}{40.0 \text{ g}} = 0.50 \text{ mol NaOH}$$

(0.500 L)(1.5 mol/L) = 0.75 mol NH_4Cl

	$NH_4^+(aq)$	+	$OH^-(aq)$	⇄	$NH_3(aq)$	+	$H_2O(l)$
before reaction (mol)	0.75		0.50		0		
change (mol)	–0.50		–0.50		+0.50		
after reaction (mol)	0.25		0		0.50		

This reaction yields a buffer solution.

$[NH_4^+]$ = 0.25 mol/0.500 L = 0.50 M

$[NH_3]$ = 0.50 mol/0.500 L = 1.0 M

$$pH = pK_a + \log\frac{[base]}{[acid]} = pK_a + \log\frac{[NH_3]}{[NH_4^+]}$$

For NH_4^+, $K_a = \dfrac{K_w}{K_b \text{ for } NH_3} = \dfrac{1.0 \times 10^{-14}}{1.8 \times 10^{-5}} = 5.6 \times 10^{-10}$; $pK_a = -\log K_a = 9.25$

$$pH = 9.25 + \log\left(\frac{1.0}{0.5}\right) = 9.55$$

16.108 For NH_4^+, $K_a = \dfrac{K_w}{K_b \text{ for } NH_3} = \dfrac{1.0 \times 10^{-14}}{1.8 \times 10^{-5}} = 5.6 \times 10^{-10}$; $pK_a = -\log K_a = 9.25$

$$pH = pK_a + \log\frac{[NH_3]}{[NH_4^+]} = 9.25 + \log\frac{(0.50)}{(0.30)} = 9.47$$

$[H_3O^+] = 10^{-pH} = 10^{-9.47} = 3.4 \times 10^{-10}$ M

For MnS, $K_{spa} = \dfrac{[Mn^{2+}][H_2S]}{[H_3O^+]^2} = 3 \times 10^{10}$

molar solubility = $[Mn^{2+}]$

$$[Mn^{2+}] = \frac{K_{spa}[H_3O^+]^2}{[H_2S]} = \frac{(3 \times 10^{10})(3.4 \times 10^{-10})^2}{(0.10)} = 3.5 \times 10^{-8} \text{ M}$$

MnS, 87.00 amu

solubility = $(3.5 \times 10^{-8}$ mol/L)(87.00 g/mol) = 3×10^{-6} g/L

17.1 (a) spontaneous (b), (c), and (d) nonspontaneous

17.2 (a) $H_2O(g) \rightarrow H_2O(l)$

A liquid is more ordered than a gas. Therefore, ΔS is negative.

(b) $I_2(g) \rightarrow 2\ I(g)$

ΔS is positive because the reaction increases the number of gaseous particles from 1 mol to 2 mol.

(c) $CaCO_3(s) \rightarrow CaO(s) + CO_2(g)$

ΔS is positive because the reaction increases the number of gaseous molecules.

(d) $Ag^+(aq) + Br^-(aq) \rightarrow AgBr(s)$

A solid is more ordered than +1 and –1 charged ions in an aqueous solution. Therefore, ΔS is negative.

17.3 (a) shuffled deck (more disorder) (b) disordered crystal

(c) 1 mole N_2 at STP (larger volume, more disorder)

(d) 1 mole N_2 at 273 K and 0.25 atm (larger volume, more disorder)

17.4 $CaCO_3(s) \rightarrow CaO(s) + CO_2(g)$

$\Delta S° = [S°(CaO) + S°(CO_2)] - S°(CaCO_3)$

$\Delta S° = [(1\ mol)(39.7\ J/(K \cdot mol)) + (1\ mol)(213.6\ J/(K \cdot mol))]$

$- (1\ mol)(92.9\ J/(K \cdot mol)) = +160.4\ J/K$

17.5 From Problem 17.4, $\Delta S_{sys} = \Delta S° = 160.4\ J/K$

$CaCO_3(s) \rightarrow CaO(s) + CO_2(g)$

$\Delta H^\circ = [\Delta H_f^\circ(CaO) + \Delta H_f^\circ(CO_2)] - \Delta H_f^\circ(CaCO_3)$

$\Delta H^\circ = [(1\ mol)(-635.1\ kJ/mol) + (1\ mol)(-393.5\ kJ/mol)]$

$- (1\ mol)(-1206.9\ kJ/mol) = +178.3\ kJ$

$$\Delta S_{surr} = \frac{-\Delta H^\circ}{T} = \frac{-178{,}300\ J}{298\ K} = -598\ J/K$$

$\Delta S_{total} = \Delta S_{sys} + \Delta S_{surr} = 160.4\ J/K + (-598\ J/K) = -438\ J/K$

Because ΔS_{total} is negative, the reaction is not spontaneous under standard-state conditions at 25°C.

17.6 (a) $\Delta G = \Delta H - T\Delta S = 57.1\ kJ - (298\ K)(0.1758\ kJ/K) = +4.7\ kJ$

Because $\Delta G > 0$, the reaction is nonspontaneous at 25°C (298 K)

(b) Set $\Delta G = 0$ and solve for T.

$0 = \Delta H - T\Delta S$

$$T = \frac{\Delta H}{\Delta S} = \frac{57.1\ kJ}{0.1758\ kJ/K} = 325\ K = 52^\circ C$$

17.7 (a) $\Delta G = \Delta H - T\Delta S = 58.5\ kJ/mol - (598\ K)[0.0929\ kJ/(K \cdot mol)]$

$\Delta G = +2.9\ kJ/mol$

Because $\Delta G > 0$, Hg does not boil at 325°C and 1 atm.

(b) The boiling point (phase change) is associated with an equilibrium. Set $\Delta G = 0$ and solve for T, the boiling point.

$$T_{bp} = \frac{\Delta H_{vap}}{\Delta S_{vap}} = \frac{58.5\ kJ/mol}{0.0929\ kJ/(K \cdot mol)} = 630\ K = 357^\circ C$$

17.8 From Problems 17.4 and 17.5:

$\Delta H^\circ = 178.3$ kJ and $\Delta S^\circ = 160.4$ J/K = 0.1604 kJ/K

(a) $\Delta G^\circ = \Delta H^\circ - T\Delta S^\circ = 178.3 \text{ kJ} - (298 \text{ K})(0.1604 \text{ kJ/K}) = +130.5 \text{ kJ}$

(b) Because $\Delta G > 0$, the reaction is nonspontaneous at 25°C (298 K).

(c) Set $\Delta G = 0$ and solve for T, the temperature above which the reaction becomes spontaneous.

$0 = \Delta H - T\Delta S$

$$T = \frac{\Delta H}{\Delta S} = \frac{178.3 \text{ kJ}}{0.1604 \text{ kJ/K}} = 1112 \text{ K} = 839^\circ\text{C}$$

17.9 (a) $CaC_2(s) + 2\ H_2O(l) \rightarrow C_2H_2(g) + Ca(OH)_2(s)$

$\Delta G^\circ = [\Delta G_f^\circ(C_2H_2) + \Delta G_f^\circ(Ca(OH)_2)] - [\Delta G_f^\circ(CaC_2) + 2\ \Delta G_f^\circ(H_2O)]$

$\Delta G^\circ = [(1 \text{ mol})(209.2 \text{ kJ/mol}) + (1 \text{ mol})(-898.6 \text{ kJ/mol})]$

$- [(1 \text{ mol})(-64.8 \text{ kJ/mol}) + (2 \text{ mol})(-237.2 \text{ kJ/mol})] = -150.2 \text{ kJ}$

This reaction can be used for the synthesis of C_2H_2 because $\Delta G < 0$.

(b) It is not possible to synthesize acetylene from solid graphite and gaseous H_2 at 25°C and 1 atm because $\Delta G_f^\circ(C_2H_2) > 0$.

17.10 $C(s) + 2\ H_2(g) \rightarrow C_2H_4(g)$

$$Q_p = \frac{P_{C_2H_4}}{(P_{H_2})^2} = \frac{(0.10)}{(100)^2} = 1.0 \times 10^{-5}$$

$\Delta G = \Delta G^\circ + 2.303RT \log Q_p$

$\Delta G = 68.1 \text{ kJ} + (2.303)(8.314 \times 10^{-3} \text{ kJ/K})(298 \text{ K})\log(1.0 \times 10^{-5}) = +39.6 \text{ kJ}$

Because $\Delta G > 0$, the reaction is spontaneous in the reverse direction.

17.11 From Problem 17.8, $\Delta G° = +130.5$ kJ

$\Delta G° = -2.303RT \log K$

$$\log K = \frac{-\Delta G°}{2.303RT} = \frac{-130.5 \text{ kJ}}{(2.303)(8.314 \times 10^{-3} \text{ kJ/K})(298 \text{ K})} = -22.9$$

$K = K_p = 10^{-22.9} = 1 \times 10^{-23}$

17.12 $H_2O(l) \rightleftarrows H_2O(g)$

$K_p = P_{H_2O}$

K_p is equal to the vapor pressure for H_2O.

$\Delta G° = \Delta G_f°(H_2O(g)) - \Delta G_f°(H_2O(l))$

$\Delta G° = (1 \text{ mol})(-228.6 \text{ kJ/mol}) - (1 \text{ mol})(-237.2 \text{ kJ/mol}) = +8.6$ kJ

$\Delta G° = -2.303RT \log K$

$$\log K = \frac{-\Delta G°}{2.303RT} = \frac{-8.6 \text{ kJ}}{(2.303)(8.314 \times 10^{-3} \text{ kJ/K})(298 \text{ K})} = -1.5$$

$K = K_p = P_{H_2O} = 10^{-1.5} = 0.03$ atm

17.13 $\Delta G° = -2.303RT \log K = -(2.303)(8.314 \times 10^{-3} \text{ kJ/K})(298 \text{ K})\log(1.0 \times 10^{-14})$

$\Delta G° = +80$ kJ

Understanding Key Concepts

1. (a)

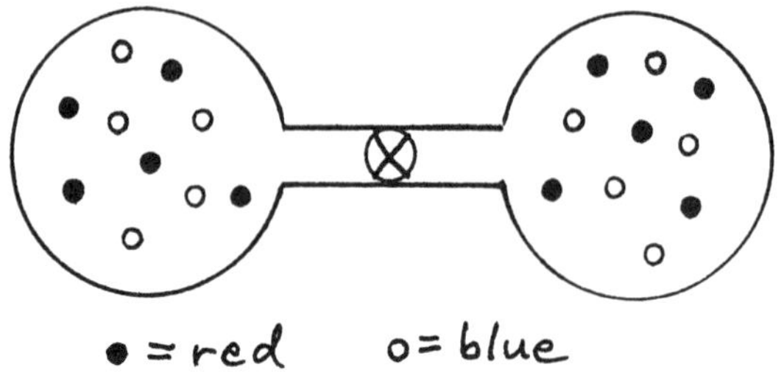

(b) $\Delta H = 0$ (no heat is gained or lost in the mixing of ideal gases)

$\Delta S = +$ (the mixture of the two gases is more disordered)

$\Delta G = -$ (the mixing of the two gases is spontaneous)

(c) For an isolated system, $\Delta S_{surr} = 0$ and $\Delta S_{sys} = \Delta S_{Total} > 0$ for the spontaneous process.

(d) $\Delta G = +$ and the process is nonspontaneous.

2. $\Delta H = +$ (heat is absorbed during sublimation)

$\Delta S = +$ (gas is more disordered than solid)

$\Delta G = -$ (the reaction is spontaneous)

3. (a) For initial state 1, $Q_p < K_p$
(more reactant (A_2) than product (A) compared to the equilibrium state)

For initial state 2, $Q_p > K_p$
(more product (A) than reactant (A_2) compared to the equilibrium state)

(b) $\Delta H = +$ (reaction involves bond breaking - endothermic)

$\Delta S = +$ (equilibrium state is more disordered than initial state 1)

$\Delta G = -$ (reaction spontaneously proceeds toward equilibrium)

(c) $\Delta H = -$ (reaction involves bond making - exothermic)

$\Delta S = -$ (equilibrium state is more ordered than initial state 2)

$\Delta G = -$ (reaction spontaneously proceeds toward equilibrium)

(d) State 1 lies to the left of the minimum in Figure 17.10. State 2 lies to the right of the minimum.

4. (a) $\Delta H^\circ = +$ (reaction involves bond breaking - endothermic)

$\Delta S^\circ = +$ (2 A's are more disordered than A_2)

(b) $\Delta S°$ is for the complete conversion of 1 mole of A_2 in its standard state to 2 moles of A in its standard state.

(c) There is not enough information to say anything about the sign of $\Delta G°$. $\Delta G°$ decreases (becomes less positive or more negative) as the temperature increases.

(d) K_p increases as the temperature increases. As the temperature increases there will be more A and less A_2.

(e) $\Delta G = 0$ at equilibrium.

Additional Problems

Spontaneous Processes

17.14 A spontaneous process is one that proceeds on its own without any external influence.

For example: $H_2O(s) \rightarrow H_2O(l)$ at 25°C

A nonspontaneous process takes place only in the presence of some continuous external influence.

For example: $2\ NaCl(s) \rightarrow 2\ Na(s) + Cl_2(g)$

17.15 (a) and (d) nonspontaneous; (b) and (c) spontaneous

17.16 (a) and (c) spontaneous; (b) and (d) nonspontaneous.

17.17 (b) and (d) spontaneous (because of the large positive K_p's)

17.18 (a) and (d) nonspontaneous (because of the small K's).

Entropy

17.19 Molecular randomness or disorder is called entropy. For the following reaction, the entropy (disorder) increases: $H_2O(s) \rightarrow H_2O(l)$ at 25°C.

17.20 Exothermic reactions can become nonspontaneous at high temperatures if ΔS is negative. Endothermic reactions can become spontaneous at high temperatures if ΔS is positive.

17.21 (a) + (solid → gas) (b) – (liquid → solid)

(c) – (aqueous ions → solid) (d) + ($CO_2(aq) \rightarrow CO_2(g)$)

17.22 (a) + (increase in moles of gas)

(b) – (decrease in moles of gas and formation of liquid)

(c) + (aqueous ions to gas)

(d) – (decrease in moles of gas)

17.23 (a) – (liquid → solid)

(b) – (decrease in number of O_2 molecules)

(c) + (gas is more disordered in larger volume)

(d) – (aqueous ions → solid)

17.24 (a) + (solid dissolved in water)

(b) + (increase in moles of gas)

(c) + (mixed gases are more disordered)

(d) + (liquid to gas)

17.25 $S = k \ln W$, $k = 1.38 \times 10^{-23}$ J/K

(a) $S = k \ln (1) = 0$

(b) $S = k \ln (1) = 0$

(c) $S = (1.38 \times 10^{-23}\text{ J/K}) \ln (2^6) = 5.74 \times 10^{-23}$ J/K

17.26 $S = k \ln W$, $k = 1.38 \times 10^{-23}$ J/K

(a) $S = k \ln (1) = 0$

(b) $S = k \ln (1) = 0$

(c) $S = (1.38 \times 10^{-23} \text{ J/K}) \ln (2^{100}) = 9.57 \times 10^{-22}$ J/K

17.27 (a) $S = (1.38 \times 10^{-23} \text{ J/K}) \ln (4^{12}) = 2.30 \times 10^{-22}$ J/K

(b) $S = (1.38 \times 10^{-23} \text{ J/K}) \ln (4^{120}) = 2.30 \times 10^{-21}$ J/K

(c) $S = (1.38 \times 10^{-23} \text{ J/K}) \ln (4^{6.02 \times 10^{23}}) = 11.5$ J/K

If all C–D bonds point in the same direction, S = 0.

17.28 $S = k \ln W$, $k = 1.38 \times 10^{-23}$ J/K

(a) W = 1; $S = k \ln (1) = 0$

(b) $W = 3^2, = 9$; $S = k \ln (3^2) = 3.03 \times 10^{-23}$ J/K

(c) W = 1; $S = k \ln (1) = 0$

(d) $W = 3^3 = 27$; $S = k \ln (3^3) = 4.55 \times 10^{-23}$ J/K

(e) W = 1; $S = k \ln (1) = 0$

(f) $W = 3^{6.02 \times 10^{23}}$; $S = k \ln (3^{6.02 \times 10^{23}}) = 9.13$ J/K

$$\Delta S = R \ln\left(\frac{V_f}{V_i}\right) = (8.314 \text{ J/K}) \ln 3 = 9.13 \text{ J/K}$$

The results are the same.

17.29 (a) H_2 at 25°C in 50 L (larger volume)

(b) O_2 at 25°C, 1 atm (larger volume)

(c) H_2 at 100°C, 1 atm (larger volume and higher T)

(d) CO_2 at 100°C, 0.1 atm (larger volume and higher T)

17.30 (a) ice at 0°C, because of the higher temperature.

(b) N_2 at STP, because it has the larger volume.

(c) Na at 0°C and 50 L, because it has the larger volume.

(d) water vapor at 150°C and 1 atm, because it has a larger volume and higher temperature.

Standard Molar Entropies and Standard Entropies of Reaction

17.31 The standard molar entropy of a substance is the entropy of 1 mol of the pure substance at 1 atm pressure and 25°C.

$\Delta S° = S°(\text{products}) - S°(\text{reactants})$

17.32 (a) $C_2H_6(g)$; more atoms/molecule

(b) $CO_2(g)$; more atoms/molecule

(c) $I_2(g)$; gas is more disordered than the solid

(d) $CH_3OH(g)$; gas is more disordered than the liquid.

17.33 (a) $P_4O_{10}(s)$ (more atoms/molecule)

(b) $SO_3(g)$ (more atoms/molecule)

(c) Hg(l) (liquid is more disordered than the solid)

(d) $CO_2(g)$ (gas is more disordered than the solid)

17.34 (a) $2\ H_2O_2(l) \rightarrow 2\ H_2O(l) + O_2(g)$

$\Delta S° = [2\ S°(H_2O(l)) + S°(O_2)] - 2\ S°(H_2O_2)$

$\Delta S° = [(2\text{ mol})(69.9\text{ J/(K}\cdot\text{mol)}) + (1\text{ mol})(205.0\text{ J/(K}\cdot\text{mol)})]$

$- (2\text{ mol})(110\text{ J/(K}\cdot\text{mol)}) = +125\text{ J/K}$ (+, because moles of gas increase)

(b) $2\ Na(s) + Cl_2(g) \rightarrow 2\ NaCl(s)$

$\Delta S° = 2\ S°(NaCl) - [2\ S°(Na) + S°(Cl_2)]$

$\Delta S° = (2\ mol)(72.1\ J/(K \cdot mol))$

$- [(2\ mol)(51.2\ J/(K \cdot mol)) + (1\ mol)(223.0\ J/(K \cdot mol))]$

$\Delta S° = -181.2\ J/K$ (–, because moles of gas decrease)

(c) $2\ O_3(g) \rightarrow 3\ O_2(g)$

$\Delta S° = 3\ S°(O_2) - 2\ S°(O_3)$

$\Delta S° = (3\ mol)(205.0\ J/(K \cdot mol)) - (2\ mol)(238.8\ J/(K \cdot mol))$

$\Delta S° = +137.4\ J/K$ (+, because moles of gas increase)

(d) $4\ Al(s) + 3\ O_2(g) \rightarrow 2\ Al_2O_3(s)$

$\Delta S° = 2\ S°(Al_2O_3) - [4\ S°(Al) + 3\ S°(O_2)]$

$\Delta S° = (2\ mol)(50.9\ J/(K \cdot mol))$

$- [(4\ mol)(28.3\ J/(K \cdot mol)) + (3\ mol)(205.0\ J/(K \cdot mol))]$

$\Delta S° = -626.4\ J/K$ (–, because moles of gas decrease)

17.35 (a) $2\ S(s) + 3\ O_2(g) \rightarrow 2\ SO_3(g)$

$\Delta S° = 2\ S°(SO_3) - [2\ S°(S) + 3\ S°(O_2)]$

$\Delta S° = (2\ mol)(256.6\ J/(K \cdot mol))$

$- [(2\ mol)(31.8\ J/(K \cdot mol)) + (3\ mol)(205.0\ J/(K \cdot mol))]$

$\Delta S° = -165.4\ J/K$ (–, because moles of gas decrease)

(b) $SO_3(g) + H_2O(l) \rightarrow H_2SO_4(aq)$

$\Delta S° = S°(H_2SO_4) - [S°(SO_3) + S°(H_2O)]$

$\Delta S° = (1\ mol)(20\ J/(K \cdot mol))$

$- [(1\ mol)(256.6\ J/(K \cdot mol)) + (1\ mol)(69.9\ J/(K \cdot mol))]$

$\Delta S° = -306$ J/K (–, because of the conversion of a gas and water to an aqueous solution)

(c) $AgCl(s) \rightarrow Ag^+(aq) + Cl^-(aq)$

$\Delta S° = [S°(Ag^+) + S°(Cl^-)] - S°(AgCl)]$

$\Delta S° = [(1\ mol)(72.7\ J/(K \cdot mol)) + (1\ mol)(56.5\ J/(K \cdot mol))]$

$- (1\ mol)(96.2\ J/(K \cdot mol))]$

$\Delta S° = +33.0$ J/K (+, because a solid is converted to ions in aqueous solution)

(d) $NH_4NO_3(s) \rightarrow N_2O(g) + 2\ H_2O(g)$

$\Delta S° = [S°(N_2O) + 2\ S°(H_2O)] - S°(NH_4NO_3)$

$\Delta S° = [(1\ mol)(219.7\ J/(K \cdot mol)) + (2\ mol)(188.7\ J/(K \cdot mol))]$

$- (1\ mol)(151.1\ J/(K \cdot mol)) = +446.0$ J/K (+, because moles of gas increase)

Entropy and the Second Law of Thermodynamics

17.36 In any spontaneous process, the total entropy of a system and its surroundings always increases.

17.37 For a spontaneous process, $\Delta S_{total} = \Delta S_{sys} + \Delta S_{surr} > 0$. For an isolated system, $\Delta S_{surr} = 0$, and so $\Delta S_{sys} > 0$ is the criterion for spontaneous change. An example of a spontaneous process in an isolated system is the mixing of two gases.

17.38 $\Delta S_{surr} = \dfrac{-\Delta H}{T}$

The temperature (T) is always positive.

(a) For an exothermic reaction, ΔH is negative and ΔS_{surr} is positive.

(b) For an endothermic reaction, ΔH is positive and ΔS_{surr} is negative.

17.39 $\Delta S_{surr} \propto \dfrac{1}{T}$

Consider the surroundings as an infinitely large constant-temperature bath to which heat can be added without changing its temperature. If the surroundings have a low temperature, they have only a small amount of disorder, in which case addition of a given quantity of heat results in a substantial increase in the amount of disorder (a relatively large value of ΔS_{surr}). If the surroundings have a high temperature, they already have a large amount of disorder, and addition of the same quantity of heat produces only a marginal increase in the amount of disorder (a relatively small value of ΔS_{surr}). Thus, we expect ΔS_{surr} to vary inversely with temperature.

17.40 $N_2(g) + 2\ O_2(g) \rightarrow N_2O_4(g)$

$\Delta H° = \Delta H_f°(N_2O_4) = 9.16\ kJ$

$\Delta S_{sys} = \Delta S° = S°(N_2O_4) - [S°(N_2) + 2\ S°(O_2)]$

$\Delta S_{sys} = (1\ mol)(304.2\ J/(K \cdot mol))$

$- [(1\ mol)(191.5\ J/(K \cdot mol)) + (2\ mol)(205.0\ J/(K \cdot mol))] = -297.3\ J/K$

$$\Delta S_{surr} = \frac{-\Delta H°}{T} = \frac{-9.16\ kJ}{298\ K} = -0.0307\ kJ/K = -30.7\ J/K$$

$\Delta S_{total} = \Delta S_{sys} + \Delta S_{surr} = -297.3\ J/K + (-30.7\ J/K) = -328.0\ J/K$

Because $\Delta S_{total} < 0$, the reaction is nonspontaneous.

17.41 $Cu_2S(s) + O_2(g) \rightarrow 2\ Cu(s) + SO_2(g)$

$\Delta H° = \Delta H_f°(SO_2) - \Delta H_f°(Cu_2S)$

$\Delta H° = (1\ mol)(-296.8\ kJ/mol) - (1\ mol)(-79.5\ kJ/mol) = -217.3\ kJ$

$\Delta S_{sys} = \Delta S° = [2\ S°(Cu) + S°(SO_2)] - [S°(Cu_2S) + S°(O_2)]$

$$\Delta S_{sys} = [(2\text{ mol})(33.1\text{ J/(K}\cdot\text{mol)}) + (1\text{ mol})(248.1\text{ J/(K}\cdot\text{mol)})]$$

$$- [(1\text{ mol})(120.9\text{ J/(K}\cdot\text{mol)}) + (1\text{ mol})(205.0\text{ J/(K}\cdot\text{mol)})] = -11.6\text{ J/K}$$

$$\Delta S_{surr} = \frac{-\Delta H^\circ}{T} = \frac{-(-217{,}300\text{ J})}{298.15\text{ K}} = +728.8\text{ J/K}$$

$$\Delta S_{total} = \Delta S_{sys} + \Delta S_{surr} = -11.6\text{ J/K} + 728.8\text{ J/K} = +717.2\text{ J/K}$$

Because ΔS_{total} is positive, the reaction is spontaneous under standard-state conditions at 25°C.

17.42 (a) $\Delta S_{surr} = \frac{-\Delta H_{vap}}{T} = \frac{-30{,}700\text{ J/mol}}{343\text{ K}} = -89.5\text{ J/(K}\cdot\text{mol)}$

$$\Delta S_{total} = \Delta S_{vap} + \Delta S_{surr} = 87.0\text{ J/(K}\cdot\text{mol)} + (-89.5\text{ J/(K}\cdot\text{mol)}) = -2.5\text{ J/(K}\cdot\text{mol)}$$

(b) $\Delta S_{surr} = \frac{-\Delta H_{vap}}{T} = \frac{-30{,}700\text{ J/mol}}{353\text{ K}} = -87.0\text{ J/(K}\cdot\text{mol)}$

$$\Delta S_{total} = \Delta S_{vap} + \Delta S_{surr} = 87.0\text{ J/(K}\cdot\text{mol)} + (-87.0\text{ J/(K}\cdot\text{mol)}) = 0$$

(c) $\Delta S_{surr} = \frac{-\Delta H_{vap}}{T} = \frac{-30{,}700\text{ J/mol}}{363\text{ K}} = -84.6\text{ J/(K}\cdot\text{mol)}$

$$\Delta S_{total} = \Delta S_{vap} + \Delta S_{surr} = 87.0\text{ J/(K}\cdot\text{mol)} + (-84.6\text{ J/(K}\cdot\text{mol)}) = +2.4\text{ J/(K}\cdot\text{mol)}$$

Benzene does not boil at 70°C (343 K) because ΔS_{total} is negative.

The normal boiling point for benzene is 80°C (353 K), where $\Delta S_{total} = 0$.

17.43 (a) $\Delta S_{surr} = \frac{-\Delta H_{fusion}}{T} = \frac{-30{,}200\text{ J/mol}}{1050\text{ K}} = -28.8\text{ J/(K}\cdot\text{mol)}$

$$\Delta S_{total} = \Delta S_{sys} + \Delta S_{surr} = 28.1\text{ J/(K}\cdot\text{mol)} + (-28.8\text{ J/(K}\cdot\text{mol)}) = -0.7\text{ J/(K}\cdot\text{mol)}$$

(b) $\Delta S_{surr} = \dfrac{-\Delta H_{fusion}}{T} = \dfrac{-30{,}200 \text{ J/mol}}{1075 \text{ K}} = -28.1 \text{ J/(K} \cdot \text{mol)}$

$\Delta S_{total} = \Delta S_{sys} + \Delta S_{surr} = 28.1 \text{ J/(K} \cdot \text{mol)} + (-28.1 \text{ J/(K} \cdot \text{mol)}) = 0$

(c) $\Delta S_{surr} = \dfrac{-\Delta H_{fusion}}{T} = \dfrac{-30{,}200 \text{ J/mol}}{1100 \text{ K}} = -27.5 \text{ J/(K} \cdot \text{mol)}$

$\Delta S_{total} = \Delta S_{sys} + \Delta S_{surr} = 28.1 \text{ J/(K} \cdot \text{mol)} + (-27.5 \text{ J/(K} \cdot \text{mol)}) = +0.6 \text{ J/(K} \cdot \text{mol)}$

NaCl melts at 1100 K because $\Delta S_{total} > 0$.

The melting point of NaCl is 1075 K, where $\Delta S_{total} = 0$.

Free Energy

17.44 The free energy (G) is that part of the system's energy that is still ordered and therefore free (available) to cause spontaneous change.

$G = H - TS$ or $\Delta G = \Delta H - T\Delta S$

17.45 If $\Delta G < 0$, the reaction is spontaneous.
If $\Delta G > 0$, the reaction is nonspontaneous.
If $\Delta G = 0$, the reaction mixture is at equilibrium.

17.46

ΔH	ΔS	$\Delta G = \Delta H - T\Delta S$	Reaction Spontaneity
–	+	–	Spontaneous at all temperatures
–	–	– or +	Spontaneous at low temperatures where $\lvert\Delta H\rvert > \lvert T\Delta S\rvert$ Nonspontaneous at high temperatures where $\lvert\Delta H\rvert < \lvert T\Delta S\rvert$
+	–	+	Nonspontaneous at all temperatures
+	+	– or +	Spontaneous at high temperatures where $T\Delta S > \Delta H$ Nonspontaneous at low temperature where $T\Delta S < \Delta H$

17.47 When ΔH and ΔS are both positive or both negative, the temperature determines the direction of spontaneous reaction. See Problem 17.46 for an explanation.

17.48 $\Delta H_{vap} = 30.7$ kJ/mol

$\Delta S_{vap} = 87.0$ J/(K · mol) = 87.0×10^{-3} kJ/(K · mol)

$\Delta G_{vap} = \Delta H_{vap} - T\Delta S_{vap}$

(a) $\Delta G_{vap} = 30.7$ kJ/mol – (343 K)(87.0×10^{-3} kJ/(K · mol)) = +0.9 kJ/mol

At 70°C (343 K), benzene does not boil because ΔG_{vap} is positive.

(b) $\Delta G_{vap} = 30.7$ kJ/mol – (353 K)(87.0×10^{-3} kJ/(K · mol)) = 0

80°C (353 K) is the boiling point for benzene because $\Delta G_{vap} = 0$

(c) $\Delta G_{vap} = 30.7$ kJ/mol – (363 K)(87.0×10^{-3} kJ/(K · mol)) = –0.9 kJ/mol

At 90°C (363 K), benzene boils because ΔG_{vap} is negative.

17.49 $\Delta H_{fusion} = 30.2$ kJ/mol; $\Delta S_{fusion} = 28.1 \times 10^{-3}$ kJ/(K · mol)

$\Delta G_{fusion} = \Delta H_{fusion} - T\Delta S_{fusion}$

(a) $\Delta G_{fusion} = 30.2$ kJ/mol – (1050 K)(28.1×10^{-3} kJ/(K · mol)) = +0.7 kJ/mol

At 1050 K, NaCl does not melt because ΔG_{fusion} is positive.

(b) $\Delta G_{fusion} = 30.2$ kJ/mol – (1075 K)(28.1×10^{-3} kJ/(K · mol)) = 0

1075 K is the melting point for NaCl because $\Delta G_{fusion} = 0$.

(c) $\Delta G_{fusion} = 30.2$ kJ/mol – (1100 K)(28.1×10^{-3} kJ/(K · mol)) = –0.7 kJ/mol

At 1100 K, NaCl does melt because ΔG_{fusion} is negative.

17.50 At the melting point (phase change), $\Delta G_{fusion} = 0$

$$\Delta G_{fusion} = \Delta H_{fusion} - T\Delta S_{fusion}$$

$$0 = \Delta H_{fusion} - T\Delta S_{fusion}$$

$$T = \frac{\Delta H_{fusion}}{\Delta S_{fusion}} = \frac{17.3 \text{ kJ/mol}}{43.8 \times 10^{-3} \text{ kJ/(K} \cdot \text{mol)}} = 395 \text{ K} = 122°\text{C}$$

17.51 At the melting point (phase change), $\Delta G_{fusion} = 0$

$$\Delta G_{fusion} = \Delta H_{fusion} - T\Delta S_{fusion}$$

$$0 = \Delta H_{fusion} - T\Delta S_{fusion}$$

$$\Delta H_{fusion} = T\Delta S_{fusion}$$

$$\Delta H_{fusion} = (401 \text{ K})(47.7 \times 10^{-3} \text{ kJ/(K} \cdot \text{mol)}) = 19.1 \text{ kJ/mol}$$

Standard Free-Energy Changes and Standard Free Energies of Formation

17.52 (a) $\Delta G°$ is the change in free energy that occurs when reactants in their standard states are converted to products in their standard states.

(b) $\Delta G_f°$ is the free-energy change for formation of one mole of a substance in its standard state from the most stable form of the constituent elements in their standard states.

17.53 The standard state of a substance (solid, liquid, or gas) is the most stable form of a pure substance at 25°C and 1 atm pressure. For solutes, the condition is 1 M at 25°C.

17.54 (a) $N_2(g) + 2\ O_2(g) \rightarrow 2\ NO_2(g)$

$$\Delta H° = 2\ \Delta H_f°(NO_2) = (2 \text{ mol})(33.2 \text{ kJ/mol}) = 66.4 \text{ kJ}$$

$$\Delta S° = 2\ S°(NO_2) - [S°(N_2) + 2\ S°(O_2)]$$

$\Delta S° = (2\ mol)(240.0\ J/(K \cdot mol))$

$- [(1\ mol)(191.5\ J/(K \cdot mol)) + (2\ mol)(205.0\ J/(K \cdot mol))]$

$\Delta S° = -121.5\ J/K = -121.5 \times 10^{-3}\ kJ/K$

$\Delta G° = \Delta H° - T\Delta S° = 66.4\ kJ - (298\ K)(-121.5 \times 10^{-3}\ kJ/K) = +102.6\ kJ$

Because $\Delta G°$ is positive, the reaction is nonspontaneous under standard-state conditions at 25°C.

(b) $2\ KClO_3(s) \rightarrow 2\ KCl(s) + 3\ O_2(g)$

$\Delta H° = 2\ \Delta H_f°(KCl) - 2\ \Delta H_f°(KClO_3)$

$\Delta H° = (2\ mol)(-436.7\ kJ/mol) - (2\ mol)(-397.7\ kJ/mol) = -78.0\ kJ$

$\Delta S° = [2\ S°(KCl) + 3\ S°(O_2)] - 2\ S°(KClO_3)$

$\Delta S° = [(2\ mol)(82.6\ J/(K \cdot mol)) + (3\ mol)(205.0\ J/(K \cdot mol))]$

$- (2\ mol)(143\ J/(K \cdot mol))$

$\Delta S° = 494.2\ J/(K \cdot mol) = 494.2 \times 10^{-3}\ kJ/(K \cdot mol)$

$\Delta G° = \Delta H° - T\Delta S° = -78.0\ kJ - (298\ K)(494.2 \times 10^{-3}\ kJ/(K \cdot mol)) = -225.3\ kJ$

Because $\Delta G°$ is negative, the reaction is spontaneous under standard-state conditions at 25°C.

(c) $CH_3CH_2OH(l) + O_2(g) \rightarrow CH_3COOH(l) + H_2O(l)$

$\Delta H° = [\Delta H_f°(CH_3COOH) + \Delta H_f°(H_2O)] - \Delta H_f°(CH_3CH_2OH)$

$\Delta H° = [(1\ mol)(-484.5\ kJ/mol) + (1\ mol)(-285.8\ kJ/mol)]$

$- (1\ mol)(-277.7\ kJ/mol) = -492.6\ kJ$

$\Delta S° = [S°(CH_3COOH) + S°(H_2O)] - [S°(CH_3CH_2OH) + S°(O_2)]$

$\Delta S° = [(1\ mol)(160\ J/(K \cdot mol)) + (1\ mol)(69.9\ J/(K \cdot mol))]$

$- [(1\ mol)(161\ J/(K \cdot mol)) + (1\ mol)(205.0\ J/(K \cdot mol))]$

$\Delta S° = -136.1\ J/(K \cdot mol) = -136.1 \times 10^{-3}\ kJ/(K \cdot mol)$

$\Delta G° = \Delta H° - T\Delta S° = -492.6\ kJ - (298\ K)(-136.1 \times 10^{-3}\ kJ/(K \cdot mol))$

$\Delta G° = -452.0\ kJ$

Because $\Delta G°$ is negative, the reaction is spontaneous under standard-state conditions at 25°C.

17.55 (a) $2\ SO_2(g) + O_2(g) \rightarrow 2\ SO_3(g)$

$\Delta H° = 2\ \Delta H_f°(SO_3) - 2\ \Delta H_f°(SO_2)$

$\Delta H° = (2\ mol)(-395.7\ kJ/mol) - (2\ mol)(-296.8\ kJ/mol) = -197.8\ kJ$

$\Delta S° = 2\ S°(SO_3) - [2\ S°(SO_2) + S°(O_2)]$

$\Delta S° = (2\ mol)(256.6\ J/(K \cdot mol))$

$- [(2\ mol)(248.1\ J/(K \cdot mol)) + (1\ mol)(205.0\ J/(K \cdot mol))]$

$\Delta S° = -188.0\ J/K = -188.0 \times 10^{-3}\ kJ/K$

$\Delta G° = \Delta H° - T\Delta S° = -197.8\ kJ - (298\ K)(-188.0 \times 10^{-3}\ kJ/K) = -141.8\ kJ$

Because $\Delta G°$ is negative, the reaction is spontaneous under standard-state conditions at 25°C.

(b) $N_2(g) + 2\ H_2(g) \rightarrow N_2H_4(l)$

$\Delta H° = \Delta H_f°(N_2H_4)$

$\Delta H° = (1\ mol)(50.6\ kJ/mol) = 50.6\ kJ$

$\Delta S° = S°(N_2H_4) - [S°(N_2) + 2\ S°(H_2)]$

$\Delta S° = (1\ mol)(121.2\ J/(K \cdot mol))$

$- [(1\ mol)(191.5\ J/(K \cdot mol)) + (2\ mol)(130.6\ J/(K \cdot mol))]$

$\Delta S° = -331.5\ J/K = -331.5 \times 10^{-3}\ kJ/K$

$\Delta G° = \Delta H° - T\Delta S° = 50.6\ kJ - (298\ K)(-331.5 \times 10^{-3}\ kJ/K) = +149.4\ kJ$

Because $\Delta G°$ is positive, the reaction is nonspontaneous under standard-state conditions at 25°C.

(c) $CH_3OH(l) + O_2(g) \rightarrow HCOOH(l) + H_2O(l)$

$\Delta H° = [\Delta H_f°(HCOOH) + \Delta H_f°(H_2O)] - \Delta H_f°(CH_3OH)$

$\Delta H° = [(1\ mol)(-424.7\ kJ/mol) + (1\ mol)(-285.8\ kJ/mol)]$

$- (1\ mol)(-238.7\ kJ/mol) = -471.8\ kJ$

$\Delta S° = [S°(HCOOH) + S°(H_2O)] - [S°(CH_3OH) + S°(O_2)]$

$\Delta S° = [(1\ mol)(129.0\ J/(K \cdot mol)) + (1\ mol)(69.9\ J/(K \cdot mol))]$

$- [(1\ mol)(127\ J/(K \cdot mol)) + (1\ mol)(205.0\ J/(K \cdot mol))]$

$\Delta S° = -133.1\ J/K = -133.1 \times 10^{-3}\ kJ/K$

$\Delta G° = \Delta H° - T\Delta S° = -471.8\ kJ - (298\ K)(-133.1 \times 10^{-3}\ kJ/K) = -432.1\ kJ$

Because $\Delta G°$ is negative, the reaction is spontaneous under standard-state conditions at 25°C.

17.56 (a) $N_2(g) + 2\ O_2(g) \rightarrow 2\ NO_2(g)$

$\Delta G° = 2\ \Delta G_f°(NO_2) = (2\ mol)(51.3\ kJ/mol) = +102.6\ kJ$

(b) $2\ KClO_3(s) \rightarrow 2\ KCl(s) + 3\ O_2(g)$

$\Delta G° = 2\ \Delta G_f°(KCl) - 2\ \Delta G_f°(KClO_3)$

$\Delta G° = (2\ mol)(-409.2\ kJ/mol) - (2\ mol)(-296.3\ kJ/mol) = -225.8\ kJ$

(c) $CH_3CH_2OH(l) + O_2(g) \rightarrow CH_3COOH(l) + H_2O(l)$

$\Delta G° = [\Delta G_f°(CH_3COOH) + \Delta G_f(H_2O)] - \Delta G_f°(CH_3CH_2OH)$

$\Delta G° = [(1\ mol)(-390\ kJ/mol) + (1\ mol)(-237.2\ kJ/mol)]$

$- (1\ mol)(-174.9\ kJ/mol) = -452\ kJ$

17.57 (a) $2\ SO_2(g) + O_2(g) \rightarrow 2\ SO_3(g)$

$\Delta G° = 2\ \Delta G_f°(SO_3) - 2\ \Delta G_f°(SO_2)$

$\Delta G° = (2\text{ mol})(-371.1\text{ kJ/mol}) - (2\text{ mol})(-300.2\text{ kJ/mol}) = -141.8\text{ kJ}$

(b) $N_2(g) + 2\ H_2(g) \rightarrow N_2H_4(l)$

$\Delta G° = \Delta G_f°(N_2H_4) = (1\text{ mol})(149.2\text{ kJ/mol}) = 149.2\text{ kJ}$

(c) $CH_3OH(l) + O_2(g) \rightarrow HCOOH(l) + H_2O(l)$

$\Delta G° = [\Delta G_f°(HCOOH) + \Delta G_f°(H_2O)] - \Delta G_f°(CH_3OH)$

$\Delta G° = [(1\text{ mol})(-361.4\text{ kJ/mol}) + (1\text{ mol})(-237.2\text{ kJ/mol})]$

$- (1\text{ mol})(-166.4\text{ kJ/mol}) = -432.2\text{ kJ}$

17.58 A compound is thermodynamically stable with respect to its constituent elements at 25°C if $\Delta G_f°$ is negative.

	$\Delta G_f°$(kJ/mol)	Feasible
(a) $BaCO_3(s)$	–1138	yes
(b) $HBr(g)$	–53.4	yes
(c) $N_2O(g)$	+104.2	no
(d) $C_2H_4(g)$	+68.1	no

17.59 A compound is thermodynamically stable with respect to its constituent elements at 25°C if $\Delta G_f°$ is negative.

	$\Delta G_f°$(kJ/mol)	Feasible
(a) $N_2O_5(g)$	+115	no
(b) $H_2S(g)$	–33.6	yes
(c) $H_2Se(g)$	+15.9	no
(d) $CCl_4(l)$	–65.3	yes

17.60 $CH_2{=}CH_2(g) + H_2O(l) \rightarrow CH_3CH_2OH(l)$

$\Delta H^\circ = \Delta H_f^\circ(CH_3CH_2OH) - [\Delta H_f^\circ(CH_2{=}CH_2) + \Delta H_f^\circ(H_2O)]$

$\Delta H^\circ = (1\ mol)(-277.7\ kJ/mol) - [(1\ mol)(52.3\ kJ/mol) + (1\ mol)(-285.8\ kJ/mol)]$

$\Delta H^\circ = -44.2\ kJ$

$\Delta S^\circ = S^\circ(CH_3CH_2OH) - [S^\circ(CH_2{=}CH_2) + S^\circ(H_2O)]$

$\Delta S^\circ = (1\ mol)(161\ J/(K \cdot mol)) - [(1\ mol)(219.5\ J/(K \cdot mol))$

$+ (1\ mol)(69.9\ J/(K \cdot mol))]$

$\Delta S^\circ = -128\ J/(K \cdot mol) = -128 \times 10^{-3}\ kJ/(K \cdot mol)$

$\Delta G^\circ = \Delta H^\circ - T\Delta S^\circ = -44.2\ kJ - (298\ K)(-128 \times 10^{-3}\ kJ/K) = -6.1\ kJ$

Because ΔG° is negative, the reaction is spontaneous under standard-state conditions at 25°C.

The reaction becomes nonspontaneous at high temperatures because ΔS° is negative.

To find the crossover temperature, set $\Delta G = 0$ and solve for T.

$$T = \frac{\Delta H^\circ}{\Delta S^\circ} = \frac{-44{,}200\ J}{-128\ J/K} = 345\ K = 72^\circ C$$

The reaction becomes nonspontaneous at 72°C.

17.61 $2\ H_2S(g) + SO_2(g) \rightarrow 3\ S(s) + 2\ H_2O(g)$

$\Delta H^\circ = 2\ \Delta H_f^\circ(H_2O) - [2\ \Delta H_f^\circ(H_2S) + \Delta H_f^\circ(SO_2)]$

$\Delta H^\circ = (2\ mol)(-241.8\ kJ/mol)$

$- [(2\ mol)(-20.6\ kJ/mol) + (1\ mol)(-296.8\ kJ/mol) = -145.6\ kJ$

$\Delta S^\circ = [3\ S^\circ(S) + 2\ S^\circ(H_2O)] - [2\ S^\circ(H_2S) + S^\circ(SO_2)]$

$\Delta S^\circ = [(3\ mol)(31.8\ J/(K \cdot mol)) + (2\ mol)(188.7\ J/(K \cdot mol))]$

$- [(2\ mol)(205.7\ J/(K \cdot mol)) + (1\ mol)(248.1\ J/(K \cdot mol))]$

$\Delta S° = -186.7\ J/K = -186.7 \times 10^{-3}\ kJ/K$

$\Delta G° = \Delta H° - T\Delta S° = -145.6\ kJ - (298\ K)(-186.7 \times 10^{-3}\ kJ/K) = -90.0\ kJ$

Because $\Delta G°$ is negative, the reaction is spontaneous under standard-state conditions at 25°C.

The reaction becomes nonspontaneous at high temperatures because $\Delta S°$ is negative.

To find the crossover temperature set $\Delta G = 0$ and solve for T.

$$T = \frac{\Delta H°}{\Delta S°} = \frac{-145{,}600\ J}{-186.7\ J/K} = 780\ K = 507°C$$

The reaction becomes nonspontaneous at 507°C.

17.62 $3\ C_2H_2(g) \rightarrow C_6H_6(l)$

$\Delta G° = \Delta G_f°(C_6H_6) - 3\ \Delta G_f°(C_2H_2)$

$\Delta G° = (1\ mol)(124.5\ kJ/mol) - (3\ mol)(209.2\ kJ/mol) = -503.1\ kJ$

Because $\Delta G°$ is negative, the reaction is possible. Look for a catalyst.

Because $\Delta G_f°$ for benzene is positive (+124.5 kJ/mol), the synthesis of benzene from graphite and gaseous H_2 at 25°C and 1 atm pressure is not possible.

17.63 $CH_2ClCH_2Cl(l) \rightarrow CH_2{=}CHCl(g) + HCl(g)$

$\Delta G° = [\Delta G_f°(CH_2{=}CHCl) + \Delta G_f°(HCl)] - \Delta G_f°(CH_2ClCH_2Cl)$

$$\Delta G° = [(1\ mol)(51.9\ kJ/mol) + (1\ mol)(-95.3\ kJ/mol)]$$
$$- (1\ mol)(-79.6\ kJ/mol) = +36.2\ kJ$$

Because $\Delta G°$ is positive, the reaction is nonspontaneous under standard-state conditions at 25°C.

$$CH_2ClCH_2Cl(l) \rightarrow CH_2{=}CHCl(g) + HCl(g)$$

Sum: $$NaOH(aq) + HCl(g) \rightarrow Na^+(aq) + Cl^-(aq) + H_2O(l)$$

$$CH_2ClCH_2Cl(l) + NaOH(aq) \rightarrow CH_2{=}CHCl(g) + Na^+(aq) + Cl^-(aq) + H_2O(l)$$

$$\Delta G^\circ = [\Delta G_f^\circ(CH_2{=}CHCl) + \Delta G_f^\circ(Na^+) + \Delta G_f^\circ(Cl^-) + \Delta G_f^\circ(H_2O)]$$

$$- [\Delta G_f^\circ(CH_2ClCH_2Cl) + \Delta G_f^\circ(NaOH)]$$

$$\Delta G^\circ = [(1\text{ mol})(51.9\text{ kJ/mol}) + (1\text{ mol})(-261.9\text{ kJ/mol})$$

$$+ (1\text{ mol})(-131.3\text{ kJ/mol}) + (1\text{ mol})(-237.2\text{ kJ/mol})]$$

$$- [(1\text{ mol})(-79.6\text{ kJ/mol}) + (1\text{ mol})(-419.2\text{ kJ/mol})] = -79.7\text{ kJ}$$

Using NaOH(aq), $\Delta G^\circ = -79.7$ kJ and the reaction is spontaneous. (More generally, base removes HCl, driving the reaction to the right.)

The synthesis of a compound from its constituent elements is thermodynamically feasible at 25°C and 1 atm pressure if ΔG_f° is negative.

Because $\Delta G_f^\circ(CH_2{=}CHCl) = +51.9$ kJ, the synthesis of vinyl chloride from its elements is not possible at 25°C and 1 atm pressure.

Free Energy, Composition, and Chemical Equilibrium

17.64 $\Delta G = \Delta G^\circ + RT \ln Q$

17.65 $\Delta G = \Delta G^\circ + RT \ln Q$

(a) If $Q < 1$, then $RT \ln Q$ is negative and $\Delta G < \Delta G^\circ$.

(b) If $Q = 1$, then $RT \ln Q = 0$ and $\Delta G = \Delta G^\circ$.

(c) If $Q > 1$, then $RT \ln Q$ is positive and $\Delta G > \Delta G^\circ$.

As Q increases the thermodynamic driving force decreases.

17.66 $\Delta G = \Delta G^\circ + 2.303RT\ \log\left[\frac{(P_{SO_3})^2}{(P_{SO_2})^2(P_{O_2})}\right]$

(a) $\Delta G = (-141.8\text{ kJ}) + (2303)(8.314 \times 10^{-3}\text{ kJ/K})(298\text{ K})\log\left[\frac{(1.0)^2}{(100)^2(100)}\right]$

$\Delta G = -176.0\text{ kJ}$

(b) $\Delta G = (-141.8\text{ kJ}) + (2.303)(8.314 \times 10^{-3}\text{ kJ/K})(298\text{ K})\log\left[\frac{(10)^2}{(2.0)^2(1.0)}\right]$

$\Delta G = -133.8\text{ kJ}$

(c) Q = 1, log Q = 0, $\Delta G = \Delta G^\circ = -141.8\text{ kJ}$

17.67 $Q = \frac{[NH_2CONH_2]}{(P_{NH_3})^2(P_{CO_2})}$

$\Delta G = \Delta G^\circ + 2.303RT\ \log Q$

(a) $\Delta G = -13.6\text{ kJ} + (2.303)(8.314 \times 10^{-3}\text{ kJ/K})(298\text{ K})\log\left(\frac{1}{(10)^2(10)}\right) = -30.7\text{ kJ}$

Because ΔG is negative, the reaction is spontaneous.

(b) $\Delta G = -13.6\text{ kJ} + (2.303)(8.314 \times 10^{-3}\text{ kJ/K})(298\text{ K})\log\left(\frac{1}{(0.10)^2(0.10)}\right) = +3.5\text{ kJ}$

Because ΔG is positive, the reaction is nonspontaneous.

17.68 $\Delta G^\circ = -RT\ \ln K = -2.303RT\ \log K$

(a) If $K > 1$, ΔG° is negative.

(b) If $K = 1$, $\Delta G^\circ = 0$.

(c) If $K < 1$, ΔG° is positive.

17.69 $K = 10^{\frac{-\Delta G^\circ}{2.303RT}}$

(a) If ΔG° is positive, K is small.

(b) If ΔG° is negative, K is large.

17.70 $\Delta G^\circ = -141.8$ kJ

$\Delta G^\circ = -2.303RT \log K$

$$\log K = \frac{-\Delta G^\circ}{2.303RT} = \frac{-(-141.8\text{ kJ})}{(2.303)(8.314 \times 10^{-3}\text{ kJ/K})(298\text{ K})} = 24.85$$

$K = 10^{24.85} = 7.1 \times 10^{24}$

17.71 $\Delta G^\circ = -2.303RT \log K$

$$\log K = \frac{-\Delta G^\circ}{2.303RT} = \frac{-(-13.6\text{ kJ})}{(2.303)(8.314 \times 10^{-3}\text{ kJ/K})(298\text{ K})} = 2.38$$

$K = 10^{2.38} = 2.4 \times 10^{2}$

17.72 $C_2H_5OH(l) \rightleftarrows C_2H_5OH(g)$

$\Delta G^\circ = \Delta G_f^\circ(C_2H_5OH(g)) - \Delta G_f^\circ(C_2H_5OH(l))$

$\Delta G^\circ = (1\text{ mol})(-168.6\text{ kJ/mol}) - (1\text{ mol})(-174.9\text{ kJ/mol}) = +6.3\text{ kJ}$

$\Delta G^\circ = -2.303RT \log K$

$$\log K = \frac{-\Delta G^\circ}{2.303RT} = \frac{-(6.3\text{ kJ})}{(2.303)(8.314 \times 10^{-3}\text{ kJ/K})(298\text{ K})} = -1.10$$

$K = 10^{-1.10} = 0.079$

$K = K_p = P_{C_2H_5OH} = 0.079$ atm

17.73 $\Delta G° = -2.303RT \log K_a$

$\Delta G° = -(2.303)(8.314 \times 10^{-3}\ kJ/K)(298\ K)\log(3.0 \times 10^{-4}) = +20.1\ kJ$

17.74 $2\ CH_2{=}CH_2(g)\ +\ O_2(g)\ \rightarrow\ 2\ C_2H_4O(g)$

$\Delta G° = 2\ \Delta G_f°(C_2H_4O) - 2\ \Delta G_f°(CH_2{=}CH_2)$

$\Delta G° = (2\ mol)(-13.1\ kJ/mol) - (2\ mol)(68.1\ kJ/mol) = -162.4\ kJ$

$\Delta G° = -2.303RT \log K$

$$\log K = \frac{-\Delta G°}{2.303RT} = \frac{-(-162.4\ kJ)}{(2.303)(8.314 \times 10^{-3}\ kJ/K)(298\ K)} = 28.46$$

$K = K_p = 10^{28.46} = 2.9 \times 10^{28}$

17.75 $CO(g)\ +\ 2\ H_2(g)\ \rightleftarrows\ CH_3OH(g)$

$\Delta G° = \Delta G_f°(CH_3OH) - \Delta G_f°(CO)$

$\Delta G° = (1\ mol)(-161.9\ kJ/mol) - (1\ mol)(-137.2\ kJ/mol) = -24.7\ kJ$

$\Delta G° = -2.303RT \log K_p$

$$\log K_p = \frac{-\Delta G°}{2.303RT} = \frac{-(-24.7\ kJ)}{(2.303)(8.314 \times 10^{-3}\ kJ/K)(298\ K)} = 4.33$$

$K_p = 10^{4.33} = 2.1 \times 10^4$

$\Delta G = \Delta G° + 2.303RT \log Q$

$$\Delta G = -24.7\ kJ + (2.303)(8.314 \times 10^{-3}\ kJ/K)(298\ K)\log\left(\frac{20}{(20)(20)^2}\right) = -39.5\ kJ$$

General Problems

17.76 A reaction is spontaneous if ΔS_{total} is positive.

$\Delta S_{total} = \Delta S_{system} + \Delta S_{surr}$

ΔS_{total} can be positive even if ΔS_{system} ($\Delta S°$) is negative as long as $\Delta S_{surr} > |\Delta S_{system}|$.

17.77 The kinetic parameters [(a), (b), and (h)] are affected by a catalyst. The thermodynamic and equilibrium parameters [(c), (d), (e), (f), and (g)] are not affected by a catalyst.

17.78 (a) $Q = 1$, $\log Q = 0$, $\Delta G = \Delta G° = +79.9$ kJ

Because ΔG is positive, the reaction is spontaneous in the reverse direction.

(b) $\Delta G = \Delta G° + 2.303RT \log Q$; $Q = [H_3O^+][OH^-] = (1.0 \times 10^{-7})^2 = 1.0 \times 10^{-14}$

$\Delta G = 79.9 \text{ kJ} + (2.303)(8.314 \times 10^{-3} \text{ kJ/K})(298 \text{ K})\log(1.0 \times 10^{-14}) = 0$

Because $\Delta G = 0$, the reaction is at equilibrium.

(c) $\Delta G = \Delta G° + 2.303RT \log Q$

$Q = [H_3O^+][OH^-] = (1.0 \times 10^{-7})(1.0 \times 10^{-10}) = 1.0 \times 10^{-17}$

$\Delta G = 79.9 \text{ kJ} + (2.303)(8.314 \times 10^{-3} \text{ kJ/K})(298 \text{ K})\log(1.0 \times 10^{-17}) = -17.1 \text{ kJ}$

Because ΔG is negative, the reaction is spontaneous in the forward direction.

The results are consistent with Le Châtelier's principle. When the $[H_3O^+]$ and $[OH^-]$ are larger than the equilibrium concentrations (a), the reverse reaction takes place. When the product of $[H_3O^+]$ and the $[OH^-]$ is less than the equilibrium value, the forward reaction is spontaneous.

$\Delta G° = -2.303RT \log K$

$$\log K = \frac{-\Delta G°}{2.303RT} = \frac{-79.9 \text{ kJ}}{(2.303)(8.314 \times 10^{-3} \text{ kJ/K})(298 \text{ K})} = -14.00$$

$K = K_a = 10^{-14.00} = 1.0 \times 10^{-14}$

17.79 At the normal boiling point, $\Delta G = 0$.

$\Delta G = \Delta H - T\Delta S$; $\qquad T = \dfrac{\Delta H_{vap}}{\Delta S_{vap}} = \dfrac{38{,}600 \text{ J}}{110 \text{ J/K}} = 351 \text{ K} = 78°C$

17.80 (a), (c), and (d) are nonspontaneous. (b) is spontaneous.

17.81 (a) Spontaneous does not mean fast, just possible.

(b) For a spontaneous reaction $\Delta S_{total} > 0$. ΔS_{sys} can be positive or negative.

(c) An endothermic reaction can be spontaneous if $\Delta S_{sys} > 0$.

(d) This statement is true because the sign of ΔG changes when the direction of a reaction is reversed.

17.82 $S = k \ln W$; $k = 1.38 \times 10^{-23}$ J/K

Point Total	Possible Ways	W	S
2	(1+1)	1	0
3	(2+1)(1+2)	2	9.57×10^{-24} J/K
4	(1+3)(2+2)(3+1)	3	1.52×10^{-23} J/K
5	(1+4)(2+3)(3+2)(4+1)	4	1.91×10^{-23} J/K
6	(1+5)(2+4)(3+3)(4+2)(5+1)	5	2.22×10^{-23} J/K
7	(1+6)(2+5)(3+4)(4+3)(5+2)(6+1)	6	2.47×10^{-23} J/K
8	(2+6)(3+5)(4+4)(5+3)(6+2)	5	2.22×10^{-23} J/K
9	(3+6)(4+5)(5+4)(6+3)	4	1.91×10^{-23} J/K
10	(4+6)(5+5)(6+4)	3	1.52×10^{-23} J/K
11	(6+5)(5+6)	2	9.57×10^{-24} J/K
12	(6+6)	1	0

Because a point total of 7 can be rolled in the most ways, it is the most probable point total.

17.83 At the normal boiling point, $\Delta G_{vap} = 0$.

$61°C = 334\ K$

$$\Delta G_{vap} = \Delta H_{vap} - T\Delta S_{vap}$$

$$\Delta S_{vap} = \frac{\Delta H_{vap}}{T} = \frac{29{,}240\ J}{334\ K} = 87.5\ J/K$$

17.84

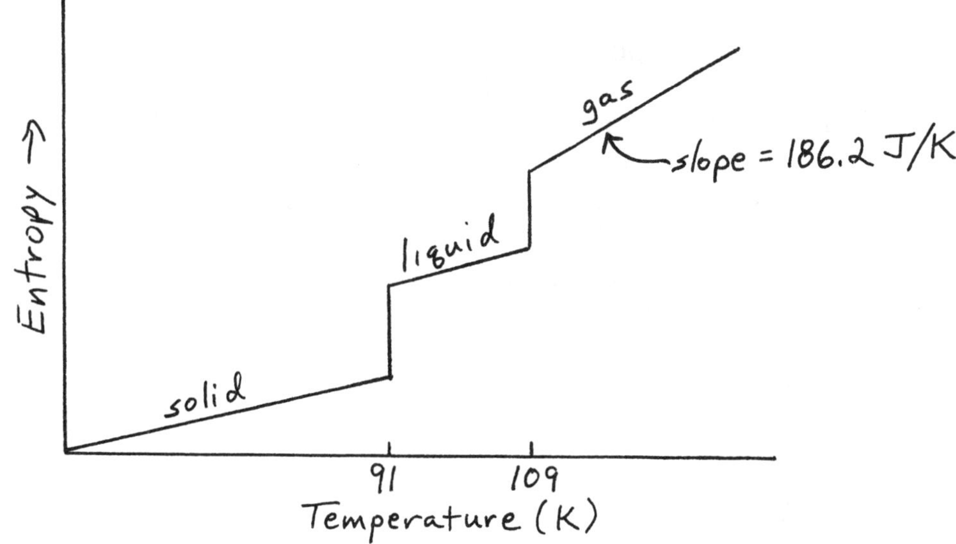

17.85 (a) $2\ Mg(s) + O_2(g) \rightarrow 2\ MgO(s)$

$\Delta H° = 2\ \Delta H_f°(MgO) = (2\ mol)(-601.7\ kJ/mol) = -1203.4\ kJ$

$\Delta S° = 2\ S°(MgO) - [2\ S°(Mg) + S°(O_2)]$

$\Delta S° = (2\ mol)(26.9\ J/(K \cdot mol))$

$- [(2\ mol)(32.7\ J/(K \cdot mol)) + (1\ mol)(205.0\ J/(K \cdot mol))]$

$\Delta S° = -216.6\ J/K = -216.6 \times 10^{-3}\ kJ/K$

$\Delta G° = \Delta H° - T\Delta S° = -1203.4\ kJ - (298\ K)(-216.6 \times 10^{-3}\ kJ/K) = -1138.8\ kJ$

Because $\Delta G°$ is negative, the reaction is spontaneous at 25°C. $\Delta G°$ becomes less negative as the temperature is raised.

(b) $MgCO_3(s) \rightarrow MgO(s) + CO_2(g)$

$\Delta H° = [\Delta H_f°(MgO) + \Delta H_f°(CO_2)] - \Delta H_f°(MgCO_3)$

$\Delta H° = [(1\ mol)(-601.1\ kJ/mol) + (1\ mol)(-393.5\ kJ/mol)]$

$- (1\ mol)(-1096\ kJ/mol) = +101\ kJ$

$\Delta S° = [S°(MgO) + S°(CO_2)] - S°(MgCO_3)$

$\Delta S° = [(1\ mol)(26.9\ J/(K \cdot mol)) + (1\ mol)(213.6\ J/(K \cdot mol))]$

$- (1\ mol)(65.7\ J/(K \cdot mol))$

$\Delta S° = 174.8\ J/K = 174.8 \times 10^{-3}\ kJ/K$

$\Delta G° = \Delta H° - T\Delta S° = 101\ kJ - (298\ K)(174.8 \times 10^{-3}\ kJ/K) = +49\ kJ$

Because $\Delta G°$ is positive, the reaction is not spontaneous at 25°C. $\Delta G°$ becomes less positive as the temperature is raised.

(c) $Fe_2O_3(s) + 2\ Al(s) \rightarrow Al_2O_3(s) + 2\ Fe(s)$

$\Delta H° = \Delta H_f°(Al_2O_3) - \Delta H_f°(Fe_2O_3)$

$\Delta H° = (1\ mol)(-1676\ kJ/mol) - (1\ mol)(-824.2\ kJ/mol) = -852\ kJ$

$\Delta S° = [S°(Al_2O_3) + 2\ S°(Fe)] - [S°(Fe_2O_3) + 2\ S°(Al)]$

$\Delta S° = [(1\ mol)(50.9\ J/(K \cdot mol)) + (2\ mol)(27.3\ J/(K \cdot mol))]$

$- [(1\ mol)(87.4\ J/(K \cdot mol)) + (2\ mol)(28.3\ J/(K \cdot mol))]$

$\Delta S° = -38.5\ J/K = -38.5 \times 10^{-3}\ kJ/K$

$\Delta G° = \Delta H° - T\Delta S° = -852\ kJ - (298\ K)(-38.5 \times 10^{-3}\ kJ/K) = -840\ kJ$

Because $\Delta G°$ is negative, the reaction is spontaneous at 25°C. $\Delta G°$ becomes less negative as the temperature is raised.

(d) $2\ NaHCO_3(s) \rightarrow Na_2CO_3(s) + CO_2(g) + H_2O(g)$

$\Delta H° = [\Delta H_f°(Na_2CO_3) + \Delta H_f°(CO_2) + \Delta H_f°(H_2O)] - 2\ \Delta H_f°(NaHCO_3)$

$\Delta H° = [(1\ mol)(-1130.7\ kJ/mol) + (1\ mol)(-393.5\ kJ/mol)$

$+ (1\ mol)(-241.8\ kJ/mol)] - (2\ mol)(-950.8\ kJ/mol) = +135.6\ kJ$

$\Delta S° = [S°(Na_2CO_3) + S°(CO_2) + S°(H_2O)] - 2\ S°(NaHCO_3)$

$\Delta S° = [(1\ mol)(135.0\ J/(K \cdot mol)) + (1\ mol)(213.6\ J/(K \cdot mol))$

$+ (1\ mol)(188.7\ J/(K \cdot mol))] - (2\ mol)(102\ J/(K \cdot mol))$

$\Delta S° = +333\ J/K = +333 \times 10^{-3}\ kJ/K$

$\Delta G° = \Delta H° - T\Delta S° = +135.6\ kJ - (298\ K)(+333 \times 10^{-3}\ kJ/K) = +36.4\ kJ$

Because $\Delta G°$ is positive, the reaction is not spontaneous at 25°C. $\Delta G°$ becomes less positive as the temperature is raised.

17.86 $NH_4NO_3(s) \rightarrow N_2O(g) + 2\ H_2O(g)$

(a) $\Delta G° = [\Delta G_f°(N_2O) + 2\ \Delta G_f°(H_2O)] - \Delta G_f°(NH_4NO_3)$

$\Delta G° = [(1\ mol)(104.2\ kJ/mol) + (2\ mol)(-228.6\ kJ/mol)] - (1\ mol)(-184.0\ kJ/mol)$

$\Delta G° = -169.0\ kJ$

Because $\Delta G°$ is negative, the reaction is spontaneous.

(b) Because the reaction increases the number of moles of gas, $\Delta S°$ is positive.

$\Delta G° = \Delta H° - T\Delta S°$

As the temperature is raised, $\Delta G°$ becomes more negative.

(c) $\Delta G° = -2.303RT \log K$

$$\log K = \frac{-\Delta G°}{2.303RT} = \frac{-(-169.0\ kJ)}{(2.303)(8.314 \times 10^{-3}\ kJ/K)(298\ K)} = 29.62$$

$K = K_p = 10^{29.62} = 4.2 \times 10^{29}$

(d) $Q = (P_{N_2O})(P_{H_2O})^2 = (30)(30)^2 = (30)^3$

$\Delta G = \Delta G° + 2.303RT \log Q$

$\Delta G = -169.0 \text{ kJ} + (2.303)(8.314 \times 10^{-3} \text{kJ/K})(298 \text{ K})\log[(30)^3] = -143.7 \text{ kJ}$

17.87

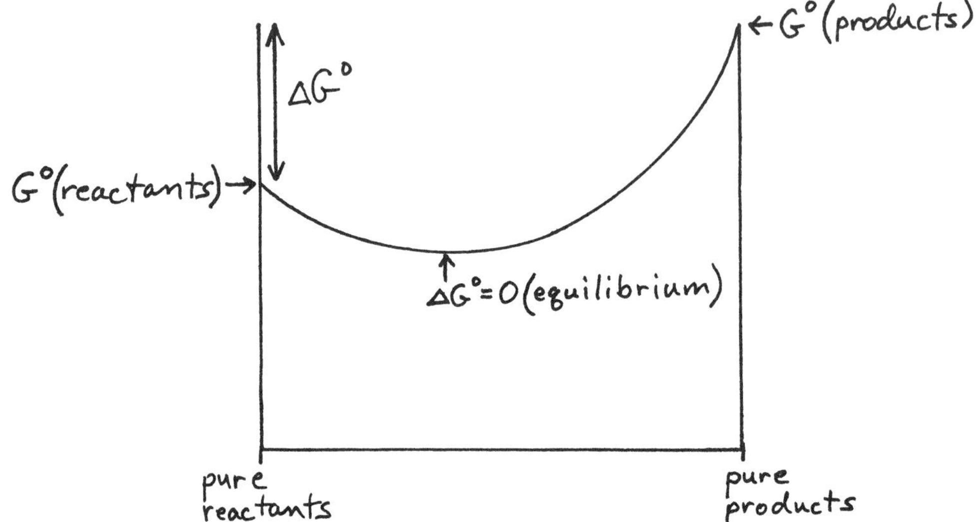

The curve has a minimum between pure reactants and pure products because the free energy decreases when pure reactants (or pure products) react to form the equilibrium mixture. The minimum is on the left side of the diagram because a positive value of $\Delta G°$ corresponds to a value of K less than 1.

17.88 For $C_6H_6(s) \rightarrow C_6H_6(l)$, both ΔH and ΔS are positive.

(a) ΔH and ΔS are positive, and because 0°C is below the melting point, ΔG is also positive.

(b) ΔH and ΔS are positive, and because 15°C is above the melting point, the reaction takes place and ΔG is negative.

17.89 (a) $\Delta H = 0$

$$\Delta S = R \ln \frac{V_{final}}{V_{initial}} = (8.314 \text{ J/K}) \ln 2 = 5.76 \text{ J/K}$$

$\Delta G = \Delta H - T\Delta S$

Because $\Delta H = 0$, $\Delta G = -T\Delta S = -(298 \text{ K})(5.76 \text{ J/K}) = -1717 \text{ J} = -1.72 \text{ kJ}$

(b) For a process in an isolated system, $\Delta S_{surr} = 0$. Therefore, $\Delta S_{total} = \Delta S_{sys} > 0$, and the process is spontaneous.

17.90 (a) $\Delta G = \Delta H - T\Delta S$; $\Delta H = \Delta G + T\Delta S$

Because ΔS is positive and ΔG is positive (the reaction is nonspontaneous), ΔH must be positive. Therefore, the reaction is endothermic.

(b) $\Delta G = \Delta H - T\Delta S$

If ΔG is set to 0, the minimum value for ΔH can be calculated.

$0 = \Delta H - T\Delta S$

$\Delta H = T\Delta S = (323 \text{ K})(104 \times 10^{-3} \text{ kJ/K}) = 33.6 \text{ kJ}$

17.91 (a)

	$\Delta H_{vap}/T_{bp}$
ammonia	120 J/K
benzene	87 J/K
carbon tetrachloride	85 J/K
chloroform	87 J/K
mercury	90 J/K

(b) All processes are the conversion of a liquid to a gas at the boiling point. They should should all have similar ΔS values. $\Delta H_{vap}/T_{bp}$ is equal to ΔS_{vap}.

(c) NH_3 deviates from Trouton's rule because of hydrogen bonding. Because $NH_3(l)$ is more ordered than the other liquids, ΔS_{vap} is larger.

17.92 $MgCO_3(s) \rightarrow MgO(s) + CO_2(g)$

From Problem 17.85(b)

$\Delta H° = +101$ kJ

$\Delta S° = 174.8$ J/K $= 174.8 \times 10^{-3}$ kJ/K

The equilibrium pressure of CO_2 is equal to $K_p = P_{CO_2}$. K_p is not affected by the quantities of $MgCO_3$ and MgO present. K_p can be calculated from $\Delta G°$.

$\Delta G° = \Delta H° - T\Delta S°$

$\Delta G° = -2.303RT \log K_p$

(a) $\Delta G° = 101 \text{ kJ} - (298 \text{ K})(174.8 \times 10^{-3} \text{ kJ/K}) = +49 \text{ kJ}$

$$\log K_p = \frac{-\Delta G°}{2.303RT} = \frac{-(49 \text{ kJ})}{(2.303)(8.314 \times 10^{-3} \text{ kJ/K})(298 \text{ K})} = -8.6$$

$K_p = P_{CO_2} = 10^{-8.6} = 3 \times 10^{-9}$ atm

(b) $\Delta G° = 101 \text{ kJ} - (553 \text{ K})(174.8 \times 10^{-3} \text{ kJ/K}) = 4.3 \text{ kJ}$

$$\log K_p = \frac{-\Delta G°}{2.303RT} = \frac{-(4.3 \text{ kJ})}{(2.303)(8.314 \times 10^{-3} \text{ kJ/K})(553 \text{ K})} = -0.4$$

$K_p = P_{CO_2} = 10^{-0.4} = 0.4$ atm

(c) $P_{CO_2} = 0.4$ atm because the temperature is the same as in (b).

17.93 $\Delta G° = -2.303RT \log K_b$

At 20 °C:

$\Delta G° = -(2.303)(8.314 \times 10^{-3} \text{ kJ/K})(293 \text{ K})\log(1.710 \times 10^{-5}) = +26.74 \text{ kJ}$

At 50 °C:

$\Delta G° = -(2.303)(8.314 \times 10^{-3}\ kJ/K)(323\ K)\log(1.892 \times 10^{-5}) = +29.21\ kJ$

$\Delta G° = \Delta H° - T\Delta S°$

$26.74 = \Delta H° - 293\Delta S°$
$29.21 = \Delta H° - 323\Delta S°$ Solve these two equations simultaneously for $\Delta H°$ and $\Delta S°$.

$26.74 + 293\Delta S° = \Delta H°$
$29.21 + 323\Delta S° = \Delta H°$ Set these two equations equal to each other.

$26.74 + 293\Delta S° = 29.21 + 323\Delta S°$

$26.74 - 29.21 = 323\Delta S° - 293\ \Delta S°$

$-2.47 = 30\Delta S°$

$\Delta S° = -2.47/30 = -0.0823 = -0.0823\ kJ/K = -82.3\ J/K$

$26.74 + 293\Delta S° = 26.74 + 293(-0.0823) = \Delta H° = +2.62\ kJ$

17.94 (a) $\Delta G° = \Delta H° - T\Delta S°$ and $\Delta G° = -RT \ln K$

Set the two equations equal to each other.

$-RT \ln K = \Delta H° - T\Delta S°$

$$\ln K = \frac{\Delta H° - T\Delta S°}{-RT}$$

$$\ln K = \frac{-\Delta H°}{RT} + \frac{T\Delta S°}{RT}$$

$$\ln K = \frac{-\Delta H°}{RT} + \frac{\Delta S°}{R}$$

$$\ln K = \frac{-\Delta H°}{R}\left(\frac{1}{T}\right) + \frac{\Delta S°}{R}$$

This is the equation for a straight line ($y = mx + b$).

$y = \ln K$

$m = -\dfrac{\Delta H^\circ}{R}$ = slope

$x = \dfrac{1}{T}$

$b = \dfrac{\Delta S^\circ}{R}$ = intercept

(b) Plot ln k versus 1/T

$\Delta H^\circ = -(R)(\text{slope})$ $\Delta S^\circ = R(\text{intercept})$

(c) For a reaction where K increases with increasing temperature, the following plot would be obtained:

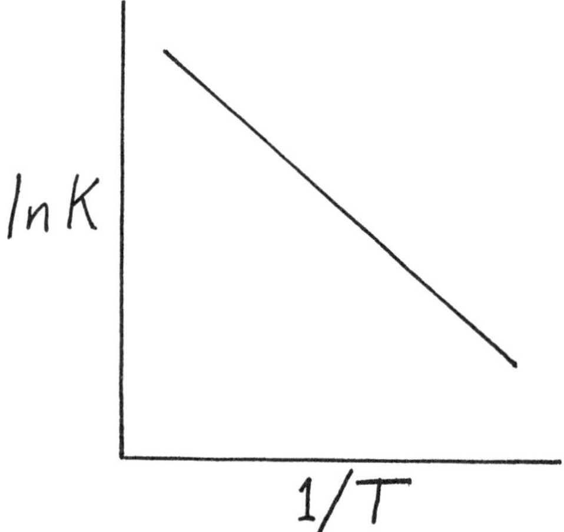

The slope is negative. Because $\Delta H^\circ = -(R)(\text{slope})$, ΔH° is positive, and the reaction is endothermic.

This prediction is in accord with LeChâtelier's principle because when you add heat (raise the temperature) for an endothermic reaction, the reaction in the forward direction takes place, the product concentrations increase and the reactant concentrations decrease. This results in an increase in K.

17.95 $Br_2(l) \rightleftarrows Br_2(g)$

$\Delta S° = S°(Br_2(g)) - S°(Br_2(l))$

$\Delta S° = (1 \text{ mol})(245.4 \text{ J/K}) - (1 \text{ mol})(152.2 \text{ J/K}) = 93.2 \text{ J/K} = 93.2 \times 10^{-3} \text{ kJ/K}$

$$T_{bp} = \frac{\Delta H_{vap}}{\Delta S_{vap}} \approx \frac{\Delta H°}{\Delta S°}$$

$\Delta H° = T_{bp}\,\Delta S° = (332 \text{ K})(93.2 \times 10^{-3} \text{ kJ/K}) = 30.9 \text{ kJ}$

$$K_p = P_{Br_2} = \left(227 \text{ mm Hg} \times \frac{1 \text{ atm}}{760 \text{ mm Hg}}\right) = 0.299 \text{ atm}$$

$\Delta G° = -2.303RT \log K_p$ and $\Delta G° = \Delta H° - T\Delta S°$
(set equations equal to each other)

$\Delta H° - T\Delta S° = -2.303RT \log K_p$ (rearrange)

$$\log K_p = \frac{-\Delta H°}{2.303R}\frac{1}{T} + \frac{\Delta S°}{2.303R} \quad \text{(solve for T)}$$

$$T = \frac{\left(\frac{-\Delta H°}{2.303R}\right)}{\left(\log K_p - \frac{\Delta S°}{2.303R}\right)} = \frac{\left(\frac{-30.9 \text{ kJ}}{(2.303)(8.314 \times 10^{-3} \text{ kJ})}\right)}{\left(\log(0.299) - \frac{93.2 \times 10^{-3} \text{ kJ}}{(2.303)(8.314 \times 10^{-3} \text{ kJ})}\right)} = 299 \text{ K} = 26°\text{C}$$

$Br_2(l)$ has a vapor pressure of 227 mm Hg at 26°C.

17.96 $N_2O_4(g) \rightleftarrows 2\,NO_2(g)$

$\Delta H° = 2\,\Delta H_f°(NO_2) - \Delta H_f°(N_2O_4)$

$\Delta H° = (2 \text{ mol})(33.2 \text{ kJ}) - (1 \text{ mol})(9.16 \text{ kJ}) = 57.2 \text{ kJ}$

$\Delta S° = 2\,S°(NO_2) - S°(N_2O_4)$

$\Delta S° = (2 \text{ mol})(240.0 \text{ J/(K} \cdot \text{mol)}) - (1 \text{ mol})(304.2 \text{ J/(K} \cdot \text{mol)})$

$\Delta S° = 175.8 \text{ J/K} = 175.8 \times 10^{-3} \text{ kJ/K}$

$\Delta G° = \Delta H° - T\Delta S° = 57.2 \text{ kJ} - (373 \text{ K})(175.8 \times 10^{-3} \text{ kJ/K}) = -8.4 \text{ kJ}$

$$K_p = \frac{(P_{NO_2})^2}{P_{N_2O_4}}$$

$\Delta G° = -2.303RT \log K_p$

$$\log K_p = \frac{-\Delta G°}{2.303RT} = \frac{-(-8.4 \text{ kJ})}{(2.303)(8.314 \times 10^{-3} \text{ kJ/K})(373 \text{ K})} = 1.2$$

$K_p = 10^{1.2} = 16$

	$N_2O_4(g)$	⇄	$2\ NO_2(g)$
initial (atm)	1		1
change (atm)	$-x$		$+2x$
equil (atm)	$1 - x$		$1 + 2x$

$$K_p = \frac{(P_{NO_2})^2}{P_{N_2O_4}} = 16 = \frac{(1+2x)^2}{(1-x)}$$

$4x^2 + 20x - 15 = 0$

Use the quadratic formula to solve for x.

$$x = \frac{(-20) \pm \sqrt{(20)^2 + (4)(4)(15)}}{2(4)} = \frac{-20 \pm 25.3}{8}$$

$x = 0.66$ and -5.7

Of the two solutions for x, only 0.66 has physical meaning because $x = -5.7$ would lead to a negative pressure for NO_2.

$P_{N_2O_4} = 1 - x = 1 - 0.66 = 0.34 \text{ atm}$

$P_{NO_2} = 1 + 2x = 1 + 2(0.66) = 2.32 \text{ atm}$

17.97 $2\ SO_2(g) + O_2(g) \rightleftarrows 2\ SO_3(g)$

$\Delta H° = 2\ \Delta H_f°(SO_3) - 2\ \Delta H_f°(SO_2)$

$\Delta H° = (2\text{ mol})(-395.7\text{ kJ/mol}) - (2\text{ mol})(-296.8\text{ kJ/mol}) = -197.8\text{ kJ}$

$\Delta S° = 2\ S°(SO_3) - [2\ S°(SO_2) + S°(O_2)]$

$\Delta S° = (2\text{ mol})(256.6\text{ J/(K}\cdot\text{mol))}$

$- [(2\text{ mol})(248.1\text{ J/(K}\cdot\text{mol))} + (1\text{ mol})(205.0\text{ J/(K}\cdot\text{mol))}]$

$\Delta S° = -188.0\text{ J/K} = -188.0 \times 10^{-3}\text{ kJ/K}$

$\Delta G° = \Delta H° - T\Delta S° = -197.8\text{ kJ} - (800\text{ K})(-188.0 \times 10^{-3}\text{ kJ/K}) = -47.4\text{ kJ}$

$\Delta G° = -2.303RT\log K_p$

$$\log K_p = \frac{-\Delta G°}{2.303RT} = \frac{-(-47.4\text{ kJ})}{(2.303)(8.314 \times 10^{-3}\text{ kJ/K})(800\text{ K})} = 3.094$$

$K_p = 10^{3.094} = 1242$

SO_2, 64.06 amu; O_2, 32.00 amu

$(192\text{ g})(1\text{ mol}/64.06\text{ g}) = 3.00\text{ mol } SO_2$

$(48.0\text{ g})(1\text{ mol}/32.00\text{ g}) = 1.50\text{ mol } O_2$

At 800 K:

$$P_{SO_2} = \frac{(3.00\text{ mol})\left(0.08206\ \frac{\text{L}\cdot\text{atm}}{\text{mol}\cdot\text{K}}\right)(800\text{ K})}{15.0\text{ L}} = 13.1\text{ atm}$$

$$P_{O_2} = \frac{(1.50\text{ mol})\left(0.08206\ \frac{\text{L}\cdot\text{atm}}{\text{mol}\cdot\text{K}}\right)(800\text{ K})}{15.0\text{ L}} = 6.57\text{ atm}$$

	$2\ SO_2(g)$	+	$O_2(g)$	⇄	$2\ SO_3(g)$
initial (atm)	13.1		6.57		0
assume complete rxn (atm)	0		0		13.1
assume a small back rxn	+2x		+x		–2x
equil (atm)	2x		x		13.1 – 2x

$$K_p = 1242 = \frac{[SO_3]^2}{[SO_2]^2[O_2]} = \frac{(13.1 - 2x)^2}{(2x)^2(x)} \approx \frac{(13.1)^2}{(2x)^2(x)}$$

Solve for x. $x^3 = 0.0345$; $x = 0.326$

Use successive approximations to solve for x because 2x is not negligible compared with 13.1.

Second approximation:

$$1242 = \frac{[13.1 - (2)(0.326)]^2}{(2x)^2(x)}$$

Solve for x. $x^3 = 0.0312$; $x = 0.315$

Third approximation:

$$1242 = \frac{[13.1 - (2)(0.315)]^2}{(2x)^2(x)}$$

Solve for x. $x^3 = 0.0313$; $x = 0.315$ (x has converged)

$P_{SO_2} = 2x = 2(0.315) = 0.63$ atm

$P_{O_2} = x = 0.32$ atm

$P_{SO_3} = 13.1 - 2x = 13.1 - 2(0.315) = 12.5$ atm

(b) The % yield of SO_3 decreases with increasing temperature because $\Delta S°$ is negative. $\Delta G°$ becomes less negative and K_p gets smaller as the temperature increases.

(c) At 1000 K:

$\Delta G^\circ = \Delta H^\circ - T\Delta S^\circ = -197.8\text{ kJ} - (1000\text{ K})(-188.0 \times 10^{-3}\text{ kJ/K}) = -9.8\text{ kJ}$

$\Delta G^\circ = -2.303RT \log K_p$

$$\log K_p = \frac{-\Delta G^\circ}{2.303RT} = \frac{-(-9.8\text{ kJ})}{(2.303)(8.314 \times 10^{-3}\text{ kJ/K})(1000\text{ K})} = 0.512$$

$K_p = 10^{0.512} = 3.25$

$$P_{SO_2} = \frac{(3.00\text{ mol})\left(0.08206\,\frac{\text{L}\cdot\text{atm}}{\text{mol}\cdot\text{K}}\right)(1000\text{ K})}{15.0\text{ L}} = 16.4\text{ atm}$$

$$P_{O_2} = \frac{(1.50\text{ mol})\left(0.08206\,\frac{\text{L}\cdot\text{atm}}{\text{mol}\cdot\text{K}}\right)(1000\text{ K})}{15.0\text{ L}} = 8.2\text{ atm}$$

	$2\ SO_2(g)$	+	$O_2(g)$	⇄	$2\ SO_3(g)$
initial (atm)	16.4		8.2		0
assume complete rxn (atm)	0		0		16.4
assume a small back rxn	+2x		+x		−2x
equil (atm)	2x		x		16.4 − 2x

$$K_p = 3.25 = \frac{[SO_3]^2}{[SO_2]^2[O_2]} = \frac{(16.4 - 2x)^2}{(2x)^2(x)} \approx \frac{(16.4)^2}{(2x)^2(x)}$$

Solve for x. $x^3 = 20.7$; $x = 2.7$

Use successive approximations to solve for x because 2x is not negligible compared with 16.4.

Second approximation:

$$3.25 = \frac{[16.4 - (2)(2.7)]^2}{(2x)^2(x)}$$

Solve for x. $x^3 = 9.31$; $x = 2.1$

Third approximation:

$$3.25 = \frac{[16.4 - (2)(2.1)]^2}{(2x)^2(x)}$$

Solve for x. $x^3 = 11.4$; $x = 2.3$

Fourth approximation:

$$3.25 = \frac{[16.4 - (2)(2.3)]^2}{(2x)^2(x)}$$

Solve for x. $x^3 = 10.7$; $x = 2.2$ (x has converged)

$P_{SO_2} = 2x = 2(2.2) = 4.4$ atm

$P_{O_2} = x = 2.2$ atm

$P_{SO_3} = 16.4 - 2x = 16.4 - 2(2.2) = 12.0$ atm

$P_{total} = P_{SO_2} + P_{O_2} + P_{SO_3} = 4.4 + 2.2 + 12.0 = 18.6$ atm

On going from 800 K to 1000 K, P_{total} increases to 18.6 atm (because K_p decreases, but P increases with temperature at constant volume).

18.1 $2\ Ag^{+}(aq) + Ni(s) \rightarrow 2\ Ag(s) + Ni^{2+}(aq)$

There is a Ni anode in an aqueous solution of Ni^{2+}, and a Ag cathode in an aqueous solution of Ag^{+}. A salt bridge connects the anode and cathode compartment. The electrodes are connected through an external circuit.

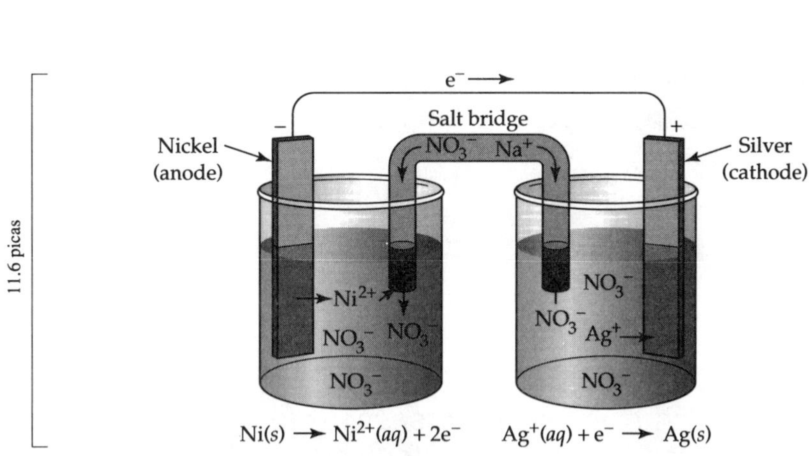

18.2 $Fe(s)\ |\ Fe^{2+}(aq)\ ||\ Sn^{2+}(aq)\ |\ Sn(s)$

18.3 $Pb(s) + Br_2(l) \rightarrow Pb^{2+}(aq) + 2\ Br^{-}(aq)$

There is a Pb anode in an aqueous solution of Pb^{2+}. The cathode is a Pt wire that dips into a pool of liquid Br_2 and an aqueous solution that is saturated with Br_2. A salt bridge connects the anode and cathode compartment. The electrodes are connected through an external circuit.

18.4 $Al(s) + Cr^{3+}(aq) \rightarrow Al^{3+}(aq) + Cr(s)$

$$\Delta G° = -nFE° = -(3 \text{ mol } e^{-})\left(\frac{96{,}500 \text{ C}}{1 \text{ mol } e^{-}}\right)(0.92 \text{ V})\left(\frac{1 \text{ J}}{1 \text{ C} \cdot \text{V}}\right)$$

$$\Delta G° = -266{,}340 \text{ J} = -270 \text{ kJ}$$

18.5

oxidation:	$Al(s) \rightarrow Al^{3+}(aq) + 3\ e^{-}$	$E° = 1.66$ V
reduction:	$\underline{Cr^{3+}(aq) + 3\ e^{-} \rightarrow Cr(s)}$	$\underline{E° = ?}$
overall	$Al(s) + Cr^{3+}(aq) \rightarrow Al^{3+}(aq) + Cr(s)$	$E° = 0.92$ V

The standard reduction potential for the Cr^{3+}/Cr half cell is:

E° = 0.92 – 1.66 = –0.74 V

18.6 (a) $Cl_2(g) + 2\ e^- \rightarrow 2\ Cl^-(aq)$ E° = 1.36 V

$Ag^+(aq) + e^- \rightarrow Ag(s)$ E° = 0.80 V

Cl_2 has the greater tendency to be reduced (larger E°). The species that has the greater tendency to be reduced is the stronger oxidizing agent. Cl_2 is the stronger oxidizing agent.

(b) $Fe^{2+}(aq) + 2\ e^- \rightarrow Fe(s)$ E° = –0.45 V

$Mg^{2+}(aq) + 2\ e^- \rightarrow Mg(s)$ E° = –2.37 V

The second half-reaction has the lesser tendency to occur in the forward direction (more negative E°) and the greater tendency to occur in the reverse direction. Therefore, Mg is the stronger reducing agent.

18.7 (a) $2\ Fe^{3+}(aq) + 2\ I^-(aq) \rightarrow 2\ Fe^{2+}(aq) + I_2(s)$

reduction: $Fe^{3+}(aq) + e^- \rightarrow Fe^{2+}(aq)$ E° = 0.77 V

oxidation: $2\ I^-(aq) \rightarrow I_2(s) + 2\ e^-$ <u>E° = –0.54 V</u>

overall E° = 0.23 V

Because E° for the overall reaction is positive, this reaction can occur under standard-state conditions.

(b) $3\ Ni(s) + 2\ Al^{3+}(aq) \rightarrow 3\ Ni^{2+}(aq) + 2\ Al(s)$

oxidation: $Ni(s) \rightarrow Ni^{2+}(aq) + 2\ e^-$ E° = 0.26 V

reduction: $Al^{3+}(aq) + 3\ e^- \rightarrow Al(s)$ <u>E° = –1.66 V</u>

overall E° = –1.40 V

Because E° for the overall reaction is negative, this reaction cannot occur under standard-state conditions. This reaction can occur in the reverse direction.

18.8 $Cu(s) + 2\ Fe^{3+}(aq) \rightarrow Cu^{2+}(aq) + 2\ Fe^{2+}(aq)$

$$E° = E°_{Cu \rightarrow Cu^{2+}} + E°_{Fe^{3+} \rightarrow Fe^{2+}} = -0.34\ V + 0.77\ V = 0.43\ V$$

$n = 2$ mol e^-

$$E = E° - \frac{0.0592}{n} \log \frac{[Cu^{2+}][Fe^{2+}]^2}{[Fe^{3+}]^2}$$

$$E = 0.43\ V - \frac{(0.0592\ V)}{2} \log \frac{(0.25)(0.20)^2}{(1.0 \times 10^{-4})^2} = 0.25\ V$$

18.9 $H_2(g) + Pb^{2+}(aq) \rightarrow 2\ H^+(aq) + Pb(s)$

$$E° = E°_{H_2 \rightarrow H^+} + E°_{Pb^{2+} \rightarrow Pb} = 0\ V + (-0.13\ V) = -0.13\ V$$

$$E = E° - \frac{0.0592}{n} \log \frac{[H_3O^+]^2}{[Pb^{2+}](P_{H_2})}$$

$$0.28\ V = -0.13\ V - \frac{(0.0592\ V)}{2} \log \frac{[H_3O^+]^2}{(1)(1)}$$

$$0.28\ V = -0.13\ V - (0.0592\ V) \log [H_3O^+]$$

$$pH = -\log [H_3O^+]$$

$$0.28\ V = -0.13\ V + (0.0592\ V)\ pH$$

$$pH = \frac{(0.28\ V + 0.13\ V)}{0.0592\ V} = 6.9$$

18.10 $4\ Fe^{2+}(aq) + O_2(g) + 4\ H^+(aq) \rightarrow 4\ Fe^{3+}(aq) + 2\ H_2O(l)$

$$E° = E°_{Fe^{2+} \rightarrow Fe^{3+}} + E°_{O_2 \rightarrow H_2O} = -0.77\ V + 1.23\ V = 0.46\ V$$

$$E° = \frac{0.0592}{n} \log K$$

$n = 4 \text{ mol } e^-$

$$\log K = \frac{nE^\circ}{0.0592} = \frac{(4)(0.46\text{ V})}{0.0592\text{ V}} = 31$$

$K = 10^{31}$ at 25°C

18.11 $E^\circ = \dfrac{0.0592}{n}\log K = \dfrac{0.0592}{2}\log(1.8 \times 10^{-5}) = -0.140\text{ V}$

18.12 (a) $Zn(s) + 2\ MnO_2(s) + 2\ NH_4^+(aq) \rightarrow$

$Zn^{2+}(aq) + Mn_2O_3(s) + 2\ NH_3(aq) + H_2O(l)$

(b) $Zn(s) + 2\ MnO_2(s) \rightarrow ZnO(s) + Mn_2O_3(s)$

(c) $Zn(s) + HgO(s) \rightarrow ZnO(s) + Hg(l)$

(d) $Cd(s) + 2\ NiO(OH)(s) + 2\ H_2O(l) \rightarrow Cd(OH)_2(s) + 2\ Ni(OH)_2(s)$

18.13 (a)

cathode reaction	$2\ H_2O(l) + 2\ e^- \rightarrow H_2(g) + 2\ OH^-(aq)$
anode reaction	$\underline{2\ Cl^-(aq) \rightarrow Cl_2(g) + 2\ e^-}$
overall reaction	$2\ Cl^-(aq) + 2\ H_2O(l) \rightarrow Cl_2(g) + H_2(g) + 2\ OH^-(aq)$

(b)

cathode reaction	$Cu^{2+}(aq) + 2\ e^- \rightarrow Cu(s)$
anode reaction	$\underline{2\ H_2O(l) \rightarrow O_2(g) + 4\ H^+(aq) + 4\ e^-}$
overall reaction	$2\ Cu^{2+}(aq) + 2\ H_2O(l) \rightarrow 2\ Cu(s) + O_2(g) + 4\ H^+(aq)$

18.14 $\text{Charge} = \left(100{,}000\ \dfrac{C}{s}\right)(8\ h)\left(\dfrac{60\ min}{h}\right)\left(\dfrac{60\ s}{min}\right) = 2.88 \times 10^9\ C$

$\text{Moles of } e^- = (2.88 \times 10^9\ C)\left(\dfrac{1\ mol\ e^-}{96{,}500\ C}\right) = 2.98 \times 10^4\ mol\ e^-$

cathode reaction: $Al^{3+} + 3\ e^- \rightarrow Al$

$$\text{mass Al} = (2.98 \times 10^4 \text{ mol e}^-) \times \frac{1 \text{ mol Al}}{3 \text{ mol e}^-} \times \frac{26.98 \text{ g Al}}{1 \text{ mol Al}} \times \frac{1 \text{ kg}}{1000 \text{ g}} = 268 \text{ kg Al}$$

18.15 $3.00 \text{ g Ag} \times \frac{1 \text{ mol Ag}}{107.9 \text{ g Ag}} = 0.0278 \text{ mol Ag}$

cathode reaction: $Ag^+(aq) + e^- \rightarrow Ag(s)$

$$\text{Charge} = (0.0278 \text{ mol Ag})\left(\frac{1 \text{ mol e}^-}{1 \text{ mol Ag}}\right)\left(\frac{96{,}500 \text{ C}}{1 \text{ mol e}^-}\right) = 2682.7 \text{ C}$$

$$\text{Time} = \frac{\text{C}}{\text{A}} = \left(\frac{2682.7 \text{ C}}{0.100 \text{ C/s}} \times \frac{1 \text{ h}}{3600 \text{ s}}\right) = 7.45 \text{ h}$$

Understanding Key Concepts

1. (a) - (d)

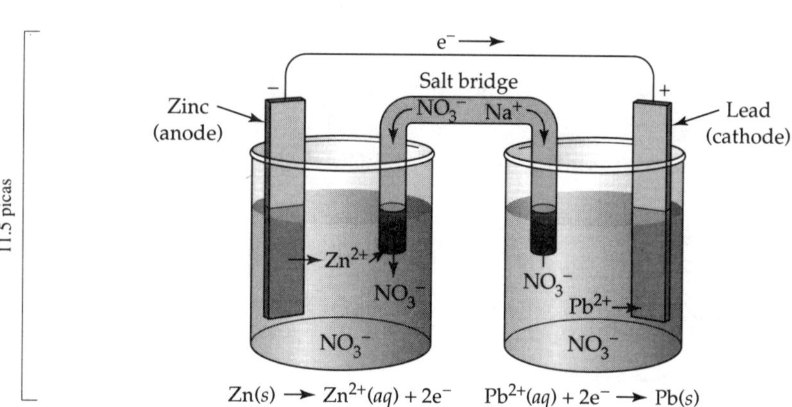

(e) anode reaction $Zn(s) \rightarrow Zn^{2+}(aq) + 2\ e^-$

cathode reaction $Pb^{2+}(aq) + 2\ e^- \rightarrow Pb(s)$

overall reaction $Zn(s) + Pb^{2+}(aq) \rightarrow Zn^{2+}(aq) + Pb(s)$

2. (a) - (b)

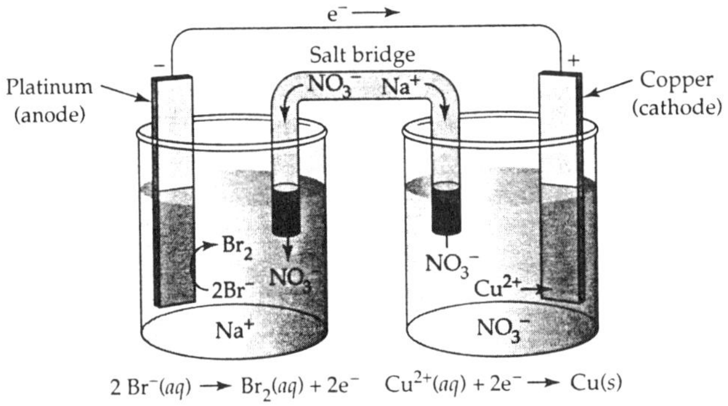

(c) anode reaction $2\ Br^-(aq) \rightarrow Br_2(aq) + 2\ e^-$

cathode reaction $Cu^{2+}(aq) + 2\ e^- \rightarrow Cu(s)$

overall reaction $Cu^{2+}(aq) + 2\ Br^-(aq) \rightarrow Cu(s) + Br_2(aq)$

3. (a) oxidizing agents: PbO_2, H^+, $Cr_2O_7^{2-}$

reducing agents: Al, Fe, Ag

(b) PbO_2 is the strongest oxidizing agent.

H^+ is the weakest oxidizing agent.

(c) Al is the strongest reducing agent.

Ag is the weakest reducing agent.

(d) oxidized by Cu^{2+}: Fe and Al

reduced by H_2O_2: PbO_2 and $Cr_2O_7^{2-}$

4. (a) From: $B + A^+ \rightarrow B^+ + A$, A^+ is reduced more easily than B^+

From: $C + A^+ \rightarrow C^+ + A$, A^+ is reduced more easily than C^+

From: $B + C^+ \rightarrow B^+ + C$, C^+ is reduced more easily than B^+

$A^+ + e^- \rightarrow A$
$C^+ + e^- \rightarrow C$
$B^+ + e^- \rightarrow B$

(b) A^+ is the strongest oxidizing agent

B is the strongest reducing agent

(c) $A^+ + B \rightarrow B^+ + A$

5. Sn is not as easily oxidized as Fe and it does not offer cathodic protection. Once the Fe is exposed (scratched) the tin can rusts.

   A galvanized can is Fe coated with Zn. The Zn offers cathodic protection even if the Fe is exposed. The Zn is sacrificially oxidized.

Additional Problems Galvanic Cells

18.16 The electrode where oxidation takes place is called the anode. For example, the lead electrode in the lead storage battery.

The electrode where reduction takes place is called the cathode. For example, the PbO_2 electrode in the lead storage battery.

18.17 The oxidizing agent gets reduced and reduction takes place at the cathode.

18.18 The cathode of a galvanic cell is considered to be the positive electrode because electrons flow through the external circuit toward the positive electrode (the cathode).

18.19 The salt bridge maintains charge neutrality in both the anode and cathode compartments of a galvanic cell.

18.20 (a) $Cd(s) + Sn^{2+}(aq) \rightarrow Cd^{2+}(aq) + Sn(s)$

21.5 picas

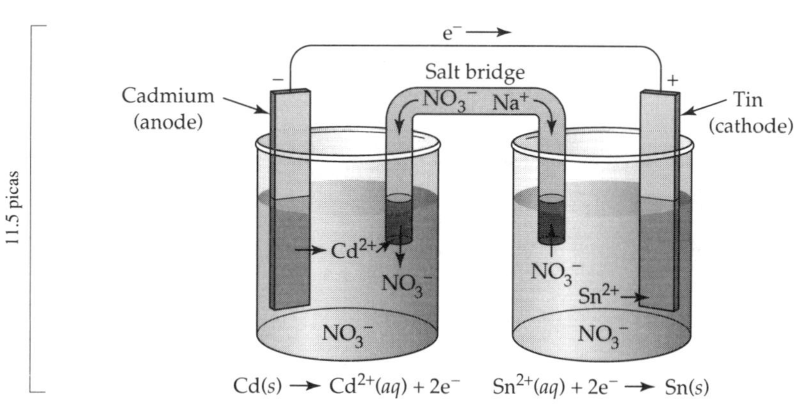

(b) $2\ Al(s)\ +\ 3\ Cd^{2+}(aq)\ \rightarrow\ 2\ Al^{3+}(aq)\ +\ 3\ Cd(s)$

23 picas

11.5 picas

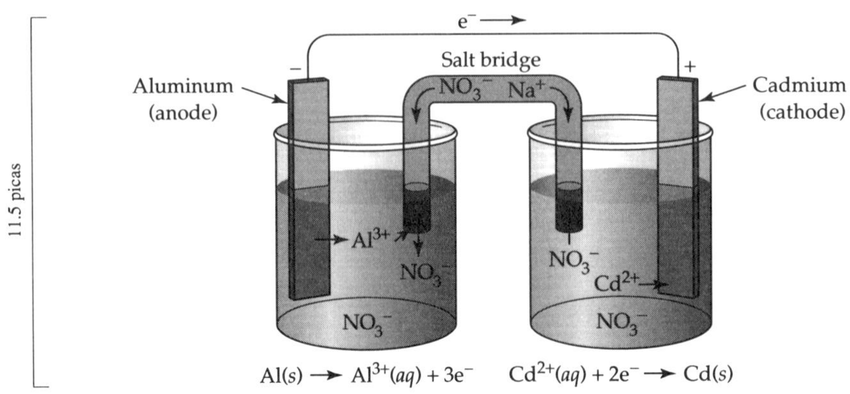

(c) $Pb(s)\ +\ 2\ H^{+}(aq)\ \rightarrow\ Pb^{2+}(aq)\ +\ H_2(g)$

23.2 picas

11.6 picas

e⁻ →

Salt bridge

NO_3^- Na^+

Lead (anode)

$H_2(g)$

Pb^{2+}

NO_3^-

NO_3^-

H^+

Standard hydrogen electrode (cathode)

NO_3^-

NO_3^-

$Pb(s) \rightarrow Pb^{2+}(aq) + 2e^-$ $\quad 2\,H^+(aq) + 2e^- \rightarrow H_2(g)$

(d) $6\ Fe^{2+}(aq) + Cr_2O_7^{2-}(aq) + 14\ H^{+}(aq)\ \rightarrow\ 6\ Fe^{3+}(aq) + 2\ Cr^{3+}(aq) + 7\ H_2O(l)$

26 picas

11.7 picas

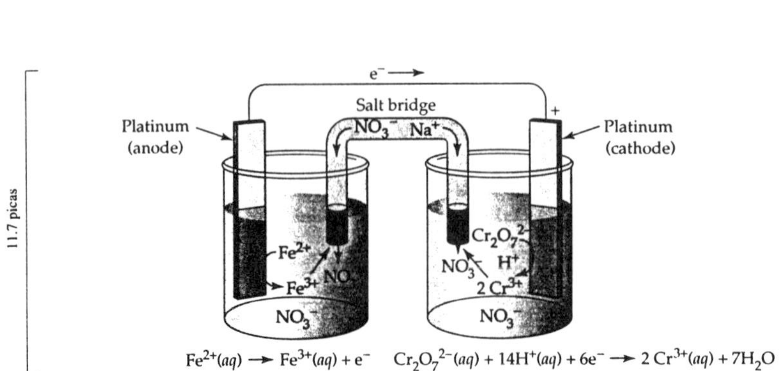

18.21 (a) $Cd(s) \mid Cd^{2+}(aq) \parallel Sn^{2+}(aq) \mid Sn(s)$

(b) $Al(s) \mid Al^{3+}(aq) \parallel Cd^{2+}(aq) \mid Cd(s)$

(c) $Pb(s) \mid Pb^{2+}(aq) \parallel H^{+}(aq) \mid H_2(g) \mid Pt(s)$

(d) $Pt(s) \mid Fe^{2+}(aq), Fe^{3+}(aq) \parallel Cr_2O_7^{2-}(aq), Cr^{3+}(aq) \mid Pt(s)$

18.22 (a)

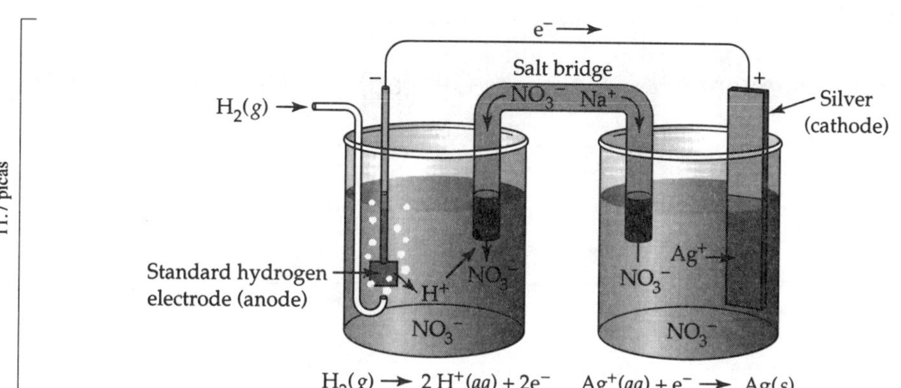

(b) anode reaction $H_2(g) \rightarrow 2\ H^{+}(aq) + 2\ e^{-}$

cathodes reaction $2\ Ag^{+}(aq) + 2\ e^{-} \rightarrow 2\ Ag(s)$

overall reaction $H_2(g) + 2\ Ag^{+}(aq) \rightarrow 2\ H^{+}(aq) + 2\ Ag(s)$

(c) $Pt(s) \mid H_2(g) \mid H^{+}(aq) \parallel Ag^{+}(aq) \mid Ag(s)$

18.23 (a)

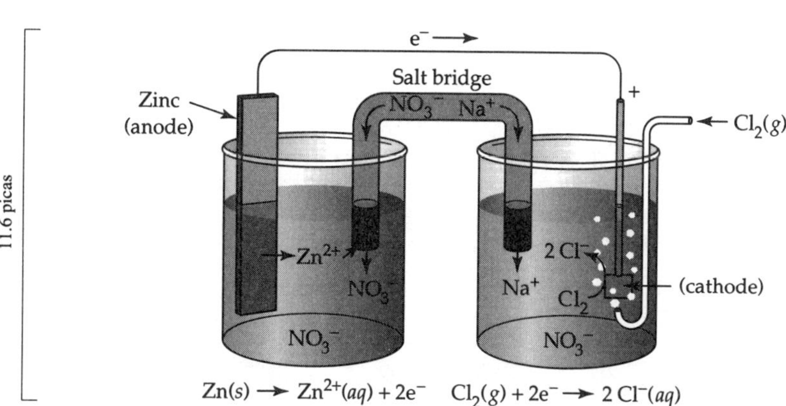

(b) anode reaction $Zn(s) \rightarrow Zn^{2+}(aq) + 2\ e^-$

cathode reaction $Cl_2(g) + 2\ e^- \rightarrow 2\ Cl^-(aq)$

overall reaction $Zn(s) + Cl_2(g) \rightarrow Zn^{2+}(aq) + 2\ Cl^-(aq)$

(c) $Zn(s)\ |\ Zn^{2+}(aq)\ \|\ Cl^-(aq)\ |\ Cl_2(g)\ |\ C(s)$

18.24 (a) anode reaction $Co(s) \rightarrow Co^{2+}(aq) + 2\ e^-$

cathode reaction $Cu^{2+}(aq) + 2\ e^- \rightarrow Cu(s)$

overall reaction $Co(s) + Cu^{2+}(aq) \rightarrow Co^{2+}(aq) + Cu(s)$

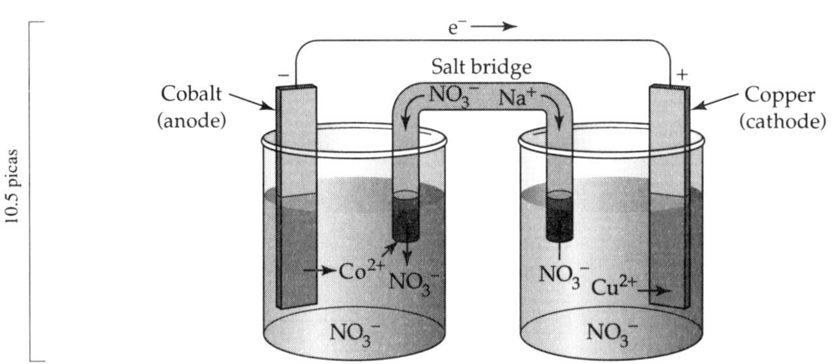

(b) anode reaction $2\ Fe(s) \rightarrow 2\ Fe^{2+}(aq) + 2\ e^-$

cathode reaction $O_2(g) + 4\ H^+(aq) + 4\ e^- \rightarrow 2\ H_2O(l)$

overall reaction $2\ Fe(s) + O_2(g) + 4\ H^+(aq) \rightarrow 2\ Fe^{2+}(aq) + 2\ H_2O(l)$

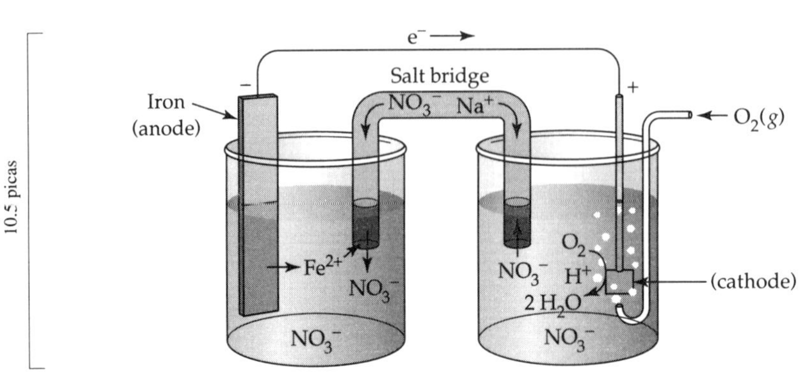

Cell Potentials and Free-Energy Changes;

Standard Reduction potentials

18.25 The SI unit of electrical potential is the volt (V).

The SI unit of charge is the coulomb (C).

The SI unit of energy is the joule (J).

$1\ J = 1\ C \cdot 1\ V$

18.26 $\Delta G = -nFE$

ΔG is the free energy for change for the cell reaction

n is the number of moles of e^-

F is the Faraday (96,500 C/mol e^-)

E is the galvanic cell potential

18.27 E is the standard cell potential (E°) when all reactants and products are in their standard states--solutes at 1 M concentrations, gases at a partial pressure of 1 atm, solids and liquids in pure form, all at 25°C.

18.28 The standard reduction potential is the potential of the reduction half reaction in a galvanic cell where the other electrode is the standard hydrogen electrode.

For a particular half reaction, the standard reduction potential and the standard oxidation potential always have the same magnitude, but they have opposite signs.

18.29 $Zn(s) + Ag_2O(s) \rightarrow ZnO(s) + 2\ Ag(s)$

n = 2 mol e^- and $1\ J = 1\ C \times 1\ V$

$$\Delta G = -nFE = -(2\ \text{mol e}^-)\left(\frac{96{,}500\ C}{1\ \text{mol e}^-}\right)(1.60\ V) = -308{,}800\ J = -309\ kJ$$

18.30 $Pb(s) + PbO_2(s) + 2\ H^+(aq) + 2\ HSO_4^-(aq) \rightarrow 2\ PbSO_4(s) + 2\ H_2O(l)$

$n = 2$ mol e^- and 1 J = 1 C x 1 V

$$\Delta G° = -nFE° = -(2 \text{ mol } e^-)\left(\frac{96{,}500 \text{ C}}{1 \text{ mol } e^-}\right)(1.924 \text{ V}) = -371{,}300 \text{ J} = -371 \text{ kJ}$$

18.31 $2\ H_2(g)\ +\ O_2(g)\ \rightarrow\ 2\ H_2O(l)$

$\Delta G° = 2\ \Delta G_f°(H_2O(l)) = (2 \text{ mol})(-237.2 \text{ kJ/mol}) = -474.4 \text{ kJ}$

$n = 4$ mol e^- and 1 V = 1 J/C

$\Delta G° = -nFE°$

$$E° = \frac{-\Delta G°}{nF} = \frac{-(-474{,}400 \text{ J})}{(4 \text{ mol } e^-)\left(\frac{96{,}500 \text{ C}}{1 \text{ mol } e^-}\right)} = +1.23 \text{ J/C} = +1.23 \text{ V}$$

18.32 $CH_4(g)\ +\ 2\ O_2(g)\ \rightarrow\ CO_2(g)\ +\ 2\ H_2O(l)$

$\Delta G° = [\Delta G_f(CO_2) + 2\ \Delta G_f°(H_2O(l))] - \Delta G_f°(CH_4)$

$\Delta G° = [(1 \text{ mol})(-394.4 \text{ kJ/mol}) + (2 \text{ mol})(-\ 237.2 \text{ kJ/mol})]$

$- (1 \text{ mol})(-50.8 \text{ kJ/mol}) = -818.0 \text{ kJ}$

$n = 8$ mol e^- and 1 V = 1 J/C

$\Delta G° = -nFE°$

$$E° = \frac{-\Delta G°}{nF} = \frac{-(-818{,}000 \text{ J})}{(8 \text{ mol } e^-)\left(\frac{96{,}500 \text{ C}}{1 \text{ mol } e^-}\right)} = +1.06 \text{ J/C} = +1.06 \text{ V}$$

18.33			
	oxidation:	$Zn(s) \rightarrow Zn^{2+}(aq) + 2\ e^-$	$E° = 0.76$ V
	reduction:	$Eu^{3+}(aq) + e^- \rightarrow Eu^{2+}(aq)$	$E° = ?$
	overall	$Zn(s) + 2\ Eu^{3+}(aq) \rightarrow Zn^{2+}(aq) + 2\ Eu^{2+}(aq)$	$E° = 0.40$ V

The standard reduction potential for the Eu^{3+}/Eu^{2+} half cell is:

E° = 0.40 – 0.76 = –0.36 V

18.34 $Cu^{2+}(aq) + 2\ Ag(s) + 2\ Br^{-}(aq) \rightarrow Cu(s) + 2\ AgBr(s)$ E° = 0.27 V

oxidation: $2\ Ag(s) + 2\ Br^{-}(aq) \rightarrow 2\ AgBr(s) + 2\ e^{-}$ E° = ?

reduction: $Cu^{2+}(aq) + 2\ e^{-} \rightarrow Cu(s)$ E° = 0.34 V

E° for the oxidation half reaction = 0.27 – 0.34 = –0.07 V

For $AgBr(s) + e^{-} \rightarrow Ag(s) + Br^{-}(aq)$, E° = –(–0.07 V) = +0.07 V

18.35 From Table 18.1, the following oxidizing agents are listed in order of increasing strength: $Cu^{2+} < O_2 < Cl_2$

18.36 From Table 18.1, the following reducing agents are listed in order of increasing strength: Ni < Zn < Mg

18.37 From Table 18.1:

MnO_4^- is the strongest oxidizing.

Fe^{2+} is the weakest oxidizing agent.

18.38 From Table 18.1:

Sn^{2+} is the strongest reducing agent.

Fe^{2+} is the weakest reducing agent.

18.39 (a) $Cd(s) + Sn^{2+}(aq) \rightarrow Cd^{2+}(aq) + Sn(s)$

oxidation: $Cd(s) \rightarrow Cd^{2+}(aq) + 2\ e^{-}$ E° = 0.40 V

reduction: $Sn^{2+}(aq) + 2\ e^{-} \rightarrow Sn(s)$ <u>E° = –0.14 V</u>

overall E° = 0.26 V

$n = 2$ mol e^- and 1 J = 1 C x 1 V

$$\Delta G° = -nFE° = -(2 \text{ mol } e^-)\left(\frac{96{,}500 \text{ C}}{1 \text{ mol } e^-}\right)(0.26 \text{ V}) = -50{,}180 \text{ J} = -50 \text{ kJ}$$

(b) $2\ Al(s) + 3\ Cd^{2+}(aq) \rightarrow 2\ Al^{3+}(aq) + 3\ Cd(s)$

oxidation:	$2\ Al(s) \rightarrow 2\ Al^{3+}(aq) + 6\ e^-$	E° = 1.66 V
reduction:	$3\ Cd^{2+}(aq) + 6\ e^- \rightarrow 3\ Cd(s)$	E° = –0.40 V
	overall	E° = 1.26 V

$n = 6$ mol e^- and 1 J = 1 C x 1 V

$$\Delta G° = -nFE° = -(6 \text{ mol } e^-)\left(\frac{96{,}500 \text{ C}}{1 \text{ mol } e^-}\right)(1.26 \text{ V}) = -729{,}540 \text{ J} = -730 \text{ kJ}$$

(c) $Pb(s) + 2\ H^+(aq) \rightarrow Pb^{2+}(aq) + H_2(g)$

oxidation:	$Pb(s) \rightarrow Pb^{2+}(aq) + 2\ e^-$	E° = 0.13 V
reduction:	$2\ H^+(aq) + 2\ e^- \rightarrow H_2(g)$	E° = 0.00 V
	overall	E° = 0.13 V

$n = 2$ mol e^- and 1 J = 1 C x 1 V

$$\Delta G° = -nFE° = -(2 \text{ mol } e^-)\left(\frac{96{,}500 \text{ C}}{1 \text{ mol } e^-}\right)(0.13 \text{ V}) = -25{,}090 \text{ J} = -25 \text{ kJ}$$

(d) $6\ Fe^{2+}(aq) + Cr_2O_7^{2-}(aq) + 14\ H^+(aq) \rightarrow 6\ Fe^{3+}(aq) + 2\ Cr^{3+}(aq) + 7\ H_2O(l)$

oxidation:	$6\ Fe^{2+}(aq) \rightarrow 6\ Fe^{3+}(aq) + 6\ e^-$	E° = –0.77 V
reduction:	$Cr_2O_7^{2-}(aq) + 14\ H^+(aq) + 6\ e^- + \rightarrow 2\ Cr^{3+}(aq) + 7\ H_2O(l)$	E° = 1.33 V
	overall	E° = 0.56 V

$n = 6$ mol e^- and 1 J = 1 C x 1 V

$$\Delta G° = -nFE° = -(6 \text{ mol e}^-)\left(\frac{96{,}500 \text{ C}}{1 \text{ mol e}^-}\right)(0.56 \text{ V}) = -324{,}240 \text{ J} = -324 \text{ kJ}$$

18.40 (a) $Co(s) + Cu^{2+}(aq) \rightarrow Co^{2+}(aq) + Cu(s)$

oxidation: $Co(s) \rightarrow Co^{2+}(aq) + 2\ e^-$ E° = 0.28 V

reduction: $Cu^{2+}(aq) + 2\ e^- \rightarrow Cu(s)$ <u>E° = 0.34 V</u>

overall E° = 0.62 V

n = 2 mol e$^-$ and 1 J = 1 C x 1 V

$$\Delta G° = -nFE° = -(2 \text{ mol e}^-)\left(\frac{96{,}500 \text{ C}}{1 \text{ mol e}^-}\right)(0.62 \text{ V}) = -119{,}660 \text{ J} = -120 \text{ kJ}$$

(b) $2\ Fe(s) + O_2(g) + 4\ H^+(aq) \rightarrow 2\ Fe^{2+}(aq) + 2\ H_2O(l)$

oxidation: $2\ Fe(s) \rightarrow 2\ Fe^{2+}(aq) + 4\ e^-$ E° = 0.45 V

reduction: $O_2(g) + 4\ H^+(aq) + 4\ e^- \rightarrow 2\ H_2O(l)$ <u>E° = 1.23 V</u>

overall E° = 1.68 V

n = 4 mol e$^-$ and 1 J = 1 C x 1 V

$$\Delta G° = -nFE° = -(4 \text{ mol e}^-)\left(\frac{96{,}500 \text{ C}}{1 \text{ mol e}^-}\right)(1.68 \text{ V}) = -648{,}480 \text{ J} = -648 \text{ kJ}$$

18.41 (a) $2\ Fe^{2+}(aq) + Pb^{2+}(aq) \rightarrow 2\ Fe^{3+}(aq) + Pb(s)$

oxidation: $2\ Fe^{2+}(aq) \rightarrow 2\ Fe^{3+}(aq) + 2\ e^-$ E° = –0.77 V

reduction: $Pb^{2+}(aq) + 2\ e^- \rightarrow Pb(s)$ <u>E° = –0.13 V</u>

overall E° = –0.90 V

Because E° is negative, this reaction is nonspontaneous.

(b) $Mg(s) + Ni^{2+}(aq) \rightarrow Mg^{2+}(aq) + Ni(s)$

oxidation: $Mg(s) \rightarrow Mg^{2+}(aq) + 2\ e^-$ E° = 2.37 V

reduction: $Ni^{2+}(aq) + 2\ e^- \rightarrow Ni(s)$ E° = –0.26 V

overall E° = 2.11 V

Because E° is positive, this reaction is spontaneous.

(c) $5\ Ag^+(aq) + Mn^{2+}(aq) + 4\ H_2O(l) \rightarrow 5\ Ag(s) + MnO_4^-(aq) + 8\ H^+(aq)$

oxidation: $Mn^{2+}(aq) + 4\ H_2O(l) \rightarrow MnO_4^-(aq) + 8\ H^+(aq) + 5\ e^-$ E° = –1.51 V

reduction: $5\ Ag^+(aq) + 5\ e^- \rightarrow 5\ Ag(s)$ E° = 0.80 V

overall E° = –0.71 V

Because E° is negative, this reaction is nonspontaneous.

(d) $2\ H_2O_2(aq) \rightarrow O_2(g) + 2\ H_2O(l)$

oxidation: $H_2O_2(aq) \rightarrow O_2(g) + 2\ H^+(aq) + 2\ e^-$ E° = –0.70 V

reduction: $H_2O_2(aq) + 2\ H^+(aq) + 2\ e^- \rightarrow 2\ H_2O(l)$ E° = 1.78 V

overall E° = 1.08 V

Because E° is positive, this reaction is spontaneous.

18.42 (a) oxidation: $Sn^{2+}(aq) \rightarrow Sn^{4+}(aq) + 2\ e^-$ E° = –0.15 V

reduction: $Br_2(l) + 2\ e^- \rightarrow 2\ Br^-(aq)$ E° = 1.09 V

overall E° = +0.94

Because the overall E° is positive, $Sn^{2+}(aq)$ can be oxidized by $Br_2(l)$.

(b) oxidation: $Sn^{2+}(aq) \rightarrow Sn^{4+}(aq) + 2\,e^-$ $E° = -0.15$ V

reduction: $Ni^{2+}(aq) + 2\,e^- \rightarrow Ni(s)$ $E° = -0.26$ V

overall $E° = -0.41$ V

Because the overall E° is negative, $Ni^{2+}(aq)$ cannot be reduced by $Sn^{2+}(aq)$.

(c) oxidation: $Ag(s) \rightarrow Ag^+(aq) + e^-$ $E° = -0.80$ V

reduction: $Pb^{2+}(aq) + 2\,e^- \rightarrow Pb(s)$ $E° = -0.13$ V

overall $E° = -0.93$ V

Because the overall E° is negative, Ag(s) cannot be oxidized by $Pb^{2+}(aq)$.

(d) oxidation: $Cu(s) \rightarrow Cu^{2+}(aq) + 2\,e^-$ $E° = -0.34$ V

reduction: $I_2(s) + 2\,e^- \rightarrow 2\,I^-(aq)$ $E° = 0.54$ V

overall $E° = +0.20$ V

Because the overall E° positive, $I_2(s)$ can be reduced by Cu(s).

18.43 (a) oxidation: $Zn(s) \rightarrow Zn^{2+}(aq) + 2\,e^-$ $E° = 0.76$ V

reduction: $Pb^{2+}(aq) + 2\,e^- \rightarrow Pb(s)$ $E° = -0.13$ V

overall $E° = 0.63$ V

$Zn(s) + Pb^{2+}(aq) \rightarrow Zn^{2+}(aq) + Pb(s)$

The reaction is spontaneous because E° is positive.

(b) oxidation: $4\,Fe^{2+}(aq) \rightarrow 4\,Fe^{3+}(aq) + 4\,e^-$ $E° = -0.77$ V

reduction: $O_2(g) + 4\,H^+(aq) + 4\,e^- \rightarrow 2\,H_2O(l)$ $E° = 1.23$ V

overall $E° = 0.46$ V

$4\ Fe^{2+}(aq) + O_2(g) + 4\ H^+(aq) \rightarrow 4\ Fe^{3+}(aq) + 2\ H_2O(l)$

The reaction is spontaneous because E° is positive.

(c) oxidation: $2\ Ag(s) \rightarrow 2\ Ag^+(aq) + 2\ e^-$ E° = –0.80 V

reduction: $Ni^{2+}(aq) + 2\ e^- \rightarrow Ni(s)$ E° = –0.26 V

overall E° = –1.06 V

There is no reaction because E° is negative.

(d) oxidation: $H_2(g) \rightarrow 2\ H^+(aq) + 2\ e^-$ E° = 0.00 V

reduction: $Cd^{2+}(aq) + 2\ e^- \rightarrow Cd(s)$ E° = –0.40 V

overall E° = –0.40 V

There is no reaction because E° is negative.

The Nernst Equation

18.44 $2\ Ag^+(aq) + Sn(s) \rightarrow 2\ Ag(s) + Sn^{2+}(aq)$

oxidation: $Sn(s) \rightarrow Sn^{2+}(aq) + 2\ e^-$ E° = 0.14 V

reduction: $2\ Ag^+(aq) + 2\ e^- \rightarrow 2\ Ag(s)$ E° = 0.80 V

overall E° = 0.94 V

$$E = E^\circ - \frac{0.0592}{n}\log\frac{[Sn^{2+}]}{[Ag^+]^2} = 0.94\ V - \frac{(0.0592\ V)}{2}\log\frac{(0.020)}{(0.010)^2} = 0.87\ V$$

18.45 $2\ Fe^{2+}(aq) + Cl_2(g) \rightarrow 2\ Fe^{3+}(aq) + 2\ Cl^-(aq)$

oxidation: $2\ Fe^{2+}(aq) \rightarrow 2\ Fe^{3+}(aq) + 2\ e^-$ E° = –0.77 V

reduction: $Cl_2(g) + 2\ e^- \rightarrow 2\ Cl^-(aq)$ E° = 1.36 V

overall E° = 0.59 V

$$E = E^\circ - \frac{0.0592}{n}\log\frac{[Fe^{3+}]^2[Cl^-]^2}{[Fe^{2+}]^2P_{Cl_2}} = 0.59\text{ V} - \frac{(0.0592\text{ V})}{2}\log\frac{(0.0010)^2(0.0030)^2}{(1.0)^2(0.50)}$$

$E = 0.91$ V

18.46 $\quad Pb(s) + Cu^{2+}(aq) \rightarrow Pb^{2+}(aq) + Cu(s)$

oxidation:	$Pb(s) \rightarrow Pb^{2+}(aq) + 2\,e^-$	$E^\circ = 0.13$ V
reduction:	$Cu^{2+}(aq) + 2\,e^- \rightarrow Cu(s)$	$\underline{E^\circ = 0.34\text{ V}}$
	overall	$E^\circ = 0.47$ V

$$E = E^\circ - \frac{0.0592}{n}\log\frac{[Pb^{2+}]}{[Cu^{2+}]} = 0.47\text{ V} - \frac{(0.0592\text{ V})}{2}\log\frac{1.0}{(1.0 \times 10^{-4})} = 0.35\text{ V}$$

When $E = 0$,

$$0 = E^\circ - \frac{0.0592}{n}\log\frac{[Pb^{2+}]}{[Cu^{2+}]} = 0.47\text{ V} - \frac{(0.0592\text{ V})}{2}\log\frac{1.0}{[Cu^{2+}]}$$

$$0 = 0.47\text{ V} + \frac{(0.0592\text{ V})}{2}\log[Cu^{2+}]$$

$$\log[Cu^{2+}] = (-0.47\text{ V})\left(\frac{2}{0.0592\text{ V}}\right) = -15.88$$

$$[Cu^{2+}] = 10^{-15.88} = 1 \times 10^{-16}\text{ M}$$

18.47 (a) $E = E^\circ - \dfrac{0.0592}{n}\log[I^-]^2 = 0.54\text{ V} - \dfrac{(0.0592\text{ V})}{2}\log(0.020)^2 = 0.64\text{ V}$

(b) $E = E^\circ - \dfrac{0.0592}{n}\log\dfrac{[Fe^{2+}]}{[Fe^{3+}]} = 0.77\text{ V} - \dfrac{(0.0592\text{ V})}{1}\log\left(\dfrac{0.10}{0.10}\right) = 0.77\text{ V}$

(c) $E = E^\circ - \dfrac{0.0592}{n}\log\dfrac{[Sn^{4+}]}{[Sn^{2+}]} = -0.15\text{ V} - \dfrac{(0.0592\text{ V})}{2}\log\left(\dfrac{0.40}{0.0010}\right) = -0.23\text{ V}$

(d) $E = E° - \frac{0.0592}{n}\log\frac{[Cr_2O_7^{2-}][H^+]^{14}}{[Cr^{3+}]^2} = -1.33\text{ V} - \frac{(0.0592\text{ V})}{6}\log\left(\frac{(1.0)(0.01)^{14}}{1.0}\right)$

$$E = -1.33\text{ V} - \frac{(0.0592\text{ V})}{6}(14)\log(0.01) = -1.05\text{ V}$$

18.48 $E = E° - \frac{0.0592}{n}\log\frac{P_{H_2}}{[H_3O^+]^2}$

$E° = 0$, $n = 2$ mol e^-, and $P_{H_2} = 1$ atm

(a) $[H_3O^+] = 1.0$ M

$$E = -\frac{0.0592}{2}\log\frac{1}{(1.0)^2} = 0$$

(b) pH = 4.00, $[H_3O^+] = 10^{-4.00} = 1.0 \times 10^{-4}$ M

$$E = -\frac{0.0592}{2}\log\frac{1}{(1.0 \times 10^{-4})^2} = -0.24\text{ V}$$

(c) $[H_3O^+] = 1.0 \times 10^{-7}$ M

$$E = -\frac{0.0592}{2}\log\frac{1}{(1.0 \times 10^{-7})^2} = -0.41\text{ V}$$

(d) $[OH^-] = 1.0$ M

$$[H_3O^+] = \frac{K_w}{[OH^-]} = \frac{1.0 \times 10^{-14}}{1.0} = 1.0 \times 10^{-14}\text{ M}$$

$$E = -\frac{0.0592}{2}\log\frac{1}{(1.0 \times 10^{-14})^2} = -0.83\text{ V}$$

18.49 $H_2(g) + Ni^{2+}(aq) \rightarrow 2\ H^+(aq) + Ni(s)$

$$E° = E°_{H_2 \rightarrow H^+} + E°_{Ni^{2+} \rightarrow Ni} = 0\ V + (-0.26\ V) = -0.26\ V$$

$$E = E° - \frac{0.0592}{n} \log \frac{[H_3O^+]^2}{[Ni^{2+}](P_{H_2})}$$

$$0.27\ V = -0.26\ V - \frac{(0.0592\ V)}{2} \log \frac{[H_3O^+]^2}{(1)(1)}$$

$$0.27\ V = -0.26\ V - (0.0592\ V) \log [H_3O^+]$$

$$pH = -\log [H_3O^+]$$

$$0.27\ V = -0.26\ V + (0.0592\ V)\ pH$$

$$pH = \frac{(0.27\ V + 0.26\ V)}{0.0592\ V} = 9.0$$

18.50 $Zn(s) + 2\ H^+(aq) \rightarrow Zn^{2+}(aq) + H_2(g)$

$$E° = E°_{H^+ \rightarrow H_2} + E°_{Zn \rightarrow Zn^{2+}} = 0\ V + 0.76\ V = 0.76\ V$$

$$E = E° - \frac{0.0592}{n} \log \frac{[Zn^{2+}](P_{H_2})}{[H_3O^+]^2}$$

$$0.58\ V = 0.76\ V - \frac{(0.0592\ V)}{2} \log \frac{(1)(1)}{[H_3O^+]^2}$$

$$0.58\ V = 0.76\ V + (0.0592\ V) \log [H_3O^+]$$

$$pH = -\log [H_3O^+]$$

$$0.58\ V = 0.76\ V - (0.0592\ V)\ pH$$

$$pH = \frac{-(0.58\ V - 0.76\ V)}{0.0592\ V} = 3.0$$

18.51 $Fe(s) + Cu^{2+}(aq) \rightarrow Fe^{2+}(aq) + Cu(s)$

oxidation: $Fe(s) \rightarrow Fe^{2+}(aq) + 2\ e^-$ $E° = 0.45\ V$

reduction: $\underline{Cu^{2+}(aq) + 2\ e^- \rightarrow Cu(s)}$ $\underline{E° = 0.34\ V}$

overall $E° = 0.79\ V$

$$E = 0.67\ V = E° - \frac{0.0592}{n} \log \frac{[Fe^{2+}]}{[Cu^{2+}]} = 0.79\ V - \frac{(0.0592\ V)}{2} \log\left(\frac{0.10}{[Cu^{2+}]}\right)$$

$$0.67\ V = 0.79\ V - \frac{(0.0592\ V)}{2} (\log (0.10) - \log [Cu^{2+}])$$

$\log [Cu^{2+}] = -5.05$; $[Cu^{2+}] = 10^{-5.05} = 8.9 \times 10^{-6}\ M$

Standard Cell Potentials and Equilibrium Constants

18.52 $Ni(s) + 2\ Ag^+(aq) \rightarrow Ni^{2+}(aq) + 2\ Ag(s)$

oxidation: $Ni(s) \rightarrow Ni^{2+}(aq) + 2\ e^-$ $E° = 0.26\ V$

reduction: $2\ Ag^+(aq) + 2\ e^- \rightarrow 2\ Ag(s)$ $\underline{E° = 0.80\ V}$

overall $E° = 1.06\ V$

$$E° = \frac{0.0592}{n} \log K$$

$$\log K = \frac{n E°}{0.0592} = \frac{(2)(1.06)}{0.0592} = 35.8$$

$K = 10^{35.8} = 6 \times 10^{35}$

18.53 $2\ MnO_4^-(aq) + 10\ Cl^-(aq) + 16\ H^+(aq) \rightarrow 2\ Mn^{2+}(aq) + 5\ Cl_2(g) + 8\ H_2O(l)$

oxidation: $10\ Cl^-(aq) \rightarrow 5\ Cl_2(g) + 10\ e^-$ $E° = -1.36\ V$

reduction: $2\ MnO_4^-(aq) + 16\ H^+(aq) + 10\ e^- \rightarrow 2\ Mn^{2+}(aq) + 8\ H_2O(l)$ $\underline{E° = 1.51\ V}$

overall $E° = 0.15\ V$

$$E^\circ = \frac{0.0592}{n}\log K$$

$$\log K = \frac{nE^\circ}{0.0592} = \frac{(10)(0.15)}{0.0592} = 25.3$$

$K = 10^{25.3} = 2 \times 10^{25}$

18.54 $\Delta G^\circ = -nFE^\circ$

Because n and F are always positive, ΔG° is negative when E° is positive because of the negative sign in the equation.

$$E^\circ = \frac{0.0592}{n}\log K$$

$$\log K = \frac{nE^\circ}{0.0592}$$

$$K = 10^{\frac{nE^\circ}{0.0592}}$$

If E° is positive, the exponent is positive (because n is positive), and K is greater than 1.

18.55 If $K < 1$, $E^\circ < 0$. When $E^\circ = 0$, $K = 1$.

18.56 E° and n are from Problem 18.39.

$$E^\circ = \frac{0.0592}{n}\log K$$

$$\log K = \frac{nE^\circ}{0.0592}$$

(a) $Cd(s) + Sn^{2+}(aq) \rightarrow Cd^{2+}(aq) + Sn(s)$

$E^\circ = 0.26$ V and n = 2 mol e^-

$\log K = \frac{(2)(0.26)}{0.0592} = 8.8$; $\quad K = 10^{8.8} = 6 \times 10^{8}$

(b) $2\ Al(s) + 3\ Cd^{2+}(aq) \rightarrow 2\ Al^{3+}(aq) + 3\ Cd(s)$

$E° = 1.26$ V and $n = 6$ mol e^-

$$\log K = \frac{(6)(1.26)}{0.0592} = 128; \qquad K = 10^{128}$$

(c) $Pb(s) + 2\ H^+(aq) \rightarrow Pb^{2+}(aq) + H_2(g)$

$E° = 0.13$ V and $n = 2$ mol e^-

$$\log K = \frac{(2)(0.13)}{0.0592} = 4.4; \qquad K = 10^{4.4} = 3 \times 10^4$$

(d) $6\ Fe^{2+}(aq) + Cr_2O_7^{2-}(aq) + 14\ H^+(aq) \rightarrow 6\ Fe^{3+}(aq) + 2\ Cr^{3+}(aq) + 7\ H_2O(l)$

$E° = 0.56$ V and $n = 6$ mol e^-

$$\log K = \frac{(6)(0.56)}{0.0592} = 57; \qquad K = 10^{57}$$

18.57 $Hg_2^{2+}(aq) \rightarrow Hg(l) + Hg^{2+}(aq)$

oxidation: $\frac{1}{2}[Hg_2^{2+}(aq) \rightarrow 2\ Hg^{2+}(aq) + 2\ e^-]$ $E° = -0.92$ V

reduction: $\frac{1}{2}[Hg_2^{2+}(aq) + 2\ e^- \rightarrow 2\ Hg(l)]$ $E° = 0.80$ V

overall $E° = -0.12$ V

$$E° = \frac{0.0592}{n} \log K$$

$$\log K = \frac{nE°}{0.0592} = \frac{(1)(-0.12)}{0.0592} = -2.027; \qquad K = 10^{-2.027} = 9 \times 10^{-3}$$

Batteries; Corrosion

18.58 (a)

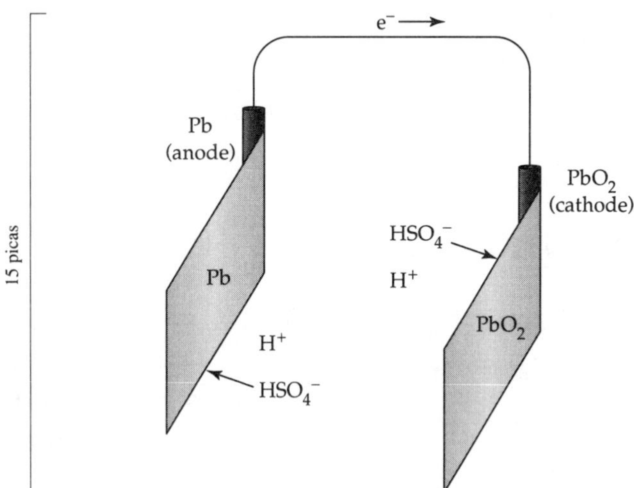

(b)

Anode: $Pb(s) + HSO_4^-(aq) \rightarrow PbSO_4(s) + H^+(aq) + 2\ e^-$ $E° = 0.296$ V

Cathode: $PbO_2(s) + 3\ H^+(aq) + HSO_4^-(aq) + 2\ e^- \rightarrow PbSO_4(s) + 2\ H_2O(l)$ $E° = 1.628$ V

Overall $Pb(s) + PbO_2(s) + 2\ H^+(aq) + 2\ HSO_4^-(aq)$

$\rightarrow 2\ PbSO_4(s) + 2\ H_2O(l)$ $E° = 1.924$ V

(c) $E° = \frac{0.0592}{n} \log K$

$\log K = \frac{nE°}{0.0592} = \frac{(2)(1.924)}{0.0592} = 65.0$; $K = 10^{65}$

(d) When the cell reaction reaches equilibrium the cell voltage = 0.

18.59 oxidation: $2\ H_2(g) + 4\ OH^-(aq) \rightarrow 4\ H_2O(l) + 4\ e^-$ $E° = 0.83$ V

reduction: $O_2(g) + 2\ H_2O(l) + 4\ e^- \rightarrow 4\ OH^-(aq)$ $E° = 0.43$ V

$2\ H_2(g) + O_2(g) \rightarrow 2\ H_2O(l)$ $E° = 1.23$ V

n = 4 mol e^- and 1 J = 1 C x 1 V

$$\Delta G° = -nFE° = -(4 \text{ mol e}^-)\left(\frac{96{,}500 \text{ C}}{1 \text{ mol e}^-}\right)(1.23 \text{ V}) = -474{,}780 \text{ J} = -475 \text{ kJ}$$

$$E° = \frac{0.0592}{n} \log K$$

$$\log K = \frac{nE°}{0.0592} = \frac{(4)(1.23)}{0.0592} = 83.1; \qquad K = 10^{83.1} = 1 \times 10^{83}$$

18.60 $Zn(s) + HgO(s) \rightarrow ZnO(s) + Hg(l)$

Zn, 65.39 amu; HgO, 216.59 amu

$$\text{mass HgO} = 2.00 \text{ g Zn} \times \frac{1 \text{ mol Zn}}{65.39 \text{ g Zn}} \times \frac{1 \text{ mol HgO}}{1 \text{ mol Zn}} \times \frac{216.59 \text{ g HgO}}{1 \text{ mol HgO}}$$

mass HgO = 6.62 g HgO

18.61 $Cd(OH)_2(s) + 2\ Ni(OH)_2(s) \rightarrow Cd(s) + 2\ NiO(OH)(s) + 2\ H_2O(l)$

$Ni(OH)_2$, 92.71 amu; Cd, 112.41 amu

$$\text{mass Cd} = 10.0 \text{ g Ni(OH)}_2 \times \frac{1 \text{ mol Ni(OH)}_2}{92.71 \text{ g Ni(OH)}_2} \times \frac{1 \text{ mol Cd}}{2 \text{ mol Ni(OH)}_2} \times \frac{112.41 \text{ g Cd}}{1 \text{ mol Cd}}$$

mass Cd = 6.06 g Cd

18.62 Rust is a hydrated form of iron(III) oxide ($Fe_2O_3 \cdot H_2O$). Rust forms from the oxidation of Fe in the presence of O_2 and H_2O. Rust can be prevented by coating Fe with Zn (galvanizing).

18.63 Cathodic protection is the attachment of a more easily oxidized metal to the metal you want to protect. This forces the metal you want to protect to be the cathode, hence the name, cathodic protection.

Zn and Al can offer cathodic protection to Fe (Ni and Sn cannot).

18.64 A sacrificial anode is a metal used for cathodic protection. It behaves as an anode and is more easily oxidized than the metal it is protecting.

An example of a sacrificial anode is Zn for protecting Fe (galvanizing).

18.65 Cr forms a protective oxide coating similar to Al.

Electrolysis

18.66 (a)

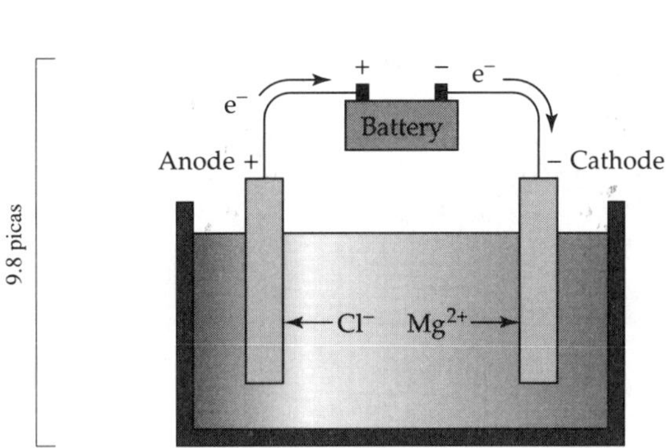

(b) anode: $2\ Cl^-(l) \rightarrow Cl_2(g) + 2\ e^-$

cathode: $Mg^{2+}(l) + 2\ e^- \rightarrow Mg(l)$

overall: $Mg^{2+}(l) + 2\ Cl^-(l) \rightarrow Mg(l) + Cl_2(g)$

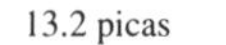

18.67 (a)

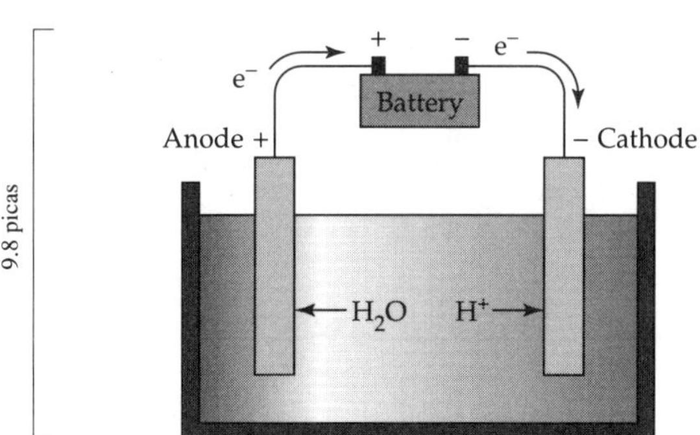

(b) anode: $2\ H_2O(l) \rightarrow O_2(g) + 4\ H^+(aq) + 4\ e^-$

cathode: $4\ H^+(aq) + 4\ e^- \rightarrow 2\ H_2(g)$

overall: $2\ H_2O(l) \rightarrow O_2(g) + 2\ H_2(g)$

18.68 possible anode reactions:

$2\ Cl^-(aq) \rightarrow Cl_2(g) + 2\ e^-$

$2\ H_2O(l) \rightarrow O_2(g) + 4\ H^+(aq) + 4\ e^-$

possible cathode reactions:

$2\ H_2O(l) + 2\ e^- \rightarrow H_2(g) + 2\ OH^-(aq)$

$Mg^{2+}(aq) + 2\ e^- \rightarrow Mg(s)$

actual reactions:

anode: $2\ Cl^-(aq) \rightarrow Cl_2(g) + 2\ e^-$

cathode: $2\ H_2O(l) + 2\ e^- \rightarrow H_2(g) + 2\ OH^-(aq)$

This anode reaction takes place instead of $2\ H_2O(l) \rightarrow O_2(g) + 4\ H^+(aq) + 4\ e^-$ because of a high overvoltage for formation of gaseous O_2.

This cathode reaction takes place instead of $Mg^{2+}(aq) + 2\ e^- \rightarrow Mg(s)$ because H_2O is easier to reduce than Mg^{2+}.

18.69 (a) $K(l)$ and $Cl_2(g)$

(b) $H_2(g)$ and $Cl_2(g)$. Solvent H_2O is reduced in preference to K^+.

18.70 (a) NaBr

anode: $2\ Br^-(aq) \rightarrow Br_2(l) + 2\ e^-$

cathode: $2\ H_2O(l) + 2\ e^- \rightarrow H_2(g) + 2\ OH^-(aq)$

overall: $2\ H_2O(l) + 2\ Br^-(aq) \rightarrow Br_2(l) + H_2(g) + 2\ OH^-(aq)$

(b) $CuCl_2$

anode: $2\ Cl^-(aq) \rightarrow Cl_2(g) + 2\ e^-$

cathode: $Cu^{2+}(aq) + 2\ e^- \rightarrow Cu(s)$

overall: $Cu^{2+}(aq) + 2\ Cl^-(aq) \rightarrow Cu(s) + Cl_2(g)$

(c) LiOH

anode: $4\ OH^-(aq) \rightarrow O_2(g) + 2\ H_2O(l) + 4\ e^-$

cathode: $4\ H_2O(l) + 4\ e^- \rightarrow 2\ H_2(g) + 4\ OH^-(aq)$

overall: $2\ H_2O(l) \rightarrow O_2(g) + 2\ H_2(g)$

(d) Ag_2SO_4

anode: $2\ H_2O(l) \rightarrow O_2(g) + 4\ H^+(aq) + 4\ e^-$

cathode: $4\ Ag^+(aq) + 4\ e^- \rightarrow 4\ Ag(s)$

overall: $4\ Ag^+(aq) + 2\ H_2O(l) \rightarrow O_2(g) + 4\ H^+(aq) + 4\ Ag(s)$

18.71 $Ag^+(aq) + e^- \rightarrow Ag(s)$; 1 A = 1 C/s

$$\text{mass Ag} = 2.40\ \frac{C}{s} \times 20.0\ \text{min} \times \frac{60\ s}{1\ \text{min}} \times \frac{1\ \text{mol}\ e^-}{96{,}500\ C} \times \frac{1\ \text{mol Ag}}{1\ \text{mol}\ e^-} \times \frac{107.87\ \text{g Ag}}{1\ \text{mol Ag}}$$

mass Ag = 3.22 g

18.72 $Cu^{2+}(aq) + 2\ e^- \rightarrow Cu(s)$

$$\text{mol}\ e^- = 100.0\ \frac{C}{s} \times 24.0\ h \times \frac{60\ \text{min}}{h} \times \frac{60\ s}{\text{min}} \times \frac{1\ \text{mol}\ e^-}{96{,}500\ C} = 89.5\ \text{mol}\ e^-$$

$$\text{mass Cu} = 89.5\ \text{mol}\ e^- \times \frac{1\ \text{mol Cu}}{2\ \text{mol}\ e^-} \times \frac{63.54\ \text{g Cu}}{1\ \text{mol Cu}} \times \frac{1\ \text{kg}}{1000\ g} = 2.84\ \text{kg Cu}$$

18.73 $2\ Na^+(l) + 2\ Cl^-(l) \rightarrow 2\ Na(l) + Cl_2(g)$

$Na^+(l) + e^- \rightarrow Na(l)$; 1 A = 1 C/s; 1000 kg = 1.000×10^6 g

$$\text{Charge} = 1.000 \times 10^6\ \text{g Na} \times \frac{1\ \text{mol Na}}{22.99\ \text{g Na}} \times \frac{1\ \text{mol}\ e^-}{1\ \text{mol Na}} \times \frac{96{,}500\ C}{1\ \text{mol}\ e^-} = 4.20 \times 10^9\ C$$

$$\text{Time} = \frac{4.20 \times 10^9\ C}{30{,}000\ C/s} \times \frac{1\ h}{3600\ s} = 38.9\ h$$

$$1.000 \times 10^6 \text{ g Na} \times \frac{1 \text{ mol Na}}{22.99 \text{ g Na}} \times \frac{1 \text{ mol } Cl_2}{2 \text{ mol Na}} = 21{,}748.6 \text{ mol } Cl_2$$

PV = nRT

$$V = \frac{nRT}{P} = \frac{(21{,}748.6 \text{ mol})\left(0.08206 \frac{\text{L}\cdot\text{atm}}{\text{mol}\cdot\text{K}}\right)(273.15 \text{ K})}{1.00 \text{ atm}} = 4.87 \times 10^5 \text{ L } Cl_2$$

18.74 $Al^{3+} + 3\ e^- \rightarrow Al$

40.0 kg = 40,000 g; 1 h = 3600 s

$$\text{Charge} = 40{,}000 \text{ g Al} \times \frac{1 \text{ mol Al}}{26.98 \text{ g Al}} \times \frac{3 \text{ mol } e^-}{1 \text{ mol Al}} \times \frac{96{,}500 \text{ C}}{1 \text{ mol } e^-} = 4.29 \times 10^8 \text{ C}$$

$$\text{Current} = \frac{4.29 \times 10^8 \text{ C}}{3600 \text{ s}} = 1.19 \times 10^5 \text{ A}$$

18.75 $PbSO_4(s) + 2\ e^- \rightarrow Pb(s)$

$$\text{mass } PbSO_4 = 10.0 \frac{\text{C}}{\text{s}} \times 1.50 \text{ h} \times \frac{3600 \text{ s}}{1 \text{ h}} \times \frac{1 \text{ mol } e^-}{96{,}500 \text{ C}} \times \frac{1 \text{ mol } PbSO_4}{2 \text{ mol } e^-} \times \frac{303.3 \text{ g } PbSO_4}{1 \text{ mol } PbSO_4}$$

mass $PbSO_4$ = 84.9 g $PbSO_4$

18.76 $Cr^{3+}(aq) + 3\ e^- \rightarrow Cr(s)$

$$\text{Charge} = 125 \text{ g Cr} \times \frac{1 \text{ mol Cr}}{52.00 \text{ g Cr}} \times \frac{3 \text{ mol } e^-}{1 \text{ mol Cr}} \times \frac{96{,}500 \text{ C}}{1 \text{ mol } e^-} = 6.96 \times 10^5 \text{ C}$$

$$\text{Time} = \frac{6.96 \times 10^5 \text{ C}}{200.0 \text{ C/s}} = 3480 \text{ s}$$

$$\text{Time} = 3480 \text{ s} \times \frac{1 \text{ min}}{60 \text{ s}} = 58.0 \text{ min}$$

General Problems

18.77 (a) oxidizing agent – a substance that causes an oxidation by gaining electrons.

reducing agent – a substance that causes a reduction by losing electrons.

(b) galvanic cell – an electrochemical cell in which a spontaneous chemical reaction generates an electric current.

electrolytic cell – an electrochemical cell in which an electric current is used to drive a nonspontaneous reaction.

(c) battery – a self-contained galvanic cell.

fuel cell – a galvanic cell in which one of the reactants is a traditional fuel such as CH_4 or H_2.

(d) anode – electrode where oxidation occurs.

cathode – electrode where reduction occurs.

18.78 (a) The Hall-Heroult process is the electrolytic process for the production of Al from Al_2O_3 in molten Na_3AlF_6.

(b) A Downs cell is used for the production of Na by electrolysis of molten NaCl.

(c) Electrorefining is a process for the purification of a metal by electrolysis.

(d) Galvanizing is the coating of Fe with Zn to prevent rusting.

18.79 (a) $2\ MnO_4^-(aq) + 16\ H^+(aq) + 5\ Sn^{2+}(aq) \rightarrow 2\ Mn^{2+}(aq) + 5\ Sn^{4+}(aq) + 8\ H_2O(l)$

(b) MnO_4^- is the oxidizing agent; Sn^{2+} is the reducing agent.

(c) $E° = 1.51\ V + (-0.15\ V) = 1.36\ V$

18.80 $2\ Mn^{3+}(aq) + 2\ H_2O(l) \rightarrow Mn^{2+}(aq) + MnO_2(s) + 4\ H^+(aq)$

$E° = 1.51\ V + (-0.95\ V) = +0.56\ V$

Because E° is positive, the disproportionation is spontaneous under standard-state conditions.

18.81 (a) Ag^+ is the strongest oxidizing agent because Ag^+ has the most positive standard reduction potential.

Pb is the strongest reducing agent because Pb has the most positive standard oxidation potential.

(b)

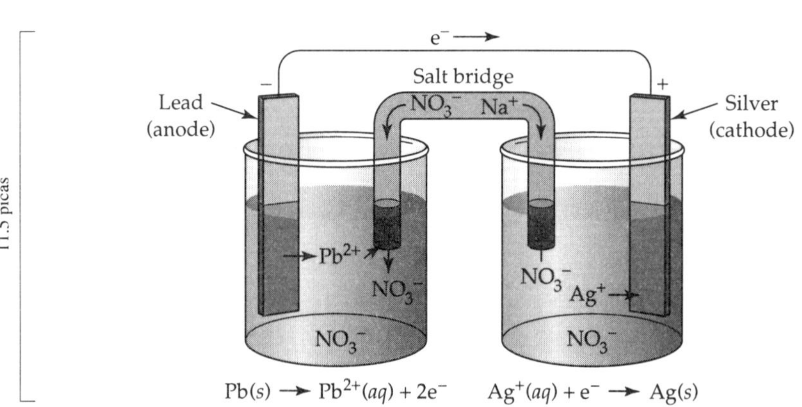

(c) $Pb(s) + 2\ Ag^+(aq) \rightarrow Pb^{2+}(aq) + 2\ Ag(s)$

$E° = E°_{anode} + E°_{cathode} = 0.13\ V + 0.80\ V = 0.93\ V$

$n = 2$ mol e^- and $1\ J = 1\ C \times 1\ V$

$$\Delta G° = -nFE° = -(2\ \text{mol e}^-)\left(\frac{96{,}500\ C}{1\ \text{mol e}^-}\right)(0.93\ V) = -179{,}490\ J = -180\ kJ$$

$$E° = \frac{0.0592}{n}\log K$$

$$\log K = \frac{nE°}{0.0592} = \frac{(2)(0.93)}{0.0592} = 31; \qquad K = 10^{31}$$

$$\text{(d) } E = E° - \frac{0.0592}{n}\log\frac{[Pb^{2+}]}{[Ag^+]^2} = 0.93\ V - \frac{0.0592}{2}\log\left(\frac{0.01}{(0.01)^2}\right) = 0.87\ V$$

18.82 From Table 18.1:

$H_2O_2(aq)$ is the strongest oxidizing agent.

$H_2O_2(aq)$ is the strongest reducing agent.

18.83

	E°(V)
$MnO_4^-(aq) + 8\ H^+(aq) + 5\ e^- \rightarrow Mn^{2+}(aq) + 4\ H_2O(l)$	1.51
$Br_2(l)\ +\ 2\ e^-\ \rightarrow\ 2\ Br^-(aq)$	1.09
$Fe^{3+}(aq)\ +\ e^-\ \rightarrow\ Fe^{2+}(aq)$	0.77
$I_2(s)\ +\ 2\ e^-\ \rightarrow\ 2\ I^-(aq)$	0.54
$Cu^{2+}(aq)\ +\ 2\ e^-\ \rightarrow\ Cu(s)$	0.34
$Pb^{2+}(aq)\ +\ 2\ e^-\ \rightarrow\ Pb(s)$	–0.13
$Fe^{2+}(aq)\ +\ 2\ e^-\ \rightarrow\ Fe(s)$	–0.45
$Al^{3+}(aq)\ +\ 3\ e^-\ \rightarrow\ Al(s)$	–1.66

Fe^{2+} can oxidize any substance with a standard reduction potential more negative than –0.45 V. Fe^{2+} can oxidize Al.

Fe^{2+} can reduce any substance with a standard reduction potential more positive than 0.77 V. Fe^{2+} can reduce $Br_2(l)$ and MnO_4^-.

18.84 For Pb^{2+}, $E = -0.13 - \dfrac{0.0592}{2} \log \dfrac{1}{[Pb^{2+}]}$

For Cd^{2+}, $E = -0.40 - \dfrac{0.0592}{2} \log \dfrac{1}{[Cd^{2+}]}$

Set these two equations for E equal to each other and solve for $[Cd^{2+}]/[Pb^{2+}]$.

$$-0.13 - \frac{0.0592}{2} \log \frac{1}{[Pb^{2+}]} = -0.40 - \frac{0.0592}{2} \log \frac{1}{[Cd^{2+}]}$$

$$0.27 = \frac{0.0592}{2}(\log[Cd^{2+}] - \log[Pb^{2+}])$$

$$0.27 = \frac{0.0592}{2}\log\frac{[Cd^{2+}]}{[Pb^{2+}]}$$

$$\log\frac{[Cd^{2+}]}{[Pb^{2+}]} = \frac{(0.27)(2)}{0.0592} = 9.1$$

$$\frac{[Cd^{2+}]}{[Pb^{2+}]} = 10^{9.1} = 1 \times 10^{9}$$

24.8 picas

18.85 (a)

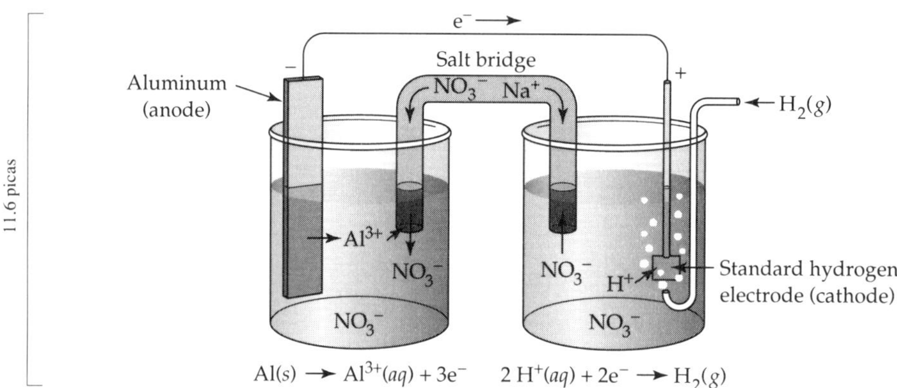

(b) $2\,Al(s) + 6\,H^+(aq) \rightarrow 2\,Al^{3+}(aq) + 3\,H_2(g)$

$E° = E°_{anode} + E°_{cathode} = 1.66\text{ V} + 0.00\text{ V} = 1.66\text{ V}$

(c)

$$E = E° - \frac{0.0592}{n}\log\frac{[Al^{3+}]^2(P_{H_2})^3}{[H^+]^6} = 1.66\text{ V} - \frac{(0.0592\text{ V})}{6}\log\left(\frac{(0.10)^2(10.0)^3}{(0.10)^6}\right) = 1.59\text{ V}$$

(d) 1 J = 1 C x 1 V

$$\Delta G° = -nFE° = -(6\text{ mol e}^-)\left(\frac{96{,}500\text{ C}}{1\text{ mol e}^-}\right)(1.66\text{ V}) = -961{,}140\text{ J} = -961\text{ kJ}$$

$$E° = \frac{0.0592}{n}\log K$$

$$\log K = \frac{n E^\circ}{0.0592} = \frac{(6)(1.66)}{0.0592} = 168; \qquad K = 10^{168}$$

(e) $$\text{mass Al} = 10.0\ \frac{C}{s} \times 25.0\ \text{min} \times \frac{60\ s}{1\ \text{min}} \times \frac{1\ \text{mol e}^-}{96{,}500\ C} \times \frac{1\ \text{mol Al}}{3\ \text{mol e}^-} \times \frac{26.98\ \text{g Al}}{1\ \text{mol Al}}$$

mass Al = 1.40 g Al

18.86 $Zn(s) \rightarrow Zn^{2+}(aq) + 2\ e^-$

mass Zn =

$$0.100\ \frac{C}{s} \times 200.0\ h \times \frac{60\ \text{min}}{h} \times \frac{60\ s}{\text{min}} \times \frac{1\ \text{mol e}^-}{96{,}500\ C} \times \frac{1\ \text{mol Zn}}{2\ \text{mol e}^-} \times \frac{65.39\ \text{g Zn}}{1\ \text{mol Zn}}$$

= 24.4 g Zn

18.87 $2\ Cl^-(aq) \rightarrow Cl_2(g) + 2\ e^-$

12 million tons = 12×10^6 tons; Cl_2, 70.91 amu

$$12 \times 10^6\ \text{tons} \times \frac{907{,}200\ g}{1\ \text{ton}} \times \frac{1\ \text{mol}\ Cl_2}{70.91\ g\ Cl_2} = 1.54 \times 10^{11}\ \text{mol}\ Cl_2$$

$$\text{Charge} = 1.54 \times 10^{11}\ \text{mol}\ Cl_2 \times \frac{2\ \text{mol e}^-}{1\ \text{mol}\ Cl_2} \times \frac{96{,}500\ C}{1\ \text{mol e}^-} = 2.97 \times 10^{16}\ C$$

1 J = 1 C x 1 V

Energy = $(2.97 \times 10^{16}\ C)(4.5\ V) = 1.34 \times 10^{17}\ J$

$$\text{kWh} = (1.34 \times 10^{17}\ J)\left(\frac{1\ \text{kWh}}{3.6 \times 10^6\ J}\right) = 3.7 \times 10^{10}\ \text{kWh}$$

18.88 (a) oxidation: $5[H_2C_2O_4(aq) \rightarrow 2\ CO_2(g) + 2\ H^+(aq) + 2\ e^-]$ $\quad E^\circ = 0.49\ V$

reduction: $2[MnO_4^-(aq) + 8\ H^+(aq) + 5\ e^- \rightarrow Mn^{2+}(aq) + 4\ H_2O(l)]$ $\quad E^\circ = 1.51\ V$

$E^\circ = 1.51\ V + 0.49\ V = 2.00\ V$

(b) 1 J = 1 C x 1 V

$$\Delta G° = -nFE° = -(10 \text{ mol } e^-)\left(\frac{96{,}500 \text{ C}}{1 \text{ mol } e^-}\right)(2.00 \text{ V}) = -1{,}930{,}000 \text{ J} = -1{,}930 \text{ kJ}$$

$$E° = \frac{0.0592}{n} \log K$$

$$\log K = \frac{nE°}{0.0592} = \frac{(10)(2.00)}{0.0592} = 338; \qquad K = 10^{338}$$

(c) $Na_2C_2O_4$, 134.0 amu

$$\text{mol } KMnO_4 = 1.200 \text{ g } Na_2C_2O_4 \times \frac{1 \text{ mol } Na_2C_2O_4}{134.0 \text{ g } Na_2C_2O_4} \times \frac{2 \text{ mol } KMnO_4}{5 \text{ mol } Na_2C_2O_4}$$

$$\text{mol } KMnO_4 = 3.582 \times 10^{-3} \text{ mol } KMnO_4$$

$$\text{molarity} = \frac{3.582 \times 10^{-3} \text{ mol}}{0.032\ 50 \text{ L}} = 0.1102 \text{ M}$$

18.89 (a) $3\ CH_3CH_2OH(aq) + 2\ Cr_2O_7^{2-}(aq) + 16\ H^+(aq) \rightarrow$

$3\ CH_3COOH(aq) + 4\ Cr^{3+}(aq) + 11\ H_2O(l)$

oxidation:

$3\ CH_3CH_2OH(aq) + 3\ H_2O(l) \rightarrow 3\ CH_3COOH(aq) + 12\ H^+(aq) + 12\ e^-$ $E° = -0.058$V

reduction:

$2\ Cr_2O_7^{2-}(aq) + 28\ H^+(aq) + 12\ e^- + \rightarrow 4\ Cr^{3+}(aq) + 14\ H_2O(l)$ $\underline{E° = 1.33 \text{ V}}$

overall $E° = 1.27$ V

(b) $$E = E° - \frac{0.0592}{n} \log \frac{[CH_3COOH]^3[Cr^{3+}]^4}{[CH_3CH_2OH]^3[Cr_2O_7^{2-}]^2[H^+]^{16}}$$

pH = 4.00, $[H^+] = 0.000\ 10$ M

$$E = 1.27\text{ V} - \frac{(0.0592\text{ V})}{12}\log\left(\frac{(1)^3(1)^4}{(1)^3(1)^2(0.000\ 10)^{16}}\right)$$

$$E = 1.27\text{ V} - \frac{(0.0592\text{ V})}{12}\log\frac{1}{(0.000\ 10)^{16}}$$

$$E = 1.27\text{ V} + \frac{(0.0592\text{ V})}{12}\log(0.000\ 10)^{16}$$

$$E = 1.27\text{ V} + \frac{(0.0592\text{ V})}{12}(16)\log(0.000\ 10) = 0.95\text{ V}$$

18.90 From Appendix D:

$AgBr(s) + e^- \rightarrow Ag(s) + Br^-(aq)$ $\quad E° = 0.07\text{ V}$

(a) oxidation: $C_6H_4(OH)_2(aq) \rightarrow C_6H_4O_2(aq) + 2\,H^+(aq) + 2\,e^-$ $\quad E° = -0.699\text{ V}$

reduction: $2[AgBr(s) + e^- \rightarrow Ag(s) + Br^-(aq)]$ $\quad E° = 0.07\text{ V}$

overall: $2\,AgBr(s) + C_6H_4(OH)_2(aq) \rightarrow$

$2\,Ag(s) + 2\,Br^-(aq) + C_6H_4O_2(aq) + 2\,H^+(aq)$

overall E° = –0.699 V + 0.07 V = –0.63 V

Because the overall E° is negative, the reaction is nonspontaneous when $[H^+] = 1M$.

(b) $$E°(\text{in 1 M } OH^-) = E°(\text{in 1 M } H^+) - \frac{0.0592}{n}\log\frac{[Br^-]^2[H^+]^2[C_6H_4O_2]}{[C_6H_4(OH)_2]}$$

$$[H^+] = \frac{K_w}{[OH^-]} = \frac{1.0 \times 10^{-14}}{1} = 1 \times 10^{-14}\text{ M}$$

$$E°(\text{in 1 M } OH^-) = -0.63\text{ V} - \frac{(0.0592\text{ V})}{2}\log\frac{(1)^2(10^{-14})^2(1)}{(1)}$$

$$E°(\text{in 1 M } OH^-) = -0.63\text{ V} + 0.83\text{ V} = +0.20\text{ V}$$

18.91 (a) $4\ CH_2{=}CHCN + 2\ H_2O \rightarrow 2\ NC(CH_2)_4CN + O_2$

(b) $\text{mol e}^- = 3000\ \text{C/s} \times 10.0\ \text{h} \times \dfrac{3600\ \text{s}}{1\ \text{h}} \times \dfrac{1\ \text{mol e}^-}{96{,}500\ \text{C}} = 1119.2\ \text{mol e}^-$

mass adiponitrile =

$$1119.2\ \text{mol e}^- \times \frac{1\ \text{mol adiponitrile}}{2\ \text{mol e}^-} \times \frac{108.14\ \text{g adiponitrile}}{1\ \text{mol adiponitrile}} \times \frac{1.0\ \text{kg}}{1000\ \text{g}}$$

$= 60.5\ \text{kg adiponitrile}$

(c) $1119.2\ \text{mol e}^- \times \dfrac{1\ \text{mol } O_2}{4\ \text{mol e}^-} = 279.8\ \text{mol } O_2$

$PV = nRT$

$$V = \frac{nRT}{P} = \frac{(279.8\ \text{mol})\left(0.08206\ \dfrac{\text{L}\cdot\text{atm}}{\text{mol}\cdot\text{K}}\right)(298\ \text{K})}{\left(740\ \text{mm Hg} \times \dfrac{1\ \text{atm}}{760\ \text{mm Hg}}\right)} = 7030\ \text{L } O_2$$

18.92 anode: $Ag(s) + Cl^-(aq) \rightarrow AgCl(s) + e^-$

cathode: $Ag^+(aq) + e^- \rightarrow Ag(s)$

overall: $Ag^+(aq) + Cl^-(aq) \rightarrow AgCl(s)$ $E° = 0.578\ V$

For $AgCl(s) \rightleftarrows Ag^+(aq) + Cl^-(aq)$ $E° = -0.578\ V$

$$E° = \frac{0.0592}{n} \log K$$

$$\log K = \frac{nE°}{0.0592} = \frac{(1)(-0.578)}{0.0592} = -9.76$$

$K = K_{sp} = 10^{-9.76} = 1.7 \times 10^{-10}$

18.93 (a) oxidation: $Cu(s) \rightarrow Cu^{2+}(aq) + 2\ e^-$ $E° = -0.34$ V

reduction: $2\ Ag^+(aq) + 2\ e^- \rightarrow 2\ Ag(s)$ $E° = 0.80$ V

$2\ Ag^+(aq) + Cu(s) \rightarrow Cu^{2+}(aq) + 2\ Ag(s)$ $E° = 0.46$ V

$$E = E° - \frac{0.0592}{n}\log\frac{[Cu^{2+}]}{[Ag^+]^2} = 0.46\ V - \frac{(0.0592\ V)}{2}\log\left(\frac{1.0}{(0.050)^2}\right) = 0.38\ V$$

(b) $$[Ag^+] = \frac{K_{sp}}{[Br^-]} = \frac{5.4 \times 10^{-13}}{1\ M} = 5.4 \times 10^{-13}\ M$$

$$E = E° - \frac{0.0592}{n}\log\frac{[Cu^{2+}]}{[Ag^+]^2} = 0.46\ V - \frac{(0.0592\ V)}{2}\log\left(\frac{1.0}{(5.4 \times 10^{-13})^2}\right) = -0.27\ V$$

The cell potential for the spontaneous reaction is E = 0.27 V.

The spontaneous reaction is:

$Cu^{2+}(aq) + 2\ Ag(s) + 2\ Br^-(aq) \rightarrow 2\ AgBr(s) + Cu(s)$

(c) $Cu^{2+}(aq) + 2\ e^- \rightarrow Cu(s)$ $E° = 0.34$ V

$2\ Ag(s) + 2\ Br^-(aq) \rightarrow 2\ AgBr(s) + 2\ e^-$ $E° = ?$

$Cu^{2+}(aq) + 2\ Ag(s) + 2\ Br^-(aq) \rightarrow 2\ AgBr(s) + Cu(s)$ $E° = 0.27$ V

$E° = ? = 0.27\ V - 0.34\ V = -0.07\ V$

For: $AgBr(s) + e^- \rightarrow Ag(s) + Br^-(aq)$

the standard reduction potential is $E° = 0.07$ V

19.1 (a) B lies above Al in group 3A, and therefore B is more nonmetallic than Al.

(b) Ge and Br are in the same row of the periodic table, but Br (group 7A) lies to the right of Ge (group 4A). Therefore, Br is more nonmetallic.

(c) Se (group 6A) has more nonmetallic than In because it lies above and to the right of In (group 3A).

19.2 (a) HNO_3

H–O–N(–O)=O ⟷ H–O–N(=O)–O

H_3PO_4

H–O–P(=O)(–O–H)–O–H

Nitrogen can form very strong pπ - pπ bonds. Phosphorus forms weaker pπ - pπ bonds, so it tends to form more single bonds. The double bond in H_3PO_4 results from pπ(O) - dπ(P) interaction.

(b) Sulfur can use empty 3d orbitals to form octahedral sp^3d^2 hybrid orbitals to bond with six fluorines in SF_6. With just 2s and 2p valence orbitals, oxygen cannot expand its valence orbitals beyond four. OF_2 has two single bonds and two lone pairs of electrons.

19.3 H–C≡N:

The carbon is sp hybridized.

19.4 $Hb\text{–}O_2 + CO \rightleftharpoons Hb\text{–}CO + O_2$

Mild cases of carbon monoxide poisoning can be treated with O_2. Le Châtelier's principle says that adding a product (O_2) will cause the reaction to proceed in the reverse direction, back to $Hb\text{–}O_2$.

19.5 $Si_3O_9^{6-}$

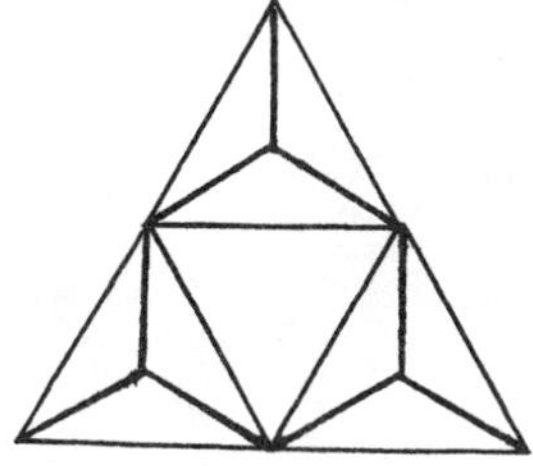

19.6 H_3PO_4

```
        O
        ||
H—O— P—O—H
        |
        O
        |
        H
```

$H_6P_4O_{13}$

```
        O      O       O       O
        ||     ||      ||      ||
H—O— P—O—P—O— P—O— P—O—H
        |      |       |       |
        O      O       O       O
        |      |       |       |
        H      H       H       H
```

$$\mathrm{H{-}O{-}P({=}O)(OH){-}O{-}H + H{-}O{-}P({=}O)(OH){-}O{-}H + H{-}O{-}P({=}O)(OH){-}O{-}H + H{-}O{-}P({=}O)(OH){-}O{-}H} \longrightarrow$$

$$\mathrm{H{-}O{-}P({=}O)(OH){-}O{-}P({=}O)(OH){-}O{-}P({=}O)(OH){-}O{-}P({=}O)(OH){-}O{-}H} \quad + \quad 3H_2O$$

$4\ H_3PO_4(aq) \rightarrow H_6P_4O_{13}(aq) + 3\ H_2O(l)$

19.7 In silicate and phosphate anions, both Si and P are surrounded by tetrahedra of O atoms, which can link together to form chains and rings.

Understanding Key Concepts

1. (a) main-group elements

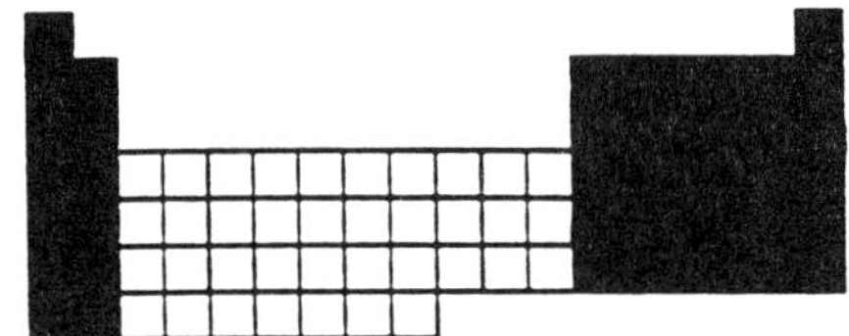

(b) s-block elements

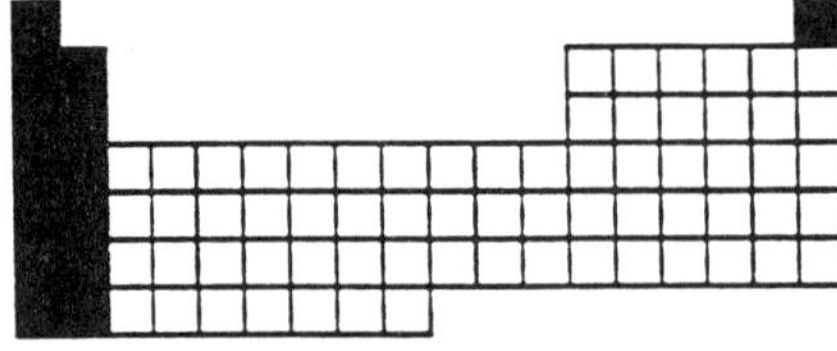

(c) p-block elements

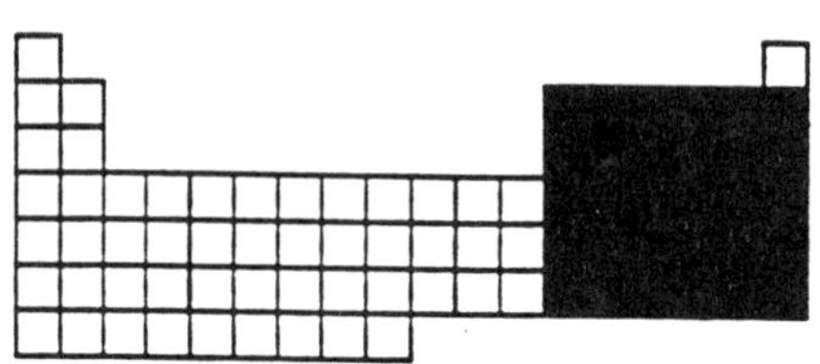

(d) main-group metals

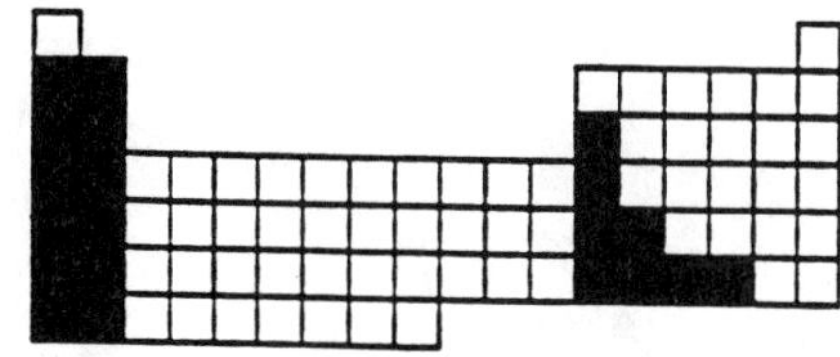

(e) nonmetals

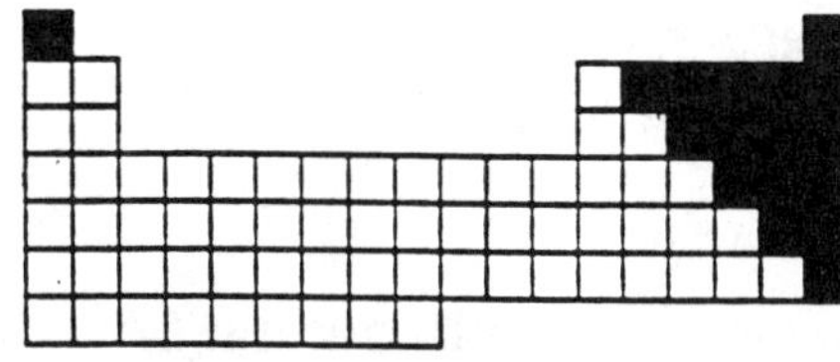

(f) semimetals

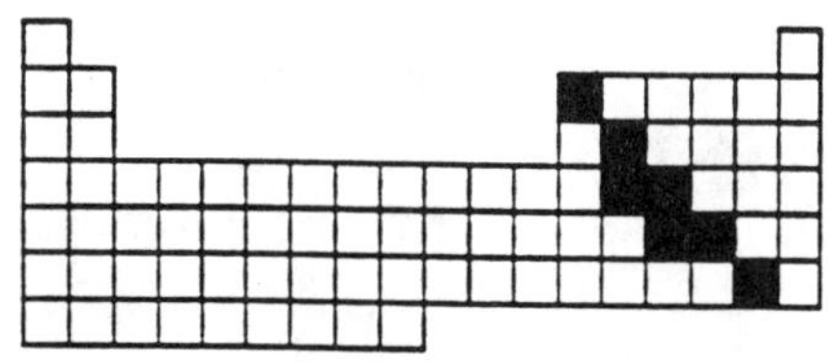

2.

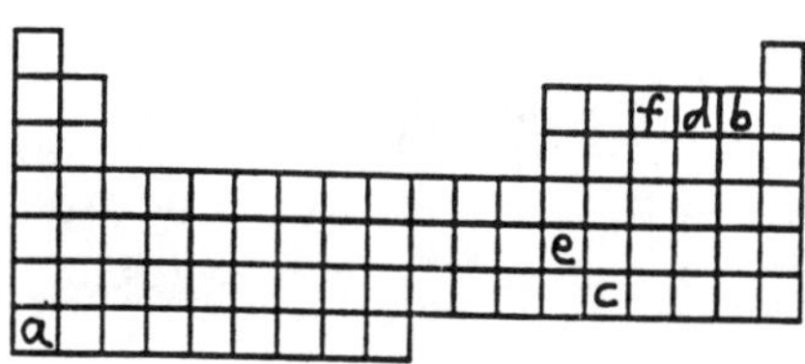

3. (a) N_2, O_2, F_2, P_4, S_8, Cl_2

(b)

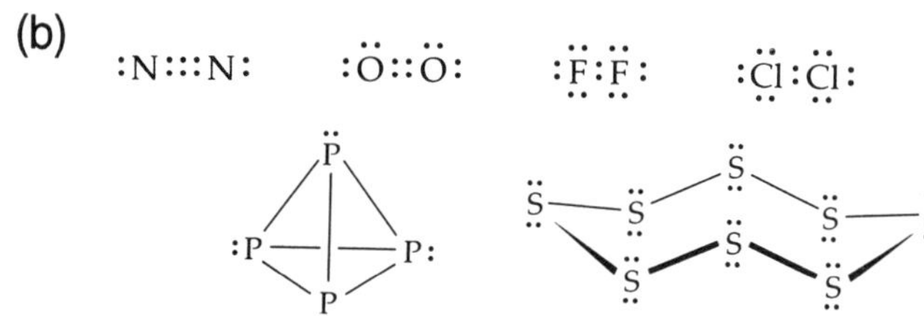

(c) The smaller N and O can form effective π bonds, whereas P and S cannot. In both F_2 and Cl_2, the atoms are joined by a single bond.

4. (a) SiO_4^{4-} (b) $Si_3O_{10}^{8-}$ (c) $Si_4O_{12}^{8-}$

Additional Problems

General Properties and Periodic Trends

19.8 The main-group elements are in groups 1A-8A. (a) Se, (b) Ca, and (d) In are main-group elements.

19.9 The p-block elements are in groups 3A - 8A. (b) P and (e) Br are p-block elements.

19.10 (a) Cl, nonmetal (b) Ga, metal (c) As, semimetal

(d) Pb, metal (e) S, nonmetal

19.11 (a) Cl (group 7A) lies to the right of S (group 6A) in the same row of the periodic table. Cl has the higher ionization energy.

(b) Si lies above Ge in group 4A. Si has the higher ionization energy.

(c) O (group 6A) lies above and to the right of In (group 3A) in the periodic table. O has the higher ionization energy.

19.12 (a) Al lies below B in group 3A. Al has the larger atomic radius.

(b) P (group 5A) lies to the left of S (group 6A) in the same row of the periodic table. P has the larger atomic radius.

(c) Pb (group 4A) lies below and to the left of Br (group 7A) in the periodic table. Pb has the larger atomic radius.

19.13 (a) I (group 7A) lies to the right of Te (group 6A) in the same row of the periodic table. I has the higher electronegativity.

(b) N lies above P in group 5A. N has the higher electronegativity.

(c) F (group 7A) lies above and to the right of In (group 3A) in the periodic table. F has the higher electronegativity.

19.14 (a) Sn lies below Si in group 4A. Sn has more metallic character.

(b) Ge (group 4A) lies to the left of Se (group 6A) in the same row of the periodic table. Ge has more metallic character.

(c) Bi (group 5A) lies below and to the left of I (group 7A) in the periodic table. Bi has more metallic character.

19.15 Na < Si < S < O (see Figure 19.2)

19.16 (a) CaH_2 (b) Ga_2O_3 (c) KCl

In each case the more ionic compound is the one formed between a metal and nonmetal.

19.17 (a) PBr_3 (b) CO (c) PH_3

In each case the more covalent molecule has the smaller electronegativity difference between bonded atoms.

19.18 Ionic (b) $BaCl_2$ (e) In_2O_3

Molecular (a) B_2H_6 (c) SO_3 (d) $GeCl_4$

19.19 (a) P_4O_{10} (b) B_2O_3 (c) SO_2

Nonmetal oxides are acidic. Metal oxides are basic or amphoteric.

19.20 (a) LiCl is an ionic compound. PCl_3 is a covalent molecular compound. The ionic compound, LiCl, has the higher melting point.

(b) Carbon forms strong π bonds with oxygen, and CO_2 is a covalent molecular compound with a low melting point. Silicon prefers to form single bonds with oxygen. SiO_2 is three dimensional extended structure with alternating silicon and oxygen singly bonded to each other. SiO_2 is a high melting solid.

(c) Nitrogen forms strong π bonds with oxygen and NO_2 is a covalent molecular compound with a low melting point. Phosphorus prefers to form single bonds with oxygen. P_4O_{10} is a larger covalent molecular compound than NO_2, with a higher metling point.

19.21 (a) Ga (b) In (c) Pb

Metals are better electrical conductors than nonmetals (S and P) or semimetals (B).

19.22 (a) Sn (b) Cl (c) Sn (d) Se (e) B

19.23 (a) F (b) Al (c) N (d) Si (e) S

19.24 Boron has only 2s and 2p valence orbitals and can form a maximum of four bonds.

Aluminum has 3d orbitals and can use octahedral sp^3d^2 hybrid orbitals to bond to six F^- ions.

19.25 In O_2 a π bond is formed by 2p orbitals on each O. S does not form strong π bonds with its 3p orbitals, which leads to the S_8 ring structure.

19.26 Carbon forms strong π bonds with oxygen. Silicon does not form strong π bonds with oxygen, and what results are chains of alternating silicon and oxygen singly bonded to each other.

The Group 3A Elements

19.27 +3 for B, Al, Ga and In

+1 for Tl

19.28 Boron is a hard semiconductor with a high melting point. Boron forms only molecular compounds and does not form an aqueous B^{3+} ion. $B(OH)_3$ is an acid.

19.29 Boron is a semimetal, whereas all the other 3A elements are metals. Boron has a much smaller atomic radius and a higher electronegativity than the other group 3A elements.

19.30 Borax is $Na_2B_4O_7 \cdot 10\ H_2O$. Borax is important because it is a mineral source for elemental boron.

19.31 $$2\ BBr_3(g) + 3\ H_2(g) \xrightarrow[1200^\circ C]{\text{Ta wire}} 2\ B(s) + 6\ HBr(g)$$

This reaction is used to produce crystalline boron.

19.32 (a) An electron deficient molecule is a molecule that doesn't have enough electrons to form a two center-two electron bond between each pair of bonded atoms. B_2H_6 is an electron deficient molecule.

(b) A three-center two-electron bond has three atoms bonded together using just two electrons. The B-H-B bridging bond in B_2H_6 is a three-center two-electron bond.

19.33

H H B H B H H H

The terminal B-H bonds in diborane are ordinary 2c-2e bonds. The bridging B-H-B bonds in diborane are 3c-2e bonds. A 3c-2e bond is longer than a 2c-2e bond because it has less electron density between each pair of adjacent atoms.

19.34 (a) Al (b) Tl (c) Ga (d) B (e) Tl (f) B

The Group 4A Elements

19.35 (a) Pb (b) C (c) Si (d) C

19.36 (a) Pb (b) Si (c) C (d) C

19.37 (a) $GeBr_4$, tetrahedral; Ge is sp^3 hybridized.

(b) CO_2, linear; C is sp hybridized.

(c) CO_3^{2-}, trigonal planar; C is sp^2 hybridized.

(d) $Sn(OH)_6^{2-}$, octahedral; Sn is sp^3d^2 hybridized.

(e) $SnCl_2$, bent; Sn is sp^2 hybridized.

19.38 Diamond is a very hard, high melting solid. It is an electrical insulator.

Diamond has a covalent network structure in which each C atom uses sp^3 hybrid orbitals to form a tetrahedral array of σ bonds. The interlocking, three-dimensional network of strong bonds makes diamond the hardest known substance with the highest melting point for an element. Because the valence electrons are localized in the σ bonds, diamond is an electrical insulator.

19.39 Graphite has a two-dimensional sheetlike structure in which each C atom uses sp^2 hybrid orbitals to form trigonal planar σ bonds to three neighboring C atoms. The carbon sheets in graphite are separated by a distance of 335 pm and are held together by weak London dispersion forces. Consequently, the sheets can easily slide over one another, thus accounting for the slippery feel of graphite and its use as a lubricant.

19.40 Graphite has a two-dimensional sheetlike structure in which each C atom uses sp^2 hybrid orbital to form trigonal planar σ bonds to three neighboring C atoms. In addition, each C atom uses its remaining p orbital, perpendicular to the plane of the sheet, to form a π bond. Because each C atom must share its π bond with its three neighbors, the π electrons are delocalized and are free to move in the plane of the sheet. As a result, the electrical conductivity of graphite in a direction parallel to the sheets is about 10^{20} times greater than the conductivity of diamond. The conductivity of graphite perpendicular to the sheets of C atoms is lower because electrons must hop from one sheet to the next.

19.41 The more dense form (diamond) is favored at high pressure because the stress of increased pressure is relieved by reducing the volume.

19.42 (a) carbon tetrachloride, CCl_4

(b) carbon monoxide, CO

(c) methane, CH_4

19.43 Some uses for CO_2 are:

(1) To provide the "bite" in soft drinks; $CO_2(aq) + H_2O(l) \rightleftharpoons H_2CO_3(aq)$

(2) CO_2 fire extinguishers; CO_2 is nonflammable and 1.5 times more dense than air.

(3) Refrigerant; dry ice, sublimes at –78°C.

19.44 $:C{\equiv}O:$ $\quad [:C{\equiv}N:]^-$

CO and CN^- are isoelectronic because they both have the same number and arrangement of valence electrons.

19.45 CO bonds to hemoglobin and prevents it from carrying O_2. CN^- bonds to cytochrome oxidase and interferes with the electron transfer associated with oxidative phosphorylation.

19.46 $CN^-(aq) + H_3O^+(aq) \rightleftharpoons HCN(aq) + H_2O(l)$

$4\,Au(s) + 8\,CN^-(aq) + O_2(g) + 2\,H_2O(l) \rightarrow 4\,Au(CN)_2^-(aq) + 4\,OH^-(aq)$

19.47 CaC_2, calcium carbide; C, –1

19.48 (a) $NaHCO_3$ (b) SnO_2 (c) $Na_2CO_3 \cdot 10\,H_2O$ (d) PbS

19.49 $\underset{\text{(sand)}}{SiO_2(l)} + 2\,C(s) \xrightarrow{\text{heat}} Si(l) + 2\,CO(g)$

Purification of silicon for semiconductor devices:

$$Si(s) + 2\ Cl_2(g) \rightarrow SiCl_4(l)$$

$$SiCl_4(g) + 2\ H_2(g) \xrightarrow{\text{heat}} Si(s) + 4\ HCl(g)$$

19.50 (a) Zone refining is a purification technique in which a heater melts a narrow zone at the top of a rod of some material and then sweeps slowly down the rod, bringing impurities with it.

(b) The silicate chain anion, $Si_2O_6^{4-}$, is an anion in which tetrahedral SiO_4 units share two O atoms to give an extended silicate chain with $Si_2O_6^{4-}$ as the repeating unit.

(c) Partial substitution of the Si^{4+} in SiO_2 with Al^{3+} gives aluminosilicates called feldspars, the most abundant of all minerals.

19.51 (a) SiO_4^{4-} (b) $Si_4O_{13}^{10-}$

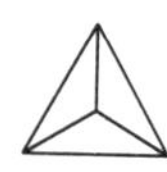

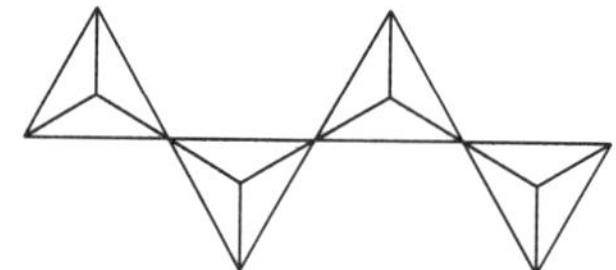

The charge on the anion is equal to the number of terminal O atoms.

19.52

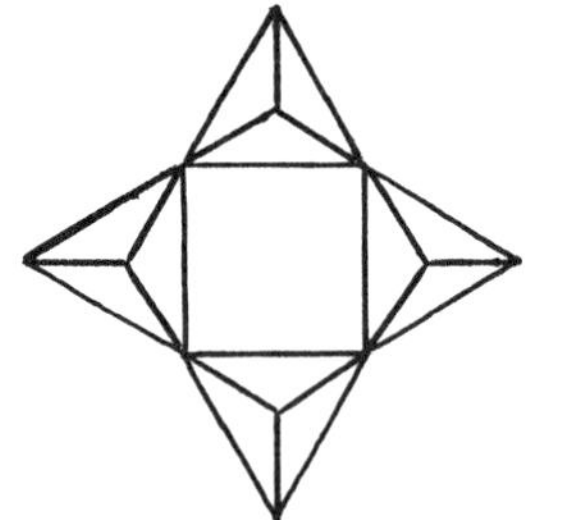

$Si_4O_{12}^{8-}$

19.53 (a) $Si_3O_{10}^{8-}$

(b) The charge on the anion is 8–. Because the Ca^{2+} to Cu^{2+} ratio is 1:1, there must be 2 Ca^{2+} and 2 Cu^{2+} ions in the formula for the mineral. There are also 2 waters. The formula of the mineral is: $Ca_2Cu_2Si_3O_{10} \cdot 2\ H_2O$

19.54 (a) spodumene, $LiAlSi_2O_6$

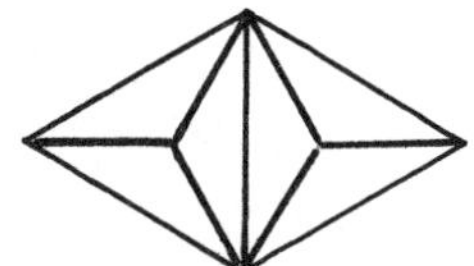

$Si_2O_6^{4-}$

(b) wollastonite, $Ca_3Si_3O_9$

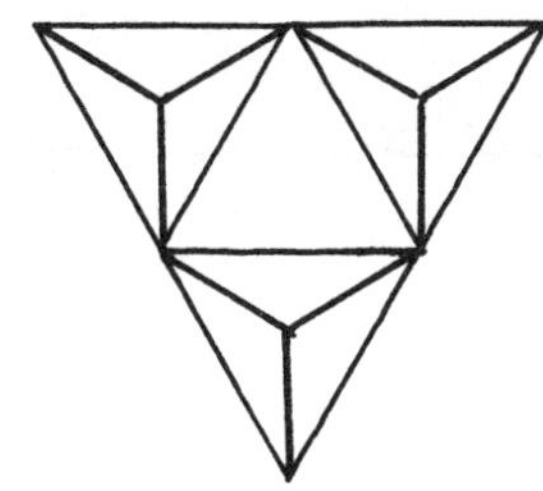

$Si_3O_9^{6-}$

(c) thortveitite, $Sc_2Si_2O_7$

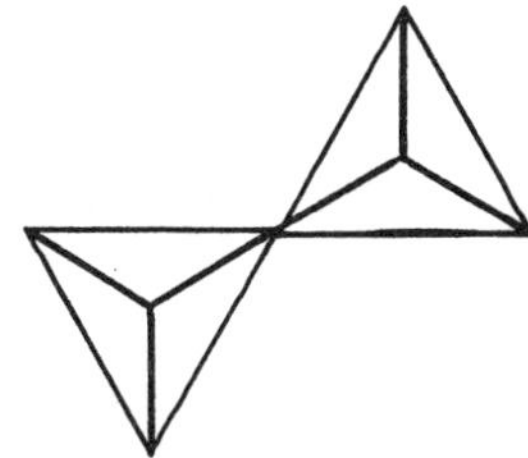

$Si_2O_7^{6-}$

(d) albite, $NaAlSi_3O_8$ - three dimensional SiO_2 structure with Al^{3+} in place of one-fourth of the Si^{4+} ions. (An equal number of Na^+ ions balance the charge.)

19.55 (a) silica sand (SiO_2)

$$SiO_2(l) + 2\ C(s) \xrightarrow{\text{heat}} Si(l) + 2\ CO(g)$$

(b) cassiterite (SnO_2)

$$SnO_2(s) + 2\ C(s) \xrightarrow{\text{heat}} Sn(l) + 2\ CO(g)$$

(c) galena (PbS)

$2\ PbS(s) + 3\ O_2(g) \rightarrow 2\ PbO(s) + 2\ SO_2(g)$

$PbO(s) + CO(g) \rightarrow Pb(l) + CO_2(g)$

19.56 Main uses for the following elements are:

(1) silicon: semiconductors, silicone oils

(2) germanium: semiconductors

(3) tin: protective coating over steel, alloys, such as bronze, pewter, solder.

(4) lead: pipes, cables, pigments, electrodes

The Group 5A Elements

19.57 (a) P (b) Sb and Bi (c) N (d) P (e) Bi

19.58 N_2 is unreactive because of the large amount of energy necessary to break the N≡N triple bond.

19.59 (a) N_2 is separated from liquid air by fractional distillation.

(b) NH_3, Haber process

$N_2(g) + 3\ H_2(g) \rightarrow 2\ NH_3(g)$

(c) HNO_3, Ostwald process

$$4\ NH_3(g) + 5\ O_2(g) \xrightarrow[\text{Pt/Rh catalyst}]{850°C} 4\ NO(g) + 6\ H_2O(g)$$

$2\ NO(g) + O_2(g) \rightarrow 2\ NO_2(g)$

$3\ NO_2(g) + H_2O(l) \rightarrow 2\ HNO_3(aq) + NO(g)$

(d) H_3PO_4

For food additive:

$P_4(l) + 5\ O_2(g) \rightarrow P_4O_{10}(s)$

$P_4O_{10}(s) + 6\ H_2O(l) \rightarrow 4\ H_3PO_4(aq)$

For fertilizers:

$Ca_3(PO_4)_2(s) + 3\ H_2SO_4(aq) \rightarrow 2\ H_3PO_4(aq) + 3\ CaSO_4(s)$

19.60 (a) N_2O, +1 (b) N_2H_4, –2 (c) Ca_3P_2, –3

(d) H_3PO_3, +3 (e) H_3AsO_4, +5

19.61 (a) NO, +2 (b) HNO_2, +3 (c) PH_3, –3

(d) P_4O_{10}, +5 (e) $H_5P_3O_{10}$, +5

19.62 (a) N_2O, linear

:N≡N—Ö: (with lone pairs above and below O)

(b) NO, linear, paramagnetic

:Ṅ=Ö:

(c) NO_2, bent, paramagnetic

:Ö—Ṅ=Ö: ⟷ :Ö=Ṅ—Ö:

19.63 (a) NO_2^-, bent; N is sp^2 hybridized.

(b) PH_3, trigonal pyramidal; P is sp^3 hybridized.

(c) PF_5, trigonal bipyramidal; P is sp^3d hybridized.

(d) N_2O, linear; central N is sp hybridized.

(e) PCl_4^+, tetrahedral; P is sp^3 hybridized.

(f) PCl_6^-, octahedral; P is sp^3d^2 hybridized.

19.64 Both NH_3 and PH_3 are colorless gases at room temperature. Both have a trigonal pyramidal geometry. NH_3 can hydrogen bond, PH_3 cannot. Aqueous solutions of NH_3 are basic. Aqueous solutions of PH_3 are neutral.

19.65 White phosphorus

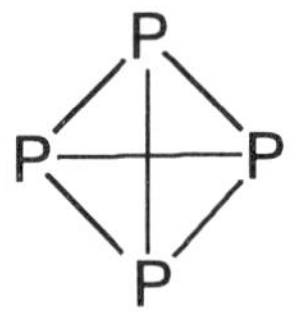

P
P P
P

Red phosphorus

P P P P
—P P — P P — P P — P P—
P P P P

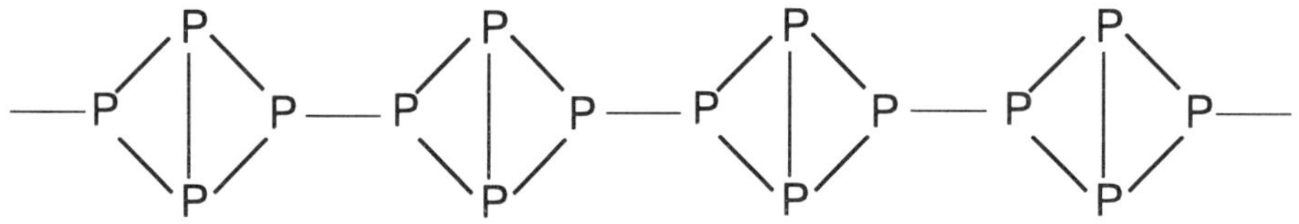

White phosphorus is reactive due to the considerable strain in the P_4 molecule.

19.66 (a) The structure for the phosphorous acid is

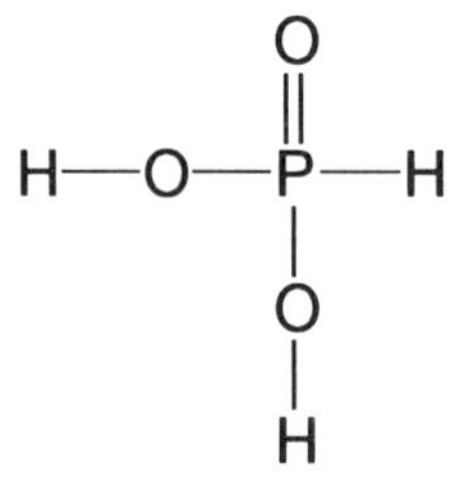

O
H—O—P—H
O
H

Only the two hydrogens bonded to oxygen are acidic.

(b) Nitric acid is a strong oxidizing agent, but phosphoric acid is not because the nitrogen atom is smaller and more electronegative than the phosphorus atom. This favors its reduction.

(c) Nitrogen forms strong π bonds, and in N_2 the nitrogen atoms are triple bonded to each other. Phosphorus does not form strong $p\pi - p\pi$ bonds, and so the P atoms are single bonded to each other in P_4

(d) P, As, and Sb can use d orbitals to become five-coordinate. Nitrogen cannot.

19.67 (a) $2\ NO(g) + O_2(g) \rightarrow 2\ NO_2(g)$

(b) $4\ HNO_3(aq) \rightarrow 4\ NO_2(aq) + O_2(g) + 2\ H_2O(l)$

(c) $3\ Ag(s) + 4\ H^+(aq) + NO_3^-(aq) \rightarrow 3\ Ag^+(aq) + NO(g) + 2\ H_2O(l)$

(d) $N_2H_4(aq) + 2\ I_2(aq) \rightarrow N_2(g) + 4\ H^+(aq) + 4\ I^-(aq)$

19.68 (a) phosphorus(III) oxide, P_4O_6

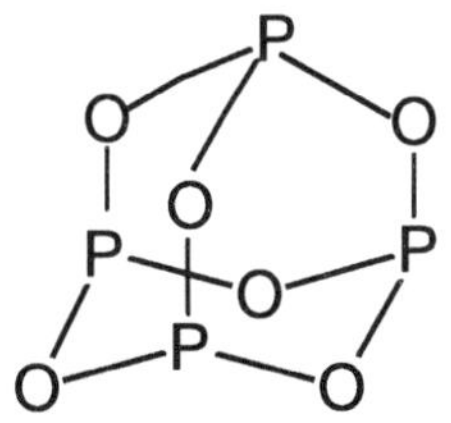

(b) phosphorus(V) oxide, P_4O_{10}

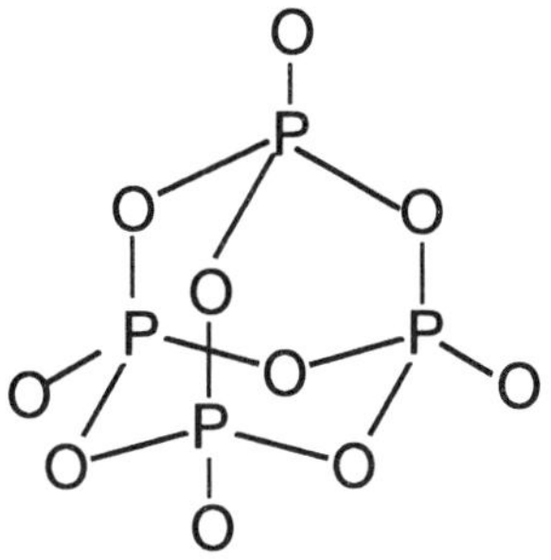

(c) phosphorous acid

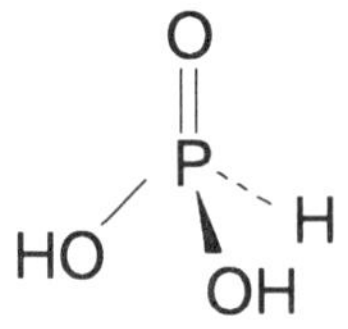

O
||
P
HO H
OH

(d) orthophosphoric acid (phosphoric)

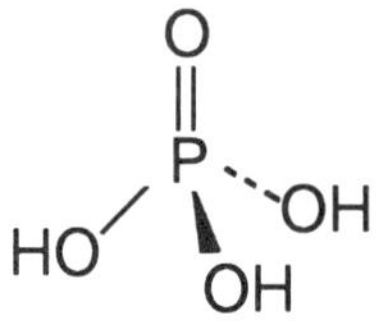

O
||
P
HO OH
OH

(e) polymetaphosphoric acid

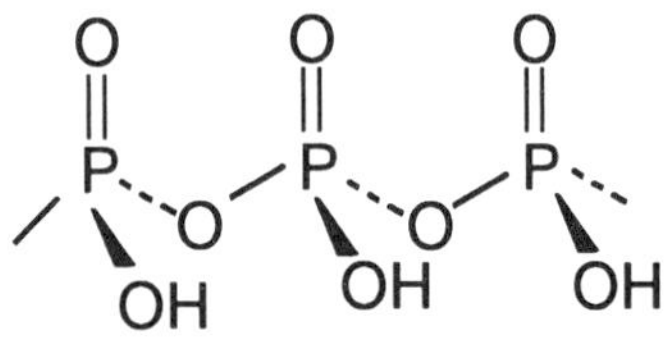

O O O
|| || ||
P O P O P
OH OH OH

19.69

H—O—P(=O)(—O—H)—O—H + H—O—P(=O)(—O—H)—O—H ⟶ H—O—P(=O)(—O—H)—O—P(=O)(—O—H)—O—H + H_2O

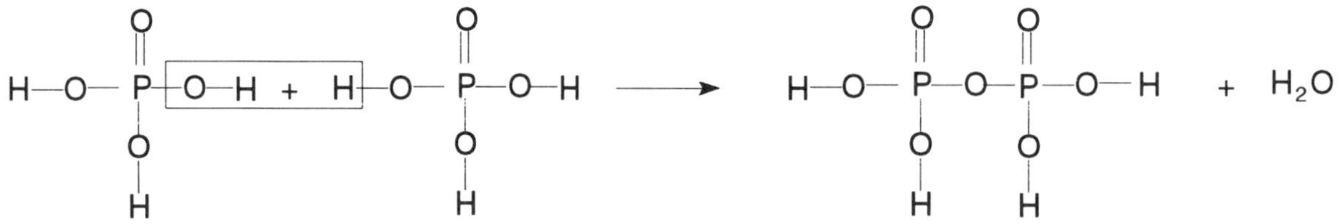

The Group 6A Elements

19.70 (a) O (b) Te (c) Po (d) S (e) O (f) O

19.71 Sulfur occurs in elemental form in large underground deposits. It is recovered from these underground deposits using superheated water under pressure to force the melted sulfur to the surface (Frasch process).

Sulfur is also present in numerous minerals, such as FeS_2, PbS, HgS, and $CaSO_4 \cdot 2\ H_2O$. Sulfur is also present in natural gas as H_2S and in crude oil as organic sulfur compounds.

The principle use for sulfur is in the manufacture of sulfuric acid.

19.72 (a) rhombic sulfur – yellow crystalline solid (mp 113°C) that contains crown-shaped S_8 rings.

(b) monoclinic sulfur – an allotrope of sulfur in which the S_8 rings pack differently in the crystal.

(c) plastic sulfur – when sulfur is cooled rapidly, the sulfur forms disordered, tangled chains, yielding an amorphous, rubbery material called plastic sulfur.

(d) Liquid sulfur between 160 and 195°C becomes dark reddish-brown and very viscous forming long polymer chains (S_n, $n > 200{,}000$).

19.73 The dramatic increase in the viscosity of molten sulfur at 160-195°C is due to opening of the S_8 rings, yielding S_8 chains that form long polymer chains, which become entangled. Above 200°C, the polymer chains begin to fragment into smaller pieces, with a decrease in viscosity.

19.74 (a) hydrogen sulfide, H_2S
lead(II) sulfide, PbS

(b) sulfur dioxide, SO_2
sulfurous acid, H_2SO_3

(c) sulfur trioxide, SO_3
sulfur hexafluoride, SF_6

19.75 (a) HgS, –2 (b) $Ca(HSO_4)_2$, +6 (c) H_2SO_3, +4

(d) FeS_2, –1 (e) SF_4, +4

19.76 Sulfuric acid (H_2SO_4) is manufactured by the contact process, a three-step reaction sequence in which (1) sulfur burns in air to give SO_2, (2) SO_2 is oxidized to SO_3 in the presence of a vanadium (V) oxide catalyst, and (3) SO_3 reacts with water to give H_2SO_4.

(1) $S(s) + O_2(g) \rightarrow SO_2(g)$

(2) $2\ SO_2(g) + O_2(g) \xrightarrow[V_2O_5\ \text{catalyst}]{\text{heat}} 2\ SO_3(g)$

(3) $SO_3(g) + H_2O$ (in conc H_2SO_4) $\rightarrow H_2SO_4(l)$

19.77 (a) $Na_2SO_3(s) + 2\ H^+(aq) \rightarrow SO_2(g) + H_2O(l) + 2\ Na^+(aq)$

(b) $CH_3C(S)NH_2(aq) + H_2O(l) \rightarrow CH_3C(O)NH_2(aq) + H_2S(aq)$

(c) $NaOH(aq) + H_2SO_4(aq) \rightarrow NaHSO_4(aq) + H_2O(l)$

19.78 (a) $Zn(s) + 2\ H_3O^+(aq) \rightarrow Zn^{2+}(aq) + H_2(g) + 2\ H_2O(l)$

(b) $BaSO_3(s) + 2\ H_3O^+(aq) \rightarrow H_2SO_3(aq) + Ba^{2+}(aq) + 2\ H_2O(l)$

(c) $Cu(s) + 2\ H_2SO_4(l) \rightarrow Cu^{2+}(aq) + SO_4^{2-}(aq) + SO_2(g) + 2\ H_2O(l)$

(d) $H_2S(aq) + I_2(aq) \rightarrow S(s) + 2\ H^+(aq) + 2\ I^-(aq)$

19.79 (a) Acid strength increases as the number of O atoms increases.

(b) In comparison with S, O is much too electronegative to form compounds of O in the +4 oxidation state. Also, S uses sp^3d hybrid orbitals for bonding in SF_4, but O doesn't have valence d orbitals and so it can't form four bonds to F.

(c) Each S is sp^3 hybridized with two lone pairs of electrons. The bond angles are therefore 109.5°. A planar ring would require bond angles of 135°.

Halogen Oxoacids and Oxoacid Salts

19.80 (a) $HBrO_3$, +5 (b) HIO, +1 (c) $NaClO_2$, +3 (d) $NaIO_4$, +7

19.81 (a) $HClO_2$, +3 (b) KIO_3, +5 (c) H_5IO_6, +7 (d) NaBrO, +1

19.82 (a) iodic acid (b) chlorous acid (c) sodium hypobromite

(d) lithium perchlorate

19.83 (a) potassium chlorite (b) metaperiodic acid (c) hypobromous acid

(d) sodium bromate

19.84 (a) HIO_3 trigonal pyramidal

O — I — O — H (with a third O bonded to I)

(b) ClO_2^- bent

[O — Cl — O]$^-$

(c) HOCl bent

O — Cl — H

(d) IO_6^{5-} octahedral

[IO_6]$^{5-}$

19.85 (a) BrO_4^- tetrahedral

```
[    ..     ]-
[   :O:     ]
[ .. .. ..  ]
[:O:Br:O:   ]
[ .. .. ..  ]
[   :O:     ]
[    ..     ]
```

(b) ClO_3^- trigonal pyramidal

```
[ ..     ]-
[:O:     ]
[ .. ..  ]
[:Cl:O:  ]
[ .. ..  ]
[:O:     ]
[ ..     ]
```

(c) HIO_4 tetrahedral

```
      ..
     :O:
   .. .. ..
 H:O:I:O:
   .. .. ..
     :O:
      ..
```

(d) HOBr bent

```
   .. ..
 H:O:Br:
   .. ..
```

19.86 Oxygen atoms are highly electronegative. Increasing the number of oxygen atoms increases the polarity of the O–H bond and increases the acid strength.

19.87 (a) $Br_2(l) + 2\ OH^-(aq) \rightarrow OBr^-(aq) + Br^-(aq) + H_2O(l)$

(b) $Cl_2(g) + H_2O(l) \rightarrow HOCl(aq) + H^+(aq) + Cl^-(aq)$

(c) $3\ Cl_2(g) + 6\ OH^-(aq) \rightarrow ClO_3^-(aq) + 5\ Cl^-(aq) + 3\ H_2O(l)$

19.88 $8\ KClO_3(s) + C_{12}H_{22}O_{11}(s) \rightarrow 8\ KCl(s) + 12\ CO_2(g) + 11\ H_2O(g)$

General Problems

19.89 (a) $Na_2B_4O_7 \cdot 10\ H_2O$ (b) $Ca_3(PO_4)_2$

(c) elemental sulfur, FeS_2, PbS, HgS, $CaSO_4 \cdot 2\ H_2O$

19.90 Carbon is a versatile element that can form millions of very stable compounds with elements such as N, O, and H. Biomolecules contain chains and rings with many C–C bonds. Si–Si bonds are much less stable and chains of Si atoms are uncommon. In addition, carbon can form very stable pπ-pπ multiple bonds. On the other hand, the chemistry of silicon (which cannot form stable pπ-pπ bonds) is dominated by structures based on the SiO_4^{4-} anion.

19.91 Gases: H, He, Ne, Ar, Kr, Xe, Rn, F, Cl, O, N

Liquid: Br

19.92 H_2SO_4 is present in acid rain.

$S(s) + O_2(g) \rightarrow SO_2(g)$ (burning of fossil fuels)

$2\ SO_2(g) + O_2(g) \rightarrow 2\ SO_3(g)$ (in the atmosphere)

$SO_3(g) + H_2O(l) \rightarrow H_2SO_4(aq)$ (rain)

19.93 Earth's crust: O, Si, Al, Fe

Human body: O, C, H, N

19.94 H_2SO_4, N_2, O_2, $CH_2{=}CH_2$, CaO, NH_3, NaOH, Cl_2, H_3PO_4, and methyl tert-butyl ether

19.95 C, Si, Ge and Sn have allotropes with the diamond structure.

Sn and Pb have metallic allotropes.

C (nonmetal), Si (semimetal), Ge (semimetal), Sn (semimetal and metal), Pb (metal)

19.96 (a) FeS_2 (b) Na_2CO_3 (c) $CaSO_4 \cdot 2H_2O$ (d) HgS

19.97 (a) $NH_4^+(aq) + OH^-(aq) \rightarrow NH_3(g) + H_2O(l)$

(b) $CO_3^{2-}(aq) + 2\ H^+(aq) \rightarrow CO_2(g) + H_2O(l)$

(c) $2\ NaBH_4 + I_2 \rightarrow B_2H_6(g) + H_2(g) + 2\ NaI$

(d) $CaC_2(s) + 2\ H_2O(l) \rightarrow C_2H_2(g) + Ca(OH)_2(aq)$

(e) $NH_4NO_3(l) \rightarrow N_2O(g) + 2\ H_2O(g)$

(f) $Cu(s) + 2\ NO_3^-(aq) + 4\ H^+(aq) \rightarrow Cu^{2+}(aq) + 2\ NO_2(g) + 2\ H_2O(l)$

19.98 (a) $H_3PO_4(aq) + H_2O(l) \rightleftharpoons H_3O^+(aq) + H_2PO_4^-(aq)$

H_3PO_4 is a Brønsted-Lowry acid.

(b) $B(OH)_3(aq) + 2\ H_2O(l) \rightleftharpoons B(OH)_4^-(aq) + H_3O^+(aq)$

$B(OH)_3$ is a Lewis acid.

19.99 (a) H_2SO_4 is used in the manufacturing of soluble phosphate and ammonium sulfate fertilizers.

(b) N_2 is used to produce NH_3 by the Haber process. NH_3 is used for fertilizers. N_2 gas is used for inert atmospheres.

(c) NH_3 is the starting material for the industrial synthesis of other important N-containing compounds, such as HNO_3. NH_3 itself is used as a fertilizer.

(d) HNO_3 is used for fertilizers, explosives, plastics, and dyes.

(e) H_3PO_4 is used for fertilizers and detergents.

19.100

+3 oxidation state	+5 oxidation state
phosphorus(III) oxide, P_4O_6	calcium phosphate, $Ca_3(PO_4)_2$
phosphorous acid, H_3PO_3	phosphorus pentachloride, PCl_5
phosphorus trichloride, PCl_3	phosphorus(V) oxide, P_4O_{10}

19.101 The angle required by P_4 is 60°. The strain would not be reduced by using sp^3 hybrid orbitals because their angle is ~109°.

19.102

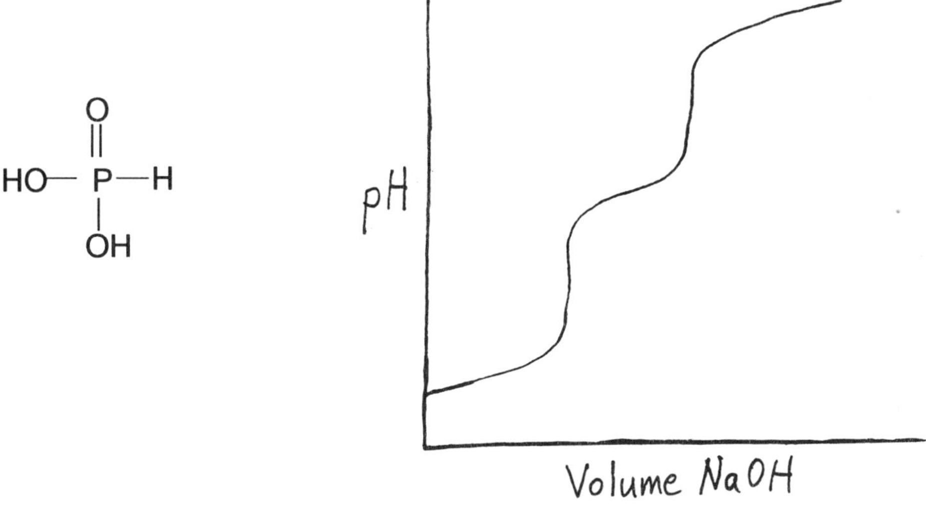

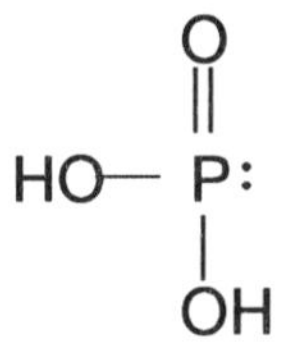

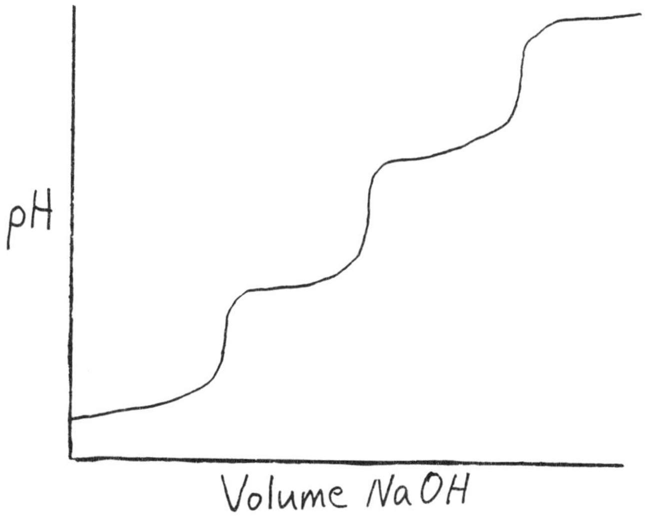

O
HO— P:
OH

pH

Volume NaOH

19.103 (a) In diamond each C is covalently bonded to four additional C atoms in a rigid three-dimensional network solid. Graphite is a two-dimensional covalent network solid of carbon sheets that can slide over each other. Both are high melting because melting requires the breaking of C–C bonds.

(b) There is some multiple bond character in terminal P–O bonds, utilizing empty d-orbitals on the P. The bridging P–O bonds are single bonds.

(c) Chlorine does not form perhalic acids of the type H_5XO_6 because its smaller size favors a tetrahedral structure over an octahedral one.

19.104 (a) C as diamond

(b) $Cl_2(g) + H_2O(l) \rightarrow HOCl(aq) + H^+(aq) + Cl^-(aq)$

(c) NO

(d) NO_2

(e) BF_3

(f) Al_2O_3

(g) Si

(h) HNO_3

(i) C as diamond, graphite, and fullerene.

20.1 (a) V, [Ar] $3d^34s^2$ (b) Co^{2+}, [Ar] $3d^7$ (c) Mn^{4+} in MnO_2, [Ar] $3d^3$

20.2 $[Cr(NH_3)_2(SCN)_4]^-$

20.3 In $Na_4[Fe(CN)_6]$ each sodium is in the +1 oxidation state (+4 total); each cyanide (CN^-) has a –1 charge (–6 total). The compound is neutral; therefore, the oxidation state of the iron is +2.

20.4 (a) tetraamminecopper(II) sulfate

(b) sodium tetrahydroxochromate(III)

(c) triglycinatocobalt(III)

(d) pentaaquathiocyanatoiron(III) ion

20.5 (a) $[Zn(NH_3)_4](NO_3)_2$ (b) $Ni(CO)_4$ (c) $K[Pt(NH_3)Cl_3]$ (d) $[Au(CN)_2]^-$

20.6 (a) Two diastereoisomers are possible.

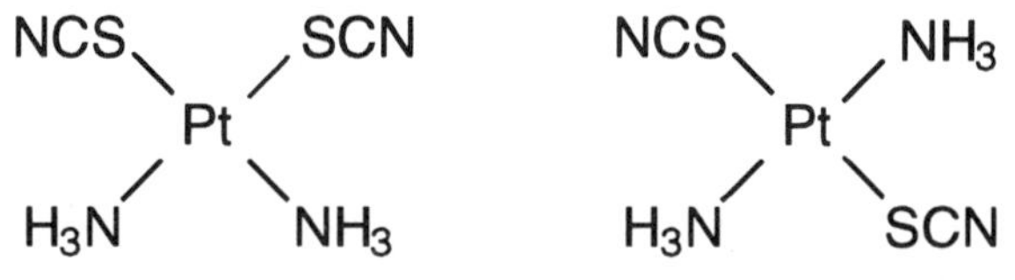

NCS, SCN, Pt, H_3N, NH_3 — cis

NCS, NH_3, Pt, H_3N, SCN — trans

(b) No isomers are possible for a tetrahedral complex of the type MA_2B_2.

(c) Two diastereoisomers are possible.

NH_3, H_3N, NH_3, Co, O_2N, NO_2, NO_2

NO_2, H_3N, NH_3, Co, O_2N, NH_3, NO_2

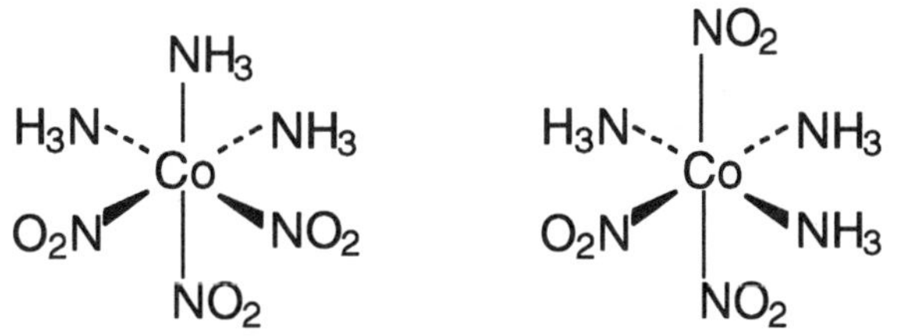

(d) No isomers are possible for a complex of this type.

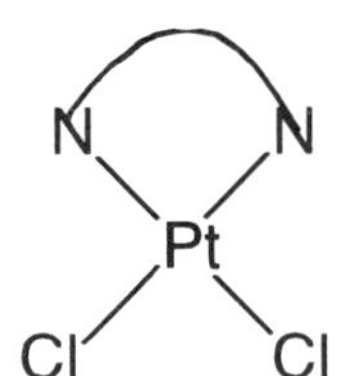

(e) Two diastereoisomers are possible.

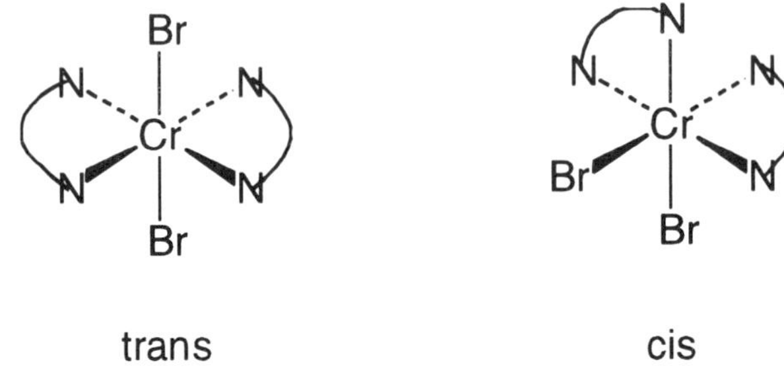

(f) No diastereoisomers are possible for a complex of this type.

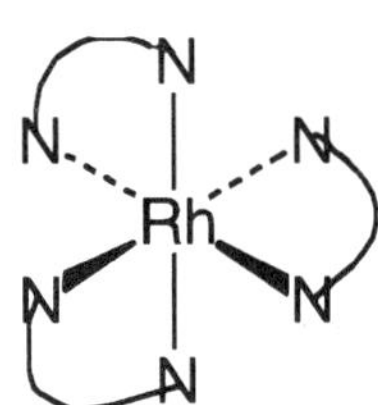

20.7 (a) glove, yes (b) baseball, no (c) chair, no

(d) foot, yes (e) pencil, no

20.8 (a) $[Fe(C_2O_4)_3]^{3-}$ can exist as enantiomers.

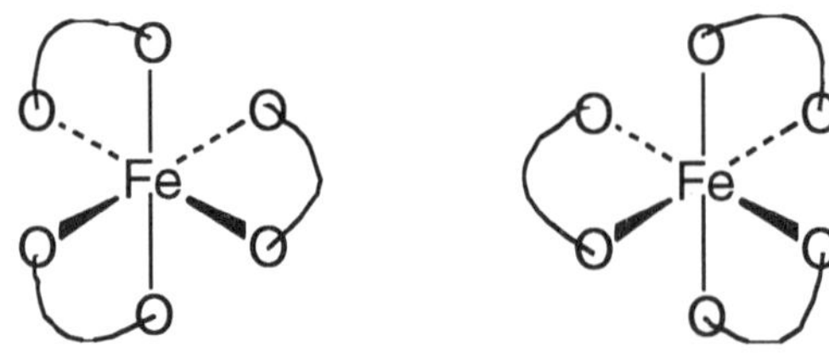

(b) $[Co(NH_3)_4en]^{3+}$ cannot exist as enantiomers.

(c) $[Co(NH_3)_2(en)_2]^{3+}$ can exist as enantiomers.

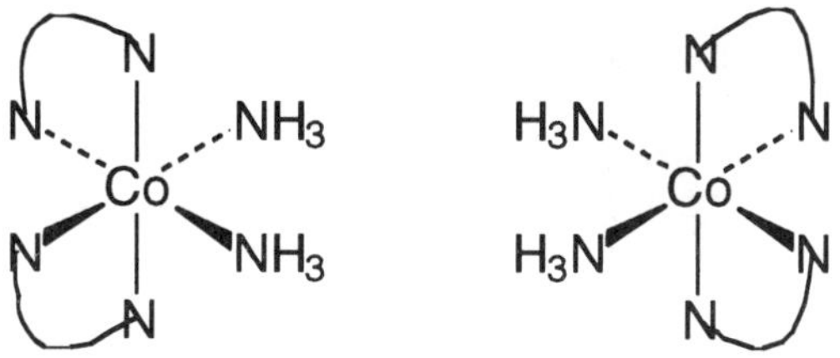

(d) $[Cr(H_2O)_4Cl_2]^+$ cannot exist as enantiomers.

20.9 (a) Fe^{3+} [Ar] ↑ ↑ ↑ ↑ ↑ (3d) _ (4s) _ _ _ (4p)

$[Fe(CN)_6]^{3-}$ [Ar] ↑↓ ↑↓ ↑ (3d) | ↑↓ ↑↓ (3d) ↑↓ (4s) ↑↓ ↑↓ ↑↓ (4p) |

d^2sp^3 1 unpaired e^-

(b) Co^{2+} [Ar] ↑↓ ↑↓ ↑ ↑ ↑ (3d) _ (4s) _ _ _ (4p)

$[Co(H_2O)_6]^{2+}$ [Ar] ↑↓ ↑↓ ↑ ↑ ↑ (3d) | ↑↓ (4s) ↑↓ ↑↓ ↑↓ (4p) ↑↓ ↑↓ (4d) | _ _ _ (4d)

sp^3d^2 3 unpaired e^-

(c) V^{3+} [Ar] ↑ ↑ _ _ _ (3d) _ (4s) _ _ _ (4p)

$[VCl_4]^-$ [Ar] ↑ ↑ _ _ _ (3d) | ↑↓ (4s) ↑↓ ↑↓ ↑↓ (4p) |

sp^3 2 unpaired e^-

(d) Pt^{2+} [Xe] ↑↓ ↑↓ ↑↓ ↑ ↑ (5d) _ (6s) _ _ _ (6p)

$[PtCl_4]^{2-}$ [Xe] ↑↓ ↑↓ ↑↓ ↑↓ (5d) | ↑↓ (5d) ↑↓ (6s) ↑↓ ↑↓ (6p) | _ (6p)

dsp^2 no unpaired e^-

20.10 (a) (b) (c)

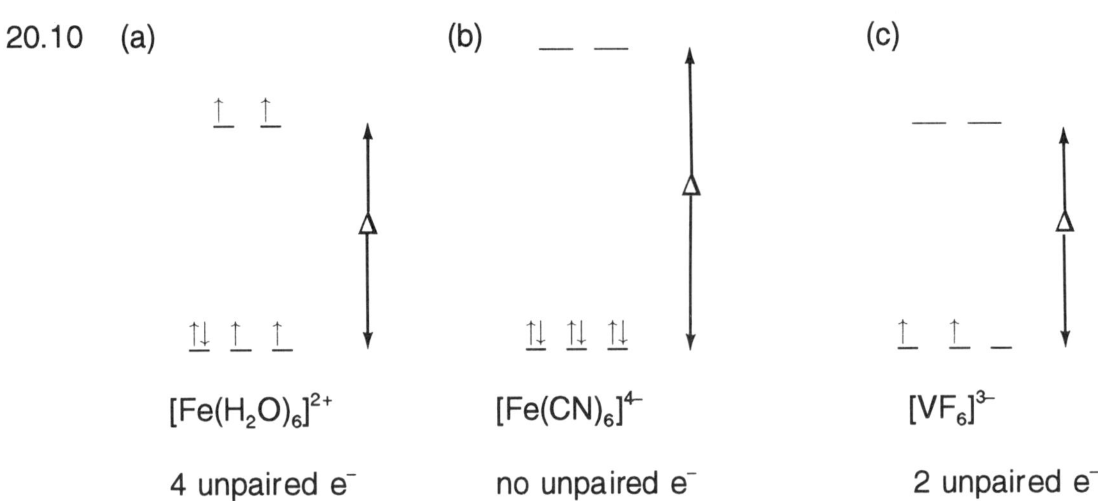

20.11 Both $[NiCl_4]^{2-}$ and $[Ni(CN)_4]^{2-}$ contain Ni^{2+} with a [Ar] $3d^8$ electron configuration.

(a) $[NiCl_4]^{2-}$ (tetrahedral)

↑↓ ↑ ↑
xy xz yz

↑↓ ↑↓
z^2 x^2-y^2 2 unpaired electrons

(b) $[Ni(CN)_4]^{2-}$ (square planar)

x^2-y^2

↑↓
xy

↑↓
z^2

↑↓ ↑↓
xz yz no unpaired electrons

Understanding Key Concepts

1. (a) Co (b) Cr (c) Zr (d) Pr

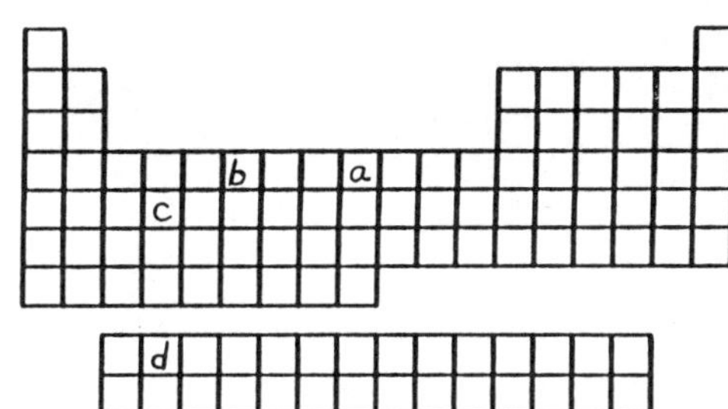

2. (a) The atomic radii decrease, at first markedly and then more gradually. Toward the end of the series, the radii increase again. The decrease in atomic radii is a result of an increase in Z_{eff}. The increase is due to electron-electron repulsions in doubly occupied d orbitals.

   (b) The densities of the transition metals are inversely related to their atomic radii. The densities initially increase from left to right and then decrease toward the end of the series.

   (c) Ionization energies generally increase from left to right across the series. The general trend correlates with an increase in Z_{eff} and a decrease in atomic radii.

   (d) The standard oxidation potentials generally decrease from left to right across the first transition series. This correlates with the general trend in ionization energies.

3. (a) **N**H_2–CH_2–CH_2–**N**H_2 is a bidentate ligand. It can form a chelate ring using the atoms indicated in bold.

   (b) CH_3–CH_2–CH_2–NH_2 is a monodentate ligand.

   (c) **N**H_2–CH_2–CH_2–**N**H–CH_2–C**O**$_2^-$ is a tridentate ligand. It can form chelate rings using the atoms indicated in bold.

   (d) NH_2–CH_2–CH_2–NH_3^+ is a monodentate ligand. The first N can coordinate to a metal.

4. (a) $Na[Au(CN)_2]$

$1\ Na^+$
$2\ CN^-$
The oxidation state of the Au is +1.

Coordination number = 2; Linear

(b) $[Co(NH_3)_5Br]SO_4$

$1\ Br^-$
$1\ SO_4^{2-}$
$5\ NH_3$ (no charge)
The oxidation state of the Co is +3.

Coordination number = 6; Octahedral

(c) $Pt(en)Cl_2$

$2\ Cl^-$
en = $NH_2–CH_2–CH_2–NH_2$ (no charge)
The oxidation state of the Pt is +2.

Coordination number = 4; Square planar

(d) $(NH_4)_2[PtCl_2(C_2O_4)_2]$

$2\ NH_4^+$
$2\ Cl^-$
$2\ C_2O_4^{2-}$
The oxidation state of the Pt is +4.

Coordination number = 6; Octahedral

5. (a) (1) cis; (2) trans; (3) trans; (4) cis

(b) (1) and (4) are the same. (2) and (3) are the same.

(c) None of the isomers exist as enantiomers because their mirror images are identical.

6. (a) (1) chiral; (2) achiral; (3) chiral; (4) chiral

(b)

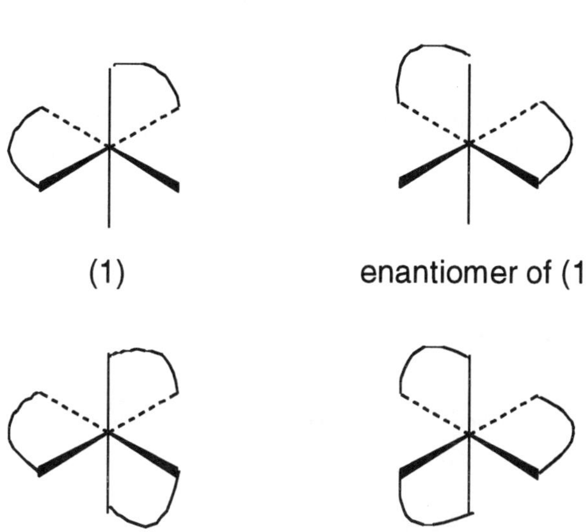

(1) enantiomer of (1)

(3) enantiomer of (3)

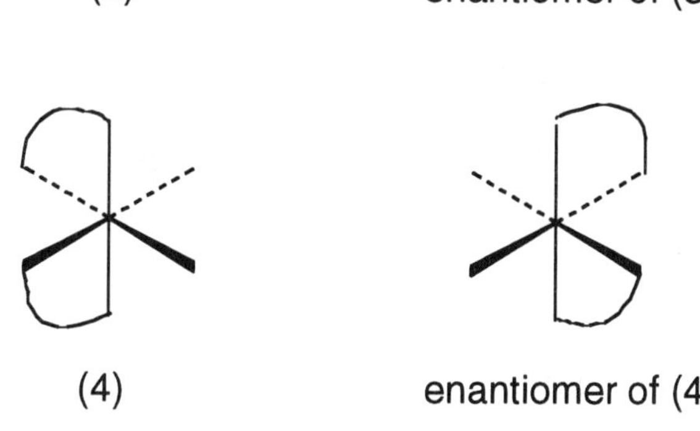

(4) enantiomer of (4)

(c) (1) and (4) are enantiomers.

7.

E

x^2-y^2

z^2

xy

xz yz

Additional Problems

Electron Configurations and Properties of Transition Elements

20.12 (a) The transition elements occupy the center portion of the periodic table, bridging the gap between groups 1A and 2A on the left and groups 3A-8A on the right. The d subshells are being filled for the elements in this region of the periodic table.

(b) the lanthanide elements (elements 58-71) are fourteen metallic elements where the 4f orbitals are progressively filled.

(c) The actinide elements (elements 90-103) are fourteen elements where the 5f orbitals are progressively filled.

(d) The d-block elements are the transition elements.

20.13 (a) Al (group 3A), p-block (b) Mo (group 6B), d-block

(c) Ba (group 2A), s-block (d) Hf (group 4B), d-block

(e) Ru (group 8B), d-block

20.14 (a) Ni, [Ar] $3d^8 4s^2$ (b) Cr, [Ar] $3d^5 4s^1$ (c) Zr, [Kr] $4d^2 5s^2$

(d) Zn, [Ar] $3d^{10} 4s^2$

20.15 (a) V, [Ar] $3d^3 4s^2$ (b) Y, [Kr] $4d^1 5s^2$ (c) Cu, [Ar] $3d^{10} 4s^1$

(d) Hg, [Xe] $4f^{14} 5d^{10} 6s^2$

20.16 (a) Co^{2+}, [Ar] $3d^7$ (b) Fe^{3+}, [Ar] $3d^5$ (c) Mo^{3+}, [Kr] $4d^3$

(d) Cr(VI), [Ar] $3d^0$

20.17 (a) V^{3+}, [Ar] $3d^2$ (b) Cu^+, [Ar] $3d^{10}$ (c) Rh^{3+}, [Kr] $4d^6$

(d) Fe(VI), [Ar] $3d^2$

20.18 (a) Cu^{2+}, [Ar] $3d^9$ ⇅ ⇅ ⇅ ⇅ ↑ (3d) — 1 unpaired e^-

(b) Ti^{2+}, [Ar] $3d^2$ ↑ ↑ _ _ _ (3d) — 2 unpaired e^-

(c) Zn^{2+}, [Ar] $3d^{10}$ ⇅ ⇅ ⇅ ⇅ ⇅ (3d) — 0 unpaired e^-

20.19 (a) Co^{3+}, [Ar] $3d^6$ ⇅ ↑ ↑ ↑ ↑ (3d) — 4 unpaired e^-

(b) Mn^{2+}, [Ar] $3d^5$ ↑ ↑ ↑ ↑ ↑ (3d) — 5 unpaired e^-

(c) Sc^{3+}, [Ar] $3d^0$ _ _ _ _ _ (3d) — 0 unpaired e^-

20.20 (a) Ti is harder than K and Ca largely because the sharing of d, as well as s, electrons results in stronger metallic bonding.

(b) Mo has a higher melting point than Y or Cd because the melting points increase as the number of unpaired d electrons available for metallic bonding increases and then decrease as the d electrons pair up and become less available for bonding.

(c) The decrease in radii with increasing atomic number is expected because the added d electrons only partially shield the added nuclear charge. As a result, Z_{eff} increases. With increasing Z_{eff}, the electrons are more strongly attracted to the nucleus, and atomic size decreases.

(d) The densities of the transition metals are inversely related to their atomic radii.

20.21 Ti > V > Cr > Mn Atomic radius decreases with increrasing Z_{eff}.

20.22 The smaller than expected sizes of the third-transition series atoms are associated with what is called the lanthanide contraction, the general decrease in atomic radii of the f-block lanthanide elements.

The lanthanide contraction is due to the increase in Z_{eff} as the 4f subshell is filled.

20.23 Zr and Hf are about the same size. The lanthanide contraction for Hf is due to the increase in Z_{eff} as the 4f subshell is filled.

20.24 Oxidation potentials generally decrease from left to right across the transition series. The general trend correlates with an increase in ionization energy, which is due in turn to an increase in Z_{eff} and a decrease in atomic radius.

20.25 (a) $Cr(s) + 2\ H^+(aq) \rightarrow Cr^{2+}(aq) + H_2(g)$

(b) $Zn(s) + 2\ H^+(aq) \rightarrow Zn^{2+}(aq) + H_2(g)$

(c) N.R.

(d) $Fe(s) + 2\ H^+(aq) \rightarrow Fe^{2+}(aq) + H_2(g)$

Oxidation States

20.26 (b) Mn (d) Cu

20.27 (c) Sc (d) Al

20.28 Sc^{3+}, Ti^{4+}, V^{5+}, Cr^{6+}, Mn^{7+}, Fe^{6+}, Co^{3+}, Ni^{2+}, Cu^{2+}, Zn^{2+}

20.29 The highest oxidation state for the group 3B-7B metals is the group number, corresponding to the loss of all valence s and d electrons. For the later transition metals, loss of all valence electrons is energetically prohibitive because of the increasing Z_{eff}. Therefore, only lower oxidation states are accessible for the later transition metals.

20.30 Cu^{2+} is a stronger oxidizing agent than Cr^{2+} because of a higher Z_{eff}.

20.31 V^{2+} is a stronger reducing agent than Co^{2+} because of a smaller Z_{eff}.

20.32 Cr^{2+} is more easily oxidized than Ni^{2+} because of a smaller Z_{eff}.

20.33 Co^{3+} is more easily reduced than Ti^{3+} because of a higher Z_{eff}.

20.34 $Mn^{2+} < MnO_2 > MnO_4^-$ because of increasing oxidation state of Mn.

20.35 $Cr_2O_7^{2-} < Cr^{3+} < Cr^{2+}$ because of decreasing oxidation state of the Cr.

Chemistry of Selected Transition Elements

20.36 (a) $Cr_2O_3(s) + 2\ Al(s) \rightarrow 2\ Cr(s) + Al_2O_3(s)$

(b) $Cu_2S(l) + O_2(g) \rightarrow 2\ Cu(l) + SO_2(g)$

20.37 (a) $Fe(s) + NO_3^-(aq) + 4\ H^+(aq) \rightarrow Fe^{3+}(aq) + NO(g) + 2\ H_2O(l)$

(b) $3\ Cu(s) + 2\ NO_3^-(aq) + 8\ H^+(aq) \rightarrow 3\ Cu^{2+}(aq) + 2\ NO(g) + 4\ H_2O(l)$

(c) $Cr(s) + NO_3^-(aq) + 4\ H^+(aq) \rightarrow Cr^{3+}(aq) + NO(g) + 2\ H_2O(l)$

20.38 (a) $Cr_2O_7^{2-}$ (b) Cr^{3+} (c) Cr^{2+} (d) Fe^{2+} (e) Cu^{2+}

20.39 (a) $Cr(OH)_2$ (b) $Cr(OH)_4^-$ (c) CrO_4^{2-} (d) $Fe(OH)_2$ (e) $Fe(OH)_3$

20.40 $Cr(OH)_2 < Cr(OH)_3 < CrO_2(OH)_2$

Acid strength increases with polarity of the O–H bond, which increases in turn with the oxidation state of Cr.

20.41 acid: $Cr(OH)_3(s) + OH^-(aq) \rightarrow Cr(OH)_4^-(aq)$

base: $Cr(OH)_3(s) + 3\ H_3O^+(aq) \rightarrow Cr^{3+}(aq) + 6\ H_2O(l)$

20.42 (c) $Cr(OH)_3$

20.43 (c) Cu^+ $2\ Cu^+(aq) \rightarrow Cu(s) + Cu^{2+}(aq)$

20.44 (a) Add excess $OH^-(aq)$ and Fe^{3+} will precipitate as $Fe(OH)_3(s)$. $Na^+(aq)$ will remain in solution.

(b) Add excess $OH^-(aq)$ and Fe^{3+} will precipitate as $Fe(OH)_3(s)$. $Cr(OH)_4^-(aq)$ will remain in solution.

(c) Add excess $NH_3(aq)$ and Fe^{3+} will precipitate as $Fe(OH)_3(s)$. $Cu(NH_3)_4^{2+}(aq)$ will remain in solution.

20.45 (a) $Cr_2O_7^{2-}(aq) + 6\ Fe^{2+}(aq) + 14\ H^+(aq) \rightarrow 2\ Cr^{3+}(aq) + 6\ Fe^{3+}(aq) + 7\ H_2O(l)$

(b) $4\ Fe^{2+}(aq) + O_2(g) + 4\ H^+(aq) \rightarrow 4\ Fe^{3+}(aq) + 2\ H_2O(l)$

(c) $Cu_2O(s) + 2\ H^+(aq) \rightarrow Cu(s) + Cu^{2+}(aq) + H_2O(l)$

(d) $Fe(s) + 2\ H^+(aq) \rightarrow Fe^{2+}(aq) + H_2(g)$

20.46 (a) $2\ CrO_4^{2-}(aq) + 2\ H_3O^+(aq) \rightarrow Cr_2O_7^{2-}(aq) + 3\ H_2O(l)$
(yellow) (orange)

(b) $[Fe(H_2O)_6]^{3+}(aq) + SCN^-(aq) \rightarrow [Fe(H_2O)_5(SCN)]^{2+}(aq) + H_2O(l)$
(red)

(c) $3\ Cu(s) + 2\ NO_3^-(aq) + 8\ H^+(aq) \rightarrow 3\ Cu^{2+}(aq) + 2\ NO(g) + 4\ H_2O(l)$
(blue)

(d) $Cr(OH)_3(s) + OH^-(aq) \rightarrow Cr(OH)_4^-(aq)$

$2\ Cr(OH)_4^-(aq) + 3\ HO_2^-(aq) \rightarrow 2\ CrO_4^{2-}(aq) + 5\ H_2O(l) + OH^-(aq)$
(yellow)

20.47 (a) $Cu^{2+}(aq) + 4\ NH_3(aq) \rightarrow [Cu(NH_3)_4]^{2+}(aq)$

(b) $Cr_2O_7^{2-}(aq) + 2\ OH^-(aq) \rightarrow 2\ CrO_4^{2-}(aq) + H_2O(l)$

(c) $Fe^{3+}(aq) + 3\ OH^-(aq) \rightarrow Fe(OH)_3(s)$

(d) $3\ CuS(s) + 8\ H^+(aq) + 2\ NO_3^-(aq) \rightarrow$

$3\ Cu^{2+}(aq) + 3\ S(s) + 2\ NO(g) + 4\ H_2O(l)$

20.48 Cr(II) $[Cr(H_2O)_6]^{2+}$

Cr(III) $Cr(OH)_4^-$

Cr(VI) CrO_4^{2-}

Fe(II) $[Fe(H_2O)_6]^{2+}$

Fe(III) $[Fe(H_2O)_5(SCN)]^{2+}$

Cu(I) $[Cu(CN)_2]^-$

Cu(II) $[Cu(NH_3)_4]^{2+}$

20.49 (a) 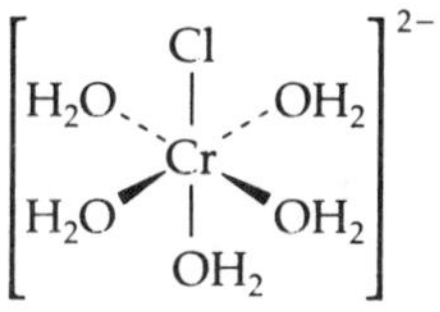 Cl, H_2O, OH_2, Cr, H_2O, OH_2, OH_2 $]^{2-}$ (b) 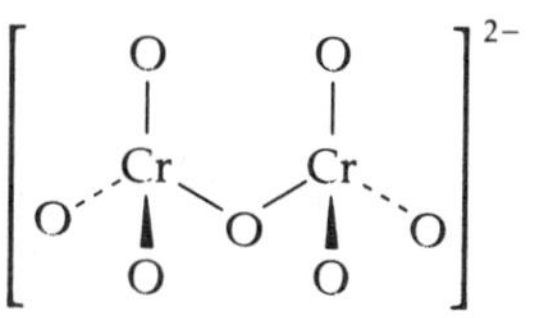 O, O, Cr, Cr, O, O, O, O, O $]^{2-}$ (c) 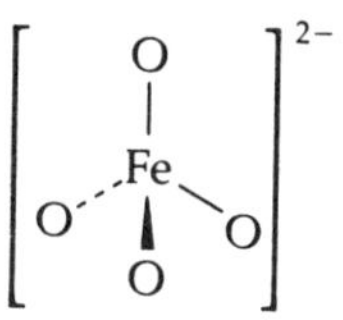 O, Fe, O, O, O $]^{2-}$

Coordination Compounds; Ligands

20.50 (a) A coordination compound is a compound in which a central metal ion is attached to a group of surrounding molecules or ions by coordinate covalent bonds. An example is $[Cu(NH_3)_4]^{2+}$.

(b) A ligand is a molecule or ion that surrounds the central metal ion in a coordination compound. An example is NH_3 in $[Cu(NH_3)_4]^{2+}$.

(c) A ligand donor atom is the ligand atom attached directly to the metal. An example is N of NH_3 $[Cu(NH_3)_4]^{2+}$.

(d) The coordination number is the number of ligand donor atoms that surround a central metal ion in a coordination compound. An example is coordination number four in $[Cu(NH_3)_4]^{2+}$.

20.51 (a) A monodentate ligand is a ligand that bonds to a metal using the electron pair of a single donor atom. An example is H_2O.

(b) A bidentate ligand is a ligand that bonds to a metal using electron pairs on two donor atoms. An example is $H_2NCH_2CH_2NH_2$. The nitrogens are the two donor atoms.

(c) Polydentate ligands are known as chelating agents because their multipoint attachment to a metal ion resembles the grasping of an object by the claws of a crab. An example is:

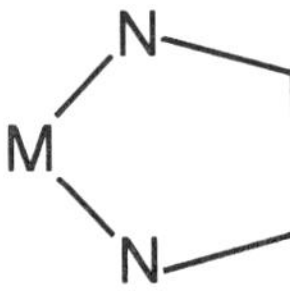

where $H_2NCH_2CH_2NH_2$ is the chelating agent.

(d) A chelate ring is the ring produced by a metal ion (M) and the chelating agent. An example is the chelate ring in (c) above.

20.52 Ni^{2+} accepts six pairs of electrons, two each from the three ethylenediamine ligands. Ni^{2+} is an electron pair acceptor, a Lewis acid. The two nitrogens in each ethylenediamine donate a pair of electrons to the Ni^{2+}. The ethylenediamine is an electron pair donor, a Lewis base. The formation of $[Ni(en)_3]^{2+}$ is a Lewis acid-base reaction.

20.53 Lewis acid (electron pair acceptor) is Fe^{3+}

Lewis base (electron pair donor) is $C_2O_4^{2-}$

20.54 (a) $[Ag(NH_3)_2]^+$ (b) $[Ni(CN)_4]^{2-}$ (c) $[Cr(H_2O)_6]^{3+}$

20.55 Coordination Number

	Coordination Number
(a) $[HgCl_4]^{2-}$	4
(b) $[Cr(H_2O)_4Cl_2]^+$	6
(c) $[Au(CN)_2]^-$	2
(d) $[ZrF_8]^{4-}$	8
(e) $[Mo(CO)_5Br]^-$	6

20.56 (a) $[HgCl_4]^{2-}$

4 Cl^-

The oxidation state of the Hg is +2.

(b) $[Cr(H_2O)_4Cl_2]^+$

4 H_2O (no charge)

2 Cl^-

The oxidation state of the Cr is +3.

(c) $[Au(CN)_2]^-$

2 CN^-

The oxidation state of the Au is +1.

(d) $[ZrF_8]^{4-}$

8 F^-

The oxidation state of the Zr is +4.

(e) $[Mo(CO)_5Br]^-$

5 CO (no charge)

1 Br^-

The oxidation state of the Mo is 0.

20.57 (a) $Ni(CO)_4$ (b) $[Ag(NH_3)_2]^+$ (c) $[Fe(CN)_6]^{3-}$ (d) $[Ni(CN)_4]^{2-}$

20.58

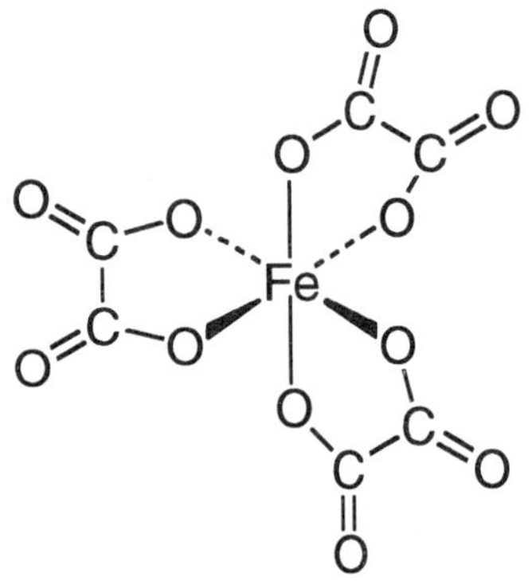

The iron is in the +3 oxidation state, and the coordination number is six. The geometry about the Fe is octahedral. The oxalate ligand is behaving as a bidentate chelate. There are three chelate rings, one formed by each oxalate ligand.

20.59 (a) $Ir(NH_3)_3Cl_3$ (b) $[Cr(H_2O)_2(C_2O_4)_2]^-$ (c) $[Pt(en)_2(SCN)_2]^{2+}$

20.60 (a) $[Ni(CN)_5]^{3-}$

5 CN^-

The oxidation state of the Ni is +2.

(b) $Ni(CO)_4$

4 CO (no charge)

The oxidation state of the Ni is 0.

(c) $[Co(en)_2(H_2O)Br]^{2+}$

2 en (NH_2–CH_2–CH_2–NH_2, no charge)
1 H_2O (no charge)
1 Br^-

The oxidation state of the Co is +3.

(d) $[Cu(H_2O)_2(C_2O_4)_2]^{2-}$

2 H_2O (no charge)
2 $C_2O_4^{2-}$

The oxidation state of the Cu is +2.

20.61 (a) $(NH_4)_3[RhCl_6]$

3 NH_4^+
6 Cl^-

The oxidation state of the Rh is +3.

(b) $[Cr(NH_3)_4(SCN)_2]Br$

4 NH_3 (no charge)
2 SCN^-
1 Br^-

The oxidation state of the Cr is +3.

(c) $[Cu(en)_2]SO_4$

2 en (NH_2–CH_2–CH_2–NH_2, no charge)
1 SO_4^{2-}

The oxidation state of the Cu is +2.

(d) $Na_2[Mn(EDTA)]$

2 Na^+
1 $EDTA^{4-}$

The oxidation state of the Mn is +2.

20.62 (a) tetrachloromanganate(II)

(b) hexaamminenickel(II)

(c) tricarbonatocobaltate(III)

(d) bis(ethylenediamine)dithiocyanatoplatinum(IV)

20.63 (a) tetrachloroaurate(III)

(b) hexacyanoferrate(II)

(c) pentaaquaisothiocyanatoiron(III)

(d) diamminedioxalatochromate(III)

20.64 (a) cesium tetrachloroferrate(III)

(b) hexaaquavanadium(III) nitrate

(c) tetraamminedibromocobalt(III) bromide

(d) diglycinatocopper(II)

20.65 (a) tetraamminecopper(II) sulfate

(b) hexacarbonylchromium(0)

(c) potassium trioxalatoferrate(III)

(d) amminecyanatobis(ethylenediamine)cobalt(III) chloride

20.66 (a) $[Pt(NH_3)_4]Cl_2$ (b) $Na_3[Fe(CN)_6]$

(c) $[Pt(en)_3](SO_4)_2$ (d) $Rh(NH_3)_3(SCN)_3$

20.67 (a) $[Ag(NH_3)_2]NO_3$ (b) $K[Co(H_2O)_2(C_2O_4)_2]$

(c) $Mo(CO)_6$ (d) $[Cr(NH_3)_2(en)_2]Cl_3$

Isomers

20.68 (a) Linkage isomers are isomers in which a ligand bonds to a metal through either of two different donor atoms. Examples are:

$[Co(NH_3)_5(NO_2)]^{2+}$ and $[Co(NH_3)_5(ONO)]^{2+}$

(b) Ionization isomers are isomers that differ in the anion bonded to the metal ion. Examples are:

$[Co(NH_3)_5Br]SO_4$ and $[Co(NH_3)_5(SO_4)]Br$

(c) Diastereoisomers are non-mirror-image stereoisomers. Examples are:

Cl Cl Pt H_3N NH_3

cis

Cl NH_3 Pt H_3N Cl

trans

(d) Enantiomers are stereoisomers that are nonidentical mirror images of each other. Examples are:

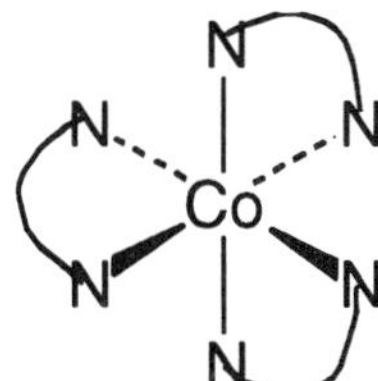

N N N Co N N N

20.69 (a) Constitutional isomers are compounds with different connections between atoms.

Stereoisomers are compounds with the same atom connections but with different spatial arrangements of atoms.

(b) A cis isomer is the isomer of a metal complex in which identical ligands are adjacent to each other. A trans isomer is the isomer of a metal complex in which identical ligands are opposite from each other.

(c) Objects that have a "handedness" are said to be chiral. Objects that lack handedness are said to be achiral.

20.70

$[Ru(NH_3)_5(NO_2)]Cl$ $[Ru(NH_3)_5(ONO)]Cl$ $[Ru(NH_3)_5Cl]NO_2$

$[Ru(NH_3)_5(NO_2)]Cl$ and $[Ru(NH_3)_5(ONO)]Cl$ are linkage isomers.

$[Ru(NH_3)_5Cl]NO_2$ is an ionization isomer of both $[Ru(NH_3)_5(NO_2)]Cl$ and $[Ru(NH_3)_5(ONO)]Cl$.

20.71

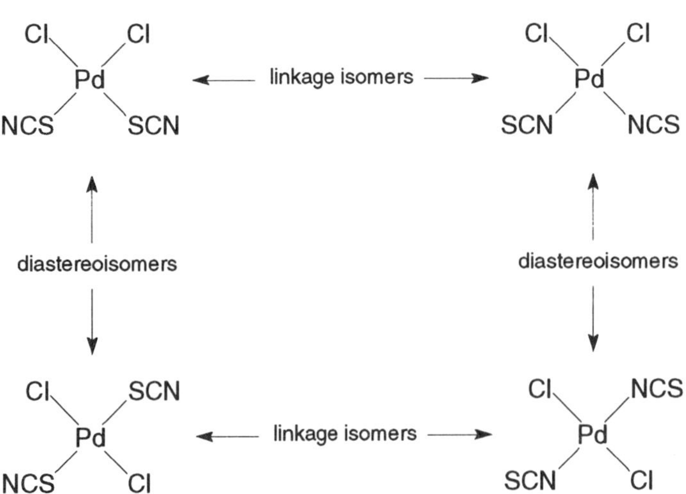

20.72 (a) $[Cr(NH_3)_2Cl_4]^-$ can exist as cis and trans diastereoisomers.

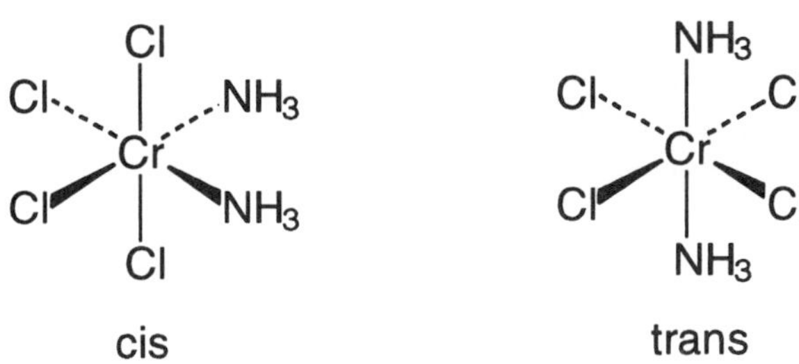

cis trans

(b) $[Co(NH_3)_5Br]^{2+}$ cannot exist as diastereoisomers.

(c) $[FeCl_2(NCS)_2]^{2-}$ (tetrahedral) cannot exist as diastereoisomers.

(d) $[PtCl_2Br_2]^{2-}$ (square planar) can exist as cis and trans diastereoisomers.

cis trans

20.73 (a) $Pt(NH_3)_2(CN)_2$ can exist as two (cis and trans) diastereoisomers.

cis trans

(b) $[Co(en)(SCN)_4]^-$ cannot exist as diastereoisomers.

(c) $[Cr(H_2O)_4Cl_2]^+$ can exist as two (cis and trans) diastereoisomers.

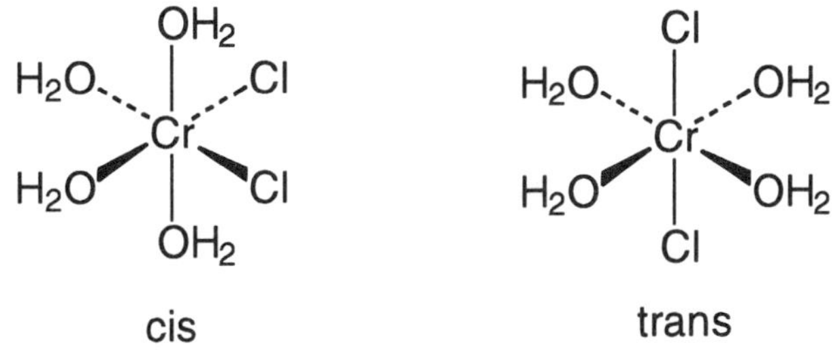

cis trans

(d) $Ru(NH_3)_3I_3$ can exist as two diastereoisomers.

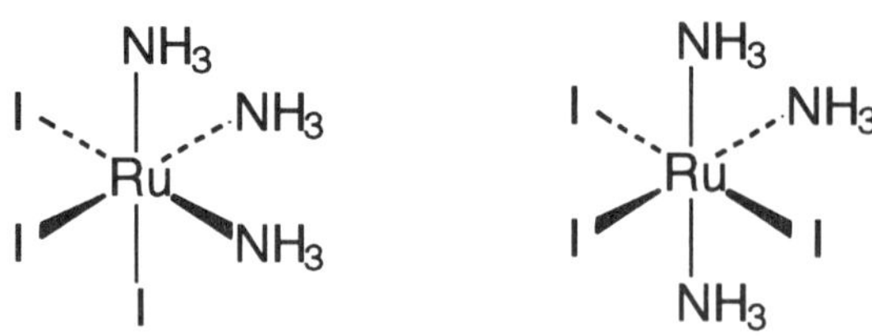

20.74 (c) cis-$[Cr(en)_2(H_2O)_2]^{3+}$ (d) $[Cr(C_2O_4)_3]^{3-}$

20.75 (a) $[Cr(en)_3]^{3+}$ can exist as enantiomers.

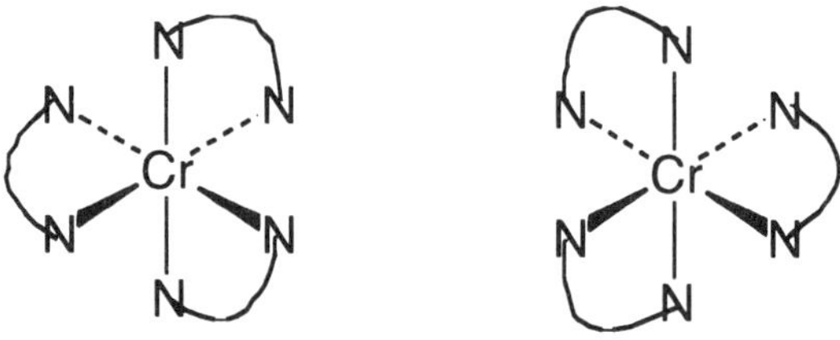

(b) cis-$[Co(en)_2(NH_3)Cl]^{2+}$ can exist as enantiomers.

(c) trans-$[Co(en)_2(NH_3)Cl]^{2+}$ cannot exist as enantiomers.

(d) $[Pt(NH_3)_3Cl_3]^+$ cannot exist as enantiomers.

20.76 (a) $Ru(NH_3)_4Cl_2$ can exist as cis and trans diastereoisomers.

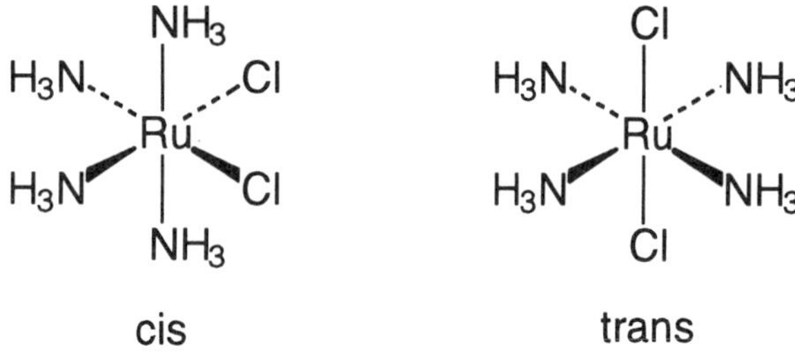

cis trans

(b) $[Pt(en)_3]^{4+}$ can exist as enantiomers.

(c) $[Pt(en)_2ClBr]^{2+}$ can exist as both diastereoisomers and enantiomers.

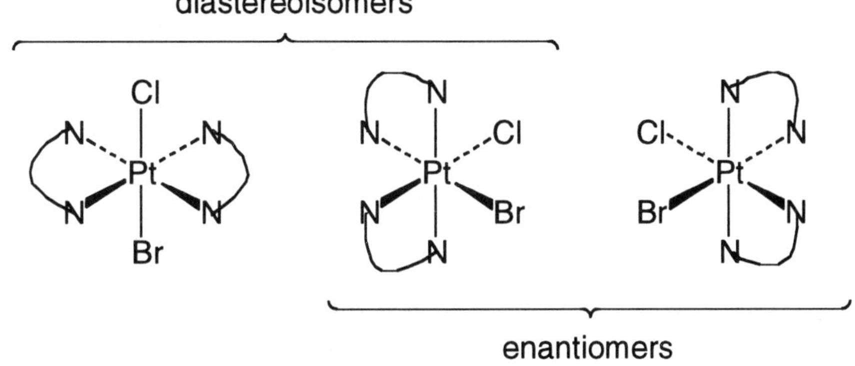

20.77 (a) $[Rh(C_2O_4)_2I_2]^{3-}$ can exist as both diastereoisomers and enantiomers.

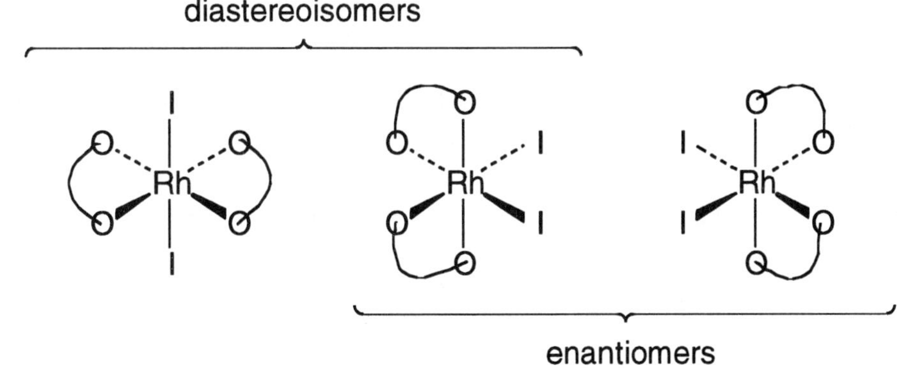

(b) $[Cr(NH_3)_2Cl_4]^-$ can exist as cis and trans diastereoisomers.

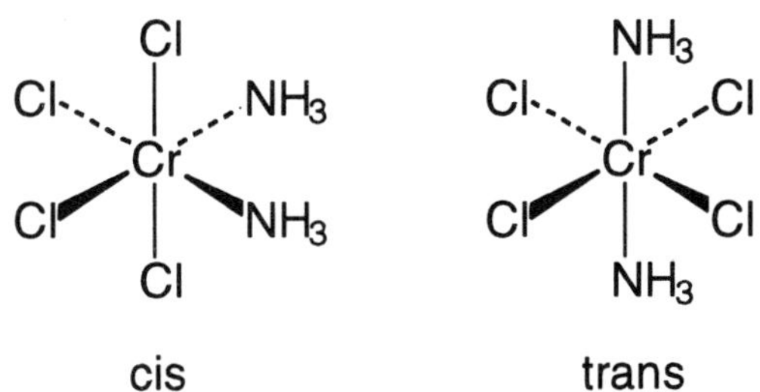

(c) $[Co(EDTA)]^-$ can exist as enantiomers.

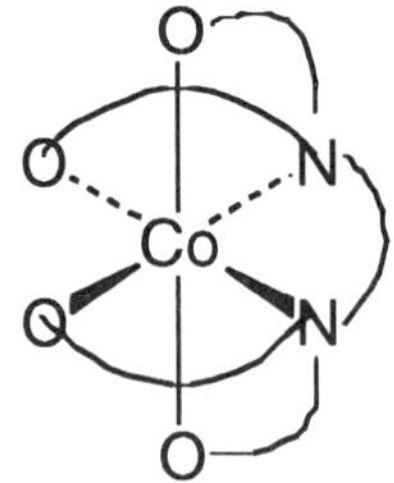

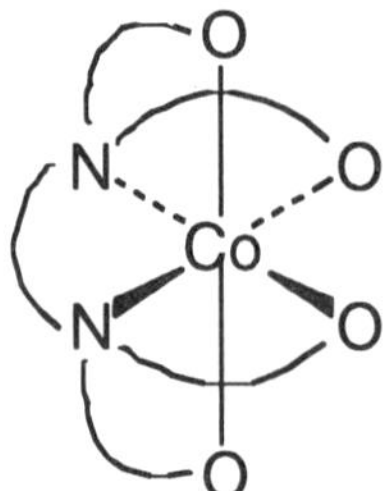

20.78 Plane-polarized light is light in which the electric vibrations of the light wave are restricted to a single plane. The following chromium complex can rotate the plane of plane-polarized light.

$[Cr(en)_3]^{3+}$

20.79 A racemic mixture is a mixture that contains equal amounts of two enantiomers. A racemic mixture does not affect plane-polarized light because the effect of one enantiomer is canceled by that of the other enantiomer.

Color and Magnetic Properties; Valence Bond and Crystal Field Theories

20.80 The measure of the amount of light absorbed by a substance is called theabsorbance, and a graph of absorbance vs wavelength is called an absorption spectrum.

If a complex absorbs at 455 nm, its color is orange (use the color wheel in Figure 20.24).

20.81 ~525 nm (use the color wheel in Figure 20.24)

20.82 (a) $[Ti(H_2O)_6]^{3+}$

		3d		4s	4p	4d
Ti^{3+}	[Ar]	↑ _ _ _ _		_	_ _ _	
$[Ti(H_2O)_6]^{3+}$	[Ar]	↑ _ _	↑↓ ↑↓	↑↓	↑↓ ↑↓ ↑↓	

d^2sp^3 1 unpaired e^-

(b) $[NiBr_4]^{2-}$

		3d	4s	4p
Ni^{2+}	[Ar]	↑↓ ↑↓ ↑↓ ↑ ↑	_	_ _ _
$[NiBr_4]^{2-}$	[Ar]	↑↓ ↑↓ ↑↓ ↑ ↑	↑↓	↑↓ ↑↓ ↑↓

sp^3 2 unpaired e^-

(c) $[Fe(CN)_6]^{3-}$ (low-spin)

		3d		4s	4p
Fe^{3+}	[Ar]	↑ ↑ ↑ ↑ ↑		_	_ _ _
$[Fe(CN)_6]^{3-}$	[Ar]	↑↓ ↑↓ ↑	↑↓ ↑↓	↑↓	↑↓ ↑↓ ↑↓

d^2sp^3 1 unpaired e^-

(d) $[MnCl_6]^{3-}$ (high-spin)

		3d	4s	4p	4d	
Mn^{3+}	[Ar]	↑ ↑ ↑ ↑ _	_	_ _ _		
$[MnCl_6]^{3-}$	[Ar]	↑ ↑ ↑ ↑ _	↑↓	↑↓ ↑↓ ↑↓	↑↓ ↑↓	_ _ _

sp^3d^2 4 unpaired e^-

20.83 (a) Au^{3+} [Xe] ↑↓ ↑↓ ↑↓ ↑ ↑ (5d) _ (6s) _ _ _ (6p)

$[AuCl_4]^-$ [Xe] ↑↓ ↑↓ ↑↓ ↑↓ (5d) | ↑↓ (5d) ↑↓ (6s) ↑↓ ↑↓ _ (6p) |

dsp² no unpaired e^-

(b) Ag^+ [Kr] ↑↓ ↑↓ ↑↓ ↑↓ ↑↓ (4d) _ (5s) _ _ _ (5p)

$[Ag(NH_3)_2]^+$ [Kr] ↑↓ ↑↓ ↑↓ ↑↓ ↑↓ (4d) | ↑↓ (5s) ↑↓ | _ _ (5p)

sp no unpaired e^-

(c) Fe^{2+} [Ar] ↑↓ ↑ ↑ ↑ ↑ (3d) _ (4s) _ _ _ (4p)

$[Fe(H_2O)_6]^{2+}$ [Ar] ↑↓ ↑ ↑ ↑ ↑ (3d) | ↑↓ (4s) ↑↓ ↑↓ ↑↓ (4p) ↑↓ ↑↓ | _ _ _ (4d)

sp^3d^2 4 unpaired e^-

(d) Fe^{2+} [Ar] ↑↓ ↑ ↑ ↑ ↑ (3d) _ (4s) _ _ _ (4p)

$[Fe(CN)_6]^{4-}$ [Ar] ↑↓ ↑↓ ↑↓ (3d) | ↑↓ ↑↓ (3d) ↑↓ (4s) ↑↓ ↑↓ ↑↓ (4p) |

d^2sp^3 no unpaired electrons

20.84 $[Ti(H_2O)_6]^{3+}$ $\quad$ Ti^{3+} $\quad$ $3d^1$

__ __ e_g
z^2 $x^2\text{–}y^2$

Δ crystal field splitting

$\uparrow$ __ __ t_{2g}
xz yz xy

$[Ti(H_2O)_6]^{3+}$ is colored because it can absorb light in the visible region, exciting the electron to the higher energy set of orbitals

20.85

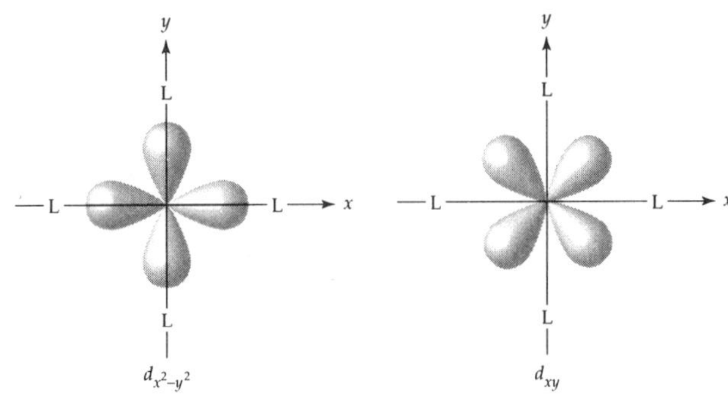

The $d_{x^2-y^2}$ orbital is higher in energy because its lobes are pointing directly at the ligands.

20.86 λ = 544 nm = 544 x 10^{-9} m

$$\Delta = \frac{hc}{\lambda} = \frac{(6.626 \times 10^{-34}\ \text{J}\cdot\text{s})(3.00 \times 10^{8}\ \text{m/s})}{(544 \times 10^{-9}\ \text{m})} = 3.65 \times 10^{-19}\ \text{J}$$

Δ = (3.65 x 10^{-19} J/ion)(6.022 x 10^{23} ion/mol) = 219,803 J/mol = 220 kJ/mol

For $[Ti(H_2O)_6]^{3+}$, Δ = 240 kJ/mol

Since $\Delta_{NCS^-} < \Delta_{H_2O}$ for the Ti complex, NCS^- is a weaker-field ligand than H_2O. If $[Ti(NCS)_6]^{3-}$ absorbs at 544 nm, its color is red (use the color wheel in Figure 20.24).

20.87 $[Cr(H_2O)_6]^{3+}$ absorbs at ~580 nm while $[Cr(CN)_6]^{3-}$ absorbs at ~415 nm. CN^- is a stronger-field ligand, and H_2O is a weaker-field ligand.

20.88 (a) $[CrF_6]^{3-}$ (b) $[V(H_2O)_6]^{3+}$ (c) $[Fe(CN)_6]^{3-}$

(a)
_ _

↑ ↑ ↑
3 unpaired e^-

(b)
_ _

↑ ↑ _
2 unpaired e^-

(c)
_ _

↑↓ ↑↓ ↑
1 unpaired e^-

20.89 (a) $[Cu(en)_3]^{2+}$ (b) $[FeF_6]^{3-}$ (c) $[Co(en)_3]^{3+}$

(a)
↑↓ ↑

↑↓ ↑↓ ↑↓
1 unpaired e^-

(b)
↑ ↑

↑ ↑ ↑
5 unpaired e^-

(c)
_ _

↑↓ ↑↓ ↑↓
0 unpaired e^-

20.90 $Ni^{2+}(aq)$ $Zn^{2+}(aq)$

$Ni^{2+}(aq)$:
↑ ↑

↑↓ ↑↓ ↑↓

$Zn^{2+}(aq)$:
↑↓ ↑↓

↑↓ ↑↓ ↑↓

$Ni^{2+}(aq)$ is green because the Ni^{2+} ion can absorb light, which promotes electrons from the filled d orbitals to the higher energy half-filled d orbitals. $Zn^{2+}(aq)$ is colorless because the d orbitals are completely filled and no electrons can be promoted so no light is absorbed.

20.91 Cr^{3+} is $3d^3$ and should be colored because absorption of light can promote electrons from the t_{2g} to the e_g orbitals. Y^{3+} is $4d^0$ and should be colorless because it can't exhibit d-d transitions.

20.92 Weak-field ligands produce a small Δ. Strong-field ligands produce a large Δ. For a metal complex with weak-field ligands, Δ < P, where P is the pairing energy, and it is easier to place an electron in either d_{z^2} or $d_{x^2-y^2}$ than to pair up electrons; high-spin complexes result. For a metal complex with strong-field ligands, Δ > P and it is easier to pair up electrons than to place them in either d_{z^2} or $d_{x^2-y^2}$; low-spin complexes result.

20.93 Ligands do not point directly at any d-orbitals, and the crystal field splitting is small. Because Δ < P, high-spin complexes result.

20.94 Square planar geometry is most common for metal ions with d^8 configurations because this configuration favors low-spin complexes in which all four lower energy d orbitals are filled and the higher energy $d_{x^2-y^2}$ orbital is vacant.

20.95 (a) $[Pt(NH_3)_4]^{2+}$ (square planar)

__ x^2-y^2

↑↓ xy

↑↓ z^2

↑↓ ↑↓ xz, yz

0 unpaired e^-

(b) $[MnCl_4]^{2-}$ (tetrahedral)

↑ ↑ ↑
xy xz yz

↑ ↑
x^2-y^2 z^2

5 unpaired e^-

(c) $[Co(NCS)_4]^{2-}$ (tetrahedral)

↑ ↑ ↑
xy xz yz

↑↓ ↑↓
x^2-y^2 z^2

3 unpaired e^-

(d) $[Cu(en)_2]^{2+}$ (square planar)

↑ x^2-y^2

↑↓ xy

↑↓ z^2

↑↓ ↑↓ xz, yz

1 unpaired e^-

General Problems

20.96 (a) Mn, [Ar] $3d^5 4s^2$ (b) Fe^{2+}, [Ar] $3d^6$

(c) V(IV), [Ar] $3d^1$ (d) Mn, [Ar] $3d^0$

20.97 (a) Sc^{3+} [Ar] _ _ _ _ _ (3d) 0 unpaired e^-, diamagnetic

(b) Co^{2+} [Ar] ↑↓ ↑↓ ↑ ↑ ↑ (3d) 3 unpaired e^-, paramagnetic

(c) V^{3+} [Ar] ↑ ↑ _ _ _ (3d) 2 unpaired e^-, paramagnetic

(d) CrO_4^{2-} [Ar] _ _ _ _ _ (3d) 0 unpaired e^-, diamagnetic

(e) MnO_4^{2-} [Ar] ↑ _ _ _ _ (3d) 1 unpaired e^-, paramagnetic

20.98 (a) $[Mn(CN)_6]^{3-}$ Mn^{3+} [Ar] $3d^4$

CN^- is a strong-field ligand. The Mn^{3+} complex is low-spin.

_ _

↑↓ ↑ ↑ 2 unpaired e^-, paramagnetic

(b) $[Zn(NH_3)_4]^{2+}$ Zn^{2+} [Ar] $3d^{10}$

$[Zn(NH_3)_4]^{2+}$ is tetrahedral.

↑↓ ↑↓ ↑↓

↑↓ ↑↓ 0 unpaired e^-, diamagnetic

(c) $[Fe(CN)_6]^{4-}$ Fe^{2+} [Ar] $3d^6$

CN^- is a strong-field ligand. The Fe^{2+} complex is low-spin.

— —

⇅ ⇅ ⇅ 0 unpaired e^-, diamagnetic

(d) $[FeF_6]^{4-}$ Fe^{2+} [Ar] $3d^6$

F^- is a weak-field ligand. The Fe^{2+} complex is high-spin.

↑ ↑

⇅ ↑ ↑ 4 unpaired e^-, paramagnetic

20.99 (a) $[Ni(H_2O)_6]^{2+}$ Ni^{2+} [Ar] $3d^8$

H_2O is a weak-field ligand.

↑ ↑

⇅ ⇅ ⇅ 2 unpaired e^-, paramagnetic

(b) $[Co(CN)_6]^{3-}$ Co^{2+} [Ar] $3d^6$

CN^- is a strong-field ligand. The Co^{3+} complex is low-spin.

— —

⇅ ⇅ ⇅ 0 unpaired e^-, diamagnetic

(c) $[HgI_4]^{2-}$ Hg^{2+} [Xe] $5d^{10}$

$[HgI_4]^{2-}$ is tetrahedral.

⇅ ⇅ ⇅

⇅ ⇅ 0 unpaired e^-, diamagnetic

(d) $[Cu(NH_3)_4]^{2+}$ Cu^{2+} [Ar] $3d^9$

$[Cu(NH_3)_4]^{2+}$ is square-planar.

↑

↑↓

↑↓

↑↓ ↑↓ 1 unpaired e^-, paramagnetic

20.100 (a) $4\,[Co^{3+}(aq) + e^- \rightarrow Co^{2+}(aq)]$

$2\,H_2O(l) \rightarrow O_2(g) + 4H^+(aq) + 4\,e^-$

$4\,Co^{3+}(aq) + 2\,H_2O(l) \rightarrow 4\,Co^{2+}(aq) + O_2(g) + 4\,H^+(aq)$

(b) $4\,Cr^{2+}(aq) + O_2(g) + 4\,H^+(aq) \rightarrow 4\,Cr^{3+}(aq) + 2\,H_2O(l)$

(c) $3\,[Cu(s) \rightarrow Cu^{2+}(aq) + 2\,e^-]$

$Cr_2O_7^{2-}(aq) + 14\,H^+(aq) + 6\,e^- \rightarrow 2\,Cr^{3+}(aq) + 7\,H_2O(l)$

$3\,Cu(s) + Cr_2O_7^{2-}(aq) + 14\,H^+(aq) \rightarrow 3\,Cu^{2+}(aq) + 2\,Cr^{3+}(aq) + 7\,H_2O(l)$

(d) $2\,CrO_4^{2-}(aq) + 2\,H^+(aq) \rightarrow Cr_2O_7^{2-}(aq) + H_2O(l)$

20.101 (a) $2\,[5\,e^- + 8\,H^+(aq) + MnO_4^-(aq) \rightarrow Mn^{2+}(aq) + 4\,H_2O(l)]$

$5\,[C_2O_4^{2-}(aq) \rightarrow 2\,CO_2(g) + 2\,e^-]$

$16\,H^+(aq) + 2\,MnO_4^-(aq) + 5\,C_2O_4^{2-}(aq) \rightarrow 2\,Mn^{2+}(aq) + 8\,H_2O(l) + 10\,CO_2(g)$

(b) $6\ [Ti^{3+}(aq) + H_2O(l) \rightarrow TiO^{2+}(aq) + 2\ H^+(aq) + e^-]$

$\underline{6\ e^- + Cr_2O_7^{2-}(aq) + 14\ H^+(aq) \rightarrow 2\ Cr^{3+}(aq) + 7\ H_2O(l)}$

$6\ Ti^{3+}(aq) + Cr_2O_7^{2-}(aq) + 2\ H^+(aq) \rightarrow 6\ TiO^{2+}(aq) + 2\ Cr^{3+}(aq) + H_2O(l)$

(c) $2\ [MnO_4^-(aq) + e^- \rightarrow MnO_4^{2-}(aq)]$

$\underline{H_2O(l) + SO_3^{2-}(aq) + 2\ OH^-(aq) \rightarrow SO_4^{2-}(aq) + 2\ H_2O(l) + 2\ e^-}$

$2\ MnO_4^-(aq) + SO_3^{2-}(aq) + 2\ OH^-(aq) \rightarrow 2\ MnO_4^{2-}(aq) + SO_4^{2-}(aq) + H_2O(l)$

(d) $4\ [H_2O(l) + Fe(OH)_2(s) \rightarrow Fe(OH)_3(s) + H^+(aq) + e^-]$

$\underline{4\ H^+(aq) + 4\ e^- + O_2(g) \rightarrow 2\ H_2O(l)}$

$2\ H_2O(l) + 4\ Fe(OH)_2(s) + O_2(g) \rightarrow 4\ Fe(OH)_3(s)$

20.102 mol Fe^{2+} = 0.400 mol/L x 0.100 L = 0.0400 mol Fe^{2+}

mol $Cr_2O_7^{2-}$ = 0.100 mol/L x 0.100 L = 0.0100 mol $Cr_2O_7^{2-}$

$6\ [Fe^{2+}(aq) \rightarrow Fe^{3+}(aq) + e^-]$

$\underline{Cr_2O_7^{2-}(aq) + 14\ H^+(aq) + 6\ e^- \rightarrow 2\ Cr^{3+}(aq) + 7\ H_2O(l)}$

$6\ Fe^{2+}(aq) + Cr_2O_7^{2-}(aq) + 14\ H^+(aq) \rightarrow 6\ Fe^{3+}(aq) + 2\ Cr^{3+}(aq) + 7\ H_2O(l)$

	Fe^{2+}	$Cr_2O_7^{2-}$	Fe^{3+}	Cr^{3+}
initial (mol)	0.0400	0.0100	0	0
change (mol)	–0.0400	$-\frac{0.0400}{6}$	+0.0400	$+\frac{0.0400}{3}$
after (mol)	0	0.00333	0.0400	0.0133

$$[Fe^{3+}] = \frac{0.0400\ \text{mol}}{0.200\ \text{L}} = 0.200\ \text{M}$$

$$[Cr^{3+}] = \frac{0.0133 \text{ mol}}{0.200 \text{ L}} = 0.0665 \text{ M}$$

$$[Cr_2O_7^{2-}] = \frac{0.00333 \text{ mol}}{0.200 \text{ L}} = 0.0166 \text{ M}$$

20.103 $2\ [Cr(OH)_4^-(aq) \rightarrow CrO_4^{2-}(aq) + 4\ H^+(aq) + 3\ e^-]$

$\underline{3\ [2\ e^- + 2\ H^+(aq) + H_2O_2(aq) \rightarrow 2\ H_2O(l)]}$

$2\ Cr(OH)_4^-(aq) + 3\ H_2O_2(aq) \rightarrow 2\ CrO_4^{2-}(aq) + 2\ H^+(aq) + 6\ H_2O(l)$

add 2 OH^- to both sides of the reaction

$2\ Cr(OH)_4^-(aq) + 3\ H_2O_2(aq) + 2\ OH^-(aq) \rightarrow 2\ CrO_4^{2-}(aq) + 8\ H_2O(l)$

mol $Cr_2(SO_4)_3$ = (0.040 L)(0.030 mol/L) = 0.0012 mol

mol $Cr(OH)_4^-$ = 2(mol $Cr_2(SO_4)_3$) = 2(0.0012 mol) = 0.0024 mol

mol H_2O_2 = (0.010 L)(0.20 mol/L) = 0.0020 mol

For 0.0024 mol $Cr(OH)_4^-$, you need (3/2)(0.0024 mol) = 0.0036 mol H_2O_2.

Because you have only 0.0020 mol H_2O_2, H_2O_2 is the limiting reactant.

$$\text{mol } CrO_4^{2-} = 0.0020 \text{ mol } H_2O_2 \times \frac{2 \text{ mol } CrO_4^{2-}}{3 \text{ mol } H_2O_2} = 0.001\ 33 \text{ mol } CrO_4^{2-}$$

$$[CrO_4^{2-}] = \frac{0.001\ 33 \text{ mol}}{0.10 \text{ L}} = 0.013 \text{ M}$$

20.104 (a) sodium aquabromodioxalatoplatinate(VI)

(b) hexaamminechromium(III) trioxalatocobaltate(III)

(c) hexaamminecobalt(III) trioxalatochromate(III)

(d) diamminebis(ethylenediamine)rhodium(III) sulfate

20.105 $[Rh(en)_2(NO_2)(SCN)]^+$

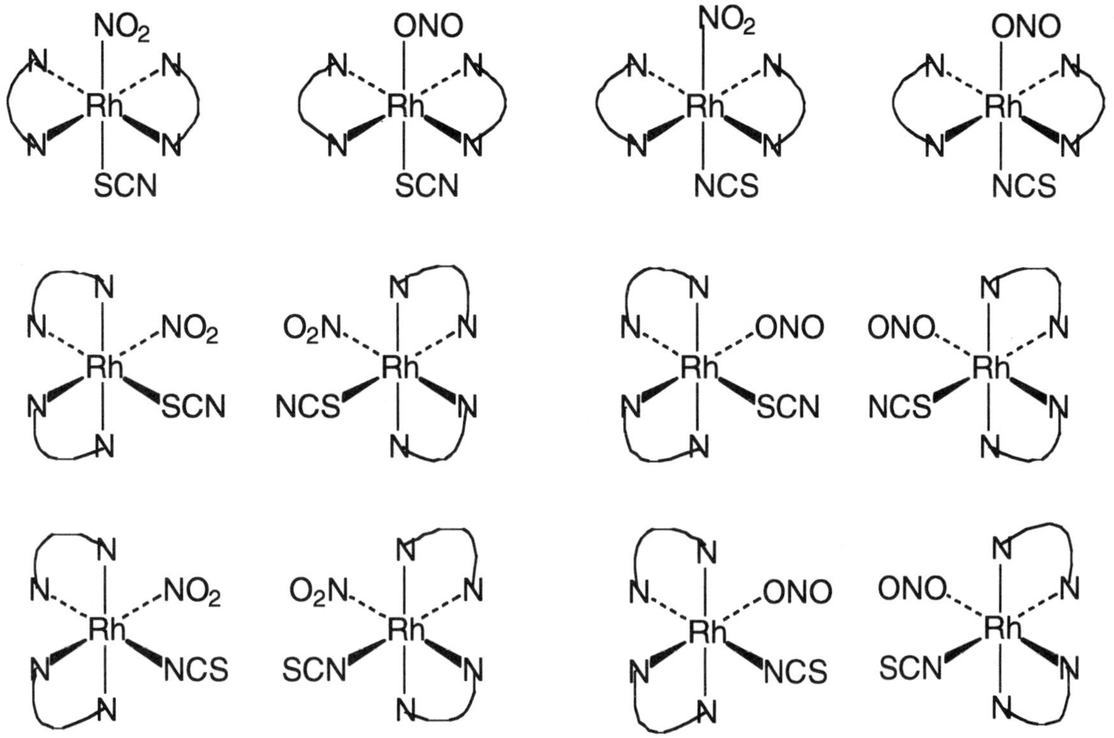

NO2 ONO NO2 ONO

Rh Rh Rh Rh

SCN SCN NCS NCS

NO2 O2N ONO ONO

Rh Rh Rh Rh

SCN NCS SCN NCS

NO2 O2N ONO ONO

Rh Rh Rh Rh

NCS SCN NCS SCN

20.106 $Co(gly)_3$

Co Co

Co Co

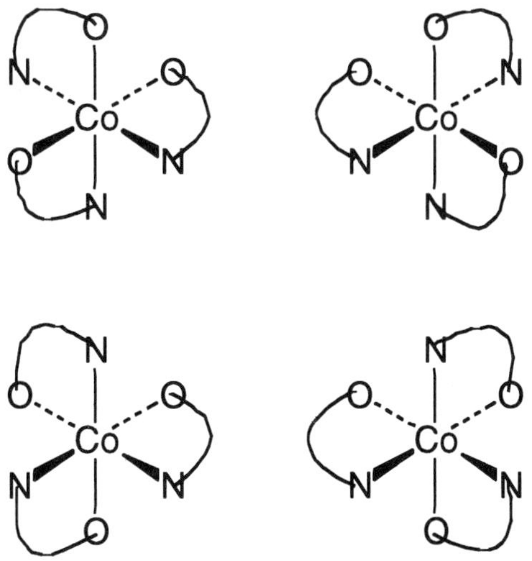

20.107 $Pt(gly)_2Cl_2$

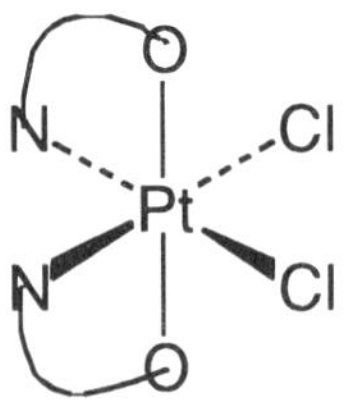

1, **2**, **3**, and **4** have a dipole moment. **1**, **2**, and **3** can exist as enantiomers.

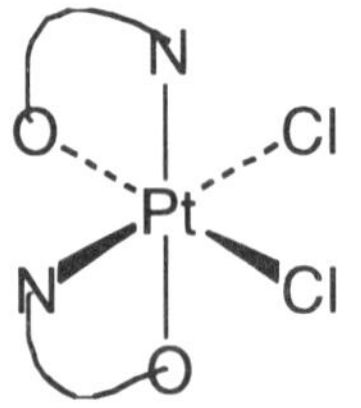

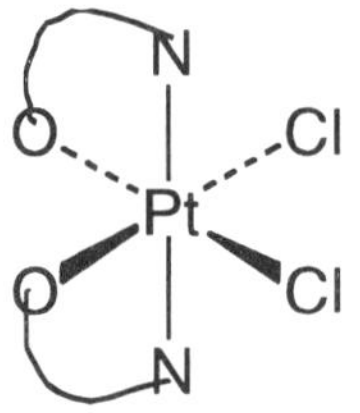

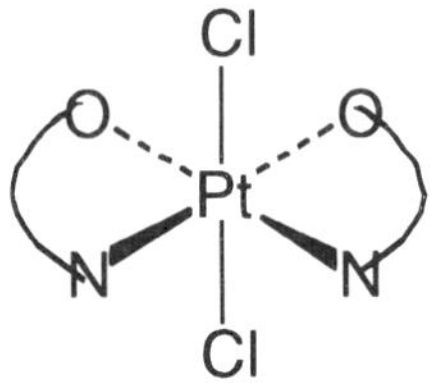

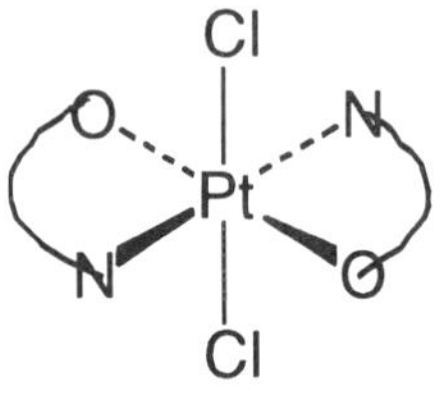

20.108 Valence bond theory describes the bonding in metal complexes in terms of two-electron, coordinate covalent bonds resulting from overlap of filled ligand orbitals with vacant metal hybrid orbitals that point in the directions of the ligands. Valence bond theory is useful in helping to visualize the bonding in metal complexes, but it doesn't account for their color or magnetic properties.

Crystal field theory assumes that the metal-ligand bonding is entirely ionic. Because of electrostatic repulsions between the d electrons and the ligands, the d orbitals are raised in energy and are differentiated by an energy separation called the crystal field splitting, Δ. The color of complexes is due to electronic transitions from one set of d orbitals to another. Crystal field theory accounts for the magnetic properties of complexes in terms of the relative values of Δ and the spin-pairing energy P. Small Δ values favor high-spin complexes and large Δ values favor low-spin complexes.

20.109 $[Mn(CN)_6]^{3-}$ Mn^{3+} $3d^4$

valence bond theory

[Ar] ↑↓ ↑ ↑ | ↑↓ ↑↓ (3d) | ↑↓ (4s) | ↑↓ ↑↓ ↑↓ (4p) — d^2sp^3

crystal field theory

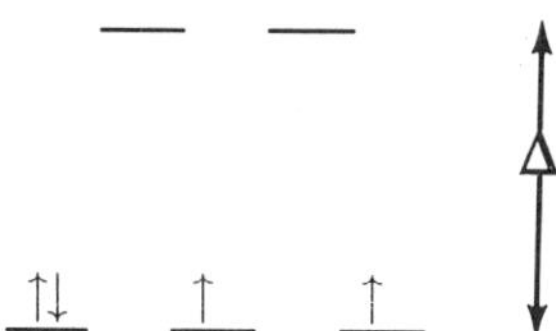

Crystal field model predicts 2 unpaired electrons.

20.110 Cl^- is a weak-field ligand, whereas CN^- is a strong-field ligand. Δ for $[Fe(CN)_6]^{3-}$ is larger than the pairing energy P; Δ for $[FeCl_6]^{3-}$ is smaller than P. Fe^{3+} has a $3d^5$ electron configuration.

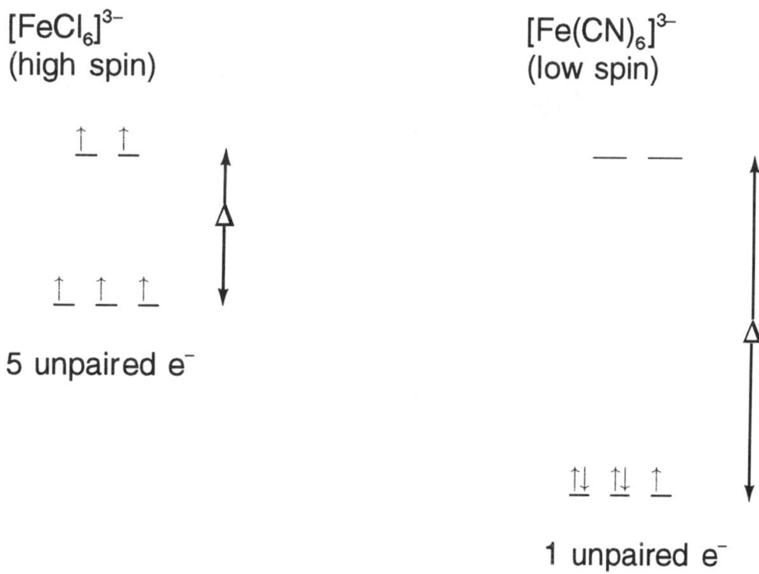

Because of the difference in Δ, $[FeCl_6]^{3-}$ is high-spin with five unpaired electrons, whereas $[Fe(CN)_6]^{3-}$ is low-spin with only one unpaired electron.

20.111 Cr^{3+} is a $3d^3$ ion. Regardless of the crystal field splitting energy, the three electrons singly occupy the three lower energy d orbitals.

20.112 A choice between high-spin and low-spin electron configurations arises only for complexes of metal ions with four to seven d electrons, the so-called d^4–d^7 complexes. For d^1–d^3 and d^8–d^{10} complexes, only one ground-state electron configuration is possible. In d^1 - d^3 complexes, all the electrons occupy the lower-energy d orbitals, independent of the value of Δ. In d^8 – d^{10} complexes, the lower-energy set of d orbitals is filled with three pairs of electrons, while the higher-energy set contains two, three, or four electrons, again independent of the value of Δ.

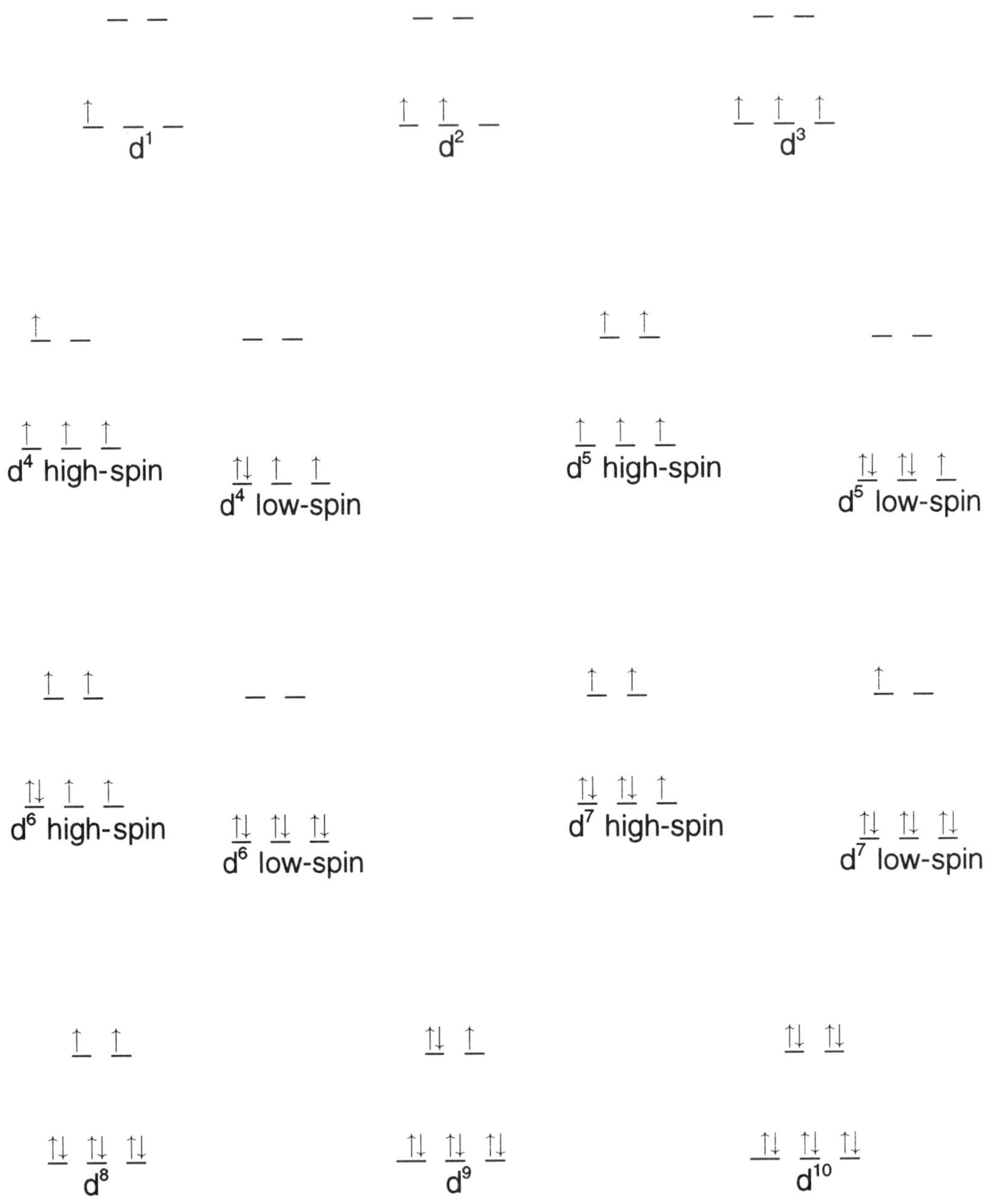

d^1

d^2

d^3

d^4 high-spin

d^4 low-spin

d^5 high-spin

d^5 low-spin

d^6 high-spin

d^6 low-spin

d^7 high-spin

d^7 low-spin

d^8

d^9

d^{10}

20.113 (a) $[Mn(H_2O)_6]^{2+}$ high-spin Mn^{2+}, $3d^5$

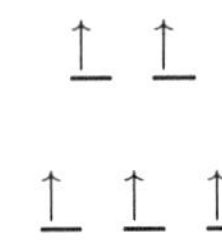

5 unpaired e^-

(b) $Pt(NH_3)_2Cl_2$ square-planar Pt^{2+}, $5d^8$

_

↑↓

↑↓

↑↓ ↑↓

0 unpaired e^-

(c) $[FeO_4]^{2-}$ tetrahedral Fe(VI), $3d^2$

_ _ _

↑ ↑

2 unpaired e^-

(d) $[Ru(NH_3)_6]^{2+}$ low-spin Ru^{2+}, $4d^6$

_ _

↑↓ ↑↓ ↑↓

0 unpaired e^-

20.114 $Fe(NO_3)_3$ is soluble in water and forms $[Fe(H_2O)_6]^{3+}$. Solutions of $Fe^{3+}(aq)$ are acidic because of the following hydrolysis reaction:

$$[Fe(H_2O)_6]^{3+}(aq) + H_2O(l) \rightleftarrows H_3O^+(aq) + [Fe(H_2O)_5(OH)]^{2+}(aq)$$

$K = 6.3 \times 10^{-3}$

20.115 $[CoCl_4]^{2-}$ is tetrahedral. $[Co(H_2O)_6]^{2+}$ is octahedral. Because $\Delta_{tet} < \Delta_{oct}$, these complexes have different colors. $[CoCl_4]^{2-}$ has absorption bands at longer wavelengths.

21.1 The electron configuration for Hg is [Xe] $4f^{14}5d^{10}6s^2$. Assuming the 5d and 6s bands overlap, the composite band can accomodate 12 valence electrons per metal atom. Weak bonding and a low melting point are expected for Hg because both the bonding and antibonding MOs are occupied.

21.2 Ge doped with As is an n-type semiconductor because As has an additional valence electron. The extra electrons are in the conduction band. The number of electrons in the conduction band of the doped Ge is much higher than for pure Ge, and the conductivity of the doped semiconductor is higher.

21.3

8 Cu at corners	8 x 1/8 = 1 Cu
8 Cu on edges	8 x 1/4 = <u>2 Cu</u>
	Total = 3 Cu

12 O on edges	12 x 1/4 = 3 O
8 O on faces	8 x 1/2 = <u>4 O</u>
	Total = 7 O

21.4 $Si(OCH_3)_4 + 4\ H_2O \rightarrow Si(OH)_4 + 4\ HOCH_3$

21.5 $Ba[OCH(CH_3)_2]_2 + Ti[OCH(CH_3)_2]_4 + 6\ H_2O \rightarrow$

$$BaTi(OH)_6(s) + 6\ HOCH(CH_3)_2$$

$$BaTi(OH)_6(s) \xrightarrow{\text{heat}} BaTiO_3(s)$$

21.6 (a) cobalt/tungsten carbide is a ceramic-metal composite.

(b) silicon carbide/zirconia is a ceramic-ceramic composite.

(c) boron nitride/epoxy is a ceramic-polymer composite.

(d) boron carbide/titanium is a ceramic-metal composite.

Understanding Key Concepts

1. A – metal oxide

   B – metal sulfide

   C – metal carbonate

   D – free metal

2. (a) (iii) electrolysis (b) (i) roasting a metal sulfide

3. (a) (1) and (4) are semiconductors; (2) is a metal; (3) is an insulator

   (b) (3) < (1) < (4) < (2). The conductivity increases with decreasing band gap.

   (c) (1) and (4) increases; (2) decreases; (3) not much change.

4. (a) (2), bonding MO's are filled.

   (b) (3), bonding and antibonding MO's are filled.

   (c) (3) < (1) < (2). Hardness increases with increasing MO bond order.

5. (a)

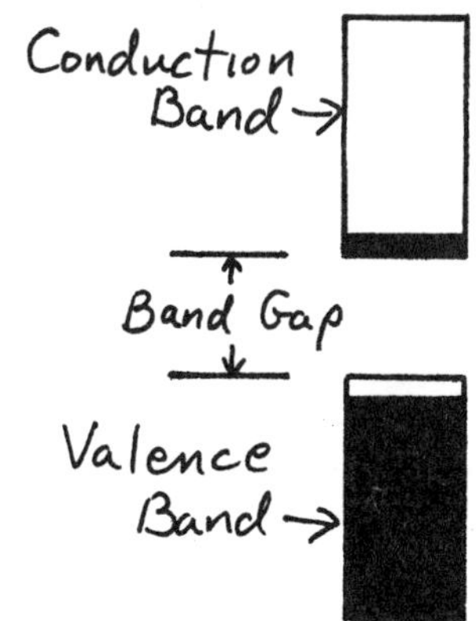

Si

(b)

Si doped with Ga
p-type semiconductor

(c) 

Si doped with As
n-type semiconductor

(d) The electrical conductivity of the doped silicon in both cases is higher than for pure silicon. Si doped with Ga is a p-type semiconductor with many more positive holes in the conduction band than in pure Si. This results in a higher conductivity. Si doped with As is an n-type semiconductor with many more electrons in the conduction band than in pure Si. This also results in a higher conductivity.

Additional Problems

Sources of the Metallic Elements

21.7 TiO_2, MnO_2, and Fe_2O_3

21.8 Pd, Pt, and Au

21.9 (a) Cu is found in nature as a sulfide.

(b) Zr is found in nature as an oxide.

(c) Pd is found in nature uncombined.

(d) Bi is found in nature as a sulfide.

21.10 (a) Ir is found in nature uncombined.

(b) Cr is found in nature as an oxide.

(c) Hg is found in nature as a sulfide.

(d) Ti is found in nature as an oxide.

21.11 The less electronegative early transition metals tend to form ionic compounds by losing electrons to highly electronegative nonmetals such as oxygen. The more electronegative late transition metals tend to form compounds with more covalent character by bonding to the less electronegative nonmetals such as sulfur.

21.12 The early transition metals, which includes Fe, generally occur as oxides because these less electronegative metals tend to form ionic compounds by losing electrons to the very electronegative oxygen. On the other hand, oxides of the s-block metals are strongly basic and far too reactive to exist in an environment that contains acidic oxdes such as CO_2 and SiO_2. Consequently, s-block metals, including Ca, are found in nature as carbonates, as silicates, and, in the case of Na and K as chlorides.

21.13 (a) Fe_2O_3, hematite (b) PbS, galena

(c) TiO_2, rutile (d) $CuFeS_2$, chalcopyrite

21.14 (a) cinnabar, HgS (b) bauxite, $Al_2O_3 \cdot x\ H_2O$

(c) sphalerite, ZnS (d) chromite, $FeCr_2O_4$

21.15 (a) Most metals occur in nature as minerals, the crystalline, inorganic constituent of the rocks that make up the earth's crust.

(b) Gangue is the economically worthless part of an ore consisting of sand, clay, and other impurities.

(c) Flotation is a metallurgical process that exploits differences in the ability of water and oil to wet the surfaces of mineral and gangue.

21.16 (a) Mineral deposits from which metals can be produced economically are called ores.

(b) Metallurgy is the science and technology of extracting metals from their ores.

(c) Roasting is a metallurgical process that involves heating a mineral in air.

21.17 The flotation process exploits the differences in the ability of water and oil to wet the surfaces of the mineral and the gangue. The gangue, which contains ionic silicates, is moistened by the polar water molecules and sinks to the bottom of the tank. The mineral particles, which contain the less polar metal sulfide, are coated by the oil and become attached to the soapy air bubbles created by the detergent. The metal sulfide particles are carried to the surface in the soapy froth, which is skimmed off at the top of the tank. This process would not work well for a metal oxide because it is too polar and will be wet by the water and sink with the gangue.

21.18 Sometimes, ores are concentrated by chemical treatment. In the Bayer process, Ga_2O_3 is separated from Fe_2O_3 by treating the ore with hot aqueous NaOH. The amphoteric Ga_2O_3 dissolves as $Ga(OH)_4^-$, but the basic Fe_2O_3 does not.

$$Ga_3O_3(s) + 2\ OH^-(aq) + 3\ H_2O(l) \rightarrow 2\ Ga(OH)_4^-(aq)$$

21.19 Since $E° < 0$ for Zn^{2+}, the reduction of Zn^{2+} is not favored.

Since $E° > 0$ for Hg^{2+}, the reduction of Hg^{2+} is favored.

The roasting of CdS should yield CdO because, like Zn^{2+}, $E° < 0$ for the reduction of Cd^{2+}.

21.20

	E°
$Mg^{2+}(aq) + 2\ e^- \rightarrow Mg(s)$	–2.37 V
$Zn^{2+}(aq) + 2\ e^- \rightarrow Zn(s)$	–0.76 V

The very negative reduction potential for Mg^{2+} indicates that Mg^{2+} it is much more difficult to reduce than the Zn^{2+}.

21.21 (a) $V_2O_5(s) + 5\ Ca(s) \rightarrow 2\ V(s) + 5\ CaO(s)$

(b) $2\ PbS(s) + 3\ O_2(g) \rightarrow 2\ PbO(s) + 2\ SO_2(g)$

(c) $MoO_3(s) + 3\ H_2(g) \rightarrow Mo(s) + 3\ H_2O(g)$

(d) $3\ MnO_2(s) + 4\ Al(s) \rightarrow Mn(s) + 2\ Al_2O_3(s)$

(e) $MgCl_2(l) \xrightarrow{\text{electrolysis}} Mg(l) + Cl_2(g)$

21.22 (a) $Fe_2O_3(s) + 3\ H_2(g) \rightarrow 2\ Fe(s) + 3\ H_2O(g)$

(b) $Cr_2O_3(s) + 2\ Al(s) \rightarrow 2\ Cr(s) + Al_2O_3(s)$

(c) $Ag_2S(s) + O_2(g) \rightarrow 2\ Ag(s) + SO_2(g)$

(d) $TiCl_4(g) + 2\ Mg(l) \rightarrow Ti(l) + 2\ MgCl_2(l)$

(e) $2\ LiCl(l) \xrightarrow{\text{electrolysis}} 2\ Li(l) + Cl_2(g)$

21.23 $2\ ZnS(s) + 3\ O_2(g) \rightarrow 2\ ZnO(s) + 2\ SO_2(g)$

$\Delta H° = [2\ \Delta H_f°(ZnO) + 2\ \Delta H_f°(SO_2)] - [2\ \Delta H_f°(ZnS)]$

$\Delta H° = [(2\text{ mol})(-348.3\text{ kJ/mol}) + (2\text{ mol})(-296.8\text{ kJ/mol})]$

$- (2\text{ mol})(-206.0\text{ kJ/mol}) = -878.2\text{ kJ}$

$\Delta G° = [2\ \Delta G_f°(ZnO) + 2\ \Delta G_f°(SO_2)] - [2\ \Delta G_f°(ZnS)]$

$\Delta G° = [(2\text{ mol})(-318.3\text{ kJ/mol}) + (2\text{ mol})(-300.2\text{ kJ/mol})]$

$- (2\text{ mol})(-201.3\text{ kJ/mol}) = -834.4\text{ kJ}$

$\Delta H°$ and $\Delta G°$ are different because of the entropy change associated with the reaction. The minus sign for $(\Delta H° - \Delta G°)$ indicates that the entropy is negative, which is consistent with a decrease in the number of moles of gas from 3 mol to 2 mol.

21.24 $HgS(s) + O_2(g) \rightarrow Hg(l) + SO_2(g)$

$\Delta H° = \Delta H_f°(SO_2) - \Delta H_f°(HgS)$

$\Delta H° = (1\text{ mol})(-296.8\text{ kJ/mol}) - (1\text{ mol})(-58.2\text{ kJ/mol}) = -238.6\text{ kJ}$

$\Delta G° = \Delta G_f°(SO_2) - \Delta G_f°(HgS)$

$\Delta G° = (1\text{ mol})(-300.2\text{ kJ/mol}) - (1\text{ mol})(-50.6\text{ kJ/mol}) = -249.6\text{ kJ}$

$\Delta H°$ and $\Delta G°$ are different because of the entropy change associated with the reaction. The plus sign for ($\Delta H° - \Delta G°$) indicates that the entropy is positive, which is consistent with a solid going to liquid and no change in the number of moles of gas.

21.25 $FeCr_2O_4(s) + 4\ C(s) \rightarrow \underbrace{Fe(s) + 2\ Cr(s)}_{\text{ferrochrome}} + 4\ CO(g)$

(a) $FeCr_2O_4$, 223.84 amu; Cr, 52.00 amu; 236 kg = 236 x 10^3 g

$$\text{mass Cr} = 236 \times 10^3\text{ g} \times \frac{1\text{ mol }FeCr_2O_4}{223.84\text{ g}} \times \frac{2\text{ mol Cr}}{1\text{ mol }FeCr_2O_4} \times \frac{52.00\text{ g Cr}}{1\text{ mol Cr}} \times \frac{1.00\text{ kg}}{1000\text{ g}}$$

mass Cr = 110 kg Cr

(b)

$$\text{mol CO} = 236 \times 10^3\text{ g} \times \frac{1\text{ mol }FeCr_2O_4}{223.84\text{ g}} \times \frac{4\text{ mol CO}}{1\text{ mol }FeCr_2O_4} = 4217.3\text{ mol CO}$$

PV = nRT

$$V = \frac{nRT}{P} = \frac{(4217.3\text{ mol})\left(0.08206\ \frac{\text{L}\cdot\text{atm}}{\text{mol}\cdot\text{K}}\right)(298\text{ K})}{\left(740\text{ mm Hg} \times \frac{1\text{ atm}}{760\text{ mm Hg}}\right)} = 1.06 \times 10^5\text{ L CO}$$

21.26 $Cu_2S(l) + O_2(g) \rightarrow 2\ Cu(l) + SO_2(g)$

Cu_2S, 159.16 amu; Cu, 63.546 amu; SO_2, 64.06 amu

(a) $\text{mass of Cu} = 8 \times 10^9\text{ kg Cu} \times \frac{1000\text{ g}}{1\text{ kg}} = 8 \times 10^{12}\text{ g Cu}$

$$\text{mass }Cu_2S = 8 \times 10^{12}\text{ g Cu} \times \frac{1\text{ mol Cu}}{63.546\text{ g Cu}} \times \frac{1\text{ mol }Cu_2S}{2\text{ mol Cu}} \times \frac{159.16\text{ g }Cu_2S}{1\text{ mol }Cu_2S} \times \frac{1\text{ kg}}{1000\text{ g}}$$

mass Cu_2S = 1 x 10^{10} kg Cu_2S

(b) $\text{mol } SO_2 = 1 \times 10^{10} \text{ kg } Cu_2S \times \frac{1000 \text{ g}}{1 \text{ kg}} \times \frac{1 \text{ mol } Cu_2S}{159.16 \text{ g } Cu_2S} \times \frac{\text{mol } SO_2}{1 \text{ mol } Cu_2S}$

$\text{mol } SO_2 = 6.3 \times 10^{10} \text{ mol } SO_2$

PV = nRT

$$V = \frac{nRT}{P} = \frac{(6.3 \times 10^{10} \text{ mol})\left(0.08206 \frac{\text{L} \cdot \text{atm}}{\text{mol} \cdot \text{K}}\right)(273 \text{ K})}{1.0 \text{ atm}} = 1.4 \times 10^{12} \text{ L}$$

$V = 1 \times 10^{12} \text{ L } SO_2$

(c) H_2SO_4, 98.08 amu

$\text{mass } H_2SO_4 = 6.3 \times 10^{10} \text{ mol } SO_2 \times \frac{1 \text{ mol } H_2SO_4}{1 \text{ mol } SO_2} \times \frac{98.08 \text{ g } H_2SO_4}{1 \text{ mol } H_2SO_4} \times \frac{1 \text{ kg}}{1000 \text{ g}}$

$\text{mass } H_2SO_4 = 6 \times 10^{9} \text{ kg } H_2SO_4$

21.27 $Ni^{2+}(aq) + 2 e^- \rightarrow Ni(s)$; 1 A = 1 C/s

$\text{mass Ni} = 52.5 \frac{\text{C}}{\text{s}} \times 8 \text{ h} \times \frac{3600 \text{ s}}{1 \text{ h}} \times \frac{1 \text{ mol } e^-}{96{,}500 \text{ C}} \times \frac{1 \text{ mol Ni}}{2 \text{ mol } e^-} \times \frac{58.69 \text{ g Ni}}{1 \text{ mol Ni}} \times \frac{1.00 \text{ kg}}{1000 \text{ g}}$

mass Ni = 0.460 kg Ni

21.28 $Cu^{2+}(aq) + 2 e^- \rightarrow Cu(s)$

$$\text{Charge} = 7.50 \text{ kg} \times \frac{1000 \text{ g}}{1 \text{ kg}} \times \frac{1 \text{ mol Cu}}{63.546 \text{ g}} \times \frac{2 \text{ mol } e^-}{1 \text{ mol Cu}} \times \frac{96{,}500 \text{ C}}{1 \text{ mol } e^-} = 2.28 \times 10^{7} \text{ C}$$

$$\text{Time} = \frac{2.28 \times 10^{7} \text{ C}}{40.0 \text{ C/s}} \times \frac{1 \text{ h}}{3600 \text{ s}} = 158 \text{ h}$$

Iron and Steel

21.29 $Fe_2O_3(s) + 3 CO(g) \rightarrow 2 Fe(l) + 3 CO_2(g)$

Fe_2O_3 is the oxidizing agent. CO is the reducing agent.

21.30 Coke (C) is converted to CO, the reducing agent used in the production of Fe.

$C(s) + CO_2(g) \rightarrow 2\ CO(g)$

$2\ C(s) + O_2(g) \rightarrow 2\ CO(g)$

21.31 Slag is a byproduct of iron production, consisting mainly of $CaSiO_3$. It is produced from the gangue in iron ore.

21.32 Limestone is added to the blast furnace in the commercial process for producing iron to remove the gangue from the iron ore. At the high temperatures of the blast furnace, the limestone decomposes to lime (CaO), a basic oxide that reacts with SiO_2 and other acidic oxides present in the gangue. The product, called slag, is a molten material consisting mainly of calcium silicate.

$CaCO_3(s) \rightarrow CaO(s) + CO_2(g)$

$CaO(s) + SiO_2(s) \rightarrow CaSiO_3(l)$ (slag)

21.33 Molten iron from a blast furnace is exposed to a jet of pure oxygen gas for about 20 minutes. The impurities are oxidized to yield a molten slag that can be poured off.

$P_4(l) + 5\ O_2(g) \rightarrow P_4O_{10}(l)$

$6\ CaO(s) + P_4O_{10}(l) \rightarrow 2\ Ca_3(PO_4)_2(l)$ (slag)

$2\ Mn(l) + O_2(g) \rightarrow 2\ MnO(s)$

$MnO(s) + SiO_2(s) \rightarrow MnSiO_3(l)$ (slag)

21.34 Slag forms in the basic oxygen process because the impurities are oxidized, and the acidic oxides that form react with the basic CaO to yield a molten slag that can be poured off. Phosphorus, for example, is oxidized to P_4O_{10}, which then reacts with CaO to give molten calcium phosphate:

$P_4(l) + 5\ O_2(g) \rightarrow P_4O_{10}(l)$

$6\ CaO(s) + P_4O_{10}(l) \rightarrow 2\ Ca_3(PO_4)_2(l)$ (slag)

Manganese also passes into the slag because its oxide is basic and reacts with added SiO_2, yielding molten manganese silicate.

$2\ Mn(l) + O_2(g) \rightarrow 2\ MnO(s)$

$MnO(s) + SiO_2(s) \rightarrow MnSiO_3(l)$ (slag)

21.35 $SiO_2(s) + 2\ C(s) \rightarrow Si(s) + 2\ CO(g)$

$Si(s) + O_2(g) \rightarrow SiO_2(s)$

$CaO(s) + SiO_2(s) \rightarrow CaSiO_3(l)$ (slag)

21.36 $CaSO_4(s) + 3\ C(s) \rightarrow CaO(s) + S(l) + 3\ CO(g)$

$2\ S(l) + 3\ O_2(g) \rightarrow 2\ SO_3(g)$

$CaO(s) + SO_3(g) \rightarrow CaSO_4(l)$ (slag)

Bonding in Metals

21.37 Metals are malleable, ductile, and good conductors of electricity.

21.38 Hardness and melting point are two properties of metals that vary considerably depending on the particular metal.

21.39

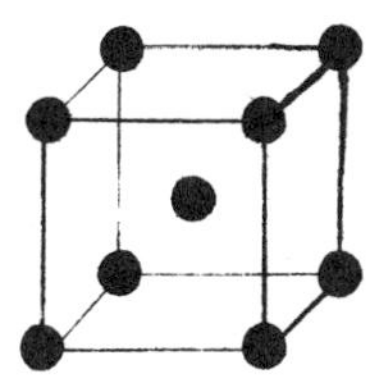

Each K has a single valence electron and has eight nearest neighbor K atoms. The valence electrons can't be localized in an electron-pair bond between any particular pair of K atoms.

21.40 Cesium has just one valence electron per Cs atom and crystallizes in a body-centered cubic structure in which each Cs atom is surrounded by eight other Cs atoms. Consequently, the valence electrons can't be localized in a bond between any particular pair of Cs atoms. Instead, the electrons are delocalized and belong to the crystal as a whole. In the electron-sea model, we visualize the crystal as a three-dimensional array of metal cations immersed in a sea of delocalized electrons that are free to move throughout the crystal. The continuum of delocalized, mobile valence electrons acts as an electrostatic glue that holds the metal cations together.

21.41 Malleability and ductility of metals follows from the fact that the delocalized bonding extends in all directions. When a metallic crystal is deformed, no localized bonds are broken. Instead, the electron sea simply adjusts to the new distribution of cations, and the energy of the deformed structure is similar to that of the original. Thus, the energy required to deform a metal is relatively small.

21.42 The electron-sea model affords a simple qualitative explanation for the electrical and thermal conductivity of metals. Because the electrons are mobile, they are free to move away from a negative electrode and toward a positive electrode when a metal is subjected to an electrical potential. The mobile electrons can also conduct heat by carrying kinetic energy from one part of the crystal to another.

21.43 The energy required to deform a transition metal like W is greater than that for Cs because W has more valence electrons and hence more electrostatic "glue".

21.44 Na has one valence electron. Mg has two valence electrons. There are more electrons per cation in the electron sea for Mg than for Na. There is more electrostatic glue for Mg, and hence the higher melting point.

21.45 The difference in energy between successive MOs in a metal decreases as the number of metal atoms increases so that the MOs merge into an almost continuous band of energy levels. Consequently, MO theory for metals is often called band theory.

21.46

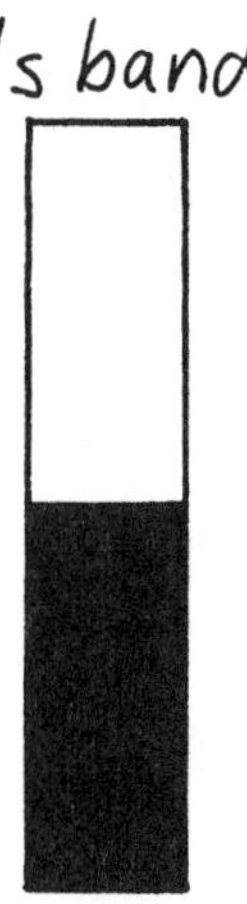

K_n

21.47 The energy levels within a band occur in degenerate pairs; one set of energy levels applies to electrons moving to the right, and the other set applies to electrons moving to the left. In the absence of an electrical potential, the two sets of levels are equally populated. As a result there is no net electric current. In the presence of an electrical potential those electrons moving to the right are accelerated, those moving to the left are slowed down, and some change direction. Thus, the two sets of energy levels are now unequally populated. The number of electrons moving to the right is now greater than the number moving to the left, and so there is a net electric current.

21.48 An electrical potential can shift electrons from one set of energy levels to the other only if the band is partially filled. If the band is completely filled, there are no available vacant energy levels to which electrons can be excited, and therefore the two sets of levels must remain equally populated, even in the presence of an electrical potential. This means that an electrical potential can't accelerate the electrons in a completely filled band. Materials that have only completely filled bands are therefore electrical insulators. By contrast, materials that have partially filled bands are metals.

21.49 (a) (b)

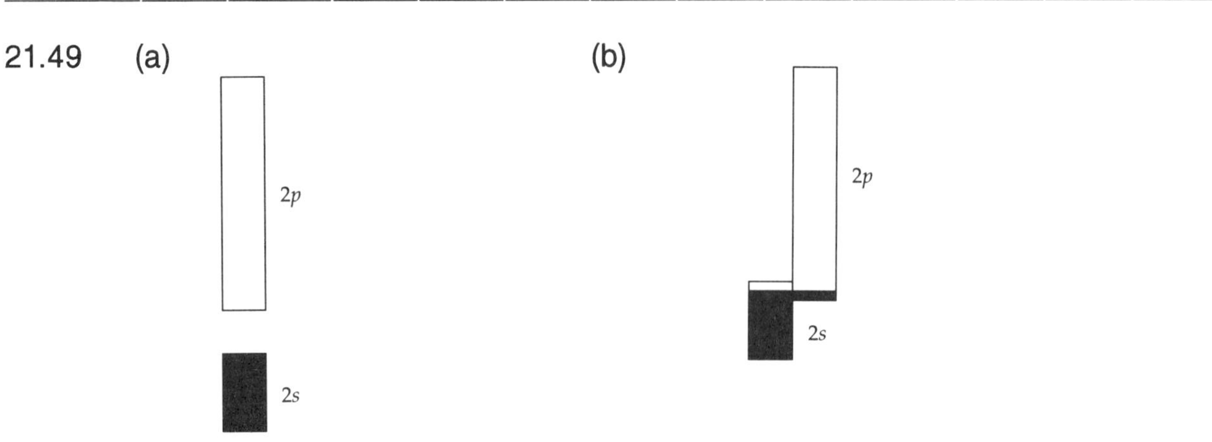

Diagram (b) shows the 2s and 2p bands overlapping in energy and the resulting composite band is only partially filled. Thus, Be is a good electrical conductor.

21.50 (a) (b)

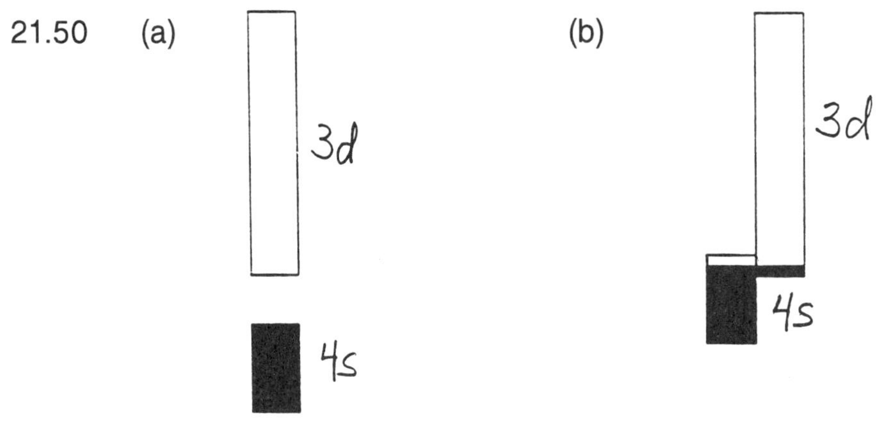

Diagram (b) shows the 4s and 3d bands overlapping in energy and the resulting composite band is only partially filled. Thus, diagram (b) agrees with the fact that Ca is a good electrical conductor.

21.51 Transition metals have a d band that can overlap the s band to give a composite band consisting of six MOs per metal atom. Half of the MOs are bonding and half are antibonding, and thus one expects maximum bonding for metals that have six valence electrons per metal atom. Accordingly, the melting points of the transition metals go through a maximum at or near group 6B.

21.52 Transition metals have a d band that can overlap the s band to give a composite band consisting of six MOs per metal atom. Half of the MOs are bonding and half are antibonding, and thus we might expect maximum bonding and hardness for metals that have six valence electrons per metal atom. The hardness will decrease as the number of valence electrons (beyond six) increases because they go into antibonding MOs. Cu has three more valence electrons than Fe, and Cu is softer.

Semiconductors

21.53 A semiconductor is a material that has an electrical conductivity intermediate between that of a metal and that of an insulator. Si, Ge, and Sn (gray) are semiconductors.

21.54 (a) The valence band is the band of bonding molecular orbitals in a semiconductor.

(b) The conduction band is the band of antibonding molecular orbitals in a semiconductor.

(c) The band gap is the energy difference between the valence band and the conduction band in a semiconductor.

(d) Doping is the addition of a small amount of impurities to increase the conductivity of a semiconductor.

21.55 (a) (b)

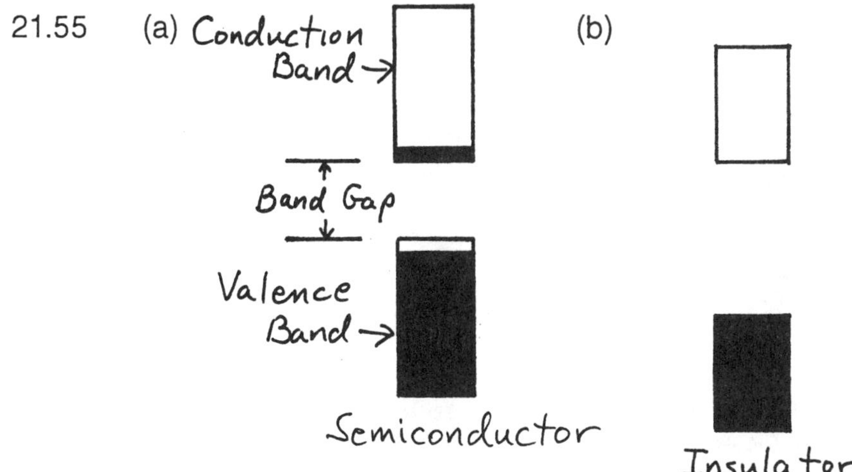

The MOs of a semiconductor are similar to those of an insulator, but the band gap in a semiconductor is smaller. As a result, a few electrons have enough energy to jump the gap and occupy the higher-energy, conduction band. The conduction band is thus partially filled, and the valence band is also partially fillled. When an electrical potential is applied to a semiconductor, it conducts a small amount of current because the potential can accelerate the electrons in the partially filled bands.

21.56 (a) (b)

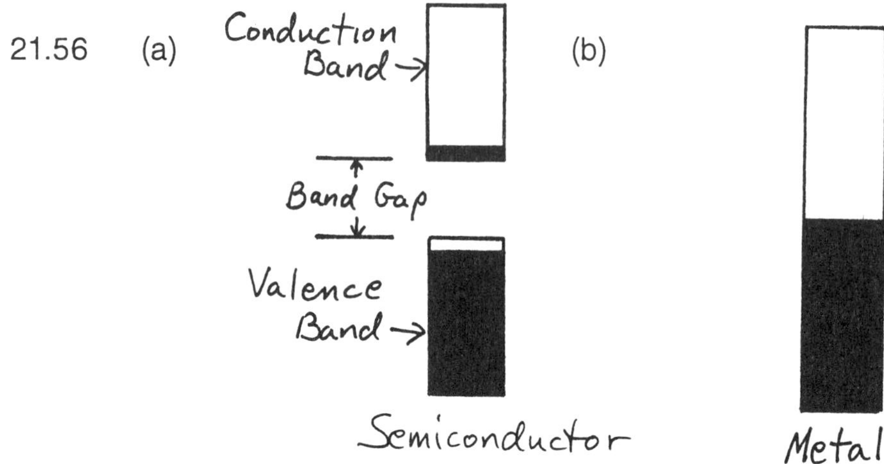

Metals have partially filled bands with no band gap between the highest occupied and lowest unoccupied MOs, which makes them conductors. Semiconductors have a moderate band gap between the valence and conduction bands. A few electrons have enough energy to jump the gap and occupy the conduction band. The result is a partially filled valence and conduction band. When an electrical potential is applied to a semiconductor, it conducts a small amount of current.

21.57 As band gap increases, electrical conductivity decreases. As the band gap increases, the number of electrons able to jump the gap and occupy the higher-energy conduction band decreases, and thus the conductivity decreases.

21.58 The electrical conductivity of a semiconductor increases with increasing temperature because the number of electrons with sufficient energy to occupy the conduction band increases as the temperature rises. At higher temperatures, there are more charge carriers (electrons) in the conduction band and more vacancies in the valence band.

21.59 An n-type semiconductor is a semiconductor doped with a substance with more valence electrons than the semiconductor itself. Si doped with P is an example.

21.60 A p-type semiconductor is a semiconductor doped with a substance with fewer valence electrons than the semiconductor itself. Si doped with B is an example.

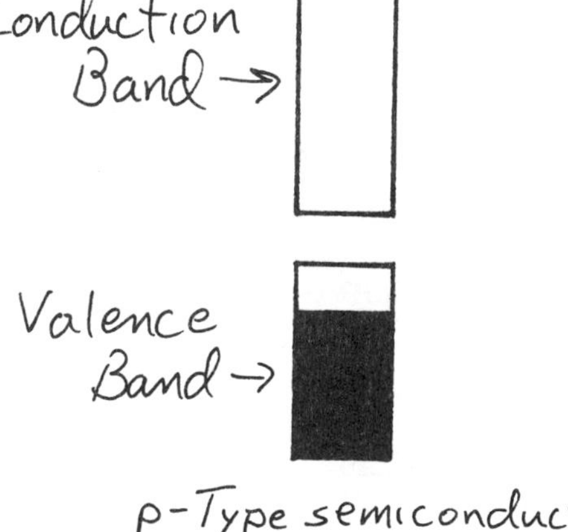

21.61 In the MO picture, the extra electrons occupy the conduction band. The number of electrons in the conduction band of the doped Ge is much greater than for pure Ge, and the conductivity of the doped semiconductor is correspondingly higher.

21.62 Si doped with Ga is a p-type semiconductor because each Ga atom has one less valence electron than Si. The valence band is thus partially filled, which accounts for the electrical conductivity. The conductivity is greater than that of pure Si because the doped Si has many more positive holes in the valence band. That is, doped Si has more vacant MOs available to which electrons can be excited by an electrical potential.

21.63 (a) p–type (In is electron deficient with respect to Si)

(b) n–type (Sb is electron rich with respect to Ge)

(c) n–type (As is electron rich with respect to gray Sn)

21.64 (a) n-type (As is electron rich with respect to Ge)

(b) p-type (B is electron deficient with respect to Ge)

(c) n-type (Sb is electron rich with respect to Si)

21.65 Al_2O_3 < Ge < Ge doped with In < Fe < Cu

21.66 NaCl < Si < gray Sn < gray Sn doped with Sb < Ag

Superconductors

21.67 A superconductor is a material that loses all electrical resistance below a certain temperature. An example is $YBa_2Cu_3O_7$.

21.68 The superconducting transition temperature (T_c) is the temperature below which a superconductor loses all electrical resistance.

21.69 (1) A superconductor is able to levitate a magnet.

(2) In a superconductor, once an electric current is started, it flows indefinitely without loss of energy. A superconductor has no electrical resistance.

21.70

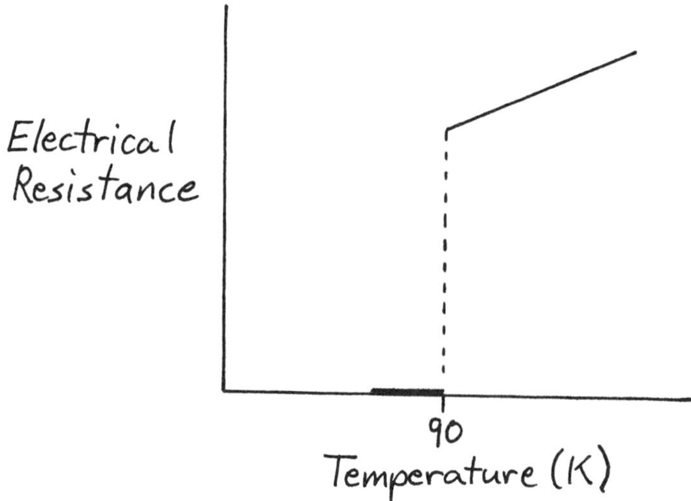

21.71 The fullerides are three-dimensional superconductors, whereas the copper oxide ceramics are two-dimensional superconductors.

21.72 In $YBa_2Cu_3O_7$

atom	coordination numbers
Cu	4 and 5
Y	8
Ba	10

Ceramics and Composites

21.73 Ceramics are inorganic, nonmetallic, nonmolecular solids, including both crystalline and amorphous materials. Ceramics have higher melting points, and they are stiffer, harder, and more resistant to wear and corrosion than are metals.

21.74 The bonding in ceramics consists of either directed covalent bonds or ionic bonds. This is very different from the electron-sea model of bonding for metals.

21.75 Ceramics have higher melting points, and they are stiffer, harder, and more wear resistant than metals because they have stronger bonding. They maintain much of their strength at high temperatures, where metals either melt or corrode because of oxidation.

21.76 Oxide ceramics don't react with oxygen because they are already fully oxidized.

21.77 The brittleness of ceramics is due to strong chemical bonding. In silicon nitride each Si atom is bonded to four N atoms and each N atom is bonded to three Si atoms. The strong, highly directional covalent bonds prevent the planes of atoms from sliding over one another when the solid is subjected to a stress. As a result, the solid can't deform to relieve the stress. It maintains its shape up to a point, but then the bonds give way suddenly and the material fails catastrophically when the stress exceeds a certain threshold value. By contrast, metals are able to deform under stress because their planes of metal cations can slide easily in the electron sea.

21.78

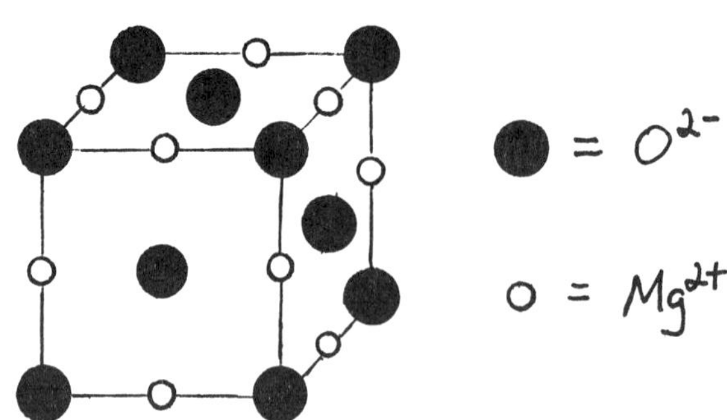

The brittleness of oxide ceramics, as well as their hardness, stiffness, and high melting points, is due to strong ionic bonding. The strong ionic bonding prevents planes of ions from sliding over one another when the solid is subjected to a stress. As a result, the solid can't deform to relieve the stress. It maintains its shape up to a point, but then the bonds give way suddenly and the material fails catastrophically. For example, if the stress displaces a plane of ions by one-half a unit-cell edge length, the resulting Mg^{2+}–Mg^{2+} and O^{2-}–O^{2-} repulsions cause the crystal to shatter. By contrast, metals are able to deform under stress because their planes of cations can slide easily in the electron sea.

21.79 Ceramic processing is the series of steps that leads from raw material to the finished ceramic object.

21.80 Sintering, which occurs below the melting point, is a process in which the particles of the powder are "welded" together without completely melting. During sintering, the crystal grains grow larger and the density of the material increases as the void spaces between particles disappear.

21.81 A sol is a colloidal dispersion of tiny particles. A gel is a more rigid gelatin-like material consisting of larger particles.

21.82 As an example, pure $Ti(OCH_2CH_3)_4$ is dissolved in an appropriate organic solvent, and water is added to bring about a hydrolysis reaction:

$$Ti(OCH_2CH_3)_4 + 4\ H_2O \rightarrow Ti(OH)_4(s) + 4\ HOCH_2CH_3$$

The $Ti(OH)_4$ forms in the liquid as a colloidal dispersion called a sol. Subsequent reactions eliminate water and form oxygen bridges between Ti atoms:

$$(HO)_3Ti\text{–}O\boxed{\text{–}H + H\text{–}O\text{–}}Ti(OH)_3 \rightarrow (HO)_3Ti\text{–}O\text{–}Ti(OH)_3 + H_2O$$

Because all of the OH groups can undergo the above reaction, the particles of the sol link together through a three-dimensional network of oxygen bridges, and the sol is converted to a more rigid, gelatin-like material called a gel.

21.83 $Zr[OCH(CH_3)_2]_4 + 4\ H_2O \rightarrow Zr(OH)_4 + 4\ HOCH(CH_3)_2$

21.84 $Zn(OCH_2CH_3)_2 + 2\ H_2O \rightarrow Zn(OH)_2 + 2\ HOCH_2CH_3$

21.85 $(HO)_3Si\text{–}O\boxed{\text{–}H + H\text{–}O\text{–}}Si(OH)_3 \rightarrow (HO)_3Si\text{–}O\text{–}Si(OH)_3 + H_2O$

Further reactions of this sort give a three-dimensional network of Si–O–Si bridges. On heating, SiO_2 is obtained.

21.86 $(HO)_2Y\text{–}O\boxed{\text{–}H + H\text{–}O\text{–}}Y(OH)_2 \rightarrow (HO)_2Y\text{–}O\text{–}Y(OH)_2 + H_2O$

On drying and sintering, Y_2O_3 is obtained.

21.87 $3\ SiCl_4(g) + 4\ NH_3(g) \rightarrow Si_3N_4(s) + 12\ HCl(g)$

21.88 $2\ BCl_3(g) + 3\ H_2(g) \xrightarrow{W} 2\ B(s) + 6\ HCl(g)$

21.89 Graphite/epoxy composites are good materials for making tennis rackets and golf clubs because of their high strength–to–weight ratios.

21.90 Silicon carbide-reinforced alumina is stronger and tougher than pure alumina. The silicon carbide whiskers have great strength along their axis because most of the chemical bonds are aligned in that direction. In addition the silicon carbide whiskers can prevent microscopic cracks from propagating.

General Problems

21.91 Pb, Hg, and Zn, which are found as PbS, HgS, and ZnS, respectively.

21.92 Al, Na, and Mg

21.93 Stainless steel contains Fe, Cr, and Ni

21.94 The chemical composition of the alkaline earth minerals is that of metal sulfates and sulfites, MSO_4 and MSO_3.

21.95 (a) A solid that forms in a liquid as a colloidal dispersion is called a sol.

(b) When the particles of the sol link together they form a more rigid, gelatin-like material called a gel.

(c) Sintering is a process in which the particles of a powder are "welded" together without completely melting.

(d) Ceramic processing is the series of steps that leads from raw material to the finished ceramic object.

21.96 Because ceramics are less dense than steel, they are attractive lightweight, high-temperature materials for replacing metal components in aircraft, space vehicles, and automotive engines. Oxide ceramics have many important uses. Alumina is the material of choice for spark-plug insulators because of its high electrical resistance, high strength, high thermal stability, and chemical inertness. Alumina is nontoxic and essentially inert in biological systems, and it is used in constructing dental crowns and the heads of artificial hips.

21.97 77 K is the boiling point of the readily available liquid N_2.

21.98 Band theory better explains how the number of valence electrons affects properties such as melting point and hardness.

21.99 W [Xe] $4f^{14}5d^4 6s^2$ Au [Xe] $4f^{14}5d^{10}6s^1$

These facts are better explained with the MO band model. Transition metals have a d band that can overlap the s band to give a composite band consisting of six MOs per metal atom. Half of the MOs are bonding and half are antibonding, and thus one expects maximum bonding for metals that have six valence electrons per metal atom. Accordingly, the hardness and melting points of the transition metals go through a maximum at or near group 6B.

21.100 V [Ar] $3d^34s^2$ Zn [Ar] $3d^{10}4s^2$

Transition metals have a d band that can overlap the s band to give a composite band consisting of six MOs per metal atom. Half of the MOs are bonding and half are antibonding. Strong bonding and a high enthalpy of vaporization are expected for V because almost all of the bonding MOs are occupied and all of the antibonding MOs are empty. Weak bonding and a low enthalpy of vaporization are expected for Zn because both the bonding and the antibonding MOs are occupied.

21.101 (a) MgO, insulator

(b) Si doped with Sb (Sb is electron rich with respect to Si), n-type semiconductor

(c) white tin, metallic conductor

(d) Ge doped with Ga (Ga is electron deficient with respect to Ge), p-type semiconductor

(e) stainless steel, metallic conductor

21.102 With a band gap of 130 kJ/mol, GaAs is a semiconductor. Because Ge lies between Ga and As in the periodic table, Ge is isoelectronic with GaAs.

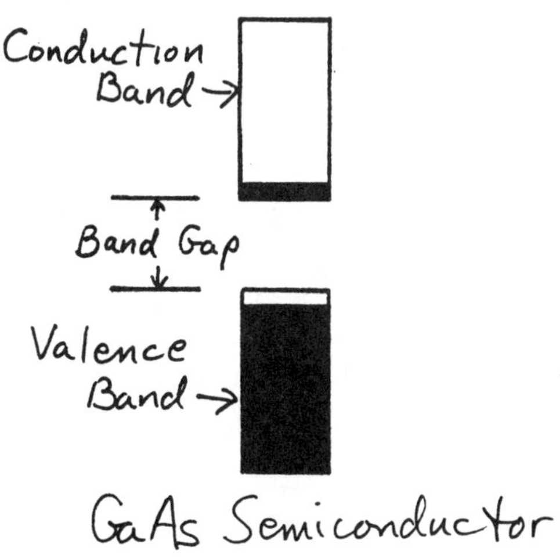

21.103 (a) $P_4(l) + 5\ O_2(g) \rightarrow P_4O_{10}(l)$

$6\ CaO(s) + P_4O_{10}(l) \rightarrow 2\ Ca_3(PO_4)_2(l)$ (slag)

(b) mass $P_4 = (0.0045)(3.4 \times 10^3\ kg) = 15.3\ kg\ P_4$

$$\text{mol } P_4O_{10} = 15.3 \times 10^3\ g \times \frac{1\ mol\ P_4}{123.9\ g} \times \frac{1\ mol\ P_4O_{10}}{1\ mol\ P_4} = 123.5\ mol\ P_4O_{10}$$

$$\text{mass CaO} = 123.5\ mol\ P_4O_{10} \times \frac{6\ mol\ CaO}{1\ mol\ P_4O_{10}} \times \frac{56.08\ g\ CaO}{1\ mol\ CaO} \times \frac{1\ kg}{1000\ g}$$

mass CaO = 42 kg CaO

21.104 $YBa_2Cu_3O_7$, 666.20 amu

$Cu(OCH_2CH_3)_2$, 153.67 amu

$Y(OCH_2CH_3)_3$, 224.09 amu

$Ba(OCH_2CH_3)_2$, 227.45 amu

$$\text{mol } Cu(OCH_2CH_3)_2 = 75.4\ g \times \frac{1\ mol}{153.67\ g} = 0.4907\ mol\ Cu(OCH_2CH_3)_2$$

mass $Y(OCH_2CH_3)_3$ = 0.4907 mol $Cu(OCH_2CH_3)_2$ x

$$\frac{1\ mol\ Y(OCH_2CH_3)_3}{3\ mol\ Cu(OCH_2CH_3)_2} \times \frac{224.09\ g\ Y(OCH_2CH_3)_3}{1\ mol\ Y(OCH_2CH_3)_3} = 36.7\ g\ Y(OCH_2CH_3)_3$$

mass $Ba(OCH_2CH_3)_2$ = 0.4907 mol $Cu(OCH_2CH_3)_2$ x

$$\frac{2\ mol\ Ba(OCH_2CH_3)_2}{3\ mol\ Cu(OCH_2CH_3)_2} \times \frac{227.45\ g\ Ba(OCH_2CH_3)_2}{1\ mol\ Ba(OCH_2CH_3)_2} = 74.4\ g\ Ba(OCH_2CH_3)_2$$

mass $YBa_2Cu_3O_7$ = 0.4907 mol $Cu(OCH_2CH_3)_2$ x

$$\frac{1 \text{ mol } YBa_2Cu_3O_7}{3 \text{ mol } Cu(OCH_2CH_3)_2} \times \frac{666.20 \text{ g } YBa_2Cu_3O_7}{1 \text{ mol } YBa_2Cu_3O_7} = 109 \text{ g } YBa_2Cu_3O_7$$

21.105 $YBa_2Cu_3O_7$, 666.20 amu

$$\% \text{ Y} = \frac{88.906 \text{ g Y}}{666.20 \text{ g } YBa_2Cu_3O_7} \times 100\% = 13.35\% \text{ Y}$$

$$\% \text{ Ba} = \frac{2(137.33 \text{ g Ba})}{666.20 \text{ g } YBa_2Cu_3O_7} \times 100\% = 41.23\% \text{ Ba}$$

$$\% \text{ Cu} = \frac{3(63.546 \text{ g Cu})}{666.20 \text{ g } YBa_2Cu_3O_7} \times 100\% = 28.62\% \text{ Cu}$$

$$\% \text{ O} = \frac{7(15.9994 \text{ g O})}{666.20 \text{ g } YBa_2Cu_3O_7} \times 100\% = 16.81\% \text{ O}$$

21.106 (a) $Al(OCH_2CH_3)_3 + 2\ Si(OCH_2CH_3)_4 + 26\ H_2O \rightarrow$

$$\underbrace{6\ Al(OH)_3(s) + 2\ Si(OH)_4(s)}_{\text{sol}} + 26\ HOCH_2CH_3$$

(b) H_2O is eliminated from the sol through a series of reactions linking the sol particles together through a three-dimensional network of O bridges to form the gel.

$$(HO)_2Al\text{–}O\text{–}H + H\text{–}O\text{–}Si(OH)_3 \rightarrow (HO)_2Al\text{–}O\text{–}Si(OH)_3 + H_2O$$

(c) The remaining H_2O and solvent are removed from the gel by heating to produce the ceramic, $3\ Al_2O_3 \cdot 2\ SiO_2$.

21.107 AgCl (transparent) $\underset{\text{dark}}{\overset{\text{sunlight}}{\rightleftarrows}}$ Ag + Cl (opaque)

21.108 $Ni(s) + 4\ CO(g) \rightleftharpoons Ni(CO)_4(g)$

$\Delta H° = -160.8$ kJ; $\Delta S° = -410$ J/K $= -410 \times 10^{-3}$ kJ/K

(a) 150°C = 423 K

$\Delta G° = \Delta H° - T\Delta S° = -160.8\text{ kJ} - (423\text{ K})(-410 \times 10^{-3}\text{ kJ/K}) = +12.6\text{ kJ}$

$\Delta G° = -2.303RT \log K$

$$\log K = \frac{-\Delta G°}{2.303RT} = \frac{-12.6\text{ kJ}}{(2.303)(8.314 \times 10^{-3}\text{ kJ/K})(423\text{ K})} = -1.56$$

$K = K_p = 10^{-1.56} = 0.028$

(b) 230°C = 503 K

$\Delta G° = \Delta H° - T\Delta S° = -160.8\text{ kJ} - (503\text{ K})(-410 \times 10^{-3}\text{ kJ/K}) = +45.4\text{ kJ}$

$\Delta G° = -2.303RT \log K$

$$\log K = \frac{-\Delta G°}{2.303RT} = \frac{-45.4\text{ kJ}}{(2.303)(8.314 \times 10^{-3}\text{ kJ/K})(503\text{ K})} = -4.71$$

$K = K_p = 10^{-4.71} = 1.9 \times 10^{-5}$

(c) $\Delta S°$ is large and negative because as the reaction proceeds in the forward direction, the number of moles of gas decrease from four to one.

Because $\Delta S°$ is negative, $-T\Delta S°$ is positive and as T increases, $\Delta G°$ becomes more positive because $\Delta G° = \Delta H° - T\Delta S°$.

(d) The reaction is exothermic because $\Delta H°$ is negative.

$Ni(s) + 4\ CO(g) \rightleftharpoons Ni(CO)_4(g) + \text{heat}$

Heat is added as the temperature is raised and the reaction proceeds in the reverse direction to relieve this stress, as predicted by Le Châtelier's principle. As the reverse reaction proceeds, the partial pressure of CO increases and the partial pressure of $Ni(CO)_4$ decreases. K_p decreases as calculated because $K_p = \dfrac{P_{Ni(CO)_4}}{(P_{CO})^4}$.

21.109 $C(s) + CO_2(g) \rightarrow 2\ CO(g)$

(a) $\Delta H^\circ = [2\ \Delta H_f^\circ(CO)] - \Delta H_f^\circ(CO_2)$

$\Delta H^\circ = (2\ mol)(-110.5\ kJ/mol) - (1\ mol)(-393.5\ kJ/mol) = 172.5\ kJ$

$\Delta S^\circ = [2\ S^\circ(CO)] - [S^\circ(C) + S^\circ(CO_2)]$

$\Delta S^\circ = (2\ mol)(197.6\ J/(K \cdot mol)) -$

$[(1\ mol)(5.7\ J/(K \cdot mol)) + (1\ mol)(213.6\ J/(K \cdot mol))]$

$\Delta S^\circ = 175.9\ J/K = 175.9 \times 10^{-3}\ kJ/K$

$\Delta G^\circ = \Delta H^\circ - T\Delta S^\circ = 172.5\ kJ - (298.15\ K)(175.9 \times 10^{-3}\ kJ/K) = 120.1\ kJ$

(b) The reaction is endothermic because $\Delta H^\circ > 0$.

(c) The number of moles of gas increases from 1 mol to 2 mol; therefore, $\Delta S^\circ > 0$.

(d) at 500°C (773 K):

$\Delta G^\circ = \Delta H^\circ - T\Delta S^\circ = 172.5\ kJ - (773\ K)(175.9 \times 10^{-3}\ kJ/K) = 36.5\ kJ$

at 1000°C (1273 K):

$\Delta G^\circ = \Delta H^\circ - T\Delta S^\circ = 172.5\ kJ - (1273\ K)(175.9 \times 10^{-3}\ kJ/K) = -51.4\ kJ$

(e) at 500°C (773 K):

$$\log K_p = \frac{-\Delta G^\circ}{2.303RT} = \frac{-36.5\ kJ}{(2.303)(8.314 \times 10^{-3}\ kJ/K)(773\ K)} = -2.47$$

$K_p = 10^{-2.47} = 3.4 \times 10^{-3}$

at 1000°C (1273 K):

$$\log K_p = \frac{-\Delta G^\circ}{2.303RT} = \frac{-(-51.4\text{ kJ})}{(2.303)(8.314 \times 10^{-3}\text{ kJ/K})(1273\text{ K})} = 2.11$$

$K_p = 10^{2.11} = 1.3 \times 10^2$

21.110 $SiO_2(s) + 2\ C(s) \rightarrow Si(s) + 2\ CO(g)$

(a) $\Delta H^\circ = 2\ \Delta H_f^\circ(CO) - \Delta H_f^\circ(SiO_2)$

ΔH° = (2 mol)(–110.5 kJ/mol) – (1 mol)(–910.9 kJ/mol) = 689.9 kJ

$\Delta S^\circ = [S^\circ(Si) + 2\ S^\circ(CO)] - [S^\circ(SiO_2) + 2\ S^\circ(C)]$

ΔS° = [(1 mol)(18.8 J/(K· mol)) + (2 mol)(197.6 J/(K· mol))]

– [(1 mol)(41.8 J/(K · mol)) + (2 mol)(5.7 J/(K · mol))]

$\Delta S^\circ = 360.8\text{ J/K} = 360.8 \times 10^{-3}\text{ kJ/K}$

$\Delta G^\circ = \Delta H^\circ - T\Delta S^\circ = 689.9\text{ kJ} - (298.15\text{ K})(360.8 \times 10^{-3}\text{ kJ/K}) = 582.3\text{ kJ}$

(b) The reaction is endothermic because $\Delta H^\circ > 0$.

(c) The number of moles of gas increases from 0 to 2 mol; therefore, $\Delta S^\circ > 0$.

(d) Because $\Delta G^\circ > 0$, the reaction is nonspontaneous at 25°C and 1 atm pressure of CO.

(e) To determine the crossover temperature, set $\Delta G^\circ = 0$ and solve for T.

$\Delta G^\circ = 0 = \Delta H^\circ - T\Delta S^\circ$

$\Delta H^\circ = T\Delta S^\circ$

$$T = \frac{\Delta H^\circ}{\Delta S^\circ} = \frac{689.9\text{ kJ}}{360.8 \times 10^{-3}\text{ kJ/K}} = 1912\text{ K} = 1639^\circ\text{C}$$

22.1 (a) In beta emission, the mass number is unchanged, and the atomic number increases by one.

$$^{106}_{44}Ru \rightarrow \, ^{0}_{-1}e + \, ^{106}_{45}Rh$$

(b) In alpha emission, the mass number decreases by four, and the atomic number decreases by two.

$$^{189}_{83}Bi \rightarrow \, ^{4}_{2}He + \, ^{185}_{81}Tl$$

(c) In electron capture, the mass number is unchanged, and the atomic number decreases by one.

$$^{204}_{84}Po + \, ^{0}_{-1}e \rightarrow \, ^{204}_{83}Bi$$

22.2 The mass number decreases by four, and the atomic number decreases by two. This is characteristic of alpha emission.

$$^{214}_{90}Th \rightarrow \, ^{210}_{88}Ra + \, ^{4}_{2}He$$

22.3 $$t_{1/2} = \frac{0.693}{k} = \frac{0.693}{1.08 \times 10^{-2}\ h^{-1}} = 64.2\ h$$

22.4 $$k = \frac{0.693}{t_{1/2}} = \frac{0.693}{5715\ y} = 1.21 \times 10^{-4}\ y^{-1}$$

22.5 $$\ln\left(\frac{N}{N_0}\right) = -0.693\left(\frac{t}{t_{1/2}}\right) = -0.693\left(\frac{16{,}230\ y}{5715\ y}\right) = -1.968$$

$$\frac{N}{N_0} = e^{-1.968} = 0.140$$

$$\frac{N}{100\%} = 0.140; \qquad N = 14.0\%$$

22.6 $\ln\left(\frac{N}{N_0}\right) = (-0.693)\left(\frac{t}{t_{1/2}}\right)$; $\frac{N}{N_0} = \frac{\text{Decay rate at time t}}{\text{Decay rate at t = 0}}$

$$\ln\left(\frac{10{,}860}{16{,}800}\right) = (-0.693)\left(\frac{28.0\text{ d}}{t_{1/2}}\right)$$

$$-0.436 = (-0.693)\left(\frac{28.0\text{ d}}{t_{1/2}}\right)$$

$$t_{1/2} = \frac{(-0.693)(28.0\text{ d})}{(-0.436)} = 44.5\text{ d}$$

22.7 For ${}^{16}_{8}O$:

First, calculate the total mass of the nucleons (8 n + 8 p)

Mass of 8 neutrons = (8)(1.008 66 amu) = 8.069 28 amu
Mass of 8 protons = (8)(1.007 28 amu) = 8.058 24 amu

Mass of 8 n + 8 p = 16.127 52 amu

Next, calculate the mass of a ${}^{16}O$ nucleus by subtracting the mass of 8 electrons from the mass of a ${}^{16}O$ atom.

Mass of ${}^{16}O$ atom = 15.994 92 amu
–Mass of 8 electrons = –(8)(5.486 x 10^{-4} amu) = –0.004 39 amu

Mass of ${}^{16}O$ nucleus = 15.990 53 amu

Then subtract the mass of the ${}^{16}O$ nucleus from the mass of the nucleons to find the mass defect:

Mass defect = mass of nucleons – mass of nucleus

= (16.127 52 amu) – (15.990 53 amu) = 0.136 99 amu

Mass defect in g/mol:
(0.136 99 amu)(1.660 54 x 10^{-24} g/amu)(6.022 x 10^{23} mol^{-1}) = 0.136 99 g/mol

Now, use the Einstein equation to convert the mass defect into the binding energy.

$\Delta E = \Delta mc^2 = (0.136\ 99\ \text{g/mol})(10^{-3}\ \text{kg/g})(3.00 \times 10^8\ \text{m/s})^2$

$\Delta E = 1.233 \times 10^{13}\ \text{J/mol} = 1.233 \times 10^{10}\ \text{kJ/mol}$

$$\Delta E = \frac{1.233 \times 10^{13}\ \text{J/mol}}{6.022 \times 10^{23}\ \text{nuclei/mol}} \times \frac{1\ \text{MeV}}{1.60 \times 10^{-13}\ \text{J}} \times \frac{1\ \text{nucleus}}{16\ \text{nucleons}} = 8.00\ \frac{\text{MeV}}{\text{nucleon}}$$

22.8 $\Delta E = -852\ \text{kJ/mol} = -852 \times 10^3\ \text{J/mol}$

$1\ \text{J} = 1\ \text{kg} \cdot \text{m}^2/\text{s}^2$

$\Delta E = \Delta mc^2$

$$\Delta m = \frac{\Delta E}{c^2} = \frac{\left(-852 \times 10^3\ \dfrac{\text{kg} \cdot \text{m}^2}{\text{s}^2 \cdot \text{mol}}\right)}{(3.00 \times 10^8\ \text{m/s})^2} = -9.47 \times 10^{-12}\ \text{kg/mol}$$

$$\Delta m = -9.47 \times 10^{-12}\ \frac{\text{kg}}{\text{mol}} \times \frac{1000\ \text{g}}{1\ \text{kg}} = -9.47 \times 10^{-9}\ \text{g/mol}$$

22.9 $^{1}_{0}\text{n} + {}^{235}_{92}\text{U} \rightarrow {}^{137}_{52}\text{Te} + {}^{97}_{40}\text{Zr} + 2\,{}^{1}_{0}\text{n}$

mass $^{235}_{92}\text{U}$	235.0439	amu
mass $^{1}_{0}\text{n}$	1.008 66	amu
–mass $^{137}_{52}\text{Te}$	–136.9254	amu
–mass $^{97}_{40}\text{Zr}$	–96.9110	amu
–mass 2 $^{1}_{0}\text{n}$	–(2)(1.008 66)	amu
mass change	0.1988	amu

$(0.1988\ \text{amu})(1.660\ 54 \times 10^{-24}\ \text{g/amu})(6.022 \times 10^{23}\ \text{mol}^{-1}) = 0.1988\ \text{g/mol}$

$\Delta E = \Delta mc^2 = (0.1988\ \text{g/mol})(10^{-3}\ \text{kg/g})(3.00 \times 10^8\ \text{m/s})^2$

$\Delta E = 1.79 \times 10^{13}\ \text{J/mol} = 1.79 \times 10^{10}\ \text{kJ/mol}$

22.10 $^{1}_{1}H + ^{2}_{1}H \rightarrow ^{3}_{2}He$

mass ^{1}H	1.007 83 amu
mass ^{2}H	2.014 10 amu
–mass ^{3}He	–3.016 03 amu
mass change	0.005 90 amu

$(0.005\ 90\ \text{amu})(1.660\ 54 \times 10^{-24}\ \text{g/amu})(6.022 \times 10^{23}\ \text{mol}^{-1}) = 0.005\ 90\ \text{g/mol}$

$\Delta E = \Delta mc^2 = (0.005\ 90\ \text{g/mol})(10^{-3}\ \text{kg/g})(3.00 \times 10^{8}\ \text{m/s})^2$

$\Delta E = 5.31 \times 10^{11}\ \text{J/mol} = 5.31 \times 10^{8}\ \text{kJ/mol}$

22.11 $^{40}_{18}Ar + ^{1}_{1}p \rightarrow ^{40}_{19}K + ^{1}_{0}n$

22.12 $^{238}_{92}U + ^{2}_{1}H \rightarrow ^{238}_{93}Np + 2\ ^{1}_{0}n$

22.13 $\dfrac{5\ \text{mrem}}{120\ \text{mrem}} \times 100\% = 4\%$

22.14 $\ln\left(\dfrac{N}{N_0}\right) = (-0.693)\left(\dfrac{t}{t_{1/2}}\right)$; $\dfrac{N}{N_0} = \dfrac{\text{Decay rate at time t}}{\text{Decay rate at time t = 0}}$

$\ln\left(\dfrac{2.4}{15.3}\right) = (-0.693)\left(\dfrac{t}{5715\ \text{y}}\right)$

$t = 1.53 \times 10^{4}\ \text{y}$

Understanding Key Concepts

1. $16\ ^{40}K \rightarrow 8\ ^{40}K \rightarrow 4\ ^{40}K$; two half-lives have passed.

2.

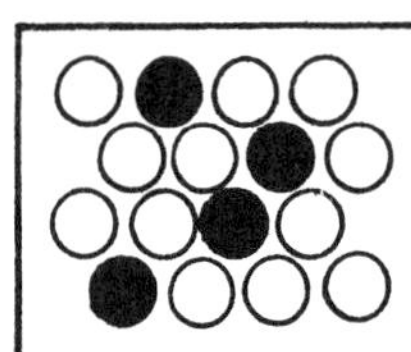

● = blue (^{40}Ar)

○ = yellow (^{40}K)

3. The isotope contains 8 neutrons and 6 protons. The isotope symbol is $^{14}_{6}C$.

4. $^{14}_{6}C$ would decay by beta emission because the n/p ratio is high.

Additional Problems

Nuclear Reactions and Radioactivity

22.15 Positron emission is the conversion of a proton in the nucleus into a neutron plus an ejected positron.

Electron capture is the process in which a proton in the nucleus captures an inner-shell electron and is thereby converted into a neutron.

22.16 An alpha particle $\left(^{4}_{2}He^{2+}\right)$ is a helium nucleus. The He atom has two electrons and is neutral.

22.17 Alpha particles move relatively slowly and can be stopped by the skin. However, inside the body, alpha particles give up their energy to the immediately surrounding tissue.

Gamma rays move at the speed of light and are very penetrating. Therefore they are equally hazardous internally and externally.

22.18 "Neutron rich" nuclides emit beta particles to decrease the number of neutrons and increase the number of protons in the nucleus.

"Neutron poor" nuclides decrease the number of protons and increase the n/p ratio by either alpha emission, positron emission, or electron capture.

22.19 There is no radioactive "neutralization" reaction like there is an acid-base neutralization reaction.

22.20 (a) $^{126}_{50}Sn \rightarrow ^{0}_{-1}e + ^{126}_{51}Sb$

(b) $^{210}_{88}Ra \rightarrow ^{4}_{2}He + ^{206}_{86}Rn$

(c) $^{77}_{37}Rb \rightarrow {}^{0}_{1}e + {}^{77}_{36}Kr$

(d) $^{76}_{36}Kr + {}^{0}_{-1}e \rightarrow {}^{76}_{35}Br$

22.21 (a) $^{90}_{38}Sr \rightarrow {}^{0}_{-1}e + {}^{90}_{39}Y$

(b) $^{247}_{100}Fm \rightarrow {}^{4}_{2}He + {}^{243}_{98}Cf$

(c) $^{49}_{25}Mn \rightarrow {}^{0}_{1}e + {}^{49}_{24}Cr$

(d) $^{37}_{18}Ar + {}^{0}_{-1}e \rightarrow {}^{37}_{17}Cl$

22.22 (a) $^{188}_{80}Hg \rightarrow {}^{188}_{79}Au + {}^{0}_{1}e$

(b) $^{218}_{85}At \rightarrow {}^{214}_{83}Bi + {}^{4}_{2}He$

(c) $^{234}_{90}Th \rightarrow {}^{234}_{91}Pa + {}^{0}_{-1}e$

22.23 (a) $^{24}_{11}Na \rightarrow {}^{24}_{12}Mg + {}^{0}_{-1}e$

(b) $^{135}_{60}Nd \rightarrow {}^{135}_{59}Pr + {}^{0}_{1}e$

(c) $^{170}_{78}Pt \rightarrow {}^{166}_{76}Os + {}^{4}_{2}He$

22.24 (a) $^{162}_{75}Re \rightarrow {}^{158}_{73}Ta + {}^{4}_{2}He$

(b) $^{138}_{62}Sm + {}^{0}_{-1}e \rightarrow {}^{138}_{61}Pm$

(c) $^{188}_{74}W \rightarrow {}^{188}_{75}Re + {}^{0}_{-1}e$

(d) $^{165}_{73}Ta \rightarrow {}^{165}_{72}Hf + {}^{0}_{1}e$

22.25 (a) $^{157}_{63}\text{Eu} \rightarrow\ ^{157}_{64}\text{Gd} +\ ^{0}_{-1}\text{e}$

(b) $^{126}_{56}\text{Ba} +\ ^{0}_{-1}\text{e} \rightarrow\ ^{126}_{55}\text{Cs}$

(c) $^{146}_{62}\text{Sm} \rightarrow\ ^{142}_{60}\text{Nd} +\ ^{4}_{2}\text{He}$

(d) $^{125}_{56}\text{Ba} \rightarrow\ ^{125}_{55}\text{Cs} +\ ^{0}_{1}\text{e}$

22.26 $^{136}_{53}\text{I}$ is neutron rich and decays by beta emission.

$^{122}_{53}\text{I}$ is neutron poor and decays by positron emission.

22.27 ^{160}W is neutron poor and decays by alpha emission. ^{185}W is neutron rich and decays by beta emission.

22.28 $^{241}_{95}\text{Am} \rightarrow\ ^{237}_{93}\text{Np} +\ ^{4}_{2}\text{He}$

$^{237}_{93}\text{Np} \rightarrow\ ^{233}_{91}\text{Pa} +\ ^{4}_{2}\text{He}$

$^{233}_{91}\text{Pa} \rightarrow\ ^{233}_{92}\text{U} +\ ^{0}_{-1}\text{e}$

$^{233}_{92}\text{U} \rightarrow\ ^{229}_{90}\text{Th} +\ ^{4}_{2}\text{He}$

$^{229}_{90}\text{Th} \rightarrow\ ^{225}_{88}\text{Ra} +\ ^{4}_{2}\text{He}$

$^{225}_{88}\text{Ra} \rightarrow\ ^{225}_{89}\text{Ac} +\ ^{0}_{-1}\text{e}$

$^{225}_{89}\text{Ac} \rightarrow\ ^{221}_{87}\text{Fr} +\ ^{4}_{2}\text{He}$

$^{221}_{87}\text{Fr} \rightarrow\ ^{217}_{85}\text{At} +\ ^{4}_{2}\text{He}$

$^{217}_{85}\text{At} \rightarrow\ ^{213}_{83}\text{Bi} +\ ^{4}_{2}\text{He}$

$^{213}_{83}\text{Bi} \rightarrow\ ^{213}_{84}\text{Po} +\ ^{0}_{-1}\text{e}$

$$^{213}_{84}Po \rightarrow ^{209}_{82}Pb + ^{4}_{2}He$$

$$^{209}_{82}Pb \rightarrow ^{209}_{83}Bi + ^{0}_{-1}e$$

22.29 $$^{222}_{86}Rn \rightarrow ^{218}_{84}Po + ^{4}_{2}He$$

$$^{218}_{84}Po \rightarrow ^{214}_{82}Pb + ^{4}_{2}He$$

$$^{214}_{82}Pb \rightarrow ^{210}_{80}Hg + ^{4}_{2}He$$

$$^{210}_{80}Hg \rightarrow ^{210}_{81}Tl + ^{0}_{-1}e$$

$$^{210}_{81}Tl \rightarrow ^{210}_{82}Pb + ^{0}_{-1}e$$

22.30 Each alpha emission decreases the mass number by four and the atomic number by two. Each beta emission increases the atomic number by one.

$$^{232}_{90}Th \rightarrow ^{208}_{82}Pb$$

$$\text{Number of } \alpha \text{ emissions} = \frac{\text{Th mass number} - \text{Pb mass number}}{4}$$

$$= \frac{232 - 208}{4} = 6\ \alpha \text{ emissions}$$

The atomic number decreases by 12 as a result of 6 alpha emissions. The resulting atomic number is (90 – 12) = 78.

Number of β emissions = Pb atomic number – 78 = 82 – 78 = 4 β emissions

Radioactive Decay Rates

22.31 If the half-life of ^{59}Fe is 44.5 d, it takes 44.5 days for half of the original amount of ^{59}Fe to decay.

22.32 The half-life is the time it takes for one-half of a radioactive sample to decay.

The decay constant is the rate constant for the first order radioactive decay.

$$k = \frac{0.693}{t_{1/2}}$$

22.33 $k = \frac{0.693}{t_{1/2}} = \frac{0.693}{2.806\ d} = 0.247\ d^{-1}$

22.34 $k = \frac{0.693}{t_{1/2}} = \frac{0.693}{78.25\ h} = 8.86 \times 10^{-3}\ h^{-1}$

22.35 $t_{1/2} = \frac{0.693}{k} = \frac{0.693}{0.227\ d^{-1}} = 3.05\ d$

22.36 $t_{1/2} = \frac{0.693}{k} = \frac{0.693}{2.88 \times 10^{-5}\ y^{-1}} = 2.41 \times 10^{4}\ y$

22.37 After 65 d:

$$\ln\left(\frac{N}{N_0}\right) = -0.693\left(\frac{t}{t_{1/2}}\right) = -0.693\left(\frac{\left[\frac{65\ d}{365\ d/y}\right]}{432.2\ y}\right) = -0.000\ 285\ 5$$

$$\frac{N}{N_0} = e^{-0.0002855} = 0.9997$$

$$\frac{N}{100\%} = 0.9997; \qquad N = 99.97\%$$

After 65 y:

$$\ln\left(\frac{N}{N_0}\right) = -0.693\left(\frac{t}{t_{1/2}}\right) = -0.693\left(\frac{65\ y}{432.2\ y}\right) = -0.1042$$

$$\frac{N}{N_0} = e^{-0.1042} = 0.9010$$

$$\frac{N}{100\%} = 0.9010; \qquad N = 90.10\%$$

After 650 y:

$$\ln\left(\frac{N}{N_0}\right) = -0.693\left(\frac{t}{t_{1/2}}\right) = -0.693\left(\frac{650\ y}{432.2\ y}\right) = -1.042$$

$$\frac{N}{N_0} = e^{-1.042} = 0.3527$$

$$\frac{N}{100\%} = 0.3527; \qquad N = 35.27\%$$

22.38 After 24 min:

$$\ln\left(\frac{N}{N_0}\right) = (-0.693)\left(\frac{t}{t_{1/2}}\right) = (-0.693)\left(\frac{24\ \text{min}}{109.8\ \text{min}}\right) = -0.1515$$

$$\frac{N}{N_0} = e^{-0.1515} = 0.8594$$

$$\frac{N}{100\%} = 0.8594; \qquad N = 85.94\%$$

After 24 h:

$$\ln\left(\frac{N}{N_0}\right) = (-0.693)\left(\frac{t}{t_{1/2}}\right) = (-0.693)\left(\frac{24\ \text{h} \times \dfrac{60\ \text{min}}{1\ \text{h}}}{109.8\ \text{min}}\right) = -9.089$$

$$\frac{N}{N_0} = e^{-9.089} = 0.000\ 113\ 0$$

$$\frac{N}{100\%} = 0.000\ 113\ 0; \qquad N = 0.011\ 30\%$$

After 24 d:

$$\ln\left(\frac{N}{N_0}\right) = (-0.693)\left(\frac{t}{t_{1/2}}\right) = (-0.693)\left(\frac{24\text{ d} \times \dfrac{24\text{ h}}{1\text{ d}} \times \dfrac{60\text{ min}}{1\text{ h}}}{109.8\text{ min}}\right) = -218.1$$

$$\frac{N}{N_0} = e^{-218.1} = 1.861 \times 10^{-95}$$

$$\frac{N}{100\%} = 1.861 \times 10^{-95}; \qquad N = 1.861 \times 10^{-93}\%$$

22.39 $\ln\left(\frac{N}{N_0}\right) = (-0.693)\left(\frac{t}{t_{1/2}}\right)$

$$\ln(0.43) = (-0.693)\left(\frac{t}{5715\text{ y}}\right)$$

$t = 6960$ y

22.40 Assume a sample of ${}^{40}_{19}K$ containing 100 atoms.

	${}^{40}_{19}K + {}^{0}_{-1}e$	$\rightarrow$	${}^{40}_{18}Ar$
before decay (atoms)	100		0
after decay (atoms)	100 – x		x

$$\frac{{}^{40}Ar}{{}^{40}K} = \frac{x}{100 - x} = 1.15$$

Solve for x. x = 53.5

$$\ln\left(\frac{N}{N_0}\right) = (-0.693)\left(\frac{t}{t_{1/2}}\right)$$

$N = 100 - x = 100 - 53.5 = 46.5$, the amount of ^{40}K at time t.

$N_0 = 100$, the original amount of ^{40}K.

$$\ln\left(\frac{46.5}{100}\right) = (-0.693)\left(\frac{t}{1.26 \times 10^9 \text{ y}}\right)$$

$t = 1.39 \times 10^9$ y

22.41 $$t_{1/2} = \frac{0.693}{k} = \frac{0.693}{7.95 \times 10^{-3} \text{ d}^{-1}} = 87.17 \text{ d}$$

$$\ln\left(\frac{N}{N_0}\right) = (-0.693)\left(\frac{t}{t_{1/2}}\right)$$

$$\ln\left(\frac{N}{N_0}\right) = (-0.693)\left(\frac{185 \text{ d}}{87.17 \text{ d}}\right) = -1.4707$$

$$\frac{N}{N_0} = e^{-1.4707} = 0.2298$$

$$\frac{N}{100\%} = 0.2298; \qquad N = 23.0\%$$

22.42 $$t_{1/2} = \frac{0.693}{k} = \frac{0.693}{2.88 \times 10^{-5} \text{ y}^{-1}} = 2.41 \times 10^4 \text{ y}$$

After 1000 y:

$$\ln\left(\frac{N}{N_0}\right) = (-0.693)\left(\frac{t}{t_{1/2}}\right) = (-0.693)\left(\frac{1000 \text{ y}}{2.41 \times 10^4 \text{ y}}\right) = -0.028\ 76$$

$$\frac{N}{N_0} = e^{-0.02876} = 0.9717$$

$$\frac{N}{100\%} = 0.9717; \qquad N = 97.17\%$$

After 25,000 y:

$$\ln\left(\frac{N}{N_0}\right) = (-0.693)\left(\frac{t}{t_{1/2}}\right) = (-0.693)\left(\frac{25{,}000\ y}{2.41 \times 10^4\ y}\right) = -0.7189$$

$$\frac{N}{N_0} = e^{-0.7189} = 0.4873$$

$$\frac{N}{100\%} = 0.4873; \quad N = 48.73\%$$

After 100,000 y:

$$\ln\left(\frac{N}{N_0}\right) = (-0.693)\left(\frac{t}{t_{1/2}}\right) = (-0.693)\left(\frac{100{,}000\ y}{2.41 \times 10^4\ y}\right) = -2.876$$

$$\frac{N}{N_0} = e^{-2.876} = 0.0564$$

$$\frac{N}{100\%} = 0.0564; \quad N = 5.64\%$$

22.43 $t_{1/2} = (105\ y)(365\ d/y)(24\ h/d)(3600\ s/h) = 3.3113 \times 10^9\ s$

$$k = \frac{0.693}{t_{1/2}} = \frac{0.693}{3.3113 \times 10^9\ s} = 2.0928 \times 10^{-10}\ s^{-1}$$

$$N = (1.0 \times 10^{-9}\ g)\left(\frac{1\ mol\ Po}{209\ g\ Po}\right)(6.022 \times 10^{23}\ atoms/mol) = 2.881 \times 10^{12}\ atoms$$

Decay rate = kN = $(2.0928 \times 10^{-10}\ s^{-1})(2.881 \times 10^{12}\ atoms) = 6.0 \times 10^2/s$

600 α particles are emitted in 1.0 s.

22.44 $t_{1/2} = (3 \times 10^5\ y)(365\ d/y)(24\ h/d)(60\ min/h) = 1.6 \times 10^{11}\ min$

$$k = \frac{0.693}{t_{1/2}} = \frac{0.693}{1.6 \times 10^{11}\ min} = 4.3 \times 10^{-12}\ min^{-1}$$

$$N = (5.0 \times 10^{-3}\ g)\left(\frac{1\ mol\ ^{36}Cl}{36\ g}\right)(6.022 \times 10^{23}\ atoms/mol) = 8.4 \times 10^{19}\ atoms$$

$$\text{Decay rate} = kN = (4.3 \times 10^{-12}\ min^{-1})(8.4 \times 10^{19}\ atoms) = 4 \times 10^{8}/min$$

$$\text{Curies} = (4 \times 10^{8}/min)\left(\frac{1\ min}{60\ s}\right)\left(\frac{1\ Ci}{3.7 \times 10^{10}/s}\right) = 2 \times 10^{-4}\ Ci$$

22.45 $\ln\left(\frac{N}{N_0}\right) = (-0.693)\left(\frac{t}{t_{1/2}}\right)$; $\frac{N}{N_0} = \frac{\text{Decay rate at time t}}{\text{Decay rate at time t = 0}}$

$$\ln\left(\frac{9.2}{15.3}\right) = (-0.693)\left(\frac{t}{5715\ y}\right)$$

$t = 4200\ y$

22.46 Decay rate = kN

$$N = (1.0 \times 10^{-3}\ g)\left(\frac{1\ mol\ ^{79}Se}{79\ g}\right)(6.022 \times 10^{23}\ atoms/mol) = 7.6 \times 10^{18}\ atoms$$

$$k = \frac{\text{Decay rate}}{N} = \frac{2.8 \times 10^{6}/s}{7.6 \times 10^{18}} = 3.7 \times 10^{-13}\ s^{-1}$$

$$t_{1/2} = \frac{0.693}{k} = \frac{0.693}{3.7 \times 10^{-13}\ s^{-1}} = 1.9 \times 10^{12}\ s$$

$$t_{1/2} = (1.9 \times 10^{12}\ s)\left(\frac{1\ h}{3600\ s}\right)\left(\frac{1\ d}{24\ h}\right)\left(\frac{1\ y}{365\ d}\right) = 6.0 \times 10^{4}\ y$$

22.47 Decay rate = kN

$$N = (1.0 \times 10^{-9}\ g)\left(\frac{1\ mol\ Ti}{44\ g\ Ti}\right)(6.022 \times 10^{23}\ atoms/mol) = 1.37 \times 10^{13}\ atoms$$

$$k = \frac{\text{Decay rate}}{N} = \frac{6.4 \times 10^3\ s^{-1}}{1.37 \times 10^{13}} = 4.68 \times 10^{-10}\ s^{-1}$$

$$k = (4.68 \times 10^{-10}\ s^{-1})(3600\ s/h)(24\ h/d)(365\ d/y) = 1.48 \times 10^{-2}\ y^{-1}$$

$$t_{1/2} = \frac{0.693}{k} = \frac{0.693}{1.48 \times 10^{-2}\ y^{-1}} = 47\ y$$

22.48 $\ln\left(\frac{N}{N_0}\right) = (-0.693)\left(\frac{t}{t_{1/2}}\right)$; $\frac{N}{N_0} = \frac{\text{Decay rate at time t}}{\text{Decay rate at time t = 0}}$

$$\ln\left(\frac{6990}{8540}\right) = (-0.693)\left(\frac{10.0\ d}{t_{1/2}}\right)$$

$$t_{1/2} = 34.6\ d$$

22.49 $\ln\left(\frac{N}{N_0}\right) = (-0.693)\left(\frac{t}{t_{1/2}}\right)$; $\frac{N}{N_0} = \frac{\text{Decay rate at time t}}{\text{Decay rate at time t = 0}}$

$$\ln\left(\frac{10{,}980}{53{,}500}\right) = (-0.693)\left(\frac{48.0\ h}{t_{1/2}}\right)$$

$$t_{1/2} = 21.0\ h$$

Energy Changes During Nuclear Reactions

22.50 The loss in mass that occurs when protons and neutrons combine to form a nucleus is called the mass defect. The lost mass is converted into the binding energy that is used to hold the nucleons together.

22.51 $E = (1.50\ \text{MeV})\left(\frac{1.60 \times 10^{-13}\ J}{1\ \text{MeV}}\right) = 2.40 \times 10^{-13}\ J$

$$\lambda = \frac{hc}{E} = \frac{(6.626 \times 10^{-34}\ J \cdot s)(3.00 \times 10^8\ m/s)}{2.40 \times 10^{-13}\ J} = 8.28 \times 10^{-13}\ m$$

22.52 $$E = (6.82\ \text{keV})\left(\frac{1\ \text{MeV}}{10^3\ \text{keV}}\right)\left(\frac{1.60 \times 10^{-13}\ \text{J}}{1\ \text{MeV}}\right) = 1.09 \times 10^{-15}\ \text{J}$$

$$\nu = \frac{E}{h} = \frac{1.09 \times 10^{-15}\ \text{J}}{6.626 \times 10^{-34}\ \text{J} \cdot \text{s}} = 1.65 \times 10^{18}/\text{s}$$

22.53 (a) For $^{52}_{26}Fe$:

First, calculate the total mass of the nucleons (26 n + 26 p)

Mass of 26 neutrons = (26)(1.008 66 amu) = 26.225 16 amu
Mass of 26 protons = (26)(1.007 28 amu) = 26.189 28 amu

Mass of 26 n + 26 p = 52.414 44 amu

Next, calculate the mass of a ^{52}Fe nucleus by subtracting the mass of 26 electrons from the mass of a ^{52}Fe atom.

Mass of ^{52}Fe atom = 51.948 11 amu
–Mass of 26 electrons = $-(26)(5.486 \times 10^{-4}$ amu) = –0.014 26 amu

Mass of ^{52}Fe nucleus = 51.933 85 amu

Then subtract the mass of the ^{52}Fe nucleus from the mass of the nucleons to find the mass defect:

Mass defect = mass of nucleons – mass of nucleus

= (52.414 44 amu) – (51.933 85 amu) = 0.480 59 amu

Mass defect in g/mol:

(0.480 59 amu)($1.660\ 54 \times 10^{-24}$ g/amu)(6.022×10^{23} mol^{-1}) = 0.480 59 g/mol

(b) For $^{92}_{42}Mo$:

First, calculate the total mass of the nucleons (50 n + 42 p)

Mass of 50 neutrons = (50)(1.008 66 amu) = 50.433 00 amu
Mass of 42 protons = (42)(1.007 28 amu) = 42.305 76 amu

Mass of 50 n + 42 p = 92.738 76 amu

Next, calculate the mass of a ^{92}Mo nucleus by subtracting the mass of 42 electrons from the mass of a ^{92}Mo atom.

Mass of ^{92}Mo atom	= 91.906 81 amu
–Mass of 42 electrons = –(42)(5.486 x 10^{-4} amu)	= –0.023 04 amu
Mass of ^{92}Mo nucleus	= 91.883 77 amu

Then subtract the mass of the ^{92}Mo nucleus from the mass of the nucleons to find the mass defect:

Mass defect = mass of nucleons – mass of nucleus

= (92.738 76 amu) – (91.883 77 amu) = 0.854 99 amu

Mass defect in g/mol:

(0.854 99 amu)(1.660 54 x 10^{-24} g/amu)(6.022 x 10^{23} mol^{-1}) = 0.854 99 g/mol

22.54 (a) For $^{32}_{16}S$:

First, calculate the total mass of the nucleons (16 n + 16 p)

Mass of 16 neutrons = (16)(1.008 66 amu)	= 16.138 56 amu
Mass of 16 protons = (16)(1.007 28 amu)	= 16.116 48 amu
Mass of 16 n + 16 p	= 32.255 04 amu

Next, calculate the mass of a ^{32}S nucleus by subtracting the mass of 16 electrons from the mass of a ^{32}S atom.

Mass of ^{32}S	= 31.972 07 amu
Mass of 16 electrons = –(16)(5.486 x 10^{-4} amu)	= –0.008 78 amu
Mass of ^{32}S nucleus	= 31.963 29 amu

Then subtract the mass of the ^{32}S nucleus from the mass of the nucleons to find the mass defect:

Mass defect = mass of nucleons – mass of nucleus

= (32.255 04 amu) – (31.963 29 amu) = 0.291 75 amu

Mass defect in g/mol:

$(0.291\ 75\ \text{amu})(1.660\ 54 \times 10^{-24}\ \text{g/amu})(6.022 \times 10^{23}\ \text{mol}^{-1}) = 0.291\ 74\ \text{g/mol}$

(b) For ${}^{40}_{20}\text{Ca}$:

First, calculate the total mass of the nucleons (20 n + 20 p)

Mass of 20 neutrons = (20)(1.008 66 amu) = 20.173 20 amu
Mass of 20 protons = (20)(1.007 28 amu) = 20.145 60 amu

Mass of 20 n + 20 p = 40.318 80 amu

Next, calculate the mass of a ^{40}Ca nucleus by subtracting the mass of 20 electrons from the mass of a ^{40}Ca atom.

Mass of ^{40}Ca = 39.962 59 amu
–Mass of 20 electrons = –(20)(5.486 x 10^{-4} amu) = –0.010 97 amu

---

Mass of ^{40}Ca nucleus = 39.951 62 amu

Then substract the mass of the ^{40}Ca nucleus from the mass of the nucleons to find the mass defect:

Mass defect = mass of nucleons – mass of nucleus

= (40.318 80 amu) – (39.951 62 amu) = 0.367 18 amu

Mass defect in g/mol:

$(0.367\ 18\ \text{amu})(1.660\ 54 \times 10^{-24}\ \text{g/amu})(6.022 \times 10^{23}\ \text{mol}^{-1}) = 0.367\ 17\ \text{g/mol}$

22.55 (a) For ${}^{58}_{28}\text{Ni}$:

First, calculate the total mass of the nucleons (30 n + 28 p)

Mass of 30 neutrons = (30)(1.008 66 amu) = 30.259 80 amu
Mass of 28 protons = (28)(1.007 28 amu) = 28.203 84 amu

Mass of 30 n + 28 p = 58.463 64 amu

Next, calculate the mass of a ^{58}Ni nucleus by subtracting the mass of 28 electrons from the mass of a ^{58}Ni atom.

Mass of ^{58}Ni atom	= 57.935 35 amu
–Mass of 28 electrons = –(28)(5.486 x 10^{-4} amu)	= –0.015 36 amu
Mass of ^{58}Ni nucleus	= 57.919 99 amu

Then subtract the mass of the ^{58}Ni nucleus from the mass of the nucleons to find the mass defect:

Mass defect = mass of nucleons – mass of nucleus

= (58.463 64 amu) – (57.919 99 amu) = 0.543 65 amu

Mass defect in g/mol:

(0.543 65 amu)(1.660 54 x 10^{-24} g/amu)(6.022 x 10^{23} mol^{-1}) = 0.543 65 g/mol

Now, use the Einstein equation to convert the mass defect into the binding energy.

$\Delta E = \Delta mc^2$ = (0.543 65 g/mol)(10^{-3} kg/g)(3.00 x 10^8 m/s)^2

ΔE = 4.893 x 10^{13} J/mol = 4.893 x 10^{10} kJ/mol

$$\Delta E = \frac{4.893 \times 10^{13}\ \text{J/mol}}{6.022 \times 10^{23}\ \text{nuclei/mol}} \times \frac{1\ \text{MeV}}{1.60 \times 10^{-13}\ \text{J}} \times \frac{1\ \text{nucleus}}{58\ \text{nucleons}} = 8.76\ \frac{\text{MeV}}{\text{nucleon}}$$

(b) For $^{84}_{36}$Kr:

First, calculate the total mass of the nucleons (48 n + 36 p)

Mass of 48 neutrons = (48)(1.008 66 amu)	= 48.415 68 amu
Mass of 36 protons = (36)(1.007 28 amu)	= 36.262 08 amu
Mass of 48 n + 36 p	= 84.677 76 amu

Next, calculate the mass of a ^{84}Kr nucleus by subtracting the mass of 36 electrons from the mass of a ^{84}Kr atom.

Mass of ^{84}Kr atom	= 83.911 51 amu
–Mass of 36 electrons = –(36)(5.486 x 10^{-4} amu)	= –0.019 75 amu
Mass of ^{84}Kr nucleus	= 83.891 76 amu

Then subtract the mass of the ^{84}Kr nucleus from the mass of the nucleons to find the mass defect:

Mass defect = mass of nucleons – mass of nucleus

= (84.677 76 amu) – (83.891 76 amu) = 0.786 00 amu

Mass defect in g/mol:

(0.786 00 amu)(1.660 54 x 10^{-24} g/mol)(6.022 x 10^{23} mol^{-1}) = 0.786 00 g/mol

Now, use the Einstein equation to convert the mass defect into the binding energy.

$\Delta E = \Delta mc^2 = (0.786\ 00\ \text{g/mol})(10^{-3}\ \text{kg/g})(3.00 \times 10^8\ \text{m/s})^2$

$\Delta E = 7.074 \times 10^{13}\ \text{J/mol} = 7.074 \times 10^{10}\ \text{kJ/mol}$

$$\Delta E = \frac{7.074 \times 10^{13}\ \text{J/mol}}{6.022 \times 10^{23}\ \text{nuclei/mol}} \times \frac{1\ \text{MeV}}{1.60 \times 10^{-13}\ \text{J}} \times \frac{1\ \text{nucleus}}{84\ \text{nucleons}} = 8.74\ \frac{\text{MeV}}{\text{nucleon}}$$

22.56 (a) For $^{63}_{29}$Cu:

First, calculate the total mass of the nucleons (34 n + 29 p)

Mass of 34 neutrons = (34)(1.008 66 amu)	= 34.294 44 amu
Mass of 29 protons = (29)(1.007 28 amu)	= 29.211 12 amu
Mass of 34 n + 29 p	= 63.505 56 amu

Next calculate the mass of a ^{63}Cu nucleus by subtracting the mass of 29 electrons from the mass of a ^{63}Cu atom.

Mass of ^{63}Cu atom	= 62.939 60 amu
–Mass of 29 electrons = –(29)(5.486 x 10^{-4}amu)	= –0.015 91 amu
Mass of ^{63}Cu nucleus	= 62.923 69 amu

Then subtract the mass of the ^{63}Cu nucleus from the mass of the nucleons to find the mass defect:

Mass defect = mass of nucleons – mass of nucleus

$$= (63.505\ 56 \text{ amu}) - (62.923\ 69 \text{ amu}) = 0.581\ 87 \text{ amu}$$

Mass defect in g/mol:

$$(0.581\ 87 \text{ amu})(1.660\ 54 \times 10^{-24} \text{ g/amu})(6.022 \times 10^{23} \text{ mol}^{-1}) = 0.581\ 86 \text{ g/mol}$$

Now, use the Einstein equation to convert the mass defect into the binding energy.

$$\Delta E = \Delta mc^2 = (0.581\ 86 \text{ g/mol})(10^{-3} \text{ kg/g})(3.00 \times 10^8 \text{ m/s})^2$$

$$\Delta E = 5.237 \times 10^{13} \text{ J/mol} = 5.237 \times 10^{10} \text{ kJ/mol}$$

$$\Delta E = \frac{5.237 \times 10^{13} \text{ J/mol}}{6.022 \times 10^{23} \text{ nuclei/mol}} \times \frac{1 \text{ MeV}}{1.60 \times 10^{-13} \text{ J}} \times \frac{1 \text{ nucleus}}{63 \text{ nucleons}} = 8.63 \frac{\text{MeV}}{\text{nucleon}}$$

(b) For $^{84}_{38}Sr$:

First, calculate the total mass of the nucleons (46 n + 38 p)

Mass of 46 neutrons = (46)(1.008 66 amu) = 46.398 36 amu
Mass of 38 protons = (38)(1.007 28 amu) = 38.276 64 amu

Mass of 46 n + 38 p = 84.675 00 amu

Next, calculate the mass of a ^{84}Sr nucleus by subtracting the mass of 38 electrons from the mass of a ^{84}Sr atom.

Mass of ^{84}Sr atom = 83.913 43 amu
–Mass of 38 electrons = $-(38)(5.486 \times 10^{-4}$ amu) = –0.020 85 amu

Mass of ^{84}Sr nucleus = 83.892 58 amu

Then subtract the mass of the ^{84}Sr nucleus from the mass of the nucleons to find the mass defect:

Mass defect = mass of nucleons – mass of nucleus

= (84.675 00 amu) – (83.892 58 amu) = 0.782 42 amu

Mass defect in g/mol:

$(0.782\ 42\ \text{amu})(1.660\ 54 \times 10^{-24}\ \text{g/amu})(6.022 \times 10^{23}\ \text{mol}^{-1}) = 0.782\ 40\ \text{g/mol}$

Now, use the Einstein equation to convert the mass defect into the binding energy.

$\Delta E = \Delta mc^2 = (0.782\ 40\ \text{g/mol})(10^{-3}\ \text{kg/g})(3.00 \times 10^8\ \text{m/s})^2$

$\Delta E = 7.042 \times 10^{13}\ \text{J/mol} = 7.042 \times 10^{10}\ \text{kJ/mol}$

$$\Delta E = \frac{7.042 \times 10^{13}\ \text{J/mol}}{6.022 \times 10^{23}\ \text{nuclei/mol}} \times \frac{1\ \text{MeV}}{1.60 \times 10^{-13}\ \text{J}} \times \frac{1\ \text{nucleus}}{84\ \text{nucleons}} = 8.70\ \frac{\text{MeV}}{\text{nucleon}}$$

22.57 $^{174}_{77}\text{Ir} \rightarrow {}^{170}_{75}\text{Re} + {}^{4}_{2}\text{He}$

mass $^{174}_{77}\text{Ir}$	173.966 66 amu
–mass $^{170}_{75}\text{Re}$	–169.958 04 amu
–mass $^{4}_{2}\text{He}$	–4.002 60 amu
mass change	0.006 02 amu

$(0.006\ 02\ \text{amu})(1.660\ 54 \times 10^{-24}\ \text{g/amu})(6.022 \times 10^{23}\ \text{mol}^{-1}) = 0.006\ 02\ \text{g/mol}$

$\Delta E = \Delta mc^2 = (0.006\ 02\ \text{g/mol})(10^{-3}\ \text{kg/g})(3.00 \times 10^8\ \text{m/s})^2$

$\Delta E = 5.42 \times 10^{11}\ \text{J/mol} = 5.42 \times 10^{8}\ \text{kJ/mol}$

22.58 $^{28}_{12}\text{Mg} \rightarrow {}^{28}_{13}\text{Al} + {}^{0}_{-1}\text{e}$

mass $^{28}_{12}\text{Mg}$	27.983 88 amu
–mass $^{28}_{13}\text{Al}$	–27.981 91 amu
–mass $^{0}_{-1}\text{e}$	–0.000 55 amu
mass change	0.001 42 amu

$(0.001\ 42 \text{ amu})(1.660\ 54 \times 10^{-24} \text{ g/amu})(6.022 \times 10^{23} \text{ mol}^{-1}) = 0.001\ 42 \text{ g/mol}$

$\Delta E = \Delta mc^2 = (0.001\ 42 \text{ g/mol})(10^{-3} \text{ kg/g})(3.00 \times 10^8 \text{ m/s})^2$

$\Delta E = 1.28 \times 10^{11} \text{ J/mol} = 1.28 \times 10^8 \text{ kJ/mol}$

22.59 $\Delta m = \dfrac{\Delta E}{c^2} = \dfrac{92.2 \times 10^3 \text{ J}}{(3.00 \times 10^8 \text{ m/s})^2} = \dfrac{92.2 \times 10^3 \text{ kg}\cdot\text{m}^2/\text{s}^2}{(3.00 \times 10^8 \text{ m/s})^2} = 1.02 \times 10^{-12} \text{ kg}$

$\Delta m = 1.02 \times 10^{-9} \text{ g}$

22.60 Mass of positron and electron

$= 2(9.109 \times 10^{-31} \text{ kg})(6.022 \times 10^{23} \text{ mol}^{-1}) = 1.097 \times 10^{-6} \text{ kg/mol}$

$\Delta E = \Delta mc^2 = (1.097 \times 10^{-6} \text{ kg/mol})(3.00 \times 10^8 \text{ m/s})^2$

$\Delta E = 9.87 \times 10^{10} \text{ J/mol} = 9.87 \times 10^7 \text{ kJ/mol}$

22.61 $2\ {}^{2}_{1}\text{H} \rightarrow {}^{3}_{2}\text{He} + {}^{1}_{0}\text{n}$

mass $2\ {}^{2}_{1}\text{H}$	2(2.0141) amu
–mass ${}^{3}_{2}\text{He}$	–3.0160 amu
–mass ${}^{1}_{0}\text{n}$	–1.008 66 amu
mass change	0.003 54 amu

$(0.003\ 54 \text{ amu})(1.660\ 54 \times 10^{-24} \text{ g/mol})(6.022 \times 10^{23} \text{ mol}^{-1}) = 0.003\ 54 \text{ g/mol}$

$\Delta E = \Delta mc^2 = (0.003\ 54 \text{ g/mol})(10^{-3} \text{ kg/g})(3.00 \times 10^8 \text{ m/s})^2$

$\Delta E = 3.2 \times 10^{11} \text{ J/mol} = 3.2 \times 10^8 \text{ kJ/mol}$

Nuclear Transmutation

22.62 (a) ${}^{109}_{47}\text{Ag} + {}^{4}_{2}\text{He} \rightarrow {}^{113}_{49}\text{In}$

(b) ${}^{10}_{5}\text{B} + {}^{4}_{2}\text{He} \rightarrow {}^{13}_{7}\text{N} + {}^{1}_{0}\text{n}$

22.63 (a) $^{235}_{92}U \rightarrow ^{160}_{62}Sm + ^{72}_{30}Zn + 3\ ^{1}_{0}n$

(b) $^{235}_{92}U \rightarrow ^{87}_{35}Br + ^{146}_{57}La + 2\ ^{1}_{0}n$

22.64 $^{209}_{83}Bi + ^{58}_{26}Fe \rightarrow ^{266}_{109}Une + ^{1}_{0}n$

22.65 $^{98}_{42}Mo + ^{1}_{0}n \rightarrow ^{99}_{42}Mo$

22.66 $^{238}_{92}U + ^{12}_{6}C \rightarrow ^{246}_{98}Cf + 4\ ^{1}_{0}n$

22.67 (a) $^{246}_{96}Cm + ^{12}_{6}C \rightarrow ^{254}_{102}No + 4\ ^{1}_{0}n$

(b) $^{253}_{99}Es + ^{4}_{2}He \rightarrow ^{256}_{101}Md + ^{1}_{0}n$

(c) $^{250}_{98}Cf + ^{11}_{5}B \rightarrow ^{257}_{103}Lr + 4\ ^{1}_{0}n$

General Problems

22.68 Decay rate = kN

$$k = \frac{0.693}{t_{1/2}} = \frac{0.693}{1.26 \times 10^9\ y} = 5.50 \times 10^{-10}\ y^{-1}$$

KCl, 74.55 amu

N = number of $^{40}K^+$ ions in a 1.00 g sample of KCl

$$N = (0.000\ 117)(1.00\ g)\left(\frac{1\ mol\ KCl}{74.55\ g}\right)\left(\frac{1\ mol\ K^+}{1\ mol\ KCl}\right)(6.022 \times 10^{23}\ mol^{-1})$$

$N = 9.45 \times 10^{17}\ ^{40}K^+$ ions

Decay rate = kN = $(5.50 \times 10^{-10}\ y^{-1})(9.45 \times 10^{17}) = 5.20 \times 10^8/y$

$$\text{Disintegration/s} = (5.20 \times 10^{8}/\text{y})\left(\frac{1\ \text{y}}{365\ \text{d}}\right)\left(\frac{1\ \text{d}}{24\ \text{h}}\right)\left(\frac{1\ \text{h}}{3600\ \text{s}}\right) = 16.5/\text{s}$$

22.69 $\ln\left(\frac{N}{N_0}\right) = (-0.693)\left(\frac{t}{t_{1/2}}\right)$

$$\ln\left(\frac{100 - 99.99}{100}\right) = (-0.693)\left(\frac{t}{1.53\ \text{s}}\right)$$

$t = 20.3$ s

22.70 $t_{1/2} = \frac{0.693}{k} = \frac{0.693}{0.063\ \text{s}^{-1}} = 11$ s

$$\ln\left(\frac{N}{N_0}\right) = (-0.693)\left(\frac{t}{t_{1/2}}\right)$$

$$\ln\left(\frac{100 - 99.99}{100}\right) = (-0.693)\left(\frac{t}{11\ \text{s}}\right)$$

$t = 150$ s

22.71 (a) For $^{50}_{24}Cr$:

First, calculate the total mass of the nucleons (26 n + 24 p)

Mass of 26 neutrons = (26)(1.008 66 amu) = 26.225 16 amu
Mass of 24 protons = (24)(1.007 28 amu) = 24.174 72 amu

Mass of 26 n + 24 p = 50.399 88 amu

Next, calculate the mass of a ^{50}Cr nucleus by subtracting the mass of 24 electrons from the mass of a ^{50}Cr atom.

Mass of ^{50}Cr atom = 49.946 05 amu
–Mass of 24 electrons = –(24)(5.486×10^{-4} amu) = –0.013 17 amu

---

Mass of ^{50}Cr nucleus = 49.932 88 amu

Then subtract the mass of the ^{50}Cr nucleus from the mass of the nucleons to find the mass defect:

Mass defect = mass of nucleons – mass of nucleus

= (50.399 88 amu) – (49.932 88 amu) = 0.467 00 amu

Mass defect in g/mol:

$(0.467\ 00 \text{ amu})(1.660\ 54 \times 10^{-24} \text{ g/amu})(6.022 \times 10^{23} \text{ mol}^{-1}) = 0.467\ 00 \text{ g/mol}$

Now, use the Einstein equation to convert the mass defect into the binding energy.

$\Delta E = \Delta mc^2 = (0.467\ 00 \text{ g/mol})(10^{-3} \text{ kg/g})(3.00 \times 10^8 \text{ m/s})^2$

$\Delta E = 4.203 \times 10^{13} \text{ J/mol} = 4.203 \times 10^{10} \text{ kJ/mol}$

$$\Delta E = \frac{4.203 \times 10^{13} \text{ J/mol}}{6.022 \times 10^{23} \text{ nuclei/mol}} \times \frac{1 \text{ MeV}}{1.60 \times 10^{-13} \text{ J}} \times \frac{1 \text{ nucleus}}{50 \text{ nucleons}} = 8.72 \frac{\text{MeV}}{\text{nucleon}}$$

(b) For $^{64}_{30}Zn$:

First, calculate the total mass of the nucleons (34 n + 30 p)

Mass of 34 neutrons	= (34)(1.008 66 amu)	= 34.294 44 amu
Mass of 30 protons	= (30)(1.007 28 amu)	= 30.218 40 amu
Mass of 34 n + 30 p		= 64.512 84 amu

Next, calculate the mass of a ^{64}Zn nucleus by subtracting the mass of 30 electrons from the mass of a ^{64}Zn atom.

Mass of ^{64}Zn atom	= 63.929 15 amu
–Mass of 30 electrons = $-(30)(5.486 \times 10^{-4}$ amu)	= –0.016 46 amu
Mass of ^{64}Zn nucleus	= 63.912 69 amu

Then subtract the mass of the ^{64}Zn nucleus from the mass of the nucleons to find the mass defect:

Mass defect = mass of nucleons – mass of nucleus

= (64.512 84 amu) – (63.912 69 amu) = 0.600 15 amu

Mass defect in g/mol:

$(0.600\ 15\ \text{amu})(1.660\ 54 \times 10^{-24}\ \text{g/amu})(6.022 \times 10^{23}\ \text{mol}^{-1}) = 0.600\ 15\ \text{g/mol}$

Now, use the Einstein equation to convert the mass defect into the binding energy.

$\Delta E = \Delta mc^2 = (0.600\ 15\ \text{g/mol})(10^{-3}\ \text{kg/g})(3.00 \times 10^8\ \text{m/s})^2$

$\Delta E = 5.401 \times 10^{13}\ \text{J/mol} = 5.401 \times 10^{10}\ \text{kJ/mol}$

$$\Delta E = \frac{5.401 \times 10^{13}\ \text{J/mol}}{6.022 \times 10^{23}\ \text{nuclei/mol}} \times \frac{1\ \text{MeV}}{1.60 \times 10^{-13}\ \text{J}} \times \frac{1\ \text{nucleus}}{64\ \text{nucleons}} = 8.76\ \frac{\text{MeV}}{\text{nucleon}}$$

The ^{64}Zn is more stable.

22.72 $\ln\left(\frac{N}{N_0}\right) = (-0.693)\left(\frac{t}{t_{1/2}}\right)$; $\frac{N}{N_0} = \frac{\text{Decay rate at time t}}{\text{Decay rate at time t = 0}}$

$$\ln\left(\frac{2.9}{15.3}\right) = (-0.693)\left(\frac{t}{5715\ \text{y}}\right)$$

$t = 1.37 \times 10^4$ y

22.73 $^{2}_{1}H + ^{3}_{2}He \rightarrow ^{4}_{2}He + ^{1}_{1}H$

mass $^{2}_{1}H$	2.0141 amu
mass $^{3}_{2}He$	3.0160 amu
–mass $^{4}_{2}He$	–4.0026 amu
–mass $^{1}_{1}H$	–1.0078 amu
mass change	0.0197 amu

$(0.0197\ \text{amu})(1.660\ 54 \times 10^{-24}\ \text{g/amu})(6.022 \times 10^{23}\ \text{mol}^{-1}) = 0.0197\ \text{g/mol}$

$\Delta E = \Delta mc^2 = (0.0197 \text{ g/mol})(10^{-3} \text{ kg/g})(3.00 \times 10^8 \text{ m/s})^2$

$\Delta E = 1.77 \times 10^{12} \text{ J/mol} = 1.77 \times 10^9 \text{ kJ/mol}$

22.74 $\;{}^{10}_{5}\text{B} + {}^{1}_{0}\text{n} \rightarrow {}^{7}_{3}\text{Li} + {}^{4}_{2}\text{He}$

22.75 $\;t_{1/2} = 1.1 \times 10^{20} \text{ y} = (1.1 \times 10^{20} \text{ y})(365 \text{ d/y}) = 4.0 \times 10^{22} \text{ d}$

$$k = \frac{0.693}{t_{1/2}} = \frac{0.693}{4.0 \times 10^{22} \text{ d}} = 1.7 \times 10^{-23} \text{ d}^{-1}$$

$N = 6.02 \times 10^{23}$ atoms

Decay rate $= kN = (1.7 \times 10^{-23} \text{ d}^{-1})(6.02 \times 10^{23} \text{ atoms}) = 10/\text{d}$

There are 10 disintegrations per day.

22.76 $\;{}^{238}_{92}\text{U} + {}^{1}_{0}\text{n} \rightarrow {}^{239}_{94}\text{Pu} + 2\,{}^{0}_{-1}\text{e}$

22.77 $\;\ln\left(\frac{N}{N_0}\right) = (-0.693)\left(\frac{t}{t_{1/2}}\right);\qquad \frac{N}{N_0} = \frac{\text{Decay rate at time t}}{\text{Decay rate at time t} = 0}$

$$\ln\left(\frac{1{,}350}{44{,}500}\right) = (-0.693)\left(\frac{5.00 \text{ min}}{t_{1/2}}\right)$$

$t_{1/2} = 0.991$ min

22.78 $\;3.9 \times 10^{23} \text{ kJ} = 3.9 \times 10^{26} \text{ J} = 3.9 \times 10^{26} \text{ kg} \cdot \text{m}^2/\text{s}^2$

$\Delta E = \Delta mc^2$

$$\Delta m = \frac{\Delta E}{c^2} = \frac{3.9 \times 10^{26} \text{ kg} \cdot \text{m}^2/\text{s}^2}{(3.00 \times 10^8 \text{ m/s})^2} = 4.3 \times 10^9 \text{ kg}$$

The sun loses mass at a rate of 4.3×10^9 kg/s.

23.1

```
    H   H   H   H   H   H   H
    |   |   |   |   |   |   |
H — C — C — C — C — C — C — C — H
    |   |   |   |   |   |   |
    H   H   H   H   H   H   H
```

23.2

```
    H   H   H   H   H   H
    |   |   |   |   |   |
H — C — C — C — C — C — C — H
    |   |   |   |   |   |
    H   H   H   H   H   H
```

```
          H
          |
      H — C — H
    H     |   H   H   H
    |     |   |   |   |
H — C  —  C — C — C — C — H
    |     |   |   |   |
    H     H   H   H   H
```

```
              H
              |
          H — C — H
    H   H     |   H   H
    |   |     |   |   |
H — C — C  —  C — C — C — H
    |   |     |   |   |
    H   H     H   H   H
```

```
          H
          |
      H — C — H
    H     |   H   H
    |     |   |   |
H — C  —  C — C — C — H
    |     |   |   |
    H     |   H   H
      H — C — H
          |
          H
```

```
          H
          |
      H — C — H
    H     |   H   H
    |     |   |   |
H — C  —  C — C — C — H
    |     |   |   |
    H     H   |   H
          H — C — H
              |
              H
```

23.3

$CH_3CH_2CH_2CH_2CH_3$

$$\begin{array}{c} \quad\quad\; CH_3 \\ \quad\quad\; | \\ CH_3CH_2CHCH_3 \end{array}$$

$$\begin{array}{c} CH_3 \\ | \\ CH_3CCH_3 \\ | \\ CH_3 \end{array}$$

23.4 Structures (a) and (c) are identical. They both contain a chain of six carbons with two $-CH_3$ branches at the fourth carbon and one $-CH_3$ branch at the second carbon. Structure (b) is different, having a chain of seven carbons.

23.5

(a) $CH_3CH_2CH_2CH_2CH_3$ pentane

$$\begin{array}{c} \quad\quad\; CH_3 \\ \quad\quad\; | \\ CH_3CH_2CHCH_3 \end{array}$$ 2-methylbutane

$$\begin{array}{c} CH_3 \\ | \\ CH_3CCH_3 \\ | \\ CH_3 \end{array}$$ 2,2-dimethylpropane

(b) 3,4-dimethylhexane

(c) 2,4-dimethylpentane

(d) 2,2,5-trimethylheptane

23.6 (a)

$$\begin{array}{l} \quad\quad\quad\quad\; CH_3 \\ \quad\quad\quad\quad\; | \\ CH_3CH_2CHCHCH_2CH_2CH_2CH_2CH_3 \\ \quad\quad\quad\quad\quad\;\; | \\ \quad\quad\quad\quad\quad\;\; CH_3 \end{array}$$

(b)

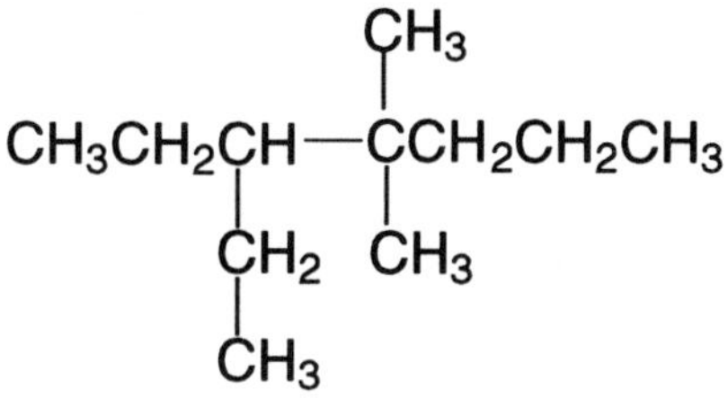

$$\begin{array}{c} \quad\quad\quad\quad CH_3 \\ CH_3CH_2CH-CCH_2CH_2CH_3 \\ CH_2\ \ CH_3 \quad\quad\quad \\ CH_3 \quad\quad\quad\quad \end{array}$$

(c)

$$\begin{array}{l} \quad\ \ CH_3\ \ CH_2CH_2CH_3 \\ CH_3CCH_2CHCH_2CH_2CH_2CH_3 \\ \quad\ \ CH_3 \end{array}$$

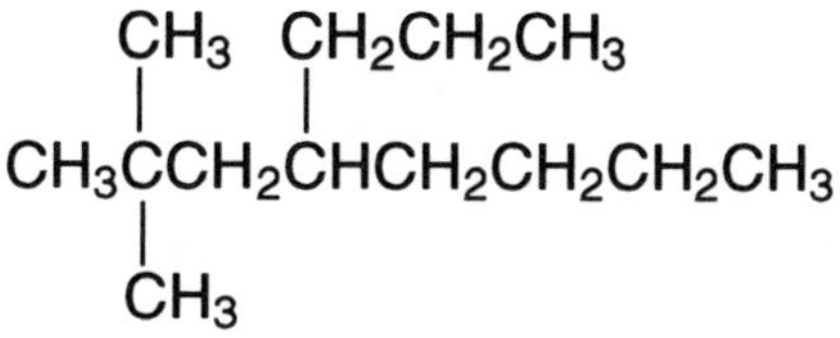

(d)

$$\begin{array}{l} \quad\ \ CH_3\ \ CH_3 \\ CH_3CCH_2CHCH_3 \\ \quad\ \ CH_3 \end{array}$$

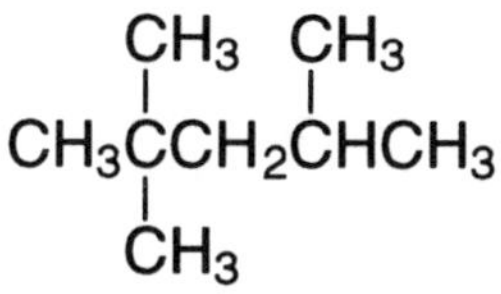

23.7 (a) 1,4-dimethylcyclohexane (b) 1-ethyl-3-methylcyclopentane

(c) isopropylcyclobutane

23.8 (a) (b) 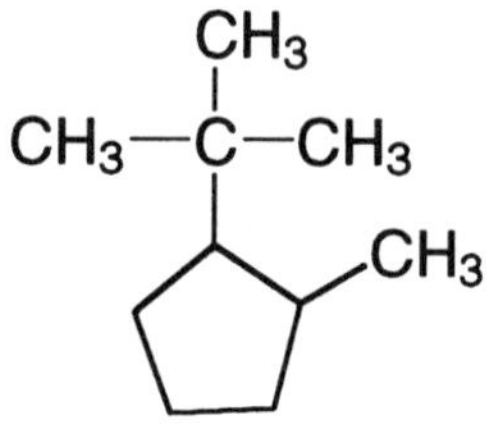 (c)

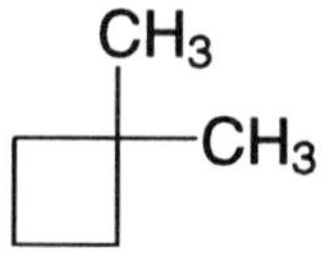

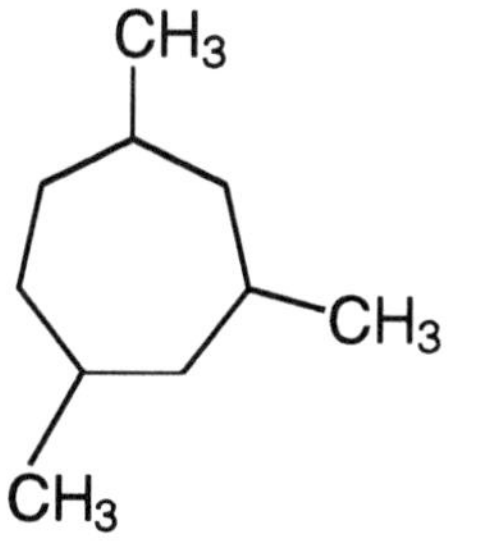

(a) cyclobutane ring with CH_3 and CH_3 on the same carbon

(b) $CH_3-\underset{|}{\overset{CH_3}{C}}-CH_3$ attached to a cyclopentane ring bearing CH_3 on the adjacent carbon

(c) cycloheptane ring with three CH_3 groups

23.9

$$\begin{array}{c} CH_3 \\ ClCH_2CHCH_2CH_3 \end{array} \quad \begin{array}{c} CH_3 \\ CH_3CCH_2CH_3 \\ Cl \end{array} \quad \begin{array}{c} CH_3 \\ CH_3CHCHCH_3 \\ Cl \end{array} \quad \begin{array}{c} CH_3 \\ CH_3CHCH_2CH_2Cl \end{array}$$

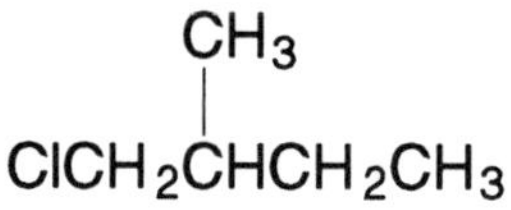

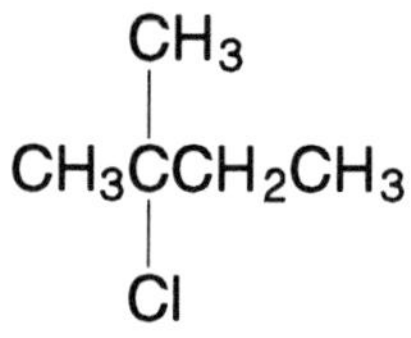

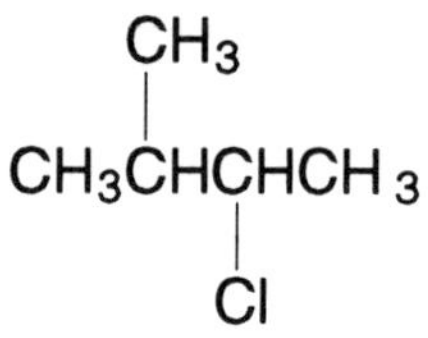

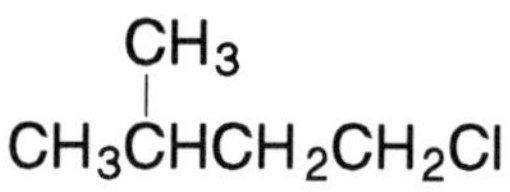

23.10 (a)

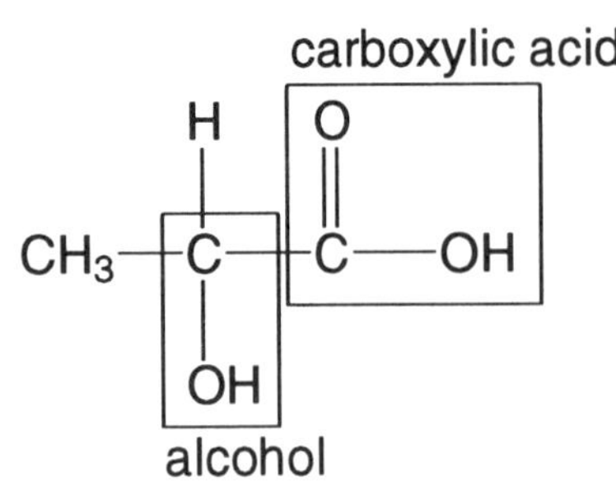

(b)

alkene

$C_6H_5-CH{=}CH_2$

arene (aromatic)

23.11 (a) $CH_3\overset{\displaystyle O}{\overset{\|}{C}}H$ (b) $CH_3CH_2\overset{\displaystyle O}{\overset{\|}{C}}OH$

23.12 (a) 3-methyl-1-butene (b) 4-methyl-3-heptene (c) 3-ethyl-1-hexyne

23.13 (a)

$$\begin{array}{l} CH_3 \\ | \\ CH_3CCH{=}CHCH_2CH_3 \\ | \\ CH_3 \end{array}$$

(b)

$$\begin{array}{l} \phantom{CH_3C{\equiv}}CH_3 \\ \phantom{CH_3C{\equiv}}| \\ \phantom{CH_3C{\equiv}}CHCH_3 \\ \phantom{CH_3C{\equiv}}| \\ CH_3C{\equiv}CCHCH_2CH_2CH_3 \end{array}$$

(c)

$$\begin{array}{ccc} CH_3CH_2 & & H \\ & C{=}C & \\ H & & CH_2CH_2CH_3 \end{array}$$

23.14 (a) $CH_3CH_2CH_2CH_3$ (b) $CH_3CHBrCHBrCH_3$ (c) $CH_3CH_2CH(OH)CH_3$

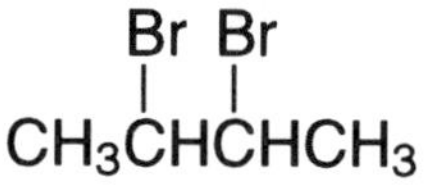

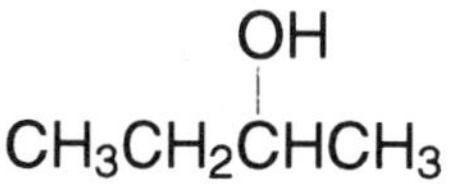

23.15

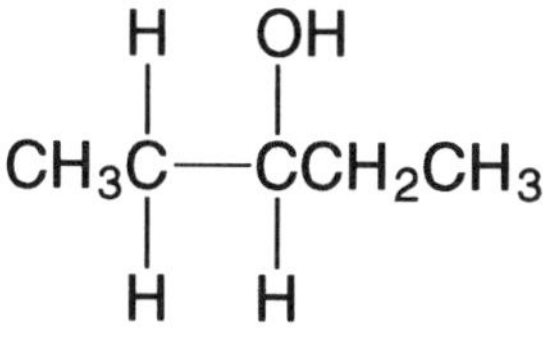

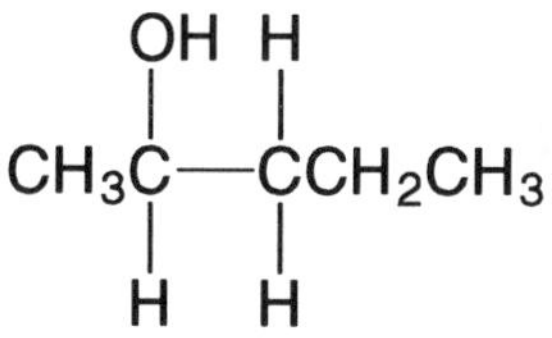

CH_3CH_2–$CH(OH)CH_2CH_3$ $CH_3CH(OH)$–$CH_2CH_2CH_3$

23.16

(a) o-dibromobenzene (Br, Br) (b) NO_2 / Cl (c) CH_2CH_3 / CH_2CH_3

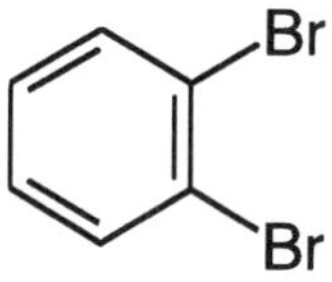

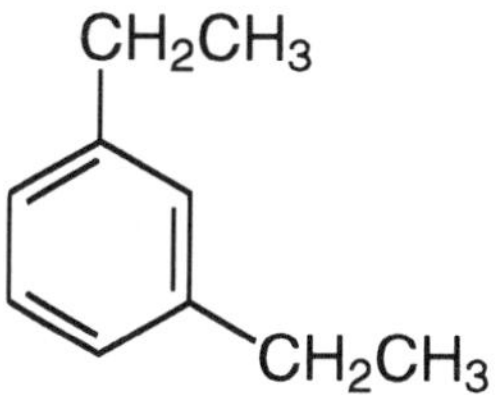

23.17

(a) CH_3, Br, CH_3 (b) CH_3, NO_2, CH_3 (c) CH_3, SO_3H, CH_3

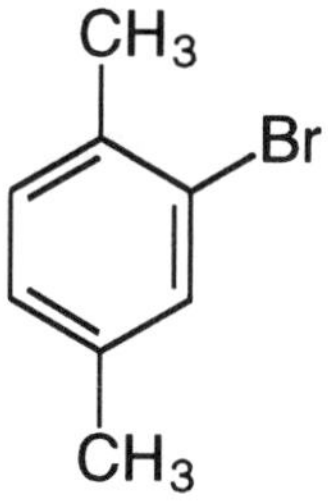

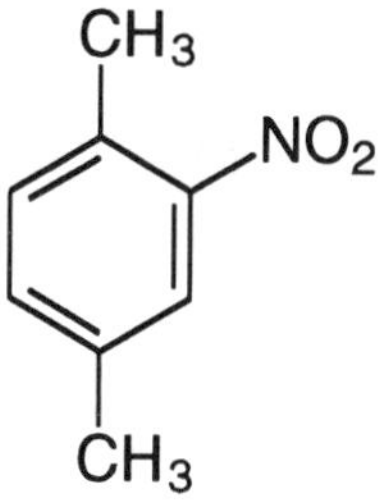

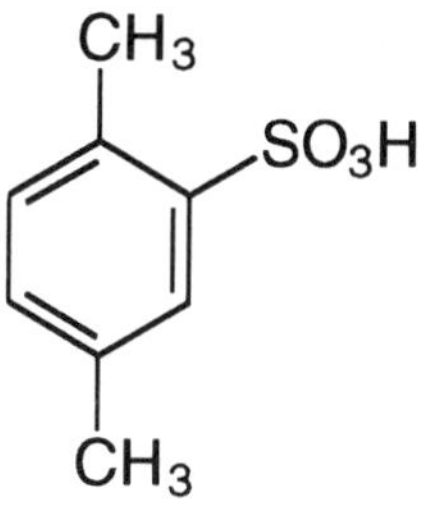

23.18

CH_3, Br CH_3, Br CH_3, Br

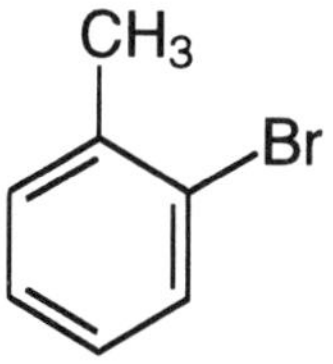

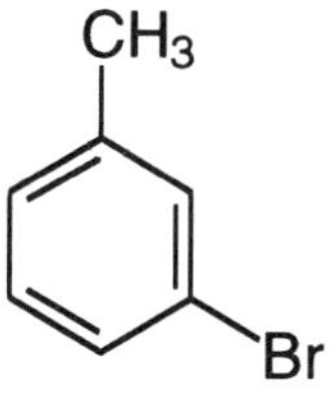

23.19

(a) 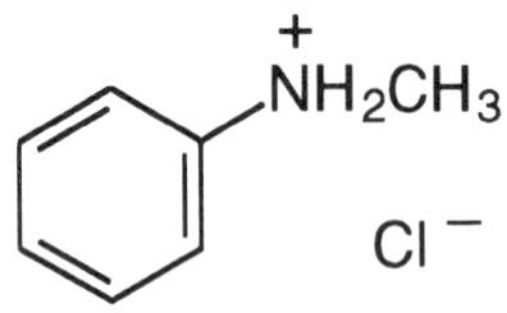 $C_6H_5\overset{+}{N}H_2CH_3$ Cl^- (b) $CH_3CH_2CH_2\overset{+}{N}H_3$ Cl^-

23.20 (a)

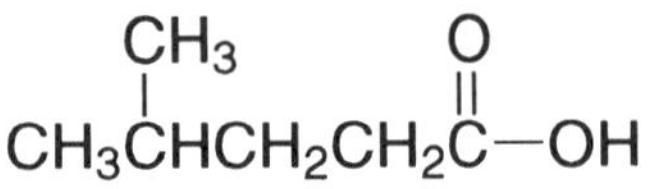

$$CH_3\underset{}{\overset{CH_3}{\overset{|}{C}}}HCH_2CH_2\overset{O}{\overset{||}{C}}-OH$$

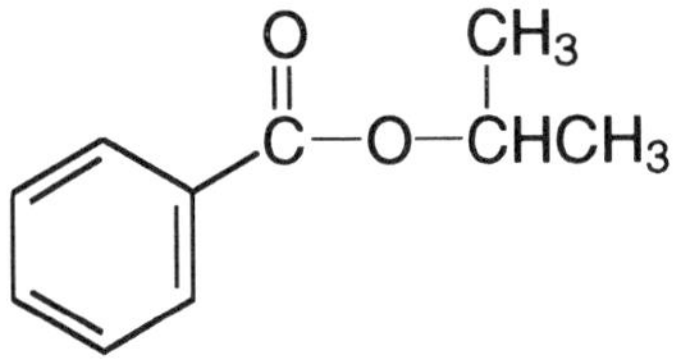

(b)

$$C_6H_5-\overset{O}{\overset{||}{C}}-O-\overset{CH_3}{\overset{|}{C}}HCH_3$$

(c)

$$CH_3CH_2\overset{O}{\overset{||}{C}}-NHCH_2CH_3$$

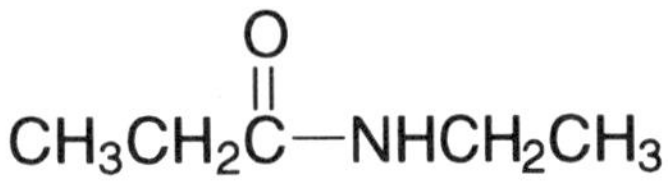

23.21

(a) 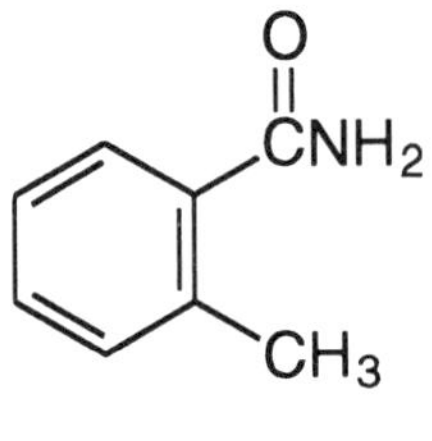 (benzene ring with $\overset{O}{\overset{||}{C}}NH_2$ and, on the adjacent carbon, CH_3)

(b)

$$CH_3\overset{Cl}{\overset{|}{C}}HCH_2\overset{O}{\overset{||}{C}}O\overset{CH_3}{\overset{|}{C}}HCH_2CH_3$$

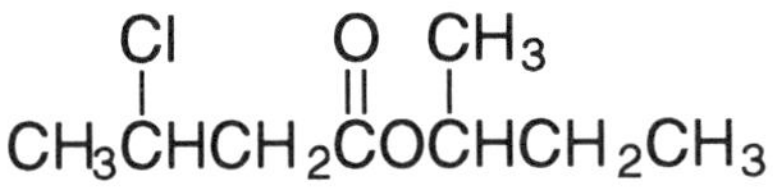

Functional Groups and Isomers

23.22 A functional group is a part of a larger molecule and is composed of an atom or group of atoms that has a characteristic chemical behavior. They are important because their chemistry controls the chemistry in molecules that contain them.

23.23

(a) CH_3CH_2OH (b) CH_3NH_2

(c) $CH_3CH_2\overset{\overset{\displaystyle O}{\|}}{C}OH$

(d) 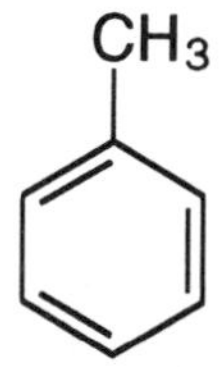 [benzene ring with CH_3 substituent]

23.24 (a) $CH_3CH_2\overset{\overset{\displaystyle O}{\|}}{C}CH_2CH_3$

(b) $CH_3CH_2CH_2\overset{\overset{\displaystyle O}{\|}}{C}{-}OCH_2CH_3$

(c) $NH_2CH_2\overset{\overset{\displaystyle O}{\|}}{C}{-}OH$

23.25

$CH_3CH_2CH_2OH$ $CH_3\overset{\overset{\displaystyle OH}{|}}{C}HCH_3$ $CH_3CH_2OCH_3$

23.26 (a) $CH_3CH_2NH_2$

(b) $CH_3CH{=}CHCH_3$

(c) $CH_3\overset{\overset{\displaystyle O}{\|}}{C}H$

(d) $H\overset{\overset{\displaystyle O}{\|}}{C}{-}OH$

23.27

(a) $CH_3CH_2CH_2CH_2OH$ $\quad$ $CH_3CH_2CH(OH)CH_3$ $\quad$ $CH_3CH(CH_3)CH_2OH$ $\quad$ $CH_3C(CH_3)_2OH$

(b) $CH_3CH_2CH_2NH_2$ $\quad$ $CH_3CH(NH_2)CH_3$ $\quad$ $CH_3CH_2NHCH_3$ $\quad$ $CH_3N(CH_3)CH_3$

(c) $CH_3CH_2CH_2C(=O)CH_3$ $\quad$ $CH_3CH_2C(=O)CH_2CH_3$ $\quad$ $CH_3CH(CH_3)C(=O)CH_3$

(d) $CH_3CH_2CH_2CH_2CH(=O)$ $\quad$ $CH_3CH(CH_3)CH_2CH(=O)$ $\quad$ $CH_3CH_2CH(CH_3)CH(=O)$ $\quad$ $CH_3C(CH_3)_2—CH(=O)$

23.28 (a) alkene and aldehyde (b) arene, alcohol, and ketone

Alkanes

23.29 In a straight-chain alkane, all the carbons are connected in a row. In a branched-chain alkane, there are branching connections of carbons along the carbon chain.

23.30 In forming alkanes, carbon uses sp^3 hybrid orbitals.

23.31 C_3H_9 contains one less H than needed for an alkane and one more H than needed for an alkene.

23.32 (a)

$\underline{C}H_3{=}CHCH_2CH_2OH$

Underlined carbon has five bonds.

(b)

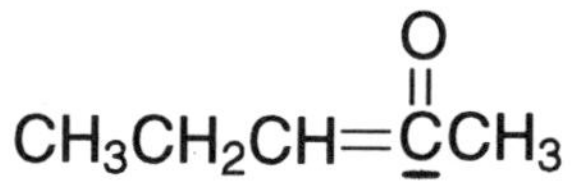

$$CH_3CH_2CH{=}\overset{\displaystyle O}{\overset{\|}{\underline{C}}}CH_3$$

Underlined carbon has five bonds.

(c)

$$CH_3CH_2C{\equiv}\underline{C}H_2CH_3$$

Underlined carbon has six bonds.

23.33 (a) 4-ethyl-3-methyloctane

(b) 4-isopropyl-2-methylheptane

(c) 2,2,6-trimethylheptane

(d) 4-ethyl-4-methyloctane

23.34 (a)

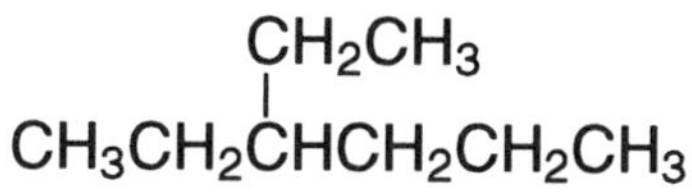

```
          CH2CH3
          |
CH3CH2CHCH2CH2CH3
```

(b)

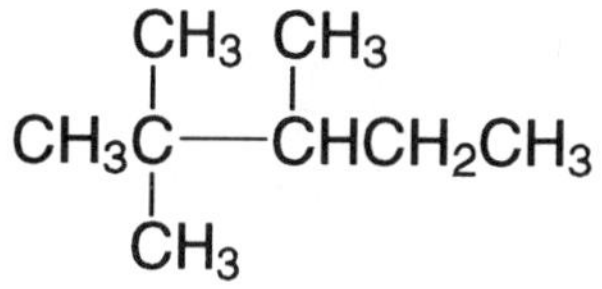

```
     CH3 CH3
     |   |
CH3C—CHCH2CH3
     |
     CH3
```

(c)

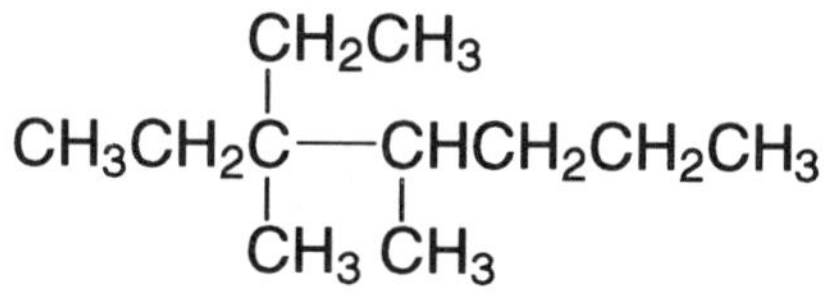

```
          CH2CH3
          |
CH3CH2C—CHCH2CH2CH3
          |     |
          CH3 CH3
```

(d)

```
                        CH3
                        |
      CH3               CHCH3
      |                 |
CH3CHCH2CH2CHCH2CH2CH3
```

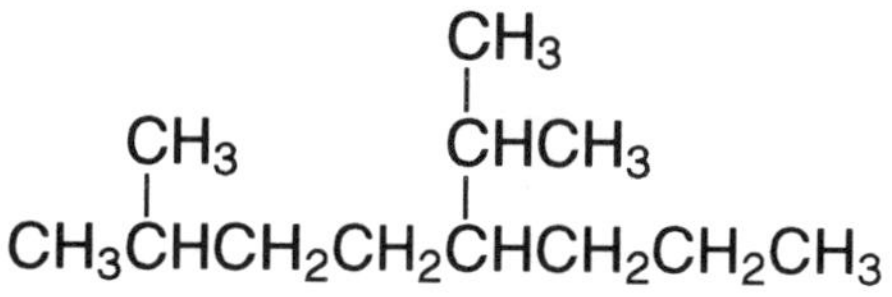

23.35 2,2,4-trimethylpentane

23.36 The structures are shown in Problem 23.2.

hexane, 2-methylpentane, 3-methylpentane, 2,2-dimethylbutane, and 2,3-dimethylbutane

23.37

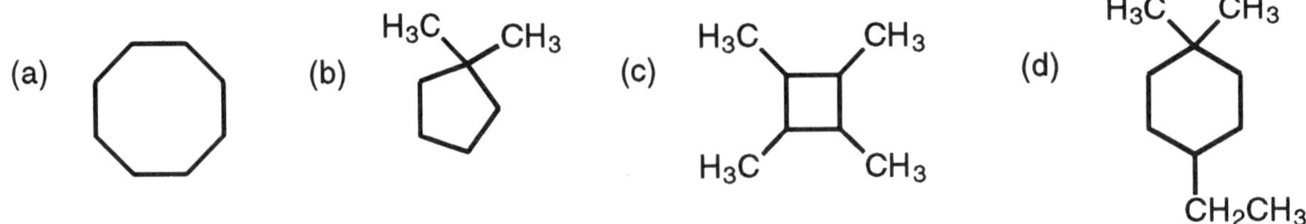

23.38 (a) 1,1-dimethylcyclopentane (b) 1-isopropyl-2-methylcyclohexane

(c) 1,2,4-trimethylcyclooctane

23.39 (a) 3,3-dimethylhexane (b) 3,5-dimethylheptane

(c) 1,3-dimethylcycloheptane

23.40

$CH_3CH_2CH_2CH_2CH_2CH_2CH_3$ heptane

$CH_3CH(CH_3)CH_2CH_2CH_2CH_3$ 2-methylhexane

$CH_3CH_2CH(CH_3)CH_2CH_2CH_3$ 3-methylhexane

$CH_3C(CH_3)_2CH_2CH_2CH_3$ 2,2-dimethylpentane

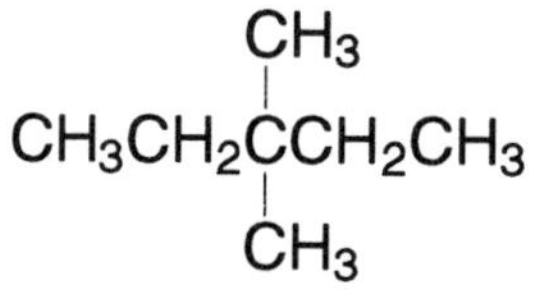

$$CH_3CH_2\overset{\overset{\displaystyle CH_3}{|}}{\underset{\underset{\displaystyle CH_3}{|}}{C}}CH_2CH_3$$

3,3-dimethylpentane

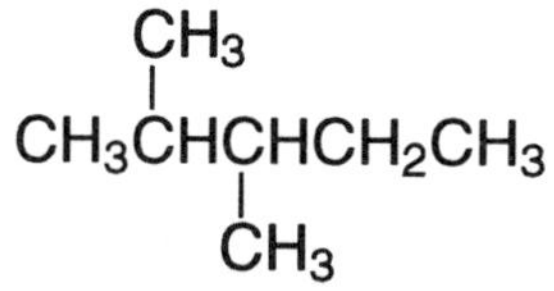

$$CH_3\overset{\overset{\displaystyle CH_3}{|}}{C}H\underset{\underset{\displaystyle CH_3}{|}}{C}HCH_2CH_3$$

2,3-dimethylpentane

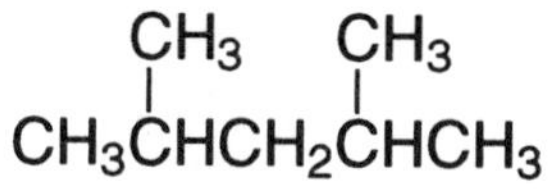

$$CH_3\overset{\overset{\displaystyle CH_3}{|}}{C}HCH_2\overset{\overset{\displaystyle CH_3}{|}}{C}HCH_3$$

2,4-dimethylpentane

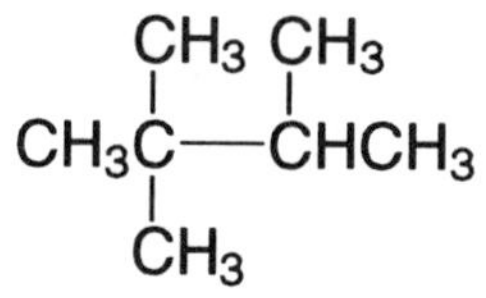

$$CH_3\overset{\overset{\displaystyle CH_3}{|}}{\underset{\underset{\displaystyle CH_3}{|}}{C}}—\overset{\overset{\displaystyle CH_3}{|}}{C}HCH_3$$

2,2,3-trimethylbutane

$$CH_3CH_2\overset{\overset{\displaystyle CH_2CH_3}{|}}{C}HCH_2CH_3$$

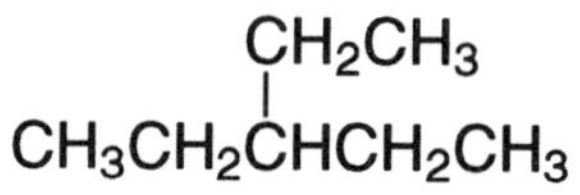

3-ethylpentane

23.41 $2\ C_4H_{10}(g) + 13\ O_2(g) \rightarrow 8\ CO_2(g) + 10\ H_2O(g)$

23.42 (a)

$ClCH_2CH_2CH_2CH_2CH_2CH_3$

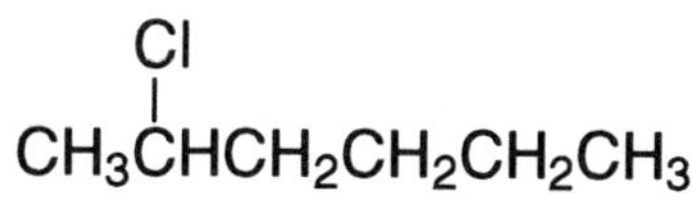

$$CH_3\overset{\overset{\displaystyle Cl}{|}}{C}HCH_2CH_2CH_2CH_3$$

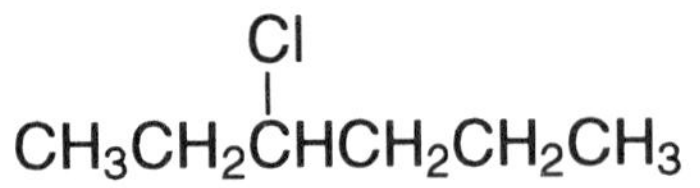

$$\begin{array}{c} Cl \\ | \\ CH_3CH_2CHCH_2CH_2CH_3 \end{array}$$

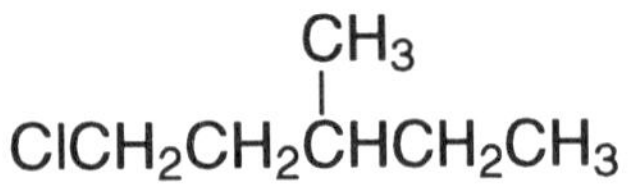

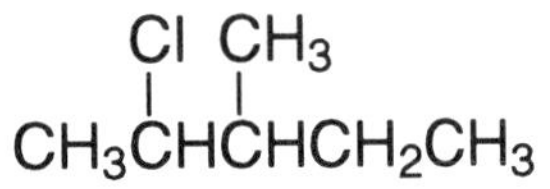

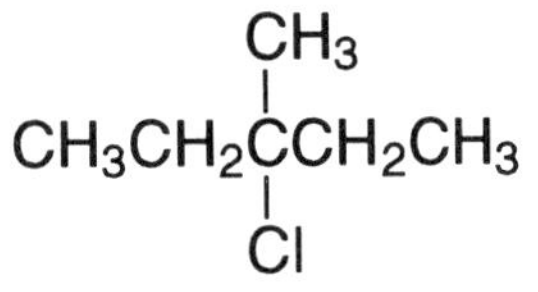

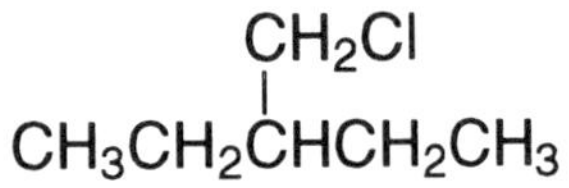

(b)

$$\begin{array}{c} CH_3 \\ | \\ ClCH_2CH_2CHCH_2CH_3 \end{array}$$

$$\begin{array}{c} Cl\;CH_3 \\ |\;\;\;| \\ CH_3CHCHCH_2CH_3 \end{array}$$

$$\begin{array}{c} CH_3 \\ | \\ CH_3CH_2CCH_2CH_3 \\ | \\ Cl \end{array}$$

$$\begin{array}{c} CH_2Cl \\ | \\ CH_3CH_2CHCH_2CH_3 \end{array}$$

(c)

CH_3 / Cl (1-chloro-2-methyl ring, adjacent positions) CH_3 / Cl (1,3 positions) CH_3 / Cl (1,4 positions) CH_2Cl Cl CH_3 (same carbon)

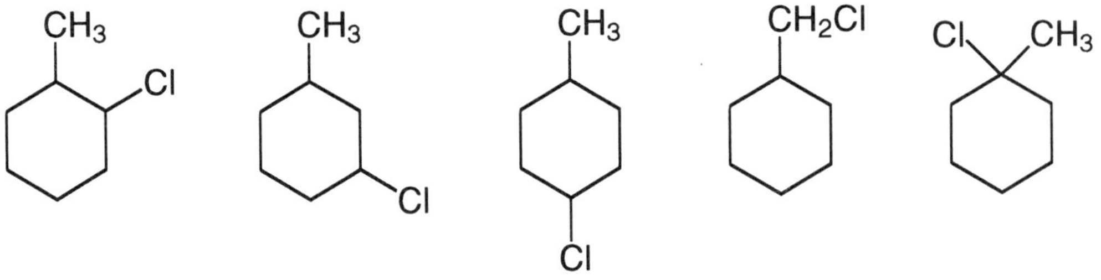

23.43 Reaction (a) is likely to have a higher yield because there is only one possible monochlorinated substitution product. Reaction (b) has four possible monochlorinated substitution products, which would result in a lower yield of the one product shown.

Alkenes, Alkynes, and Aromatic Compounds

23.44 (a) sp^2 (b) sp (c) sp^2

23.45 Alkenes, alkynes, and aromatic compounds are said to be unsaturated because they do not contain as many hydrogens as their alkane analogs.

23.46 Today the term "aromatic" refers to the class of compounds containing a six-membered ring with three double bonds, not to the fragrance of a compound.

23.47 An addition reaction is the reaction of an XY molecule with an alkene or alkyne.

$$>C{=}C< \quad + \quad XY \longrightarrow -\overset{X}{\underset{|}{\overset{|}{C}}}-\overset{Y}{\underset{|}{\overset{|}{C}}}-$$

23.48 (a) $CH_3CH{=}CHCH_2CH_3$ (b) $HC{\equiv}CCH_2CH_3$

(c) benzene ring with CH_3 and CH_3 (meta)

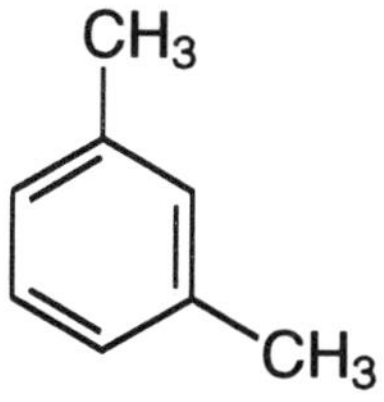

23.49

$CH_2{=}C{=}CHCH_2CH_3$ $CH_2{=}CHCH{=}CH_2$ $CH_2{=}CHCH_2CH{=}CH_2$

$CH_3CH{=}C{=}CHCH_3$ $CH_2{=}\overset{CH_3}{\overset{|}{C}}CH{=}CH_2$ $CH_2{=}C{=}\overset{CH_3}{\overset{|}{C}}CH_3$

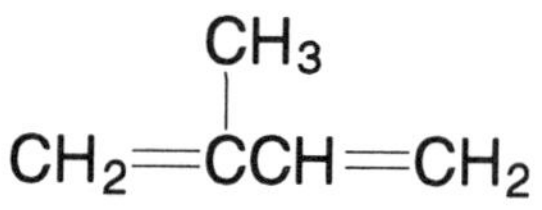

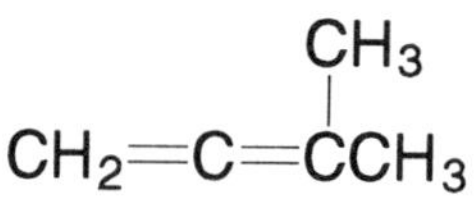

23.50 (a) 4-methyl-2-pentene (b) 3-methyl-1-pentene

(c) 1,2-dichlorobenzene, or o-dichlorobenzene

(d) 2-methyl-2-butene (e) 7-methyl-3-octyne

23.51

(a)

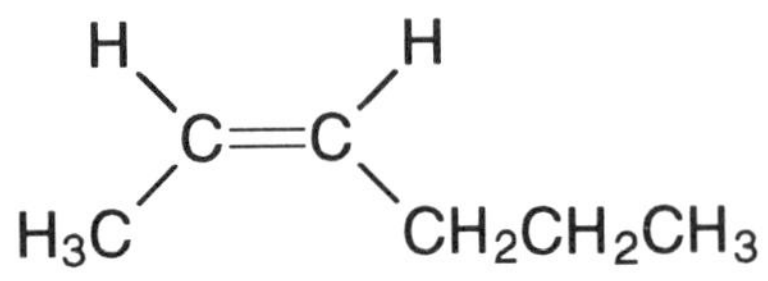

```
  H           H
   \         /
    C ═ C
   /         \
H3C           CH2CH2CH3
```

(b)

```
       CH3
        |
CH3CHCH═CHCH2CH3
```

(c)

```
       CH3
        |
CH2═CCH═CH2
```

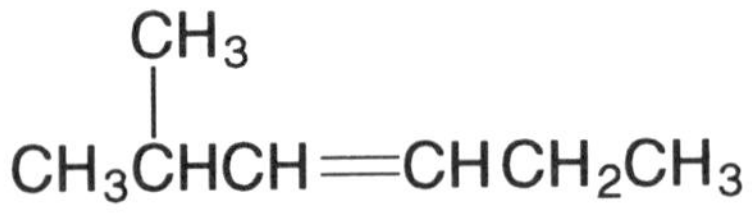

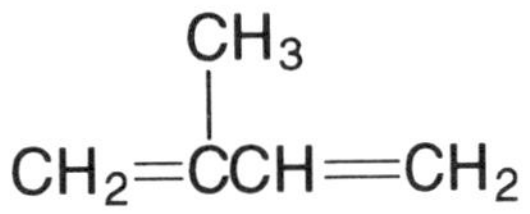

23.52

$CH_2{=}CHCH_2CH_2CH_3$ 1-pentene

$CH_3CH{=}CHCH_2CH_3$ 2-pentene

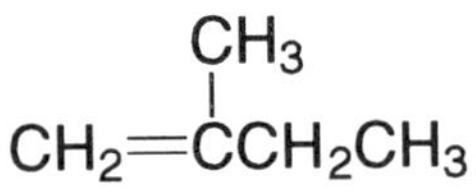

```
        CH3
         |
CH2═CCH2CH3
```

2-methyl-1-butene

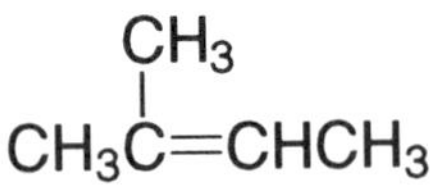

```
     CH3
      |
CH3C═CHCH3
```

2-methyl-2-butene

```
            CH3
             |
CH2═CHCHCH3
```

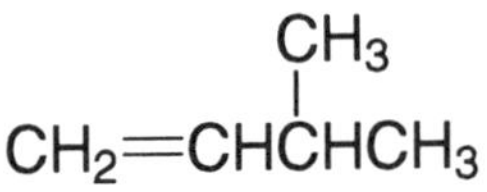

3-methyl-1-butene

23.53 2-pentene

23.54

$HC{\equiv}CCH_2CH_2CH_3$ 1-pentyne

$CH_3C{\equiv}CCH_2CH_3$ 2-pentyne

$HC{\equiv}CCH(CH_3)CH_3$ 3-methyl-1-butyne

23.55

(a) $CH_3CH(CH_3)CH_2CH{=}CHCH_3$ 5-methyl-2-hexene

(b) $CH_3CH_2C{\equiv}CC(CH_3)_2CH_3$ 2,2-dimethyl-3-hexyne

(c) $CH_2{=}C(CH_3)CH_2CH_2CH_2CH_3$ 2-methyl-1-hexene

(d) 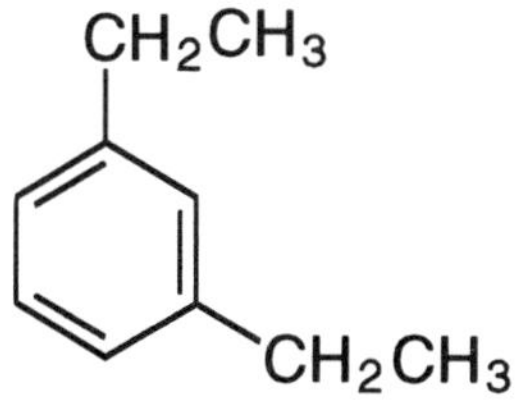 $C_6H_4(CH_2CH_3)_2$ (benzene ring with CH_2CH_3 groups at positions 1 and 3) 1,3-diethylbenzene

23.56 Cis-trans isomers are possible for substituted alkenes because of the lack of rotation about the carbon-carbon double bond. Alkanes and alkynes cannot form cis-trans isomers because alkanes have free rotation about carbon-carbon single bonds and alkynes are linear about the carbon-carbon triple bond.

23.57 (a) $CH_2{=}CHCH_2CH_2CH_2CH_3$

This compound cannot form cis-trans isomers.

(b) $CH_3CH{=}CHCH_2CH_2CH_3$

This compound can form cis-trans isomers because of the different groups on each double bond C.

(c) $CH_3CH_2CH{=}CHCH_2CH_3$

This compound can form cis-trans isomers because of the different groups on each double bond C.

23.58 (a)

```
  CH3
  |
CH3CHCH=CHCH3
```

This compound can form cis-trans isomers because of the different groups on each double bond C.

(b)

```
        CH=CH2
        |
CH3CH2CHCH3
```

This compound cannot form cis-trans isomers.

(c)

```
          Cl
          |
CH3CH=CHCHCH2CH3
```

This compound can form cis-trans isomers because of the different groups on each double bond C.

23.59

(a)

```
      H           H
       \         /
        C=====C
       /         \
CH3CH2            CH2CH2CH3
```

(b)

```
   H         H
    \       /
     C=====C
    /       \
 H3C         CHCH3
             |
             CH3
```

(c)

```
      CH3
      |
   CH3CH         H
        \       /
         C=====C
        /       \
       H         CHCH3
                 |
                 CH3
```

23.60 Small-ring cycloalkenes don't exist as cis-trans isomers because the trans isomer could not close the carbon-carbon chain back on itself to form a ring.

23.61 (a)

```
    CH3                                  CH3 CH3
    |                       Pd           |   |
CH3C=CCH3      +     H2    ---->    CH3C---CCH3
      |                                  |   |
      CH3                                H   H
```

(b)

```
    CH3                                  CH3 CH3
    |                                    |   |
CH3C=CCH3      +     Br2   ---->    CH3C---CCH3
      |                                  |   |
      CH3                                Br  Br
```

(c)

```
    CH3                                  CH3 CH3
    |                      H2SO4         |   |
CH3C=CCH3      +     H2O   ---->    CH3C---CCH3
      |                                  |   |
      CH3                                OH  H
```

23.62 (a)

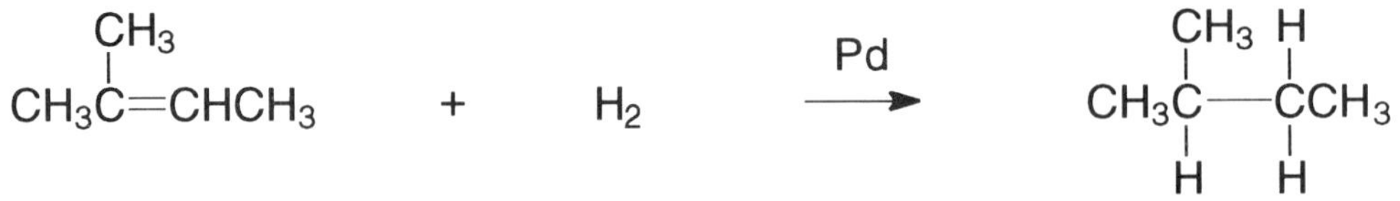

$$CH_3C(CH_3){=}CHCH_3 + H_2 \xrightarrow{Pd} CH_3CH(CH_3){-}CH_2CH_3$$

(b)

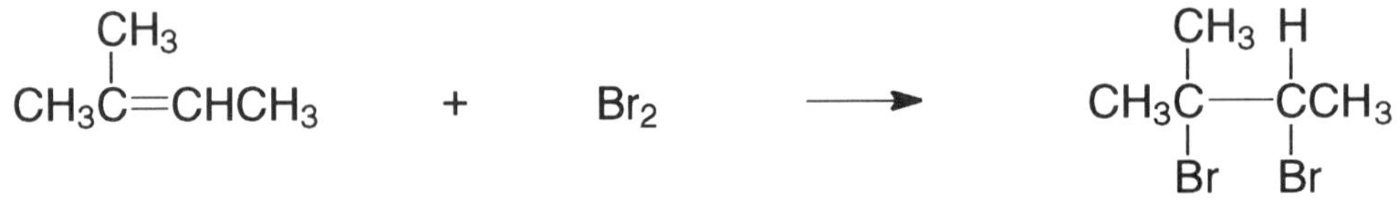

$$CH_3C(CH_3){=}CHCH_3 + Br_2 \longrightarrow CH_3CBr(CH_3){-}CHBrCH_3$$

(c)

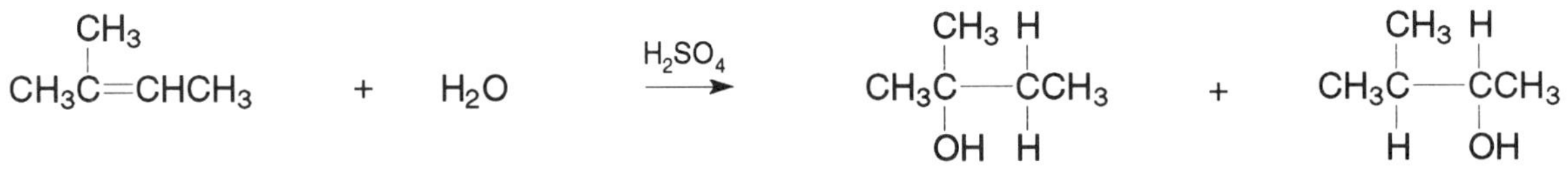

$$CH_3C(CH_3){=}CHCH_3 + H_2O \xrightarrow{H_2SO_4} CH_3C(OH)(CH_3){-}CH_2CH_3 + CH_3CH(CH_3){-}CH(OH)CH_3$$

23.63 (a)

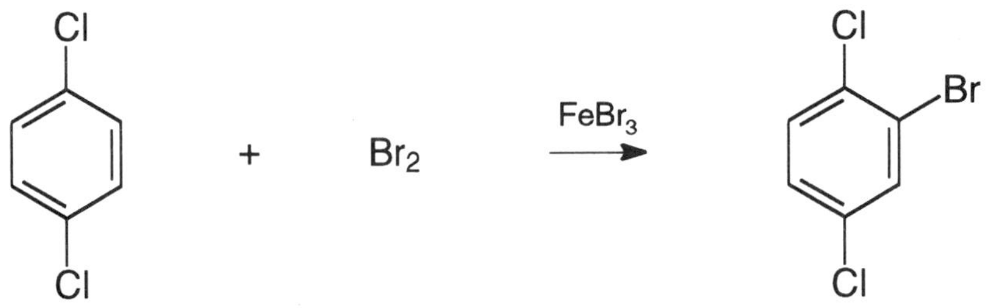

$$C_6H_4Cl_2 \text{ (1,4-dichlorobenzene)} + Br_2 \xrightarrow{FeBr_3} C_6H_3Cl_2Br \text{ (2-bromo-1,4-dichlorobenzene)}$$

(b)

$$C_6H_4Cl_2 \text{ (1,4-dichlorobenzene)} + HNO_3 \xrightarrow{H_2SO_4} C_6H_3Cl_2NO_2 \text{ (1,4-dichloro-2-nitrobenzene)}$$

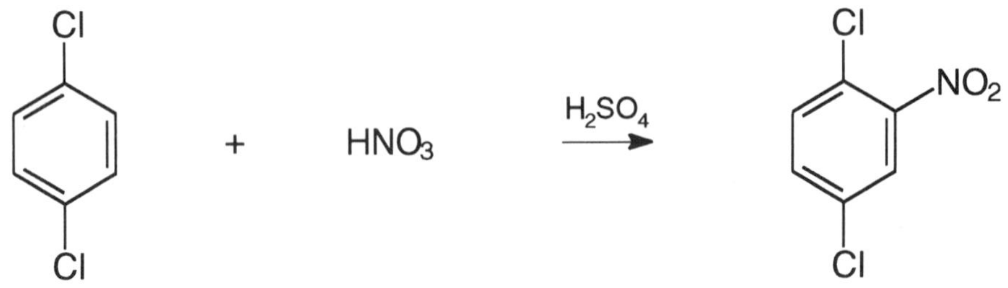

(c)

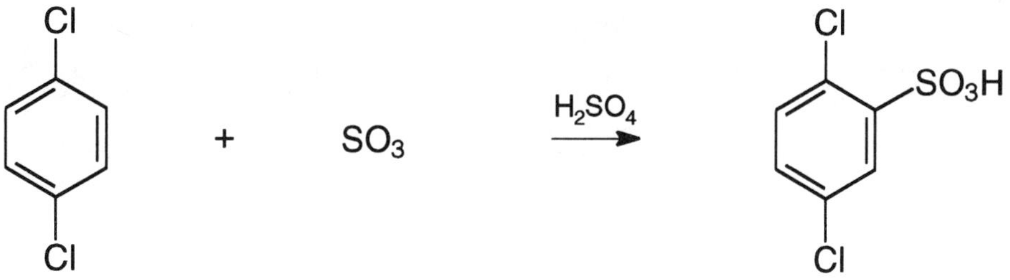

1,4-dichlorobenzene + SO_3 $\xrightarrow{H_2SO_4}$ 2,5-dichlorobenzenesulfonic acid (Cl, SO_3H, Cl)

(d)

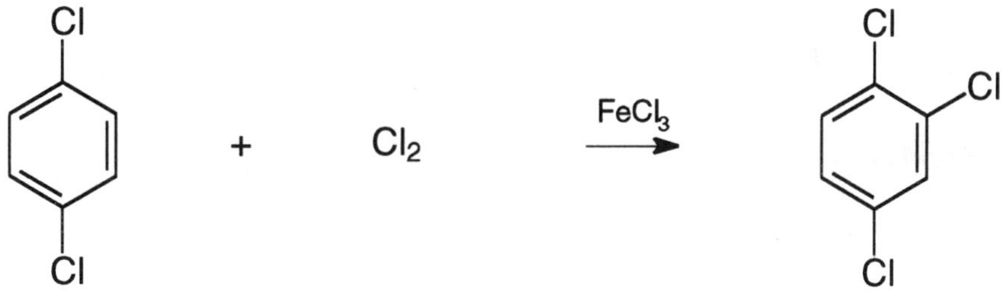

1,4-dichlorobenzene + Cl_2 $\xrightarrow{FeCl_3}$ 1,2,4-trichlorobenzene (Cl, Cl, Cl)

23.64

cyclohexane

23.65

CH_3, O_2N, NO_2, NO_2

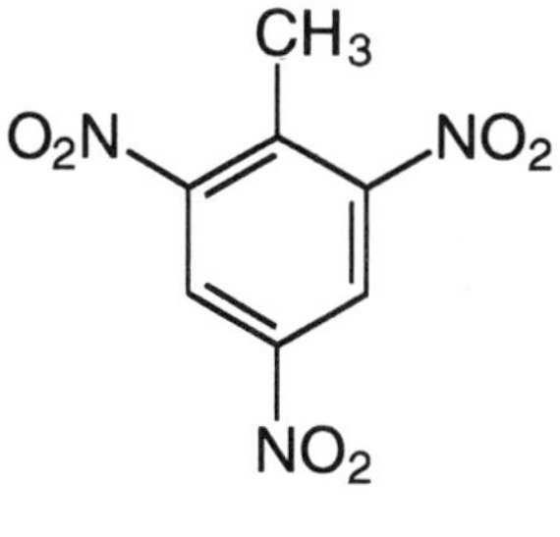

Alcohols, Amines, and Carbonyl Compounds

23.66 (a)

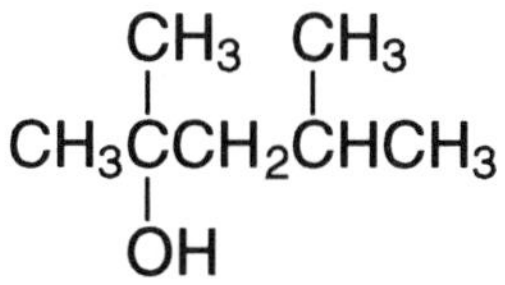

$$\mathrm{CH_3\overset{\displaystyle CH_3}{\underset{\displaystyle OH}{C}}CH_2\overset{\displaystyle CH_3}{C}HCH_3}$$

(b)

OH, CH_3, CH_3

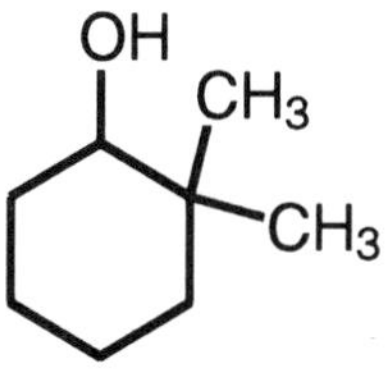

(c)

$$\begin{array}{c} \quad\quad\quad\quad\quad\quad\quad\quad\quad CH_2CH_3 \\ \quad\quad\quad\quad\quad\quad\quad\quad\quad | \\ HOCH_2CH_2CH_2CH_2CCH_2CH_3 \\ \quad\quad\quad\quad\quad\quad\quad\quad\quad | \\ \quad\quad\quad\quad\quad\quad\quad\quad\quad CH_2CH_3 \end{array}$$

(d)

$$\begin{array}{c} \quad CH_2CH_3 \\ \quad | \\ CH_3CH_2CCH_2CH_2CH_3 \\ | \\ OH \end{array}$$

23.67 (a) $CH_3CH_2CH_2NH_2$ (b) $(CH_3CH_2)_2NH$ (c) $CH_3CH_2CH_2NHCH_3$

23.68 (a) $CH_3CH_2CH_2NH_3^+\ Cl^-$

(b) $(CH_3CH_2)_2NH_2^+\ Cl^-$

(c) $CH_3CH_2CH_2\overset{+}{N}H_2CH_3\ Cl^-$

23.69 Quinine, a base will dissolve in aqueous acid, but menthol is insoluble.

23.70 Pentanoic acid will react with aqueous $NaHCO_3$ to yield CO_2, but methyl butanoate will not.

23.71 An aldehyde has a terminal carbonyl group. A ketone has the carbonyl group located between two carbon atoms.

23.72 In aldehydes and ketones, the carbonyl-group carbon is bonded to atoms (H and C) that don't attract electrons strongly. In carboxylic acids, esters, and amides, the carbonyl-group carbon is bonded to an atom (O or N) that does attract electrons strongly.

23.73 Carboxylic acids, esters, and amides undergo carbonyl-group substitution reactions, in which a group -Y substitutes for the –OH, –OC, or –N group of the starting material.

23.74 (a) ketone (b) aldehyde (c) ketone (d) amide (e) ester

23.75 $C_6H_5COOH(aq) + H_2O(l) \rightleftharpoons H_3O^+(aq) + C_6H_5COO^-(aq)$

	C_6H_5COOH		H_3O^+	$C_6H_5COO^-$
initial (M)	1.0		~0	0
change (M)	–x		+x	+x
equil (M)	1.0 – x		x	x

$$K_a = \frac{[H_3O^+][C_6H_5COO^-]}{[C_6H_5COOH]} = 6.5 \times 10^{-5} = \frac{x^2}{1.0 - x} \approx \frac{x^2}{1.0}$$

$x = [H_3O^+] = [C_6H_5COOH]_{diss} = 0.0081\ M$

$$\%\ \text{dissociation} = \frac{[C_6H_5COOH]_{diss}}{[C_6H_5COOH]_{initial}} \times 100\% = \frac{0.0081\ M}{1.0\ M} \times 100\% = 0.81\%$$

23.76

benzoic acid (COOH on benzene ring), $C_7H_6O_2$, 122.12 amu + NaOH 40.00 amu → COONa on benzene ring + H_2O

$$\text{mass NaOH} = 5.0\text{ g } C_7H_6O_2 \times \frac{1\text{ mol } C_7H_6O_2}{122.12\text{ g}} \times \frac{1\text{ mol NaOH}}{1\text{ mol } C_7H_6O_2} \times \frac{40.00\text{ g}}{1\text{ mol NaOH}}$$

mass NaOH = 1.6 g NaOH

23.77

(a) $CH_3\overset{O}{\overset{\|}{C}}OH$ (b) $CH_3\overset{O}{\overset{\|}{C}}OH$ (c) $CH_3\overset{O}{\overset{\|}{C}}O^-\ Na^+$

23.78 (a)

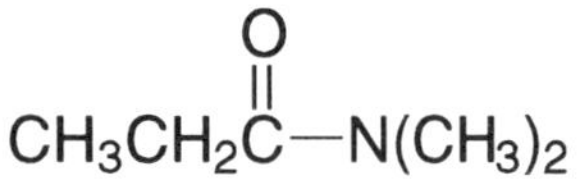

$$CH_3CH_2\overset{\overset{\displaystyle O}{\|}}{C}-N(CH_3)_2$$ N,N-dimethylpropanamide

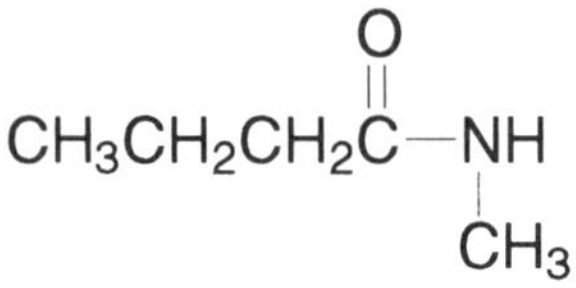

$$CH_3CH_2CH_2\overset{\overset{\displaystyle O}{\|}}{C}-\underset{\underset{\displaystyle CH_3}{|}}{NH}$$ N-methylbutanamide

$$CH_3CH_2CH_2CH_2\overset{\overset{\displaystyle O}{\|}}{C}-NH_2$$ pentanamide

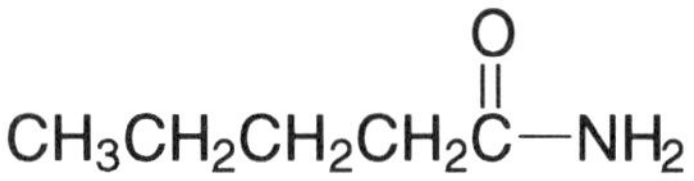

(b)

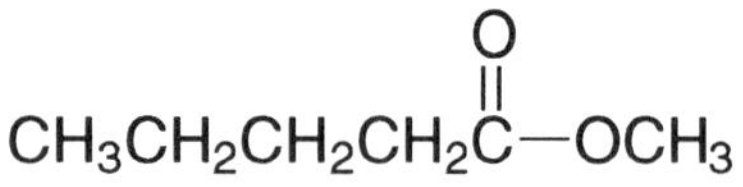

$$CH_3CH_2CH_2CH_2\overset{\overset{\displaystyle O}{\|}}{C}-OCH_3$$ methyl pentanoate

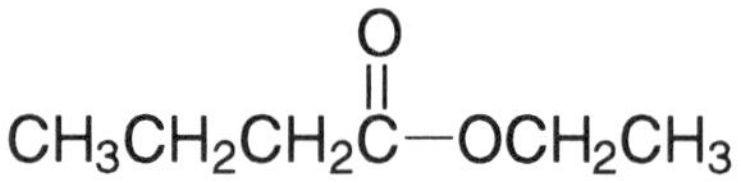

$$CH_3CH_2CH_2\overset{\overset{\displaystyle O}{\|}}{C}-OCH_2CH_3$$ ethyl butanoate

$$CH_3CH_2\overset{\overset{\displaystyle O}{\|}}{C}-OCH_2CH_2CH_3$$ propyl propanoate

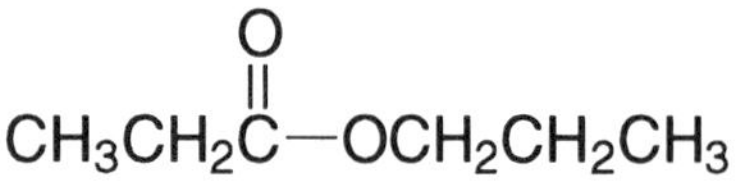

23.79 (a) methyl 4-methylpentanoate (b) 4,4-dimethylpentanoic acid

(c) 2-methylpentanamide

23.80 (a)

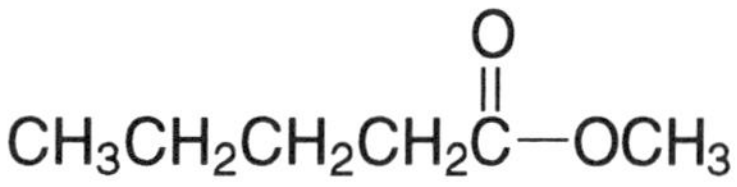

$$CH_3CH_2CH_2CH_2\overset{\overset{\displaystyle O}{\|}}{C}-OCH_3$$

(b)

$$CH_3CH_2\underset{\underset{\displaystyle CH_3}{|}}{C}H\overset{\overset{\displaystyle O}{\|}}{C}-OCH(CH_3)_2$$

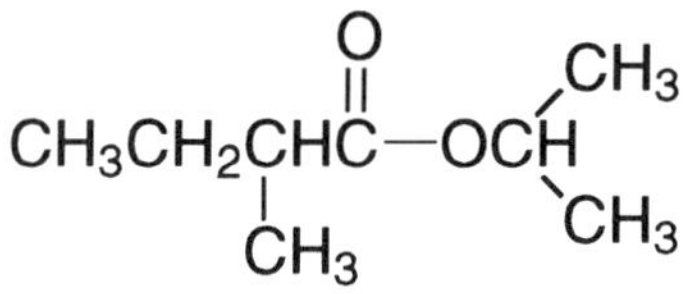

(c)

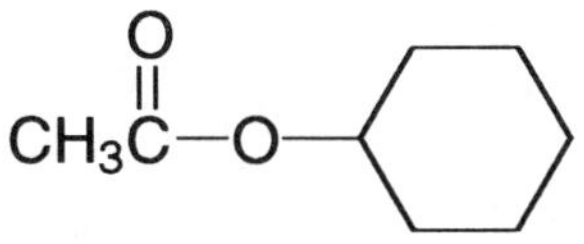

$CH_3C(=O)–O–C_6H_{11}$ (cyclohexyl)

23.81 (a)

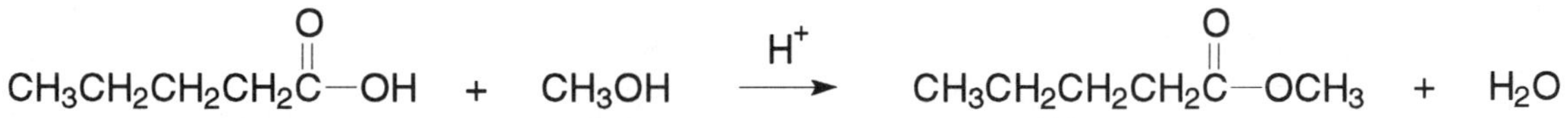

$$CH_3CH_2CH_2CH_2C(=O)–OH + CH_3OH \xrightarrow{H^+} CH_3CH_2CH_2CH_2C(=O)–OCH_3 + H_2O$$

(b)

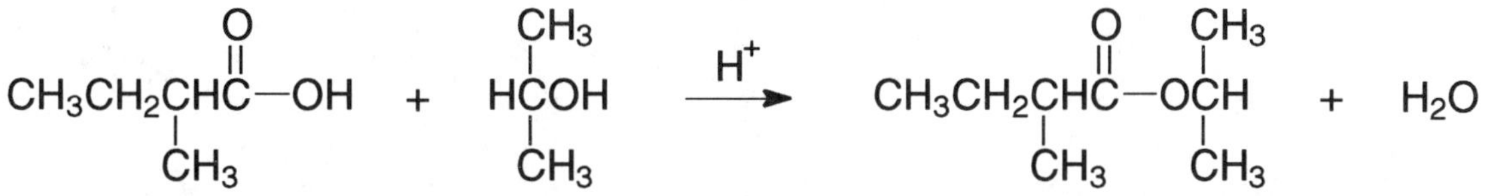

$$CH_3CH_2CH(CH_3)C(=O)–OH + HC(CH_3)_2OH \xrightarrow{H^+} CH_3CH_2CH(CH_3)C(=O)–OCH(CH_3)_2 + H_2O$$

(c)

$$CH_3C(=O)–OH + HO–C_6H_{11} \xrightarrow{H^+} CH_3C(=O)–O–C_6H_{11} + H_2O$$

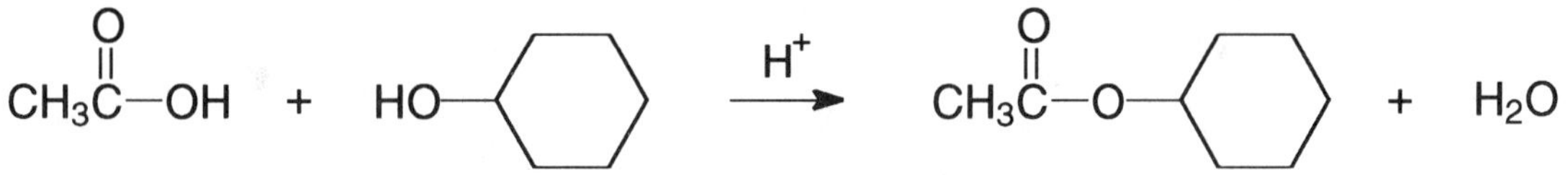

23.82 (a)

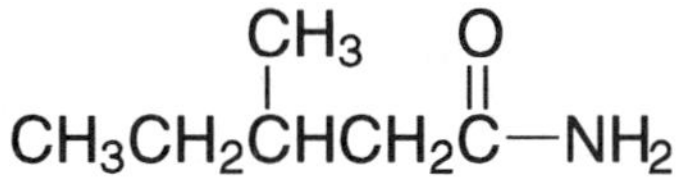

$CH_3CH_2CH(CH_3)CH_2C(=O)–NH_2$

(b)

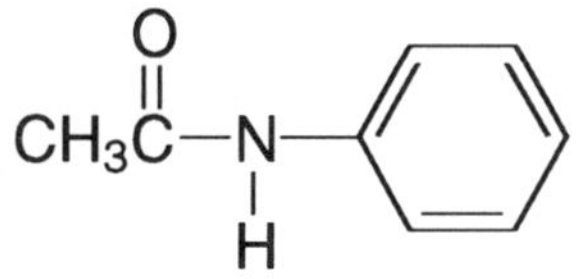

$CH_3C(=O)–N(H)–C_6H_5$

(c)

$C_6H_5–C(=O)–N(CH_2CH_3)(CH_3)$

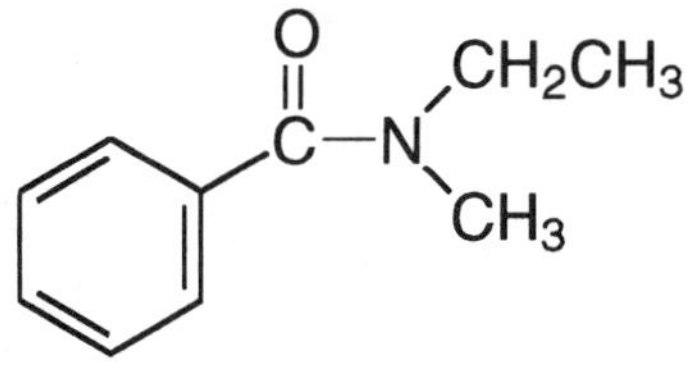

23.83 (a)

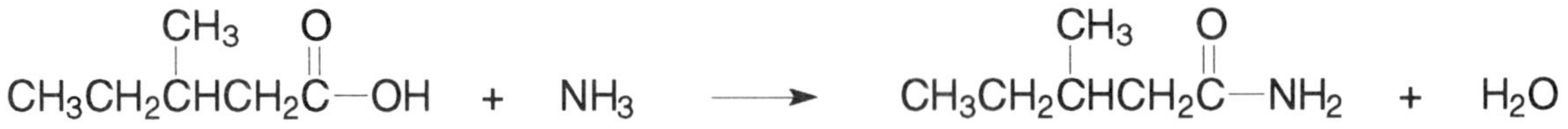

$$CH_3CH_2CH(CH_3)CH_2C(=O)\text{—}OH + NH_3 \longrightarrow CH_3CH_2CH(CH_3)CH_2C(=O)\text{—}NH_2 + H_2O$$

(b)

$$CH_3C(=O)\text{—}OH + NH_2\text{—}C_6H_5 \longrightarrow CH_3C(=O)\text{—}N(H)\text{—}C_6H_5 + H_2O$$

(c)

$$C_6H_5\text{—}C(=O)\text{—}OH + HN(CH_3)(CH_2CH_3) \longrightarrow C_6H_5\text{—}C(=O)\text{—}N(CH_3)(CH_2CH_3) + H_2O$$

23.84 amine, arene, and ester

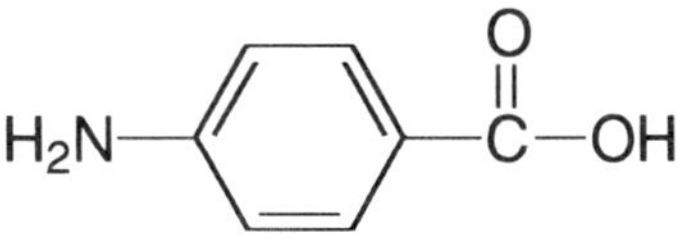

$$H_2N\text{—}C_6H_4\text{—}C(=O)\text{—}OH$$

carboxylic acid

$$HOCH_2CH_2N(CH_2CH_3)CH_2CH_3$$

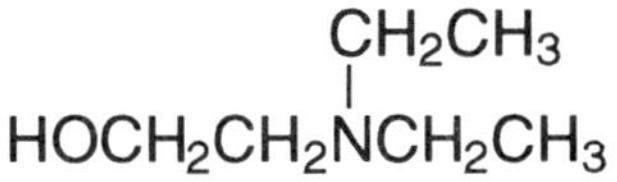

alcohol

23.85

$$CH_3(CH_2)_{14}C(=O)\text{—}OK$$

$$CH_3(CH_2)_7CH{=}CH(CH_2)_7C(=O)\text{—}OK$$

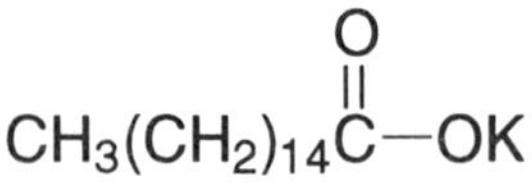

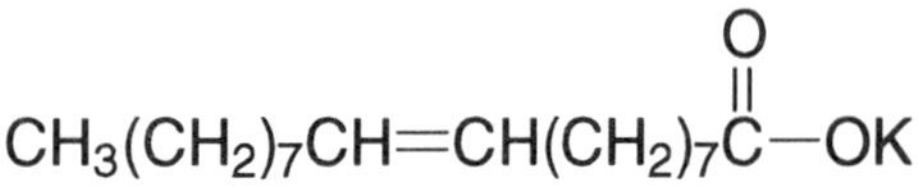

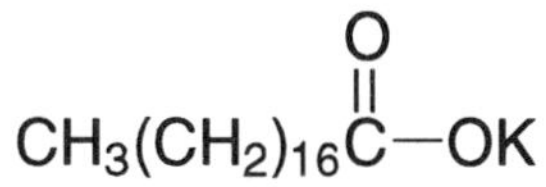

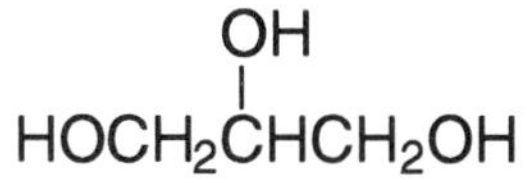

$$CH_3(CH_2)_{16}\overset{\overset{\large O}{||}}{C}{-}OK$$

$$HOCH_2\overset{\overset{\large OH}{|}}{C}HCH_2OH$$

23.86 Polymers are large molecules formed by the repetitive bonding together of many smaller molecules, called monomers.

23.87

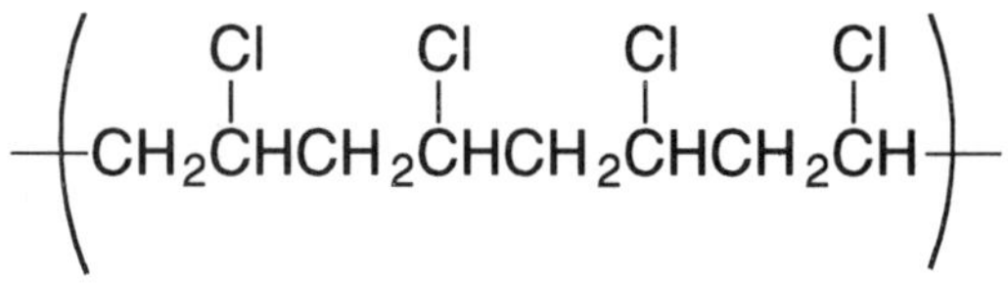

$$\left(CH_2\overset{\overset{\large Cl}{|}}{C}HCH_2\overset{\overset{\large Cl}{|}}{C}HCH_2\overset{\overset{\large Cl}{|}}{C}HCH_2\overset{\overset{\large Cl}{|}}{C}H\right)$$

23.88 (a)

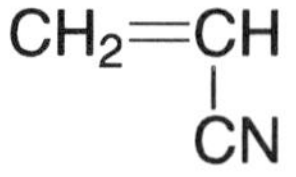

$$CH_2{=}\underset{\underset{\large CN}{|}}{C}H$$

(b)

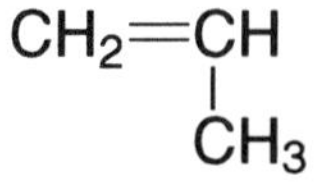

$$CH_2{=}\underset{\underset{\large CH_3}{|}}{C}H$$

(c)

$$CH_2{=}CCl_2$$

23.89 (a)

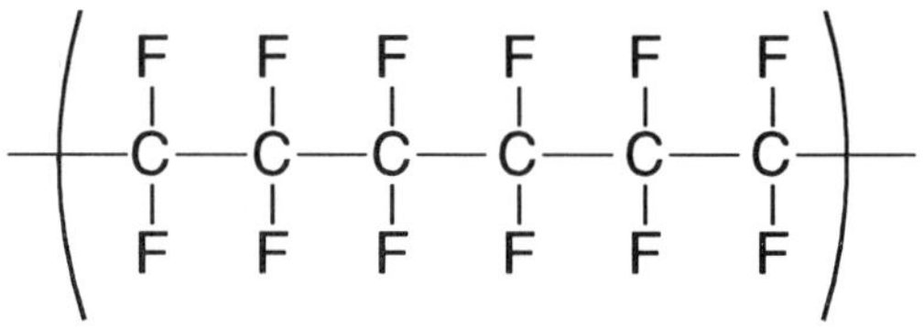

$$\left(\underset{\underset{\large F}{|}}{\overset{\overset{\large F}{|}}{C}}-\underset{\underset{\large F}{|}}{\overset{\overset{\large F}{|}}{C}}-\underset{\underset{\large F}{|}}{\overset{\overset{\large F}{|}}{C}}-\underset{\underset{\large F}{|}}{\overset{\overset{\large F}{|}}{C}}-\underset{\underset{\large F}{|}}{\overset{\overset{\large F}{|}}{C}}-\underset{\underset{\large F}{|}}{\overset{\overset{\large F}{|}}{C}}\right)$$

(b)

$$\left(CH_2-\overset{\overset{\overset{\large O}{||}}{COCH_3}}{\overset{|}{C}H}-CH_2-\overset{\overset{\overset{\large O}{||}}{COCH_3}}{\overset{|}{C}H}-CH_2-\overset{\overset{\overset{\large O}{||}}{COCH_3}}{\overset{|}{C}H}\right)$$

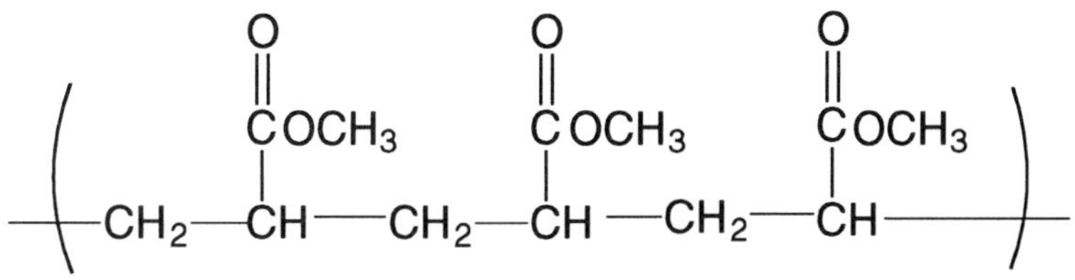

23.90

$$\left(\overset{O}{\overset{\|}{C}}-C_6H_4-\overset{O}{\overset{\|}{C}}-\overset{H}{\overset{|}{N}}-C_6H_4-\overset{H}{\overset{|}{N}}-\overset{O}{\overset{\|}{C}}-C_6H_4-\overset{O}{\overset{\|}{C}}-\overset{H}{\overset{|}{N}}-C_6H_4-\overset{H}{\overset{|}{N}} \right)$$

repeating unit

23.91

$$\left(OCH_2CH_2O\overset{O}{\overset{\|}{C}}CH_2CH_2\overset{O}{\overset{\|}{C}}OCH_2CH_2O\overset{O}{\overset{\|}{C}}CH_2CH_2\overset{O}{\overset{\|}{C}} \right)$$

repeating unit

General Problems

23.92 (a)

$$CH_3\underset{}{\overset{CH_3}{\overset{|}{C}}}HCH_2CH_2CH_2CH_2CH_3$$

(b)

$$CH_3\overset{CH_3}{\overset{|}{C}}HCH_2\overset{CH_2CH_3}{\overset{|}{C}}HCH_2CH_3$$

(c)

$$CH_3CH_2\overset{CH_3}{\overset{|}{C}}H-\underset{CH_3}{\underset{|}{\overset{CH_2CH_3}{\overset{|}{C}}}}CH_2CH_2CH_2CH_3$$

(d)

$$CH_3\overset{CH_3}{\overset{|}{C}}HCH_2\underset{CH_3}{\underset{|}{\overset{CH_3}{\overset{|}{C}}}}CH_2CH_2CH_3$$

(e)

1,1-dimethylcyclopentane: a cyclopentane ring with two CH_3 groups on the same carbon.

(f)

$$CH_3CH_2\overset{CH_3}{\overset{|}{C}}H\underset{\underset{CH_3}{\underset{|}{CHCH_3}}}{\underset{|}{C}}HCH_2CH_2CH_3$$

23.93 (a) 2,3-dimethylhexane

(b) 4-isopropyloctane

(c) 4-ethyl-2,4-dimethylhexane

(d) 3,3-diethylpentane

23.94 Cyclohexene will react with Br_2 and decolorize it. Cyclohexane will not react.

23.95 Cyclohexene will react with Br_2 and decolorize it. Benzene will not react with Br_2 without a catalyst.

23.96 (a)

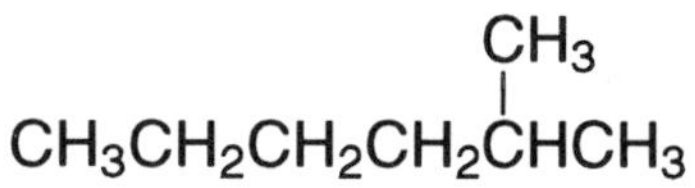

$$CH_3CH_2CH_2CH_2\underset{}{\overset{CH_3}{\overset{|}{C}}}HCH_3$$

(b)

Br

NO_2

Br

(c)

CH_3

CH_3

OH

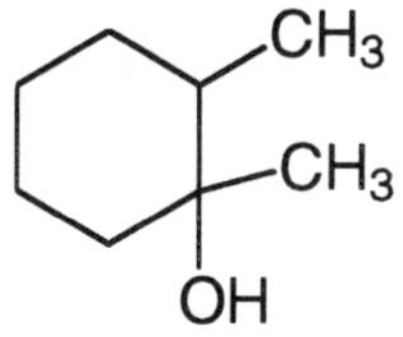

23.97

$$\left(\!-NHCH_2CH_2CH_2CH_2CH_2\overset{O}{\overset{||}{C}}NHCH_2CH_2CH_2CH_2CH_2\overset{O}{\overset{||}{C}}-\right)$$

repeating unit

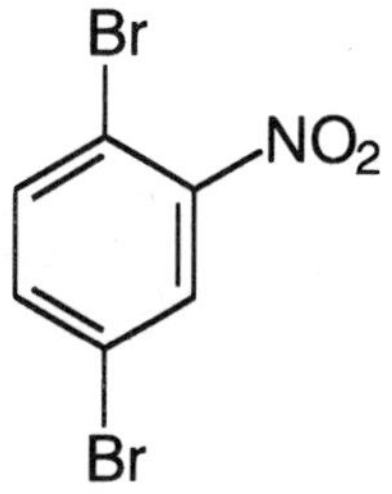

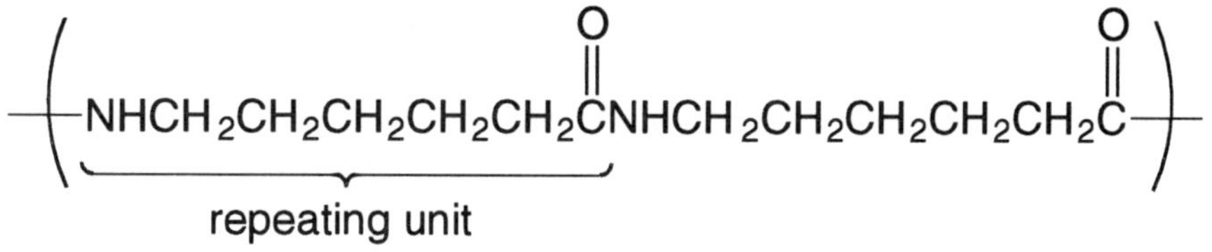

24.1

$$HOCH_2CH(OH)CH_2OH + ATP \longrightarrow HOCH_2CH(OH)CH_2OPO_3^{2-} + ADP$$

24.2 Amino acids that contain an aromatic ring:

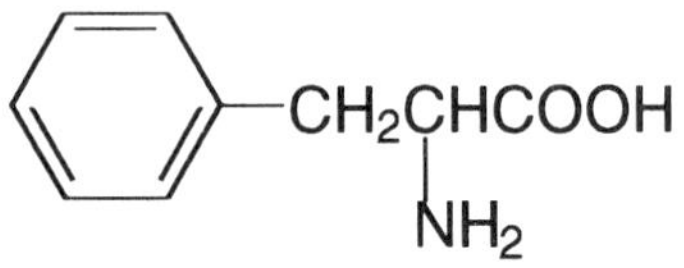

$C_6H_5CH_2CH(NH_2)COOH$ phenylalanine

(indol-3-yl)$CH_2CH(NH_2)COOH$ tryptophan

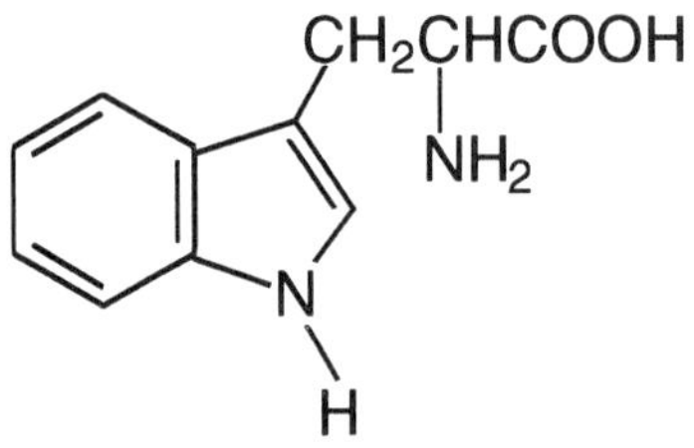

$HO{-}C_6H_4{-}CH_2CH(NH_2)COOH$ tyrosine

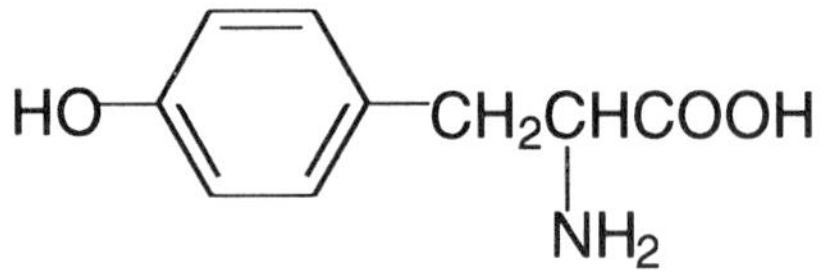

Amino acids that contain sulfur:

$CH_3SCH_2CH_2CH(NH_2)COOH$ methionine

$HSCH_2CH(NH_2)COOH$ cysteine

Amino acids that are alcohols:

$$HOCH_2\underset{\displaystyle NH_2}{\underset{|}{C}}HCOOH$$ serine

$$CH_3\overset{\displaystyle OH}{\overset{|}{C}}H\underset{\displaystyle NH_2}{\underset{|}{C}}HCOOH$$ threonine

$$HO-C_6H_4-CH_2\underset{\displaystyle NH_2}{\underset{|}{C}}HCOOH$$ tyrosine

Amino acids that have alkyl-group side chains:

$$CH_3\underset{\displaystyle NH_2}{\underset{|}{C}}HCOOH$$ alanine

$$CH_3\overset{\displaystyle CH_3}{\overset{|}{C}}H\underset{\displaystyle NH_2}{\underset{|}{C}}HCOOH$$ valine

$$CH_3CH_2\overset{\displaystyle CH_3}{\overset{|}{C}}H\underset{\displaystyle NH_2}{\underset{|}{C}}HCOOH$$ isoleucine

$$CH_3\overset{\displaystyle CH_3}{\overset{|}{C}}HCH_2\underset{\displaystyle NH_2}{\underset{|}{C}}HCOOH$$ leucine

24.3

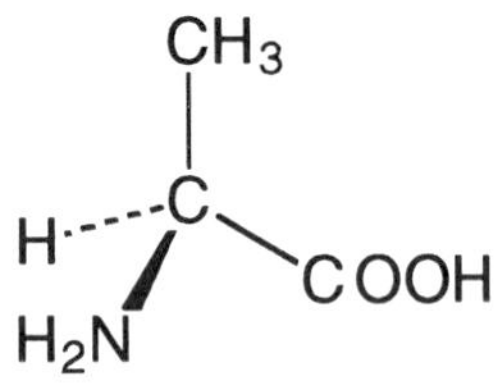

$$\begin{array}{c} CH_3 \\ | \\ H\cdots C \diagdown COOH \\ H_2N \end{array}$$

24.4 (a) a glove and (c) a screw are chiral

24.5 2-aminopropane

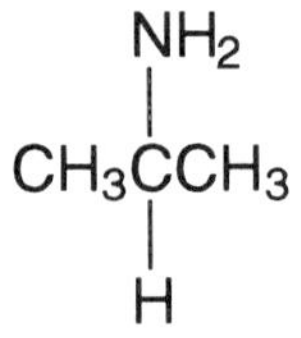

$$\begin{array}{c} NH_2 \\ | \\ CH_3CCH_3 \\ | \\ H \end{array}$$

No carbon in 2-aminopropane has four different groups attached to it so the molecule is achiral.

2-aminobutane

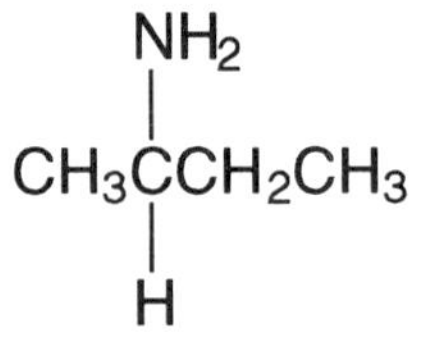

$$\begin{array}{c} NH_2 \\ | \\ CH_3CCH_2CH_3 \\ | \\ H \end{array}$$

The second carbon in 2-aminobutane has four different groups attached to it so the molecule is chiral.

24.6 (a) 3-chloropentane

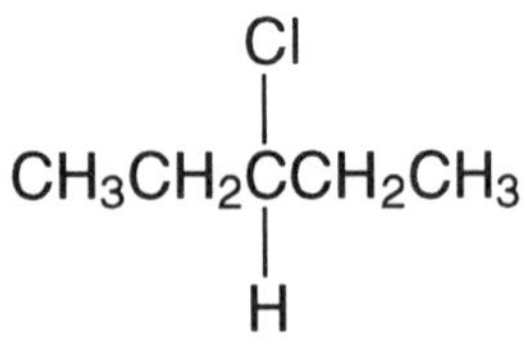

$$\begin{array}{c} Cl \\ | \\ CH_3CH_2CCH_2CH_3 \\ | \\ H \end{array}$$

No carbon in 3-chloropentane has four different groups attached to it so the molecule is achiral.

(b) 2-chloropentane

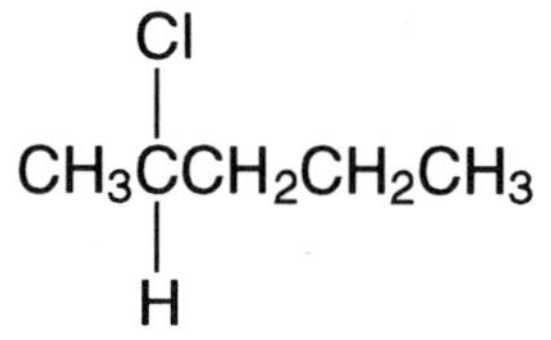

```
   Cl
   |
CH3CCH2CH2CH3
   |
   H
```

The second carbon in 2-chloropentane has four different groups attached to it so the molecule is chiral.

(c)

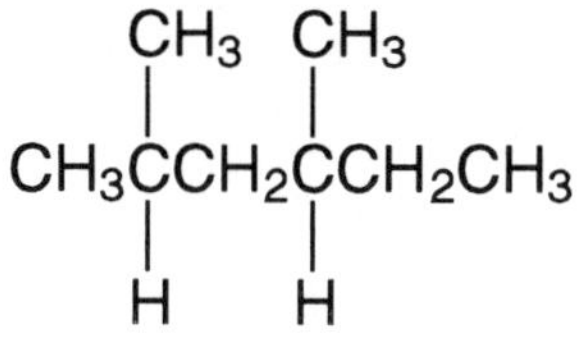

```
    CH3  CH3
     |    |
CH3CCH2CCH2CH3
     |    |
     H    H
```

The fourth carbon in the straight chain has four different groups attached to it so the molecule is chiral.

24.7 isoleucine and threonine

24.8 Val-Cys

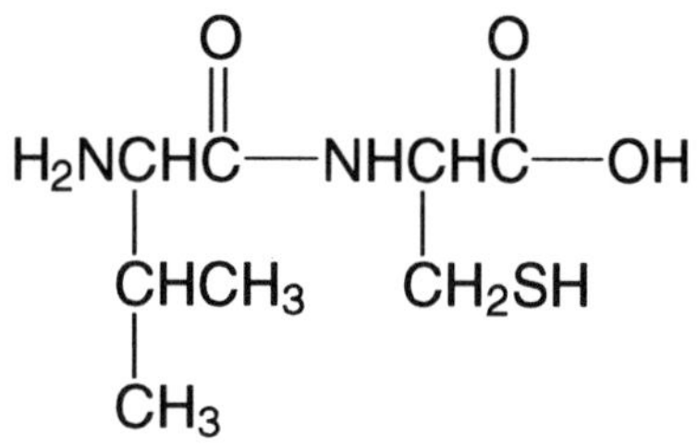

```
          O         O
          ||        ||
H2NCHC—NHCHC—OH
      |         |
      CHCH3     CH2SH
      |
      CH3
```

Cys-Val

```
          O         O
          ||        ||
H2NCHC—NHCHC—OH
      |         |
      CH2SH     CHCH3
                |
                CH3
```

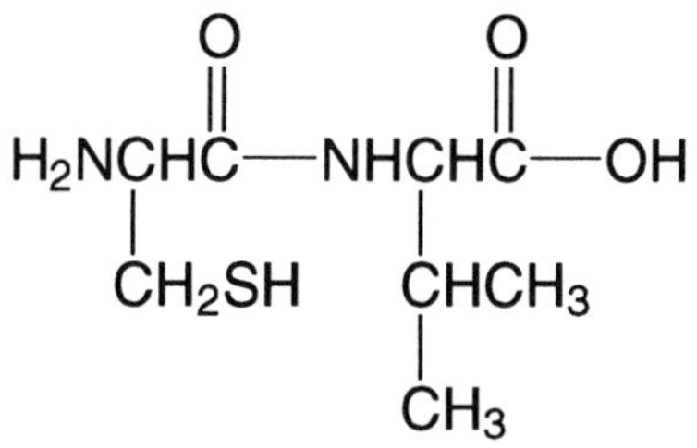

24.9 Val–Tyr–Gly; Val–Gly–Tyr; Tyr–Gly–Val;

Tyr–Val–Gly; Gly–Tyr–Val; Gly–Val–Tyr

24.10 (a) aldopentose (b) ketotriose (c) aldotetrose

24.11

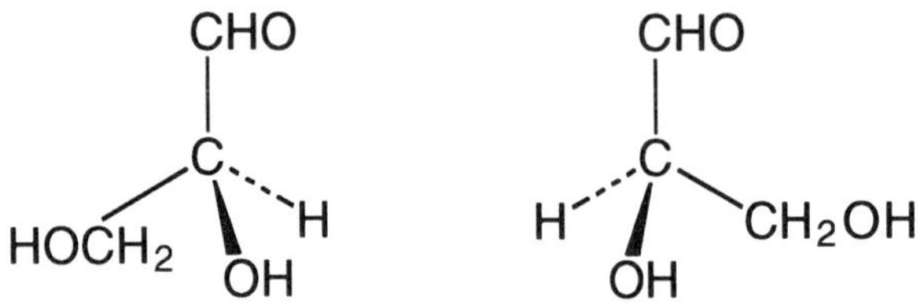

CHO | C (HOCH$_2$, H, OH)   CHO | C (H, CH$_2$OH, OH)

24.12 In general, a compound with n chiral carbon atoms has a maximum of 2^n possible forms.

Since ribose has three chiral carbon atoms, the maximum number of ribose forms = 2^3 = 8.

24.13

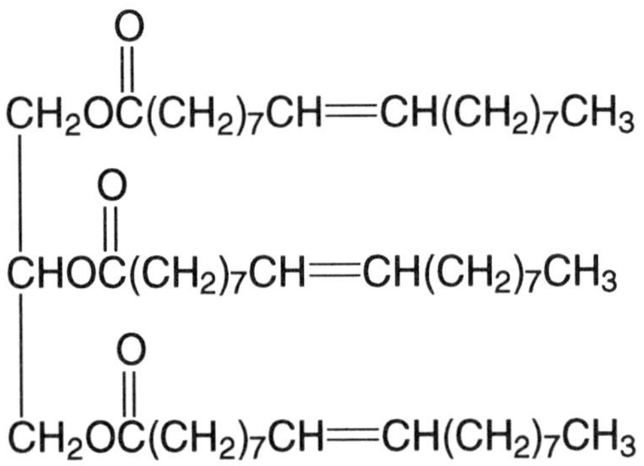

$CH_2OC(=O)(CH_2)_7CH{=}CH(CH_2)_7CH_3$
|
$CHOC(=O)(CH_2)_7CH{=}CH(CH_2)_7CH_3$
|
$CH_2OC(=O)(CH_2)_7CH{=}CH(CH_2)_7CH_3$

24.14 DNA dinucleotide A – G.

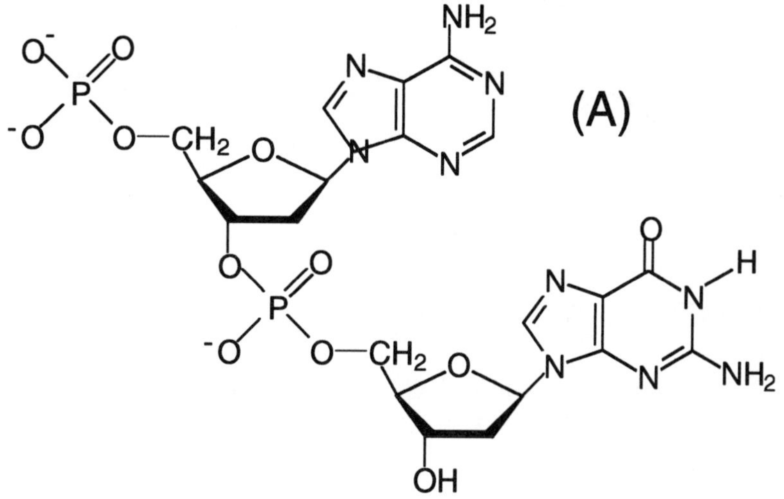

24.15 RNA dinucleotide U – A.

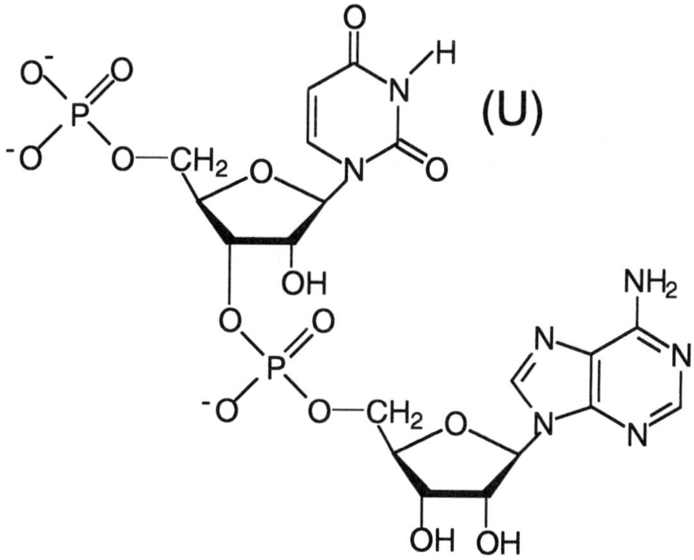

24.16 Original: G–G–C–C–C–G–T–A–A–T

Complement: C–C–G–G–G–C–A–T–T–A

24.17

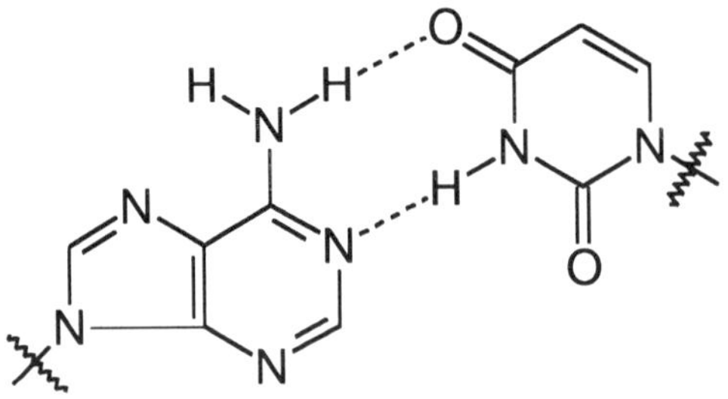

24.18 C–G–T–G–A–T–T–A–C–A (DNA)

G–C–A–C–U–A–A–U–G–U (RNA)

24.19 U–G–C–A–U–C–G–A–G–U (RNA)

A–C–G–T–A–G–C–T–C–A (DNA)

24.20 The "α" in α-amino acids means that the amino group in each is connected to the carbon atom alpha to (next to) the carboxylic acid group.

24.21 The naturally occurring amino acids are called L-amino acids because they are chiral molecules in the "left-handed" L-form (from levo, Latin for "left").

24.22 (a) serine (b) threonine (c) proline

(d) phenylalanine (e) cysteine

24.23 (a)

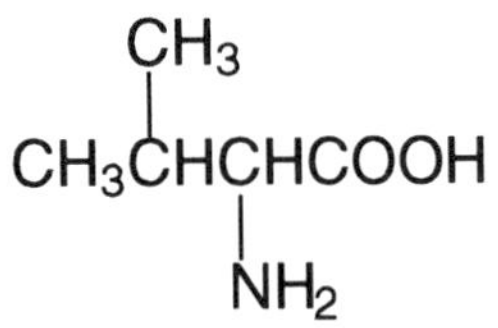

$CH_3CH(CH_3)CH(NH_2)COOH$ valine

$CH_3CH(CH_3)CH_2CH(NH_2)COOH$ leucine

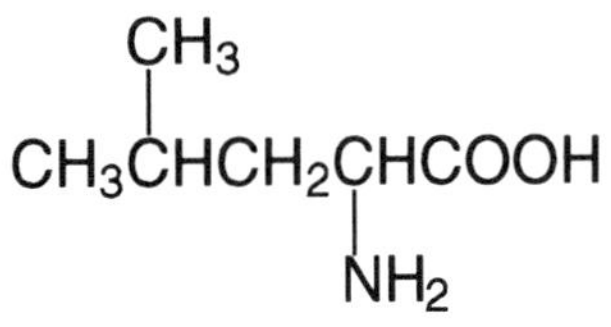

(b)

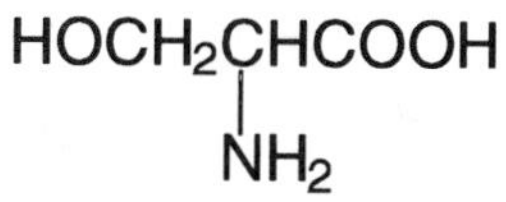

$HOCH_2CHCOOH$ (with NH_2 on the α-carbon) serine

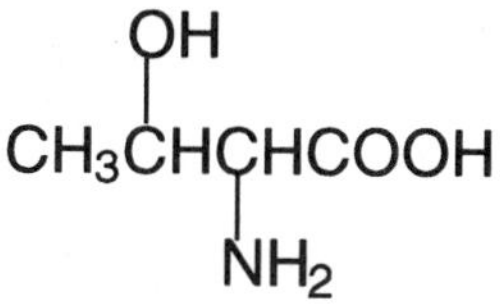

$CH_3CH(OH)CH(NH_2)COOH$ threonine

$HO-C_6H_4-CH_2CHCOOH$ (with NH_2 on the α-carbon) tyrosine

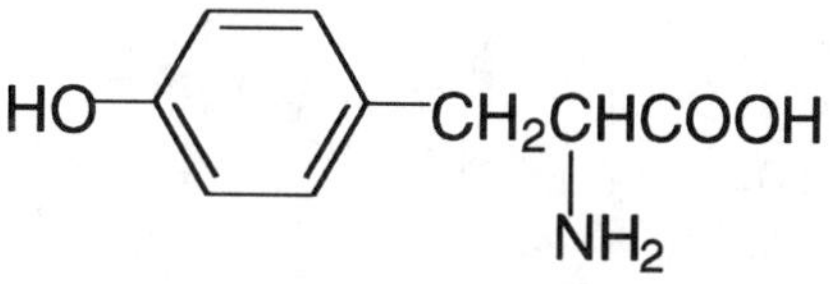

(c)

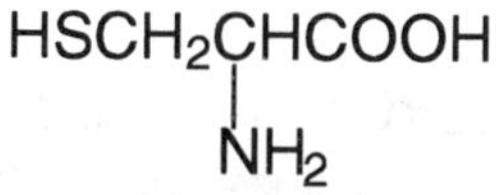

$HSCH_2CHCOOH$ (with NH_2 on the α-carbon) cysteine

(d)

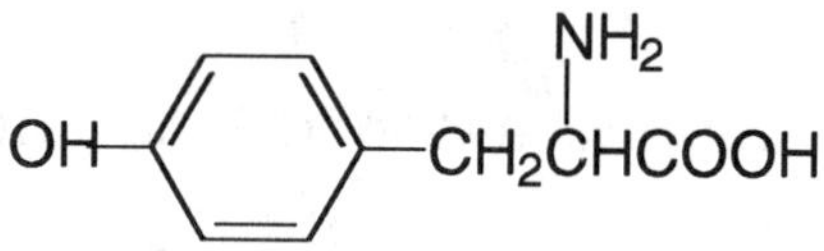

$OH-C_6H_4-CH_2CHCOOH$ (with NH_2 on the α-carbon) tyrosine

Indole ring (N–H) with $CH_2CHCOOH$ (with NH_2 on the α-carbon) tryptophan

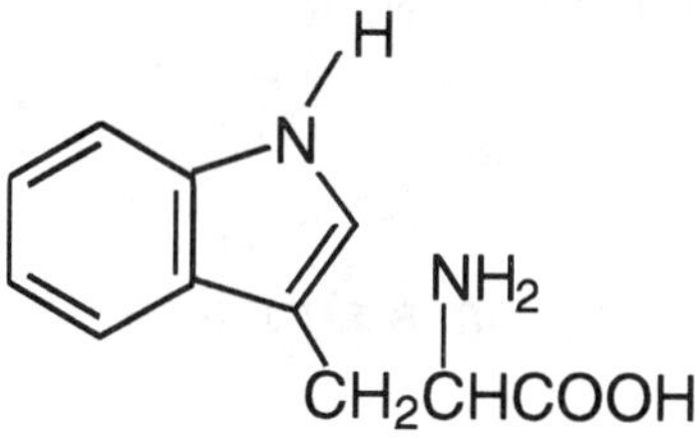

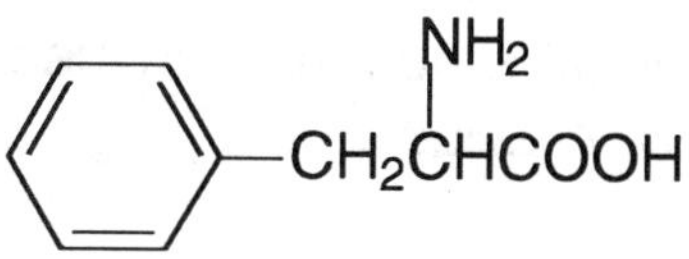

$C_6H_5-CH_2CHCOOH$ (with NH_2 on the α-carbon) phenylalanine

24.24 (a)

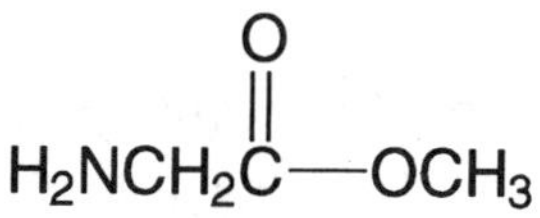

$H_2NCH_2\overset{O}{\overset{\|}{C}}-OCH_3$

(b)

$HO\overset{O}{\overset{\|}{C}}-CH_2\overset{+}{N}H_3\ Cl^-$

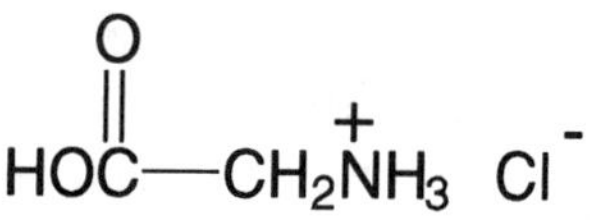

24.25 Val–Ser–Phe–Met–Thr–Ala

24.26 The N-terminal amino acid is aspartic acid.

The C-terminal amino acid is phenylalanine.

24.27 (a) The primary structure of a protein is the sequence in which amino acids are linked together.

(b) The secondary structure of a protein is the orientation of segments of a protein chain into a regular pattern.

(c) The tertiary structure of a protein is the way in which a protein chain folds into a specific three-dimensional shape.

24.28 Proteins are classified according to their three-dimensional shape as either fibrous or globular. Fibrous proteins consist of polypeptide chains arranged side by side in long filaments. Globular proteins are usually coiled into compact, nearly spherical shapes.

24.29 Several biological functions that proteins have in the body are (see Table 24.1):

(1) enzymes – to catalyze biological processes

(2) hormones – to regulate body processes

(3) structural proteins – to form an organism's structure

24.30 A protein's tertiary structure is stabilized by amino acid side-chain interactions, which include hydrophobic interactions, covalent disulfide bridges, electrostatic salt bridges, and hydrogen bonds.

24.31 Hydrogen bonding stabilizes helical and β-pleated-sheet secondary protein structure.

24.32 Cysteine is an important amino acid for defining the tertiary structure of proteins because the nearby cysteine residues can link together forming a disulfide bridge.

24.33 outside: (c) and (d), hydrophilic

inside: (a) and (b), hydrophobic

24.34 Met–Ile–Lys, Met–Lys–Ile, Ile–Met–Lys,

Ile–Lys–Met, Lys–Met–Ile, Lys–Ile–Met

24.35

O O

$H_2NCHCNHCHCOH$

CH_2 CH_2CH_2COOH

O O

$H_2NCHCNHCHCOH$

CH_2 CH_2

CH_2COOH

24.36 Aspartic acid and phenylalanine are the amino acids in aspartame.

Digestion products:

O O

$HO—CCH_2CHC—OH$ aspartic acid

NH_2

O

$—CH_2CHC—OH$ phenylalanine

NH_2

CH_3OH methanol

24.37 $$\frac{55.847 \text{ amu Fe}}{\text{mol wt cyt c}} \times 100\% = 0.43\%$$

mol wt cyt c = 13,000 amu

Molecular Handedness

24.38 (a) a shoe and (c) a light bulb are chiral.

24.39 chiral: screw, glove; achiral: pencil, tire tread pattern

24.40 (a) 2,4-dimethylheptane

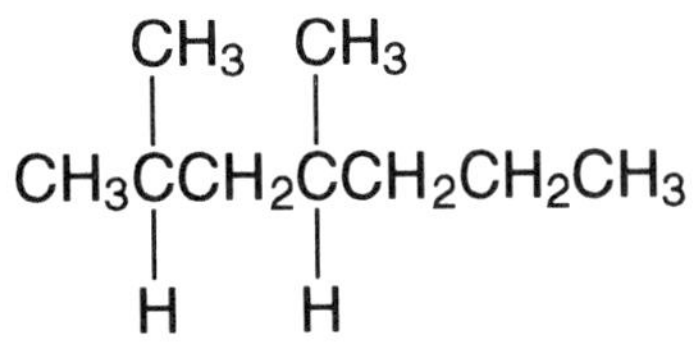

```
    CH3  CH3
    |    |
CH3CCH2CCH2CH2CH3
    |    |
    H    H
```

The fourth carbon in the straight chain has four different groups attached to it so the molecule is chiral.

(b) 5–ethyl–3,3–dimethylheptane

```
       CH3 CH2CH3
       |   |
CH3CH2CCH2CCH2CH3
       |   |
       CH3 H
```

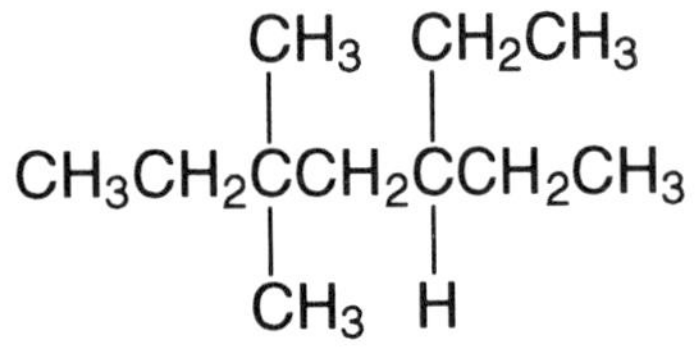

No carbon in 5–ethyl–3,3–dimethylheptane has four different groups attached to it so the molecule is achiral.

24.41 (a)

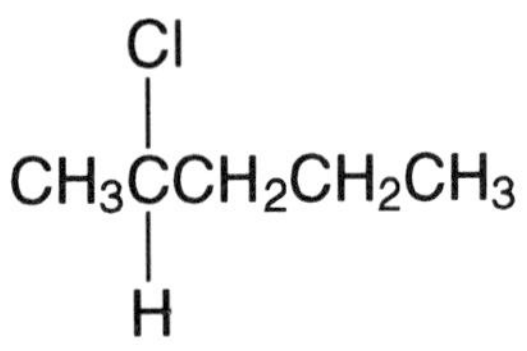

```
   Cl
   |
CH3CCH2CH2CH3
   |
   H
```

(b)

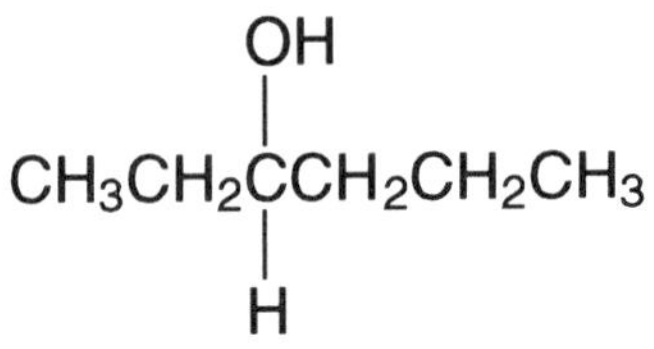

```
      OH
      |
CH3CH2CCH2CH2CH3
      |
      H
```

(c)

$$\begin{array}{c} CH_3 \\ | \\ CH_3CH_2CCH{=}CH_2 \\ | \\ H \end{array}$$

(d)

$$\begin{array}{c} CH_3 \\ | \\ CH_3CH_2CCH_2CH_2CH_2CH_3 \\ | \\ H \end{array}$$

24.42

$$HOCH_2CH_2CH_2CH_2CH_3$$

$$\begin{array}{c} OH \\ | \\ CH_3CH_2CHCH_2CH_3 \end{array}$$

$$\begin{array}{c} CH_3 \\ | \\ CH_3CCH_2CH_3 \\ | \\ OH \end{array}$$

$$\begin{array}{c} CH_3 \\ | \\ CH_3CHCH_2CH_2OH \end{array}$$

$$\begin{array}{c} CH_3 \\ | \\ CH_3CCH_2OH \\ | \\ CH_3 \end{array}$$

$$\begin{array}{c} OH \\ | \\ CH_3CCH_2CH_2CH_3 \\ | \\ H \end{array}$$ (chiral)

$$\begin{array}{c} CH_3 \\ | \\ HOCH_2CCH_2CH_3 \\ | \\ H \end{array}$$ (chiral)

$$\begin{array}{c} CH_3 \quad OH \\ | \qquad | \\ CH_3CH{-}CCH_3 \\ | \\ H \end{array}$$ (chiral)

24.43 (a)

$$CH_3CH_2\overset{\displaystyle OH}{\overset{|}{C}}HCH_3$$

(b)

$$CH_3CH_2\underset{\displaystyle CH_3}{\underset{|}{C}}H\overset{\displaystyle O}{\overset{\|}{C}}H$$

(c)

$$CH_3\overset{\displaystyle OH}{\overset{|}{C}}H\underset{\displaystyle OH}{\underset{|}{C}}HCH_2CH_3$$

Carbohydrates

24.44 An aldose contains the aldehyde functional group while a ketose contains the ketone functional group.

24.45 (a) aldopentose (b) aldohexose (c) ketohexose

24.46 Structurally, starch differs from cellulose in that it contains α- rather than β-glucose units.

Cellulose is the fibrous substance used by plants as a structural material in grasses, leaves, and stems. It consists of several thousand β-glucose molecules joined together by 1,4 links to form an immense polysaccharide.

Starch is a polymer made up of α-glucose units. Unlike cellulose, starch is edible. The starch in beans, rice, and potatoes is an essential part of the human diet.

24.47 There are two types of starch, amylose and amylopectin. Amylose, which accounts for about 20% of starch, consists of several hundred to 1000 α-glucose units joined together in a long chain by 1,4 links. Amylopectin, which accounts for about 80% of starch, is much larger than amylose (up to 100,000 glucose units per molecule) and has branches approximately every 25 units along its chain.

24.48 Structurally, glycogen is similar to amylopectin in being a long polymer of α-glucose units with branch points in its chain. Glycogen has many more branches than amylopectin, however, and is much larger–up to 1 million glucose units per molecule.

Glycogen, sometimes called animal starch, serves the same food-storage role in animals that starch serves in plants.

24.49

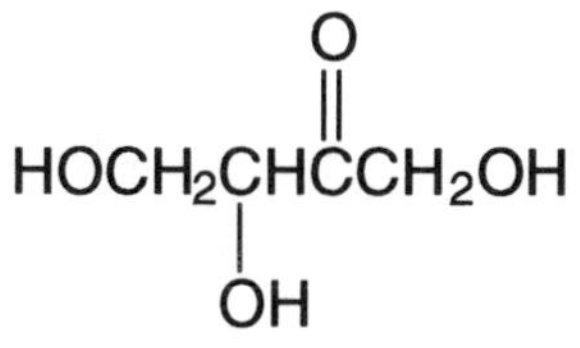

$HOCH_2CH(OH)C(=O)CH_2OH$

24.50

$HOCH_2CH(OH)CH_2CHO$

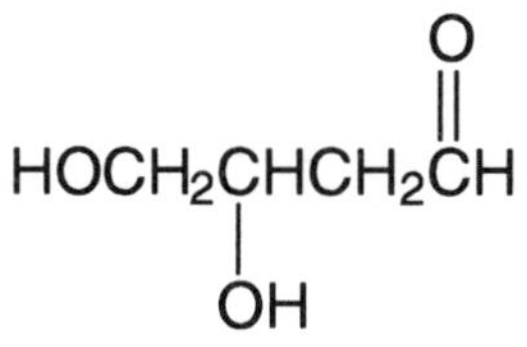

24.51

α

β

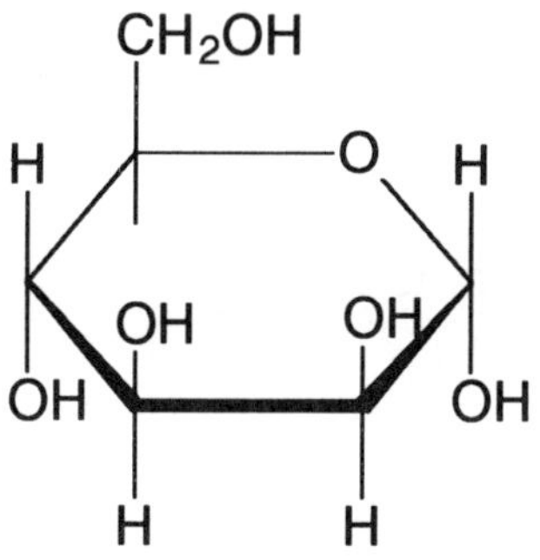

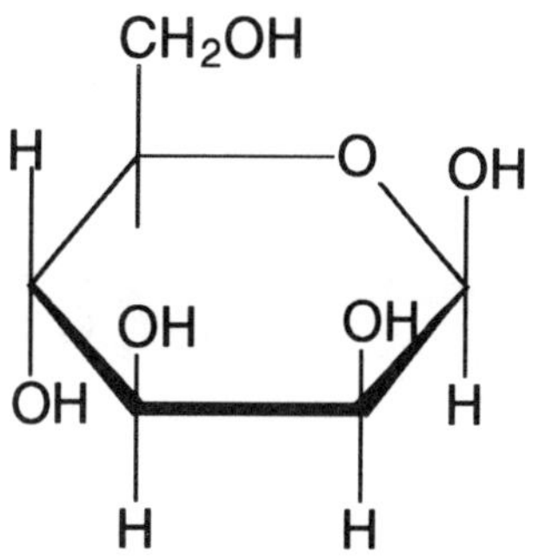

24.52

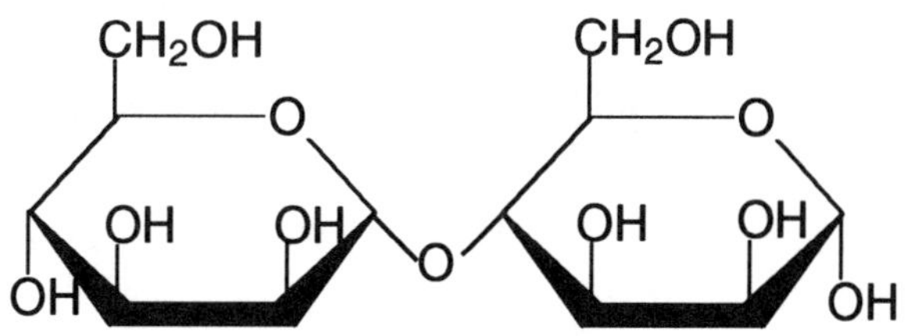

Lipids

24.53 Long-chain carboxylic acids are called fatty acids. Fatty acids are usually unbranched and have an even number of carbon atoms in the range of 12-22.

24.54 All fats and oils are triesters of glycerol (1,2,3-propanetriol) with three long-chain carboxylic acids, hence they are called triacylglycerols, or triglycerides.

24.55

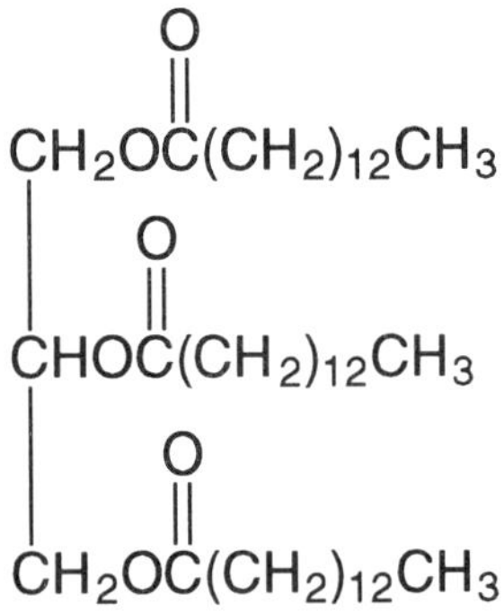

$$\begin{array}{l} CH_2O\overset{\overset{O}{\|}}{C}(CH_2)_{12}CH_3 \\ | \\ CHO\overset{\overset{O}{\|}}{C}(CH_2)_{12}CH_3 \\ | \\ CH_2O\overset{\overset{O}{\|}}{C}(CH_2)_{12}CH_3 \end{array}$$

24.56

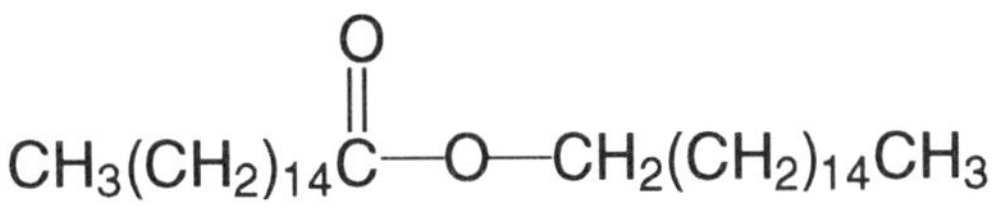

$$CH_3(CH_2)_{14}\overset{\overset{O}{\|}}{C}-O-CH_2(CH_2)_{14}CH_3$$

24.57

$$\begin{array}{l} CH_2O\overset{\overset{O}{\|}}{C}(CH_2)_{16}CH_3 \\ | \\ CHO\overset{\overset{O}{\|}}{C}(CH_2)_{16}CH_3 \\ | \\ CH_2O\overset{\overset{O}{\|}}{C}(CH_2)_{14}CH_3 \end{array} \qquad \begin{array}{l} CH_2O\overset{\overset{O}{\|}}{C}(CH_2)_{16}CH_3 \\ | \\ CHO\overset{\overset{O}{\|}}{C}(CH_2)_{14}CH_3 \\ | \\ CH_2O\overset{\overset{O}{\|}}{C}(CH_2)_{16}CH_3 \end{array}$$

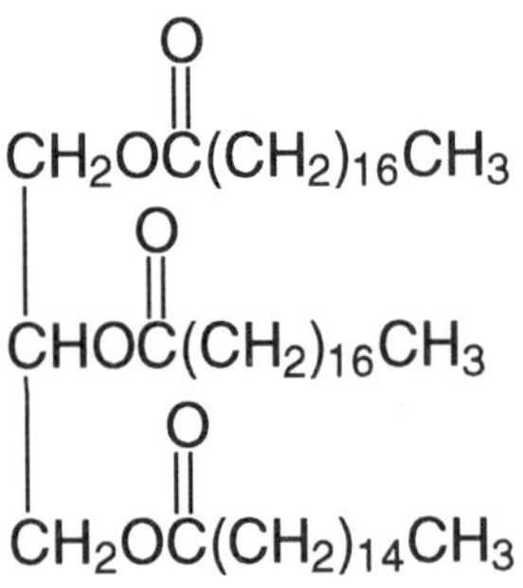

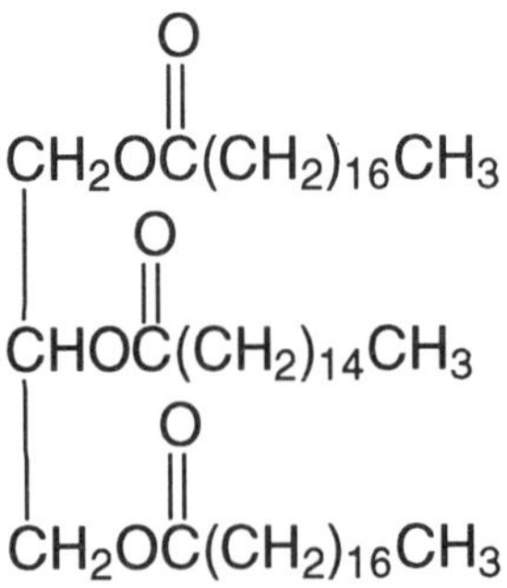

The two fat molecules differ from each other depending on where the palmitic acid chain is located. In the first fat molecule, palmitic acid is on an end and in the second it is in the middle.

24.58

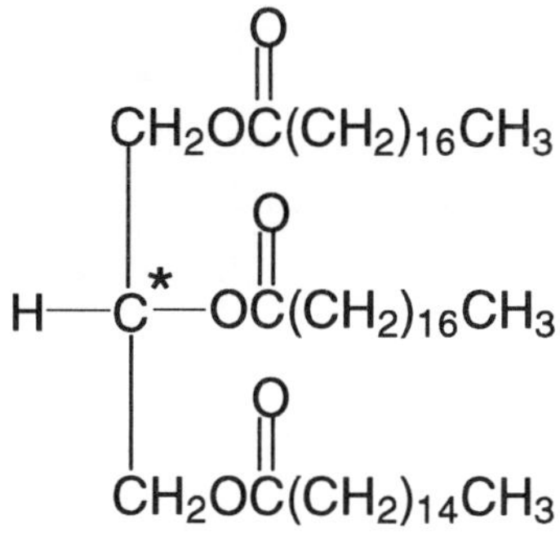

$$
\begin{array}{l}
\quad\quad\quad\;\; O \\
\quad\quad\quad\;\; \| \\
\quad CH_2OC(CH_2)_{16}CH_3 \\
\quad | \quad\quad\; O \\
\quad | \quad\quad\; \| \\
H-\overset{*}{C}-OC(CH_2)_{16}CH_3 \\
\quad | \quad\quad\; O \\
\quad | \quad\quad\; \| \\
\quad CH_2OC(CH_2)_{14}CH_3
\end{array}
$$

The molecule shown above is chiral. The carbon marked with an asterisk has four different groups attached to it.

24.59 potassium stearate $CH_3(CH_2)_{16}\overset{O}{\overset{\|}{C}}O^- K^+$

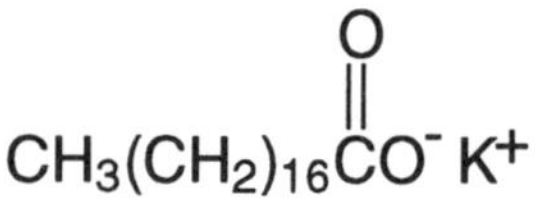

potassium oleate $CH_3(CH_2)_7CH{=}CH(CH_2)_7\overset{O}{\overset{\|}{C}}O^- K^+$

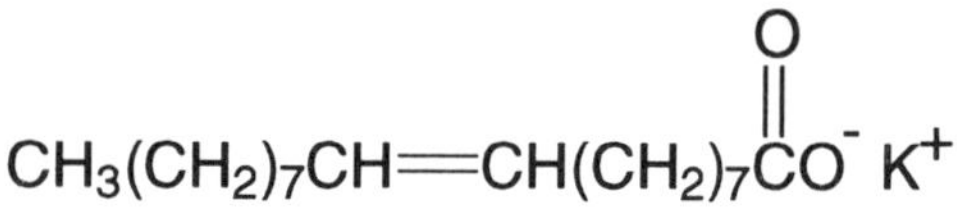

potassium linolenate

$$CH_3CH_2CH{=}CHCH_2CH{=}CHCH_2CH{=}CH(CH_2)_7\overset{O}{\overset{\|}{C}}O^- K^+$$

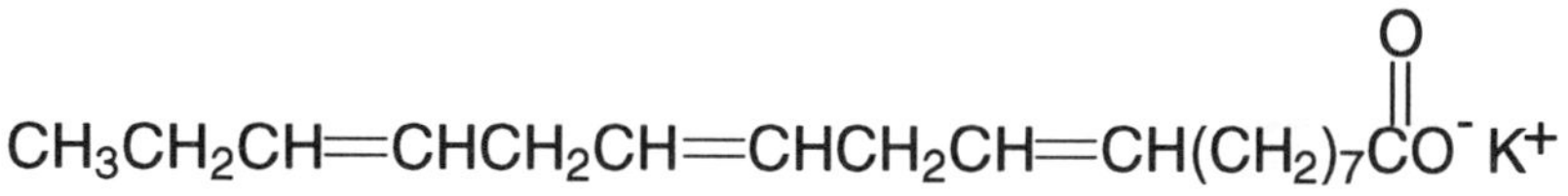

glycerol $HOCH_2\overset{OH}{\overset{|}{C}}HCH_2OH$

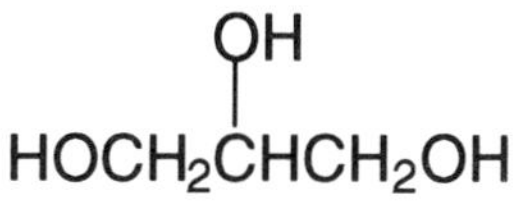

24.60 Glyceryl tristearate

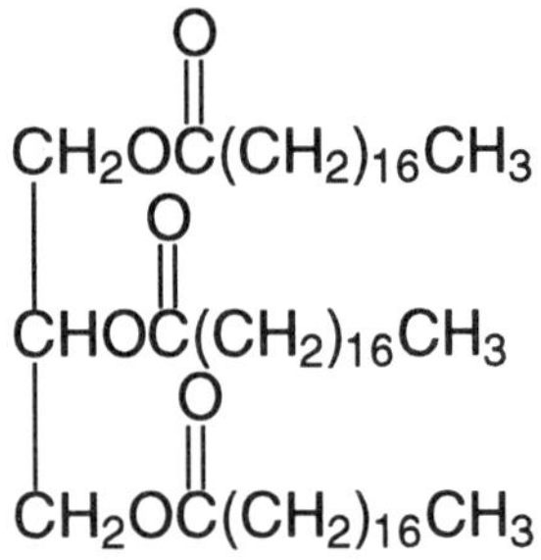

$$\begin{array}{l} \overset{\displaystyle O}{\|} \\ CH_2O\overset{}{C}(CH_2)_{16}CH_3 \\ | \quad \overset{\displaystyle O}{\|} \\ CHOC(CH_2)_{16}CH_3 \\ | \quad \overset{\displaystyle O}{\|} \\ CH_2OC(CH_2)_{16}CH_3 \end{array}$$

24.61 The hydrogenation product would have a higher melting point because it is saturated.

24.62 (a)

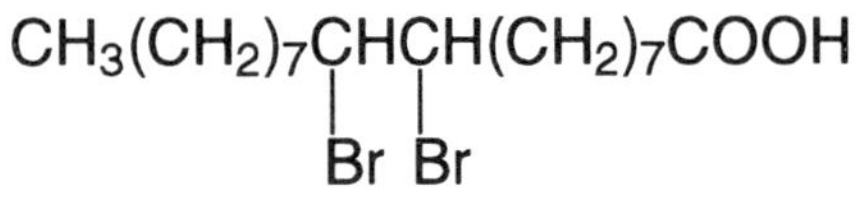

$$CH_3(CH_2)_7\underset{\displaystyle Br}{\underset{|}{C}}H\underset{\displaystyle Br}{\underset{|}{C}}H(CH_2)_7COOH$$

(b)

$$CH_3(CH_2)_{16}COOH$$

(c)

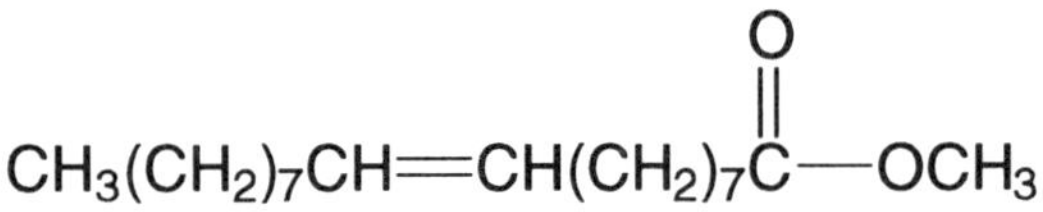

$$CH_3(CH_2)_7CH{=}CH(CH_2)_7\overset{\displaystyle O}{\overset{\|}{C}}{-}OCH_3$$

24.63

$$CH_3(CH_2)_{18}\overset{\displaystyle O}{\overset{\|}{C}}O(CH_2)_{31}CH_3$$

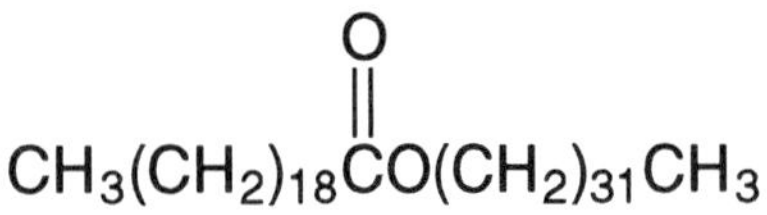

Nucleic Acids

24.64 Just as proteins are polymers made of amino acid units, nucleic acids are polymers made up of nucleotide units linked together to form a long chain. Each nucleotide contains a phosphate group, an aldopentose sugar, and an amine base.

24.65 The sugar component in RNA is ribose, and the sugar in DNA is 2-deoxyribose (2-deoxy means that oxygen is missing from C2 of ribose).

24.66 Most DNA of higher organisms is found in the nucleus of cells.

24.67 (a) Base pairing in nucleic acids means that the amine bases are found in specific pairs. In DNA, for example, adenine and thymine form hydrogen bonds to each other, as do cytosine and guanine.

(b) Replication is the process by which identical copies of DNA are made.

(c) Translation is the process by which RNA builds proteins.

(d) Transcription is the process by which information in DNA is transferred to RNA.

24.68 A chromosome is a threadlike strand of DNA in the cell nucleus. A gene is a segment of a DNA chain that contains the instructions necessary to make a specific protein.

24.69 A single gene contains the instructions necessary to make a specific protein.

24.70

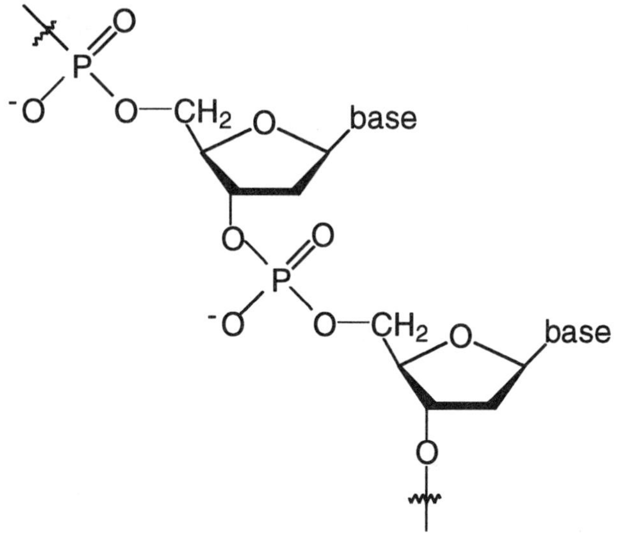

24.71

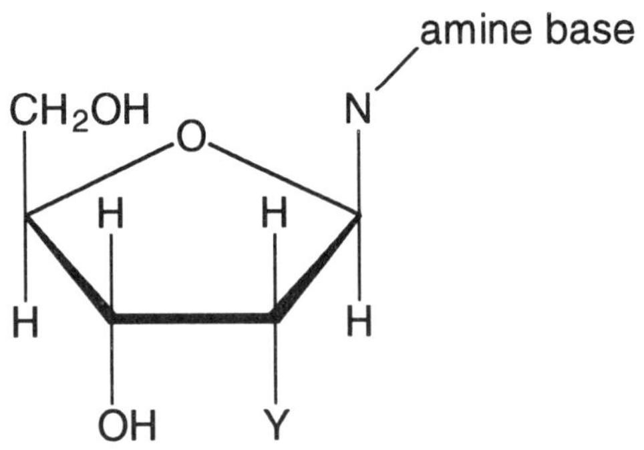

amine base

CH_2OH O N H H H H OH Y

24.72

NH_2 N N O O⁻ O P ⁻O O—CH_2 O OH

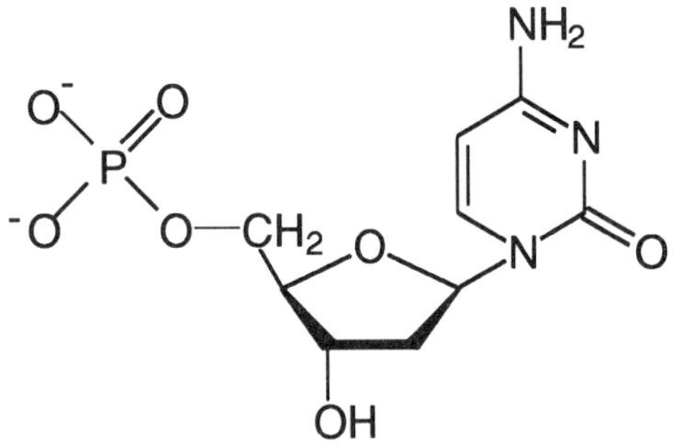

24.73 32% A and T because they are complementary.

18% G and C because they are complementary.

24.74 Original: T–A–C–C–G–A

Complement: A–T–G–G–C–T

24.75 T–A–C–C–G–A (DNA)

A–U–G–G–C–U (mRNA)

24.76 It takes three nucleotides to code for a specific amino acid. In insulin the 21 amino acid chain would require (3 x 21) = 63 nucleotides to code for it, and the 30 amino acid chain would require (3 x 30) = 90 nucleotides to code for it.

General Problems

24.77 ΔG° = + 2870 kJ/mol, because photosynthesis is the reverse reaction.

24.78

$$H_2N\underset{\underset{\underset{\text{C}_6\text{H}_4\text{OH}}{|}}{CH_2}}{\underset{|}{C}}H\overset{O}{\overset{\|}{C}}NH\underset{H}{\underset{|}{C}}H\overset{O}{\overset{\|}{C}}NH\underset{H}{\underset{|}{C}}H\overset{O}{\overset{\|}{C}}NH\underset{\underset{\underset{\text{C}_6\text{H}_5}{|}}{CH_2}}{\underset{|}{C}}H\overset{O}{\overset{\|}{C}}NH\underset{\underset{\underset{SCH_3}{|}}{\underset{CH_2}{|}}}{\underset{CH_2}{\underset{|}{C}}}H\overset{O}{\overset{\|}{C}}OH$$

24.79

(a)

$$H_2N\underset{\underset{\underset{CH_3}{|}}{CHCH_3}}{\underset{|}{C}}H\overset{O}{\overset{\|}{C}}NH\underset{\underset{\underset{\text{C}_6\text{H}_5}{|}}{CH_2}}{\underset{|}{C}}H\overset{O}{\overset{\|}{C}}NH\underset{CH_2SH}{\underset{|}{C}}H\overset{O}{\overset{\|}{C}}OH$$

(b)

$$H_2N\underset{\underset{\underset{CH_2COOH}{|}}{CH_2}}{\underset{|}{C}}H\overset{O}{\overset{\|}{C}}-N(\text{pyrrolidine ring})-\overset{O}{\overset{\|}{C}}NH\underset{\underset{\underset{CH_2CH_3}{|}}{HCCH_3}}{\underset{|}{C}}H\overset{O}{\overset{\|}{C}}NH\underset{\underset{\underset{CH_3}{|}}{CH_2CHCH_3}}{\underset{|}{C}}H\overset{O}{\overset{\|}{C}}OH$$

24.80 proline containing segment of a protein chain:

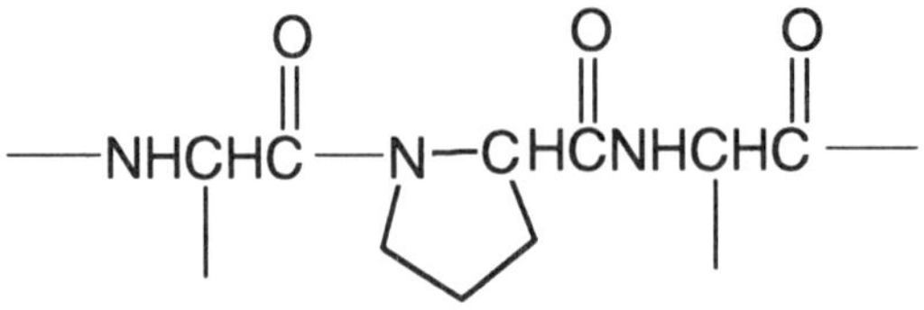

$$-\mathrm{NHCH}\overset{\overset{O}{\|}}{\mathrm{C}}-\mathrm{N}-\mathrm{CH}\overset{\overset{O}{\|}}{\mathrm{C}}\mathrm{NHCH}\overset{\overset{O}{\|}}{\mathrm{C}}-$$

The α-helix is stabilized by the formation of hydrogen bonds between the N–H group of one amino acid and the C=O group of another amino acid four residues away. The nitrogen in proline has no attached hydrogen and cannot hydrogen bond. Hence proline is never encountered in a protein α-helix.

24.81 Diabetics must receive insulin by injection rather than by mouth because insulin is a protein and the peptide (amide) bonds would be hydrolyzed (broken) by stomach acid.

24.82

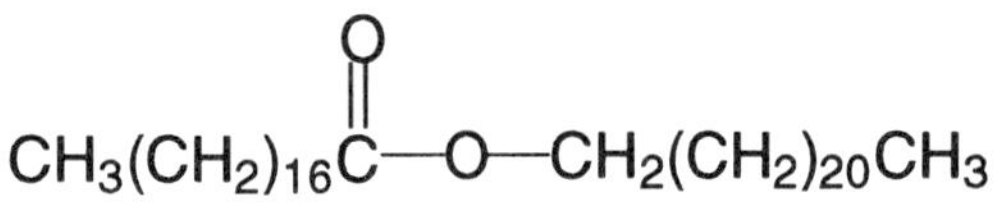

$$CH_3(CH_2)_{16}\overset{\overset{O}{\|}}{C}-O-CH_2(CH_2)_{20}CH_3$$

24.83 (a) a fat

$$\begin{array}{l} CH_2O\overset{\overset{O}{\|}}{C}(CH_2)_{16}CH_3 \\ | \\ CHO\overset{\overset{O}{\|}}{C}(CH_2)_{16}CH_3 \\ | \\ CH_2O\overset{\overset{O}{\|}}{C}(CH_2)_{14}CH_3 \end{array}$$

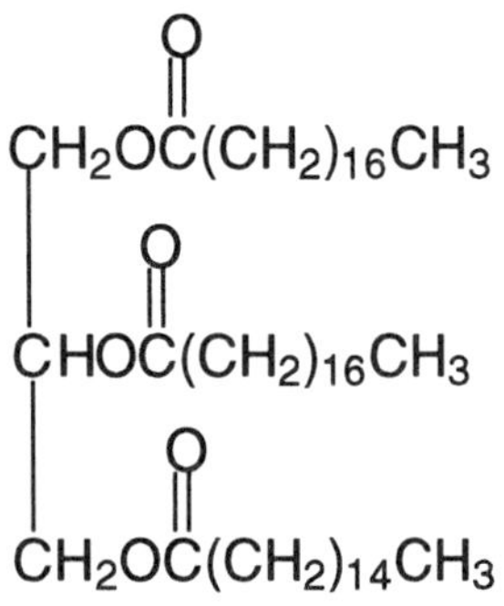

(b) a vegetable oil

$$\begin{array}{l} CH_2OC(=O)(CH_2)_7CH{=}CH(CH_2)_7CH_3 \\ | \\ CHOC(=O)(CH_2)_7CH{=}CH(CH_2)_7CH_3 \\ | \\ CH_2OC(=O)(CH_2)_7CH{=}CHCH_2CH{=}CH(CH_2)_4CH_3 \end{array}$$

(c) A steroid

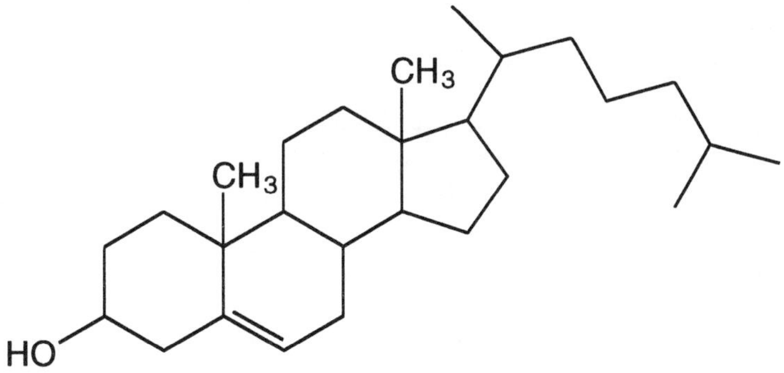

CH$_3$ CH$_3$ HO

24.84 Original: A–G–T–T–C–A–T–C–G

Complement: T–C–A–A–G–T–A–G–C